Texts in Computer Science

Series Editors

David Gries, Department of Computer Science, Cornell University, Ithaca, NY, USA

Orit Hazzan, Faculty of Education in Technology and Science, Technion—Israel Institute of Technology, Haifa, Israel

More information about this series at http://www.springer.com/series/3191

Andreas Antoniou • Wu-Sheng Lu

Practical Optimization

Algorithms and Engineering Applications

Second Edition

 Springer

Andreas Antoniou
Department of Electrical and Computer
Engineering
University of Victoria
Victoria, BC, Canada

Wu-Sheng Lu
Department of Electrical and Computer
Engineering
University of Victoria
Victoria, BC, Canada

ISSN 1868-0941   ISSN 1868-095X   (electronic)
Texts in Computer Science
ISBN 978-1-0716-0841-8   ISBN 978-1-0716-0843-2   (eBook)
https://doi.org/10.1007/978-1-0716-0843-2

This Springer imprint is published by the registered company Springer Science+Business Media, LLC part of Springer Nature.
The registered company address is: 1 New York Plaza, New York, NY 10004, U.S.A.

To
Lynne
and
Chi-Tang Catherine
with our love

Preface to the Second Edition

Optimization methods and algorithms continue to evolve at a tremendous rate and are providing solutions to many problems that could not be solved before in economics, finance, geophysics, molecular modeling, computational systems biology, operations research, and all branches of engineering (see the following link for details: https://en.wikipedia.org/wiki/Mathematical_optimization#Molecular_modeling).

The growing demand for optimization methods and algorithms has been addressed in the second edition by updating some material, adding more examples, and introducing some recent innovations, techniques, and methodologies. The emphasis continues to be on practical methods and efficient algorithms that work.

Chapters 1–8 continue to deal with the basics of optimization. Chapter 5 now includes two increasingly popular line search techniques, namely, the so-called *two-point* and *backtracking* line searches. In Chap. 6, a new section has been added that deals with the application of the conjugate-gradient method for the solution of linear systems of equations.

In Chap. 9, some state-of-the art applications of unconstrained optimization to machine learning and source localization are added. The first application is in the area of character recognition and it is a method for classifying handwritten digits using a regression technique known as *softmax*. The method is based on an accelerated gradient descent algorithm. The second application is in the area of communications and it deals of the problem formulation and solution methods for identifying the location of a radiating source given the distances between the source and several sensors.

The contents of Chaps. 10–12 are largely unchanged except for some editorial changes whereas Chap. 13 combines the material found in Chaps. 13 and 14 of the first edition.

Chapter 14 is a new chapter that presents additional concepts and properties of convex functions that are not covered in Chapter 2. It also describes several algorithms for the solution of general convex problems and includes a detailed exposition of the so-called *alternating direction method of multipliers (ADMM)*.

Chapter 15 is a new chapter that focuses on sequential convex programming, sequential quadratic programming, and convex-concave procedures for general nonconvex problems. It also includes a section on heuristic ADMM techniques for nonconvex problems.

In Chap. 16, we have added some new state-of-the art applications of constrained optimization for the design of Finite-Duration Impulse Response (FIR) and Infinite-Duration Impulse Response (IIR) digital filters, also known as nonrecursive and recursive filters, respectively, using second-order cone programming. Digital filters that would satisfy multiple specifications such as maximum passband gain, minimum stopband gain, maximum transition-band gain, and maximum pole radius, can be designed with these methods.

The contents of Appendices A and B are largely unchanged except for some editorial changes.

Many of our past students at the University of Victoria have helped a great deal in improving the first edition and some of them, namely, Drs. M. L. R. de Campos, Sunder Kidambi, Rajeev C. Nongpiur, Ana Maria Sevcenco, and Ioana Sevcenco have provided meaningful help in the evolution of the second edition as well. We would also like to thank Drs. Z. Dong, T. Hinamoto, Y. Q. Hu, and W. Xu for useful discussions on optimization theory and its applications, Catherine Chang for typesetting the first draft of the second edition, and to Lynne Barrett for checking the entire second edition for typographical errors.

Victoria, Canada   Andreas Antoniou
                                                                                          Wu-Sheng Lu

Preface to the First Edition

The rapid advancements in the efficiency of digital computers and the evolution of reliable software for numerical computation during the past three decades have led to an astonishing growth in the theory, methods, and algorithms of numerical optimization. This body of knowledge has, in turn, motivated widespread applications of optimization methods in many disciplines, e.g., engineering, business, and science, and led to problem solutions that were considered intractable not too long ago.

Although excellent books are available that treat the subject of optimization with great mathematical rigor and precision, there appears to be a need for a book that provides a practical treatment of the subject aimed at a broader audience ranging from college students to scientists and industry professionals. This book has been written to address this need. It treats unconstrained and constrained optimization in a unified manner and places special attention on the algorithmic aspects of optimization to enable readers to apply the various algorithms and methods to specific problems of interest. To facilitate this process, the book provides many solved examples that illustrate the principles involved, and includes, in addition, two chapters that deal exclusively with applications of unconstrained and constrained optimization methods to problems in the areas of pattern recognition, control systems, robotics, communication systems, and the design of digital filters. For each application, enough background information is provided to promote the understanding of the optimization algorithms used to obtain the desired solutions.

Chapter 1 gives a brief introduction to optimization and the general structure of optimization algorithms. Chapters 2 to 9 are concerned with unconstrained optimization methods. The basic principles of interest are introduced in Chapter 2. These include the first-order and second-order necessary conditions for a point to be a local minimizer, the second-order sufficient conditions, and the optimization of convex functions. Chapter 3 deals with general properties of algorithms such as the concepts of descent function, global convergence, and rate of convergence. Chapter 4 presents several methods for one-dimensional optimization, which are commonly referred to as line searches. The chapter also deals with inexact line-search methods that have been found to increase the efficiency in many optimization algorithms. Chapter 5 presents several basic gradient methods that include the steepest-descent, Newton, and Gauss-Newton methods. Chapter 6 presents a class of methods based on the concept of conjugate directions such as the conjugate-gradient, Fletcher-Reeves, Powell, and Partan methods. An important class of unconstrained optimization methods known as quasi-Newton methods is presented in Chapter 7. Representative methods of this class such as the Davidon-Fletcher-Powell and Broydon-Fletcher-Goldfarb-Shanno methods and their properties are investigated. The chapter also includes a practical, efficient, and reliable quasi-Newton algorithm that eliminates some problems associated with the basic quasi-Newton method. Chapter 8 presents minimax methods that are used in many applications including the design of digital filters. Chapter 9 presents three case studies in which several of the unconstrained optimization methods described in Chapters 4 to 8 are applied to point pattern matching, inverse kinematics for robotic manipulators, and the design of digital filters.

Chapters 10 to 16 are concerned with constrained optimization methods. Chapter 10 introduces the fundamentals of constrained optimization. The concept of Lagrange multipliers, the first-order necessary conditions known as Karush-Kuhn-Tucker conditions, and the duality principle of convex programming are addressed in detail and are illustrated by many examples. Chapters 11 and 12 are concerned with linear programming (LP) problems. The general properties of LP and the simplex method for standard LP problems are addressed in Chapter 11. Several interior-point methods including the primal affine-scaling, primal Newton-barrier, and primal-dual path-following methods are presented in Chapter 12. Chapter 13 deals with quadratic and general convex programming. The so-called active-set methods and several interior-point methods for convex quadratic programming are investigated. The chapter also includes the so-called cutting plane and ellipsoid algorithms for general convex programming problems. Chapter 14 presents two special classes of convex programming known as semidefinite and second-order cone programming, which have found interesting applications in a variety of disciplines. Chapter 15 treats general constrained optimization problems that do not belong to the class of convex programming; special emphasis is placed on several sequential quadratic programming methods that are enhanced through the use of efficient line searches and approximations of the Hessian matrix involved. Chapter 16, which concludes the book, examines several applications of constrained optimization for the design of digital filters, for the control of dynamic systems, for evaluating the force distribution in robotic systems, and in multiuser detection for wireless communication systems.

The book also includes two appendices, A and B, which provide additional support material. Appendix A deals in some detail with the relevant parts of linear algebra to consolidate the understanding of the underlying mathematical principles involved whereas Appendix B provides a concise treatment of the basics of digital filters to enhance the understanding of the design algorithms included in Chaps. 8, 9, and 16.

The book can be used as a text for a sequence of two one-semester courses on optimization. The first course comprising Chaps. 1 to 7, 9, and part of Chap. 10 may be offered to senior undergraduate or first-year graduate students. The prerequisite knowledge is an undergraduate mathematics background of calculus and linear algebra. The material in Chaps. 8 and 10 to 16 may be used as a text for an advanced graduate course on minimax and constrained optimization. The prerequisite knowledge for this course is the contents of the first optimization course.

The book is supported by online solutions of the end-of-chapter problems under password as well as by a collection of MATLAB programs for free access by the readers of the book, which can be used to solve a variety of optimization problems. These materials can be downloaded from book's website: https://www.ece.uvic.ca/ optimization/.

We are grateful to many of our past students at the University of Victoria, in particular, Drs. M. L. R. de Campos, S. Netto, S. Nokleby, D. Peters, and Mr. J. Wong who took our optimization courses and have helped improve the manuscript in one way or another; to Chi-Tang Catherine Chang for typesetting the first draft of the manuscript and for producing most of the illustrations; to R. Nongpiur for checking a large part of the index; and to P. Ramachandran for proofreading the entire manuscript. We would also like to thank Professors M. Ahmadi, C. Charalambous, P. S. R. Diniz, Z. Dong, T. Hinamoto, and P. P. Vaidyanathan for useful discussions on optimization theory and practice; Tony Antoniou of Psicraft Studios for designing the book cover; the Natural Sciences and Engineering Research Council of Canada for supporting the research that led to some of the new results described in Chapters 8, 9, and 16; and last but not least the University of Victoria for supporting the writing of this book over a number of years.

Andreas Antoniou
Wu-Sheng Lu

Contents

1 The Optimization Problem   1
 1.1   Introduction   1
 1.2   The Basic Optimization Problem   4
 1.3   General Structure of Optimization Algorithms.   8
 1.4   Constraints   9
 1.5   The Feasible Region   17
 1.6   Branches of Mathematical Programming.   20
       1.6.1   Linear Programming   21
       1.6.2   Integer Programming   22
       1.6.3   Quadratic Programming   22
       1.6.4   Nonlinear Programming   23
       1.6.5   Dynamic Programming   23
 Problems   24
 References   25

2 Basic Principles   27
 2.1   Introduction   27
 2.2   Gradient Information   27
 2.3   The Taylor Series   29
 2.4   Types of Extrema   31
 2.5   Necessary and Sufficient Conditions For Local Minima
       and Maxima   33
       2.5.1   First-Order Necessary Conditions.   34
       2.5.2   Second-Order Necessary Conditions.   36
 2.6   Classification of Stationary Points   40
 2.7   Convex and Concave Functions   50
 2.8   Optimization of Convex Functions   57
 Problems   59
 References   62

3 General Properties of Algorithms   63
 3.1   Introduction   63
 3.2   An Algorithm as a Point-to-Point Mapping.   63

3.3   An Algorithm as a Point-to-Set Mapping   65
3.4   Closed Algorithms   66
3.5   Descent Functions   68
3.6   Global Convergence   69
3.7   Rates of Convergence   73
Problems   75
References   76

4   **One-Dimensional Optimization**   77
4.1   Introduction   77
4.2   Dichotomous Search   78
4.3   Fibonacci Search   79
4.4   Golden-Section Search   88
4.5   Quadratic Interpolation Method   91
        4.5.1   Two-Point Interpolation   94
4.6   Cubic Interpolation   94
4.7   Algorithm of Davies, Swann, and Campey   97
4.8   Inexact Line Searches   101
Problems   110
References   113

5   **Basic Multidimensional Gradient Methods**   115
5.1   Introduction   115
5.2   Steepest-Descent Method   116
        5.2.1   Ascent and Descent Directions   116
        5.2.2   Basic Method   117
        5.2.3   Orthogonality of Directions   119
        5.2.4   Step-Size Estimation for Steepest-Descent
                 Method   120
        5.2.5   Step-Size Estimation Using the Barzilai–Borwein
                 Two-Point Formulas   122
        5.2.6   Convergence   124
        5.2.7   Scaling   126
5.3   Newton Method   126
        5.3.1   Modification of the Hessian   128
        5.3.2   Computation of the Hessian   135
        5.3.3   Newton Decrement   135
        5.3.4   Backtracking Line Search   135
        5.3.5   Independence of Linear Changes in Variables   136
5.4   Gauss–Newton Method   137
Problems   140
References   144

6   Conjugate-Direction Methods 145
   6.1   Introduction 145
   6.2   Conjugate Directions 146
   6.3   Basic Conjugate-Directions Method 148
   6.4   Conjugate-Gradient Method 152
   6.5   Minimization of Nonquadratic Functions 157
   6.6   Fletcher–Reeves Method 158
   6.7   Powell's Method 161
   6.8   Partan Method 169
   6.9   Solution of Systems of Linear Equations 173
   Problems 175
   References 178

7   Quasi-Newton Methods 179
   7.1   Introduction 179
   7.2   The Basic Quasi-Newton Approach 180
   7.3   Generation of Matrix S_k 181
   7.4   Rank-One Method 185
   7.5   Davidon–Fletcher–Powell Method 190
      7.5.1   Alternative Form of DFP Formula 196
   7.6   Broyden–Fletcher–Goldfarb–Shanno Method 197
   7.7   Hoshino Method 199
   7.8   The Broyden Family 200
      7.8.1   Fletcher Switch Method 200
   7.9   The Huang Family 201
   7.10   Practical Quasi-Newton Algorithm 202
   Problems 206
   References 208

8   Minimax Methods 211
   8.1   Introduction 211
   8.2   Problem Formulation 211
   8.3   Minimax Algorithms 213
   8.4   Improved Minimax Algorithms 219
   Problems 235
   References 236

9   Applications of Unconstrained Optimization 239
   9.1   Introduction 239
   9.2   Classification of Handwritten Digits 240
      9.2.1   Handwritten-Digit Recognition Problem 240
      9.2.2   Histogram of Oriented Gradients 240
      9.2.3   Softmax Regression for Use in Multiclass
             Classification 243

        9.2.4   Use of Softmax Regression for the Classification
              of Handwritten Digits 249
  9.3   Inverse Kinematics for Robotic Manipulators 253
        9.3.1   Position and Orientation of a Manipulator 253
        9.3.2   Inverse Kinematics Problem 256
        9.3.3   Solution of Inverse Kinematics Problem. 257
  9.4   Design of Digital Filters 261
        9.4.1   Weighted Least-Squares Design of FIR Filters 262
        9.4.2   Minimax Design of FIR Filters 267
  9.5   Source Localization. 275
        9.5.1   Source Localization Based on Range
              Measurements. 275
        9.5.2   Source Localization Based on Range-Difference
              Measurements. 279
  Problems 282
  References 283

10 Fundamentals of Constrained Optimization 285
  10.1   Introduction 285
  10.2   Constraints 286
        10.2.1   Notation and Basic Assumptions 286
        10.2.2   Equality Constraints 286
        10.2.3   Inequality Constraints 290
  10.3   Classification of Constrained Optimization Problems 292
        10.3.1   Linear Programming 293
        10.3.2   Quadratic Programming 294
        10.3.3   Convex Programming 295
        10.3.4   General Constrained Optimization Problem 295
  10.4   Simple Transformation Methods. 296
        10.4.1   Variable Elimination 296
        10.4.2   Variable Transformations 300
  10.5   Lagrange Multipliers 303
        10.5.1   Equality Constraints 305
        10.5.2   Tangent Plane and Normal Plane 308
        10.5.3   Geometrical Interpretation 310
  10.6   First-Order Necessary Conditions 312
        10.6.1   Equality Constraints 312
        10.6.2   Inequality Constraints 314
  10.7   Second-Order Conditions. 319
        10.7.1   Second-Order Necessary Conditions. 320
        10.7.2   Second-Order Sufficient Conditions 323
  10.8   Convexity. 326
  10.9   Duality 328

Problems 332
References 338

11 Linear Programming Part I: The Simplex Method 339
   11.1 Introduction 339
   11.2 General Properties 339
       11.2.1 Formulation of LP Problems 339
       11.2.2 Optimality Conditions 341
       11.2.3 Geometry of an LP Problem 346
       11.2.4 Vertex Minimizers 360
   11.3 Simplex Method 363
       11.3.1 Simplex Method for Alternative-Form LP
           Problem 363
       11.3.2 Simplex Method for Standard-Form LP
           Problems 374
       11.3.3 Tabular Form of the Simplex Method 383
       11.3.4 Computational Complexity 385
   Problems 387
   References 391

12 Linear Programming Part II: Interior-Point Methods 393
   12.1 Introduction 393
   12.2 Primal-Dual Solutions and Central Path 394
       12.2.1 Primal-Dual Solutions 394
       12.2.2 Central Path 396
   12.3 Primal Affine Scaling Method 398
   12.4 Primal Newton Barrier Method 402
       12.4.1 Basic Idea 402
       12.4.2 Minimizers of Subproblem 402
       12.4.3 A Convergence Issue 403
       12.4.4 Computing a Minimizer of the Problem
           in Eqs. (12.26a) and (12.26b) 404
   12.5 Primal-Dual Interior-Point Methods 407
       12.5.1 Primal-Dual Path-Following Method 407
       12.5.2 A Nonfeasible-Initialization Primal-Dual
           Path-Following Method 412
       12.5.3 Predictor-Corrector Method 415
   Problems 419
   References 424

**13 Quadratic, Semidefinite, and Second-Order Cone
Programming** 425
   13.1 Introduction 425
   13.2 Convex QP Problems with Equality Constraints 426

13.3   Active-Set Methods for Strictly Convex QP Problems   429
        13.3.1   Primal Active-Set Method   430
        13.3.2   Dual Active-Set Method   434
13.4   Interior-Point Methods for Convex QP Problems   435
        13.4.1   Dual QP Problem, Duality Gap, and Central Path . . .   435
        13.4.2   A Primal-Dual Path-Following Method
                  for Convex QP Problems   437
        13.4.3   Nonfeasible-initialization Primal-Dual
                  Path-Following Method for Convex QP Problems . . .   439
        13.4.4   Linear Complementarity Problems   442
13.5   Primal and Dual SDP Problems   445
        13.5.1   Notation and Definitions   445
        13.5.2   Examples   447
13.6   Basic Properties of SDP Problems   450
        13.6.1   Basic Assumptions   450
        13.6.2   Karush-Kuhn-Tucker Conditions   450
        13.6.3   Central Path   451
        13.6.4   Centering Condition   452
13.7   Primal-Dual Path-Following Method   453
        13.7.1   Reformulation of Centering Condition   453
        13.7.2   Symmetric Kronecker Product   454
        13.7.3   Reformulation of Eqs. (13.79a)–(13.79c)   455
        13.7.4   Primal-Dual Path-Following Algorithm   457
13.8   Predictor-Corrector Method   460
13.9   Second-Order Cone Programming   465
        13.9.1   Notation and Definitions   465
        13.9.2   Relations Among LP, QP, SDP, and SOCP
                  Problems   466
        13.9.3   Examples   468
13.10   A Primal-Dual Method for SOCP Problems   472
        13.10.1   Assumptions and KKT Conditions   472
        13.10.2   A Primal-Dual Interior-Point Algorithm   473
Problems   476
References   480

14   **Algorithms for General Convex Problems**   483
14.1   Introduction   483
14.2   Concepts and Properties of Convex Functions   483
        14.2.1   Subgradient   484
        14.2.2   Convex Functions with Lipschitz-Continuous
                  Gradients   488
        14.2.3   Strongly Convex Functions   491

	14.2.4	Conjugate Functions	494
	14.2.5	Proximal Operators	498
14.3	Extension of Newton Method to Convex Constrained and Unconstrained Problems		500
	14.3.1	Minimization of Smooth Convex Functions Without Constraints	500
	14.3.2	Minimization of Smooth Convex Functions Subject to Equality Constraints	502
	14.3.3	Newton Algorithm for Problem in Eq. (14.34) with a Nonfeasible x_0	504
	14.3.4	A Newton Barrier Method for General Convex Programming Problems	507
14.4	Minimization of Composite Convex Functions		512
	14.4.1	Proximal-Point Algorithm	512
	14.4.2	Fast Algorithm For Solving the Problem in Eq. (14.56)	514
14.5	Alternating Direction Methods		519
	14.5.1	Alternating Direction Method of Multipliers	519
	14.5.2	Application of ADMM to General Constrained Convex Problem	526
	14.5.3	Alternating Minimization Algorithm (AMA)	529
Problems			530
References			537

15 Algorithms for General Nonconvex Problems ... 539
15.1	Introduction		539
15.2	Sequential Convex Programming		540
	15.2.1	Principle of SCP	540
	15.2.2	Convex Approximations for $f(x)$ and $c_j(x)$ and Affine Approximation of $a_i(x)$	541
	15.2.3	Exact Penalty Formulation	544
	15.2.4	Alternating Convex Optimization	546
15.3	Sequential Quadratic Programming		550
	15.3.1	Basic SQP Algorithm	552
	15.3.2	Positive Definite Approximation of Hessian	554
	15.3.3	Robustness and Solvability of QP Subproblem of Eqs. (15.16a)–(15.16c)	555
	15.3.4	Practical SQP Algorithm for the Problem of Eq. (15.1)	556
15.4	Convex-Concave Procedure		557
	15.4.1	Basic Convex-Concave Procedure	558
	15.4.2	Penalty Convex-Concave Procedure	559
15.5	ADMM Heuristic Technique for Nonconvex Problems		562

Problems 565
References 568

16 Applications of Constrained Optimization. 571
    16.1 Introduction 571
    16.2 Design of Digital Filters 572
        16.2.1 Design of Linear-Phase FIR Filters Using QP 572
        16.2.2 Minimax Design of FIR Digital Filters
                Using SDP 574
        16.2.3 Minimax Design of IIR Digital Filters
                Using SDP 578
        16.2.4 Minimax Design of FIR and IIR Digital Filters
                Using SOCP 586
        16.2.5 Minimax Design of IIR Digital Filters Satisfying
                Multiple Specifications 587
    16.3 Model Predictive Control of Dynamic Systems 591
        16.3.1 Polytopic Model for Uncertain Dynamic Systems . . . 591
        16.3.2 Introduction to Robust MPC 592
        16.3.3 Robust Unconstrained MPC by Using SDP 594
        16.3.4 Robust Constrained MPC by Using SDP 597
    16.4 Optimal Force Distribution for Robotic Systems with Closed
          Kinematic Loops 602
        16.4.1 Force Distribution Problem in Multifinger Dextrous
                Hands 602
        16.4.2 Solution of Optimal Force Distribution Problem
                by Using LP 608
        16.4.3 Solution of Optimal Force Distribution Problem
                by Using SDP 611
    16.5 Multiuser Detection in Wireless Communication Channels . . . 614
        16.5.1 Channel Model and ML Multiuser Detector 615
        16.5.2 Near-Optimal Multiuser Detector Using SDP
                Relaxation 617
        16.5.3 A Constrained Minimum-BER Multiuser
                Detector 625
    Problems 631
    References 633

Appendix A: Basics of Linear Algebra 635

Appendix B: Basics of Digital Filters 673

Index 691

About the Authors

Andreas Antoniou received the B.Sc. and Ph.D. degrees in Electrical Engineering from the University of London, UK, in 1963 and 1966, respectively. He is a Life Member of the Association of Professional Engineers and Geoscientists of British Columbia, Canada, a Fellow of the Institution of Engineering and Technology, and a Life Fellow of the Institute of Electrical and Electronic Engineers. He served as the founding Chair of the Department of Electrical and Computer Engineering at the University of Victoria, BC, Canada and is now Professor Emeritus. He is the author of Digital Filters: Analysis, Design, and Signal Processing Applications published by McGraw-Hill in 2018 (previous editions of the book were published by McGraw-Hill under slightly different titles in 1979, 1993, and 2005). He served as Associate Editor/Editor of IEEE Transactions on Circuits and Systems from June 1983 to May 1987, as a Distinguished Lecturer of the IEEE Signal Processing Society in 2003, as General Chair of the 2004 International Symposium on Circuits and Systems, and as a Distinguished Lecturer of the IEEE Circuits and Systems Society during 2006–2007. He received the Ambrose Fleming Premium for 1964 from the IEE (best paper award), the CAS Golden Jubilee Medal from the IEEE Circuits and Systems Society in recognition of outstanding achievements in the area of circuits and systems, the BC Science Council Chairman's Award for Career Achievement both in 2000, the Doctor Honoris Causa degree by the Metsovio National Technical University, Athens, Greece, in 2002, the IEEE Circuits and Systems Society Technical Achievement Award for 2005, the IEEE Canada Outstanding Engineering Educator Silver Medal for 2008, the IEEE Circuits and Systems Society Education Award for 2009, the Craigdarroch Gold Medal for Career Achievement for 2011, and the Legacy Award for 2011 both from the University of Victoria.

Wu-Sheng Lu received the B.Sc. degree in Mathematics from Fudan University, Shanghai, China, in 1964, the M.S. degree in Electrical Engineering, and the Ph.D. degree in Control Science both from the University of Minnesota, Minneapolis, in 1983 and 1984, respectively. He is a Member of the Association of Professional Engineers and Geoscientists of British Columbia, Canada, a Fellow of the Engineering Institute of Canada, and a Fellow of the Institute of Electrical and Electronics Engineers. He has been teaching and carrying out research in the areas of digital signal processing and application of optimization methods at the

University of Victoria, BC, Canada, since 1987. He is the co-author with A. Antoniou of Two-Dimensional Digital Filters published by Marcel Dekker in 1992. He served as an Associate Editor of the Canadian Journal of Electrical and Computer Engineering in 1989, and Editor of the same journal from 1990 to 1992. He served as an Associate Editor for the IEEE Transactions on Circuits and Systems, Part II, from 1993 to 1995 and for Part I of the same journal from 1999 to 2001 and 2004 to 2005. He received two best paper awards from the IEEE Asia Pacific Conference on Circuits and Systems in 2006 and 2014, the Outstanding Teacher Award of the Engineering Institute of Canada, Vancouver Island Branch, for 1988 and 1990, and the University of Victoria Alumni Association Award for Excellence in Teaching for 1991.

Abbreviations

ADMM	Alternating direction methods of multipliers
AMA	Alternating minimization algorithms
AWGN	Additive white Gaussian noise
BER	Bit-error rate
BFGS	Broyden-Fletcher-Goldfarb-Shanno
CCP	Convex-concave procedure
CDMA	Code-division multiple access
CMBER	Constrained minimum BER
CP	Convex programming
DFP	Davidon-Fletcher-Powell
DH	Denavit-Hartenberg
DNB	Dual Newton barrier
DS-CDMA	Direct-sequence CDMA
FDMA	Frequency-division multiple access
FIR	Finite-duration impulse response
FISTA	Fast iterative shrinkage-thresholding algorithm
FR	Fletcher-Reeves
GCO	General constrained optimization
GN	Gauss-Newton
HOG	Histogram of oriented gradient
HWDR	Handwritten digits recognition
IIR	Infinite-duration impulse response
IP	Integer programming
KKT	Karush-Kuhn-Tucker
LMI	Linear matrix inequality
LP	Linear programming
LSQI	Least-squares minimization with quadratic inequality
LU	Lower-upper
MAI	Multiple access interference
ML	Maximum-likelihood
MLE	Maximum-likelihood estimation
MNIST	Modified National Institute for Standards and Technology
MPC	Model predictive control
NAG	Nesterov's accelerated gradient

PAS	Primal affine scaling
PCCP	Penalty convex-concave procedure
PCM	Predictor-corrector method
PNB	Primal Newton barrier
QP	Quadratic programming
SCP	Sequential convex programming
SD	Steepest-descent
SDP	Semidefinite programming
SDPR-D	SDP relaxation dual
SDPR-P	SDP relaxation primal
SNR	Signal-to-noise ratio
SOCP	Second-order cone programming
SQP	Sequential quadratic programming
SVD	Singular-value decomposition
TDMA	Time-division multiple access

The Optimization Problem

1

1.1 Introduction

Throughout the ages, man has continuously been involved with the process of optimization. In its earliest form, optimization consisted of unscientific rituals and prejudices like pouring libations and sacrificing animals to the gods, consulting the oracles, observing the positions of the stars, and watching the flight of birds. When the circumstances were appropriate, the timing was thought to be auspicious (or optimum) for planting the crops or embarking on a war.

As the ages advanced and the age of reason prevailed, unscientific rituals were replaced by rules of thumb and later, with the development of mathematics, mathematical calculations began to be applied.

Interest in the process of optimization has taken a giant leap with the advent of the digital computer in the early fifties. In recent years, optimization techniques advanced rapidly and considerable progress has been achieved. At the same time, digital computers became faster, more versatile, and more efficient. As a consequence, it is now possible to solve complex optimization problems which were thought intractable only a few years ago.

The process of optimization is the process of obtaining the '*best*', if it is possible to measure and change what is 'good' or 'bad'. In practice, one wishes the 'most' or 'maximum' (e.g., salary) or the 'least' or 'minimum' (e.g., expenses). Therefore, the word '*optimum*' is taken to mean '*maximum*' or '*minimum*' depending on the circumstances; 'optimum' is a technical term which implies quantitative measurement and is a stronger word than 'best' which is more appropriate for everyday use. Likewise, the word '*optimize*', which means to achieve an optimum, is a stronger word than 'improve'. Optimization theory is the branch of mathematics encompassing the quantitative study of optima and methods for finding them. Optimization practice, on the other hand, is the collection of techniques, methods, procedures, and algorithms that can be used to find the optima.

Optimization problems occur in most disciplines like engineering, physics, mathematics, economics, administration, commerce, social sciences, and even politics.

© Springer Science+Business Media, LLC, part of Springer Nature 2021
A. Antoniou and W.-S. Lu, *Practical Optimization*, Texts in Computer Science,
https://doi.org/10.1007/978-1-0716-0843-2_1

Optimization problems abound in the various fields of engineering like electrical, mechanical, civil, chemical, and building engineering. Typical areas of application are modeling, characterization, and design of devices, circuits, and systems; design of tools, instruments, and equipment; design of structures and buildings; process control; approximation theory, curve fitting, solution of systems of equations; forecasting, production scheduling, quality control; maintenance and repair; inventory control, accounting, budgeting, etc. Some recent innovations rely almost entirely on optimization theory, for example, neural networks and adaptive systems. Most real-life problems have several solutions and occasionally an infinite number of solutions may be possible. Assuming that the problem at hand admits more than one solution, optimization can be achieved by finding the best solution of the problem in terms of some performance criterion. If the problem admits only one solution, that is, only a unique set of parameter values is acceptable, then optimization cannot be applied.

Several general approaches to optimization are available, as follows:

1. Analytical methods
2. Graphical methods
3. Experimental methods
4. Numerical methods

Analytical methods are based on the classical techniques of differential calculus. In these methods the maximum or minimum of a performance criterion is determined by finding the values of parameters $x_1, x_2, \ldots, x_n$ that cause the derivatives of $f(x_1, x_2, \ldots, x_n)$ with respect to $x_1, x_2, \ldots, x_n$ to assume zero values. The problem to be solved must obviously be described in mathematical terms before the rules of calculus can be applied. The method need not entail the use of a digital computer. However, it cannot be applied to highly nonlinear problems or to problems where the number of independent parameters exceeds two or three.

A graphical method can be used to plot the function to be maximized or minimized if the number of variables does not exceed two. If the function depends on only one variable, say, x_1, a plot of $f(x_1)$ versus x_1 will immediately reveal the maxima and/or minima of the function. Similarly, if the function depends on only two variables, say, x_1 and x_2, a set of contours can be constructed. A *contour* is a set of points in the (x_1, x_2) plane for which $f(x_1, x_2)$ is constant, and so a contour plot, like a topographical map of a specific region, will reveal readily the peaks and valleys of the function. For example, the contour plot of $f(x_1, x_2)$ depicted in Fig. 1.1 shows that the function has a minimum at point A. Unfortunately, the graphical method is of limited usefulness since in most practical applications the function to be optimized depends on several variables, usually in excess of four.

The optimum performance of a system can sometimes be achieved by direct experimentation. In this method, the system is set up and the process variables are adjusted one by one and the performance criterion is measured in each case. This method may lead to optimum or near optimum operating conditions. However, it can lead to unreliable results since in certain systems, two or more variables interact with

Fig. 1.1 Contour plot of $f(x_1, x_2)$

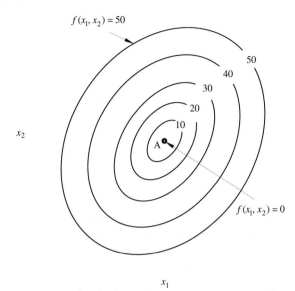

each other, and must be adjusted simultaneously to yield the optimum performance criterion.

The most important general approach to optimization is based on numerical methods. In this approach, iterative numerical procedures are used to generate a series of progressively improved solutions to the optimization problem, starting with an initial estimate for the solution. The process is terminated when some convergence criterion is satisfied. For example, when changes in the independent variables or the performance criterion from iteration to iteration become insignificant.

Numerical methods can be used to solve highly complex optimization problems of the type that cannot be solved analytically. Furthermore, they can be readily programmed on the digital computer. Consequently, they have all but replaced most other approaches to optimization.

The discipline encompassing the theory and practice of numerical optimization methods has come to be known as *mathematical programming* [1–5]. During the past 40 years, several branches of mathematical programming have evolved, as follows:

1. Linear programming
2. Integer programming
3. Quadratic programming
4. Nonlinear programming
5. Dynamic programming

Each one of these branches of mathematical programming is concerned with a specific class of optimization problems. The differences among them will be examined in Sect. 1.6.

1.2 The Basic Optimization Problem

Before optimization is attempted, the problem at hand must be properly formulated. A performance criterion F must be derived in terms of n parameters $x_1, x_2, \ldots, x_n$ as

$$F = f(x_1, x_2, \ldots, x_n)$$

F is a scalar quantity which can assume numerous forms. It can be the cost of a product in a manufacturing environment or the difference between the desired performance and the actual performance in a system. Variables $x_1, x_2, \ldots, x_n$ are the parameters that influence the product cost in the first case or the actual performance in the second case. They can be independent variables, like time, or control parameters that can be adjusted.

The most basic optimization problem is to adjust variables $x_1, x_2, \ldots, x_n$ in such a way as to minimize quantity F. This problem can be stated mathematically as

$$\text{minimize} \quad F = f(x_1, x_2, \ldots, x_n) \tag{1.1}$$

Quantity F is usually referred to as the *objective* or *cost function*.

The objective function may depend on a large number of variables, sometimes as many as 100 or more. To simplify the notation, matrix notation is usually employed. If $\mathbf{x}$ is a column vector with elements $x_1, x_2, \ldots, x_n$, the transpose of $\mathbf{x}$, namely, $\mathbf{x}^T$, can be expressed as the row vector

$$\mathbf{x}^T = [x_1 \; x_2 \; \cdots \; x_n]$$

In this notation, the basic optimization problem of Eq. (1.1) can be expressed as

$$\text{minimize} \quad F = f(\mathbf{x}) \qquad \text{for } \mathbf{x} \in E^n$$

where E^n represents the *n-dimensional Euclidean space*.

On many occasions, the optimization problem consists of finding the maximum of the objective function. Since

$$\max[f(\mathbf{x})] = -\min[-f(\mathbf{x})]$$

the maximum of F can be readily obtained by finding the minimum of the negative of F and then changing the sign of the minimum. Consequently, in this and subsequent chapters we focus our attention on minimization without loss of generality.

In many applications, a number of distinct functions of $\mathbf{x}$ need to be optimized simultaneously. For example, if the system of nonlinear simultaneous equations

$$f_i(\mathbf{x}) = 0 \qquad \text{for } i = 1, 2, \ldots, m$$

needs to be solved, a vector $\mathbf{x}$ is sought which will reduce all $f_i(\mathbf{x})$ to zero simultaneously. In such a problem, the functions to be optimized can be used to construct a vector

$$\mathbf{F}(\mathbf{x}) = [f_1(\mathbf{x}) \; f_2(\mathbf{x}) \; \cdots \; f_m(\mathbf{x})]^T$$

The problem can be solved by finding a point $\mathbf{x} = \mathbf{x}^*$ such that $\mathbf{F}(\mathbf{x}^*) = \mathbf{0}$. Very frequently, a point $\mathbf{x}^*$ that reduces all the $f_i(\mathbf{x})$ to zero simultaneously may not exist but an approximate solution, i.e., $\mathbf{F}(\mathbf{x}^*) \approx \mathbf{0}$, may be available which could be entirely satisfactory in practice.

A similar problem arises in scientific or engineering applications when the function of $\mathbf{x}$ that needs to be optimized is also a function of a continuous independent parameter (e.g., time, position, speed, frequency) that can assume an infinite set of values in a specified range. The optimization might entail adjusting variables $x_1, x_2, \ldots, x_n$ so as to optimize the function of interest over a given range of the independent parameter. In such an application, the function of interest can be sampled with respect to the independent parameter, and a vector of the form

$$\mathbf{F}(\mathbf{x}) = [f(\mathbf{x}, t_1) \ f(\mathbf{x}, t_2) \ \cdots \ f(\mathbf{x}, t_m)]^T$$

can be constructed, where t is the independent parameter. Now if we let

$$f_i(\mathbf{x}) \equiv f(\mathbf{x}, t_i)$$

we can write

$$\mathbf{F}(\mathbf{x}) = [f_1(\mathbf{x}) \ f_2(\mathbf{x}) \ \cdots \ f_m(\mathbf{x})]^T$$

A solution of such a problem can be obtained by optimizing functions $f_i(\mathbf{x})$ for $i = 1, 2, \ldots, m$ simultaneously. Such a solution would, of course, be approximate because any variations in $f(\mathbf{x}, t)$ between sample points are ignored. Nevertheless, reasonable solutions can be obtained in practice by using a sufficiently large number of sample points. This approach is illustrated by the following example.

Example 1.1 The step response $y(\mathbf{x}, t)$ of an nth-order control system is required to satisfy the specification

$$y_0(\mathbf{x}, t) = \begin{cases} t & \text{for } 0 \le t < 2 \\ 2 & \text{for } 2 \le t < 3 \\ -t + 5 & \text{for } 3 \le t < 4 \\ 1 & \text{for } 4 \le t \end{cases}$$

as closely as possible. Construct a vector $\mathbf{F}(\mathbf{x})$ that can be used to obtain a function $f(\mathbf{x}, t)$ such that

$$y(\mathbf{x}, t) \approx y_0(\mathbf{x}, t) \qquad \text{for } 0 \le t \le 5$$

Solution The difference between the actual and specified step responses, which constitutes the *approximation error*, can be expressed as

$$f(\mathbf{x}, t) = y(\mathbf{x}, t) - y_0(\mathbf{x}, t)$$

and if $f(\mathbf{x}, t)$ is sampled at $t = 0, 1, \ldots, 5$, we obtain

$$\mathbf{F}(\mathbf{x}) = [f_1(\mathbf{x}) \ f_2(\mathbf{x}) \ \cdots \ f_6(\mathbf{x})]^T$$

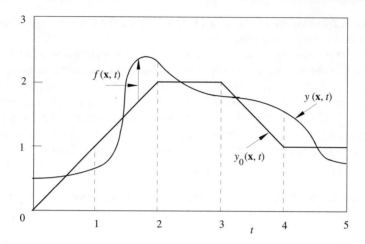

Fig. 1.2 Graphical construction for Example 1.1

where

$$f_1(\mathbf{x}) = f(\mathbf{x}, \ 0) = y(\mathbf{x}, \ 0)$$
$$f_2(\mathbf{x}) = f(\mathbf{x}, \ 1) = y(\mathbf{x}, \ 1) - 1$$
$$f_3(\mathbf{x}) = f(\mathbf{x}, \ 2) = y(\mathbf{x}, \ 2) - 2$$
$$f_4(\mathbf{x}) = f(\mathbf{x}, \ 3) = y(\mathbf{x}, \ 3) - 2$$
$$f_5(\mathbf{x}) = f(\mathbf{x}, \ 4) = y(\mathbf{x}, \ 4) - 1$$
$$f_6(\mathbf{x}) = f(\mathbf{x}, \ 5) = y(\mathbf{x}, \ 5) - 1$$

The problem is illustrated in Fig. 1.2. It can be solved by finding a point $\mathbf{x} = \mathbf{x}^*$ such that $\mathbf{F}(\mathbf{x}^*) \approx \mathbf{0}$. Evidently, the quality of the approximation obtained for the step response of the system will depend on the density of the sampling points and the higher the density of points, the better the approximation. ∎

Problems of the type just described can be solved by defining a suitable objective function in terms of the element functions of $\mathbf{F}(\mathbf{x})$. The objective function must be a *scalar quantity* and its optimization must lead to the simultaneous optimization of the element functions of $\mathbf{F}(\mathbf{x})$ in some sense. Consequently, a norm of some type must be used. An objective function can be defined in terms of the L_p *norm* as

$$F \equiv L_p = \left\{ \sum_{i=1}^{m} |f_i(\mathbf{x})|^p \right\}^{1/p}$$

where p is an integer.[1]

---

[1] See Sect. A.8 for more details on vector and matrix norms. Appendix A also deals with other aspects of linear algebra that are important to optimization.

Several special cases of the L_p norm are of particular interest. If $p = 1$

$$F \equiv L_1 = \sum_{i=1}^{m} |f_i(\mathbf{x})|$$

and, therefore, in a minimization problem like that in Example 1.1, the sum of the magnitudes of the individual element functions is minimized. This is called an L_1 *problem*.

If $p = 2$, the *Euclidean norm*

$$F \equiv L_2 = \left\{ \sum_{i=1}^{m} |f_i(\mathbf{x})|^2 \right\}^{1/2}$$

is minimized, and if the square root is omitted, the sum of the squares is minimized. Such a problem is commonly referred to as a *least-squares problem*.

In the case where $p = \infty$, if we assume that there is a unique maximum of $|f_i(\mathbf{x})|$ designated $\hat{F}$ such that

$$\hat{F} = \max_{1 \leq i \leq m} |f_i(\mathbf{x})|$$

then we can write

$$F \equiv L_\infty = \lim_{p \to \infty} \left\{ \sum_{i=1}^{m} |f_i(\mathbf{x})|^p \right\}^{1/p}$$

$$= \hat{F} \lim_{p \to \infty} \left\{ \sum_{i=1}^{m} \left[\frac{|f_i(\mathbf{x})|}{\hat{F}} \right]^p \right\}^{1/p}$$

Since all the terms in the summation except one are less than unity, they tend to zero when raised to a large positive power. Therefore, we obtain

$$F = \hat{F} = \max_{1 \leq i \leq m} |f_i(\mathbf{x})|$$

Evidently, if the L_∞ *norm* is used in Example 1.1, the maximum approximation error is minimized and the problem is said to be a *minimax problem*.

Often the individual element functions of $\mathbf{F}(\mathbf{x})$ are modified by using constants $w_1, w_2, \ldots, w_m$ as *weights*. For example, the least-squares objective function can be expressed as

$$F = \sum_{i=1}^{m} [w_i f_i(\mathbf{x})]^2$$

so as to emphasize important or critical element functions and de-emphasize unimportant or uncritical ones. If F is minimized, the residual errors in $w_i f_i(\mathbf{x})$ at the end of the minimization would tend to be of the same order of magnitude, i.e.,

$$\text{error in } |w_i f_i(\mathbf{x})| \approx \varepsilon$$

and so

$$\text{error in } |f_i(\mathbf{x})| \approx \frac{\varepsilon}{|w_i|}$$

Consequently, if a large positive weight w_i is used with $f_i(\mathbf{x})$, a small residual error is achieved in $|f_i(\mathbf{x})|$.

1.3   General Structure of Optimization Algorithms

Most of the available optimization algorithms entail a series of steps which are executed sequentially. A typical pattern is as follows:

Algorithm 1.1 General optimization algorithm
Step 1
(*a*) Set $k = 0$ and initialize $\mathbf{x}_0$.
(*b*) Compute $F_0 = f(\mathbf{x}_0)$.
Step 2
(*a*) Set $k = k + 1$.
(*b*) Compute the changes in $\mathbf{x}_k$ given by column vector $\mathbf{\Delta x}_k$ where

$$\mathbf{\Delta x}_k^T = [\Delta x_1 \; \Delta x_2 \; \cdots \; \Delta x_n]$$

by using an appropriate procedure.
(*c*) Set $\mathbf{x}_k = \mathbf{x}_{k-1} + \mathbf{\Delta x}_k$
(*d*) Compute $F_k = f(\mathbf{x}_k)$ and $\Delta F_k = F_{k-1} - F_k$.
Step 3
Check if convergence has been achieved by using an appropriate criterion, e.g., by checking ΔF_k and/or $\Delta \mathbf{x}_k$. If this is the case, continue to Step 4; otherwise, go to Step 2.
Step 4
(*a*) Output $\mathbf{x}^* = \mathbf{x}_k$ and $F^* = f(\mathbf{x}^*)$.
(*b*) Stop.

In Step 1, vector $\mathbf{x}_0$ is initialized by estimating the solution using knowledge about the problem at hand. Often the solution cannot be estimated and an arbitrary solution may be assumed, say, $\mathbf{x}_0 = \mathbf{0}$. Steps 2 and 3 are then executed repeatedly until convergence is achieved. Each execution of Steps 2 and 3 constitutes one iteration, that is, k is the number of iterations.

When convergence is achieved, Step 4 is executed. In this step, column vector

$$\mathbf{x}^* = [x_1^* \; x_2^* \; \cdots \; x_n^*]^T = \mathbf{x}_k$$

and the corresponding value of F, namely,

$$F^* = f(\mathbf{x}^*)$$

are output. The column vector $\mathbf{x}^*$ is said to be the *optimum, minimum, solution point*, or simply the *minimizer*, and F^* is said to be the optimum or minimum value of the objective function. The pair $\mathbf{x}^*$ and F^* constitute the solution of the optimization problem.

Convergence can be checked in several ways, depending on the optimization problem and the optimization technique used. For example, one might decide to stop the algorithm when the reduction in F_k between any two iterations has become insignificant, that is,

$$|\Delta F_k| = |F_{k-1} - F_k| < \varepsilon_F \tag{1.2}$$

where ε_F is an *optimization tolerance* for the objective function. Alternatively, one might decide to stop the algorithm when the changes in all variables have become insignificant, that is,

$$|\Delta x_i| < \varepsilon_x \quad \text{for } i = 1, 2, \ldots, n \tag{1.3}$$

where ε_x is an optimization tolerance for variables $x_1, x_2, \ldots, x_n$. A third possibility might be to check if both criteria given by Eqs. (1.2) and (1.3) are satisfied simultaneously.

There are numerous algorithms for the minimization of an objective function. However, we are primarily interested in algorithms that entail the minimum amount of effort. Therefore, we shall focus our attention on algorithms that are simple to apply, are reliable when applied to a diverse range of optimization problems, and entail a small amount of computation. A reliable algorithm is often referred to as a '*robust*' algorithm in the terminology of mathematical programming.

1.4 Constraints

In many optimization problems, the variables are interrelated by physical laws like the conservation of mass or energy, Kirchhoff's voltage and current laws, and other system equalities that must be satisfied. In effect, in these problems certain equality constraints of the form

$$a_i(\mathbf{x}) = 0 \quad \text{for } \mathbf{x} \in E^n$$

where $i = 1, 2, \ldots, p$ must be satisfied before the problem can be considered solved. In other optimization problems a collection of inequality constraints might be imposed on the variables or parameters to ensure physical realizability, reliability, compatibility, or even to simplify the modeling of the problem. For example, the power dissipation might become excessive if a particular current in a circuit exceeds a given upper limit or the circuit might become unreliable if another current is reduced below a lower limit, the mass of an element in a specific chemical reaction must be positive, and so on. In these problems, a collection of inequality constraints of the form

$$c_j(\mathbf{x}) \geq 0 \quad \text{for } \mathbf{x} \in E^n$$

where $j = 1, 2, \ldots, q$ must be satisfied before the optimization problem can be considered solved.

An optimization problem may entail a set of equality constraints and possibly a set of inequality constraints. If this is the case, the problem is said to be a *constrained optimization problem*. The most general constrained optimization problem can be expressed mathematically as

$$\text{minimize } f(\mathbf{x}) \quad \text{for } \mathbf{x} \in E^n \tag{1.4a}$$

$$\text{subject to: } a_i(\mathbf{x}) = 0 \quad \text{for } i = 1, 2, \ldots, p \tag{1.4b}$$

$$c_j(\mathbf{x}) \geq 0 \quad \text{for } j = 1, 2, \ldots, q \tag{1.4c}$$

Fig. 1.3 The double inverted
pendulum

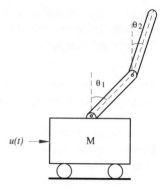

A problem that does not entail any equality or inequality constraints is said to be an
unconstrained optimization problem.

Constrained optimization is usually much more difficult than unconstrained optimization, as might be expected. Consequently, the general strategy that has evolved
in recent years towards the solution of constrained optimization problems is to reformulate constrained problems as unconstrained optimization problems. This can
be done by redefining the objective function such that the constraints are simultaneously satisfied when the objective function is minimized. Some real-life constrained
optimization problems are given as Examples 1.2 to 1.4 below.

Example 1.2 Consider a control system that comprises a double inverted pendulum
as depicted in Fig. 1.3. The objective of the system is to maintain the pendulum in the
upright position using the minimum amount of energy. This is achieved by applying
an appropriate control force to the car to damp out any displacements $\theta_1(t)$ and $\theta_2(t)$.
Formulate the problem as an optimization problem.

Solution The dynamic equations of the system are nonlinear and the standard practice
is to apply a linearization technique to these equations to obtain a small-signal linear
model of the system as [6]

$$\dot{\mathbf{x}}(t) = \mathbf{A}\mathbf{x}(t) + \mathbf{f}u(t) \tag{1.5}$$

where

$$\mathbf{x}(t) = \begin{bmatrix} \theta_1(t) \\ \dot{\theta}_1(t) \\ \theta_2(t) \\ \dot{\theta}_2(t) \end{bmatrix}, \quad \mathbf{A} = \begin{bmatrix} 0 & 1 & 0 & 0 \\ \alpha & 0 & -\beta & 0 \\ 0 & 0 & 0 & 1 \\ -\alpha & 0 & \alpha & 0 \end{bmatrix}, \quad \mathbf{f} = \begin{bmatrix} 0 \\ -1 \\ 0 \\ 0 \end{bmatrix}$$

with $\alpha > 0$, $\beta > 0$, and $\alpha \neq \beta$. In the above equations, $\dot{\mathbf{x}}(t)$, $\dot{\theta}_1(t)$, and $\dot{\theta}_2(t)$
represent the first derivatives of $\mathbf{x}(t)$, $\theta_1(t)$, and $\theta_2(t)$, respectively, with respect
to time, $\ddot{\theta}_1(t)$ and $\ddot{\theta}_2(t)$ would be the second derivatives of $\theta_1(t)$ and $\theta_2(t)$, and
parameters α and β depend on system parameters such as the length and weight of
each pendulum, the mass of the car, etc. Suppose that at instant $t = 0$ small nonzero

displacements $\theta_1(t)$ and $\theta_2(t)$ occur, which would call for immediate control action in order to steer the system back to the equilibrium state $\mathbf{x}(t) = \mathbf{0}$ at time $t = T_0$. In order to develop a digital controller, the system model in Eq. (1.5) is discretized to become

$$\mathbf{x}(k+1) = \Phi\mathbf{x}(k) + \mathbf{g}u(k) \tag{1.6}$$

where $\Phi = \mathbf{I} + \Delta t\mathbf{A}$, $\mathbf{g} = \Delta t\mathbf{f}$, Δt is the sampling interval, and $\mathbf{I}$ is the identity matrix. Let $\mathbf{x}(0) \neq \mathbf{0}$ be given and assume that T_0 is a multiple of Δt, i.e., $T_0 = K\Delta t$ where K is an integer. We seek to find a sequence of control actions $u(k)$ for $k = 0, 1, \ldots, K-1$ such that the zero equilibrium state is achieved at $t = T_0$, i.e., $\mathbf{x}(T_0) = \mathbf{0}$.

Let us assume that the energy consumed by these control actions, namely,

$$J = \sum_{k=0}^{K-1} u^2(k)$$

needs to be minimized. This optimal control problem can be formulated analytically as

$$\text{minimize } J = \sum_{k=0}^{K-1} u^2(k) \tag{1.7a}$$

$$\text{subject to: } \mathbf{x}(K) = \mathbf{0} \tag{1.7b}$$

From Eq. (1.6), we know that the state of the system at $t = K\Delta t$ is determined by the initial value of the state and system model in Eq. (1.6) as

$$\mathbf{x}(K) = \Phi^K\mathbf{x}(0) + \sum_{k=0}^{K-1} \Phi^{K-k-1}\mathbf{g}u(k)$$

$$\equiv -\mathbf{h} + \sum_{k=0}^{K-1} \mathbf{g}_k u(k)$$

where $\mathbf{h} = -\Phi^K\mathbf{x}(0)$ and $\mathbf{g}_k = \Phi^{K-k-1}\mathbf{g}$. Hence the constraint in Eq. (1.7b) is equivalent to

$$\sum_{k=0}^{K-1} \mathbf{g}_k u(k) = \mathbf{h} \tag{1.8}$$

If we define $\mathbf{u} = [u(0)\ u(1)\ \cdots\ u(K-1)]^T$ and $\mathbf{G} = [\mathbf{g}_0\ \mathbf{g}_1\ \cdots\ \mathbf{g}_{K-1}]$, then the constraint in Eq. (1.8) can be expressed as $\mathbf{G}\mathbf{u} = \mathbf{h}$, and the optimal control problem at hand can be formulated as the problem of finding a $\mathbf{u}$ that solves the minimization problem

$$\text{minimize } \mathbf{u}^T \mathbf{u} \tag{1.9a}$$

$$\text{subject to: } \mathbf{a}(\mathbf{u}) = \mathbf{0} \tag{1.9b}$$

where $\mathbf{a}(\mathbf{u}) = \mathbf{G}\mathbf{u} - \mathbf{h}$. In practice, the control actions cannot be made arbitrarily large in magnitude. Consequently, additional constraints are often imposed on $|u(i)|$, for instance,

$$|u(i)| \le m \quad \text{for } i = 0, 1, \ldots, K - 1$$

These constraints are equivalent to

$$m + u(i) \ge 0$$
$$m - u(i) \ge 0$$

Hence if we define

$$\mathbf{c}(\mathbf{u}) = \begin{bmatrix} m + u(0) \\ m - u(0) \\ \vdots \\ m + u(K-1) \\ m - u(K-1) \end{bmatrix}$$

then the magnitude constraints can be expressed as

$$\mathbf{c}(\mathbf{u}) \ge \mathbf{0} \tag{1.9c}$$

Obviously, the problem in Eqs. (1.9a)–(1.9c) fits nicely into the standard form of optimization problems given by Eqs. (1.4a)–(1.4c). ∎

Example 1.3 High performance in modern optical instruments depends on the quality of components like lenses, prisms, and mirrors. These components have reflecting or partially reflecting surfaces, and their performance is limited by the reflectivities of the materials of which they are made. The surface reflectivity can, however, be altered by the deposition of a thin transparent film. In fact, this technique facilitates the control of losses due to reflection in lenses and makes possible the construction of mirrors with unique properties [7,8].

As is depicted in Fig. 1.4, a typical N-layer thin-film system consists of N layers of thin films of certain transparent media deposited on a glass substrate. The thickness and refractive index of the ith layer are denoted as x_i and n_i, respectively. The refractive index of the medium above the first layer is denoted as n_0. If ϕ_0 is the angle of incident light, then the transmitted ray in the $(i - 1)$th layer is refracted at an angle ϕ_i which is given by Snell's law, namely,

$$n_i \sin \phi_i = n_0 \sin \phi_0$$

Given angle ϕ_0 and the wavelength of light, λ, the energy of the light reflected from the film surface and the energy of the light transmitted through the film surface are usually measured by the reflectance R and transmittance T, which satisfy the relation

$$R + T = 1$$

Fig. 1.4 An N-layer
thin-film system

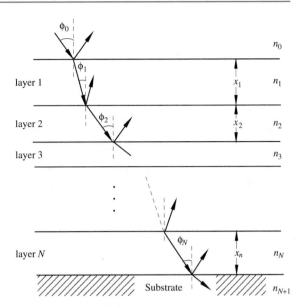

For an N-layer system, R is given by (see [9] for details)

$$R(x_1, \ldots, x_N, \lambda) = \left| \frac{\eta_0 - y}{\eta_0 + y} \right|^2$$

$$y = \frac{c}{b}$$

$$\begin{bmatrix} b \\ c \end{bmatrix} = \left\{ \prod_{k=1}^{N} \begin{bmatrix} \cos \delta_k & (j \sin \delta_k)/\eta_k \\ j\eta_k \sin \delta_k & \cos \delta_k \end{bmatrix} \right\} \begin{bmatrix} 1 \\ \eta_{N+1} \end{bmatrix}$$

where $j = \sqrt{-1}$ and

$$\delta_k = \frac{2\pi n_k x_k \cos \phi_k}{\lambda}$$

$$\eta_k = \begin{cases} n_k / \cos \phi_k & \text{for light polarized with the electric} \\ & \text{vector lying in the plane of incidence} \\ \\ n_k \cos \phi_k & \text{for light polarized with the electric} \\ & \text{vector perpendicular to the} \\ & \text{plane of incidence} \end{cases}$$

The design of a multilayer thin-film system can now be accomplished as follows: Given a range of wavelengths $\lambda_l \leq \lambda \leq \lambda_u$ and an angle of incidence ϕ_0, find $x_1, x_2, \ldots, x_N$ such that the reflectance $R(\mathbf{x}, \lambda)$ best approximates a desired reflectance $R_d(\lambda)$ for $\lambda \in [\lambda_l, \lambda_u]$. Formulate the design problem as an optimization problem.

Solution In practice, the desired reflectance is specified at grid points $\lambda_1, \lambda_2, \ldots, \lambda_K$ in the interval $[\lambda_l, \ \lambda_u]$; hence the design may be carried out by selecting x_i such that the objective function

$$J = \sum_{i=1}^{K} w_i [R(\mathbf{x}, \ \lambda_i) - R_d(\lambda_i)]^2 \tag{1.10}$$

is minimized, where

$$\mathbf{x} = [x_1 \ x_2 \ \cdots \ x_N]^T$$

and $w_i > 0$ is a weight to reflect the importance of term $[R(\mathbf{x}, \ \lambda_i) - R_d(\lambda_i)]^2$ in Eq. (1.10). If we let $\boldsymbol{\eta} = [1 \ \eta_{N+1}]^T$, $\mathbf{e}_+ = [\eta_0 \ 1]^T$, $\mathbf{e}_- = [\eta_0 \ -1]^T$, and

$$\mathbf{M}(\mathbf{x}, \ \lambda) = \prod_{k=1}^{N} \begin{bmatrix} \cos \delta_k & (j \sin \delta_k)/\eta_k \\ j\eta_k \sin \delta_k & \cos \delta_k \end{bmatrix}$$

then $R(\mathbf{x}, \ \lambda)$ can be expressed as

$$R(\mathbf{x}, \ \lambda) = \left| \frac{b\eta_0 - c}{b\eta_0 + c} \right|^2 = \left| \frac{\mathbf{e}_-^T \mathbf{M}(\mathbf{x}, \ \lambda)\boldsymbol{\eta}}{\mathbf{e}_+^T \mathbf{M}(\mathbf{x}, \ \lambda)\boldsymbol{\eta}} \right|^2$$

Finally, we note that the thickness of each layer cannot be made arbitrarily thin or arbitrarily large and, therefore, constraints must be imposed on the elements of $\mathbf{x}$ as

$$d_{il} \leq x_i \leq d_{iu} \quad \text{for } i = 1, \ 2, \ \ldots, \ N$$

The design problem can now be formulated as the constrained minimization problem

$$\text{minimize } J = \sum_{i=1}^{K} w_i \left| \frac{\mathbf{e}_-^T \mathbf{M}(\mathbf{x}, \ \lambda_i)\boldsymbol{\eta}}{\mathbf{e}_+^T \mathbf{M}(\mathbf{x}, \ \lambda_i)\boldsymbol{\eta}} \right|^2 - R_d(\lambda_i)^2$$

$$\text{subject to: } \quad x_i - d_{il} \geq 0 \quad \text{for } i = 1, \ 2, \ \ldots, \ N$$

$$d_{iu} - x_i \geq 0 \quad \text{for } i = 1, \ 2, \ \ldots, \ N$$

∎

Example 1.4 Quantities $q_1, q_2, \ldots, q_m$ of a certain product are produced by m manufacturing divisions of a company, which are at distinct locations. The product is to be shipped to n destinations that require quantities $b_1, b_2, \ldots, b_n$. Assume that the cost of shipping a unit from manufacturing division i to destination j is c_{ij} with $i = 1, 2, \ldots, m$ and $j = 1, 2, \ldots, n$. Find the quantity x_{ij} to be shipped from division i to destination j so as to minimize the total cost of transportation, i.e.,

$$\text{minimize } C = \sum_{i=1}^{m} \sum_{j=1}^{n} c_{ij} x_{ij}$$

This is known as the *transportation problem*. Formulate the problem as an optimization problem.

Solution Note that there are several constraints on variables x_{ij}. First, each division can provide only a fixed quantity of the product, hence

$$\sum_{j=1}^{n} x_{ij} = q_i \qquad \text{for } i = 1, 2, \ldots, m$$

Second, the quantity to be shipped to a specific destination has to meet the need of that destination and so

$$\sum_{i=1}^{m} x_{ij} = b_j \qquad \text{for } j = 1, 2, \ldots, n$$

In addition, the variables x_{ij} are nonnegative and thus, we have

$$x_{ij} \geq 0 \qquad \text{for } i = 1, 2, \ldots, m \quad \text{and} \quad j = 1, 2, \ldots, n$$

If we let

$$\mathbf{c} = [c_{11} \; \cdots \; c_{1n} \; c_{21} \; \cdots \; c_{2n} \; \cdots \; c_{m1} \; \cdots \; c_{mn}]^T$$

$$\mathbf{x} = [x_{11} \; \cdots \; x_{1n} \; x_{21} \; \cdots \; x_{2n} \; \cdots \; x_{m1} \; \cdots \; x_{mn}]^T$$

$$\mathbf{A} = \begin{bmatrix} 1 \; 1 & \cdots \; 1 & 0 \; 0 & \cdots \; 0 & \cdots \cdots \cdots \cdots \\ 0 \; 0 & \cdots \; 0 & 1 \; 1 & \cdots \; 1 & \cdots \cdots \cdots \cdots \\ \cdots \cdots \cdots \cdots \cdots \cdots \cdots \cdots \cdots \\ 0 \; 0 & \cdots \; 0 & 0 \; 0 & \cdots \; 0 & \cdots \; 1 \; 1 & \cdots \; 1 \\ 1 \; 0 & \cdots \; 0 & 1 \; 0 & \cdots \; 0 & \cdots \; 1 \; 0 & \cdots \; 0 \\ 0 \; 1 & \cdots \; 0 & 0 \; 1 & \cdots \; 0 & \cdots \; 0 \; 1 & \cdots \; 0 \\ \cdots \cdots \cdots \cdots \cdots \cdots \cdots \cdots \cdots \\ 0 \; 0 & \cdots \; 1 & 0 \; 0 & \cdots \; 1 & \cdots \; 0 \; 0 & \cdots \; 1 \end{bmatrix}$$

$$\mathbf{b} = [q_1 \; \cdots \; q_m \; b_1 \; \cdots \; b_n]^T$$

then the minimization problem can be stated as

$$\text{minimize } C = \mathbf{c}^T \mathbf{x} \tag{1.11a}$$

$$\text{subject to: } \mathbf{A}\mathbf{x} = \mathbf{b} \tag{1.11b}$$

$$\mathbf{x} \geq \mathbf{0} \tag{1.11c}$$

where $\mathbf{c}^T \mathbf{x}$ is the inner product of $\mathbf{c}$ and $\mathbf{x}$. The above problem, like those in Examples 1.2 and 1.3, fits into the standard optimization problem in Eqs. (1.4a)–(1.4c). Since both the objective function in Eq. (1.11a) and the constraints in Eqs. (1.11b) and (1.11c) are linear, the problem is known as a *linear programming (LP) problem* (see Sect. 1.6.1). ∎

Example 1.5 An investment portfolio is a collection of a variety of securities owned by an individual whereby each security is associated with a unique possible return and a corresponding risk. Design an optimal investment portfolio that would minimize the risk involved subject to an acceptable return for a portfolio that comprises n securities.

Solution Some background theory on portfolio selection can be found in [10]. Assume that the amount of resources to be invested is normalized to unity (e.g., 1 million dollars) and let x_i represent the return of security i at some specified time in the future, say, in one month's time. The return x_i can be assumed to be a random variable and hence three quantities pertaining to security i can be evaluated, namely, the expected return

$$\mu_i = E[x_i]$$

the variance of the return

$$\sigma_i^2 = E[(x_i - \mu_i)^2]$$

and the correlation between the returns of the ith and jth securities

$$\rho_{i,j} = \frac{E[(x_i - \mu_i)(x_j - \mu_j)]}{\sigma_i \sigma_j} \quad \text{for } i, j = 1, 2, \ldots, n$$

With these quantities known, constructing a portfolio amounts to allocating a fraction w_i of the available resources to security i, for $i = 1, 2, \ldots, n$. This leads to the constraints

$$0 \le w_i \le 1 \quad \text{for } i = 1, 2, \ldots, n \quad \text{and} \quad \sum_{i=1}^{n} w_i = 1$$

Given a set of investment allocations $\{w_i, \ i = 1, 2, \ldots, n\}$, the expected return of the portfolio can be deduced as

$$E\left[\sum_{i=1}^{n} w_i x_i\right] = \sum_{i=1}^{n} w_i \mu_i$$

The variance for the portfolio, which measures the risk of the investment, can be evaluated as

$$E\left[\sum_{i=1}^{n} w_i x_i - E\left(\sum_{i=1}^{n} w_i x_i\right)\right]^2 = E\left\{\left[\sum_{i=1}^{n} w_i (x_i - \mu_i)\right]\left[\sum_{j=1}^{n} w_j (x_j - \mu_j)\right]\right\}$$

$$= E\left[\sum_{i=1}^{n} \sum_{j=1}^{n} (x_i - \mu_i)(x_j - \mu_j) w_i w_j\right]$$

$$= \sum_{i=1}^{n} \sum_{j=1}^{n} E[(x_i - \mu_i)(x_j - \mu_j)] w_i w_j$$

$$= \sum_{i=1}^{n} \sum_{j=1}^{n} (\sigma_i \sigma_j \rho_{ij}) w_i w_j$$

$$= \sum_{i=1}^{n} \sum_{j=1}^{n} q_{ij} w_i w_j$$

where

$$q_{ij} = \sigma_i \sigma_j \rho_{ij}$$

The portfolio can be optimized in several ways. One possibility would be to minimize the investment risk subject to an acceptable expected return μ_*, namely,

$$\underset{w_i,\, 1 \le i \le n}{\text{minimize}} \quad f(\mathbf{w}) = \sum_{i=1}^{n}\sum_{j=1}^{n} q_{ij} w_i w_j \tag{1.12a}$$

$$\text{subject to:} \quad \sum_{i=1}^{n} \mu_i w_i \ge \mu_* \tag{1.12b}$$

$$w_i \ge 0 \quad \text{for } 1 \le i \le n \tag{1.12c}$$

$$\sum_{i=1}^{n} w_i = 1 \tag{1.12d}$$

Another possibility would be to maximize the expected return subject to an acceptable risk σ_*^2, namely

$$\underset{w_i,\, 1 \le i \le n}{\text{maximize}} \quad F(\mathbf{w}) = \sum_{i=1}^{n} \mu_i w_i$$

$$\text{subject to:} \quad \sum_{i=1}^{n}\sum_{j=1}^{n} q_{ij} w_i w_j \le \sigma_*^2$$

$$w_i \ge 0 \quad \text{for } 1 \le i \le n$$

$$\sum_{i=1}^{n} w_i = 1$$

∎

1.5 The Feasible Region

Any point $\mathbf{x}$ that satisfies both the equality as well as the inequality constraints is said to be a *feasible point* of the optimization problem. The set of all points that satisfy the constraints constitutes the *feasible domain region* of $f(\mathbf{x})$. Evidently, the constraints define a subset of E^n. Therefore, the feasible region can be defined as a set[2]

$$\mathcal{R} = \{\mathbf{x} : a_i(\mathbf{x}) = 0 \text{ for } i = 1,\ 2,\ \ldots,\ p \text{ and } c_j(\mathbf{x}) \ge 0 \text{ for } j = 1,\ 2,\ \ldots,\ q\}$$

where $\mathcal{R} \subset E^n$.

The optimum point $\mathbf{x}^*$ must be located in the feasible region, and so the general constrained optimization problem can be stated as

$$\text{minimize } f(\mathbf{x}) \quad \text{for } \mathbf{x} \in \mathcal{R}$$

Any point $\mathbf{x}$ not in $\mathcal{R}$ is said to be a *nonfeasible point*.

If the constraints in an optimization problem are all inequalities, the constraints divide the points in the E^n space into three types of points, as follows:

----

[2]The above notation for a set will be used consistently throughout the book.

1. Interior points
2. Boundary points
3. Exterior points

An *interior point* is a point for which $c_j(\mathbf{x}) > 0$ for all j. A *boundary point* is a point for which at least one $c_j(\mathbf{x}) = 0$, and an *exterior point* is a point for which at least one $c_j(\mathbf{x}) < 0$. Interior points are feasible points, boundary points may or may not be feasible points, whereas exterior points are nonfeasible points.

If a constraint $c_m(\mathbf{x})$ is zero during a specific iteration, the constraint is said to be *active*, and if $c_m(\mathbf{x}^*)$ is zero when convergence is achieved, the optimum point $\mathbf{x}^*$ is located on the boundary. In such a case, the optimum point is said to be constrained. If the constraints are all equalities, the feasible points must be located on the intersection of all the hypersurfaces corresponding to $a_i(\mathbf{x}) = 0$ for $i = 1, 2, \ldots, p$. The above definitions and concepts are illustrated by the following two examples.

Example 1.6 By using a graphical method, solve the following optimization problem

$$\text{minimize } f(\mathbf{x}) = x_1^2 + x_2^2 - 4x_1 + 4$$
$$\text{subject to: } c_1(\mathbf{x}) = x_1 - 2x_2 + 6 \geq 0$$
$$c_2(\mathbf{x}) = -x_1^2 + x_2 - 1 \geq 0$$
$$c_3(\mathbf{x}) = x_1 \geq 0$$
$$c_4(\mathbf{x}) = x_2 \geq 0$$

Solution The equation defining the objective function can be expressed as

$$(x_1 - 2)^2 + x_2^2 = f(\mathbf{x})$$

Hence the contours of $f(\mathbf{x})$ in the (x_1, x_2) plane are concentric circles with radius $\sqrt{f(\mathbf{x})}$ centered at $x_1 = 2$, $x_2 = 0$. Constraints $c_1(\mathbf{x})$ and $c_2(\mathbf{x})$ dictate that

$$x_2 \leq \tfrac{1}{2}x_1 + 3$$

and

$$x_2 \geq x_1^2 + 1$$

respectively, while constraints $c_3(\mathbf{x})$ and $c_4(\mathbf{x})$ dictate that x_1 and x_2 be positive. The contours of $f(\mathbf{x})$ and the boundaries of the constraints can be constructed as shown in Fig. 1.5.

The feasible region for this problem is the shaded region in Fig. 1.5. The solution is located at point A on the boundary of constraint $c_2(\mathbf{x})$. In effect, the solution is a constrained optimum point. Consequently, if this problem is solved by means of mathematical programming, constraint $c_2(\mathbf{x})$ will be active when the solution is reached.

In the absence of constraints, the minimization of $f(\mathbf{x})$ would yield point B as the solution.   ∎

Fig. 1.5 Graphical
construction for Example 1.6

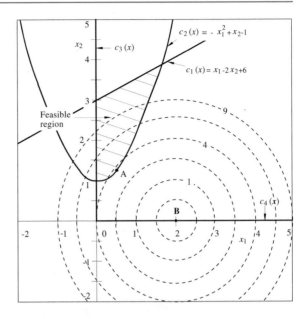

Example 1.7 By using a graphical method, solve the optimization problem

$$\text{minimize } f(\mathbf{x}) = x_1^2 + x_2^2 + 2x_2$$
$$\text{subject to: } a_1(\mathbf{x}) = x_1^2 + x_2^2 - 1 = 0$$
$$c_1(\mathbf{x}) = x_1 + x_2 - 0.5 \geq 0$$
$$c_2(\mathbf{x}) = x_1 \geq 0$$
$$c_3(\mathbf{x}) = x_2 \geq 0$$

Solution The equation defining the objective function can be expressed as

$$x_1^2 + (x_2 + 1)^2 = f(\mathbf{x}) + 1$$

Hence the contours of $f(\mathbf{x})$ in the (x_1, x_2) plane are concentric circles with radius $\sqrt{f(\mathbf{x}) + 1}$, centered at $x_1 = 0$, $x_2 = -1$. Constraint $a_1(\mathbf{x})$ is a circle centered at the origin with radius 1. On the other hand, constraint $c_1(\mathbf{x})$ is a straight line since it is required that

$$x_2 \geq -x_1 + 0.5$$

The last two constraints dictate that x_1 and x_2 be nonnegative. Hence the required construction can be obtained as depicted in Fig. 1.6.

In this case, the feasible region is the arc of circle $a_1(\mathbf{x}) = 0$ located in the first quadrant of the (x_1, x_2) plane. The solution, which is again a constrained optimum point, is located at point A. There are two active constraints in this example, namely, $a_1(\mathbf{x})$ and $c_3(\mathbf{x})$.

In the absence of constraints, the solution would be point B in Fig. 1.6. ∎

Fig. 1.6 Graphical
construction for Example 1.7

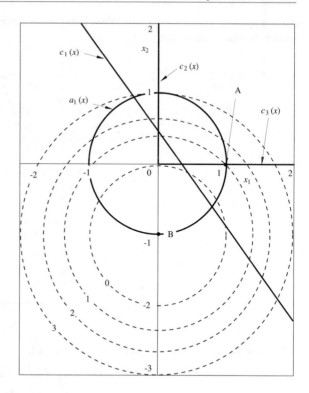

In the above examples, the set of points comprising the feasible region are simply connected as depicted in Fig. 1.7a. Sometimes the feasible region may consist of two or more disjoint sub-regions, as depicted in Fig. 1.7b. If this is the case, the following difficulty may arise. A typical optimization algorithm is an iterative numerical procedure that will generate a series of progressively improved solutions, starting with an initial estimate for the solution. Therefore, if the feasible region consists of two sub-regions, say, A and B, an initial estimate for the solution in sub-region A is likely to yield a solution in sub-region A, and a better solution in sub-region B may be missed. Fortunately, however, in most real-life optimization problems, this difficulty can be avoided by formulating the problem carefully.

1.6   Branches of Mathematical Programming

Several branches of mathematical programming were enumerated in Sect. 1.1, namely, linear, integer, quadratic, nonlinear, and dynamic programming. Each one of these branches of mathematical programming consists of the theory and application of a collection of optimization techniques that are suited to a specific class of optimization problems. The differences among the various branches of mathematical

Fig. 1.7 Examples of simply connected and disjoint feasible regions

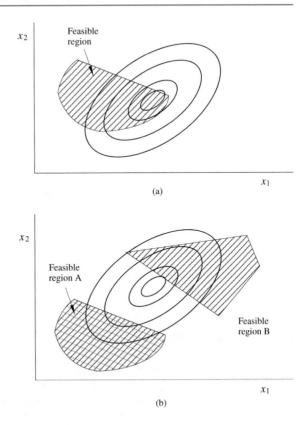

programming are closely linked to the structure of the optimization problem and to the mathematical nature of the objective and constraint functions. A brief description of each branch of mathematical programming is as follows.

1.6.1 Linear Programming

If the objective and constraint functions are linear and the variables are constrained to be positive, as in Example 1.4, the general optimization problem assumes the form

$$\text{minimize } f(\mathbf{x}) = \sum_{i=1}^{n} \alpha_i x_i$$

$$\text{subject to: } a_j(\mathbf{x}) = \sum_{i=1}^{n} \beta_{ij} x_i - \mu_j = 0 \quad \text{for } j = 1, 2, \ldots, p$$

$$c_j(\mathbf{x}) = \sum_{i=1}^{n} \gamma_{ij} x_i - \nu_j \geq 0 \quad \text{for } j = 1, 2, \ldots, q$$

$$x_i \geq 0 \quad \text{for } i = 1, 2, \ldots, n$$

where α_i, β_{ij}, γ_{ij}, μ_j, and ν_j are constants. For example,

$$\text{minimize} \ \ f(\mathbf{x}) = -2x_1 + 4x_2 + 7x_3 + x_4 + 5x_5$$

$$\text{subject to:} \ \ a_1(\mathbf{x}) = -x_1 + x_2 + 2x_3 + x_4 + 2x_5 - 7 = 0$$
$$a_2(\mathbf{x}) = -x_1 + 2x_2 + 3x_3 + x_4 + x_5 - 6 = 0$$
$$a_3(\mathbf{x}) = -x_1 + x_2 + x_3 + 2x_4 + x_5 - 4 = 0$$
$$x_i \geq 0 \quad \text{for} \ \ i = 1, \ 2, \ \dots, \ 5$$

or

$$\text{minimize} \ \ f(\mathbf{x}) = 3x_1 + 4x_2 + 5x_3$$

$$\text{subject to:} \ \ c_1(\mathbf{x}) = x_1 + 2x_2 + 3x_3 - 5 \geq 0$$
$$c_2(\mathbf{x}) = 2x_1 + 2x_2 + x_3 - 6 \geq 0$$
$$x_1 \geq 0, \ x_2 \geq 0, \ x_3 \geq 0$$

Optimization problems like the above occur in many disciplines. Their solution can be readily achieved by using some powerful LP algorithms as will be shown in Chaps. 11 and 12.

1.6.2   Integer Programming

In certain linear programming problems, at least some of the variables are required to assume only integer values. This restriction renders the programming problem nonlinear. Nevertheless, the problem is referred to as linear since the objective and constraint functions are linear [11].

1.6.3   Quadratic Programming

If the optimization problem assumes the form

$$\text{minimize} \ \ f(\mathbf{x}) = \alpha_0 + \boldsymbol{\gamma}^T \mathbf{x} + \mathbf{x}^T \mathbf{Q} \, \mathbf{x}$$

$$\text{subject to:} \ \ \boldsymbol{\alpha}^T \mathbf{x} \geq \boldsymbol{\beta}$$

where

$$\boldsymbol{\alpha} = \begin{bmatrix} \alpha_{11} & \alpha_{22} & \dots & \alpha_{1q} \\ \alpha_{21} & \alpha_{22} & \dots & \alpha_{2q} \\ \vdots & \vdots & & \vdots \\ \alpha_{n1} & \alpha_{n2} & \dots & \alpha_{nq} \end{bmatrix}$$

$$\boldsymbol{\beta}^T = \begin{bmatrix} \beta_1 & \beta_2 & \cdots & \beta_q \end{bmatrix}$$
$$\boldsymbol{\gamma}^T = \begin{bmatrix} \gamma_1 & \gamma_2 & \dots & \gamma_n \end{bmatrix}$$

and $\mathbf{Q}$ is a positive definite or semidefinite symmetric square matrix, then the constraints are linear and the objective function is quadratic. Such an optimization problem is said to be a quadratic programming (QP) problem (see Chap. 10 of [5]). A typical example of this type of problem is as follows:

$$\text{minimize } f(\mathbf{x}) = \tfrac{1}{2}x_1^2 + \tfrac{1}{2}x_2^2 - x_1 - 2x_2$$

$$\text{subject to: } c_1(\mathbf{x}) = 6 - 2x_1 - 3x_2 \geq 0$$

$$c_2(\mathbf{x}) = 5 - x_1 - 4x_2 \geq 0$$

$$c_3(\mathbf{x}) = x_1 \geq 0$$

$$c_4(\mathbf{x}) = x_2 \geq 0$$

1.6.4   Nonlinear Programming

In nonlinear programming problems, the objective function and usually the constraint functions are nonlinear. Typical examples were given earlier as Examples 1.1 to 1.3. This is the most general branch of mathematical programming and, in effect, LP and QP can be considered as special cases of nonlinear programming. Although it is possible to solve linear or quadratic programming problems by using nonlinear programming algorithms, the specialized algorithms developed for linear or quadratic programming should be used for these problems since they are usually much more efficient.

The choice of optimization algorithm depends on the mathematical behavior and structure of the objective function. Most of the time, the objective function is a well-behaved nonlinear function and all that is necessary is a general-purpose, robust, and efficient algorithm. For certain applications, however, specialized algorithms exist which are often more efficient than general-purpose ones. These are often referred to by the type of norm minimized, for example, an algorithm that minimizes an L_1, L_2, or L_∞ norm is said to be an L_1, L_2, or *minimax algorithm*.

1.6.5   Dynamic Programming

In many applications, a series of decisions must be made in sequence, where subsequent decisions are influenced by earlier ones. In such applications, a number of optimizations have to be performed in sequence and a general strategy may be required to achieve an overall optimum solution. For example, a large system which cannot be optimized owing to the size and complexity of the problem can be partitioned into a set of smaller sub-systems that can be optimized individually. Often individual sub-systems interact with each other and, consequently, a general solution strategy is required if an overall optimum solution is to be achieved. Dynamic programming is a collection of techniques that can be used to develop general solution strategies for problems of the type just described. It is usually based on the use of linear, integer, quadratic, or nonlinear optimization algorithms.

Problems

1.1 (a) Solve the following minimization problem by using a graphical method:

$$\text{minimize } f(\mathbf{x}) = x_1^2 + x_2 + 4$$
$$\text{subject to: } c_1(\mathbf{x}) = -x_1^2 - (x_2 + 4)^2 + 16 \geq 0$$
$$c_2(\mathbf{x}) = x_1 - x_2 - 6 \geq 0$$

Note: An explicit numerical solution is required.
(b) Indicate the feasible region.
(c) Is the optimum point constrained?

1.2 Repeat Prob. 1(a) to (c) for the problem

$$\text{minimize } f(\mathbf{x}) = x_2 - \frac{8}{x_1}$$
$$\text{subject to: } c_1(\mathbf{x}) = \tfrac{1}{5}x_1 - x_2 \geq 0$$
$$c_2(\mathbf{x}) = 16 - (x_1 - 5)^2 - x_2^2 \geq 0$$

Note: Obtain an accurate solution by using MATLAB.

1.3 Repeat Prob. 1(a) to (c) for the problem

$$\text{minimize } f(\mathbf{x}) = (x_1 - 12)x_1 + (x_2 - 6)x_2 + 45$$
$$\text{subject to: } c_1(\mathbf{x}) = \tfrac{7}{5}x_1 - x_2 - \tfrac{7}{5} \geq 0$$
$$c_2(\mathbf{x}) = x_2 + \tfrac{7}{5}x_1 - \tfrac{77}{5} \geq 0$$
$$c_3(\mathbf{x}) = x_2 \geq 0$$

1.4 Repeat Prob. 1(a) to (c) for the problem

$$\text{minimize } f(\mathbf{x}) = \tfrac{1}{4}(x_1 - 6)^2 + (x_2 - 4)^2$$
$$\text{subject to: } a_1(\mathbf{x}) = x_1 - 3 = 0$$
$$c_1(\mathbf{x}) = \tfrac{80}{7} - x_2 - \tfrac{8}{7}x_1 \geq 0$$
$$c_2(\mathbf{x}) = x_2 \geq 0$$

1.5 Develop a method to determine the coordinates of point A in Example 1.6 based on the following observation: From Fig. 1.5, we see that there will be no intersection points between the contour of $f(\mathbf{x}) = r^2$ and constraint $c_2(\mathbf{x}) = 0$ if radius r is smaller than the distance A to B and there will be two distinct intersection points between them if r is larger than the distance A to B. Therefore, the solution point A can be identified by determining the value of r for which the distance between the two intersection points is sufficiently small.

1.6 Solve the constrained minimization problem

$$\text{minimize} \ \ f(\mathbf{x}) = 3x_1 + 2x_2 + x_3$$
$$\text{subject to:} \ \ a_1(\mathbf{x}) = 2x_1 + 3x_2 + x_3 - 30 = 0$$
$$c_1(\mathbf{x}) = x_1 \geq 0$$
$$c_2(\mathbf{x}) = x_2 \geq 0$$
$$c_3(\mathbf{x}) = x_3 \geq 0$$

Hint: (i) Use the equality constraint to eliminate variable x_3, and (ii) use $x = \hat{x}^2$ to eliminate constraint $x \geq 0$.

1.7 Consider the constrained minimization problem

$$\text{minimize} \ \ f(\mathbf{x}) = -5\sin(x_1 + x_2) + (x_1 - x_2)^2 - x_1 - 2x_2$$
$$\text{subject to:} \ \ c_1(\mathbf{x}) = 5 - x_1 \geq 0$$
$$c_2(\mathbf{x}) = 5 - x_2 \geq 0$$

(a) Plot a dense family of contours for $f(\mathbf{x})$ over the region $D = \{(x_1, x_2) : -5 < x_1 < 5, \ -5 < x_2 < 5\}$ to identify all local minimizers and local maximizers of $f(\mathbf{x})$ in D.
(b) Convert the problem in part (a) into an unconstrained minimization problem by eliminating the inequality constraints. Hint: A constraint $x \leq a$ can be eliminated by using the variable substitution $x = a - \hat{x}^2$.

1.8 Consider the portfolio optimization problem in Eqs. (1.12a)–(1.12d) with $n = 2$, $\mu_1 = 5$, $\mu_2 = 12$, $\sigma_1^2 = 9$, $\sigma_2^2 = 196$, $\rho_{12} = \rho_{21} = -0.5$, and $\mu_* = 7$.

(a) Specify the problem in Eqs. (1.12a)–(1.12d) with the data given above.
(b) Plot the feasible region of the problem.
(c) Solve the problem without using software.
(d) Evaluate and compare the expected returns and associated risks for three instances: (i) Invest in security 1 only; (ii) invest in security 2 only; and (iii) invest in both securities 1 and 2 using the solution obtained in part (c).
(e) Repeat parts (a)–(c) for the same data set but with μ_* changed to $\mu_* = 9$. Compare the solution obtained with that obtained in part (c) in terms of expected return and associated risk.

References

1. G. B. Dantzig, *Linear Programming and Extensions*. Princeton, NJ: Princeton University Press, 1963.
2. D. M. Himmelblau, *Applied Nonlinear Programming*. New York: McGraw-Hill, 1972.

3. P. E. Gill, W. Murray, and M. H. Wright, *Practical Optimization*. New York: Academic Press, 1981.
4. D. G. Luenberger and Y. Ye, *Linear and Nonlinear Programming*, 4th ed. New York: Springer, 2008.
5. R. Fletcher, *Practical Methods of Optimization*, 2nd ed. New York: Wiley, 1987.
6. B. C. Kuo, *Automatic Control Systems*, 5th ed. Englewood Cliffs, NJ: Prentice-Hall, 1987.
7. K. D. Leaver and B. N. Chapman, *Thin Films*. London: Wykeham, 1971.
8. O. S. Heavens, *Thin Film Physics*. London: Methuen, 1970.
9. Z. Knittl, *Optics of Thin Films, An Optical Multilayer Theory*. New York: Wiley, 1976.
10. H. M. Markowitz, "Portfolio selection," *The Journal of Finance*, vol. 7, no. 1, pp. 77–91, 1952.
11. G. L. Nemhauser and L. A. Wolsey, *Integer and Combinatorial Optimization*. New York: Wiley, 1988.

Basic Principles

2

2.1 Introduction

Nonlinear programming is based on a collection of definitions, theorems, and principles that must be clearly understood if the available nonlinear programming methods are to be used effectively. This chapter begins with the definition of the gradient vector, the Hessian matrix, and the various types of extrema (maxima and minima). The conditions that must hold at the solution point are then discussed and techniques for the characterization of the extrema are described. Subsequently, the classes of convex and concave functions are introduced. These provide a natural formulation for the theory of global convergence.

Throughout the chapter, we focus our attention on the nonlinear optimization problem

$$\text{minimize} \ \ f = f(\mathbf{x})$$
$$\text{subject to:} \ \ \mathbf{x} \in \mathcal{R}$$

where $f(\mathbf{x})$ is a real-valued function and $\mathcal{R} \subset E^n$ is the feasible region.

2.2 Gradient Information

In many optimization methods, gradient information pertaining to the objective function is required. This information consists of the first and second derivatives of $f(\mathbf{x})$ with respect to the n variables.

If $f(\mathbf{x}) \in C^1$, that is, if $f(\mathbf{x})$ has continuous first-order partial derivatives, the *gradient* of $f(\mathbf{x})$ is defined as

© Springer Science+Business Media, LLC, part of Springer Nature 2021
A. Antoniou and W.-S. Lu, *Practical Optimization*, Texts in Computer Science,
https://doi.org/10.1007/978-1-0716-0843-2_2

$$\mathbf{g}(\mathbf{x}) = \left[\frac{\partial f}{\partial x_1} \; \frac{\partial f}{\partial x_2} \; \cdots \; \frac{\partial f}{\partial x_n} \right]^T$$
$$= \nabla f(\mathbf{x}) \tag{2.1}$$

where

$$\nabla = \left[\frac{\partial}{\partial x_1} \; \frac{\partial}{\partial x_2} \; \cdots \; \frac{\partial}{\partial x_n} \right]^T \tag{2.2}$$

If $f(\mathbf{x}) \in C^2$, that is, if $f(\mathbf{x})$ has continuous second-order partial derivatives, the *Hessian*[1] of $f(\mathbf{x})$ is defined as

$$\mathbf{H}(\mathbf{x}) = \nabla \mathbf{g}^T = \nabla \{ \nabla^T f(\mathbf{x}) \} \tag{2.3}$$

Hence Eqs. (2.1)–(2.3) give

$$\mathbf{H}(\mathbf{x}) = \begin{bmatrix}
\frac{\partial^2 f}{\partial x_1^2} & \frac{\partial^2 f}{\partial x_1 \partial x_2} & \cdots & \frac{\partial^2 f}{\partial x_1 \partial x_n} \\[2mm]
\frac{\partial^2 f}{\partial x_2 \partial x_1} & \frac{\partial^2 f}{\partial x_2^2} & \cdots & \frac{\partial^2 f}{\partial x_2 \partial x_n} \\[2mm]
\vdots & \vdots & & \vdots \\[2mm]
\frac{\partial^2 f}{\partial x_n \partial x_1} & \frac{\partial^2 f}{\partial x_n \partial x_2} & \cdots & \frac{\partial^2 f}{\partial x_n^2}
\end{bmatrix}$$

For a function $f(\mathbf{x}) \in C^2$

$$\frac{\partial^2 f}{\partial x_i \partial x_j} = \frac{\partial^2 f}{\partial x_j \partial x_i}$$

since differentiation is a linear operation and hence $\mathbf{H}(\mathbf{x})$ is an $n \times n$ square symmetric matrix.

The gradient and Hessian at a point $\mathbf{x} = \mathbf{x}_k$ are represented by $\mathbf{g}(\mathbf{x}_k)$ and $\mathbf{H}(\mathbf{x}_k)$ or by the simplified notation $\mathbf{g}_k$ and $\mathbf{H}_k$, respectively. Sometimes, when confusion is not likely to arise, $\mathbf{g}(\mathbf{x})$ and $\mathbf{H}(\mathbf{x})$ are simplified to $\mathbf{g}$ and $\mathbf{H}$.

The gradient and Hessian tend to simplify the optimization process considerably. Nevertheless, in certain applications it may be uneconomic, time-consuming, or impossible to deduce and compute the partial derivatives of $f(\mathbf{x})$. For these applications, methods are preferred that do not require gradient information.

Gradient methods, namely, methods based on gradient information, may use only $\mathbf{g}(\mathbf{x})$ or both $\mathbf{g}(\mathbf{x})$ and $\mathbf{H}(\mathbf{x})$. In the latter case, the inversion of matrix $\mathbf{H}(\mathbf{x})$ may be required which tends to introduce numerical inaccuracies and is time-consuming. Such methods are often avoided.

---

[1] For the sake of simplicity, the gradient vector and Hessian matrix will be referred to as the gradient and Hessian, respectively, henceforth.

2.3 The Taylor Series

Some of the nonlinear programming procedures and methods utilize linear or quadratic approximations for the objective function and the equality and inequality constraints, namely, $f(\mathbf{x})$, $a_i(\mathbf{x})$, and $c_j(\mathbf{x})$ in Eqs. (1.4a)–(1.4c). Such approximations can be obtained by using the Taylor series. If $f(\mathbf{x})$ is a function of two variables x_1 and x_2 such that $f(\mathbf{x}) \in C^P$ where $P \to \infty$, that is, $f(\mathbf{x})$ has continuous partial derivatives of all orders, then the value of function $f(\mathbf{x})$ at point $[x_1 + \delta_1, \ x_2 + \delta_2]$ is given by the Taylor series as

$$f(x_1 + \delta_1, \ x_2 + \delta_2) = f(x_1, \ x_2) + \frac{\partial f}{\partial x_1}\delta_1 + \frac{\partial f}{\partial x_2}\delta_2$$
$$+ \frac{1}{2}\left(\frac{\partial^2 f}{\partial x_1^2}\delta_1^2 + \frac{2\partial^2 f}{\partial x_1 \partial x_2}\delta_1\delta_2 + \frac{\partial^2 f}{\partial x_2^2}\delta_2^2\right)$$
$$+ O(\|\boldsymbol{\delta}\|_2^3) \tag{2.4a}$$

where

$$\boldsymbol{\delta} = [\delta_1 \ \delta_2]^T$$

$O(\|\boldsymbol{\delta}\|_2^3)$ is the *remainder*, and $\|\boldsymbol{\delta}\|_2$ is the Euclidean norm of $\boldsymbol{\delta}$ given by

$$\|\boldsymbol{\delta}\|_2 = \sqrt{\boldsymbol{\delta}^T \boldsymbol{\delta}}$$

The notation $\phi(x) = O(x)$ *denotes that* $\phi(x)$ *approaches zero at least as fast as x approaches zero*, that is, there exists a constant $K \geq 0$ such that

$$\left|\frac{\phi(x)}{x}\right| \leq K \quad \text{as } x \to 0$$

The remainder term in Eq. (2.4a) can also be expressed as $o(\|\boldsymbol{\delta}\|_2^2)$ where the notation $\phi(x) = o(x)$ *denotes that* $\phi(x)$ *approaches zero faster than x as x approaches zero*, that is,

$$\left|\frac{\phi(x)}{x}\right| \to 0 \quad \text{as } x \to 0$$

If $f(\mathbf{x})$ is a function of n variables, then the Taylor series of $f(\mathbf{x})$ at point $[x_1 + \delta_1, \ x_2 + \delta_2, \ \ldots]$ is given by

$$f(x_1 + \delta_1, \ x_2 + \delta_2, \ \ldots) = f(x_1, \ x_2, \ \ldots) + \sum_{i=1}^{n} \frac{\partial f}{\partial x_i}\delta_i$$
$$+ \frac{1}{2}\sum_{i=1}^{n}\sum_{j=1}^{n}\delta_i \frac{\partial^2 f}{\partial x_i \partial x_j}\delta_j$$
$$+ o(\|\boldsymbol{\delta}\|_2^2) \tag{2.4b}$$

Alternatively, on using matrix notation

$$f(\mathbf{x} + \boldsymbol{\delta}) = f(\mathbf{x}) + \mathbf{g}(\mathbf{x})^T \boldsymbol{\delta} + \tfrac{1}{2} \boldsymbol{\delta}^T \mathbf{H}(\mathbf{x}) \boldsymbol{\delta} + o(\|\boldsymbol{\delta}\|_2^2) \tag{2.4c}$$

where $\mathbf{g}(\mathbf{x})$ is the gradient, and $\mathbf{H}(\mathbf{x})$ is the Hessian at point $\mathbf{x}$.

As $\|\boldsymbol{\delta}\|_2 \to 0$, second- and higher-order terms can be neglected and can be obtained for $f(\mathbf{x} + \boldsymbol{\delta})$ as

$$f(\mathbf{x} + \boldsymbol{\delta}) \approx f(\mathbf{x}) + \mathbf{g}(\mathbf{x})^T \boldsymbol{\delta} \tag{2.4d}$$

Similarly, a *quadratic approximation* for $f(\mathbf{x} + \boldsymbol{\delta})$ can be obtained as

$$f(\mathbf{x} + \boldsymbol{\delta}) \approx f(\mathbf{x}) + \mathbf{g}(\mathbf{x})^T \boldsymbol{\delta} + \tfrac{1}{2} \boldsymbol{\delta}^T \mathbf{H}(\mathbf{x}) \boldsymbol{\delta} \tag{2.4e}$$

Another form of the Taylor series, which includes an expression for the remainder term, is

$$f(\mathbf{x} + \boldsymbol{\delta}) = f(\mathbf{x})$$
$$+ \sum_{1 \le k_1 + k_2 + \cdots + k_n \le P} \frac{\partial^{k_1 + k_2 + \cdots + k_n} f(\mathbf{x})}{\partial x_1^{k_1} \partial x_2^{k_2} \cdots \partial x_n^{k_n}} \prod_{i=1}^{n} \frac{\delta_i^{k_i}}{k_i!}$$
$$+ \sum_{k_1 + k_2 + \cdots + k_n = P+1} \frac{\partial^{P+1} f(\mathbf{x} + \alpha \boldsymbol{\delta})}{\partial x_1^{k_1} \partial x_2^{k_2} \cdots \partial x_n^{k_n}} \prod_{i=1}^{n} \frac{\delta_i^{k_i}}{k_i!} \tag{2.4f}$$

where $0 \le \alpha \le 1$ and

$$\sum_{1 \le k_1 + k_2 + \cdots + k_n \le P} \frac{\partial^{k_1 + k_2 + \cdots + k_n} f(\mathbf{x})}{\partial x_1^{k_1} \partial x_2^{k_2} \cdots \partial x_n^{k_n}} \prod_{i=1}^{n} \frac{\delta_i^{k_i}}{k_i!}$$

is the sum of terms taken over all possible combinations of $k_1, k_2, \ldots, k_n$ that add up to a number in the range 1 to P. (See Chap. 4 of Protter and Morrey [1] for proof.) This representation of the Taylor series is completely general and, therefore, it can be used to obtain cubic and higher-order approximations for $f(\mathbf{x} + \boldsymbol{\delta})$. Furthermore, it can be used to obtain linear, quadratic, cubic, and higher-order *exact closed-form* expressions for $f(\mathbf{x} + \boldsymbol{\delta})$. If $f(\mathbf{x}) \in C^1$ and $P = 0$, Eq. (2.4f) gives

$$f(\mathbf{x} + \boldsymbol{\delta}) = f(\mathbf{x}) + \mathbf{g}(\mathbf{x} + \alpha \boldsymbol{\delta})^T \boldsymbol{\delta} \tag{2.4g}$$

and if $f(\mathbf{x}) \in C^2$ and $P = 1$, then

$$f(\mathbf{x} + \boldsymbol{\delta}) = f(\mathbf{x}) + \mathbf{g}(\mathbf{x})^T \boldsymbol{\delta} + \tfrac{1}{2} \boldsymbol{\delta}^T \mathbf{H}(\mathbf{x} + \alpha \boldsymbol{\delta}) \boldsymbol{\delta} \tag{2.4h}$$

where $0 \le \alpha \le 1$. Equation (2.4g) is usually referred to as the *mean-value theorem for differentiation*.

Yet another form of the Taylor series can be obtained by regrouping the terms in Eq. (2.4f) as

$$f(\mathbf{x} + \boldsymbol{\delta}) = f(\mathbf{x}) + \mathbf{g}(\mathbf{x})^T \boldsymbol{\delta} + \tfrac{1}{2} \boldsymbol{\delta}^T \mathbf{H}(\mathbf{x}) \boldsymbol{\delta} + \tfrac{1}{3!} D^3 f(\mathbf{x})$$
$$+ \cdots + \frac{1}{(r-1)!} D^{r-1} f(\mathbf{x}) + \cdots \tag{2.4i}$$

where

$$D^r f(\mathbf{x}) = \sum_{i_1=1}^{n} \sum_{i_2=1}^{n} \cdots \sum_{i_r=1}^{n} \left\{ \delta_{i_1} \delta_{i_2} \cdots \delta_{i_r} \frac{\partial^r f(\mathbf{x})}{\partial x_{i_1} \partial x_{i_2} \cdots \partial x_{i_r}} \right\}$$

2.4 Types of Extrema

The *extrema* of a function are its minima and maxima. Points at which a function has minima (maxima) are said to be *minimizers* (*maximizers*). Several types of minimizers (maximizers) can be distinguished, namely, local or global and weak or strong.

Definition 2.1 *Weak local minimizer* A point $\mathbf{x}^* \in \mathcal{R}$, where $\mathcal{R}$ is the feasible region, is said to be a *weak local minimizer* of $f(\mathbf{x})$ if there exists a distance $\varepsilon > 0$ such that

$$f(\mathbf{x}) \geq f(\mathbf{x}^*) \tag{2.5}$$

if

$$\mathbf{x} \in \mathcal{R} \quad \text{and} \quad \|\mathbf{x} - \mathbf{x}^*\|_2 < \varepsilon$$

∎

Definition 2.2 *Weak global minimizer* A point $\mathbf{x}^* \in \mathcal{R}$ is said to be a *weak global minimizer* of $f(\mathbf{x})$ if

$$f(\mathbf{x}) \geq f(\mathbf{x}^*) \tag{2.6}$$

for all $\mathbf{x} \in \mathcal{R}$. ∎

If Definition 2.2 is satisfied at $\mathbf{x}^*$, then Definition 2.1 is also satisfied at $\mathbf{x}^*$, and so a global minimizer is also a local minimizer.

Definition 2.3 *Strong local (or global) minimizer* If Eq. (2.5) in Definition 2.1 or Eq. (2.6) in Definition 2.2 is replaced by

$$f(\mathbf{x}) > f(\mathbf{x}^*) \tag{2.7}$$

$\mathbf{x}^*$ is said to be a *strong local* (or *global*) *minimizer*. ∎

The minimum at a weak local, weak global, etc. minimizer is called a weak local, weak global, etc. minimum.

A strong global minimum in E^2 is depicted in Fig. 2.1.

Weak or strong and local or global maximizers can similarly be defined by reversing the inequalities in Eqs. (2.5)–(2.7).

Example 2.1 The function of Fig. 2.2 has a feasible region defined by the set

$$\mathcal{R} = \{x : x_1 \leq x \leq x_2\}$$

Classify its minimizers.

Fig. 2.1 A strong global
minimizer

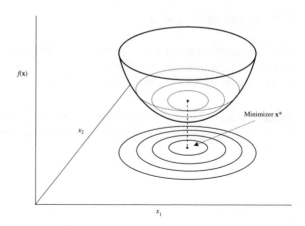

Fig. 2.2 Types of minima
(Example 2.1)

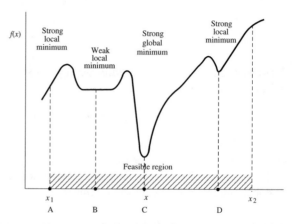

Solution The function has a weak local minimum at point B, strong local minima at
points A, C, and D, and a strong global minimum at point C. ∎

In the general optimization problem, we are in principle seeking the global minimum (or maximum) of $f(\mathbf{x})$. In practice, an optimization problem may have two or
more local minima. Since optimization algorithms in general are iterative procedures
which start with an initial estimate of the solution and converge to a single solution,
one or more local minima may be missed. If the global minimum is missed, a suboptimal solution will be achieved, which may or may not be acceptable. This problem
can to some extent be overcome by performing the optimization several times using
a different initial estimate for the solution in each case in the hope that several distinct local minima will be located. If this approach is successful, the best minimizer,
namely, the one yielding the lowest value for the objective function, can be selected.
Although such a solution could be acceptable from a practical point of view, usually
there is no guarantee that the global minimum will be achieved. Therefore, for the
sake of convenience, the term 'minimize $f(\mathbf{x})$' in the general optimization problem
will be interpreted as 'find a local minimum of $f(\mathbf{x})$'.

In a specific class of problems where function $f(\mathbf{x})$ and set $\mathcal{R}$ satisfy certain convexity properties, any local minimum of $f(\mathbf{x})$ is also a global minimum of $f(\mathbf{x})$. In this class of problems an optimal solution can be assured. These problems will be examined in Sect. 2.7.

2.5 Necessary and Sufficient Conditions For Local Minima and Maxima

The gradient $\mathbf{g}(\mathbf{x})$ and the Hessian $\mathbf{H}(\mathbf{x})$ must satisfy certain conditions at a local minimizer $\mathbf{x}^*$, (see [2, Chap. 7]). Two sets of conditions will be discussed, as follows:

1. Conditions which are satisfied at a local minimizer $\mathbf{x}^*$. These are the necessary conditions.
2. Conditions which guarantee that $\mathbf{x}^*$ is a local minimizer. These are the sufficient conditions.

The necessary and sufficient conditions can be described in terms of a number of theorems. A concept that is used extensively in these theorems is the concept of a feasible direction.

Definition 2.4 *Feasible direction* Let $\delta = \alpha\mathbf{d}$ be a change in $\mathbf{x}$ where α is a positive constant and $\mathbf{d}$ is a direction vector. If $\mathcal{R}$ is the feasible region and a constant $\hat{\alpha} > 0$ exists such that

$$\mathbf{x} + \alpha\mathbf{d} \in \mathcal{R}$$

for all α in the range $0 \leq \alpha \leq \hat{\alpha}$, then $\mathbf{d}$ is said to be a *feasible direction* at point $\mathbf{x}$. ■

In effect, if a point $\mathbf{x}$ remains in $\mathcal{R}$ after it is moved a finite distance in a direction $\mathbf{d}$, then $\mathbf{d}$ is a feasible direction vector at $\mathbf{x}$.

Example 2.2 The feasible region in an optimization problem is given by

$$\mathcal{R} = \{\mathbf{x} : x_1 \geq 2,\ x_2 \geq 0\}$$

as depicted in Fig. 2.3. Which of the vectors $\mathbf{d}_1 = [-2\ 2]^T$, $\mathbf{d}_2 = [0\ 2]^T$, $\mathbf{d}_3 = [2\ 0]^T$ are feasible directions at points $\mathbf{x}_1 = [4\ 1]^T$, $\mathbf{x}_2 = [2\ 3]^T$, and $\mathbf{x}_3 = [1\ 4]^T$?

Solution Since

$$\mathbf{x}_1 + \alpha\mathbf{d}_1 \in \mathcal{R}$$

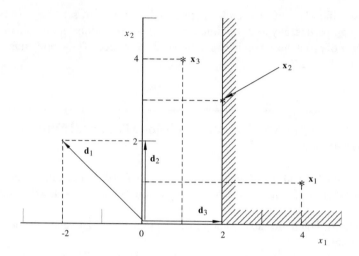

Fig. 2.3 Graphical construction for Example 2.2

for all α in the range $0 \le \alpha \le \hat{\alpha}$ for $\hat{\alpha} = 1$, $\mathbf{d}_1$ is a feasible direction at point $\mathbf{x}_1$; for any range $0 \le \alpha \le \hat{\alpha}$

$$\mathbf{x}_1 + \alpha\mathbf{d}_2 \in \mathcal{R} \quad \text{and} \quad \mathbf{x}_1 + \alpha\mathbf{d}_3 \in \mathcal{R}$$

Hence $\mathbf{d}_2$ and $\mathbf{d}_3$ are feasible directions at $\mathbf{x}_1$.

Since no constant $\hat{\alpha} > 0$ can be found such that

$$\mathbf{x}_2 + \alpha\mathbf{d}_1 \in \mathcal{R} \quad \text{for } 0 \le \alpha \le \hat{\alpha}$$

$\mathbf{d}_1$ is not a feasible direction at $\mathbf{x}_2$. On the other hand, a positive constant $\hat{\alpha}$ exists such that

$$\mathbf{x}_2 + \alpha\mathbf{d}_2 \in \mathcal{R} \quad \text{and} \quad \mathbf{x}_2 + \alpha\mathbf{d}_3 \in \mathcal{R}$$

for $0 \le \alpha \le \hat{\alpha}$, and so $\mathbf{d}_2$ and $\mathbf{d}_3$ are feasible directions at $\mathbf{x}_2$.

Since $\mathbf{x}_3$ is not in $\mathcal{R}$, no $\hat{\alpha} > 0$ exists such that

$$\mathbf{x}_3 + \alpha\mathbf{d} \in \mathcal{R} \quad \text{for } 0 \le \alpha \le \hat{\alpha}$$

for any $\mathbf{d}$. Hence $\mathbf{d}_1$, $\mathbf{d}_2$, and $\mathbf{d}_3$ are not feasible directions at $\mathbf{x}_3$. ∎

2.5.1 First-Order Necessary Conditions

The objective function must satisfy two sets of conditions in order to have a minimum, namely, first- and second-order conditions. The first-order conditions are in terms of the first derivatives, i.e., the gradient.

Theorem 2.1 *First-order necessary conditions for a minimum*

(a) *If* $f(\mathbf{x}) \in C^1$ *and* $\mathbf{x}^*$ *is a local minimizer, then*

$$\mathbf{g}(\mathbf{x}^*)^T \mathbf{d} \geq 0$$

for every feasible direction $\mathbf{d}$ *at* $\mathbf{x}^*$.
(b) *If* $\mathbf{x}^*$ *is located in the interior of* $\mathcal{R}$, *then*

$$\mathbf{g}(\mathbf{x}^*) = 0$$

Proof (a) If $\mathbf{d}$ is a feasible direction at $\mathbf{x}^*$, then from Definition 2.4

$$\mathbf{x} = \mathbf{x}^* + \alpha \mathbf{d} \in \mathcal{R} \qquad \text{for } 0 \leq \alpha \leq \hat{\alpha}$$

From the Taylor series

$$f(\mathbf{x}) = f(\mathbf{x}^*) + \alpha \mathbf{g}(\mathbf{x}^*)^T \mathbf{d} + o(\alpha \|\mathbf{d}\|_2)$$

If

$$\mathbf{g}(\mathbf{x}^*)^T \mathbf{d} < 0$$

then as $\alpha \to 0$

$$\alpha \mathbf{g}(\mathbf{x}^*)^T \mathbf{d} + o(\alpha \|\mathbf{d}\|_2) < 0$$

and so

$$f(\mathbf{x}) < f(\mathbf{x}^*)$$

This contradicts the assumption that $\mathbf{x}^*$ is a minimizer. Therefore, a necessary condition for $\mathbf{x}^*$ to be a minimizer is

$$\mathbf{g}(\mathbf{x}^*)^T \mathbf{d} \geq 0$$

(b) If $\mathbf{x}^*$ is in the interior of $\mathcal{R}$, vectors exist in all directions which are feasible. Thus from part (a), a direction $\mathbf{d} = \mathbf{d}_1$ yields

$$\mathbf{g}(\mathbf{x}^*)^T \mathbf{d}_1 \geq 0$$

Similarly, for a direction $\mathbf{d} = -\mathbf{d}_1$

$$-\mathbf{g}(\mathbf{x}^*)^T \mathbf{d}_1 \geq 0$$

Therefore, in this case, a necessary condition for $\mathbf{x}^*$ to be a local minimizer is

$$\mathbf{g}(\mathbf{x}^*) = 0$$

∎

2.5.2 Second-Order Necessary Conditions

The second-order necessary conditions involve the first as well as the second derivatives or, equivalently, the gradient and the Hessian.

Definition 2.5 *Quadratic forms*

(a) Let $\mathbf{d}$ be an arbitrary direction vector at point $\mathbf{x}$. The quadratic form $\mathbf{d}^T \mathbf{H}(\mathbf{x})\mathbf{d}$ is said to be *positive definite, positive semidefinite, negative semidefinite, negative definite* if $\mathbf{d}^T \mathbf{H}(\mathbf{x})\mathbf{d} > 0, \ \geq 0, \ \leq 0, \ < 0$, respectively, for all $\mathbf{d} \neq \mathbf{0}$ at $\mathbf{x}$. If $\mathbf{d}^T \mathbf{H}(\mathbf{x})\mathbf{d}$ can assume positive as well as negative values, it is said to be *indefinite*.
(b) If $\mathbf{d}^T \mathbf{H}(\mathbf{x})\mathbf{d}$ is positive definite, positive semidefinite, etc., then matrix $\mathbf{H}(\mathbf{x})$ is said to be positive definite, positive semidefinite, etc.   ∎

Theorem 2.2 *Second-order necessary conditions for a minimum*

(a) *If $f(\mathbf{x}) \in C^2$ and $\mathbf{x}^*$ is a local minimizer, then for every feasible direction $\mathbf{d}$ at $\mathbf{x}^*$*

    (i) $\mathbf{g}(\mathbf{x}^*)^T \mathbf{d} \geq 0$
    (ii) *If* $\mathbf{g}(\mathbf{x}^*)^T \mathbf{d} = 0$, *then* $\mathbf{d}^T \mathbf{H}(\mathbf{x}^*)\mathbf{d} \geq 0$

(b) *If $\mathbf{x}^*$ is a local minimizer in the interior of $\mathcal{R}$, then*

    (i) $\mathbf{g}(\mathbf{x}^*) = \mathbf{0}$
    (ii) $\mathbf{d}^T \mathbf{H}(\mathbf{x}^*)\mathbf{d} \geq 0$ *for all $\mathbf{d} \neq \mathbf{0}$*

Proof Conditions $(a)(i)$ and $(b)(i)$ are the same as in Theorem 2.1.
Condition (ii) of part (a) can be proved by letting $\mathbf{x} = \mathbf{x}^* + \alpha\mathbf{d}$, where $\mathbf{d}$ is a feasible direction. The Taylor series gives

$$f(\mathbf{x}) = f(\mathbf{x}^*) + \alpha \mathbf{g}(\mathbf{x}^*)^T \mathbf{d} + \tfrac{1}{2}\alpha^2 \mathbf{d}^T \mathbf{H}(\mathbf{x}^*)\mathbf{d} + o(\alpha^2 \|\mathbf{d}\|_2^2)$$

Now if condition (i) is satisfied with the equal sign, then

$$f(\mathbf{x}) = f(\mathbf{x}^*) + \tfrac{1}{2}\alpha^2 \mathbf{d}^T \mathbf{H}(\mathbf{x}^*)\mathbf{d} + o(\alpha^2 \|\mathbf{d}\|_2^2)$$

If

$$\mathbf{d}^T \mathbf{H}(\mathbf{x}^*)\mathbf{d} < 0$$

then as $\alpha \to 0$

$$\tfrac{1}{2}\alpha^2 \mathbf{d}^T \mathbf{H}(\mathbf{x}^*)\mathbf{d} + o(\alpha^2 \|\mathbf{d}\|_2^2) < 0$$

and so

$$f(\mathbf{x}) < f(\mathbf{x}^*)$$

This contradicts the assumption that $\mathbf{x}^*$ is a minimizer. Therefore, if $\mathbf{g}(\mathbf{x}^*)^T\mathbf{d} = 0$, then

$$\mathbf{d}^T\mathbf{H}(\mathbf{x}^*)\mathbf{d} \geq 0$$

If $\mathbf{x}^*$ is a local minimizer in the interior of $\mathcal{R}$, then all vectors $\mathbf{d}$ are feasible directions and, therefore, condition (ii) of part (b) holds. This condition is equivalent to stating that $\mathbf{H}(\mathbf{x}^*)$ is positive semidefinite, according to Definition 2.5.   ∎

Example 2.3 Point $\mathbf{x}^* = [\frac{1}{2}\ 0]^T$ is a local minimizer of the problem

$$\text{minimize}\ \ f(x_1,\ x_2) = x_1^2 - x_1 + x_2 + x_1 x_2$$

$$\text{subject to} :\ \ x_1 \geq 0,\ x_2 \geq 0$$

Show that the necessary conditions for $\mathbf{x}^*$ to be a local minimizer are satisfied.

Solution The partial derivatives of $f(x_1,\ x_2)$ are

$$\frac{\partial f}{\partial x_1} = 2x_1 - 1 + x_2, \quad \frac{\partial f}{\partial x_2} = 1 + x_1$$

Hence if $\mathbf{d} = [d_1\ d_2]^T$ is a feasible direction, we obtain

$$\mathbf{g}(\mathbf{x})^T\mathbf{d} = (2x_1 - 1 + x_2)d_1 + (1 + x_1)d_2$$

At $\mathbf{x} = \mathbf{x}^*$

$$\mathbf{g}(\mathbf{x}^*)^T\mathbf{d} = \tfrac{3}{2}d_2$$

and since $d_2 \geq 0$ for $\mathbf{d}$ to be a feasible direction, we have

$$\mathbf{g}(\mathbf{x}^*)^T\mathbf{d} \geq 0$$

Therefore, the first-order necessary conditions for a minimum are satisfied.
   Now

$$\mathbf{g}(\mathbf{x}^*)^T\mathbf{d} = 0$$

if $d_2 = 0$. The Hessian is

$$\mathbf{H}(\mathbf{x}^*) = \begin{bmatrix} 2 & 1 \\ 1 & 0 \end{bmatrix}$$

and so

$$\mathbf{d}^T\mathbf{H}(\mathbf{x}^*)\mathbf{d} = 2d_1^2 + 2d_1 d_2$$

For $d_2 = 0$, we obtain

$$\mathbf{d}^T\mathbf{H}(\mathbf{x}^*)\mathbf{d} = 2d_1^2 \geq 0$$

for every feasible value of d_1. Therefore, the second-order necessary conditions for a minimum are satisfied.   ∎

Example 2.4 Points $\mathbf{p}_1 = [0\ 0]^T$ and $\mathbf{p}_2 = [6\ 9]^T$ are possible minimizers for the problem

$$\text{minimize } f(x_1, x_2) = x_1^3 - x_1^2 x_2 + 2x_2^2$$

$$\text{subject to}: \quad x_1 \geq 0, \ x_2 \geq 0$$

Check whether the necessary conditions of Theorems 2.1 and 2.2 are satisfied.

Solution The partial derivatives of $f(x_1, x_2)$ are

$$\frac{\partial f}{\partial x_1} = 3x_1^2 - 2x_1 x_2, \quad \frac{\partial f}{\partial x_2} = -x_1^2 + 4x_2$$

Hence if $\mathbf{d} = [d_1\ d_2]^T$, we obtain

$$\mathbf{g(x)}^T \mathbf{d} = (3x_1^2 - 2x_1 x_2) d_1 + (-x_1^2 + 4x_2) d_2$$

At points $\mathbf{p}_1$ and $\mathbf{p}_2$

$$\mathbf{g(x)}^T \mathbf{d} = 0$$

i.e., the first-order necessary conditions are satisfied. The Hessian is

$$\mathbf{H(x)} = \begin{bmatrix} 6x_1 - 2x_2 & -2x_1 \\ -2x_1 & 4 \end{bmatrix}$$

and if $\mathbf{x} = \mathbf{p}_1$, then

$$\mathbf{H(p_1)} = \begin{bmatrix} 0 & 0 \\ 0 & 4 \end{bmatrix}$$

and so

$$\mathbf{d}^T \mathbf{H(p_1)} \mathbf{d} = 4d_2^2 \geq 0$$

Hence the second-order necessary conditions are satisfied at $\mathbf{x} = \mathbf{p}_1$, and $\mathbf{p}_1$ can be a local minimizer.

If $\mathbf{x} = \mathbf{p}_2$, then

$$\mathbf{H(p_2)} = \begin{bmatrix} 18 & -12 \\ -12 & 4 \end{bmatrix}$$

and

$$\mathbf{d}^T \mathbf{H(p_2)} \mathbf{d} = 18d_1^2 - 24d_1 d_2 + 4d_2^2$$

Since $\mathbf{d}^T \mathbf{H(p_2)} \mathbf{d}$ is indefinite, the second-order necessary conditions are violated, that is, $\mathbf{p}_2$ cannot be a local minimizer. ∎

Analogous conditions hold for the case of a local maximizer as stated in the following theorem.

Theorem 2.3 *Second-order necessary conditions for a maximum*

(a) *If* $f(\mathbf{x}) \in C^2$ *and* $\mathbf{x}^*$ *is a local maximizer, then for every feasible direction* $\mathbf{d}$ *at* $\mathbf{x}^*$

    (i) $\mathbf{g}(\mathbf{x}^*)^T\mathbf{d} \le 0$
    (ii) *If* $\mathbf{g}(\mathbf{x}^*)^T\mathbf{d} = 0$*, then* $\mathbf{d}^T\mathbf{H}(\mathbf{x}^*)\mathbf{d} \le 0$

(b) *If* $\mathbf{x}^*$ *is a local maximizer in the interior of* $\mathcal{R}$ *then*

    (i) $\mathbf{g}(\mathbf{x}^*) = \mathbf{0}$
    (ii) $\mathbf{d}^T\mathbf{H}(\mathbf{x}^*)\mathbf{d} \le 0$ *for all* $\mathbf{d} \ne \mathbf{0}$

■

Condition (ii) of part (b) is equivalent to stating that $\mathbf{H}(\mathbf{x}^*)$ is negative semidefinite.

The conditions considered are necessary but not sufficient for a point to be a local extremum point, that is, a point may satisfy these conditions without being a local extremum point. We now focus our attention on a set of stronger conditions that are *sufficient* for a point to be a local extremum. We consider conditions that are applicable in the case where $\mathbf{x}^*$ is located in the interior of the feasible region. Sufficient conditions that are applicable to the case where $\mathbf{x}^*$ is located on a boundary of the feasible region are somewhat more difficult to deduce and will be considered in Chap. 10.

Theorem 2.4 *Second-order sufficient conditions for a minimum*   *If* $f(\mathbf{x}) \in C^2$ *and* $\mathbf{x}^*$ *is located in the interior of* $\mathcal{R}$*, then the conditions*

(a) $\mathbf{g}(\mathbf{x}^*) = \mathbf{0}$
(b) $\mathbf{H}(\mathbf{x}^*)$ *is positive definite*

are sufficient for $\mathbf{x}^*$ *to be a strong local minimizer.*

Proof   For any direction $\mathbf{d}$, the Taylor series yields

$$f(\mathbf{x}^* + \mathbf{d}) = f(\mathbf{x}^*) + \mathbf{g}(\mathbf{x}^*)^T\mathbf{d} + \tfrac{1}{2}\mathbf{d}^T\mathbf{H}(\mathbf{x}^*)\mathbf{d} + o(\|\mathbf{d}\|_2^2)$$

and if condition (a) is satisfied, we have

$$f(\mathbf{x}^* + \mathbf{d}) = f(\mathbf{x}^*) + \tfrac{1}{2}\mathbf{d}^T\mathbf{H}(\mathbf{x}^*)\mathbf{d} + o(\|\mathbf{d}\|_2^2)$$

Now if condition (b) is satisfied, then

$$\tfrac{1}{2}\mathbf{d}^T\mathbf{H}(\mathbf{x}^*)\mathbf{d} + o(\|\mathbf{d}\|_2^2) > 0 \qquad \text{as } \|\mathbf{d}\|_2 \to 0$$

Therefore,

$$f(\mathbf{x}^* + \mathbf{d}) > f(\mathbf{x}^*)$$

that is, $\mathbf{x}^*$ is a strong local minimizer.   ■

Analogous conditions hold for a maximizer as stated in Theorem 2.5 below.

Theorem 2.5 *Second-order sufficient conditions for a maximum If $f(\mathbf{x}^*) \in C^2$
and $\mathbf{x}^*$ is located in the interior of $\mathcal{R}$, then the conditions*

(a) $\mathbf{g(x)} = \mathbf{0}$
(b) $\mathbf{H(x^*)}$ *is negative definite*

are sufficient for $\mathbf{x}^*$ to be a strong local maximizer.   ■

2.6 Classification of Stationary Points

If the extremum points of the type considered so far, namely, minimizers and maximizers, are located in the interior of the feasible region, they are called *stationary
points* since $\mathbf{g(x)} = \mathbf{0}$ at these points. Another type of stationary point of interest is
the saddle point.

Definition 2.6 *Saddle point* A point $\bar{\mathbf{x}} \in \mathcal{R}$, where $\mathcal{R}$ is the feasible region, is said
to be a *saddle point* if

(a) $\mathbf{g(\bar{x})} = \mathbf{0}$
(b) point $\bar{\mathbf{x}}$ is neither a maximizer nor a minimizer.   ■

A saddle point in E^2 is illustrated in Fig. 2.4.

At a point $\mathbf{x} = \bar{\mathbf{x}} + \alpha \mathbf{d} \in \mathcal{R}$ in the neighborhood of a saddle point $\bar{\mathbf{x}}$, the Taylor
series gives

$$f(\mathbf{x}) = f(\bar{\mathbf{x}}) + \tfrac{1}{2}\alpha^2 \mathbf{d}^T \mathbf{H}(\bar{\mathbf{x}})\mathbf{d} + o(\alpha^2 \|\mathbf{d}\|_2^2)$$

since $\mathbf{g(\bar{x})} = \mathbf{0}$. From the definition of a saddle point, directions $\mathbf{d}_1$ and $\mathbf{d}_2$ must exist
such that

$$f(\bar{\mathbf{x}} + \alpha \mathbf{d}_1) < f(\bar{\mathbf{x}}) \quad \text{and} \quad f(\bar{\mathbf{x}} + \alpha \mathbf{d}_2) > f(\bar{\mathbf{x}})$$

Since $\bar{\mathbf{x}}$ is neither a minimizer nor a maximizer, then as $\alpha \to 0$ we have

$$\mathbf{d}_1^T \mathbf{H}(\bar{\mathbf{x}})\mathbf{d}_1 < 0 \quad \text{and} \quad \mathbf{d}_2^T \mathbf{H}(\bar{\mathbf{x}})\mathbf{d}_2 > 0$$

Therefore, matrix $\mathbf{H}(\bar{\mathbf{x}})$ must be indefinite.

Stationary points can be located and classified as follows:

Fig. 2.4 A saddle point in E^2

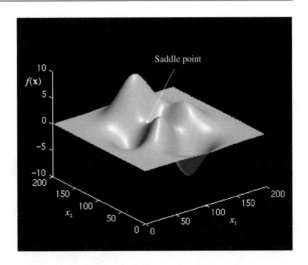

1. Find the points $\mathbf{x}_i$ at which $\mathbf{g}(\mathbf{x}_i) = \mathbf{0}$.
2. Obtain the Hessian $\mathbf{H}(\mathbf{x}_i)$.
3. Determine the character of $\mathbf{H}(\mathbf{x}_i)$ for each point $\mathbf{x}_i$.

If $\mathbf{H}(\mathbf{x}_i)$ is positive (or negative) definite, $\mathbf{x}_i$ is a minimizer (or maximizer); if $\mathbf{H}(\mathbf{x}_i)$ is indefinite, $\mathbf{x}_i$ is a saddle point. If $\mathbf{H}(\mathbf{x}_i)$ is positive (or negative) semidefinite, $\mathbf{x}_i$ can be a minimizer (or maximizer); in the special case where $\mathbf{H}(\mathbf{x}_i) = \mathbf{0}$, $\mathbf{x}_i$ can be a minimizer or maximizer since the necessary conditions are satisfied in both cases. Evidently, if $\mathbf{H}(\mathbf{x}_i)$ is semidefinite, insufficient information is available for the complete characterization of a stationary point and further work is, therefore, necessary in such a case. A possible approach would be to deduce the third partial derivatives of $f(\mathbf{x})$ and then calculate the fourth term in the Taylor series, namely, term $D^3 f(\mathbf{x})/3!$ in Eq. (2.4i). If the fourth term is zero, then the fifth term needs to be calculated and so on. An alternative and more practical approach would be to compute $f(\mathbf{x}_i + \mathbf{e}_j)$ and $f(\mathbf{x}_i - \mathbf{e}_j)$ for $j = 1, \ 2, \ \ldots, \ n$ where $\mathbf{e}_j$ is a vector with elements

$$
e_{jk} = \begin{cases} 0 & \text{for } k \neq j \\ \varepsilon & \text{for } k = j \end{cases}
$$

for some small positive value of ε and then check whether the definition of a minimizer or maximizer is satisfied.

Example 2.5 Find and classify the stationary points of

$$
f(\mathbf{x}) = (x_1 - 2)^3 + (x_2 - 3)^3
$$

Solution The first-order partial derivatives of $f(\mathbf{x})$ are

$$\frac{\partial f}{\partial x_1} = 3(x_1 - 2)^2$$

$$\frac{\partial f}{\partial x_2} = 3(x_2 - 3)^2$$

If $\mathbf{g} = \mathbf{0}$, then

$$3(x_1 - 2)^2 = 0 \quad \text{and} \quad 3(x_2 - 3)^2 = 0$$

and so there is a stationary point at

$$\mathbf{x} = \mathbf{x}_1 = [2\ 3]^T$$

The Hessian is given by

$$\mathbf{H} = \begin{bmatrix} 6(x_1 - 2) & 0 \\ 0 & 6(x_2 - 3) \end{bmatrix}$$

and at $\mathbf{x} = \mathbf{x}_1$

$$\mathbf{H} = \mathbf{0}$$

Since $\mathbf{H}$ is semidefinite, more work is necessary in order to determine the type of stationary point.

The third derivatives are all zero except for $\partial^3 f/\partial x_1^3$ and $\partial^3 f/\partial x_2^3$ which are both equal to 6. For point $\mathbf{x}_1 + \boldsymbol{\delta}$, the fourth term in the Taylor series is given by

$$\frac{1}{3!}\left(\delta_1^3 \frac{\partial^3 f}{\partial x_1^3} + \delta_2^3 \frac{\partial^3 f}{\partial x_2^3}\right) = \delta_1^3 + \delta_2^3$$

and is positive for $\delta_1,\ \delta_2 > 0$ and negative for $\delta_1,\ \delta_2 < 0$. Hence

$$f(\mathbf{x}_1 + \boldsymbol{\delta}) > f(\mathbf{x}_1) \qquad \text{for } \delta_1,\ \delta_2 > 0$$

and

$$f(\mathbf{x}_1 + \boldsymbol{\delta}) < f(\mathbf{x}_1) \qquad \text{for } \delta_1,\ \delta_2 < 0$$

that is, $\mathbf{x}_1$ is neither a minimizer nor a maximizer. Therefore, $\mathbf{x}_1$ is a saddle point. ∎

From the preceding discussion, it follows that the problem of classifying the stationary points of function $f(\mathbf{x})$ reduces to the problem of characterizing the Hessian. This problem can be solved by using the following theorems.

Theorem 2.6 *Characterization of symmetric matrices A real symmetric $n \times n$ matrix $\mathbf{H}$ is positive definite, positive semidefinite, etc., if and only if for every nonsingular matrix $\mathbf{B}$ of the same order, the $n \times n$ matrix $\hat{\mathbf{H}}$ given by*

$$\hat{\mathbf{H}} = \mathbf{B}^T \mathbf{H} \mathbf{B}$$

is positive definite, positive semidefinite, etc.

Proof   The "if" part is proved by considering the case where **B** is taken to be the identity matrix so that the condition given by the theorem immediately implies that **H** is positive definite, positive semidefinite, etc.

Conversely, if **H** is positive definite, positive semidefinite, etc., then for all $\mathbf{d} \neq \mathbf{0}$

$$\mathbf{d}^T \hat{\mathbf{H}} \mathbf{d} = \mathbf{d}^T (\mathbf{B}^T \mathbf{H} \mathbf{B}) \mathbf{d}$$
$$= (\mathbf{d}^T \mathbf{B}^T) \mathbf{H} (\mathbf{B} \mathbf{d})$$
$$= (\mathbf{B} \mathbf{d})^T \mathbf{H} (\mathbf{B} \mathbf{d})$$

Since **B** is nonsingular, $\mathbf{B}\mathbf{d} = \hat{\mathbf{d}}$ is a nonzero vector and thus

$$\mathbf{d}^T \hat{\mathbf{H}} \mathbf{d} = \hat{\mathbf{d}}^T \mathbf{H} \hat{\mathbf{d}} > 0, \; \geq 0, \text{ etc.}$$

for all $\mathbf{d} \neq \mathbf{0}$. Therefore,

$$\hat{\mathbf{H}} = \mathbf{B}^T \mathbf{H} \mathbf{B}$$

is positive definite, positive semidefinite, etc.   ∎

Theorem 2.7   *Characterization of symmetric matrices via diagonalization*

(a) *If the $n \times n$ matrix **B** is nonsingular and*
$$\hat{\mathbf{H}} = \mathbf{B}^T \mathbf{H} \mathbf{B}$$
*is a diagonal matrix with diagonal elements $\hat{h}_1$, $\hat{h}_2$, ..., $\hat{h}_n$ then **H** is positive definite, positive semidefinite, negative semidefinite, negative definite, if $\hat{h}_i > 0, \; \geq 0, \; \leq 0, \; < 0$ for $i = 1, \; 2, \; ..., \; n$. Otherwise, if some $\hat{h}_i$ are positive and some are negative, **H** is indefinite.*

(b) *The converse of part (a) is also true, that is, if **H** is positive definite, positive semidefinite, etc., then $\hat{h}_i > 0, \; \geq 0$, etc., and if **H** is indefinite, then some $\hat{h}_i$ are positive and some are negative.*

Proof   (a) For all $\mathbf{d} \neq \mathbf{0}$
$$\mathbf{d}^T \hat{\mathbf{H}} \mathbf{d} = d_1^2 \hat{h}_1 + d_2^2 \hat{h}_2 + \cdots + d_n^2 \hat{h}_n$$

Therefore, if $\hat{h}_i > 0, \; \geq 0$, etc. for $i = 1, \; 2, \; ..., \; n$, then
$$\mathbf{d}^T \hat{\mathbf{H}} \mathbf{d} > 0, \; \geq 0, \text{ etc.}$$

that is, $\hat{\mathbf{H}}$ is positive definite, positive semidefinite, etc. If some $\hat{h}_i$ are positive and some are negative, a vector **d** can be found which will yield a positive or negative $\mathbf{d}^T \hat{\mathbf{H}} \mathbf{d}$ and then $\hat{\mathbf{H}}$ is indefinite. Now since $\hat{\mathbf{H}} = \mathbf{B}^T \mathbf{H} \mathbf{B}$, it follows from Theorem 2.6 that if $\hat{h}_i > 0, \; \geq 0$, etc. for $i = 1, \; 2, \; ..., \; n$, then **H** is positive definite, positive semidefinite, etc.

(b) Suppose that **H** is positive definite, positive semidefinite, etc. Since $\hat{\mathbf{H}} = \mathbf{B}^T \mathbf{H} \mathbf{B}$, it follows from Theorem 2.6 that $\hat{\mathbf{H}}$ is positive definite, positive semidefinite, etc. If **d** is a vector with element d_k given by

$$d_k = \begin{cases} 0 & \text{for } k \neq i \\ 1 & \text{for } k = i \end{cases}$$

then

$$\mathbf{d}^T \hat{\mathbf{H}} \mathbf{d} = \hat{h}_i > 0, \geq 0, \text{ etc.} \quad \text{for } i = 1, 2, \ldots, n$$

If $\mathbf{H}$ is indefinite, then from Theorem 2.6 it follows that $\hat{\mathbf{H}}$ is indefinite, and, therefore, some $\hat{h}_i$ must be positive and some must be negative. ∎

A diagonal matrix $\hat{\mathbf{H}}$ can be obtained by performing row and column operations on $\mathbf{H}$, like adding k times a given row to another row or adding m times a given column to another column. For a symmetric matrix, these operations can be carried out by applying *elementary transformations*, that is, $\hat{\mathbf{H}}$ can be formed as

$$\hat{\mathbf{H}} = \cdots \mathbf{E}_3 \mathbf{E}_2 \mathbf{E}_1 \mathbf{H} \mathbf{E}_1^T \mathbf{E}_2^T \mathbf{E}_3^T \cdots \tag{2.8}$$

where $\mathbf{E}_1, \mathbf{E}_2, \cdots$ are elementary matrices. Typical elementary matrices are

$$\mathbf{E}_a = \begin{bmatrix} 1 & 0 & 0 \\ 0 & 1 & 0 \\ 0 & k & 1 \end{bmatrix}$$

and

$$\mathbf{E}_b = \begin{bmatrix} 1 & m & 0 & 0 \\ 0 & 1 & 0 & 0 \\ 0 & 0 & 1 & 0 \\ 0 & 0 & 0 & 1 \end{bmatrix}$$

If E_a premultiplies a 3×3 matrix, it will cause k times the second row to be added to the third row, and if $\mathbf{E}_b$ postmultiplies a 4×4 matrix it will cause m times the first column to be added to the second column. If

$$\mathbf{B} = \mathbf{E}_1^T \mathbf{E}_2^T \mathbf{E}_3^T \cdots$$

then

$$\mathbf{B}^T = \cdots \mathbf{E}_3 \mathbf{E}_2 \mathbf{E}_1$$

and so Eq. (2.8) can be expressed as

$$\hat{\mathbf{H}} = \mathbf{B}^T \mathbf{H} \mathbf{B}$$

Since elementary matrices are nonsingular, $\mathbf{B}$ is nonsingular, and hence $\hat{\mathbf{H}}$ is positive definite, positive semidefinite, etc., if $\mathbf{H}$ is positive definite, positive semidefinite, etc.

Therefore, the characterization of $\mathbf{H}$ can be achieved by diagonalizing $\mathbf{H}$, through the use of appropriate elementary matrices, and then using Theorem 2.7.

Example 2.6 Diagonalize the matrix

$$\mathbf{H} = \begin{bmatrix} 1 & -2 & 4 \\ -2 & 2 & 0 \\ 4 & 0 & -7 \end{bmatrix}$$

and then characterize it.

Solution Add 2 times the first row to the second row as

$$
\begin{bmatrix} 1\,0\,0 \\ 2\,1\,0 \\ 0\,0\,1 \end{bmatrix}
\begin{bmatrix} 1 & -2 & 4 \\ -2 & 2 & 0 \\ 4 & 0 & -7 \end{bmatrix}
\begin{bmatrix} 1\,2\,0 \\ 0\,1\,0 \\ 0\,0\,1 \end{bmatrix}
=
\begin{bmatrix} 1 & 0 & 4 \\ 0 & -2 & 8 \\ 4 & 8 & -7 \end{bmatrix}
$$

Add -4 times the first row to the third row as

$$
\begin{bmatrix} 1 & 0\,0 \\ 0 & 1\,0 \\ -4 & 0\,1 \end{bmatrix}
\begin{bmatrix} 1 & 0 & 4 \\ 0 & -2 & 8 \\ 4 & 8 & -7 \end{bmatrix}
\begin{bmatrix} 1\,0 & -4 \\ 0\,1 & 0 \\ 0\,0 & 1 \end{bmatrix}
=
\begin{bmatrix} 1 & 0 & 0 \\ 0 & -2 & 8 \\ 0 & 8 & -23 \end{bmatrix}
$$

Now add 4 times the second row to the third row as

$$
\begin{bmatrix} 1\,0\,0 \\ 0\,1\,0 \\ 0\,4\,1 \end{bmatrix}
\begin{bmatrix} 1 & 0 & 0 \\ 0 & -2 & 8 \\ 0 & 8 & -23 \end{bmatrix}
\begin{bmatrix} 1\,0\,0 \\ 0\,1\,4 \\ 0\,0\,1 \end{bmatrix}
=
\begin{bmatrix} 1 & 0 & 0 \\ 0 & -2 & 0 \\ 0 & 0 & 9 \end{bmatrix}
$$

Since $\hat{h}_1 = 1, \hat{h}_2 = -2, \hat{h}_3 = 9$, **H** is indefinite.   ∎

Example 2.7 Diagonalize the matrix

$$
\mathbf{H} = \begin{bmatrix} 4 & -2 & 0 \\ -2 & 3 & 0 \\ 0 & 0 & 50 \end{bmatrix}
$$

and determine its characterization.

Solution Add 0.5 times the first row to the second row as

$$
\begin{bmatrix} 1 & 0\,0 \\ 0.5 & 1\,0 \\ 0 & 0\,1 \end{bmatrix}
\begin{bmatrix} 4 & -2 & 0 \\ -2 & 3 & 0 \\ 0 & 0 & 50 \end{bmatrix}
\begin{bmatrix} 1 & 0.5 & 0 \\ 0 & 1 & 0 \\ 0 & 0 & 1 \end{bmatrix}
=
\begin{bmatrix} 4 & 0 & 0 \\ 0 & 2 & 0 \\ 0 & 0 & 50 \end{bmatrix}
$$

Hence **H** is positive definite.   ∎

Another theorem that can be used to characterize the Hessian is as follows.

Theorem 2.8 (*Eigendecomposition of symmetric matrices*) *If* **H** *is a real symmetric matrix, then*

(a) *there exists a real unitary (or orthogonal) matrix* **U** *such that*

$$
\Lambda = \mathbf{U}^T \mathbf{H} \mathbf{U}
$$

is a diagonal matrix whose diagonal elements are the eigenvalues of **H**.

(b) *The eigenvalues of* **H** *are real.*   ∎

(See Chap. 4 of Horn and Johnson [3] for proofs.)

For a real unitary matrix, we have $\mathbf{U}^T \mathbf{U} = \mathbf{I}_n$ where

$$\mathbf{I}_n = \begin{bmatrix} 1 & 0 & \cdots & 0 \\ 0 & 1 & \cdots & 0 \\ \vdots & \vdots & & \vdots \\ 0 & 0 & \cdots & 1 \end{bmatrix}$$

is the $n \times n$ identity matrix, and hence det $\mathbf{U} = \pm 1$, that is, $\mathbf{U}$ is nonsingular. From Theorem 2.6, $\mathbf{\Lambda}$ is positive definite, positive semidefinite, etc. if $\mathbf{H}$ is positive definite, positive semidefinite, etc. Therefore, $\mathbf{H}$ can be characterized by deducing its eigenvalues and then checking their signs as in Theorem 2.7.

Another approach for the characterization of a square matrix $\mathbf{H}$ is based on the evaluation of the so-called *principal minors* and *leading principal minors* of $\mathbf{H}$, which are described in Sect. A.6. The details of this approach are summarized in terms of the following theorem.

Theorem 2.9 *Properties of matrices*

(a) *If* $\mathbf{H}$ *is positive semidefinite or positive definite, then*
$$\det \mathbf{H} \geq 0 \text{ or } > 0$$

(b) $\mathbf{H}$ *is positive definite if and only if all its leading principal minors are positive, i.e.,* det $\mathbf{H}_i > 0$ *for* $i = 1, 2, \ldots, n$.

(c) $\mathbf{H}$ *is positive semidefinite if and only if all its principal minors are nonnegative, i.e.,* det $(\mathbf{H}_i^{(l)}) \geq 0$ *for all possible selections of* $l_1, l_2, \ldots, l_i$ *for* $i = 1, 2, \ldots, n$.

(d) $\mathbf{H}$ *is negative definite if and only if all the leading principal minors of* $-\mathbf{H}$ *are positive, i.e.,* det $(-\mathbf{H}_i) > 0$ *for* $i = 1, 2, \ldots, n$.

(e) $\mathbf{H}$ *is negative semidefinite if and only if all the principal minors of* $-\mathbf{H}$ *are nonnegative, i.e.,* det $(-\mathbf{H}_i^{(l)}) \geq 0$ *for all possible selections of* $l_1, l_2, \ldots, l_i$ *for* $i = 1, 2, \ldots, n$.

(f) $\mathbf{H}$ *is indefinite if neither (c) nor (e) holds.*

Proof (a) Elementary transformations do not change the determinant of a matrix and hence

$$\det \mathbf{H} = \det \hat{\mathbf{H}} = \prod_{i=1}^{n} \hat{h}_i$$

where $\hat{\mathbf{H}}$ is a diagonalized version of $\mathbf{H}$ with diagonal elements $\hat{h}_i$. If $\mathbf{H}$ is positive semidefinite or positive definite, then $\hat{h}_i \geq 0$ or > 0 from Theorem 2.7 and, therefore,

$$\det \mathbf{H} \geq 0 \text{ or } > 0$$

(b) If

$$\mathbf{d} = [d_1 \ d_2 \ \cdots \ d_i \ 0 \ 0 \cdots 0]^T$$

and $\mathbf{H}$ is positive definite, then

$$\mathbf{d}^T \mathbf{H} \mathbf{d} = \mathbf{d}_0^T \mathbf{H}_i \mathbf{d}_0 > 0$$

for all $\mathbf{d}_0 \neq \mathbf{0}$ where

$$\mathbf{d}_0 = [d_1 \ d_2 \ \cdots \ d_i]^T$$

and $\mathbf{H}_i$ is the ith leading principal submatrix of $\mathbf{H}$. The preceding inequality holds for $i = 1, \ 2, \ \ldots, \ n$, and hence $\mathbf{H}_i$ is positive definite for $i = 1, \ 2, \ \ldots, \ n$. From part ($a$)

$$\det \mathbf{H}_i > 0 \quad \text{for } i = 1, \ 2, \ \ldots, \ n$$

Now we prove the sufficiency of the theorem by induction. If $n = 1$, then $\mathbf{H} = a_{11}$, and $\det (\mathbf{H}_1) = a_{11} > 0$ implies that $\mathbf{H}$ is positive definite. We assume that the sufficiency is valid for matrix $\mathbf{H}$ of size $(n - 1)$ by $(n - 1)$ and we shall show that the sufficiency is also valid for matrix $\mathbf{H}$ of size n by n. First, we write $\mathbf{H}$ as

$$\mathbf{H} = \begin{bmatrix} \mathbf{H}_{n-1} & \mathbf{h} \\ \mathbf{h}^T & h_{nn} \end{bmatrix}$$

where

$$\mathbf{H}_{n-1} = \begin{bmatrix} h_{11} & h_{12} & \cdots & h_{1,n-1} \\ h_{21} & h_{22} & \cdots & h_{2,n-1} \\ \vdots & \vdots & & \vdots \\ h_{n-1,1} & h_{n-1,2} & \cdots & h_{n-1,n-1} \end{bmatrix}, \quad \mathbf{h} = \begin{bmatrix} h_{1n} \\ h_{2n} \\ \vdots \\ h_{n-1,n} \end{bmatrix}$$

By assumption $\mathbf{H}_{n-1}$ is positive definite; hence there exists an $\mathbf{R}$ such that

$$\mathbf{R}^T \mathbf{H}_{n-1} \mathbf{R} = \mathbf{I}_{n-1}$$

where $\mathbf{I}_{n-1}$ is the $(n - 1) \times (n - 1)$ identity matrix. If we let

$$\mathbf{S} = \begin{bmatrix} \mathbf{R} & \mathbf{0} \\ \mathbf{0} & 1 \end{bmatrix}$$

we obtain

$$\mathbf{S}^T \mathbf{H} \mathbf{S} = \begin{bmatrix} \mathbf{R}^T & \mathbf{0} \\ \mathbf{0} & 1 \end{bmatrix} \begin{bmatrix} \mathbf{H}_{n-1} & \mathbf{h} \\ \mathbf{h}^T & h_{nn} \end{bmatrix} \begin{bmatrix} \mathbf{R} & \mathbf{0} \\ \mathbf{0} & 1 \end{bmatrix} = \begin{bmatrix} \mathbf{I}_{n-1} & \mathbf{R}^T \mathbf{h} \\ \mathbf{h}^T \mathbf{R} & h_{nn} \end{bmatrix}$$

If we define

$$\mathbf{T} = \begin{bmatrix} \mathbf{I}_{n-1} & -\mathbf{R}^T \mathbf{h} \\ \mathbf{0} & 1 \end{bmatrix}$$

then

$$\mathbf{T}^T \mathbf{S}^T \mathbf{H} \mathbf{S} \mathbf{T} = \begin{bmatrix} \mathbf{I}_{n-1} & \mathbf{0} \\ -\mathbf{h}^T \mathbf{R} & 1 \end{bmatrix} \begin{bmatrix} \mathbf{I}_{n-1} & \mathbf{R}^T \mathbf{h} \\ \mathbf{h}^T \mathbf{R} & h_{nn} \end{bmatrix} \begin{bmatrix} \mathbf{I}_{n-1} & -\mathbf{R}^T \mathbf{h} \\ \mathbf{0} & 1 \end{bmatrix}$$

$$= \begin{bmatrix} \mathbf{I}_{n-1} & \mathbf{0} \\ \mathbf{0} & h_{nn} - \mathbf{h}^T \mathbf{R} \mathbf{R}^T \mathbf{h} \end{bmatrix}$$

So if we let $\mathbf{U} = \mathbf{ST}$ and $\alpha = h_{nn} - \mathbf{h}^T \mathbf{RR}^T \mathbf{h}$, then

$$\mathbf{U}^T \mathbf{HU} = \begin{bmatrix} 1 & & & \\ & \ddots & & \\ & & 1 & \\ & & & \alpha \end{bmatrix}$$

which implies that

$$(\det \mathbf{U})^2 \det \mathbf{H} = \alpha$$

As $\det \mathbf{H} > 0$, we obtain $\alpha > 0$ and, therefore, $\mathbf{U}^T \mathbf{HU}$ is positive definite which implies the positive definiteness of $\mathbf{H}$.

(c) The proof of the necessity is similar to the proof of part (b). If

$$\mathbf{d} = [0 \ \cdots \ 0 \ d_{l_1} \ 0 \ \cdots \ 0 \ d_{l_2} \ 0 \ \cdots \ 0 \ d_{l_i} \ 0 \ \cdots \ 0]^T$$

and $\mathbf{H}$ is positive semidefinite, then

$$\mathbf{d}^T \mathbf{Hd} = \mathbf{d}_0^T \mathbf{H}_i^{(l)} \mathbf{d}_0 \geq 0$$

for all $\mathbf{d}_0 \neq \mathbf{0}$ where

$$\mathbf{d}_0 = [d_{l_1} \ d_{l_2} \ \cdots \ d_{l_i}]^T$$

and $\mathbf{H}_i^{(l)}$ is an $i \times i$ principal submatrix. Hence $\mathbf{H}_i^{(l)}$ is positive semidefinite for all possible selections of rows (and columns) from the set $l = \{l_1, l_2, \ldots, l_i,\}$ with $1 \leq l_1 \leq l_2 < \ldots < l_i \leq n\}$ and $i = 1, 2, \ldots, n$. Now from part (a)

$$\det (\mathbf{H}_i^l) \geq 0 \qquad \text{for } 1, 2, \ldots, n.$$

The proof of sufficiency is rather lengthy and is omitted. The interested reader is referred to Chap. 7 of [3].

(d) If $\mathbf{H}_i$ is negative definite, then $-\mathbf{H}_i$ is positive definite by definition and hence the proof of part (b) applies to part (d).

(e) If $\mathbf{H}_i^{(l)}$ is negative semidefinite, then $-\mathbf{H}_i^{(l)}$ is positive semidefinite by definition and hence the proof of part (c) applies to part (e).

(f) If neither part (c) nor part (e) holds, then $\mathbf{d}^T \mathbf{Hd}$ can be positive or negative and hence $\mathbf{H}$ is indefinite.   ∎

Example 2.8 Characterize the Hessian matrices in Examples 2.6 and 2.7 by using the determinant method.

Solution Let

$$\Delta_i = \det (\mathbf{H}_i)$$

be the leading principal minors of $\mathbf{H}$. From Example 2.6, we have

$$\Delta_1 = 1, \quad \Delta_2 = -2, \quad \Delta_3 = -18$$

and if $\Delta_i' = \det(-\mathbf{H}_i)$, then

$$\Delta_1' = -1, \quad \Delta_2' = -2, \quad \Delta_3' = 18$$

since

$$\det(-\mathbf{H}_i) = (-1)^i \det(\mathbf{H}_i)$$

Hence $\mathbf{H}$ is indefinite.

From Example 2.7, we get

$$\Delta_1 = 4, \quad \Delta_2 = 8, \quad \Delta_3 = 400$$

Hence $\mathbf{H}$ is positive definite.   ∎

Example 2.9 Find and classify the stationary points of

$$f(\mathbf{x}) = x_1^2 + 2x_1 x_2 + 2x_2^2 + 2x_1 + x_2$$

Solution The first partial derivatives of $f(\mathbf{x})$ are

$$\frac{\partial f}{\partial x_1} = 2x_1 + 2x_2 + 2$$

$$\frac{\partial f}{\partial x_2} = 2x_1 + 4x_2 + 1$$

If $\mathbf{g} = \mathbf{0}$, then

$$2x_1 + 2x_2 + 2 = 0$$

$$2x_1 + 4x_2 + 1 = 0$$

and so there is a stationary point at

$$\mathbf{x} = \mathbf{x}_1 = [-\tfrac{3}{2} \ \tfrac{1}{2}]^T$$

The Hessian is deduced as

$$\mathbf{H} = \begin{bmatrix} 2 & 2 \\ 2 & 4 \end{bmatrix}$$

and since $\Delta_1 = 2$ and $\Delta_2 = 4$, $\mathbf{H}$ is positive definite. Therefore, $\mathbf{x}_1$ is a minimizer. ∎

Example 2.10 Find and classify the stationary points of function

$$f(\mathbf{x}) = x_1^2 - x_2^2 + x_3^2 - 2x_1 x_3 - x_2 x_3 + 4x_1 + 12$$

Solution The first-order partial derivatives of $f(\mathbf{x})$ are

$$\frac{\partial f}{\partial x_1} = 2x_1 - 2x_3 + 4$$

$$\frac{\partial f}{\partial x_2} = -2x_2 - x_3$$

$$\frac{\partial f}{\partial x_3} = -2x_1 - x_2 + 2x_3$$

On equating the gradient to zero and then solving the simultaneous equations obtained, the stationary point $\mathbf{x}_1 = [-10 \ 4 \ -8]^T$ can be deduced. The Hessian is

$$\mathbf{H} = \begin{bmatrix} 2 & 0 & -2 \\ 0 & -2 & -1 \\ -2 & -1 & 2 \end{bmatrix}$$

and since $\Delta_1 = 2$, $\Delta_2 = -4$, $\Delta_3 = -2$, and $\Delta_1' = -2$, $\Delta_2' = -4$, $\Delta_3' = 2$, $\mathbf{H}$ is indefinite. Therefore, point $\mathbf{x}_1 = [-10 \ 4 \ -8]^T$ is a saddle point. The solution can be readily checked by diagonalizing $\mathbf{H}$ as

$$\hat{\mathbf{H}} = \begin{bmatrix} 2 & 0 & 0 \\ 0 & -2 & 0 \\ 0 & 0 & 2\frac{1}{2} \end{bmatrix}$$

∎

2.7 Convex and Concave Functions

Usually, in practice, the function to be minimized has several extremum points and, consequently, uncertainty arises as to whether the extremum point located by an optimization algorithm is the global one. In a specific class of functions referred to as convex and concave functions, any local extremum point is also a global extremum point. Therefore, if the objective function is convex or concave, optimality can be assured. The basic principles relating to convex and concave functions entail a collection of definitions and theorems.

Definition 2.7 *Convex sets* A set $\mathcal{R}_c \in E^n$ is said to be *convex* if for every pair of points $\mathbf{x}_1$, $\mathbf{x}_2 \in \mathcal{R}_c$ and for every real number α in the range $0 < \alpha < 1$, the point

$$\mathbf{x} = \alpha \mathbf{x}_1 + (1 - \alpha)\mathbf{x}_2$$

is located in $\mathcal{R}_c$, i.e., $\mathbf{x} \in \mathcal{R}_c$. ∎

In effect, if any two points $\mathbf{x}_1$, $\mathbf{x}_2 \in \mathcal{R}_c$ are connected by a straight line, then $\mathcal{R}_c$ is convex if every point on the line segment between $\mathbf{x}_1$ and $\mathbf{x}_2$ is a member of $\mathcal{R}_c$.

Fig. 2.5 Convexity in sets

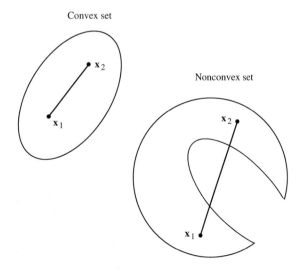

If some points on the line segment between $\mathbf{x}_1$ and $\mathbf{x}_2$ are not in $\mathcal{R}_c$, the set is said to be *nonconvex*. Convexity in sets is illustrated in Fig. 2.5.

The concept of convexity can also be applied to functions.

Definition 2.8 *Convex functions*

(a) A function $f(\mathbf{x})$ defined over a convex set $\mathcal{R}_c$ is said to be convex if for every pair of points $\mathbf{x}_1$, $\mathbf{x}_2 \in \mathcal{R}_c$ and every real number α in the range $0 < \alpha < 1$, the inequality

$$f[\alpha\mathbf{x}_1 + (1 - \alpha)\mathbf{x}_2] \leq \alpha f(\mathbf{x}_1) + (1 - \alpha)f(\mathbf{x}_2) \tag{2.9}$$

holds. If $\mathbf{x}_1 \neq \mathbf{x}_2$ and

$$f[\alpha\mathbf{x}_1 + (1 - \alpha)\mathbf{x}_2] < \alpha f(\mathbf{x}_1) + (1 - \alpha)f(\mathbf{x}_2)$$

then $f(\mathbf{x})$ is said to be *strictly convex*.
(b) If $\phi(\mathbf{x})$ is defined over a convex set $\mathcal{R}_c$ and $f(\mathbf{x}) = -\phi(\mathbf{x})$ is convex, then $\phi(\mathbf{x})$ is said to be *concave*. If $f(\mathbf{x})$ is strictly convex, $\phi(\mathbf{x})$ is *strictly concave*. ∎

In the left-hand side of Eq. (2.9), function $f(\mathbf{x})$ is evaluated on the line segment joining points $\mathbf{x}_1$ and $\mathbf{x}_2$ whereas in the right-hand side of Eq. (2.9) an approximate value is obtained for $f(\mathbf{x})$ based on linear interpolation. Thus a function is convex if linear interpolation between any two points overestimates the value of the function. The functions shown in Fig. 2.6a, b are convex whereas that in Fig. 2.6c is nonconvex.

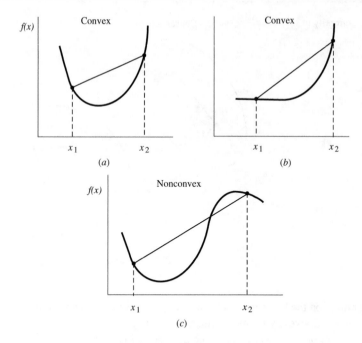

Fig. 2.6 Convexity in functions

Theorem 2.10 *Convexity of linear combination of convex functions If*

$$f(\mathbf{x}) = af_1(\mathbf{x}) + bf_2(\mathbf{x})$$

where a, $b \geq 0$ and $f_1(\mathbf{x})$, $f_2(\mathbf{x})$ are convex functions on the convex set $\mathcal{R}_c$, then $f(\mathbf{x})$ is convex on the set $\mathcal{R}_c$.

Proof Since $f_1(\mathbf{x})$ and $f_2(\mathbf{x})$ are convex, and a, $b \geq 0$, then for $\mathbf{x} = \alpha\mathbf{x}_1 + (1-\alpha)\mathbf{x}_2$ we have

$$af_1[\alpha\mathbf{x}_1 + (1 - \alpha)\mathbf{x}_2] \leq a[\alpha f_1(\mathbf{x}_1) + (1 - \alpha)f_1(\mathbf{x}_2)]$$

$$bf_2[\alpha\mathbf{x}_1 + (1 - \alpha)\mathbf{x}_2] \leq b[\alpha f_2(\mathbf{x}_1) + (1 - \alpha)f_2(\mathbf{x}_2)]$$

where $0 < \alpha < 1$. Hence

$$f(\mathbf{x}) = af_1(\mathbf{x}) + bf_2(\mathbf{x})$$

$$\begin{aligned}
f[\alpha\mathbf{x}_1 + (1 - \alpha)\mathbf{x}_2] &= af_1[\alpha\mathbf{x}_1 + (1 - \alpha)\mathbf{x}_2] + bf_2[\alpha\mathbf{x}_1 + (1 - \alpha)\mathbf{x}_2] \\
&\leq \alpha[af_1(\mathbf{x}_1) + bf_2(\mathbf{x}_1)] + (1 - \alpha)[af_1(\mathbf{x}_2) \\
&\quad + bf_2(\mathbf{x}_2)]
\end{aligned}$$

Since

$$af_1(\mathbf{x}_1) + bf_2(\mathbf{x}_1) = f(\mathbf{x}_1)$$
$$af_1(\mathbf{x}_2) + bf_2(\mathbf{x}_2) = f(\mathbf{x}_2)$$

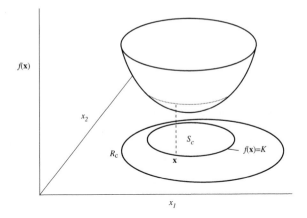

Fig. 2.7 Graphical construction for Theorem 2.11

the above inequality can be expressed as

$$f[\alpha \mathbf{x}_1 + (1 - \alpha)\mathbf{x}_2] \leq \alpha f(\mathbf{x}_1) + (1 - \alpha)f(\mathbf{x}_2)$$

that is, $f(\mathbf{x})$ is convex. ∎

Theorem 2.11 *Relation between convex functions and convex sets If $f(\mathbf{x})$ is a convex function on a convex set $\mathcal{R}_c$, then the set*

$$\mathcal{S}_c = \{\mathbf{x} : \mathbf{x} \in \mathcal{R}_c, \ f(\mathbf{x}) \leq K\}$$

is convex for every real number K.

Proof If $\mathbf{x}_1, \ \mathbf{x}_2 \in \mathcal{S}_c$, then $f(\mathbf{x}_1) \leq K$ and $f(\mathbf{x}_2) \leq K$ from the definition of $\mathcal{S}_c$. Since $f(\mathbf{x})$ is convex

$$f[\alpha \mathbf{x}_1 + (1 - \alpha)\mathbf{x}_2] \leq \alpha f(\mathbf{x}_1) + (1 - \alpha)f(\mathbf{x}_2)$$
$$\leq \alpha K + (1 - \alpha)K$$

or

$$f(\mathbf{x}) \leq K \quad \text{for } \mathbf{x} = \alpha \mathbf{x}_1 + (1 - \alpha)\mathbf{x}_2 \quad \text{and} \quad 0 < \alpha < 1$$

Therefore

$$\mathbf{x} \in \mathcal{S}_c$$

that is, $\mathcal{S}_c$ is convex by virtue of Definition 2.7. ∎

Theorem 2.11 is illustrated in Fig. 2.7, where set $\mathcal{S}_c$ is convex if $f(\mathbf{x})$ is a convex function on convex set $\mathcal{R}_c$.

An alternative view of convexity can be generated by examining some theorems which involve the gradient and Hessian of $f(\mathbf{x})$.

Theorem 2.12 *Property of convex functions relating to gradient* If $f(\mathbf{x}) \in C^1$, *then $f(\mathbf{x})$ is convex over a convex set $\mathcal{R}_c$ if and only if*

$$f(\mathbf{x}_1) \geq f(\mathbf{x}) + \mathbf{g}(\mathbf{x})^T (\mathbf{x}_1 - \mathbf{x})$$

for all $\mathbf{x}$ and $\mathbf{x}_1 \in \mathcal{R}_c$, where $\mathbf{g}(\mathbf{x})$ is the gradient of $f(\mathbf{x})$.

Proof   The proof of this theorem consists of two parts. First we prove that if $f(\mathbf{x})$ is convex, the inequality holds. Then we prove that if the inequality holds, $f(\mathbf{x})$ is convex. The two parts constitute the necessary and sufficient conditions of the theorem. If $f(\mathbf{x})$ is convex, then for all α in the range $0 < \alpha < 1$

$$f[\alpha\mathbf{x}_1 + (1 - \alpha)\mathbf{x}] \leq \alpha f(\mathbf{x}_1) + (1 - \alpha)f(\mathbf{x})$$

or

$$f[\mathbf{x} + \alpha(\mathbf{x}_1 - \mathbf{x})] - f(\mathbf{x}) \leq \alpha[f(\mathbf{x}_1) - f(\mathbf{x})]$$

As $\alpha \to 0$, the Taylor series of $f[\mathbf{x} + \alpha(\mathbf{x}_1 - \mathbf{x})]$ yields

$$f(\mathbf{x}) + \mathbf{g}(\mathbf{x})^T \alpha(\mathbf{x}_1 - \mathbf{x}) - f(\mathbf{x}) \leq \alpha[f(\mathbf{x}_1) - f(\mathbf{x})]$$

and so

$$f(\mathbf{x}_1) \geq f(\mathbf{x}) + \mathbf{g}(\mathbf{x})^T (\mathbf{x}_1 - \mathbf{x}) \tag{2.10}$$

Now if this inequality holds at points $\mathbf{x}$ and $\mathbf{x}_2 \in \mathcal{R}_c$, then

$$f(\mathbf{x}_2) \geq f(\mathbf{x}) + \mathbf{g}(\mathbf{x})^T (\mathbf{x}_2 - \mathbf{x}) \tag{2.11}$$

Hence Eqs. (2.10) and (2.11) yield

$$\alpha f(\mathbf{x}_1) + (1 - \alpha)f(\mathbf{x}_2) \geq \alpha f(\mathbf{x}) + \alpha\mathbf{g}(\mathbf{x})^T (\mathbf{x}_1 - \mathbf{x}) + (1 - \alpha)f(\mathbf{x})$$
$$+ (1 - \alpha)\mathbf{g}(\mathbf{x})^T (\mathbf{x}_2 - \mathbf{x})$$

or

$$\alpha f(\mathbf{x}_1) + (1 - \alpha)f(\mathbf{x}_2) \geq f(\mathbf{x}) + \mathbf{g}^T(\mathbf{x})[\alpha\mathbf{x}_1 + (1 - \alpha)\mathbf{x}_2 - \mathbf{x}]$$

With the substitution

$$\mathbf{x} = \alpha\mathbf{x}_1 + (1 - \alpha)\mathbf{x}_2$$

we obtain

$$f[\alpha\mathbf{x}_1 + (1 - \alpha)\mathbf{x}_2] \leq \alpha f(\mathbf{x}_1) + (1 - \alpha)f(\mathbf{x}_2)$$

for $0 < \alpha < 1$. Therefore, from Definition 2.8 $f(\mathbf{x})$ is convex.   ∎

Theorem 2.12 states that a linear approximation of $f(\mathbf{x})$ at point $\mathbf{x}_1$ based on the derivatives of $f(\mathbf{x})$ at $\mathbf{x}$ underestimates the value of the function. This fact is illustrated in Fig. 2.8.

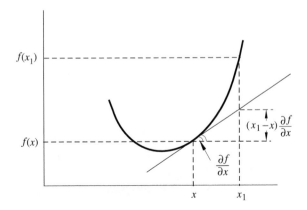

Fig. 2.8 Graphical construction for Theorem 2.12

Theorem 2.13 *Property of convex functions relating to the Hessian* *A function* $f(\mathbf{x}) \in C^2$ *is convex over a convex set* $\mathcal{R}_c$ *if and only if the Hessian* $H(\mathbf{x})$ *of* $f(\mathbf{x})$ *is positive semidefinite for* $\mathbf{x} \in \mathcal{R}_c$.

Proof If $\mathbf{x}_1 = \mathbf{x} + \mathbf{d}$ where $\mathbf{x}_1$ and $\mathbf{x}$ are arbitrary points in $\mathcal{R}_c$, then the Taylor series yields

$$f(\mathbf{x}_1) = f(\mathbf{x}) + \mathbf{g}(\mathbf{x})^T (\mathbf{x}_1 - \mathbf{x}) + \tfrac{1}{2}\mathbf{d}^T H(\mathbf{x} + \alpha\mathbf{d})\mathbf{d} \qquad (2.12)$$

where $0 \le \alpha \le 1$ (see Eq. (2.4h)). Now if $H(\mathbf{x})$ is positive semidefinite everywhere in $\mathcal{R}_c$, then

$$\tfrac{1}{2}\mathbf{d}^T H(\mathbf{x} + \alpha\mathbf{d})\mathbf{d} \ge 0$$

and so

$$f(\mathbf{x}_1) \ge f(\mathbf{x}) + \mathbf{g}(\mathbf{x})^T (\mathbf{x}_1 - \mathbf{x})$$

Therefore, from Theorem 2.12, $f(\mathbf{x})$ is convex.

If $H(\mathbf{x})$ is not positive semidefinite everywhere in $\mathcal{R}_c$, then a point $\mathbf{x}$ and at least a $\mathbf{d}$ exist such that

$$\mathbf{d}^T H(\mathbf{x} + \alpha\mathbf{d})\mathbf{d} < 0$$

and so Eq. (2.12) yields

$$f(\mathbf{x}_1) < f(\mathbf{x}) + \mathbf{g}(\mathbf{x})^T (\mathbf{x}_1 - \mathbf{x})$$

and $f(\mathbf{x})$ is nonconvex from Theorem 2.12. Therefore, $f(\mathbf{x})$ is convex if and only if $H(\mathbf{x})$ is positive semidefinite everywhere in $\mathcal{R}_c$.   ∎

For a strictly convex function, Theorems 2.11–2.13 are modified as follows.

Theorem 2.14 *Properties of strictly convex functions*

(a) If $f(\mathbf{x})$ is a strictly convex function on a convex set $\mathcal{R}_c$, then the set

$$S_c = \{\mathbf{x} : \mathbf{x} \in \mathcal{R}_c \ \text{for} \ f(\mathbf{x}) < K\}$$

is convex for every real number K.
(b) If $f(\mathbf{x}) \in C^1$, then $f(\mathbf{x})$ is strictly convex over a convex set if and only if

$$f(\mathbf{x}_1) > f(\mathbf{x}) + \mathbf{g}(\mathbf{x})^T(\mathbf{x}_1 - \mathbf{x})$$

for all $\mathbf{x}$ and $\mathbf{x}_1 \in \mathcal{R}_c$ where $\mathbf{g}(\mathbf{x})$ is the gradient of $f(\mathbf{x})$.
(c) A function $f(\mathbf{x}) \in C^2$ is strictly convex over a convex set $\mathcal{R}_c$ if and only if the Hessian $\mathbf{H}(\mathbf{x})$ is positive definite for $\mathbf{x} \in \mathcal{R}_c$.   ∎

If the second-order sufficient conditions for a minimum hold at $\mathbf{x}^*$ as in Theorem 2.4, in which case $\mathbf{x}^*$ is a strong local minimizer, then from Theorem 2.14(c), $f(\mathbf{x})$ must be strictly convex in the neighborhood of $\mathbf{x}^*$. Consequently, convexity assumes considerable importance even though the class of convex functions is quite restrictive.

If $\phi(\mathbf{x})$ is defined over a convex set $\mathcal{R}_c$ and $f(\mathbf{x}) = -\phi(\mathbf{x})$ is strictly convex, then $\phi(\mathbf{x})$ is strictly concave and the Hessian of $\phi(\mathbf{x})$ is negative definite. Conversely, if the Hessian of $\phi(\mathbf{x})$ is negative definite, then $\phi(\mathbf{x})$ is strictly concave.

Example 2.11 Check the following functions for convexity:

(a) $f(\mathbf{x}) = e^{x_1} + x_2^2 + 5$
(b) $f(\mathbf{x}) = 3x_1^2 - 5x_1x_2 + x_2^2$
(c) $f(\mathbf{x}) = \frac{1}{4}x_1^4 - x_1^2 + x_2^2$
(d) $f(\mathbf{x}) = 50 + 10x_1 + x_2 - 6x_1^2 - 3x_2^2$

Solution In each case the problem reduces to the derivation and characterization of the Hessian $\mathbf{H}$.

(a) The Hessian can be obtained as

$$\mathbf{H} = \begin{bmatrix} e^{x_1} & 0 \\ 0 & 2 \end{bmatrix}$$

For $-\infty < x_1 < \infty$, $\mathbf{H}$ is positive definite and $f(\mathbf{x})$ is strictly convex.

(b) In this case, we have

$$\mathbf{H} = \begin{bmatrix} 6 & -5 \\ -5 & 2 \end{bmatrix}$$

Since $\Delta_1 = 6$, $\Delta_2 = -13$ and $\Delta_1' = -6$, $\Delta_2' = -13$, where $\Delta_i = \det(\mathbf{H}_i)$ and $\Delta_i' = \det(-\mathbf{H}_i)$, $\mathbf{H}$ is indefinite. Thus $f(\mathbf{x})$ is neither convex nor concave.

(c) For this example, we get

$$\mathbf{H} = \begin{bmatrix} 3x_1^2 - 2 & 0 \\ 0 & 2 \end{bmatrix}$$

For $x_1 \leq -\sqrt{2/3}$ and $x_1 \geq \sqrt{2/3}$, $\mathbf{H}$ is positive semidefinite and $f(\mathbf{x})$ is convex; for $x_1 < -\sqrt{2/3}$ and $x_1 > \sqrt{2/3}$, $\mathbf{H}$ is positive definite and $f(\mathbf{x})$ is strictly convex; for $-\sqrt{2/3} < x_1 < \sqrt{2/3}$, $\mathbf{H}$ is indefinite, and $f(\mathbf{x})$ is neither convex nor concave.

(d) As before

$$\mathbf{H} = \begin{bmatrix} -12 & 0 \\ 0 & -6 \end{bmatrix}$$

In this case $\mathbf{H}$ is negative definite, and $f(\mathbf{x})$ is strictly concave.   ∎

2.8   Optimization of Convex Functions

The above theorems and results can now be used to deduce the following three important theorems.

Theorem 2.15   Relation between local and global minimizers in convex functions
If $f(\mathbf{x})$ is a convex function defined on a convex set $\mathcal{R}_c$, then

(a) *the set of points $\mathcal{S}_c$ where $f(\mathbf{x})$ is minimum is convex;*
(b) *any local minimizer of $f(\mathbf{x})$ is a global minimizer.*

Proof   (a) If F^* is a minimum of $f(\mathbf{x})$, then $\mathcal{S}_c = \{\mathbf{x} : f(\mathbf{x}) \leq F^*, \mathbf{x} \in \mathcal{R}_c\}$ is convex by virtue of Theorem 2.11.
  (b) If $\mathbf{x}^* \in \mathcal{R}_c$ is a local minimizer but there is another point $\mathbf{x}^{**} \in \mathcal{R}_c$ which is a global minimizer such that

$$f(\mathbf{x}^{**}) < f(\mathbf{x}^*)$$

then on line $\mathbf{x} = \alpha \mathbf{x}^{**} + (1 - \alpha)\mathbf{x}^*$

$$f[\alpha \mathbf{x}^{**} + (1 - \alpha)\mathbf{x}^*] \leq \alpha f(\mathbf{x}^{**}) + (1 - \alpha)f(\mathbf{x}^*)$$
$$< \alpha f(\mathbf{x}^*) + (1 - \alpha)f(\mathbf{x}^*)$$

or

$$f(\mathbf{x}) < f(\mathbf{x}^*) \qquad \text{for all } \alpha$$

This contradicts the fact that $\mathbf{x}^*$ is a local minimizer and so

$$f(\mathbf{x}) \geq f(\mathbf{x}^*)$$

for all $\mathbf{x} \in \mathcal{R}_c$. Therefore, any local minimizers are located in a convex set, and all are global minimizers.   ∎

Theorem 2.16 *Existence of a global minimizer in convex functions* If $f(\mathbf{x}) \in C^1$ *is a convex function on a convex set* $\mathcal{R}_c$ *and there is a point* $\mathbf{x}^*$ *such that*

$$\mathbf{g}(\mathbf{x}^*)^T \mathbf{d} \geq 0 \quad \text{where } \mathbf{d} = \mathbf{x}_1 - \mathbf{x}^*$$

for all $\mathbf{x}_1 \in \mathcal{R}_c$, *then* $\mathbf{x}^*$ *is a global minimizer of* $f(\mathbf{x})$.

Proof From Theorem 2.12

$$f(\mathbf{x}_1) \geq f(\mathbf{x}^*) + \mathbf{g}(\mathbf{x}^*)^T(\mathbf{x}_1 - \mathbf{x}^*)$$

where $\mathbf{g}(\mathbf{x}^*)$ is the gradient of $f(\mathbf{x})$ at $\mathbf{x} = \mathbf{x}^*$. Since

$$\mathbf{g}(\mathbf{x}^*)^T(\mathbf{x}_1 - \mathbf{x}^*) \geq 0$$

we have

$$f(\mathbf{x}_1) \geq f(\mathbf{x}^*)$$

and so $\mathbf{x}^*$ is a local minimizer. By virtue of Theorem 2.15, $\mathbf{x}^*$ is also a global minimizer.

Similarly, if $f(\mathbf{x})$ is a strictly convex function and

$$\mathbf{g}(\mathbf{x}^*)^T \mathbf{d} \geq 0$$

then $\mathbf{x}^*$ is a strong global minimizer. ∎

The above theorem states, in effect, that if $f(\mathbf{x})$ is convex, then the first-order necessary conditions become sufficient for $\mathbf{x}^*$ to be a global minimizer.

Since a convex function of one variable is in the form of the letter U whereas a convex function of two variables is in the form of a bowl, there are no theorems analogous to Theorems 2.15 and 2.16 pertaining to the maximization of a convex function. However, the following theorem, which is intuitively plausible, is sometimes useful.

Theorem 2.17 *Location of maximum of a convex function* If $f(\mathbf{x})$ *is a convex function defined on a bounded, closed, convex set* $\mathcal{R}_c$, *then if* $f(\mathbf{x})$ *has a single maximum over* $\mathcal{R}_c$, *it occurs at the boundary of* $\mathcal{R}_c$.

Proof If point $\mathbf{x}$ is in the interior of $\mathcal{R}_c$, a line can be drawn through $\mathbf{x}$ which intersects the boundary at two points, say, $\mathbf{x}_1$ and $\mathbf{x}_2$, since $\mathcal{R}_c$ is bounded and closed. Since $f(\mathbf{x})$ is convex, some α exists in the range $0 < \alpha < 1$ such that

$$\mathbf{x} = \alpha\mathbf{x}_1 + (1 - \alpha)\mathbf{x}_2$$

and

$$f(\mathbf{x}) \leq \alpha f(\mathbf{x}_1) + (1 - \alpha)f(\mathbf{x}_2)$$

If $f(\mathbf{x}_1) > f(\mathbf{x}_2)$, we have

$$f(\mathbf{x}) < \alpha f(\mathbf{x}_1) + (1 - \alpha)f(\mathbf{x}_1)$$
$$= f(\mathbf{x}_1)$$

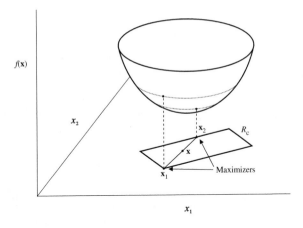

Fig. 2.9 Graphical construction for Theorem 2.17

If

$$f(\mathbf{x}_1) < f(\mathbf{x}_2)$$

we obtain

$$f(\mathbf{x}) < \alpha f(\mathbf{x}_2) + (1 - \alpha) f(\mathbf{x}_2)$$
$$= f(\mathbf{x}_2)$$

Now if

$$f(\mathbf{x}_1) = f(\mathbf{x}_2)$$

the result

$$f(\mathbf{x}) \le f(\mathbf{x}_1) \quad \text{and} \quad f(\mathbf{x}) \le f(\mathbf{x}_2)$$

is obtained. Evidently, in all possibilities a maximizer occurs on the boundary of $\mathcal{R}_c$. ∎

This theorem is illustrated in Fig. 2.9.

Problems

2.1 (*a*) Obtain a quadratic approximation for the function
$$f(\mathbf{x}) = 2x_1^3 + x_2^2 + x_1^2 x_2^2 + 4x_1 x_2 + 3$$
at point $\mathbf{x} + \delta$ if $\mathbf{x}^T = [1 \ 1]$.
(*b*) Now obtain a linear approximation.

2.2 An *n*-variable quadratic function is given by
$$f(\mathbf{x}) = a + \mathbf{b}^T \mathbf{x} + \tfrac{1}{2} \mathbf{x}^T \mathbf{Q} \mathbf{x}$$

where $\mathbf{Q}$ is an $n \times n$ symmetric matrix. Show that the gradient and Hessian of $f(\mathbf{x})$ are given by

$$\mathbf{g} = \mathbf{b} + \mathbf{Q}\mathbf{x} \quad \text{and} \quad \nabla^2 f(\mathbf{x}) = \mathbf{Q}$$

respectively.

2.3 Point $\mathbf{x}_a = [2\ 4]^T$ is a possible minimizer of the problem

$$\text{minimize } f(\mathbf{x}) = \tfrac{1}{4}[x_1^2 + 4x_2^2 - 4(3x_1 + 8x_2) + 100]$$

$$\text{subject to: } x_1 = 2, \ x_2 \geq 0$$

(a) Find the feasible directions.
(b) Check if the second-order necessary conditions are satisfied.

2.4 Points $\mathbf{x}_a = [0\ 3]^T$, $\mathbf{x}_b = [4\ 0]^T$, $\mathbf{x}_c = [4\ 3]^T$ are possible maximizers of the problem

$$\text{maximize } f(\mathbf{x}) = 2(4x_1 + 3x_2) - (x_1^2 + x_2^2 + 25)$$

$$\text{subject to: } x_1 \geq 0, \ x_2 \geq 0$$

(a) Find the feasible directions.
(b) Check if the second-order necessary conditions are satisfied.

2.5 Point $\mathbf{x}_a = [4\ -1]^T$ is a possible minimizer of the problem

$$\text{minimize } f(\mathbf{x}) = \tfrac{16}{x_1} - x_2$$

$$\text{subject to: } x_1 + x_2 = 3, \ x_1 \geq 0$$

(a) Find the feasible directions.
(b) Check if the second-order necessary conditions are satisfied.

2.6 Classify the following matrices as positive definite, positive semidefinite, etc. by using the diagonalization technique described in Example 2.6:

(a) $\mathbf{H} = \begin{bmatrix} 5 & 3 & 1 \\ 3 & 4 & 2 \\ 1 & 2 & 6 \end{bmatrix}$

(b) $\mathbf{H} = \begin{bmatrix} -5 & 1 & 1 \\ 1 & -2 & 2 \\ 1 & 2 & -4 \end{bmatrix}$

(c) $\mathbf{H} = \begin{bmatrix} -1 & 2 & -3 \\ 2 & 4 & 5 \\ -3 & 5 & -20 \end{bmatrix}$

2.7 Check the results in Problem 2.6 by using the determinant method.

2.8 Classify the following matrices by using the eigenvalue method described in Theorems 2.7 and 2.8:

(a) $\mathbf{H} = \begin{bmatrix} 2 & 3 \\ 3 & 4 \end{bmatrix}$

(b) $\mathbf{H} = \begin{bmatrix} 10 & 4 \\ 0 & 2 & 0 \\ 4 & 0 & 18 \end{bmatrix}$

2.9 Consider a Hessian matrix $\mathbf{H}$.

(a) Show that $\mathbf{H}$ cannot be positive or negative definite if there is at least one zero element on the diagonal of $\mathbf{H}$.

(b) Show that $\mathbf{H}$ is indefinite if there is at least one positive element and one negative element on the diagonal of $\mathbf{H}$.

2.10 One of the points $\mathbf{x}_a = [1 \ -1]^T$, $\mathbf{x}_b = [0 \ 0]^T$, $\mathbf{x}_c = [1 \ 1]^T$ minimizes the function

$$f(\mathbf{x}) = 100(x_2 - x_1^2)^2 + (1 - x_1)^2$$

By using appropriate tests, identify the minimizer.

2.11 An optimization algorithm has given a solution $\mathbf{x}_a = [0.6959 \ -1.3479]^T$ for the problem

$$\text{minimize} \ \ f(\mathbf{x}) = x_1^4 + x_1 x_2 + (1 + x_2)^2$$

(a) Classify the general Hessian of $f(\mathbf{x})$ (i.e., positive definite, ..., etc.).

(b) Determine whether $\mathbf{x}_a$ is a minimizer, maximizer, or saddle point.

2.12 Find and classify the stationary points for the function

$$f(\mathbf{x}) = x_1^2 - x_2^2 + x_3^2 - 2x_1 x_3 - x_2 x_3 + 4x_1 + 12$$

2.13 Find and classify the stationary points for the following functions:

(a) $f(\mathbf{x}) = 2x_1^2 + x_2^2 - 2x_1 x_2 + 2x_1^3 + x_1^4$

(b) $f(\mathbf{x}) = x_1^2 x_2^2 - 4x_1^2 x_2 + 4x_1^2 + 2x_1 x_2^2 + x_2^2 - 8x_1 x_2 + 8x_1 - 4x_2$

2.14 Show that

$$f(\mathbf{x}) = (x_2 - x_1^2)^2 + x_1^5$$

has only one stationary point which is neither a minimizer nor a maximizer.

2.15 Investigate the following functions and determine whether they are convex or concave:

(a) $f(\mathbf{x}) = x_1^2 + \cosh x_2$
(b) $f(\mathbf{x}) = x_1^2 + 2x_2^2 + 2x_3^2 + x_4^2 - x_1x_2 + x_1x_3 - 2x_2x_4 + x_1x_4$
(c) $f(\mathbf{x}) = x_1^2 - 2x_2^2 - 2x_3^2 + x_4^2 - x_1x_2 + x_1x_3 - 2x_2x_4 + x_1x_4$

2.16 A given quadratic function $f(\mathbf{x})$ is known to be convex for $\|\mathbf{x}\|_2 < \varepsilon$. Show that it is convex for all $\mathbf{x} \in E^n$.

2.17 Two functions $f_1(\mathbf{x})$ and $f_2(\mathbf{x})$ are convex over a convex set $\mathcal{R}_c$. Show that

$$f(\mathbf{x}) = \alpha f_1(\mathbf{x}) + \beta f_2(\mathbf{x})$$

where α and β are nonnegative scalars is convex over $\mathcal{R}_c$.

2.18 Assume that functions $f_1(\mathbf{x})$ and $f_2(\mathbf{x})$ are convex and let

$$f(\mathbf{x}) = \max\{f_1(\mathbf{x}),\ f_2(\mathbf{x})\}$$

Show that $f(\mathbf{x})$ is a convex function.

2.19 Let $\gamma(t)$ be a single-variable convex function which is monotonic non-decreasing, i.e., $\gamma(t_1) \geq \gamma(t_2)$ for $t_1 > t_2$. Show that the compound function $\gamma[f(\mathbf{x})]$ is convex if $f(\mathbf{x})$ is convex [2].

2.20 Let $f(\mathbf{x})$ be a smooth convex function where $\mathbf{x} \in R^{n \times 1}$ is related to another variable $\mathbf{u}$ by the equation $\mathbf{x} = A\mathbf{u} + \mathbf{b}$ with $A \in R^{n \times m}$ and $\mathbf{b} \in R^{n \times 1}$ and define function $F(\mathbf{u}) = f(A\mathbf{u} + \mathbf{b})$. Show that

(a) $\nabla F(\mathbf{u}) = \mathbf{A}^T \cdot \nabla f(\mathbf{x})\big|_{\mathbf{x}=A\mathbf{u}+\mathbf{b}}$
(b) $\nabla^2 F(\mathbf{u}) = \mathbf{A}^T \cdot \nabla^2 f(\mathbf{x})\big|_{\mathbf{x}=A\mathbf{u}+\mathbf{b}} \cdot \mathbf{A}$
(c) Function $F(\mathbf{u})$ is convex with respect to $\mathbf{u}$.
(d) If $m \leq n$, $\mathbf{A}$ is of full column rank, and $f(\mathbf{x})$ is strictly convex, then $F(\mathbf{u})$ is strictly convex with respect to $\mathbf{u}$.

References

1. M. H. Protter and C. B. Morrey, Jr., *Modern Mathematical Analysis*. Reading, MA: Addison-Wesley, 1964.
2. D. G. Luenberger and Y. Ye, *Linear and Nonlinear Programming*, 4th ed. New York: Springer, 2008.
3. R. A. Horn and C. R. Johnson, *Matrix Analysis*, 2nd ed. New York: Cambridge University Press, 2013.

General Properties of Algorithms

3

3.1 Introduction

In Chap. 1, an optimization algorithm has been informally introduced as a sequence of steps that can be executed repeatedly in order to obtain a series of progressively improved solutions, starting with an initial estimate of the solution. In this chapter, a more formal and mathematical description of an algorithm will be supplied and some fundamental concepts pertaining to all algorithms in general will be studied.

The chapter includes a discussion on the principle of global convergence. Specifically, a general theorem that enumerates the circumstances and conditions under which convergence can be assured in any given algorithm is proved [1, Chap. 4] [2, Chap. 7] [3, Chap. 7].

The chapter concludes with a quantitative discussion relating to the speed of convergence of an optimization algorithm. In particular, quantitative criteria are described that can be used to compare the efficiency of different types of algorithms.

3.2 An Algorithm as a Point-to-Point Mapping

There are numerous algorithms that can be used for the solution of nonlinear programming problems ranging from some simple to some highly complex algorithms. Although different algorithms differ significantly in their structure, mathematical basis, and range of applications, they share certain common properties that can be regarded as universal. The two most fundamental common properties of nonlinear programming algorithms are

1. They are iterative algorithms.
2. They are descent algorithms.

© Springer Science+Business Media, LLC, part of Springer Nature 2021
A. Antoniou and W.-S. Lu, *Practical Optimization*, Texts in Computer Science,
https://doi.org/10.1007/978-1-0716-0843-2_3

Fig. 3.1 Block diagram for
an iterative algorithm

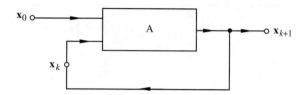

An algorithm is *iterative* if the solution is obtained by calculating a series of points
in sequence, starting with an initial estimate of the solution. On the other hand, an
algorithm is a *descent* algorithm if each new point generated by the algorithm yields
a reduced value of some function, possibly the objective function.

In mathematical terms, an algorithm can be regarded as a *point-to-point mapping*
where a point $\mathbf{x}_k$ in some space, normally a subspace of the E^n vector space, is
mapped onto another point $\mathbf{x}_{k+1}$ in the same space. The value of $\mathbf{x}_{k+1}$ is governed by
some rule of correspondence. In effect, if point $\mathbf{x}_k$ is used as input to the algorithm,
a point $\mathbf{x}_{k+1}$ is obtained as output. An algorithm can thus be represented by a block
diagram as depicted in Fig. 3.1. In this representation, $\mathbf{x}_0$ is an initial estimate of the
solution and the feedback line denotes the iterative nature of the algorithm. The rule
of correspondence between $\mathbf{x}_{k+1}$ and $\mathbf{x}_k$, which might range from a simple expression
to a large number of formulas, can be represented by the relation

$$\mathbf{x}_{k+1} = A(\mathbf{x}_k)$$

When applied iteratively to successive points, an algorithm will generate a series
(or sequence) of points $\{\mathbf{x}_0, \mathbf{x}_1, \ldots, \mathbf{x}_k, \ldots\}$ in space X, as depicted in Fig. 3.2. If
the sequence converges to a limit $\hat{\mathbf{x}}$, then $\hat{\mathbf{x}}$ is the required solution.

A sequence $\{\mathbf{x}_0, \mathbf{x}_1, \ldots, \mathbf{x}_k, \ldots\}$ is said to *converge to a limit* $\hat{\mathbf{x}}$ if for any given
$\varepsilon > 0$, an integer K exists such that

$$\|\mathbf{x}_k - \hat{\mathbf{x}}\|_2 < \varepsilon \qquad \text{for all } k \geq K$$

where $\| \cdot \|_2$ denotes the Euclidean norm. Such a sequence can be represented by the
notation $\{\mathbf{x}_k\}_{k=0}^{\infty}$ and its limit as $k \to \infty$ by $\mathbf{x}_k \to \hat{\mathbf{x}}$. If the sequence converges, it
has a unique limit point.

Later on in this chapter, reference will be made to subsequences of a given se-
quence. A subsequence of $\{\mathbf{x}_k\}_{k=0}^{\infty}$, denoted as $\{\mathbf{x}_k\}_{k \in I}$, where I is a set of positive
integers, can be obtained by deleting certain elements in $\{\mathbf{x}_k\}_{k=0}^{\infty}$. For example, if
$I = \{k : k \geq 10\}$ then $\{\mathbf{x}_k\}_{k \in I} = \{\mathbf{x}_{10}, \mathbf{x}_{11}, \mathbf{x}_{12}, \ldots\}$, if $I = \{k : k$ even and
greater than zero$\}$ then $\{\mathbf{x}_k\}_{k \in I} = \{\mathbf{x}_2, \mathbf{x}_4, \mathbf{x}_6, \ldots\}$, and if $I = \{k : 0 < k \leq 100\}$,
then $\{\mathbf{x}_k\}_{k \in I} = \{\mathbf{x}_1, \mathbf{x}_2, \ldots, \mathbf{x}_{100}\}$. In our notation $S = \{k : \mathcal{P}\}$, S is the set of
elements such that k has property $\mathcal{P}$.

If the sequence of points generated by an algorithm A converges to a limit $\hat{\mathbf{x}}$ as
described above, then algorithm A is said to be *continuous*.

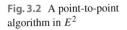

Fig. 3.2 A point-to-point
algorithm in E^2

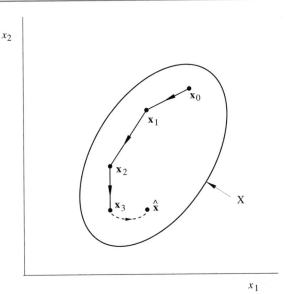

3.3 An Algorithm as a Point-to-Set Mapping

In the above discussion, an algorithm was considered as a point-to-point mapping
in that for any given point x_k a corresponding unique point x_{k+1} is generated. In
practice, this is the true nature of an algorithm only if a specific version of the
algorithm is implemented on a specific computer. Since different implementations
of an algorithm by different programmers on different computers are very likely
to give slightly different results, owing to the accumulation of roundoff errors, it
is advantageous to consider an algorithm as a point-to-set mapping. In this way, if
any general properties of an algorithm are deduced, they will hold for all possible
implementations of the algorithm. Furthermore, they may hold for similar algorithms.
For these reasons, the following more general definition of an algorithm will be used
throughout the rest of this chapter.

Definition 3.1 *An algorithm as a point-to-set mapping* An algorithm is a *point-to-
set mapping* on space X that assigns a subset of X to every point $\mathbf{x} \in X$.   ■

According to this definition, an algorithm A will generate a sequence of points
$\{x_k\}_{k=1}^{\infty}$ by assigning a set X_1 which is a subset of X, i.e., $X_1 \subset X$, to a given initial
point $x_0 \in X$. Then an arbitrary point $x_1 \in X_1$ is selected and a set $X_2 \subset X$ is
assigned to it, and so on, as depicted in Fig. 3.3. The rule of correspondence between
x_{k+1} and x_k is, therefore, of the form

$$x_{k+1} \in A(x_k)$$

where $A(x_k)$ is the set of all possible outputs if x_k is the input.

Clearly, the above definition encompasses all possible implementations of an
algorithm and it would encompass a class of algorithms that are based on a similar

Fig. 3.3 A point-to-set
algorithm in E^2

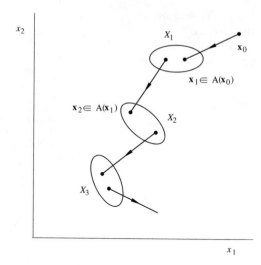

mathematical structure. The concept of the point-to-set algorithm can be visualized
by noting that in a typical algorithm

$$\mathbf{x}_{k+1} = A(\mathbf{x}_k) + \boldsymbol{\varepsilon}_q$$

where $\boldsymbol{\varepsilon}_q$ is a random vector due to the quantization of numbers. Since the quantization error tends to depend heavily on the sequence in which arithmetic operations
are performed and on the precision of the computer used, the exact location of $\mathbf{x}_{k+1}$
is not known. Nevertheless, it is known that $\mathbf{x}_{k+1}$ is a member of a small subset of
X.

3.4   Closed Algorithms

In the above discussion, reference was made to the continuity of a point-to-point
algorithm. A more general property which is applicable to point-to-point as well
as to point-to-set algorithms is the property of closedness. This property reduces to
continuity in a point-to-point algorithm.

Definition 3.2 *Closed point-to-set algorithm*

(*a*) A point-to-set algorithm A, from space X to space X_1 is said to be *closed at
point* $\hat{\mathbf{x}} \in X$ if the assumptions

$$\mathbf{x}_k \rightarrow \hat{\mathbf{x}} \qquad \text{for } \mathbf{x}_k \in X$$
$$\mathbf{x}_{k+1} \rightarrow \hat{\mathbf{x}}_1 \qquad \text{for } \mathbf{x}_{k+1} \in A(\mathbf{x}_k)$$

imply that

$$\hat{\mathbf{x}}_1 \in A(\hat{\mathbf{x}})$$

The notation $\mathbf{x}_k \rightarrow \hat{\mathbf{x}}$ denotes that the sequence $\{\mathbf{x}_k\}_{k=0}^{\infty}$ converges to a limit $\hat{\mathbf{x}}$.

Fig. 3.4 Definition of a closed algorithm in E^2

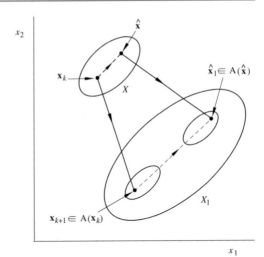

(*b*) A point-to-set algorithm A is said to be *closed on X* if it is closed at each point of X.　■

This definition is illustrated in Fig. 3.4. It states that algorithm A is closed at point $\hat{\mathbf{x}}$ if a solid line can be drawn between $\hat{\mathbf{x}}$ and $\hat{\mathbf{x}}_1$, and if a solid line can be drawn for all $\hat{\mathbf{x}} \in X$, then A is closed on X.

Example 3.1 An algorithm A is defined by

$$x_{k+1} = A(x_k) = \begin{cases} \frac{1}{2}(x_k + 2) & \text{for } x_k > 1 \\ \frac{1}{4}x_k & \text{for } x_k \leq 1 \end{cases}$$

(see Fig. 3.5). Show that the algorithm is not closed at $\hat{x} = 1$.

Solution Let sequence $\{x_k\}_{k=0}^{\infty}$ be defined by

$$x_k = 1 + \frac{1}{2^{k+1}}$$

The sequence can be obtained as

$$\{x_k\}_{k=0}^{\infty} = \{1.5, \ 1.25, \ 1.125 \ldots, \ 1\}$$

and hence

$$x_k \rightarrow \hat{x} = 1$$

The corresponding sequence $\{x_{k+1}\}_{k=0}^{\infty}$ is given by

$$x_{k+1} = A(x_k) = \frac{1}{2}(x_k + 2)$$

Fig. 3.5 Graph for
Example 3.1

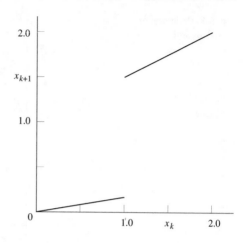

and so

$$\{x_{k+1}\}_{k=0}^{\infty} = \{1.75, \ 1.625, \ 1.5625, \ \ldots, \ 1.5\}$$

Thus

$$x_{k+1} \rightarrow \hat{x}_1 = 1.5$$

Now

$$A(\hat{x}) = \tfrac{1}{4}$$

and since $\hat{x}_1 = 1.5$, we have

$$\hat{x}_1 \neq A(\hat{x})$$

Therefore, A is not closed at $\hat{x} = 1$. The problem is due to the discontinuity of $A(x_k)$ at $x_k = 1$. ■

Example 3.2 An algorithm A is defined by

$$x_{k+1} = A(x_k) = x_k^2 \quad \text{for} \ -\infty < x_k < \infty$$

Show that A is closed.

Solution Let $\{x_k\}$ be a sequence converging to $\hat{x}$, i.e., $x_k \rightarrow \hat{x}$. Then $\{x_{k+1}\} = \{A(x_k)\} = \{x_k^2\}$ is a sequence that converges to $\hat{x}^2$, i.e., $x_k^2 \rightarrow \hat{x}_1 = \hat{x}^2$. Since $\hat{x}_1 = A(\hat{x})$, we conclude that for all $\hat{x}$ in the range $-\infty < \hat{x} < \infty$, A is closed. ■

3.5 Descent Functions

In any descent algorithm, a specific function $D(\mathbf{x})$ is utilized, which is reduced continuously throughout the optimization until convergence is achieved. $D(\mathbf{x})$ may be the objective function itself or some related function, and it is referred to as the descent function. A formal definition summarizing the required specifications for a function to be a descent function is as follows. This will be used later in Theorem 3.1.

Definition 3.3 *Descent function* Let $S \subset X$ be the set containing the solution points, and assume that A is an algorithm on X. A continuous real-valued function $D(\mathbf{x})$ on X is said to be a *descent function* for S and A if it satisfies the following specifications:

(a) if $\mathbf{x}_k \notin S$, then $D(\mathbf{x}_{k+1}) < D(\mathbf{x}_k)$ for all $\mathbf{x}_{k+1} \in A(\mathbf{x}_k)$
(b) if $\mathbf{x}_k \in S$, then $D(\mathbf{x}_{k+1}) \leq D(\mathbf{x}_k)$ for all $\mathbf{x}_{k+1} \in A(\mathbf{x}_k)$ ■

Example 3.3 Obtain a descent function for the algorithm

$$x_{k+1} = A(x_k) = \tfrac{1}{4} x_k$$

Solution For an arbitrary point x_0, the sequence

$$\{x_k\}_{k=0}^{\infty} = \{x_0, \frac{x_0}{4}, \frac{x_0}{4^2}, \ldots, 0\}$$

is generated. Therefore, $D(x_k) = |x_k|$ satisfies condition (a). The solution set is a single point at $x_\infty = 0$. Therefore, condition (b) is satisfied. Hence $D(x) = |x|$ is a descent function for the algorithm. ■

3.6 Global Convergence

If an algorithm has the important property that an arbitrary initial point $\mathbf{x}_0 \in X$ will lead to a converging sequence of points $\{\mathbf{x}_k\}_{k=0}^{\infty}$, then the algorithm is said to be *globally convergent*. In practice, even the most efficient algorithms are likely to fail if certain conditions are violated. For example, an algorithm may generate sequences that do not converge or may converge to points that are not solutions. There are several factors that are likely to cause failure in an algorithm. However, if they are clearly understood, certain precautions can be taken which will circumvent the cause of failure. Consequently, the study of global convergence is of particular interest not only to the theorist but also to the practitioner.

A large segment of the theory of global convergence deals with the circumstances and conditions that will guarantee global convergence. An important theorem in this area is as follows.

Theorem 3.1 *Convergence of an algorithm* Let A be an algorithm on X and assume that an initial point $\mathbf{x}_0$ will yield an infinite sequence $\{\mathbf{x}_k\}_{k=0}^{\infty}$ where

$$\mathbf{x}_{k+1} \in A(\mathbf{x}_k)$$

If a solution set S and a descent function $D(\mathbf{x}_k)$ exist for the algorithm such that

(a) all points $\mathbf{x}_k$ are contained in a compact subset of X,
(b) $D(\mathbf{x}_k)$ satisfies the specifications of Definition 3.3, and
(c) the mapping of A is closed at all points outside S,

then the limit of any convergent subsequence of $\{\mathbf{x}_k\}_{k=0}^{\infty}$ is a solution point.

Proof   The proof of this important theorem consists of two parts. In part (a), we suppose that $\hat{\mathbf{x}}$ is the limit of any subsequence of $\{\mathbf{x}_k\}_{k=0}^{\infty}$, say, $\{\mathbf{x}_k\}_{k\in I}$, where I is a set of integers, and show that $D(\mathbf{x}_k)$ converges with respect to the infinite sequence $\{\mathbf{x}_k\}_{k=0}^{\infty}$. In part (b), we show that $\hat{\mathbf{x}}$ is in the solution set S.

The second part of the proof relies heavily on the Weierstrass theorem (see [4, Chap. 9]), which states that if W is a compact set, then the sequence $\{\mathbf{x}_k\}_{k=0}^{\infty}$, where $\mathbf{x}_k \in W$, has a limit point in W. A set W is *compact* if it is *closed*. A set W is *closed* if all points on the boundary of W belong to W. A set W is *bounded* if it can be circumscribed by a hypersphere of finite radius. A consequence of the Weierstrass theorem is that a subsequence $\{\mathbf{x}_k\}_{k\in I}$ of $\{\mathbf{x}_k\}_{k=0}^{\infty}$ has a limit point in set $\bar{W} = \{\mathbf{x}_k : k \in I\}$ since $\bar{W}$ is a subset of W and is, therefore, compact.

(a) Since $D(\mathbf{x}_k)$ is continuous on X and $\hat{\mathbf{x}}$ is assumed to be the limit of $\{\mathbf{x}_k\}_{k\in I}$, a positive number and an integer K exist such that

$$D(\mathbf{x}_k) - D(\hat{\mathbf{x}}) < \varepsilon \tag{3.1}$$

for $k \geq K$ with $k \in I$. Hence $D(\mathbf{x}_k)$ converges with respect to the subsequence $\{\mathbf{x}_k\}_{k\in I}$. We must show, however, that $D(\mathbf{x}_k)$ converges with respect to the infinite sequence $\{\mathbf{x}_k\}_{k=0}^{\infty}$.

For any $k \geq K$, we can write

$$D(\mathbf{x}_k) - D(\hat{\mathbf{x}}) = [D(\mathbf{x}_k) - D(\mathbf{x}_K)] + [D(\mathbf{x}_K) - D(\hat{\mathbf{x}})] \tag{3.2}$$

If $k = K$ in Eq. (3.1)

$$D(\mathbf{x}_K) - D(\hat{\mathbf{x}}) < \varepsilon \tag{3.3}$$

and if $k \geq K$, then $D(\mathbf{x}_k) \leq D(\mathbf{x}_K)$ from Definition 3.3 and hence

$$D(\mathbf{x}_k) - D(\mathbf{x}_K) \leq 0 \tag{3.4}$$

Now from Eqs. (3.2)–(3.4), we have

$$D(\mathbf{x}_k) - D(\hat{\mathbf{x}}) < \varepsilon$$

for *all* $k \geq K$. Therefore,

$$\lim_{k\to\infty} D(\mathbf{x}_k) = D(\hat{\mathbf{x}}) \tag{3.5}$$

that is, $D(\mathbf{x}_k)$ converges with respect to the infinite series, as $\mathbf{x}_k \to \hat{\mathbf{x}}$.

(b) Let us assume that $\hat{\mathbf{x}}$ is not in the solution set. Since the elements of subsequence $\{\mathbf{x}_{k+1}\}_{k\in I}$ belong to a compact set according to condition (a), a compact subset $\{\mathbf{x}_{k+1} : k \in \bar{I} \subset I\}$ exists such that $\mathbf{x}_{k+1}$ converges to some limit $\bar{\mathbf{x}}$ by virtue of the Weierstrass theorem. As in part (a), we can show that

$$\lim_{k\to\infty} D(\mathbf{x}_{k+1}) = D(\bar{\mathbf{x}}) \tag{3.6}$$

Therefore, from Eqs. (3.5) and (3.6)

$$D(\bar{\mathbf{x}}) = D(\hat{\mathbf{x}}) \tag{3.7}$$

On the other hand,

$$\mathbf{x}_k \to \hat{\mathbf{x}} \qquad \text{for } k \in \bar{I} \quad \text{(from part } (a)\text{)}$$

$$\mathbf{x}_{k+1} \to \bar{\mathbf{x}} \qquad \text{for } \mathbf{x}_{k+1} \in A(\mathbf{x})$$

and since $\hat{\mathbf{x}} \notin S$ by supposition, and A is closed at points outside S according to condition (c), we have

$$\bar{\mathbf{x}} \in A(\hat{\mathbf{x}})$$

Consequently,

$$D(\bar{\mathbf{x}}) < D(\hat{\mathbf{x}}) \tag{3.8}$$

On comparing Eqs. (3.7) and (3.8), a contradiction is observed and, in effect, our assumption that point $\hat{\mathbf{x}}$ is not in the solution set S is not valid. That is, the limit of any convergent subsequence of $\{\mathbf{x}_k\}_{k=0}^{\infty}$ is a solution point. ∎

In simple terms, the above theorem states that if

(a) the points that can be generated by the algorithm are located in the *finite* E^n space,
(b) a *descent function* can be found that satisfies the strict requirements stipulated, and
(c) the algorithm is *closed* outside the neighborhood of the solution,

then the algorithm is globally convergent. Further, a very close approximation to the solution can be obtained in a finite number of iterations, since the limit of any convergent finite subsequence of $\{\mathbf{x}_k\}_{k=0}^{\infty}$ is a solution.

A corollary of Theorem 3.1 which is of some significance is as follows.

Corollary If, under the conditions of Theorem 3.1, the solution set S consists of a single point $\hat{\mathbf{x}}$, then the sequence $\{\mathbf{x}_k\}_{k=0}^{\infty}$ converges to $\hat{\mathbf{x}}$.

Proof If we suppose that there is a subsequence $\{\mathbf{x}_k\}_{k \in I}$ that does not converge to $\hat{\mathbf{x}}$, then

$$\|\mathbf{x}_k - \hat{\mathbf{x}}\|_2 > \varepsilon \tag{3.9}$$

for all $k \in I$ and $\varepsilon > 0$. Now set $\{\mathbf{x}_k : k \in I' \subset I\}$ is compact and hence $\{\mathbf{x}_k\}_{k \in I'}$ converges to a limit point, say, $\mathbf{x}'$, by virtue of the Weierstrass theorem. From Theorem 3.1,

$$\|\mathbf{x}_k - \mathbf{x}'\|_2 < \varepsilon \tag{3.10}$$

for all $k \geq K$. Since the solution set consists of a single point, we have $\mathbf{x}' = \hat{\mathbf{x}}$. Under these circumstances, Eqs. (3.9) and (3.10) are contradictory and, in effect, our supposition is false. That is, any subsequence of $\{\mathbf{x}_k\}_{k=0}^{\infty}$, including the sequence itself, converges to $\hat{\mathbf{x}}$. ∎

If one or more of the conditions in Theorem 3.1 are violated, an algorithm may fail to converge. The possible causes of failure are illustrated in terms of the following examples.

Example 3.4 A possible algorithm for the problem

$$\text{minimize } f(x) = |x|$$

is

$$x_{k+1} = A(x_k) = \begin{cases} \frac{1}{2}(x_k + 2) & \text{for } x_k > 1 \\ \frac{1}{4}x_k & \text{for } x_k \leq 1 \end{cases}$$

Show that the algorithm is not globally convergent and explain why.

Solution If $x_0 = 4$, the algorithm will generate the sequence

$$\{x_k\}_{k=0}^{\infty} = \{4, \ 3, \ 2.5, \ 2.25, \ \ldots, \ 2\}$$

and if $x_0 = -4$, we have

$$\{x_k\}_{k=0}^{\infty} = \{-4, \ -1, \ -0.25, \ -0.0625, \ \ldots, \ 0\}$$

Since two distinct initial points lead to different limit points, the algorithm is not globally convergent. The reason is that the algorithm is not closed at point $x_k = 1$ (see Example 3.1), i.e., condition (c) of Theorem 3.1 is violated. ∎

Example 3.5 A possible algorithm for the problem

$$\text{minimize } f(x) = x^3$$

is

$$x_{k+1} = A(x_k) = -(x_k^2 + 1)$$

Show that the algorithm is not globally convergent and explain why.

Solution For an initial point x_0 the solution sequence is

$$\{x_k\}_{k=0}^{\infty} = \{x_0, \ -x_0^2 + 1, \ -(x_0^2 + 1)^2 + 1, \ [(x_0^2 + 1)^2 + 1]^2 + 1, \ \ldots, \ -\infty\}$$

Hence the sequence does not converge, and its elements are not in a compact set. Therefore, the algorithm is not globally convergent since condition (a) of Theorem 3.1 is violated. ∎

Example 3.6 A possible algorithm for the problem

$$\text{minimize } f(x) = |x - 1|$$

$$\text{subject to: } x > 0$$

is

$$x_{k+1} = A(x_k) = \sqrt{x_k}$$

Show that the algorithm is globally convergent for $0 < x_0 < \infty$.

Solution For any initial point x_0 in the range $0 < x_0 < \infty$, we have

$$\{x_k\}_{k=0}^{\infty} = \{x_0,\ x_0^{1/2},\ x_0^{1/4},\ \ldots,\ 1\}$$
$$\{x_k\}_{k=0}^{\infty} = \{x_0^{1/2},\ x_0^{1/4},\ x_0^{1/8},\ \ldots,\ 1\}$$

Thus

$$x_k \to \hat{x} = 1, \quad x_{k+1} \to \hat{x}_1 = 1$$

Evidently, all points x_k belong to a compact set and so condition (a) is satisfied. The objective function $f(x)$ is a descent function since

$$|x_{k+1} - 1| < |x_k - 1| \quad \text{for all } k < \infty$$

and so condition (b) is satisfied.

Since

$$x_k \to \hat{x} \quad \text{for } x_k > 0$$
$$x_{k+1} \to \hat{x}_1 \quad \text{for } x_{k+1} = A(x_k)$$

and

$$\hat{x}_1 = A(\hat{x})$$

the algorithm is closed, and so condition (c) is satisfied. The algorithm is, therefore, globally convergent. ∎

3.7 Rates of Convergence

The many available algorithms differ significantly in their computational efficiency. An efficient or fast algorithm is one that requires only a small number of iterations to converge to a solution and the amount of computation will be small. Economical reasons dictate that the most efficient algorithm for the application be chosen and, therefore, quantitative measures or criteria that can be used to measure the rate of convergence in a set of competing algorithms are required.

The most basic criterion in this area is the order of convergence of a sequence. If $\{x_k\}_{k=0}^{\infty}$ is a sequence of real numbers, its *order of convergence* is the largest nonnegative integer p that will satisfy the relation

$$0 \le \beta < \infty$$

where

$$\beta = \lim_{k \to \infty} \frac{|x_{k+1} - \hat{x}|}{|x_k - \hat{x}|^p} \tag{3.11}$$

and $\hat{x}$ is the limit of the sequence as $k \to \infty$. Parameter β is called the *convergence ratio*.

Example 3.7 Find the order of convergence and convergence ratio of the sequence $\{x_k\}_{k=0}^{\infty}$ if

(a) $x_k = \gamma^k$ for $0 < \gamma < 1$
(b) $x_k = \gamma^{2^k}$ for $0 < \gamma < 1$

Solution (a) Since $\hat{x} = 0$, Eq. (3.11) gives

$$\beta = \lim_{k \to \infty} \gamma^{k(1-p)+1}$$

Hence for $p = 0,\ 1,\ 2$ we have $\beta = 0,\ \gamma,\ \infty$. Thus $p = 1$ and $\beta = \gamma$.

(b) In this case

$$\beta = \lim_{k \to \infty} \frac{\gamma^{2^{(k+1)}}}{\gamma^{2^k p}} = \lim_{k \to \infty} \{\gamma^{2^k(2-p)}\}$$

Hence for $p = 0,\ 1,\ 2,\ 3$, we have $\beta = 0,\ 0,\ 1,\ \infty$. Thus $p = 2$ and $\beta = 1$. ∎

If the limit in Eq. (3.11) exists, then

$$\lim_{k \to \infty} |x_k - \hat{x}| = \varepsilon$$

where $\varepsilon < 1$. As a result

$$\lim_{k \to \infty} |x_{k+1} - \hat{x}| = \beta \varepsilon^p$$

Therefore, the rate of convergence is increased if p is increased and β is reduced. If $\gamma = 0.8$ in Example 3.7, the sequences in parts (a) and (b) will be

$$\{x_k\}_{k=0}^{\infty} = \{1,\ 0.8,\ 0.64,\ 0.512,\ 0.409,\ \ldots,\ 0\}$$

and

$$\{x_k\}_{k=0}^{\infty} = \{1,\ 0.64,\ 0.409,\ 0.167,\ 0.023,\ \ldots,\ 0\}$$

respectively. The rate of convergence in the second sequence is much faster since $p = 2$.

If $p = 1$ and $\beta < 1$, the sequence is said to have *linear convergence*. If $p = 1$ and $\beta = 0$ or $p \geq 2$ the sequence is said to have *superlinear convergence*.

Most of the available nonlinear programming algorithms have linear convergence and hence their comparison is based on the value of β.

Another measure of the rate of convergence of a sequence is the so-called *average order of convergence*. This is the lowest nonnegative integer that will satisfy the relation

$$\gamma = \lim_{k \to \infty} |x_k - \hat{x}|^{1/(p+1)^k} = 1$$

If no $p > 0$ can be found, then the order of convergence is infinity.

Example 3.8 Find the average order of convergence of the sequence $\{x_k\}_{k=0}^{\infty}$ for the following sequences:

(a) $x_k = \tau^k$ for $0 < \tau < 1$
(b) $x_k = \tau^{2^k}$ for $0 < \tau < 1$

Solution (*a*) Since $\hat{x} = 0$,

$$\gamma = \lim_{k \to \infty} (\tau^k)^{1/(p+1)^k} = \lim_{k \to \infty} \tau^{k/(p+1)^k}$$

Hence for $p = 0, 1, 2$, we have $\gamma = 0, 1, 1$. Thus $p = 1$.

(*b*) In this case,

$$\gamma = \lim_{k \to \infty} (\tau^{2^k})^{1/(p+1)^k} = \lim_{k \to \infty} \tau^{2^k/(p+1)^k}$$

Hence for $p = 0, 1, 2, 3$, we have $\gamma = 0, \tau, 1, 1$. Thus $p = 2$. ■

If the average order of convergence is unity, then the sequence is said to have an *average linear convergence*. The average convergence ratio can be defined as

$$\gamma = \lim_{k \to \infty} |x_k - \hat{x}|^{1/k}$$

In the above discussion, the convergence of a sequence of numbers has been considered. Such a sequence might consist of the values of the objective function as the solution is approached. In such a case, we are measuring the rate at which the objective function is approaching its minimum. Alternatively, if we desire to know how fast the variables of the problem approach their optimum values, a sequence of numbers can be generated by considering the magnitudes or the square magnitudes of the vectors $\mathbf{x}_k - \hat{\mathbf{x}}$, namely, $\|\mathbf{x}_k - \hat{\mathbf{x}}\|_2$ or $\|\mathbf{x}_k - \hat{\mathbf{x}}\|_2^2$, as the solution is approached.

In the above measures of the rate of convergence, the emphasis is placed on the efficiency of an algorithm in the neighborhood of the solution. Usually in optimization a large percentage of the computation is used in the neighborhood of the solution and, consequently, the above measures are quite meaningful. Occasionally, however, a specific algorithm may be efficient in the neighborhood of the solution and very inefficient elsewhere. In such a case, the use of the above criteria would lead to misleading results and, therefore, other criteria should also be employed.

Problems

3.1 Let A be a point-to-set algorithm from space E^1 to space E^1. The *graph* of A is defined as the set

$$\{(x, y) : x \in E^1, \ y \in A(x)\}$$

(*a*) Show that algorithm A defined by

$$A(x) = \{y : x/4 \le y \le x/2\}$$

is closed on E^1.

(*b*) Plot the graph of A.

3.2 Examine whether or not the following point-to-set mappings from E^1 to E^1 are closed:

(a)

$$A(x) = \begin{cases} \frac{1}{x} & \text{if } x \neq 0 \\ x & \text{if } x = 0 \end{cases}$$

(b)

$$A(x) = \begin{cases} \frac{1}{x} & \text{if } x \neq 0 \\ 1 & \text{if } x = 0 \end{cases}$$

(c)

$$A(x) = \begin{cases} x & \text{if } x \neq 0 \\ 1 & \text{if } x = 0 \end{cases}$$

3.3 Define the point-to-set mapping on E^n by

$$A(\mathbf{x}) = \{\mathbf{y} : \ \mathbf{y}^T \mathbf{x} \geq 1\}$$

Is A closed on E^n?

3.4 Let $\{b_k, \ k = 0, \ 1, \ \ldots\}$ and $\{c_k, \ k = 0, \ 1, \ \ldots\}$ be sequences of real numbers, where $b_k \to 0$ superlinearly in the sense that $p = 1$ and $\beta = 0$ (see Eq. (3.11)) and $c \leq c_k \leq C$ with $c > 0$. Show that $\{b_k c_k, \ k = 0, \ 1, \ \ldots\}$ converges to zero superlinearly.

References

1. W. I. Zangwill, *Nonlinear Programming: A Unified Approach*. Englewood Cliffs, NJ: Prentice-Hall, 1969.
2. M. S. Bazaraa and C. M. Shetty, *Nonlinear Programming*. New York: Wiley, 1979.
3. D. G. Luenberger and Y. Ye, *Linear and Nonlinear Programming*, 4th ed. New York: Springer, 2008.
4. H. M. Edwards, *Advanced Calculus*. Boston, MA: Houghton Mifflin, 1969.

One-Dimensional Optimization

4

4.1 Introduction

Three general classes of nonlinear optimization problems can be identified, as follows:

1. One-dimensional unconstrained problems
2. Multidimensional unconstrained problems
3. Multidimensional constrained problems

Problems of the first class are the easiest to solve whereas those of the third class are the most difficult. In practice, multidimensional constrained problems are usually reduced to multidimensional unconstrained problems which, in turn, are reduced to one-dimensional unconstrained problems. In effect, most of the available nonlinear programming algorithms are based on the minimization of a function of a single variable without constraints. Therefore, *efficient* one-dimensional optimization algorithms are required, if efficient multidimensional unconstrained and constrained algorithms are to be constructed.

The one-dimensional optimization problem is

$$\text{minimize } F = f(x)$$

where $f(x)$ is a function of one variable. This problem has a solution if $f(x)$ is *unimodal* in some range of x, i.e., $f(x)$ has only one minimum in some range $x_L \leq x \leq x_U$, where x_L and x_U are the lower and upper limits of the minimizer x^*.

Two general classes of one-dimensional optimization methods are available, namely, *search methods* and *approximation methods*.

In search methods, an interval $[x_L, \ x_U]$ containing x^*, known as a *bracket*, is established and is then repeatedly reduced on the basis of function evaluations until a reduced bracket $[x_{L,k}, \ x_{U,k}]$ is obtained which is sufficiently small. The minimizer can be assumed to be at the center of interval $[x_{L,k}, \ x_{U,k}]$. These methods can be applied to any function and differentiability of $f(x)$ is not essential.

© Springer Science+Business Media, LLC, part of Springer Nature 2021
A. Antoniou and W.-S. Lu, *Practical Optimization*, Texts in Computer Science,
https://doi.org/10.1007/978-1-0716-0843-2_4

In approximation methods, an approximation of the function in the form of a low-order polynomial, usually a second- or third-order polynomial, is assumed. This is then analyzed using elementary calculus and an approximate value of x^* is deduced. The interval $[x_L, x_U]$ is then reduced and the process is repeated several times until a sufficiently precise value of x^* is obtained. In these methods, $f(x)$ is required to be continuous and differentiable, i.e., $f(x) \in C^1$.

Several one-dimensional optimization approaches will be examined in this chapter, as follows [1–8]:

1. Dichotomous search
2. Fibonacci search
3. Golden-section search
4. Quadratic interpolation method
5. Cubic interpolation method
6. The Davies, Swann, and Campey method

The first three are search methods, the fourth and fifth are approximation methods, and the sixth is a practical and useful method that combines a search method with an approximation method.

The chapter will also deal with a so-called *inexact line search* due to Fletcher [9,10], which offers certain important advantages such as reduced computational effort in some optimization methods.

4.2   Dichotomous Search

Consider a unimodal function which is known to have a minimum in the interval $[x_L, x_U]$. This interval is said to be the *range of uncertainty*. The minimizer x^* of $f(x)$ can be located by reducing progressively the range of uncertainty until a sufficiently small range is obtained. In search methods, this can be achieved by using the values of $f(x)$ at suitable points.

If the value of $f(x)$ is known at a single point x_a in the range $x_L < x_a < x_U$, point x^* is equally likely to be in the range x_L to x_a or x_a to x_U as depicted in Fig. 4.1a. Consequently, the information available is not sufficient to allow the reduction of the range of uncertainty. However, if the value of $f(x)$ is known at two points, say, x_a and x_b, an immediate reduction is possible. Three possibilities may arise, namely,

(a) $f(x_a) < f(x_b)$
(b) $f(x_a) > f(x_b)$
(c) $f(x_a) = f(x_b)$

In case (a), x^* may be located in range $x_L < x^* < x_a$ or $x_a < x^* < x_b$, that is, $x_L < x^* < x_b$, as illustrated in Fig. 4.1a. The possibility $x_b < x^* < x_U$ is definitely ruled out since this would imply that $f(x)$ has two minima: one to the left of x_b and one to the right of x_b. Similarly, for case (b), we must have $x_a < x^* < x_U$ as in Fig. 4.1b. For case (c), we must have $x_a < x^* < x_b$, that is, both inequalities $x_L < x^* < x_b$ and $x_a < x^* < x_U$ must be satisfied as in Fig. 4.1c.

A rudimentary strategy for reducing the range of uncertainty is the so-called *dichotomous search*. In this method, $f(x)$ is evaluated at two points $x_a = x_1 - \varepsilon/2$ and $x_b = x_1 + \varepsilon/2$ where ε is a small positive number. Then depending on whether $f(x_a) < f(x_b)$ or $f(x_a) > f(x_b)$, range x_L to $x_1 + \varepsilon/2$ or $x_1 - \varepsilon/2$ to x_U can be selected and if $f(x_a) = f(x_b)$ either will do fine. If we assume that $x_1 - x_L = x_U - x_1$, i.e., $x_1 = (x_L + x_U)/2$, the region of uncertainty is immediately reduced by half. The same procedure can be repeated for the reduced range, that is, $f(x)$ can be evaluated at $x_2 - \varepsilon/2$ and $x_2 + \varepsilon/2$ where x_2 is located at the center of the reduced range, and so on. For example, if the dichotomous search is applied to the function of Fig. 4.2 the range of uncertainty will be reduced from $0 < x^* < 1$ to $9/16 + \varepsilon/2 < x^* < 5/8 - \varepsilon/2$ in four iterations.

Each iteration reduces the range of uncertainty by half and, therefore, after k iterations, the interval of uncertainty reduces to

$$I_k = (\tfrac{1}{2})^k I_0$$

where $I_0 = x_U - x_L$. For example, after 7 iterations the range of uncertainty would be reduced to less than 1% of the initial interval. The corresponding computational effort would be 14 function evaluations since two evaluations are required for each iteration.

4.3 Fibonacci Search

Consider an interval of uncertainty

$$I_k = [x_{L,k}, \ x_{U,k}]$$

and assume that two points $x_{a,k}$ and $x_{b,k}$ are located in I_k, as depicted in Fig. 4.3. As in Sect. 4.2, the values of $f(x)$ at $x_{a,k}$ and $x_{b,k}$, namely, $f(x_{a,k})$ and $f(x_{b,k})$, can be used to select the left interval

$$I_{k+1}^L = [x_{L,k}, \ x_{b,k}]$$

if $f(x_{a,k}) < f(x_{b,k})$, the right interval

$$I_{k+1}^R = [x_{a,k}, \ x_{U,k}]$$

if $f(x_{a,k}) > f(x_{b,k})$, or either of I_{k+1}^R and I_{k+1}^L if

$$f(x_a, k) = f(x_b, k)$$

If the right interval I_{k+1}^R is selected, it contains the minimizer and, in addition, the value of $f(x)$ is known at one interior point of I_{k+1}^R, namely, at point $x_{b,k}$. If $f(x)$ is evaluated at one more interior point of I_{k+1}^R, say, at point $x_{b,k+1}$, sufficient

Fig. 4.1 Reduction of range
of uncertainty: **a** case (a),
$f(x_a) < f(x_b)$, **b** case (b),
$f(x_a) > f(x_b)$. Reduction
of range of uncertainty: **c**
case $(c), f(x_a) = f(x_b)$

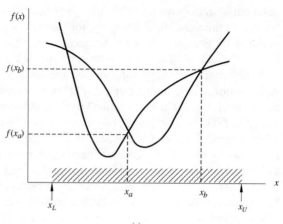

(a)

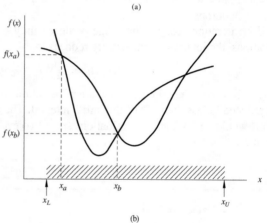

(b)

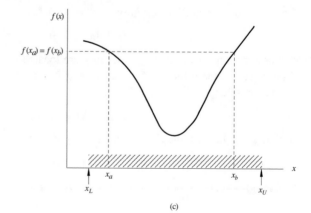

(c)

Fig. 4.2 Construction for dichotomous search

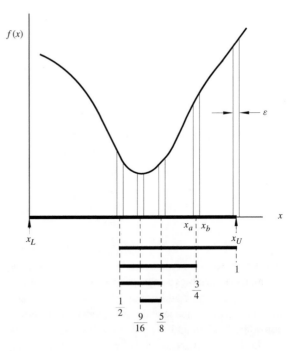

Fig. 4.3 Reduction of range of uncertainty

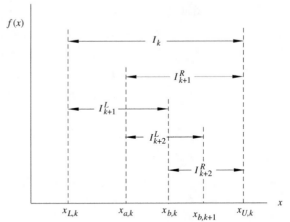

information is available to allow a further reduction in the region of uncertainty, and the above cycle of events can be repeated. One of the two new sub-intervals I_{k+2}^L and I_{k+2}^R, shown in Fig. 4.3, can be selected as before, and so on. In this way, only one function evaluation is required per iteration, and the amount of computation will be reduced relative to that required in the dichotomous search.

From Fig. 4.3

$$I_k = I_{k+1}^L + I_{k+2}^R \tag{4.1}$$

and if, for the sake of convenience, we assume equal intervals, then

$$I_{k+1}^L = I_{k+1}^R = I_{k+1}$$
$$I_{k+2}^L = I_{k+2}^R = I_{k+2}$$

Equation (4.1) gives the recursive relation

$$I_k = I_{k+1} + I_{k+2} \tag{4.2}$$

If the above procedure is repeated a number of times, a sequence of intervals $\{I_1, I_2, \ldots, I_n\}$ will be generated as follows:

$$I_1 = I_2 + I_3$$
$$I_2 = I_3 + I_4$$
$$\vdots$$
$$I_n = I_{n+1} + I_{n+2}$$

In the above set of n equations, there are $n+2$ variables and if I_1 is the given initial interval, $n+1$ variables remain. Therefore, an infinite set of sequences can be generated by specifying some additional rule. Two specific sequences of particular interest are the Fibonacci sequence and the golden-section sequence. The Fibonacci sequence is considered below and the golden-section sequence is considered in Sect. 4.4.

The Fibonacci sequence is generated by assuming that the interval for iteration $n + 2$ vanishes, that is, $I_{n+2} = 0$. If we let $k = n$ in Eq. (4.2), we can write

$$I_{n+1} = I_n - I_{n+2} = I_n \equiv F_0 I_n$$
$$I_n = I_{n+1} + I_{n+2} = I_n \equiv F_1 I_n$$
$$I_{n-1} = I_n + I_{n+1} = 2I_n \equiv F_2 I_n$$
$$I_{n-2} = I_{n-1} + I_n = 3I_n \equiv F_3 I_n$$
$$I_{n-3} = I_{n-2} + I_{n-1} = 5I_n \equiv F_4 I_n$$
$$I_{n-4} = I_{n-3} + I_{n-2} = 8I_n \equiv F_5 I_n$$
$$\vdots \qquad\qquad \vdots$$
$$I_k = I_{k+1} + I_{k+2} = F_{n-k+1} I_n \tag{4.3a}$$
$$\vdots \qquad\qquad \vdots$$
$$I_1 = I_2 + I_3 = F_n I_n \tag{4.3b}$$

The sequence generated, namely,

$$\{1, \ 1, \ 2, \ 3, \ 5, \ 8, \ 13, \ \ldots\} = \{F_0, \ F_1, \ F_2, \ F_3, \ F_4, \ F_5, \ F_6 \ \ldots\}$$

is the well-known *Fibonacci sequence*, which occurs in various branches of mathematics. It can be generated by using the recursive relation

$$F_k = F_{k-1} + F_{k-2} \quad \text{for } k \geq 2 \tag{4.4}$$

Fig. 4.4 Fibonacci search for $n = 6$

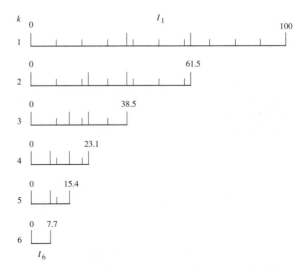

where $F_0 = F_1 = 1$. Its application in one-dimensional optimization gives rise to the *Fibonacci search method*. The method is illustrated in Fig. 4.4 for $n = 6$ and $I_1 = 100$ for the case where the left interval is consistently selected, i.e., the minimum occurs in the neighborhood of $x = 0$.

If the number of iterations is assumed to be n, then from Eq. (4.3b) the Fibonacci search reduces the interval of uncertainty to

$$I_n = \frac{I_1}{F_n} \tag{4.5}$$

From Eq. (4.3a), it follows that $I_n = I_{n-1}/2$. Consequently, in order to obtain an approximate minimizer in the range I_n, we only need to carry out $n - 2$ Fibonacci iterations to obtain the region of uncertainty I_{n-2} and then take the middle point of interval I_{n-1} as the solution. The Fibonacci search, therefore, requires $n - 2$ iterations. Since one function evaluation is required per iteration except the first iteration, which requires two function evaluations, the Fibonacci search requires a total of $n - 1$ function evaluations. For example, if $n = 11$, then $F_n = 144$ and so I_n is reduced to a value of less than 1% of the value of I_1. This would entail 9 iterations and hence 10 function evaluations as opposed to 14 function evaluations required by the dichotomous search to achieve the same precision. In effect, the Fibonacci search is more efficient than the dichotomous search. Indeed, it can be shown that it achieves the largest interval reduction relative to the other search methods and it is, therefore, the most efficient in terms of computational effort required.

The Fibonacci sequence of intervals can be generated only if n is known. If the objective of the optimization is to find x^* to within a prescribed tolerance, the required n can be readily deduced by using Eq. (4.5). However, if the objective is to determine the minimum of $f(x)$ to within a prescribed tolerance, difficulty will be experienced in determining the required n without solving the problem. The only available information is that n will be low if the minimum of $f(x)$ is shallow and high if $f(x)$ varies rapidly in the neighborhood of the solution.

The above principles can be used to implement the Fibonacci search. Let us assume that the initial bounds of the minimizer, namely, $x_{L,1}$ and $x_{U,1}$, and the value of n are given, and a mathematical description of $f(x)$ is available. The implementation consists of computing the successive intervals, evaluating $f(x)$, and selecting the appropriate intervals.

At the kth iteration, the quantities $x_{L,k}$, $x_{a,k}$, $x_{b,k}$, $x_{U,k}$, I_{k+1} and

$$f_{a,k} = f(x_{a,k}), \quad f_{b,k} = f(x_{b,k})$$

are known, and the quantities $x_{L,k+1}$, $x_{a,k+1}$, $x_{b,k+1}$, $x_{U,k+1}$, I_{k+2}, $f_{a,k+1}$, and $f_{b,k+1}$ are required. Interval I_{k+2} can be obtained from Eq. (4.3a) as

$$I_{k+2} = \frac{F_{n-k-1}}{F_{n-k}} I_{k+1} \tag{4.6}$$

The remaining quantities can be computed as follows.

If $f_{a,k} > f_{b,k}$, then x^* is in interval $[x_{a,k}, x_{U,k}]$ and so the new bounds of x^* can be updated as

$$x_{L,k+1} = x_{a,k} \tag{4.7}$$
$$x_{U,k+1} = x_{U,k} \tag{4.8}$$

Similarly, the two interior points of the new interval, namely, $x_{a,k+1}$ and $x_{b,k+1}$ will be $x_{b,k}$ and $x_{L,k+1} + I_{k+2}$, respectively. We can thus assign

$$x_{a,k+1} = x_{b,k} \tag{4.9}$$
$$x_{b,k+1} = x_{L,k+1} + I_{k+2} \tag{4.10}$$

as illustrated in Fig. 4.5. The value $f_{b,k}$ is retained as the value of $f(x)$ at $x_{a,k+1}$, and the value of $f(x)$ at $x_{b,k+1}$ is calculated, i.e.,

$$f_{a,k+1} = f_{b,k} \tag{4.11}$$
$$f_{b,k+1} = f(x_{b,k+1}) \tag{4.12}$$

On the other hand, if $f_{a,k} < f_{b,k}$, then x^* is in interval $[x_{L,k}, x_{b,k}]$. In this case, we assign

$$x_{L,k+1} = x_{L,k} \tag{4.13}$$
$$x_{U,k+1} = x_{b,k} \tag{4.14}$$
$$x_{a,k+1} = x_{U,k+1} - I_{k+2} \tag{4.15}$$
$$x_{b,k+1} = x_{a,k} \tag{4.16}$$
$$f_{b,k+1} = f_{a,k} \tag{4.17}$$

and calculate

$$f_{a,k+1} = f(x_{a,k+1}) \tag{4.18}$$

as depicted in Fig. 4.6. In the unlikely event that $f_{a,k} = f_{b,k}$, either of the above sets of assignments can be used since x^* is contained by both intervals $[x_{L,k}, x_{b,k}]$ and $[x_{a,k}, x_{U,k}]$.

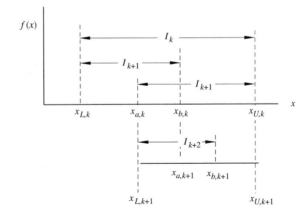

Fig. 4.5 Assignments in kth iteration of the Fibonacci search if $f_{a,k} > f_{b,k}$

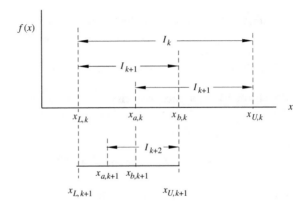

Fig. 4.6 Assignments in kth iteration of the Fibonacci search if $f_{a,k} < f_{b,k}$

The above procedure is repeated until $k = n - 2$ in which case

$$I_{k+2} = I_n$$

and

$$x^* = x_{a,k+1} = x_{b,k+1}$$

as depicted in Fig. 4.7. Evidently, the minimizer is determined to within a tolerance $\pm 1/F_n$.

The error in x^* can be divided by two by applying one stage of the dichotomous search. This is accomplished by evaluating $f(x)$ at point $x = x_{a,k+1} + \varepsilon$ where $|\varepsilon| < 1/F_n$ and then assigning

$$x^* = \begin{cases} x_{a,k+1} + \frac{1}{2F_n} & \text{if } f(x_{a,k+1} + \varepsilon) < f(x_{a,k+1}) \\ x_{a,k+1} + \frac{\varepsilon}{2} & \text{if } f(x_{a,k+1} + \varepsilon) = f(x_{a,k+1}) \\ x_{a,k+1} - \frac{1}{2F_n} & \text{if } f(x_{a,k+1} + \varepsilon) > f(x_{a,k+1}) \end{cases}$$

If n is very large, the difference between $x_{a,k}$ and $x_{b,k}$ can become very small, and it is possible for $x_{a,k}$ to exceed $x_{b,k}$, owing to roundoff errors. If this happens, unreliable results will be obtained. In such applications, checks should be incorporated in the

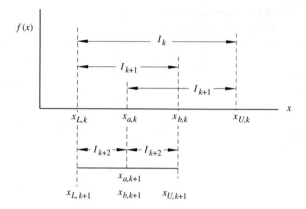

Fig. 4.7 Assignments in iteration $n - 2$ of the Fibonacci search if $f_{a,k} < f_{b,k}$

algorithm for the purpose of eliminating the problem, if it occurs. One possibility would be to terminate the algorithm since, presumably, sufficient precision has been achieved if $x_{a,k} \approx x_{b,k}$.

The above principles can be used to construct the following algorithm.

Algorithm 4.1 Fibonacci search
Step 1
Input $x_{L,1}$, $x_{U,1}$, and n.
Step 2
Compute F_1, F_2, $\ldots$, F_n using Eq. (4.4).
Step 3
Assign $I_1 = x_{U,1} - x_{L,1}$ and compute

$$I_2 = \frac{F_{n-1}}{F_n} I_1 \quad \text{(see Eq. (4.6))}$$
$$x_{a,1} = x_{U,1} - I_2, \quad x_{b,1} = x_{L,1} + I_2$$
$$f_{a,1} = f(x_{a,1}), \quad f_{b,1} = f(x_{b,1})$$

Set $k = 1$.
Step 4
Compute I_{k+2} using Eq. (4.6).
If $f_{a,k} \geq f_{b,k}$, then update $x_{L,k+1}$, $x_{U,k+1}$, $x_{a,k+1}$, $x_{b,k+1}$, $f_{a,k+1}$, and $f_{b,k+1}$ using Eqs. (4.7)–(4.12). Otherwise, if $f_{a,k} < f_{b,k}$, update information using Eqs. (4.13)–(4.18).
Step 5
If $k = n - 2$ or $x_{a,k+1} > x_{b,k+1}$, output $x^* = x_{a,k+1}$ and $f^* = f(x^*)$, and stop. Otherwise, set $k = k + 1$ and repeat from Step 4.

The condition $x_{a,k+1} > x_{b,k+1}$ implies that $x_{a,k+1} \approx x_{b,k+1}$ within the precision of the computer used, as was stated earlier, or that there is an error in the algorithm. It is thus used as an alternative stopping criterion.

Fig. 4.8 Plot of $f(x)$ for Example 4.1

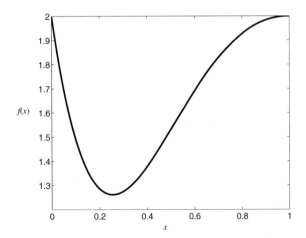

Example 4.1 Applying the Fibonacci search, find an approximate solution $\hat{x}$ of the problem

$$\text{minimize} \quad f(x) = -2x^5 + 11.5x^4 - 24x^3 + 21.5x^2 - 7x + 2$$
$$0 \le x \le 1$$

with error $|\hat{x} - x^*| \le 10^{-4}$ where x^* denotes the true minimizer of $f(x)$ in the interval $[0, 1]$.

Solution From Fig. 4.8, it can be seen that $f(x)$ is a unimodal function over the interval $[0, 1]$. By setting its first-order derivative to zero, namely,

$$f'(x) = -10x^4 + 46x^3 - 72x^2 + 43x - 7 = 0$$

four roots can be found, with one in the interval $[0, 1]$ given by

$$x^* = 0.25474927$$

The second-order derivative of $f(x)$ at point x^* is found to be $f''(x^*) = 14.6106 > 0$ and hence x^* is the only minimizer of $f(x)$ in the interval $[0, 1]$.

The number of Fibonacci iterations can be estimated by noting that

$$I_n = \frac{I_1}{F_n} = \frac{1}{F_n} \le 10^{-4}$$

which yields $F_n \ge 10^4$. Since $F_{19} = 6765$ and $F_{20} = 10946$, we have $n = 20$. By applying Algorithm 4.1 to the problem at hand with $x_{L,1} = 0$, $x_{U,1} = 1$, and $n = 20$, we obtain the solution

$$\hat{x} = 0.25465924$$

The magnitude of the error in $\hat{x}$ is $|\hat{x} - x^*| = 9.0030 \times 10^{-5}$. ∎

4.4   Golden-Section Search

The main disadvantage of the Fibonacci search is that the number of iterations must be supplied as input. A search method in which iterations can be performed until the desired accuracy in either the minimizer or the minimum value of the objective function is achieved is the so-called *golden-section search*. In this approach, as in the Fibonacci search, a sequence of intervals $\{I_1, I_2, I_3, \ldots\}$ is generated as illustrated in Fig. 4.9 by using the recursive relation of Eq. (4.2). The rule by which the lengths of successive intervals are generated is that the ratio of any two adjacent intervals is constant, that is

$$\frac{I_k}{I_{k+1}} = \frac{I_{k+1}}{I_{k+2}} = \frac{I_{k+2}}{I_{k+3}} = \cdots = K \qquad (4.19)$$

so that

$$\frac{I_k}{I_{k+2}} = K^2 \qquad (4.20)$$

$$\frac{I_k}{I_{k+3}} = K^3$$

and so on.

Upon dividing I_k in Eq. (4.2) by I_{k+2}, we obtain

$$\frac{I_k}{I_{k+2}} = \frac{I_{k+1}}{I_{k+2}} + 1 \qquad (4.21)$$

and from Eqs. (4.19)–(4.21)

$$K^2 = K + 1 \qquad (4.22)$$

Now solving for K, we get

$$K = \frac{1 \pm \sqrt{5}}{2} \qquad (4.23)$$

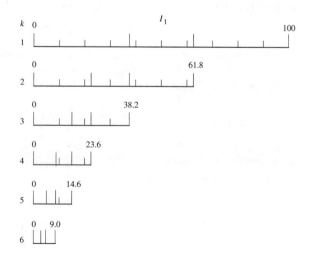

Fig. 4.9   Golden-section search

The negative value of K is irrelevant and so $K = 1.618034$. This constant is known as the *golden ratio*. The terminology has arisen from the fact that in classical Greece, a rectangle with sides bearing a ratio $1 : K$ was considered the most pleasing rectangle and hence it came to be known as the golden rectangle. In turn, the sequence $\{I_1, \; I_1/K, \; I_1/K^2, \; \ldots, \; I_1/K^{n-1}\}$ came to be known as the *golden-section sequence*.

The golden-section search is illustrated in Fig. 4.9 for the case where the left interval is consistently selected. As can be seen, this search resembles the Fibonacci search in most respects. The two exceptions are:

1. Successive intervals are independent of n. Consequently, iterations can be performed until the range of uncertainty or the change in the value of the objective function is reduced below some tolerance ε.
2. The ratio between successive intervals, namely, F_{n-k-1}/F_{n-k}, is replaced by the ratio $1/K$ where

$$\frac{1}{K} = K - 1 = 0.618034$$

according to Eqs. (4.22) and (4.23).

The efficiency of the golden-section search can be easily compared with that of the Fibonacci search. A known relation between F_n and K which is applicable for large values of n is

$$F_n \approx \frac{K^{n+1}}{\sqrt{5}} \qquad (4.24)$$

(e.g., if $n = 11$, $F_n = 144$ and $K^{n+1}/\sqrt{5} \approx 144.001$). Thus Eqs. (4.5) and (4.24) give the region of uncertainty for the Fibonacci search as

$$\Lambda_F = I_n = \frac{I_1}{F_n} \approx \frac{\sqrt{5}}{K^{n+1}} I_1$$

Similarly, for the golden-section search

$$\Lambda_{GS} = I_n = \frac{I_1}{K^{n-1}}$$

and hence

$$\frac{\Lambda_{GS}}{\Lambda_F} = \frac{K^2}{\sqrt{5}} \approx 1.17$$

Therefore, if the number of iterations is the same in the two methods, the region of uncertainty in the golden-section search is larger by about 17% relative to that in the Fibonacci search. Alternatively, the golden-section search will require more iterations to achieve the same precision as the Fibonacci search. However, this disadvantage is offset by the fact that the total number of iterations need not be supplied at the start of the optimization.

An implementation of the golden-section search is as follows:

Algorithm 4.2 Golden-section search
Step 1
Input $x_{L,1}$, $x_{U,1}$, and ε.
Step 2
Assign $I_1 = x_{U,1} - x_{L,1}$, $K = 1.618034$ and compute

$$I_2 = I_1/K$$
$$x_{a,1} = x_{U,1} - I_2, \quad x_{b,1} = x_{L,1} + I_2$$
$$f_{a,1} = f(x_{a,1}), \quad f_{b,1} = f(x_{b,1})$$

Set $k = 1$.
Step 3
Compute

$$I_{k+2} = I_{k+1}/K$$

If $f_{a,k} \geq f_{b,k}$, then update $x_{L,k+1}$, $x_{U,k+1}$, $x_{a,k+1}$, $x_{b,k+1}$, $f_{a,k+1}$, and $f_{b,k+1}$
using Eqs. (4.7)–(4.12). Otherwise, if $f_{a,k} < f_{b,k}$, update information using
Eqs. (4.13)–(4.18).
Step 4
If $I_k < \varepsilon$ or $x_{a,k+1} > x_{b,k+1}$, then do:
  If $f_{a,k+1} > f_{b,k+1}$, compute

$$x^* = \tfrac{1}{2}(x_{b,k+1} + x_{U,k+1})$$

  If $f_{a,k+1} = f_{b,k+1}$, compute

$$x^* = \tfrac{1}{2}(x_{a,k+1} + x_{b,k+1})$$

  If $f_{a,k+1} < f_{b,k+1}$, compute

$$x^* = \tfrac{1}{2}(x_{L,k+1} + x_{a,k+1})$$

Compute $f^* = f(x^*)$.
Output x^* and f^*, and stop.
Step 5
Set $k = k + 1$ and repeat from Step 3.

Example 4.2 Applying the golden-section search to the problem in Example 4.1,
obtain an approximate solution $\hat{x}$ with magnitude error $|\hat{x} - x^*| \leq 10^{-4}$.

Solution The number of golden-section iterations can be estimated by using the
equation

$$I_n = \frac{I_1}{K^{n-1}} \leq 10^{-4}$$

which gives

$$n \geq 1 + \frac{\log_{10}(I_1 \cdot 10^4)}{\log_{10}(K)} = 20.1399$$

Hence $n = 21$, i.e., 20 iterations are required. By applying Algorithm 4.2 with $x_{L,1} = 0$, $x_{U,1} = 1$ and $\varepsilon = 10^{-4}$, we obtain the solution

$$\hat{x} = 0.25477561$$

The magnitude of the error of the solution can be obtained as $|\hat{x} - x^*| = 2.6340 \times 10^{-5}$. ∎

4.5   Quadratic Interpolation Method

In the approximation approach to one-dimensional optimization, an approximate expression for the objective function is assumed, usually in the form of a low-order polynomial. If a second-order polynomial of the form

$$p(x) = a_0 + a_1 x + a_2 x^2 \tag{4.25}$$

is assumed, where a_0, a_1, and a_2 are constants, a quadratic interpolation method is obtained.

Let

$$p(x_i) = a_0 + a_1 x_i + a_2 x_i^2 = f(x_i) = f_i \tag{4.26}$$

for $i = 1$, 2, and 3 where $[x_1, x_3]$ is a bracket on the minimizer x^* of $f(x)$. Assuming that the values f_i are known, the three constants a_0, a_1, and a_2 can be deduced by solving the three simultaneous equations in Eq. (4.26). Thus a polynomial $p(x)$ can be deduced which is an approximation for $f(x)$. Under these circumstances, the plots of $p(x)$ and $f(x)$ will assume the form depicted in Fig. 4.10. As can be seen, the minimizer $\bar{x}$ of $p(x)$ is close to x^*, and if $f(x)$ can be accurately represented by a second-order polynomial, then $\bar{x} \approx x^*$. If $f(x)$ is a quadratic function, then $p(x)$ becomes an exact representation of $f(x)$ and $\bar{x} = x^*$.

The first derivative of $p(x)$ with respect to x is obtained from Eq. (4.25) as

$$p'(x) = a_1 + 2a_2 x$$

and if

$$p'(x) = 0$$

and $a_2 \neq 0$, then the minimizer of $p(x)$ can be deduced as

$$\bar{x} = -\frac{a_1}{2a_2} \tag{4.27}$$

By solving the simultaneous equations in Eq. (4.26), we find that

$$a_1 = -\frac{(x_2^2 - x_3^2)f_1 + (x_3^2 - x_1^2)f_2 + (x_1^2 - x_2^2)f_3}{(x_1 - x_2)(x_1 - x_3)(x_2 - x_3)} \tag{4.28}$$

$$a_2 = \frac{(x_2 - x_3)f_1 + (x_3 - x_1)f_2 + (x_1 - x_2)f_3}{(x_1 - x_2)(x_1 - x_3)(x_2 - x_3)} \tag{4.29}$$

Fig. 4.10 Quadratic
interpolation method

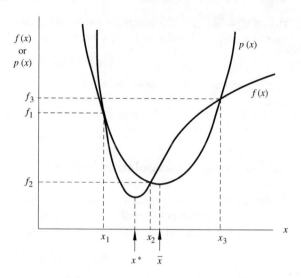

and from Eqs. (4.27)–(4.29), we have

$$\bar{x} = \frac{(x_2^2 - x_3^2)f_1 + (x_3^2 - x_1^2)f_2 + (x_1^2 - x_2^2)f_3}{2[(x_2 - x_3)f_1 + (x_3 - x_1)f_2 + (x_1 - x_2)f_3]} \tag{4.30}$$

The above approach constitutes one iteration of the quadratic interpolation method. If $f(x)$ cannot be represented accurately by a second-order polynomial, a number of such iterations can be performed. The appropriate strategy is to attempt to reduce the interval of uncertainty in each iteration as was done in the search methods of Sects. 4.2–4.4. This can be achieved by rejecting either x_1 or x_3 and then using the two remaining points along with point $\bar{x}$ for a new interpolation.

After a number of iterations, the three points will be in the neighborhood of x^*. Consequently, the second-order polynomial $p(x)$ will be an accurate representation of $f(x)$ by virtue of the Taylor series, and x^* can be determined to within any desired accuracy.

An algorithm based on the above principles is as follows:

Algorithm 4.3 Quadratic interpolation search
Step 1
Input x_1, x_3, and ε.
Set $\bar{x}_0 = 10^{99}$.
Step 2
Compute
$x_2 = \frac{1}{2}(x_1 + x_3)$ and $f_i = f(x_i)$ for $i = 1, 2, 3$.

Step 3
Compute $\bar{x}$ using Eq. (4.30) and $\bar{f} = f(\bar{x})$.
If $|\bar{x} - \bar{x}_0| < \varepsilon$, then output $x^* = \bar{x}$ and $f(x^*) = \bar{f}$, and stop.
Step 4
If $x_1 < \bar{x} < x_2$, then do:
   If $\bar{f} \le f_2$, assign $x_3 = x_2$, $f_3 = f_2$, $x_2 = \bar{x}$, $f_2 = \bar{f}$;
   otherwise, if $\bar{f} > f_2$, assign $x_1 = \bar{x}$, $f_1 = \bar{f}$.
If $x_2 < \bar{x} < x_3$, then do:
   If $\bar{f} \le f_2$, assign $x_1 = x_2$, $f_1 = f_2$, $x_2 = \bar{x}$, $f_2 = \bar{f}$;
   otherwise, if $\bar{f} > f_2$, assign $x_3 = \bar{x}$, $f_3 = \bar{f}$.
Set $\bar{x}_0 = \bar{x}$, and repeat from Step 3.

In Step 4, the bracket on x^* is reduced judiciously to $[x_1, x_2]$ or $[\bar{x}, x_3]$ if $x_1 < \bar{x} < x_2$; or to $[x_2, x_3]$ or $[x_1, \bar{x}]$ if $x_2 < \bar{x} < x_3$ by using the principles developed in Sect. 4.2. The algorithm entails one function evaluation per iteration (see Step 3) except for the first iteration in which three additional function evaluations are required in Step 2.

An implicit assumption in the above algorithm is that interval $[x_1, x_3]$ is a bracket on x^*. If it is not, one can be readily established by varying x in the direction of decreasing $f(x)$ until $f(x)$ begins to increase.

A simplified version of the interpolation formula in Eq. (4.30) can be obtained by assuming that points x_1, x_2, and x_3 are equally spaced. If we let

$$x_1 = x_2 - \delta \quad \text{and} \quad x_3 = x_2 + \delta$$

then Eq. (4.30) becomes

$$\bar{x} = x_2 + \frac{(f_1 - f_3)\delta}{2(f_1 - 2f_2 + f_3)} \tag{4.31}$$

Evidently, this formula involves less computation than that in Eq. (4.30) and, if equal spacing is allowed, it should be utilized. The minimum of the function can be deduced as

$$f_{\min} = f_2 - \frac{(f_1 - f_3)^2}{8(f_1 - 2f_2 + f_3)}$$

(see Problem 4.10).

Example 4.3 Applying the quadratic interpolation search to the problem in Example 4.1, obtain an approximate solution $\hat{x}$ with magnitude error $|\hat{x} - x^*| \le 10^{-4}$ and compare the results obtained with those obtained with the Fibonacci and golden-section searches.

Solution With $x_{L,1} = 0, x_{U,1} = 1$, and $\varepsilon = 10^{-4}$, it took Algorithm 4.3 15 iterations to obtain the solution

$$\hat{x} = 0.25477273$$

The magnitude of the error of the solution was obtained as 2.3460×10^{-5}. On comparing the above solution with the solutions obtained in Example 4.1 using the

Fibonacci search and in Example 4.2 using the golden-section search, the quadratic interpolation search gave a solution of comparable accuracy requiring the least number of function evaluations.   ∎

4.5.1 Two-Point Interpolation

The interpolation formulas in Eqs. (4.30) and (4.31) are said to be *three-point formulas* since they entail the values of $f(x)$ at three distinct points. *Two-point interpolation formulas* can be obtained by assuming that the values of $f(x)$ and its first derivatives are available at two distinct points. If the values of $f(x)$ at $x = x_1$ and $x = x_2$ and the first derivative of $f(x)$ at $x = x_1$ are available, we can write

$$p(x_1) = a_0 + a_1 x_1 + a_2 x_1^2 = f(x_1) \equiv f_1$$
$$p(x_2) = a_0 + a_1 x_2 + a_2 x_2^2 = f(x_2) \equiv f_2$$
$$p'(x_1) = a_1 + 2a_2 x_1 = f'(x_1) \equiv f_1'$$

The solution of these equations gives a_1 and a_2, and thus from Eq. (4.27), the two-point interpolation formula

$$\bar{x} = x_1 + \frac{f_1'(x_2 - x_1)^2}{2[f_1 - f_2 + f_1'(x_2 - x_1)]}$$

can be obtained.

An alternative two-point interpolation formula of the same class can be generated by assuming that the first derivative of $f(x)$ is known at two points x_1 and x_2. If we let

$$p'(x_1) = a_1 + 2a_2 x_1 = f'(x_1) \equiv f_1'$$
$$p'(x_2) = a_1 + 2a_2 x_2 = f'(x_2) \equiv f_2'$$

we deduce

$$\bar{x} = \frac{x_2 f_1' - x_1 f_2'}{f_1' - f_2'} = \frac{x_2 f_1' - x_2 f_2' + x_2 f_2' - x_1 f_2'}{f_1' - f_2'}$$
$$= x_2 + \frac{(x_2 - x_1) f_2'}{f_1' - f_2'}$$

4.6 Cubic Interpolation

Another one-dimensional optimization method which is sometimes quite useful is the *cubic interpolation method*. This is based on the third-order polynomial

$$p(x) = a_0 + a_1 x + a_2 x^2 + a_3 x^3 \tag{4.32}$$

As in the quadratic interpolation method, the coefficients a_i can be determined such that $p(x)$ and/or its derivatives at certain points are equal to $f(x)$ and/or its derivatives. Since there are four coefficients in Eq. (4.32), four equations are needed for

Fig. 4.11 Possible forms of
third-order polynomial

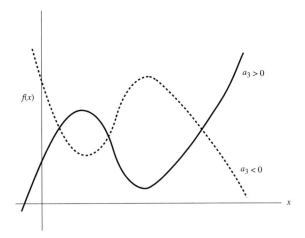

the complete characterization of $p(x)$. These equations can be chosen in a number
of ways and several cubic interpolation formulas can be generated.

The plot of $p(x)$ can assume either of the forms depicted in Fig. 4.11 and, in effect,
$p(x)$ can have a maximum as well as a minimum. By equating the first derivative of
$p(x)$ to zero, that is,

$$p'(x) = a_1 + 2a_2x + 3a_3x^2 = 0 \tag{4.33}$$

and then solving for x, the extremum points of $p(x)$ can be determined as

$$x = \frac{1}{3a_3}\left(-a_2 \pm \sqrt{a_2^2 - 3a_1a_3}\right) \tag{4.34}$$

At the minimizer $\bar{x}$, the second derivative of $p(x)$ is positive, and thus Eq. (4.33)
gives

$$p''(\bar{x}) = 2a_2 + 6a_3\bar{x} > 0$$

or

$$\bar{x} > -\frac{a_2}{3a_3} \tag{4.35}$$

Thus, the solution in Eq. (4.34) that corresponds to the minimizer of $p(x)$ can be
readily selected.

Polynomial $p(x)$ will be an approximation for $f(x)$ if four independent equations
are chosen which interrelate $p(x)$ with $f(x)$. One of many possibilities is to let

$$p(x_i) = a_0 + a_1x_i + a_2x_i^2 + a_3x_i^3 = f(x_i)$$

for $i = 1,\ 2,$ and 3 and

$$p'(x_1) = a_1 + 2a_2x_1 + 3a_3x_1^2 = f'(x_1)$$

By solving this set of equations, coefficients a_1 and a_3 can be determined as

$$a_3 = \frac{\beta - \gamma}{\theta - \psi} \tag{4.36}$$

$$a_2 = \beta - \theta a_3 \tag{4.37}$$

$$a_1 = f'(x_1) - 2a_2 x_1 - 3a_3 x_1^2 \tag{4.38}$$

where

$$\beta = \frac{f(x_2) - f(x_1) + f'(x_1)(x_1 - x_2)}{(x_1 - x_2)^2} \tag{4.39}$$

$$\gamma = \frac{f(x_3) - f(x_1) + f'(x_1)(x_1 - x_3)}{(x_1 - x_3)^2} \tag{4.40}$$

$$\theta = \frac{2x_1^2 - x_2(x_1 + x_2)}{(x_1 - x_2)} \tag{4.41}$$

$$\psi = \frac{2x_1^2 - x_3(x_1 + x_3)}{(x_1 - x_3)} \tag{4.42}$$

The minimizer $\bar{x}$ can now be obtained by using Eqs. (4.34) and (4.35).

An implementation of the cubic interpolation method is as follows:

Algorithm 4.4   Cubic interpolation search
Step 1
Input x_1, x_2, x_3, and initialize the tolerance ε.
Step 2
Set $\bar{x}_0 = 10^{99}$.
Compute $f_1' = f'(x_1)$ and $f_i = f(x_i)$ for $i = 1, 2, 3$.
Step 3
Compute constants β, γ, θ, and ψ using Eqs. (4.39)–(4.42).
Compute constants a_3, a_2, and a_1 using Eqs. (4.36)–(4.38).
Compute the extremum points of $p(x)$ using Eq. (4.34), and select the minimizer $\bar{x}$ using Eq. (4.35).
Compute $\bar{f} = f(\bar{x})$.
Step 4
If $|\bar{x} - \bar{x}_0| < \varepsilon$, then output $x^* = \bar{x}$ and $f(x^*) = \bar{f}$, and stop.
Step 5
Find m such that $f_m = \max (f_1, f_2, f_3)$.
Set $\bar{x}_0 = \bar{x}$, $x_m = \bar{x}$, $f_m = \bar{f}$.
If $m = 1$, compute $f_1' = f'(\bar{x})$.
Repeat from Step 3.

In this algorithm, a bracket is maintained on x^* by replacing the point that yields the largest value in $f(x)$ by the new estimate of the minimizer $\bar{x}$ in Step 5. If the point that is replaced is x_1, the first derivative $f'(x_1)$ is computed since it is required for the calculation of a_1, β, and γ.

As can be seen in Eqs. (4.36)–(4.42), one iteration of cubic interpolation entails a lot more computation than one iteration of quadratic interpolation. Nevertheless, the former can be more efficient. The reason is that a third-order polynomial is a more accurate approximation for $f(x)$ than a second-order one and, as a result, convergence will be achieved in a smaller number of iterations. For the same reason, the method is more tolerant to an inadvertent loss of the bracket.

4.7 Algorithm of Davies, Swann, and Campey

The methods described so far are either search methods or approximation methods. A method due to Davies, Swann, and Campey [8] will now be described, which combines a search method with an approximation method. The search method is used to establish and maintain a bracket on x^*, whereas the approximation method is used to generate estimates of x^*.

In this method, $f(x)$ is evaluated for increasing or decreasing values of x until x^* is bracketed. Then the quadratic interpolation formula for equally-spaced points is used to predict x^*. This procedure is repeated several times until sufficient accuracy in the solution is achieved, as in previous methods.

The input to the algorithm consists of an initial point $x_{0,1}$, an initial increment δ_1, a scaling constant K, and the optimization tolerance ε.

At the kth iteration, an initial point $x_{0,k}$ and an initial increment δ_k are available, and a new initial point $x_{0,k+1}$ as well as a new increment δ_{k+1} are required for the next iteration.

Initially, $f(x)$ is evaluated at points $x_{0,k} - \delta_k$, $x_{0,k}$, and $x_{0,k} + \delta_k$. Three possibilities can arise, namely,

(a) $f(x_{0,k} - \delta_k) > f(x_{0,k}) > f(x_{0,k} + \delta_k)$
(b) $f(x_{0,k} - \delta_k) < f(x_{0,k}) < f(x_{0,k} + \delta_k)$
(c) $f(x_{0,k} - \delta_k) \geq f(x_{0,k}) \leq f(x_{0,k} + \delta_k)$

In case (a), the minimum of $f(x)$ is located in the positive direction and so $f(x)$ is evaluated for increasing values of x until a value of $f(x)$ is obtained which is larger than the previous one. If this occurs on the nth function evaluation, the interval $[x_{0,k}, x_{n,k}]$ is a bracket on x^*. The interval between successive points is increased geometrically, and so this procedure will yield the sequence of points

$$x_{0,k}$$
$$x_{1,k} = x_{0,k} + \delta_k$$
$$x_{2,k} = x_{1,k} + 2\delta_k$$
$$x_{3,k} = x_{2,k} + 4\delta_k$$
$$\vdots$$
$$x_{n,k} = x_{n-1,k} + 2^{n-1}\delta_k \tag{4.43}$$

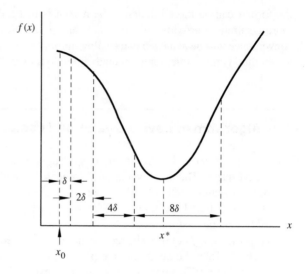

Fig. 4.12 Search method used in the Davies, Swann, and Campey algorithm

as illustrated in Fig. 4.12. Evidently, the most recent interval is twice as long as the previous one and if it is divided into two equal sub-intervals at point

$$x_{m,k} = x_{n-1,k} + 2^{n-2}\delta_k \tag{4.44}$$

then four equally-spaced points are available which bracket the minimizer.

If $f(x)$ is evaluated at point $x_{m,k}$, the function values

$$f_{n-2,k} \equiv f(x_{n-2,k}) \tag{4.45}$$

$$f_{n-1,k} \equiv f(x_{n-1,k}) \tag{4.46}$$

$$f_{m,k} \equiv f(x_{m,k}) \tag{4.47}$$

$$f_{n,k} \equiv f(x_{n,k}) \tag{4.48}$$

will be available. If $f_{m,k} \geq f_{n-1,k}$, x^* is located in the interval $[x_{n-2,k}, x_{m,k}]$ (see Fig. 4.13) and so the use of Eq. (4.31) and Eqs. (4.45)–(4.48) yields an estimate for x^* as

$$x_{0,k+1} = x_{n-1,k} + \frac{2^{n-2}\delta_k(f_{n-2,k} - f_{m,k})}{2(f_{n-2,k} - 2f_{n-1,k} + f_{m,k})} \tag{4.49}$$

Similarly, if $f_{m,k} < f_{n-1,k}$, x^* is located in the interval $[x_{n-1,k}, x_{n,k}]$ (see Fig. 4.14) then an estimate for x^* is

$$x_{0,k+1} = x_{m,k} + \frac{2^{n-2}\delta_k(f_{n-1,k} - f_{n,k})}{2(f_{n-1,k} - 2f_{m,k} + f_{n,k})} \tag{4.50}$$

In case (b), x^* is located in the negative direction, and so x is decreased in steps δ_k, $2\delta_k$, ... until the minimum of $f(x)$ is located. The procedure is as in case (a) except that δ_k is negative in Eqs. (4.49) and (4.50).

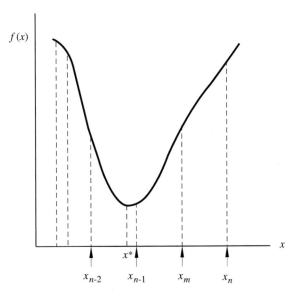

Fig. 4.13 Reduction of range of uncertainty in Davies, Swann, and Campey algorithm if $f_m \geq f_{n-1}$

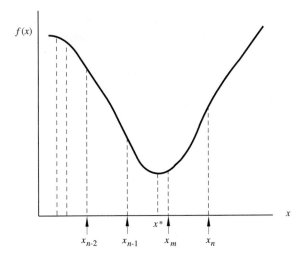

Fig. 4.14 Reduction of range of uncertainty in Davies, Swann, and Campey algorithm if $f_m < f_{n-1}$

In case (c), x^* is bracketed by $x_{0,k} - \delta_k$ and $x_{0,k} + \delta_k$ and if

$$f_{-1,k} = f(x_{0,k} - \delta_k)$$
$$f_{0,k} = f(x_{0,k})$$
$$f_{1,k} = f(x_{0,k} + \delta_k)$$

Equation (4.31) yields an estimate for x^* as

$$x_{0,k+1} = x_{0,k} + \frac{\delta_k(f_{-1,k} - f_{1,k})}{2(f_{-1,k} - 2f_{0,k} + f_{1,k})} \qquad (4.51)$$

The kth iteration is completed by defining a new increment

$$\delta_{k+1} = K\delta_k$$

where K is a constant in the range 0–1. The motivation for this scaling is that as the solution is approached, a reduced range of x will be searched and, therefore, the resolution of the algorithm needs to be increased. A suitable value for K might be 0.1.

The above principles can be used to construct the following algorithm:

Algorithm 4.5 Davies, Swann, and Campey search
Step 1
Input $x_{0,1}$, δ_1, K, and initialize the tolerance ε.
Set $k = 0$.
Step 2
Set $k = k + 1$, $x_{-1,k} = x_{0,k} - \delta_k$, $x_{1,k} = x_{0,k} + \delta_k$.
Compute $f_{0,k} = f(x_{0,k})$ and $f_{1,k} = f(x_{1,k})$.
Step 3
If $f_{0,k} > f_{1,k}$, set $p = 1$ and go to Step 4; otherwise, compute $f_{-1,k} = f(x_{-1,k})$.
If $f_{-1,k} < f_{0,k}$, set $p = -1$ and go to Step 4.
Otherwise, if $f_{-1,k} \geq f_{0,k} \leq f_{1,k}$ go to Step 7.
Step 4
For $n = 1, 2, \ldots$ compute $f_{n,k} = f(x_{n-1,k} + 2^{n-1}p\delta_k)$ until $f_{n,k} > f_{n-1,k}$.
Step 5
Compute $f_{m,k} = f(x_{n-1,k} + 2^{n-2}p\delta_k)$.
Step 6
If $f_{m,k} \geq f_{n-1,k}$, compute

$$x_{0,k+1} = x_{n-1,k} + \frac{2^{n-2}p\delta_k(f_{n-2,k} - f_{m,k})}{2(f_{n-2,k} - 2f_{n-1,k} + f_{m,k})}$$

Otherwise, if $f_{m,k} < f_{n-1,k}$, compute

$$x_{0,k+1} = x_{m,k} + \frac{2^{n-2}p\delta_k(f_{n-1,k} - f_{n,k})}{2(f_{n-1,k} - 2f_{m,k} + f_{n,k})}$$

(see Eqs. (4.49) and (4.50))
If $2^{n-2}\delta_k \leq \varepsilon$ go to Step 8; otherwise, set $\delta_{k+1} = K\delta_k$ and repeat from Step 2.
Step 7 Compute

$$x_{0,k+1} = x_{0,k} + \frac{\delta_k(f_{-1,k} - f_{1,k})}{2(f_{-1,k} - 2f_{0,k} + f_{1,k})}$$

(see Eq. (4.51)).
If $\delta_k \leq \varepsilon$ go to Step 8; otherwise, set $\delta_{k+1} = K\delta_k$ and repeat from Step 2.
Step 8
Output $x^* = x_{0,k+1}$ and $f(x^*) = f_{0,k+1}$, and stop.

Parameter δ_1 is a small positive constant that would depend on the problem, say, $0.1x_{0,1}$. Constant p in Steps 3–6, which can be 1 or -1, is used to render the formulas in Eqs. (4.49) and (4.50) applicable for increasing as well as decreasing values of x. Constant ε in Step 1 determines the precision of the solution. If ε is very small, say, less than 10^{-6}, then as the solution is approached, we have

$$f_{n-2,k} \approx f_{n-1,k} \approx f_{m,k} \approx f_{n,k}$$

Consequently, the distinct possibility of dividing by zero may arise in the evaluation of $x_{0,k+1}$. However, this problem can be easily prevented by using appropriate checks in Steps 6 and 7.

An alternative form of the above algorithm can be obtained by replacing the quadratic interpolation formula for equally-spaced points by the general formula of Eq. (4.30). If this is done, the mid-interval function evaluation of Step 5 is unnecessary. Consequently, if the additional computation required by Eq. (4.31) is less than one complete evaluation of $f(x)$, then the modified algorithm is likely to be more efficient.

Another possible modification is to use the cubic interpolation of Sect. 4.6 instead of quadratic interpolation. Such an algorithm is likely to reduce the number of function evaluations. However, the amount of computation could increase owing to the more complex formulation in the cubic interpolation.

4.8   Inexact Line Searches

In the multidimensional algorithms to be studied, most of the computational effort is spent in performing function and gradient evaluations in the execution of line searches. Consequently, the amount of computation required tends to depend on the efficiency and precision of the line searches used. If high-precision line searches are necessary, the amount of computation will be large and if inexact line searches do not affect the convergence of an algorithm, a small amount of computation might be sufficient.

Many optimization methods have been found to be quite tolerant to line-search imprecision and, for this reason, inexact line searches are usually used in these methods.

Let us assume that

$$\mathbf{x}_{k+1} = \mathbf{x}_k + \alpha \mathbf{d}_k$$

where $\mathbf{d}_k$ is a given direction vector and α is an independent search parameter, and that function $f(\mathbf{x}_{k+1})$ has a unique minimum for some positive value of α. The linear approximation of the Taylor series in Eq. (2.4d) gives

$$f(\mathbf{x}_{k+1}) = f(\mathbf{x}_k) + \mathbf{g}_k^T \mathbf{d}_k \alpha \tag{4.52}$$

where

$$\mathbf{g}_k^T \mathbf{d}_k = \left. \frac{df(\mathbf{x}_k + \alpha \mathbf{d}_k)}{d\alpha} \right|_{\alpha=0}$$

Equation (4.52) represents line A shown in Fig. 4.15a. The equation

$$f(\mathbf{x}_{k+1}) = f(\mathbf{x}_k) + \rho \mathbf{g}_k^T \mathbf{d}_k \alpha \tag{4.53}$$

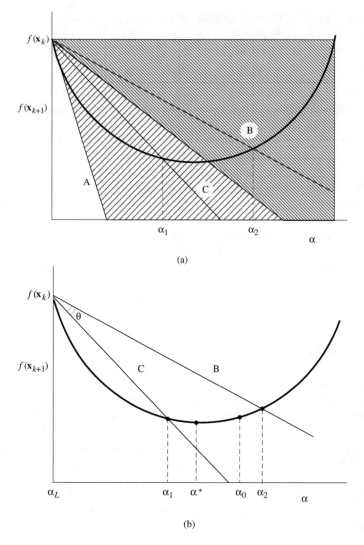

Fig. 4.15 **a** The Goldstein tests. **b** Goldstein tests satisfied

where $0 \leq \rho < \frac{1}{2}$ represents line B in Fig. 4.15a whose slope ranges from 0 to $\frac{1}{2}\mathbf{g}_k^T\mathbf{d}_k$ depending on the value of ρ, as depicted by shaded area B in Fig. 4.15a. On the other hand, the equation

$$f(\mathbf{x}_{k+1}) = f(\mathbf{x}_k) + (1 - \rho)\mathbf{g}_k^T\mathbf{d}_k\alpha \tag{4.54}$$

represents line C in Fig. 4.15a whose slope ranges from $\mathbf{g}_k^T\mathbf{d}_k$ to $\frac{1}{2}\mathbf{g}_k^T\mathbf{d}_k$ as depicted by shaded area C in Fig. 4.15a. The angle between lines C and B, designated as θ, is given by

$$\theta = \tan^{-1}\left[\frac{-(1-2\rho)\mathbf{g}_k^T\mathbf{d}_k}{1+\rho(1-\rho)(\mathbf{g}_k^T\mathbf{d}_k)^2}\right]$$

as illustrated in Fig. 4.15b. Evidently by adjusting ρ in the range $0-\frac{1}{2}$, the slope of θ can be varied in the range $-\mathbf{g}_k^T\mathbf{d}_k$ to 0. By fixing ρ at some value in the permissible range, two values of α are defined by the intercepts of the lines in Eqs. (4.53) and (4.54) and the curve for $f(\mathbf{x}_{k+1})$, say, α_1 and α_2, as depicted in Fig. 4.15b.

Let α_0 be an estimate of the value of α that minimizes $f(\mathbf{x}_k + \alpha\mathbf{d}_k)$. If $f(\mathbf{x}_{k+1})$ for $\alpha = \alpha_0$ is equal to or less than the corresponding value of $f(\mathbf{x}_{k+1})$ given by Eq. (4.53), and is equal to or greater than the corresponding value of $f(\mathbf{x}_{k+1})$ given by Eq. (4.54), that is, if

$$f(\mathbf{x}_{k+1}) \leq f(\mathbf{x}_k) + \rho\mathbf{g}_k^T\mathbf{d}_k\alpha_0 \tag{4.55}$$

and

$$f(\mathbf{x}_{k+1}) \geq f(\mathbf{x}_k) + (1-\rho)\mathbf{g}_k^T\mathbf{d}_k\alpha_0 \tag{4.56}$$

then α_0 may be deemed to be an acceptable estimate of α^* in that it will yield a sufficient reduction in $f(\mathbf{x})$. Under these circumstances, we have $\alpha_1 \leq \alpha_0 \leq \alpha_2$, as depicted in Fig. 4.15b, i.e., α_1 and α_2 constitute a bracket of the estimated minimizer α_0. Equations (4.55) and (4.56), which are often referred to as the *Goldstein conditions*, form the basis of a class of *inexact line searches*. In these methods, an estimate α_0 is generated by some means, based on available information, and the conditions in Eqs. (4.55) and (4.56) are checked. If both conditions are satisfied, then the reduction in $f(\mathbf{x}_{k+1})$ is deemed to be acceptable, and the procedure is terminated. On the other hand, if either Eq. (4.55) or Eq. (4.56) is violated, the reduction in $f(\mathbf{x}_{k+1})$ is deemed to be insufficient and an improved estimate of α^*, say, $\breve{\alpha}_0$, can be obtained. If Eq. (4.55) is violated, then $\alpha_0 > \alpha_2$ as depicted in Fig. 4.16a and since $\alpha_L < \alpha^* < \alpha_0$, the new estimate $\breve{\alpha}_0$ can be determined by using interpolation. On the other hand, if Eq. (4.56) is violated, $\alpha_0 < \alpha_1$ as depicted in Fig. 4.16b, and since α_0 is likely to be in the range $\alpha_L < \alpha_0 < \alpha^*$, $\breve{\alpha}_0$ can be determined by using extrapolation.

If the value of $f(\mathbf{x}_k + \alpha\mathbf{d}_k)$ and its derivative with respect to α are known for $\alpha = \alpha_L$ and $\alpha = \alpha_0$, then for $\alpha_0 > \alpha_2$ a good estimate for $\breve{\alpha}_0$ can be deduced by using the interpolation formula

$$\breve{\alpha}_0 = \alpha_L + \frac{(\alpha_0 - \alpha_L)^2 f_L'}{2[f_L - f_0 + (\alpha_0 - \alpha_L)f_L']} \tag{4.57}$$

and for $\alpha_0 < \alpha_1$ the extrapolation formula

$$\breve{\alpha}_0 = \alpha_0 + \frac{(\alpha_0 - \alpha_L)f_0'}{(f_L' - f_0')} \tag{4.58}$$

can be used where

$$f_L = f(\mathbf{x}_k + \alpha_L\mathbf{d}_k), \quad f_L' = f'(\mathbf{x}_k + \alpha_L\mathbf{d}_k) = \mathbf{g}(\mathbf{x}_k + \alpha_L\mathbf{d}_k)^T\mathbf{d}_k$$
$$f_0 = f(\mathbf{x} + \alpha_0\mathbf{d}_k), \quad f_0' = f'(\mathbf{x}_k + \alpha_0\mathbf{d}_k) = \mathbf{g}(\mathbf{x}_k + \alpha_0\mathbf{d}_k)^T\mathbf{d}_k$$

(see Sect. 4.5).

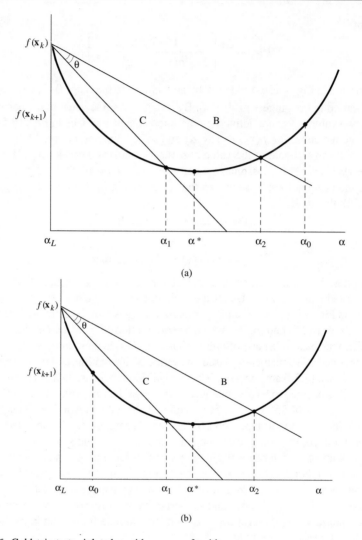

Fig. 4.16 Goldstein tests violated: **a** with $\alpha_0 > \alpha_2$, **b** with $\alpha_0 < \alpha_1$

Repeated application of the above procedure will eventually yield a value of $\breve{\alpha}_0$ such that $\alpha_1 < \breve{\alpha}_0 < \alpha_2$ and the inexact line search is terminated.

A useful theorem relating to the application of the Goldstein tests in an inexact line search is as follows.

Theorem 4.1 *Convergence of inexact line search*   *If*

(a) $f(\mathbf{x}_k)$ has a lower bound,
(b) $\mathbf{g}_k$ is uniformly continuous on set $\{\mathbf{x} : f(\mathbf{x}) < f(\mathbf{x}_0)\}$,
(c) directions $\mathbf{d}_k$ are not orthogonal to $-\mathbf{g}_k$ for all k,

then a descent algorithm using an inexact line search based on Eqs. (4.55) and (4.56) will converge to a stationary point as $k \to \infty$.

The proof of this theorem is given by Fletcher [9]. The theorem does not guarantee that a descent algorithm will converge to a minimizer since a saddle point is also a stationary point. Nevertheless, the theorem is of importance since it demonstrates that inaccuracies due to the inexactness of the line search are not detrimental to convergence.

Conditions (*a*) and (*b*) of Theorem 4.1 are normally satisfied but condition (*c*) may be violated. Nevertheless, the problem can be avoided in practice by changing direction $\mathbf{d}_k$. For example, if θ_k is the angle between $\mathbf{d}_k$ and $-\mathbf{g}_k$ and

$$\theta_k = \cos^{-1} \frac{-\mathbf{g}_k^T \mathbf{d}_k}{\|\mathbf{g}_k\|_2 \, \|\mathbf{d}_k\|_2} = \frac{\pi}{2}$$

then $\mathbf{d}_k$ can be modified slightly to ensure that

$$\theta_k = \frac{\pi}{2} - \mu$$

where $\mu > 0$.

The Goldstein conditions sometimes lead to the situation illustrated in Fig. 4.17, where α^* is not in the range $[\alpha_1, \ \alpha_2]$. Evidently, in such a case a value α_0 in the interval $[\alpha^*, \ \alpha_1]$ will not terminate the line search even though the reduction in $f(\mathbf{x}_k)$ would be larger than that for any α_0 in the interval $[\alpha_1, \ \alpha_2]$. Although the problem is not serious, since convergence is assured by Theorem 4.1, the amount of computation may be increased. The problem can be eliminated by replacing the second Goldstein condition, namely, Eq. (4.56), by the condition

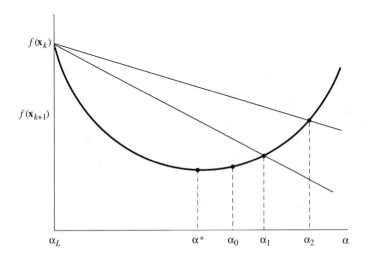

Fig. 4.17 Goldstein tests violated with $\alpha^* < \alpha_1$

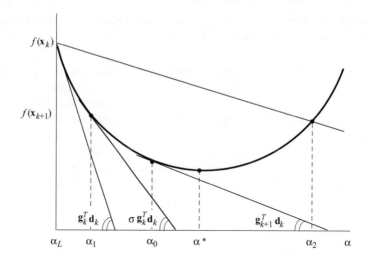

Fig. 4.18 Fletcher's modification of the Goldstein tests

$$\mathbf{g}_{k+1}^T \mathbf{d}_k \geq \sigma \mathbf{g}_k^T \mathbf{d}_k \qquad (4.59)$$

where $0 < \sigma < 1$ and $\sigma \geq \rho$. This modification to the second Goldstein condition was proposed by Fletcher [10]. It is illustrated in Fig. 4.18. The scalar $\mathbf{g}_k^T \mathbf{d}_k$ is the derivative of $f(\mathbf{x}_k + \alpha \mathbf{d}_k)$ at $\alpha = 0$, and since $0 < \sigma < 1$, $\sigma \mathbf{g}_k^T \mathbf{d}_k$ is the derivative of $f(\mathbf{x}_k + \alpha \mathbf{d}_k)$ at some value of α, say, α_1, such that $\alpha_1 < \alpha^*$. Now if the condition in Eq. (4.59) is satisfied at some point

$$\mathbf{x}_{k+1} = \mathbf{x}_k + \alpha_0 \mathbf{d}_k$$

then the slope of $f(\mathbf{x}_k + \alpha \mathbf{d}_k)$ at $\alpha = \alpha_0$ is less negative (more positive) than the slope at $\alpha = \alpha_1$ and, consequently, we conclude that $\alpha_1 \leq \alpha_0$. Now if Eq. (4.55) is also satisfied, then we must have $\alpha_1 < (\alpha^* \text{ or } \alpha_0) < \alpha_2$, as depicted in Fig. 4.18.

The precision of a line search based on Eqs. (4.55) and (4.59) can be increased by reducing the value of σ. While $\sigma = 0.9$ results in a somewhat imprecise line search, the value $\sigma = 0.1$ results in a fairly precise line search. Note, however, that a more precise line search could slow down the convergence.

A disadvantage of the condition in Eq. (4.59) is that it does not lead to an exact line search as $\sigma \to 0$. An alternative condition that eliminates this problem is obtained by modifying the condition in Eq. (4.59) as

$$|\mathbf{g}_{k+1}^T \mathbf{d}_k| \leq -\sigma \mathbf{g}_k^T \mathbf{d}_k$$

In order to demonstrate that an exact line search can be achieved with the above condition, let us assume that $\mathbf{g}_k^T \mathbf{d}_k < 0$. If $\mathbf{g}_{k+1}^T \mathbf{d}_k < 0$, the line search will not terminate until

$$-|\mathbf{g}_{k+1}^T \mathbf{d}_k| \geq \sigma \mathbf{g}_k^T \mathbf{d}_k$$

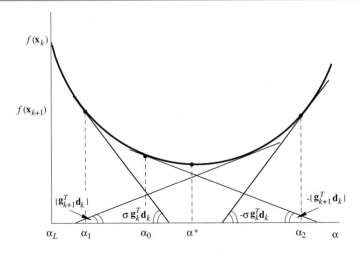

Fig. 4.19   Conversion of inexact line search into an exact line search

and if $\mathbf{g}_{k+1}^T \mathbf{d}_k > 0$, the line search will not terminate until

$$|\mathbf{g}_{k+1}^T \mathbf{d}_k| \le -\sigma \mathbf{g}_k^T \mathbf{d}_k \tag{4.60}$$

Now if $\sigma \mathbf{g}_k^T \mathbf{d}_k$, $\mathbf{g}_{k+1}^T \mathbf{d}_k$, and $-\sigma \mathbf{g}_k^T \mathbf{d}_k$ are the derivatives of $f(\mathbf{x}_k + \alpha \mathbf{d}_k)$ at points $\alpha = \alpha_1, \alpha = \alpha_0$, and $\alpha = \alpha_2$, respectively, we have $\alpha_1 \le \alpha_0 \le \alpha_2$ as depicted in Fig. 4.19. In effect, Eq. (4.60) overrides both of the Goldstein conditions in Eqs. (4.55) and (4.56). Since interval $[\alpha_1, \ \alpha_2]$ can be reduced as much as desired by reducing σ, it follows that α^* can be determined as accurately as desired, and as $\sigma \to 0$, the line search becomes exact. In such a case, the amount of computation would be comparable to that required by any other exact line search and the computational advantage of using an inexact line search would be lost.

An inexact line search based on Eqs. (4.55) and (4.59) due to Fletcher [10] is as follows:

Algorithm 4.6   Inexact line search
Step 1
Input $\mathbf{x}_k$, $\mathbf{d}_k$, and compute $\mathbf{g}_k$.
Initialize algorithm parameters ρ, σ, τ, and χ.
Set $\alpha_L = 0$ and $\alpha_U = 10^{99}$.
Step 2
Compute $f_L = f(\mathbf{x}_k + \alpha_L \mathbf{d}_k)$.
Compute $f_L' = \mathbf{g}(\mathbf{x}_k + \alpha_L \mathbf{d}_k)^T \mathbf{d}_k$.
Step 3
Estimate α_0.

Step 4

Compute $f_0 = f(\mathbf{x}_k + \alpha_0 \mathbf{d}_k)$.

Step 5 (Interpolation)

If $f_0 > f_L + \rho(\alpha_0 - \alpha_L)f'_L$, then do:

   a. If $\alpha_0 < \alpha_U$, then set $\alpha_U = \alpha_0$.

   b. Compute $\breve{\alpha}_0$ using Eq. (4.57).

   c. If $\breve{\alpha}_0 < \alpha_L + \tau(\alpha_U - \alpha_L)$, then set $\breve{\alpha}_0 = \alpha_L + \tau(\alpha_U - \alpha_L)$.

   d. If $\breve{\alpha}_0 > \alpha_U - \tau(\alpha_U - \alpha_L)$, then set $\breve{\alpha}_0 = \alpha_U - \tau(\alpha_U - \alpha_L)$.

   e. Set $\alpha_0 = \breve{\alpha}_0$ and go to Step 4.

Step 6

Compute $f'_0 = \mathbf{g}(\mathbf{x}_k + \alpha_0 \mathbf{d}_k)^T \mathbf{d}_k$.

Step 7 (Extrapolation)

If $f'_0 < \sigma f'_L$, then do:

   a. Compute $\Delta\alpha_0 = (\alpha_0 - \alpha_L)f'_0/(f'_L - f'_0)$ (see Eq. (4.58)).

   b. If $\Delta\alpha_0 < \tau(\alpha_0 - \alpha_L)$, then set $\Delta\alpha_0 = \tau(\alpha_0 - \alpha_L)$.

   c. If $\Delta\alpha_0 > \chi(\alpha_0 - \alpha_L)$, then set $\Delta\alpha_0 = \chi(\alpha_0 - \alpha_L)$.

   d. Compute $\breve{\alpha}_0 = \alpha_0 + \Delta\alpha_0$.

   e. Set $\alpha_L = \alpha_0$, $\alpha_0 = \breve{\alpha}_0$, $f_L = f_0$, $f'_L = f'_0$, and go to Step 4.

Step 8

Output α_0 and $f_0 = f(\mathbf{x}_k + \alpha_0 \mathbf{d}_k)$, and stop.

   The precision to which the minimizer is determined depends on the values of ρ and σ. Small values like $\rho = \sigma = 0.1$ will yield a relatively precise line search whereas values like $\rho = 0.3$ and $\sigma = 0.9$ will yield a somewhat imprecise line search. The values $\rho = 0.1$ and $\sigma = 0.7$ give good results.

   An estimate of α_0 in Step 3 can be determined by assuming that $f(\mathbf{x})$ is a convex quadratic function and using $\alpha_0 = \|\mathbf{g}_0\|_2^2/(\mathbf{g}_0^T \mathbf{H}_0 \mathbf{g}_0)$ which is the minimum point for a convex quadratic function.

   In Step 5, $\breve{\alpha}_0$ is checked and if necessary it is adjusted through a series of interpolations to ensure that $\alpha_L < \breve{\alpha}_0 < \alpha_U$. A suitable value for τ is 0.1. This assures that $\breve{\alpha}_0$ is no closer to α_L or α_U than 10% of the permissible range. A similar check is applied in the case of extrapolation, as can be seen in Step 7. The value for χ suggested by Fletcher is 9.

   The algorithm maintains a running bracket (or range of uncertainty) $[\alpha_L, \alpha_U]$ that contains the minimizer and is initially set to $[0, 10^{99}]$ in Step 1. This is gradually reduced by reducing α_U in Step 5a and increasing α_L in Step 7e.

   In Step 7e, known data that can be used in the next iteration are saved, i.e., α_0, f_0, and f'_0 become α_L, f_L, and f'_L, respectively. This keeps the amount of computation to a minimum.

   Note that the Goldstein condition in Eq. (4.55) is modified as in Step 5 to take into account the fact that α_L assumes a value greater than zero when extrapolation is applied at least once.

Example 4.4 Applying the inexact line search, minimize the function $F(\alpha) = f(\mathbf{x} + \alpha\mathbf{d})$ where

$$f(\mathbf{x}) = 0.6x_2^4 + 5x_1^2 - 7x_2^2 + \sin(x_1 x_2) - 5x_2$$

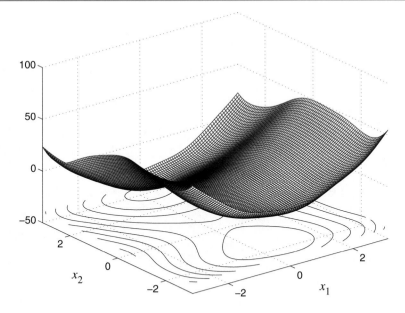

Fig. 4.20 Plot of $f(\mathbf{x})$ for Example 4.4

$\mathbf{x}_0 = [-\pi \ \ -\pi]^T$ and examine the two search directions $\mathbf{d}_0 = [1 \ 1.01]^T$ and $\hat{\mathbf{d}}_0 = [1 \ 0.85]^T$.

Solution A plot of function $f(\mathbf{x})$ over the region $-\pi \leq x_1 \leq \pi$, $-\pi \leq x_2 \leq \pi$ is shown in Fig. 4.20. By inspection, we observe that at point $\mathbf{x}_0 = [-\pi \ \ -\pi]^T$ both $\mathbf{d}_0$ and $\hat{\mathbf{d}}_0$ are descent directions.

The gradient $\nabla f(\mathbf{x})$ is given by

$$\nabla f(\mathbf{x}) = \begin{bmatrix} 10x_1 + x_2\cos(x_1x_2) \\ 2.4x_2^3 - 14x_2 + x_1\cos(x_1x_2) - 5 \end{bmatrix}$$

and thus

$$\nabla f(\mathbf{x}_0) = \begin{bmatrix} -28.5801 \\ -32.5969 \end{bmatrix}$$

With the above data and $\rho = 0.1$, $\sigma = 0.1$, $\tau = 0.1$, and $\chi = 0.75$ at $\mathbf{x}_0$ with search direction $\mathbf{d}_0$, Algorithm 4.6 yields $\alpha_0 = 5.0123$ and hence the next iterate is given by

$$\mathbf{x}_1 = \mathbf{x}_0 + \alpha_0\mathbf{d}_0 = \begin{bmatrix} 1.8707 \\ 1.9208 \end{bmatrix}$$

Function $f(\mathbf{x})$ at $\mathbf{x}_0$ and $\mathbf{x}_1$ can be evaluated as

$$f(\mathbf{x}_0) = 53.9839 \quad \text{and} \quad f(\mathbf{x}_1) = -10.2022$$

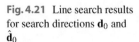

Fig. 4.21 Line search results for search directions $\mathbf{d}_0$ and $\hat{\mathbf{d}}_0$

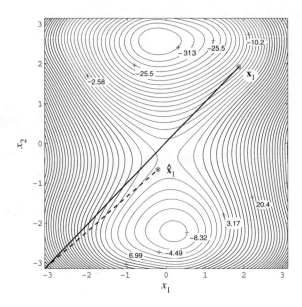

respectively. By applying Algorithm 4.6 at point $\mathbf{x}_0$ with search direction $\hat{\mathbf{d}}_0$, we obtain $\hat{\alpha}_0 = 2.9286$ and hence the next iterate is given by

$$\hat{\mathbf{x}}_1 = \mathbf{x}_0 + \hat{\alpha}_0 \hat{\mathbf{d}}_0 = \begin{bmatrix} -0.2130 \\ -0.6523 \end{bmatrix}$$

Function $f(\mathbf{x})$ at point $\hat{\mathbf{x}}_1$ can be evaluated as $f(\hat{\mathbf{x}}_1) = 0.7570$. The results obtained are depicted in Fig. 4.21 where the solid line represents search direction $\mathbf{d}_0$ and the dashed line represents search direction $\hat{\mathbf{d}}_0$.

This example demonstrates the importance of using a good search direction as a slightly different search direction can lead to quite a different iterate. ∎

Problems

4.1 (*a*) Assuming that the ratio of two consecutive Fibonacci numbers, F_{k-1}/F_k, converges to a finite limit α, use Eq. (4.4) to show that

$$\lim_{k \to \infty} \frac{F_{k-1}}{F_k} = \alpha = \frac{2}{\sqrt{5}+1} \approx 0.6180$$

(*b*) Use MATLAB to verify the value of α in part (*a*).

4.2 The 5th-order polynomial

$$f(x) = -5x^5 + 4x^4 - 12x^3 + 11x^2 - 2x + 1$$

is known to be a unimodal function on interval $[-0.5, \ 0.5]$.

(a) Use the dichotomous search to find the minimizer of $f(x)$ on $[-0.5, 0.5]$ with the range of uncertainty less than 10^{-5}.

(b) Solve the line search problem in part (a) using the Fibonacci search.

(c) Solve the line search problem in part (a) using the golden-section search.

(d) Solve the line search problem in part (a) using the quadratic interpolation method of Sect. 4.5.

(e) Solve the line search problem in part (a) using the cubic interpolation method of Sect. 4.6.

(f) Solve the line search problem in part (a) using the algorithm of Davies, Swann, and Campey.

(g) Compare the computational efficiency of the methods in $(a) - (f)$ in terms of number of function evaluations.

4.3 The function[1]

$$f(x) = \ln^2(x - 2) + \ln^2(10 - x) - x^{0.2}$$

is known to be a unimodal function on $[6, 9.9]$. Repeat Problem 4.2 for the above function.

4.4 The function

$$f(x) = -3x \sin 0.75x + e^{-2x}$$

is known to be a unimodal function on $[0, \ 2\pi]$. Repeat Problem 4.2 for the above function.

4.5 The function

$$f(x) = e^{3x} + 5e^{-2x}$$

is known to be a unimodal function on $[0, 1]$. Repeat Problem 4.2 for the above function.

4.6 The function

$$f(x) = 0.2x \ln x + (x - 2.3)^2$$

is known to be a unimodal function on $[0.5, 2.5]$. Repeat Problem 4.2 for the above function.

4.7 Let $f_1(x)$ and $f_2(x)$ be two convex functions such that $f_1(-0.4) = 0.36$, $f_1(0.6) = 2.56$, $f_2(-0.4) = 3.66$, and $f_2(1) = 2$ and define the function

$$f(x) = \max\{f_1(x), \ f_2(x)\}$$

Identify the smallest interval in which the minimizer of $f(x)$ is guaranteed to exist.

---

[1] Here and in the rest of the book, the logarithms of x to the base e and 10 will be denoted as $\ln(x)$ and $\log_{10}(x)$, respectively.

4.8 The values of a function $f(x)$ at points $x = x_1$ and $x = x_2$ are f_1 and f_2, respectively, and the derivative of $f(x)$ at point x_1 is f_1'. Show that

$$\bar{x} = x_1 + \frac{f_1'(x_2 - x_1)^2}{2[f_1 - f_2 + f_1'(x_2 - x_1)]}$$

is an estimate of the minimizer of $f(x)$.

4.9 By letting $x_1 = x_2 - \delta$ and $x_3 = x_2 + \delta$ in Eq. (4.30), show that the minimizer $\bar{x}$ can be computed using Eq. (4.31).

4.10 A convex quadratic function $f(x)$ assumes the values f_1, f_2, and f_3 at $x = x_1$, x_2, and x_3, respectively, where $x_1 = x_2 - \delta$ and $x_3 = x_2 + \delta$. Show that the minimum of the function is given by

$$f_{\min} = f_2 - \frac{(f_1 - f_3)^2}{8(f_1 - 2f_2 + f_3)}$$

4.11 Line search can be performed in closed form for convex quadratic functions. Consider the objective function

$$f(\mathbf{x}) = \tfrac{1}{2}\mathbf{x}^T \mathbf{Q}\mathbf{X} + \mathbf{b}^T\mathbf{x} + c$$

where $\mathbf{Q}$ is positive definite.

(a) For given $\mathbf{x}_k$ and $\mathbf{d}_k$, derive an explicit expression of $f(\mathbf{x}_k + \alpha\mathbf{d}_k)$ as a second-order polynomial of α.

(b) Use the expression of $f(\mathbf{x}_k + \alpha\mathbf{d}_k)$ from part (a) to show that $f(\mathbf{x}_k + \alpha\mathbf{d}_k)$ is convex with respect to α and its minimizer is given by

$$\alpha_k = -\frac{\mathbf{d}_k^T(\mathbf{Q}\mathbf{x}_k + \mathbf{b})}{\mathbf{d}_k^T\mathbf{Q}\mathbf{d}_k} = -\frac{\mathbf{d}_k^T\mathbf{g}_k}{\mathbf{d}_k^T\mathbf{Q}\mathbf{d}_k}$$

where $\mathbf{g}_k = \mathbf{Q}\mathbf{x}_k + \mathbf{b}$ is the gradient of $f(\mathbf{x})$ at $\mathbf{x}_k$.

4.12 (a) Use MATLAB to plot

$$f(\mathbf{x}) = 0.7x_1^4 - 8x_1^2 + 6x_2^2 + \cos(x_1 x_2) - 8x_1$$

over the region $-\pi \le x_1, x_2 \le \pi$. A MATLAB command for plotting the surface of a two-variable function is mesh.

(b) Use MATLAB to generate a contour plot of $f(\mathbf{x})$ over the same region as in (a) and 'hold' it.

(c) Compute the gradient of $f(\mathbf{x})$, and prepare MATLAB function files to evaluate $f(\mathbf{x})$ and its gradient.

(d) Use Fletcher's inexact line search algorithm to update point $\mathbf{x}_0$ along search direction $\mathbf{d}_0$ by solving the problem

$$\underset{\alpha \geq 0}{\text{minimize}} \quad f(\mathbf{x}_0 + \alpha \mathbf{d}_0)$$

where

$$\mathbf{x}_0 = \begin{bmatrix} -\pi \\ \pi \end{bmatrix}, \quad \mathbf{d}_0 = \begin{bmatrix} 1.0 \\ -1.3 \end{bmatrix}$$

This can be done in several steps:

- Record the numerical values of α^* obtained.
- Record the updated point $\mathbf{x}_1 = \mathbf{x}_0 + \alpha^* \mathbf{d}_0$.
- Evaluate $f(\mathbf{x}_1)$ and compare it with $f(\mathbf{x}_0)$.
- Plot the line search result on the contour plot generated in (b).
- Plot $f(\mathbf{x}_0 + \alpha \mathbf{d}_0)$ as a function of α over the interval [0, 4.8332]. Based on the plot, comment on the precision of Fletcher's inexact line search.

(e) Repeat Part (d) for

$$\mathbf{x}_0 = \begin{bmatrix} -\pi \\ \pi \end{bmatrix}, \quad \mathbf{d}_0 = \begin{bmatrix} 1.0 \\ -1.1 \end{bmatrix}$$

The interval of α for plotting $f(\mathbf{x}_0 + \alpha \mathbf{d}_0)$ in this case is [0, 5.7120].

References

1. D. M. Himmelblau, *Applied Nonlinear Programming*. New York: McGraw-Hill, 1972.
2. B. S. Gottfried and J. Weisman, *Introduction to Optimization Theory*. Englewood Cliffs, NJ: Prentice-Hall, 1973.
3. P. R. Adby and M. A. H. Dempster, *Introduction to Optimization Methods*. London, UK: Chapman and Hall, 1974.
4. C. S. Beightler, D. T. Phillips, and D. J. Wilde, *Foundations of Optimization*. Englewood Cliffs, NJ: Prentice-Hall, 1979.
5. M. S. Bazaraa and C. M. Shetty, *Nonlinear Programming*. New York: Wiley, 1979.
6. P. E. Gill, W. Murray, and M. H. Wright, *Practical Optimization*. New York: Academic Press, 1981.
7. G. P. McCormick, *Nonlinear Programming*. New York: Wiley, 1983.
8. M. J. Box, D. Davies, and W. H. Swann, *Nonlinear Optimization Techniques*. London, UK: Oliver and Boyd, 1969.
9. R. Fletcher, *Practical Methods of Optimization*, 2nd ed. New York: Wiley, 1987.
10. R. Fletcher, *Practical Methods of Optimization*, vol. 1. New York: Wiley, 1980.

Basic Multidimensional Gradient Methods

5

5.1 Introduction

In Chap. 4, several methods were considered that can be used for the solution of one-dimensional unconstrained problems. In this chapter, we consider the solution of multidimensional unconstrained problems.

As for one-dimensional optimization, there are two general classes of multidimensional methods, namely, search methods and gradient methods. In search methods, the solution is obtained by using only function evaluations. The general approach is to explore the parameter space in an organized manner in order to find a trajectory that leads progressively to reduced values of the objective function. A rudimentary method of this class might be to adjust all the parameters at a specific starting point, one at a time, and then select a new point by comparing the calculated values of the objective function. The same procedure can then be repeated at the new point, and so on. Multidimensional search methods are thus analogous to their one-dimensional counterparts and, like the latter, they are not very efficient. As a result, their application is restricted to problems where gradient information is unavailable or difficult to obtain, for example, in applications where the objective function is not continuous.

Gradient methods are based on gradient information. They can be grouped into two classes, first-order and second-order methods. First-order methods are based on the linear approximation of the Taylor series, and hence they entail the gradient $\mathbf{g}$. Second-order methods, on the other hand, are based on the quadratic approximation of the Taylor series. They entail the gradient $\mathbf{g}$ as well as the Hessian $\mathbf{H}$.

Gradient methods range from some simple to some highly sophisticated methods. In this chapter, we focus our attention on the most basic ones, which are as follows:

1. Steepest-descent method
2. Newton method
3. Gauss–Newton method

Some more advanced gradient methods will be considered later in Chaps. 6 and 7.

© Springer Science+Business Media, LLC, part of Springer Nature 2021
A. Antoniou and W.-S. Lu, *Practical Optimization*, Texts in Computer Science,
https://doi.org/10.1007/978-1-0716-0843-2_5

5.2 Steepest-Descent Method

Consider the optimization problem

$$\text{minimize } F = f(\mathbf{x}) \quad \text{for } \mathbf{x} \in E^n$$

From the Taylor series

$$F + \Delta F = f(\mathbf{x} + \boldsymbol{\delta}) \approx f(\mathbf{x}) + \mathbf{g}^T \boldsymbol{\delta} + \tfrac{1}{2} \boldsymbol{\delta}^T \mathbf{H} \boldsymbol{\delta}$$

and as $\|\boldsymbol{\delta}\|_2 \to 0$, the change in F due to change $\boldsymbol{\delta}$ is obtained as

$$\Delta F \approx \mathbf{g}^T \boldsymbol{\delta}$$

The product at the right-hand side is the *scalar* or *dot product* of vectors $\mathbf{g}$ and $\boldsymbol{\delta}$. If

$$\mathbf{g} = [g_1 \ g_2 \ \cdots \ g_n]^T$$

and

$$\boldsymbol{\delta} = [\delta_1 \ \delta_2 \ \cdots \ \delta_n]^T$$

then

$$\Delta F \approx \sum_{i=1}^{n} g_i \delta_i = \|\mathbf{g}\|_2 \, \|\boldsymbol{\delta}\|_2 \cos \theta$$

where θ is the angle between vectors $\mathbf{g}$ and $\boldsymbol{\delta}$, and

$$\|\mathbf{g}\|_2 = (\mathbf{g}^T \mathbf{g})^{1/2} = \left(\sum_{i=1}^{n} g_i^2 \right)^{1/2}$$

5.2.1 Ascent and Descent Directions

Consider the contour plot of Fig. 5.1. If $\mathbf{x}$ and $\mathbf{x} + \boldsymbol{\delta}$ are adjacent points on contour A, then as $\|\boldsymbol{\delta}\|_2 \to 0$

$$\Delta F \approx \|\mathbf{g}\|_2 \, \|\boldsymbol{\delta}\|_2 \, \cos \theta = 0$$

since F is constant on a contour. We thus conclude that the angle θ between vectors $\mathbf{g}$ and $\boldsymbol{\delta}$ is equal to $90°$. In effect, the gradient at point $\mathbf{x}$ is orthogonal to contour A, as depicted in Fig. 5.1. Now for any vector $\boldsymbol{\delta}$, ΔF assumes a maximum positive value if $\theta = 0$, that is, $\boldsymbol{\delta}$ must be in the direction $\mathbf{g}$. On the other hand, ΔF assumes a maximum negative value if $\theta = \pi$, that is, $\boldsymbol{\delta}$ must be in the direction $-\mathbf{g}$. The gradient $\mathbf{g}$ and its negative $-\mathbf{g}$ are thus said to be the *steepest-ascent* and *steepest-descent directions*, respectively. These concepts are illustrated in Figs. 5.1 and 5.2.

Fig. 5.1 Steepest-descent and steepest-ascent directions

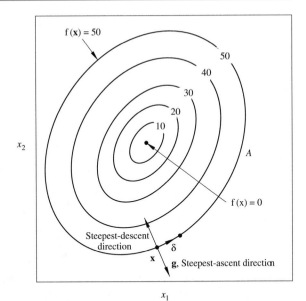

Fig. 5.2 Construction for steepest-descent method

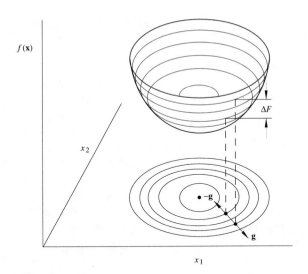

5.2.2 Basic Method

Assume that a function $f(\mathbf{x})$ is continuous in the neighborhood of point $\mathbf{x}$. If $\mathbf{d}$ is the steepest-descent direction at point $\mathbf{x}$, i.e.,

$$\mathbf{d} = -\mathbf{g}$$

then a change δ in $\mathbf{x}$ given by

$$\delta = \alpha \mathbf{d}$$

Fig. 5.3 Line search in
steepest-descent direction

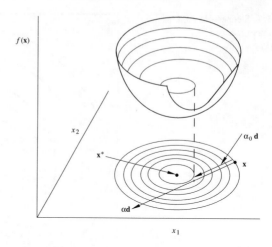

where α is a small positive constant, will decrease the value of $f(\mathbf{x})$. Maximum reduction in $f(\mathbf{x})$ can be achieved by solving the one-dimensional optimization problem

$$\underset{\alpha}{\text{minimize}} \quad F = f(\mathbf{x} + \alpha\mathbf{d}) \tag{5.1}$$

as depicted in Fig. 5.3.

If the steepest-descent direction at point $\mathbf{x}$ happens to point towards the minimizer $\mathbf{x}^*$ of $f(\mathbf{x})$, then a value of α exists that minimizes $f(\mathbf{x} + \alpha\mathbf{d})$ with respect to α and $f(\mathbf{x})$ with respect to $\mathbf{x}$. Consequently, in such a case the multidimensional problem can be solved by solving the one-dimensional problem in Eq. (5.1) once. Unfortunately, in most real-life problems, $\mathbf{d}$ does not point in the direction of $\mathbf{x}^*$ and, therefore, an iterative procedure must be used for the solution. Starting with an initial point $\mathbf{x}_0$, a direction $\mathbf{d} = \mathbf{d}_0 = -\mathbf{g}$ can be calculated, and the value of α that minimizes $f(\mathbf{x}_0 + \alpha\mathbf{d}_0)$, say, α_0, can be determined. Thus a point $\mathbf{x}_1 = \mathbf{x}_0 + \alpha_0\mathbf{d}_0$ is obtained. The minimization can be performed by using one of the methods of Chap. 4 as a line search. The same procedure can then be repeated at points

$$\mathbf{x}_{k+1} = \mathbf{x}_k + \alpha_k\mathbf{d}_k \tag{5.2}$$

for $k = 1, 2, \ldots$ until convergence is achieved. The procedure can be terminated when $\|\alpha_k\mathbf{d}_k\|_2$ becomes insignificant or if $\alpha_k \leq K\alpha_0$ where K is a sufficiently small positive constant. A typical solution trajectory for the steepest-descent method is illustrated in Fig. 5.4. A corresponding algorithm is as follows.

Algorithm 5.1 Steepest-descent algorithm
Step 1
Input $\mathbf{x}_0$ and initialize the tolerance ε.
Set $k = 0$.
Step 2
Calculate gradient $\mathbf{g}_k$ and set $\mathbf{d}_k = -\mathbf{g}_k$.
Step 3
Find α_k, the value of α that minimizes $f(\mathbf{x}_k + \alpha\mathbf{d}_k)$, using a line search.
Step 4
Set $\mathbf{x}_{k+1} = \mathbf{x}_k + \alpha_k\mathbf{d}_k$ and calculate $f_{k+1} = f(\mathbf{x}_{k+1})$.
Step 5
If $\|\alpha_k\mathbf{d}_k\|_2 < \varepsilon$, then do:
    Output $\mathbf{x}^* = \mathbf{x}_{k+1}$ and $f(\mathbf{x}^*) = f_{k+1}$, and stop.
Otherwise, set $k = k + 1$ and repeat from Step 2.

5.2.3 Orthogonality of Directions

In the steepest-descent method, the trajectory to the solution follows a zig-zag pattern, as can be seen in Fig. 5.4. If α is chosen such that $f(\mathbf{x}_k + \alpha\mathbf{d}_k)$ is minimized in each iteration, then successive directions are orthogonal. To demonstrate this fact, we note that

$$\frac{df(\mathbf{x}_k + \alpha\mathbf{d}_k)}{d\alpha} = \sum_{i=1}^{n} \frac{\partial f(\mathbf{x}_k + \alpha\mathbf{d}_k)}{\partial x_{ki}} \frac{d(x_{ki} + \alpha d_{ki})}{d\alpha}$$

$$= \sum_{i=1}^{n} g_i(\mathbf{x}_k + \alpha\mathbf{d}_k)d_{ki}$$

$$= \mathbf{g}(\mathbf{x}_k + \alpha\mathbf{d}_k)^T \mathbf{d}_k$$

where $\mathbf{g}(\mathbf{x}_k + \alpha\mathbf{d}_k)$ is the gradient at point $\mathbf{x}_k + \alpha\mathbf{d}_k$. If α^* is the value of α that minimizes $f(\mathbf{x}_k + \alpha\mathbf{d}_k)$, then

$$\mathbf{g}(\mathbf{x}_k + \alpha^*\mathbf{d}_k)^T \mathbf{d}_k = 0$$

or

$$\mathbf{d}_{k+1}^T\mathbf{d}_k = 0$$

where

$$\mathbf{d}_{k+1} = -\mathbf{g}(\mathbf{x}_k + \alpha^*\mathbf{d}_k)$$

is the steepest-descent direction at point $\mathbf{x}_k + \alpha^*\mathbf{d}_k$. In effect, successive directions $\mathbf{d}_k$ and $\mathbf{d}_{k+1}$ are orthogonal as depicted in Fig. 5.4.

Fig. 5.4 Typical solution trajectory in steepest-descent algorithm

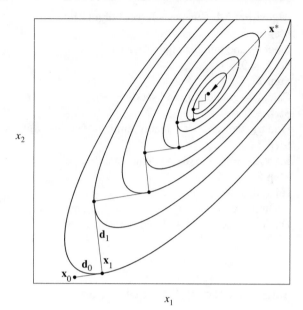

5.2.4 Step-Size Estimation for Steepest-Descent Method

If the Hessian of $f(\mathbf{x})$ is available, the value of α, α_k, that minimizes $f(\mathbf{x}_k + \alpha \mathbf{d})$ can be determined by using an analytical method. If $\boldsymbol{\delta}_k = \alpha \mathbf{d}_k$, the Taylor series yields

$$f(\mathbf{x}_k + \boldsymbol{\delta}_k) \approx f(\mathbf{x}_k) + \boldsymbol{\delta}_k^T \mathbf{g}_k + \tfrac{1}{2}\boldsymbol{\delta}_k^T \mathbf{H}_k \boldsymbol{\delta}_k \tag{5.3}$$

and if $\mathbf{d}_k$ is the steepest-descent direction, i.e.,

$$\boldsymbol{\delta}_k = -\alpha \mathbf{g}_k$$

we obtain

$$f(\mathbf{x}_k - \alpha \mathbf{g}_k) \approx f(\mathbf{x}_k) - \alpha \mathbf{g}_k^T \mathbf{g}_k + \tfrac{1}{2}\alpha^2 \mathbf{g}_k^T \mathbf{H}_k \mathbf{g}_k \tag{5.4}$$

By differentiating and setting the result to zero, we get

$$\frac{df(\mathbf{x}_k - \alpha \mathbf{g}_k)}{d\alpha} \approx -\mathbf{g}_k^T \mathbf{g}_k + \alpha \mathbf{g}_k^T \mathbf{H}_k \mathbf{g}_k = 0$$

or

$$\alpha = \alpha_k \approx \frac{\mathbf{g}_k^T \mathbf{g}_k}{\mathbf{g}_k^T \mathbf{H}_k \mathbf{g}_k} \tag{5.5}$$

Note that the value of α_k is positive if $\mathbf{g}_k^T \mathbf{H}_k \mathbf{g}_k > 0$ which is guaranteed to hold if $\mathbf{H}_k$ is positive definite.

Now if we assume that $\alpha = \alpha_k$ minimizes $f(\mathbf{x}_k + \alpha \mathbf{d}_k)$, Eq. (5.2) can be expressed as

$$\mathbf{x}_{k+1} = \mathbf{x}_k - \frac{\mathbf{g}_k^T \mathbf{g}_k}{\mathbf{g}_k^T \mathbf{H}_k \mathbf{g}_k} \mathbf{g}_k$$

The accuracy of α_k will depend heavily on the magnitude of $\boldsymbol{\delta}_k$ since the quadratic approximation of the Taylor series is valid only in the neighborhood of point $\mathbf{x}_k$. At the start of the optimization, $\|\boldsymbol{\delta}_k\|_2$ will be relatively large and so α_k will be inaccurate. Nevertheless, reduction will be achieved in $f(\mathbf{x})$ since $f(\mathbf{x}_k + \alpha\mathbf{d}_k)$ is minimized in the steepest-descent direction. As the solution is approached, $\|\boldsymbol{\delta}_k\|_2$ is decreased and, consequently, the accuracy of α_k will progressively be improved, and the maximum reduction in $f(\mathbf{x})$ will eventually be achieved in each iteration. Convergence will thus be achieved. For quadratic functions, Eq. (5.3) is satisfied with the equal sign and hence $\alpha = \alpha_k$ yields the maximum reduction in $f(\mathbf{x})$ in every iteration.

If the Hessian is not available, the value of α_k can be determined by calculating $f(\mathbf{x})$ at points $\mathbf{x}_k$ and $\mathbf{x}_k - \hat{\alpha}\mathbf{g}_k$ where $\hat{\alpha}$ is an estimate of α_k. If

$$f_k = f(\mathbf{x}_k) \quad \text{and} \quad \hat{f} = f(\mathbf{x}_k - \hat{\alpha}\mathbf{g}_k)$$

Equation (5.4) gives

$$\hat{f} \approx f_k - \hat{\alpha}\mathbf{g}_k^T\mathbf{g}_k + \tfrac{1}{2}\hat{\alpha}^2\mathbf{g}_k^T\mathbf{H}_k\mathbf{g}_k$$

or

$$\mathbf{g}_k^T\mathbf{H}_k\mathbf{g}_k \approx \frac{2(\hat{f} - f_k + \hat{\alpha}\mathbf{g}_k^T\mathbf{g}_k)}{\hat{\alpha}^2} \tag{5.6}$$

Now from Eqs. (5.5) and (5.6), we get

$$\alpha_k \approx \frac{\mathbf{g}_k^T\mathbf{g}_k\hat{\alpha}^2}{2(\hat{f} - f_k + \hat{\alpha}\mathbf{g}_k^T\mathbf{g}_k)} \tag{5.7}$$

Note that Eq. (5.7) yields a positive α_k only if $\hat{f} - f + \hat{\alpha}\mathbf{g}_k^T\mathbf{g}_k$ is positive. From Eq. (5.6) this is guaranteed to be true if $\mathbf{H}_K$ is positive definite. A suitable value for $\hat{\alpha}$ is the optimum value of α in the previous iteration, namely, α_{k-1}. For the first iteration, the value $\hat{\alpha} = 1$ can be used.

An algorithm that uses an estimated step size instead of a line search is as follows:

Algorithm 5.2 Steepest-descent algorithm without line search
Step 1
Input $\mathbf{x}_1$ and initialize the tolerance ε.
Set $k = 1$ and $\alpha_0 = 1$.
Compute $f_1 = f(\mathbf{x}_1)$.
Step 2
Compute $\mathbf{g}_k$.
Step 3
Set $\mathbf{d}_k = -\mathbf{g}_k$ and $\hat{\alpha} = \alpha_{k-1}$.
Compute $\hat{f} = f(\mathbf{x}_k - \hat{\alpha}\mathbf{g}_k)$.
Compute α_k from Eq. (5.7).
Step 4
Set $\mathbf{x}_{k+1} = \mathbf{x}_k + \alpha_k\mathbf{d}_k$ and calculate $f_{k+1} = f(\mathbf{x}_{k+1})$.
Step 5
If $\|\alpha_k\mathbf{d}_k\|_2 < \varepsilon$, then do:
    Output $\mathbf{x}^* = \mathbf{x}_{k+1}$ and $f(\mathbf{x}^*) = f_{k+1}$, and stop.
Otherwise, set $k = k + 1$ and repeat from Step 2.

The value of α_k in Step 3 is an accurate estimate of the value of α that minimizes $f(\mathbf{x}_k + \alpha \mathbf{d}_k)$ to the extent that the quadratic approximation of the Taylor series is an accurate representation of $f(\mathbf{x})$. Thus, as was argued earlier, the reduction in $f(\mathbf{x})$ per iteration tends to approach the maximum possible as $\mathbf{x}^*$ is approached, and if $f(\mathbf{x})$ is quadratic the maximum possible reduction is achieved in every iteration.

5.2.5 Step-Size Estimation Using the Barzilai–Borwein Two-Point Formulas

An alternative method for the estimation of the step size that is particularly suitable for the steepest-descent algorithm is the so-called *two-point method* proposed by Barzilai and Borwein [1]. The analysis involved is detailed below.

If the Hessian in Eq. (5.3) is positive definite, then the convex quadratic function at the right-hand side of Eq. (5.3) has a unique minimizer which can be obtained as

$$\boldsymbol{\delta}_k = -\mathbf{H}_k^{-1}\mathbf{g}_k \tag{5.8}$$

by setting the gradient of the quadratic function to zero. On comparing the $\boldsymbol{\delta}_k$ in Eq. (5.8) with that obtained in the steepest-descent method, i.e.,

$$\boldsymbol{\delta}_k = -\alpha_k \mathbf{g}_k \tag{5.9}$$

we note that α_k in Eq. (5.9) corresponds to $\mathbf{H}_k^{-1}$ in Eq. (5.8). Thus we can approximate the inverse Hessian matrix $\mathbf{H}_k^{-1}$ by $\alpha_k \mathbf{I}$ or, equivalently, we can approximate $\mathbf{H}_k$ by $\beta_k \mathbf{I}$ with $\beta_k = 1/\alpha_k$. Hence we can write

$$\mathbf{g}_k \approx \mathbf{g}_{k-1} + \mathbf{H}_k \boldsymbol{\delta}_{k-1} \tag{5.10}$$

which implies that

$$\mathbf{H}_k^{-1}\boldsymbol{\gamma}_{k-1} \approx \boldsymbol{\delta}_{k-1} \tag{5.11a}$$

and

$$\mathbf{H}_k \boldsymbol{\delta}_{k-1} \approx \boldsymbol{\gamma}_{k-1} \tag{5.11b}$$

where

$$\boldsymbol{\delta}_{k-1} = \mathbf{x}_k - \mathbf{x}_{k-1}, \quad \boldsymbol{\gamma}_{k-1} = \mathbf{g}_k - \mathbf{g}_{k-1}$$

From Eq. (5.11a), it follows that $\alpha_k \mathbf{I}$ approximates $\mathbf{H}_k^{-1}$ if $\alpha = \alpha_k$ is a solution of the problem

$$\underset{\alpha}{\text{minimize}} \quad \|\alpha \boldsymbol{\gamma}_{k-1} - \boldsymbol{\delta}_{k-1}\|_2 \tag{5.12}$$

which is a single-variable least-squares problem that involves the objective function

$$\|\boldsymbol{\gamma}_{k-1}\|_2^2 \alpha^2 - 2\boldsymbol{\delta}_{k-1}^T \boldsymbol{\gamma}_{k-1}\alpha + \|\boldsymbol{\delta}_{k-1}\|_2^2$$

By setting the derivative of the above function with respect to α to zero, the minimizer of Eq. (5.12) is obtained as

$$\alpha_k = \frac{\boldsymbol{\delta}_{k-1}^T \boldsymbol{\gamma}_{k-1}}{\boldsymbol{\gamma}_{k-1}^T \boldsymbol{\gamma}_{k-1}} \tag{5.13}$$

Alternatively, from Eq. (5.11b) it follows that $\beta_k \mathbf{I}$ approximates $\mathbf{H}_k$ if $\beta = \beta_k$ is a solution of the single-variable least-squares optimization problem

$$\underset{\beta}{\text{minimize}} \quad \|\beta \boldsymbol{\delta}_{k-1} - \boldsymbol{\gamma}_{k-1}\|_2 \tag{5.14}$$

The minimizer of the problem in Eq. (5.14) is given by

$$\beta_k = \frac{\boldsymbol{\delta}_{k-1}^T \boldsymbol{\gamma}_{k-1}}{\boldsymbol{\delta}_{k-1}^T \boldsymbol{\delta}_{k-1}}$$

and, in effect, an alternative formula for α_k is obtained as

$$\alpha_k = \frac{\boldsymbol{\delta}_{k-1}^T \boldsymbol{\delta}_{k-1}}{\boldsymbol{\delta}_{k-1}^T \boldsymbol{\gamma}_{k-1}} \tag{5.15}$$

In summary, we have derived two formulas for α_k, given by Eqs. (5.13) and (5.15), which can be used to implement Step 3 of Algorithm 5.2. Since these formulas involve $\boldsymbol{\delta}_{k-1}$ and $\boldsymbol{\gamma}_{k-1}$ which are the differences between the two points $\mathbf{x}_k$ and $\mathbf{x}_{k-1}$ and the gradients at these two points, respectively, they are usually referred to as the *two-point Barzilai–Borwein step-size estimation formulas* [1]. It should be mentioned that the solution produced by a steepest-descent algorithm using the Barzilai–Borwein formulas is not guaranteed to be monotonically decreasing. However, convergence analyses reported in [1–3] as well as extensive use of these formulas for the minimization of general nonlinear objective functions have shown that steepest-descent algorithms using these formulas are often significantly more efficient relative to steepest-descent algorithms using conventional line searches or the step-size estimate in Eq. (5.7).

Example 5.1 The unconstrained optimization problem

$$\text{minimize } f(\mathbf{x}) = \tfrac{1}{2}\mathbf{x}^T \mathbf{Q}\mathbf{x} + \mathbf{b}^T \mathbf{x}$$

where

$$\mathbf{Q} = \begin{bmatrix} 30 & 0 & 0 & 0 \\ 0 & 10 & 0 & 0 \\ 0 & 0 & 3 & 0 \\ 0 & 0 & 0 & 1 \end{bmatrix}, \quad \mathbf{b} = -\begin{bmatrix} 1 \\ 1 \\ 1 \\ 1 \end{bmatrix}$$

is to be solved by using Algorithm 5.2.

(a) Solve the problem by using the formula in Eq. (5.5) in Step 3 of the algorithm. Start with the initial point $\mathbf{x}_0 = [0\ 0\ 0\ 0]^T$ and assume a termination tolerance $\varepsilon = 10^{-9}$.
(b) Repeat part (a) using the Barzilai–Borwein formula in Eq. (5.13).
(c) Repeat part (a) using the Barzilai–Borwein formula in Eq. (5.15).
(d) Compare the computational efficiencies of the different versions of Algorithm 5.2 by estimating the average CPU time required by each version of the algorithm.

Solution (a) Since $\mathbf{Q}$ is positive definite, the objective function $f(\mathbf{x})$ is a strictly convex quadratic function; hence the formula in Eq. (5.5) yields the exact minimizer for $f(\mathbf{x}_k - \alpha\mathbf{g}_k)$ as

$$\alpha_k = \frac{\mathbf{g}_k^T \mathbf{g}_k}{\mathbf{g}_k^T \mathbf{Q}\mathbf{g}_k}$$

where

$$\mathbf{g}_k = \mathbf{Q}\mathbf{x}_k + \mathbf{b}$$

It took Algorithm 5.2 243 iterations to converge to the solution

$$\mathbf{x}^* = \begin{bmatrix} 0.033333 \\ 0.100000 \\ 0.333333 \\ 0.999999 \end{bmatrix}$$

(b) In this case, it took Algorithm 5.2 24 iterations to converge to the solution obtained in part (a).

(c) In this case, it took Algorithm 5.2 21 iterations to converge to the solution obtained in part (a).

(d) The average CPU time required by each algorithm can be estimated by using the following pseudocode:

1. Set Initial_CPU_time = cputime.
2. Run Algorithm 5.2 for $i = 1 : 1000$.
3. Set Final_CPU_time = cputime.
4. Compute the average CPU time as:
   Average_CPU_time = (Final_CPU_time − Initial_CPU_time)/1000

Using the above pseudocode, the average CPU times required by the three versions of Algorithm 5.2, normalized with respect to the largest CPU time, were found to be 1.0, 0.1121, and 0.0972, respectively. In effect, the step-size estimations produced by the Barzilai–Borwein formulas have reduced the amount of computational effort by a factor of the order of 10 relative to that required by the use of the formula in Eq. (5.5) with the formula in Eq. (5.15) offering an additional 10 percent reduction relative to the formula in Eq. (5.13) in the present example. ∎

5.2.6 Convergence

If a function $f(\mathbf{x}) \in C^2$ has a local minimizer $\mathbf{x}^*$ and its Hessian is positive definite at $\mathbf{x} = \mathbf{x}^*$, then it can be shown that if $\mathbf{x}_k$ is sufficiently close to $\mathbf{x}^*$, we have

$$f(\mathbf{x}_{k+1}) - f(\mathbf{x}^*) \le \left(\frac{1-r}{1+r}\right)^2 [f(\mathbf{x}_k) - f(\mathbf{x}^*)] \tag{5.16}$$

where

$$r = \frac{\text{smallest eigenvalue of } \mathbf{H}_k}{\text{largest eigenvalue of } \mathbf{H}_k}$$

Furthermore, if $f(\mathbf{x})$ is a quadratic function then the inequality in Eq. (5.16) holds for all k (see [4, Chap. 8] for proof). In effect, subject to the conditions stated, the steepest-descent method converges linearly (see Sect. 3.7) with a convergence ratio

$$\beta = \left(\frac{1-r}{1+r}\right)^2$$

Evidently, the rate of convergence is high if the eigenvalues of $\mathbf{H}_k$ are all nearly equal, or low if at least one eigenvalue is small relative to the largest eigenvalue.

The eigenvalues of $\mathbf{H}$, namely, λ_i for $i = 1, 2, \ldots, n$, determine the geometry of the surface

$$\mathbf{x}^T \mathbf{H} \mathbf{x} = \text{constant}$$

This equation gives the contours of $\mathbf{x}^T \mathbf{H} \mathbf{x}$ and if $\mathbf{H}$ is positive definite, the contours are ellipsoids with axes proportional to $1/\sqrt{\lambda_i}$. If the number of variables is two, the contours are ellipses with axes proportional to $1/\sqrt{\lambda_1}$ and $1/\sqrt{\lambda_2}$. Consequently, if the steepest-descent method is applied to a two-dimensional problem, convergence will be fast if the contours are nearly circular, as is to be expected, and if they are circular, i.e., $r = 1$, convergence will be achieved in one iteration. On the other hand, if the contours are elongated ellipses or if the function exhibits long narrow valleys, progress will be very slow, in particular as the solution is approached. The influence of r on convergence can be appreciated by comparing Figs. 5.4 and 5.5.

The steepest-descent method attempts, in effect, to reduce the gradient to zero. Since at a saddle point the gradient is zero, it might be questioned whether such a point is a likely solution. It turns out that such a solution is highly unlikely, in practice, for two reasons. First, the probability of locating a saddle point exactly as

Fig. 5.5 Solution trajectory in steepest-descent algorithm if $r \approx 1$

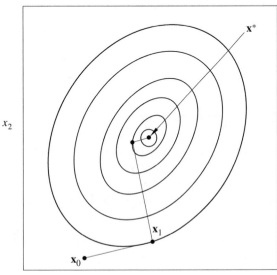

the next iteration point is infinitesimal. Second, there is always a descent direction in the neighborhood of a saddle point.

5.2.7 Scaling

The eigenvalues of $\mathbf{H}$ in a specific optimization problem and, in turn, the performance of the steepest-descent method tend to depend to a large extent on the choice of variables. For example, in one and the same two-dimensional problem, the contours may be nearly circular or elliptical depending on the choice of units. Consequently, the rate of convergence can often be improved by scaling the variables through variable transformations.

A possible approach to scaling might be to let

$$\mathbf{x} = \mathbf{T}\mathbf{y}$$

where $\mathbf{T}$ is an $n \times n$ diagonal matrix, and then solve the problem

$$\underset{\mathbf{y}}{\text{minimize}} \quad h(\mathbf{y}) = f(\mathbf{x})\big|_{\mathbf{x}=\mathbf{T}\mathbf{y}}$$

The gradient and Hessian of the new problem are

$$\mathbf{g}_h = \mathbf{T}^T \mathbf{g}_x \quad \text{and} \quad \mathbf{H}_h = \mathbf{T}^T \mathbf{H}\mathbf{T}$$

respectively, and, therefore, both the steepest-descent direction as well as the eigenvalues associated with the problem are changed. Unfortunately, the choice of $\mathbf{T}$ tends to depend heavily on the problem at hand and, as a result, no general rules can be stated. As a rule of thumb, we should strive to as far as possible equalize the second derivatives

$$\frac{\partial^2 f}{\partial x_i^2} \quad \text{for } i = 1, \ 2, \ \dots, \ n.$$

5.3 Newton Method

The steepest-descent method is a first-order method since it is based on the linear approximation of the Taylor series. A second-order method known as the Newton (also known as the Newton–Raphson) method can be developed by using the quadratic approximation of the Taylor series. If $\boldsymbol{\delta}$ is a change in $\mathbf{x}$, $f(\mathbf{x} + \boldsymbol{\delta})$ is given by

$$f(\mathbf{x} + \boldsymbol{\delta}) \approx f(\mathbf{x}) + \sum_{j=1}^{n} \frac{\partial f}{\partial x_i} \delta_i + \frac{1}{2} \sum_{i=1}^{n} \sum_{j=1}^{n} \frac{\partial^2 f}{\partial x_i \partial x_j} \delta_i \delta_j \qquad (5.17)$$

Assuming that this is an accurate representation of the function at point $\mathbf{x} + \boldsymbol{\delta}$, then differentiating $f(\mathbf{x}+\boldsymbol{\delta})$ with respect to δ_k for $k = 1, 2, \ldots, n$ and setting the result to zero will give the values of δ_k that minimize $f(\mathbf{x} + \boldsymbol{\delta})$. This approach yields

$$\frac{\partial f}{\partial x_k} + \sum_{i=1}^{n} \frac{\partial^2 f}{\partial x_i \partial x_k} \delta_i = 0 \quad \text{for } k = 1, 2, \ldots, n$$

or in matrix form

$$\mathbf{g} = -\mathbf{H}\boldsymbol{\delta}$$

Therefore, the optimum change in $\mathbf{x}$ is

$$\boldsymbol{\delta} = -\mathbf{H}^{-1}\mathbf{g} \tag{5.18}$$

This solution exists if and only if the following conditions hold:

(a) The Hessian is nonsingular.
(b) The approximation in Eq. (5.17) is valid.

Assuming that the second-order sufficiency conditions for a minimum hold at point $\mathbf{x}^*$, then $\mathbf{H}$ is positive definite at $\mathbf{x}^*$ and also in the neighborhood of the solution i.e., for $\|\mathbf{x}-\mathbf{x}^*\|_2 < \varepsilon$. This means that $\mathbf{H}$ is nonsingular and has an inverse for $\|\mathbf{x}-\mathbf{x}^*\|_2 < \varepsilon$. Since any function $f(x) \in C^2$ can be accurately represented in the neighborhood of $\mathbf{x}^*$ by the quadratic approximation of the Taylor series, the solution in Eq. (5.18) exists. Furthermore, for any point $\mathbf{x}$ such that $\|\mathbf{x} - \mathbf{x}^*\|_2 < \varepsilon$ one iteration will yield $\mathbf{x} \approx \mathbf{x}^*$.

Any quadratic function has a Hessian which is constant for any $\mathbf{x} \in E^n$, as can be readily demonstrated. If the function has a minimum, and the second-order sufficiency conditions for a minimum hold, then $\mathbf{H}$ is positive definite and, therefore, nonsingular at any point $\mathbf{x} \in E^n$. Since any quadratic function is represented exactly by the quadratic approximation of the Taylor series, the solution in Eq. (5.18) exists. Furthermore, for any point $\mathbf{x} \in E^n$ one iteration will yield the solution.

If a general nonquadratic function is to be minimized and an arbitrary point $\mathbf{x}$ is assumed, condition (a) and/or condition (b) may be violated. If condition (a) is violated, Eq. (5.18) may have an infinite number of solutions or no solution at all. If, on the other hand, condition (b) is violated, then $\boldsymbol{\delta}$ may not yield the solution in one iteration and, if $\mathbf{H}$ is not positive definite, $\boldsymbol{\delta}$ may not even yield a reduction in the objective function.

The first problem can be overcome by forcing $\mathbf{H}$ to become positive definite by means of some manipulation. The second problem, on the other hand, can be overcome by using an iterative procedure which incorporates a line search for the calculation of the change in $\mathbf{x}$. The iterative procedure will counteract the fact that one iteration will not yield the solution, and the line search can be used to achieve the maximum reduction in $f(\mathbf{x})$ along the direction predicted by Eq. (5.18). This approach can be implemented by selecting the next point $\mathbf{x}_{k+1}$ as

$$\mathbf{x}_{k+1} = \mathbf{x}_k + \boldsymbol{\delta}_k = \mathbf{x}_k + \alpha_k \mathbf{d}_k \tag{5.19}$$

where

$$\mathbf{d}_k = -\mathbf{H}_k^{-1}\mathbf{g}_k \tag{5.20}$$

and α_k is the value of α that minimizes $f(\mathbf{x}_k + \alpha\mathbf{d}_k)$. The vector $\mathbf{d}_k$ is referred to as the *Newton direction* at point $\mathbf{x}_k$. In the case where conditions (*a*) and (*b*) are satisfied, the first iteration will yield the solution with $\alpha_k = 1$.

At the start of the minimization, progress might be slow for certain types of functions. Nevertheless, continuous reduction in $f(\mathbf{x})$ will be achieved through the choice of α. As the solution is approached, however, both conditions (*a*) and (*b*) will be satisfied and, therefore, convergence will be achieved. The order of convergence can be shown to be two (see [4, Chap. 8]). In effect, the Newton method has convergence properties that are complementary to those of the steepest-descent method, namely, it can be slow away from the solution and fast close to the solution.

The above principles lead readily to the basic Newton algorithm summarized below.

Algorithm 5.3 Basic Newton algorithm
Step 1
Input $\mathbf{x}_0$ and initialize the tolerance ε.
Set $k = 0$.
Step 2
Compute $\mathbf{g}_k$ and $\mathbf{H}_k$.
If $\mathbf{H}_k$ is not positive definite, force it to become positive definite.
Step 3
Compute $\mathbf{H}_k^{-1}$ and $\mathbf{d}_k = -\mathbf{H}_k^{-1}\mathbf{g}_k$.
Step 4
Find α_k, the value of α that minimizes $f(\mathbf{x}_k + \alpha\mathbf{d}_k)$, using a line search.
Step 5
Set $\mathbf{x}_{k+1} = \mathbf{x}_k + \alpha_k\mathbf{d}_k$.
Compute $f_{k+1} = f(\mathbf{x}_{k+1})$.
Step 6
If $\|\alpha_k\mathbf{d}_k\|_2 < \varepsilon$, then output $\mathbf{x}^* = \mathbf{x}_{k+1}$ and $f(\mathbf{x}^*) = f_{k+1}$, and stop; otherwise, set $k = k + 1$ and repeat from Step 2.

5.3.1 Modification of the Hessian

If the Hessian is not positive definite in any iteration of Algorithm 5.3, it is forced to become positive definite in Step 2 of the algorithm. This modification of $\mathbf{H}_k$ can be accomplished in one of several ways.

One approach proposed by Goldfeld, Quandt, and Trotter [5] is to replace matrix $\mathbf{H}_k$ by the $n \times n$ identity matrix $\mathbf{I}_n$ whenever it becomes nonpositive definite. Since $\mathbf{I}_n$ is positive definite, the problem of a nonsingular $\mathbf{H}_k$ is eliminated. Another approach would be to let

$$\hat{\mathbf{H}}_k = \frac{1}{1 + \beta}(\mathbf{H}_k + \beta\mathbf{I}_n) \tag{5.21}$$

where β is a positive scalar that is slightly larger than the absolute value of the most negative eigenvalue of $\mathbf{H}_k$ so as to assure the positive definiteness of $\mathbf{H}_k$ in Eq. (5.21).

If β is large, then

$$\hat{\mathbf{H}}_k \approx \mathbf{I}_n$$

and from Eq. (5.20)

$$\mathbf{d}_k \approx -\mathbf{g}_k$$

In effect, the modification in Eq. (5.21) converts the Newton method into the steepest-descent method.

A nonpositive definite $\mathbf{H}_k$ is likely to arise at points far from the solution where the steepest-descent method is most effective in reducing the value of $f(\mathbf{x})$. Therefore, the modification in Eq. (5.21) leads to an algorithm that combines the complementary convergence properties of the Newton and steepest-descent methods.

A second possibility due to Zwart [6] is to form a modified matrix

$$\hat{\mathbf{H}}_k = \mathbf{U}^T \mathbf{H}_k \mathbf{U} + \boldsymbol{\varepsilon}$$

where $\mathbf{U}$ is a real unitary matrix (i.e., $\mathbf{U}^T \mathbf{U} = \mathbf{I}_n$) and $\boldsymbol{\varepsilon}$ is a diagonal $n \times n$ matrix with diagonal elements ε_i. It can be shown that a matrix $\mathbf{U}$ exists such that $\mathbf{U}^T \mathbf{H}_k \mathbf{U}$ is diagonal with diagonal elements λ_i for $i = 1, 2, \ldots, n$, where λ_i are the eigenvalues of $\mathbf{H}_k$ (see Theorem 2.8). In effect, $\hat{\mathbf{H}}_k$ is diagonal with elements $\lambda_i + \varepsilon_i$. Therefore, if

$$\varepsilon_i = \begin{cases} 0 & \text{if } \lambda_i > 0 \\ \delta - \lambda_i & \text{if } \lambda_i \leq 0 \end{cases}$$

where δ is a positive constant, then $\hat{\mathbf{H}}_k$ will be positive definite. With this modification, changes in the components of $\mathbf{x}_k$ in Eq. (5.20) due to negative eigenvalues are ignored. Matrix $\mathbf{U}^T \mathbf{H}_k \mathbf{U}$ can be formed by solving the equation

$$\det(\mathbf{H}_k - \lambda \mathbf{I}_n) = 0 \tag{5.22}$$

This method entails minimal disturbance of $\mathbf{H}_k$, and hence the convergence properties of the Newton method are largely preserved. Unfortunately, however, the solution of Eq. (5.22) involves the determination of the n roots of the characteristic polynomial of $\mathbf{H}_k$ and is, therefore, time-consuming.

A third method for the manipulation of $\mathbf{H}_k$, which is attributed to Matthews and Davies [7], is based on Gaussian elimination. This method leads simultaneously to the modification of $\mathbf{H}_k$ and the computation of the Newton direction $\mathbf{d}_k$ and is, therefore, one of the most practical to use. As was shown in Sect. 2.6, given a matrix $\mathbf{H}_k$, a diagonal matrix $\mathbf{D}$ can be deduced as

$$\mathbf{D} = \mathbf{L} \mathbf{H}_k \mathbf{L}^T \tag{5.23}$$

where

$$\mathbf{L} = \mathbf{E}_{n-1} \cdots \mathbf{E}_2 \mathbf{E}_1$$

is a unit lower triangular matrix, and $\mathbf{E}_1, \mathbf{E}_2, \ldots$ are elementary matrices. If $\mathbf{H}_k$ is positive definite, then $\mathbf{D}$ is positive definite and vice versa (see Theorem 2.7). If $\mathbf{D}$ is not positive definite, then a positive definite diagonal matrix $\hat{\mathbf{D}}$ can be formed by replacing each zero or negative element in $\mathbf{D}$ by a positive element. In this way a positive definite matrix $\hat{\mathbf{H}}_k$ can be formed as

$$\hat{\mathbf{H}}_k = \mathbf{L}^{-1}\hat{\mathbf{D}}(\mathbf{L}^T)^{-1} \tag{5.24}$$

Now from Eq. (5.20)

$$\hat{\mathbf{H}}_k\mathbf{d}_k = -\mathbf{g}_k \tag{5.25}$$

and hence Eqs. (5.24) and (5.25) yield

$$\mathbf{L}^{-1}\hat{\mathbf{D}}(\mathbf{L}^T)^{-1}\mathbf{d}_k = -\mathbf{g}_k \tag{5.26}$$

If we let

$$\hat{\mathbf{D}}(\mathbf{L}^T)^{-1}\mathbf{d}_k = \mathbf{y}_k \tag{5.27}$$

then Eq. (5.26) can be expressed as

$$\mathbf{L}^{-1}\mathbf{y}_k = -\mathbf{g}_k$$

Therefore,

$$\mathbf{y}_k = -\mathbf{L}\mathbf{g}_k \tag{5.28}$$

and from Eq. (5.27)

$$\mathbf{d}_k = \mathbf{L}^T\hat{\mathbf{D}}^{-1}\mathbf{y}_k \tag{5.29}$$

The computation of $\mathbf{d}_k$ can thus be carried out by generating the unit lower triangular matrix $\mathbf{L}$ and the corresponding positive definite diagonal matrix $\hat{\mathbf{D}}$. If

$$\mathbf{H}_k = \begin{bmatrix} h_{11} & h_{12} & \cdots & h_{1n} \\ h_{21} & h_{22} & \cdots & h_{2n} \\ \vdots & \vdots & & \vdots \\ h_{n1} & h_{n2} & \cdots & h_{nn} \end{bmatrix}$$

then

$$\mathbf{L} = \begin{bmatrix} l_{11} & 0 & \cdots & 0 \\ l_{21} & l_{22} & \cdots & 0 \\ \vdots & \vdots & & \vdots \\ l_{n1} & l_{n2} & \cdots & l_{nn} \end{bmatrix}$$

and

$$\hat{\mathbf{D}} = \begin{bmatrix} \hat{d}_{11} & 0 & \cdots & 0 \\ 0 & \hat{d}_{22} & \cdots & 0 \\ \vdots & \vdots & & \vdots \\ 0 & 0 & \cdots & \hat{d}_{nn} \end{bmatrix}$$

can be computed by using the following algorithm.

Algorithm 5.4 Matthews and Davies algorithm
Step 1
Input $\mathbf{H}_k$ and n.
Set $\mathbf{L} = \mathbf{0}, \hat{\mathbf{D}} = \mathbf{0}$.
If $h_{11} > 0$, then set $h_{00} = h_{11}$, else set $h_{00} = 1$.
Step 2
For $k = 2, 3, \ldots, n$ do:
  Set $m = k - 1$, $l_{mm} = 1$.
  If $h_{mm} \leq 0$, set $h_{mm} = h_{00}$.
  Step 2.1
  For $i = k, k + 1, \ldots, n$ do:
    Set $l_{im} = -h_{im}/h_{mm}$, $h_{im} = 0$.
    Step 2.1.1
    For $j = k, k + 1, \ldots, n$ do:
      Set $h_{ij} = h_{ij} + l_{im}h_{mj}$
  If $0 < h_{kk} < h_{00}$, set $h_{00} = h_{kk}$.
Step 3
Set $l_{nn} = 1$. If $h_{nn} \leq 0$, set $h_{nn} = h_{00}$.
For $i = 1, 2, \ldots, n$ set $\hat{d}_{ii} = h_{ii}$.
Stop.

This algorithm will convert $\mathbf{H}$ into an upper triangular matrix with positive diagonal elements, and will then assign the diagonal elements obtained to $\hat{\mathbf{D}}$. Any zero or negative elements of $\mathbf{D}$ are replaced by the most recent lowest positive element of $\hat{\mathbf{D}}$, except if the first element is zero or negative, which is replaced by unity.

If $\mathbf{H}_k$ is a 4×4 matrix, k and m are initially set to 2 and 1, respectively, and l_{11} is set to unity; h_{11} is checked and if it is zero or negative it is changed to unity. The execution of Step 2.1 yields

$$\mathbf{L} = \begin{bmatrix} 1 & 0 & 0 & 0 \\ -\dfrac{h_{21}}{h_{11}} & 0 & 0 & 0 \\ -\dfrac{h_{31}}{h_{11}} & 0 & 0 & 0 \\ -\dfrac{h_{41}}{h_{11}} & 0 & 0 & 0 \end{bmatrix}$$

In addition, the elements in column 1 of $\mathbf{H}_k$ other than h_{11} are set to zero and the elements of rows 2–4 and columns 2–4 are updated to give

$$\mathbf{H}_k = \begin{bmatrix} h_{11} & h_{12} & h_{13} & h_{14} \\ 0 & h'_{22} & h'_{23} & h'_{24} \\ 0 & h'_{32} & h'_{33} & h'_{34} \\ 0 & h'_{42} & h'_{43} & h'_{44} \end{bmatrix}$$

If $0 < h'_{22} < h_{00}$, then h'_{22} is used to update h_{00}. Indices k and m are then set to 3 and 2, respectively, and l_{22} is set to unity. If $h'_{22} \leq 0$, it is replaced by the most recent value of h_{00}. The execution of Step 2.1 yields

$$
\mathbf{L} = \begin{bmatrix}
1 & 0 & 0 & 0 \\
-\dfrac{h_{21}}{h_{11}} & 1 & 0 & 0 \\
-\dfrac{h_{31}}{h_{11}} & -\dfrac{h'_{32}}{h'_{22}} & 0 & 0 \\
-\dfrac{h_{41}}{h_{11}} & -\dfrac{h'_{42}}{h'_{22}} & 0 & 0
\end{bmatrix}
$$

and

$$
\mathbf{H}_k = \begin{bmatrix}
h_{11} & h_{12} & h_{13} & h_{14} \\
0 & h'_{22} & h'_{23} & h'_{24} \\
0 & 0 & h''_{33} & h''_{34} \\
0 & 0 & h''_{43} & h''_{44}
\end{bmatrix}
$$

If $0 < h''_{33} < h_{00}$, h''_{33} is assigned to h_{00}, and so on. In Step 3, h'''_{44} is checked and is changed to h_{00} if found to be zero or negative, and l_{44} is set to unity. Then the diagonal elements of $\mathbf{H}_k$ are assigned to $\hat{\mathbf{D}}$.

With $\hat{\mathbf{D}}$ known, $\hat{\mathbf{D}}^{-1}$ can be readily obtained by replacing the diagonal elements of $\hat{\mathbf{D}}$ by their reciprocals. The computation of $\mathbf{y}_k$ and $\mathbf{d}_k$ can be completed by using Eqs. (5.28) and (5.29). Algorithm 5.4 is illustrated by the following example.

Example 5.2 Compute $\mathbf{L}$ and $\hat{\mathbf{D}}$ for the 4×4 matrix

$$
\mathbf{H}_k = \begin{bmatrix}
h_{11} & h_{12} & h_{13} & h_{14} \\
h_{21} & h_{22} & h_{23} & h_{24} \\
h_{31} & h_{32} & h_{33} & h_{34} \\
h_{41} & h_{42} & h_{43} & h_{44}
\end{bmatrix}
$$

using Algorithm 5.4.

Solution The elements of $\mathbf{L}$ and $\hat{\mathbf{D}}$, namely, l_{ij} and $\hat{d}_{ii}$, can be computed as follows:
Step 1
Input $\mathbf{H}_k$ and set $n = 4$. Initialize $\mathbf{L}$ and $\hat{\mathbf{D}}$ as

$$
\mathbf{L} = \begin{bmatrix}
0 & 0 & 0 & 0 \\
0 & 0 & 0 & 0 \\
0 & 0 & 0 & 0 \\
0 & 0 & 0 & 0
\end{bmatrix}, \quad
\hat{\mathbf{D}} = \begin{bmatrix}
0 & 0 & 0 & 0 \\
0 & 0 & 0 & 0 \\
0 & 0 & 0 & 0 \\
0 & 0 & 0 & 0
\end{bmatrix}
$$

Step 2
If $h_{11} > 0$, then set $h_{00} = h_{11}$, else set $h_{00} = 1$.
$k = 2$;
   $m = 1, l_{11} = 1$;
   if $h_{11} \le 0$, set $h_{11} = h_{00}$;
   Step 2.1
   $i = 2$;

$l_{21} = -h_{21}/h_{11}, \ h_{21} = 0;$
Step 2.1.1
$\quad j = 2;$
$\qquad h_{22} = h_{22} + l_{21}h_{12} = h_{22} - h_{21}h_{12}/h_{11} \ (= h'_{22});$
$\quad j = 3;$
$\qquad h_{23} = h_{23} + l_{21}h_{13} = h_{23} - h_{21}h_{13}/h_{11} \ (= h'_{23});$
$\quad j = 4;$
$\qquad h_{24} = h_{24} + l_{21}h_{14} = h_{24} - h_{21}h_{14}/h_{11} \ (= h'_{24});$
$i = 3;$
$\quad l_{31} = -h_{31}/h_{11}, \ h_{31} = 0;$
$\quad j = 2;$
$\qquad h_{32} = h_{32} + l_{31}h_{12} = h_{32} - h_{31}h_{12}/h_{11} \ (= h'_{32});$
$\quad j = 3;$
$\qquad h_{33} = h_{33} + l_{31}h_{13} = h_{33} - h_{31}h_{13}/h_{11} \ (= h'_{33});$
$\quad j = 4;$
$\qquad h_{34} = h_{34} + l_{31}h_{14} = h_{34} - h_{31}h_{14}/h_{11} \ (= h'_{34});$
$i = 4;$
$\quad l_{41} = -h_{41}/h_{11}, \ h_{41} = 0;$
$\quad j = 2;$
$\qquad h_{42} = h_{42} + l_{41}h_{12} = h_{42} - h_{41}h_{12}/h_{11} \ (= h'_{42});$
$\quad j = 3;$
$\qquad h_{43} = h_{43} + l_{41}h_{13} = h_{43} - h_{41}h_{13}/h_{11} \ (= h'_{43});$
$\quad j = 4;$
$\qquad h_{44} = h_{44} + l_{41}h_{14} = h_{44} - h_{41}h_{14}/h_{11} \ (= h'_{44});$
if $0 < h_{22} < h_{00}$, set $h_{00} = h_{22};$
$k = 3;$
$\quad m = 2, l_{22} = 1;$
$\quad$if $h_{22} < 0$, set $h_{22} = h_{00};$
$\quad i = 3;$
$\qquad l_{32} = -h_{32}/h_{22}, \ h_{32} = 0;$
$\qquad j = 3;$
$\qquad\quad h_{33} = h_{33} + l_{32}h_{23} = h_{33} - h_{32}h_{23}/h_{22} \ (= h''_{33});$
$\qquad j = 4;$
$\qquad\quad h_{34} = h_{34} + l_{32}h_{24} = h_{34} - h_{32}h_{24}/h_{22} \ (= h''_{34});$
$\qquad i = 4;$
$\qquad l_{42} = -h_{42}/h_{22}, \ h_{42} = 0;$
$\qquad j = 3;$
$\qquad\quad h_{43} = h_{43} + l_{42}h_{23} = h_{43} - h_{42}h_{23}/h_{22} \ (= h''_{43});$
$\qquad j = 4;$
$\qquad\quad h_{44} = h_{44} + l_{42}h_{24} = h_{43} - h_{42}h_{24}/h_{22} \ (= h''_{44});$
$\quad$if $0 < h_{33} < h_{00}$, set $h_{00} = h_{33}.$
$k = 4;$
$\quad m = 3, l_{33} = 1;$
$\quad$if $h_{33} \le 0$, set $h_{33} = h_{00};$
$\quad i = 4;$

$l_{34} = -h_{43}/h_{33}, h_{43} = 0;$
$j = 4;$
$\quad h_{44} = h_{44} + l_{44}h_{34} = h_{44} - h_{43}h_{34}/h_{33} = h_{44}'''.$

Step 3
$l_{44} = 1;$
if $h_{44} \le 0$, set $h_{44} = h_{00};$
set $\hat{d}_{ii} = h_{ii}$ for $i = 1, 2, \ldots, n.$ ∎

Example 5.3 The gradient and Hessian are given by

$$\mathbf{g}_k^T = [-\tfrac{1}{5}\ 2], \quad \mathbf{H}_k = \begin{bmatrix} 3 & -6 \\ -6 & \frac{59}{5} \end{bmatrix}$$

Deduce the Newton direction $\mathbf{d}_k$.

Solution If

$$\mathbf{L} = \begin{bmatrix} 1 & 0 \\ 2 & 1 \end{bmatrix}$$

Equation (5.23) gives

$$\mathbf{D} = \begin{bmatrix} 1 & 0 \\ 2 & 1 \end{bmatrix} \begin{bmatrix} 3 & -6 \\ -6 & \frac{59}{5} \end{bmatrix} \begin{bmatrix} 1 & 2 \\ 0 & 1 \end{bmatrix} = \begin{bmatrix} 3 & 0 \\ 0 & -\frac{1}{5} \end{bmatrix}$$

A positive definite diagonal matrix is

$$\hat{\mathbf{D}} = \begin{bmatrix} 3 & 0 \\ 0 & \frac{1}{5} \end{bmatrix}$$

Hence

$$\hat{\mathbf{D}}^{-1} = \begin{bmatrix} \frac{1}{3} & 0 \\ 0 & 5 \end{bmatrix}$$

From Eq. (5.28), we get

$$\mathbf{y}_k = -\begin{bmatrix} 1 & 0 \\ 2 & 1 \end{bmatrix} \begin{bmatrix} -\frac{1}{5} \\ 2 \end{bmatrix} = \begin{bmatrix} \frac{1}{5} \\ -\frac{8}{5} \end{bmatrix}$$

Therefore, from Eq. (5.29)

$$\mathbf{d}_k = \begin{bmatrix} 1 & 2 \\ 0 & 1 \end{bmatrix} \begin{bmatrix} \frac{1}{3} & 0 \\ 0 & 5 \end{bmatrix} \begin{bmatrix} \frac{1}{5} \\ -\frac{8}{5} \end{bmatrix} = \begin{bmatrix} -\frac{239}{15} \\ -8 \end{bmatrix} \quad ∎$$

5.3.2   Computation of the Hessian

The main disadvantage of the Newton method is that the second derivatives of the function are required so that the Hessian may be computed. If exact formulas are unavailable or are difficult to obtain, the second derivatives can be computed by using the numerical formulas

$$\frac{\partial f}{\partial x_1} = \lim_{\delta \to 0} \frac{f(\mathbf{x} + \boldsymbol{\delta}_1) - f(\mathbf{x})}{\delta} = f'(\mathbf{x}) \quad \text{with } \boldsymbol{\delta}_1 = [\delta \; 0 \; 0 \; \cdots \; 0]^T$$

$$\frac{\partial^2 f}{\partial x_1 \partial x_2} = \lim_{\delta \to 0} \frac{f'(\mathbf{x} + \boldsymbol{\delta}_2) - f'(\mathbf{x})}{\delta} \quad \text{with } \boldsymbol{\delta}_2 = [0 \; \delta \; 0 \; \cdots \; 0]^T$$

5.3.3   Newton Decrement

The so-called *Newton decrement* of a smooth and strictly convex objective function $f(\mathbf{x})$ at point $\mathbf{x}$ is defined as

$$\lambda(\mathbf{x}) = [\mathbf{g}(\mathbf{x})^T \, \mathbf{H}(\mathbf{x})^{-1} \mathbf{g}(\mathbf{x})]^{1/2}$$

where $\mathbf{g}(\mathbf{x})$ and $\mathbf{H}(\mathbf{x})$ are the gradient and Hessian of $f(\mathbf{x})$, respectively [8]. It can be used to derive an approximate formula for the error in the objective function of the Newton algorithm after convergence. It can be shown that $\lambda(\mathbf{x})^2/2$ is an estimate of $f(\mathbf{x}) - f^*$ where $f^* = \min f(\mathbf{x})$ (see Problem 5.24 for proof). Thus a stopping criterion of the form

$$f(\mathbf{x}) - f^* < \varepsilon$$

where ε is a termination tolerance can be formulated.

5.3.4   Backtracking Line Search

When iterate $\mathbf{x}_k$ is near a local minimizer, the objective function is well approximated by a convex quadratic function. Since the Newton algorithm is developed on the basis of a quadratic approximation, it follows from Eqs. (5.9) and (5.10) that the value of α obtained in Step 3 of Algorithm 5.3, namely, α_k, will approach unity as $\mathbf{x}_k$ approaches a solution point. For this reason, a line search known as *backtracking* [9] is especially suitable for the Newton algorithm. The principles involved are as follows.

With α set to unity, the inequality

$$f(\mathbf{x}_k + \alpha \mathbf{d}_k) \leq f(\mathbf{x}_k) + \rho \mathbf{g}_k^T \mathbf{d}_k \alpha \tag{5.30}$$

where ρ is in the range 0–0.5 is checked and if it is satisfied, then by moving from point $\mathbf{x}_k$ to point $\mathbf{x}_k + \alpha \mathbf{d}_k$, the objective function would be reduced since $\mathbf{d}_k$ is a descent direction and, as a result, both $-\mathbf{g}_k^T \mathbf{d}_k$ and $-\rho \mathbf{g}_k^T \mathbf{d}_k \alpha$ would be positive. Consequently, the objective function would be reduced at least by an amount $-\rho \mathbf{g}_k^T \mathbf{d}_k \alpha$. Under these circumstances, this value of α is deemed acceptable and the search is terminated.

Note that the inequality in Eq. (5.30) with the equal sign is actually Eq. (4.53) which is represented by line B in Fig. 4.15a. If the inequality in Eq. (5.30) is satisfied, then the value of the objective function would be in the area below line B in Fig. 4.15a and, as can be seen in Fig. 4.15a, a significant reduction in the objective function would be achieved. Note that the above principles are also part of Fletcher's inexact line search.

Now if the inequality in Eq. (5.30) is not satisfied, then

$$f(\mathbf{x}_k + \alpha \mathbf{d}_k) > f(\mathbf{x}_k) + \rho \mathbf{g}_k^T \mathbf{d}_k \alpha \tag{5.31}$$

and, in effect, the value of the objective function at the new point $\mathbf{x}_k + \alpha \mathbf{d}_k$ is larger than the value of the right-hand side of the inequality in Eq. (5.31) i.e., it is located in the area above line B in Fig. 4.15a and any reduction in the objective function is not considered significant and, therefore, it is not deemed acceptable. In this case, α is reduced (or *backtracked*) and the procedure is repeated from the start until a value of α is found that satisfies Eq. (5.30).

An algorithm based on the above principles is as follows.

Algorithm 5.5 Backtracking line search
Step 1
Input $\mathbf{x}_k$ and $\mathbf{d}_k$.
Select a $\rho \in (0, 0.5)$ and a $\gamma \in (0, 1)$, say, $\rho = 0.1$ and $\gamma = 0.5$.
Set $\alpha = 1$.
Step 2
If Eq. (5.31) holds, output $\alpha_k = \alpha$ and stop.
Step 3
Set $\alpha = \gamma * \alpha$ and repeat from Step 2.

5.3.5 Independence of Linear Changes in Variables

Unlike a steepest-descent direction, a Newton direction is independent of linear changes in the variables [8]. Suppose that variables $\mathbf{x}$ and $\mathbf{y}$ are interrelated in terms of a linear transformation $\mathbf{T}$ as $\mathbf{x} = \mathbf{T}\mathbf{y}$. With such a linear variable transformation, the objective function in terms of variable $\mathbf{y}$ becomes

$$\hat{f}(\mathbf{y}) = f(\mathbf{x}) = f(\mathbf{T}\mathbf{y})$$

Hence the gradient and Hessian of $\hat{f}(\mathbf{y})$ with respect to $\mathbf{y}$ are given by

$$\nabla \hat{f}(\mathbf{y}) = \mathbf{T}^T \nabla f(\mathbf{x}) \quad \text{and} \quad \nabla^2 \hat{f}(\mathbf{y}) = \mathbf{T}^T \nabla^2 f(\mathbf{x}) \mathbf{T}$$

respectively. Consequently, the Newton step for $\hat{f}(\mathbf{y})$, denoted as $\Delta \mathbf{y}$, is related to the Newton step of $f(\mathbf{x})$, denoted as $\Delta \mathbf{x}$, by the relation

$$\begin{aligned}
\Delta \mathbf{y} &= -\nabla^2 \hat{f}(\mathbf{y})^{-1} \nabla \hat{f}(\mathbf{y}) \\
&= -(\mathbf{T}^T \nabla^2 f(\mathbf{x}) \mathbf{T})^{-1} \mathbf{T}^T \nabla f(\mathbf{x}) \\
&= -\mathbf{T}^{-1} \nabla^2 f(\mathbf{x})^{-1} \nabla f(\mathbf{x}) \\
&= \mathbf{T}^{-1} \Delta \mathbf{x}
\end{aligned}$$

that is,

$$\Delta \mathbf{x} = \mathbf{T} \Delta \mathbf{y}$$

On comparing the above relation with $\mathbf{x} = \mathbf{T}\mathbf{y}$, we note that the Newton directions for $f(\mathbf{x})$ and $\hat{f}(\mathbf{y})$ are interrelated by the same linear transformation as the variables $\mathbf{x}$ and $\mathbf{y}$ themselves.

5.4   Gauss–Newton Method

In many optimization problems, the objective function is in the form of a vector of functions given by

$$\mathbf{f} = [f_1(\mathbf{x}) \; f_2(\mathbf{x}) \; \cdots \; f_m(\mathbf{x})]^T$$

where $f_p(\mathbf{x})$ for $p = 1, 2, \ldots, m$ are independent functions of $\mathbf{x}$ (see Sect. 1.2). The solution sought is a point $\mathbf{x}$ such that all $f_p(\mathbf{x})$ are reduced to zero simultaneously.

In problems of this type, a real-valued function can be formed as

$$F = \sum_{p=1}^{m} f_p(\mathbf{x})^2 = \mathbf{f}^T \mathbf{f} \tag{5.32}$$

If F is minimized by using a multidimensional unconstrained algorithm, then the individual functions $f_p(\mathbf{x})$ are minimized in the least-squares sense (see Sect. 1.2).

A method for the solution of the above class of problems, known as the *Gauss–Newton method*, can be readily developed by applying the Newton method of Sect. 5.3.

Since there are a number of functions $f_p(\mathbf{x})$ and each one depends on x_i for $i = 1, 2, \ldots, n$ a gradient matrix, referred to as the *Jacobian*, can be formed as

$$\mathbf{J} = \begin{bmatrix} \frac{\partial f_1}{\partial x_1} & \frac{\partial f_1}{\partial x_2} & \cdots & \frac{\partial f_1}{\partial x_n} \\[2mm] \frac{\partial f_2}{\partial x_1} & \frac{\partial f_2}{\partial x_2} & \cdots & \frac{\partial f_2}{\partial x_n} \\[2mm] \vdots & \vdots & & \vdots \\[2mm] \frac{\partial f_m}{\partial x_1} & \frac{\partial f_m}{\partial x_2} & \cdots & \frac{\partial f_m}{\partial x_n} \end{bmatrix}$$

The number of functions m may exceed the number of variables n, that is, the Jacobian need not be a square matrix.

By differentiating F in Eq. (5.32) with respect to x_i, we obtain

$$\frac{\partial F}{\partial x_i} = \sum_{p=1}^{m} 2 f_p(\mathbf{x}) \frac{\partial f_p}{\partial x_i} \tag{5.33}$$

for $i = 1, 2, \ldots, n$. Alternatively, in matrix form

$$
\begin{bmatrix} \frac{\partial F}{\partial x_1} \\ \frac{\partial F}{\partial x_2} \\ \vdots \\ \frac{\partial F}{\partial x_n} \end{bmatrix} = 2 \begin{bmatrix} \frac{\partial f_1}{\partial x_1} & \frac{\partial f_2}{\partial x_1} & \cdots & \frac{\partial f_m}{\partial x_1} \\ \frac{\partial f_1}{\partial x_2} & \frac{\partial f_2}{\partial x_2} & \cdots & \frac{\partial f_m}{\partial x_2} \\ \vdots & \vdots & & \vdots \\ \frac{\partial f_1}{\partial x_n} & \frac{\partial f_2}{\partial x_n} & \cdots & \frac{\partial f_m}{\partial x_n} \end{bmatrix} \begin{bmatrix} f_1(\mathbf{x}) \\ f_2(\mathbf{x}) \\ \vdots \\ f_m(\mathbf{x}) \end{bmatrix}
$$

Hence the gradient of F, designated by $\mathbf{g}_F$, can be expressed as

$$
\mathbf{g}_F = 2\mathbf{J}^T \mathbf{f} \tag{5.34}
$$

Assuming that $f_p(\mathbf{x}) \in C^2$, Eq. (5.33) yields

$$
\frac{\partial^2 F}{\partial x_i \partial x_j} = 2 \sum_{p=1}^{m} \frac{\partial f_p}{\partial x_i} \frac{\partial f_p}{\partial x_j} + 2 \sum_{p=1}^{m} f_p(\mathbf{x}) \frac{\partial^2 f_p}{\partial x_i \partial x_j}
$$

for $i, j = 1, 2, \ldots, n$. If the second derivatives of $f_p(\mathbf{x})$ are neglected, we have

$$
\frac{\partial^2 F}{\partial x_i \partial x_j} \approx 2 \sum_{p=1}^{m} \frac{\partial f_p}{\partial x_i} \frac{\partial f_p}{\partial x_j}
$$

Thus the Hessian of F, designated by $\mathbf{H}_F$, can be deduced as

$$
\mathbf{H}_F \approx 2\mathbf{J}^T \mathbf{J} \tag{5.35}
$$

Since the gradient and Hessian of F are now known, the Newton method can be applied for the solution of the problem. The necessary recursive relation is given by Eqs. (5.19), (5.20), (5.34), and (5.35) as

$$
\begin{aligned}
\mathbf{x}_{k+1} &= \mathbf{x}_k - \alpha_k (2\mathbf{J}^T \mathbf{J})^{-1} (2\mathbf{J}^T \mathbf{f}) \\
&= \mathbf{x}_k - \alpha_k (\mathbf{J}^T \mathbf{J})^{-1} (\mathbf{J}^T \mathbf{f})
\end{aligned}
$$

where α_k is the value of α that minimizes $F(\mathbf{x}_k + \alpha \mathbf{d}_k)$. As k is increased, successive line searches bring about reductions in F_k and $\mathbf{x}_k$ approaches $\mathbf{x}^*$. When $\mathbf{x}_k$ is in the neighborhood of $\mathbf{x}^*$, functions $f_p(\mathbf{x}_k)$ can be accurately represented by the linear approximation of the Taylor series, the matrix in Eq. (5.35) becomes an accurate representation of the Hessian of F_k, and the method converges very rapidly. If functions $f_p(\mathbf{x})$ are linear, F is quadratic, the matrix in Eq. (5.35) is the Hessian, and the problem is solved in one iteration.

The method breaks down if $\mathbf{H}_F$ becomes singular, as in the case of the Newton method. However, the remedies described in Sect. 5.3 can also be applied to the Gauss–Newton method.

An algorithm based on the above principles is as follows:

Algorithm 5.6 Gauss–Newton algorithm
Step 1
Input $\mathbf{x}_0$ and initialize the tolerance ε.
Set $k = 0$.
Step 2
Compute $f_{pk} = f_p(\mathbf{x}_k)$ for $p = 1, 2, \ldots, m$ and F_k.
Step 3
Compute $\mathbf{J}_k$, $\mathbf{g}_k = 2\mathbf{J}_k^T\mathbf{f}_k$, and $\mathbf{H}_k = 2\mathbf{J}_k^T\mathbf{J}_k$.
Step 4
Compute $\mathbf{L}_k$ and $\hat{\mathbf{D}}_k$ using Algorithm 5.4.
Compute $\mathbf{y}_k = -\mathbf{L}_k\mathbf{g}_k$ and $\mathbf{d}_k = \mathbf{L}_k^T\hat{\mathbf{D}}_k^{-1}\mathbf{y}_k$.
Step 5
Find α_k, the value of α that minimizes $F(\mathbf{x}_k + \alpha\mathbf{d}_k)$.
Step 6
Set $\mathbf{x}_{k+1} = \mathbf{x}_k + \alpha_k\mathbf{d}_k$.
Compute $f_{p(k+1)}$ for $p = 1, 2, \ldots, m$ and F_{k+1}.
Step 7
If $|F_{k+1} - F_k| < \varepsilon$, then output $\mathbf{x}^* = \mathbf{x}_{k+1}$, $f_{p(k+1)}(\mathbf{x}^*)$ for $p = 1, 2, \ldots, m$,
and F_{k+1} and stop; otherwise, set $k = k + 1$ and repeat from Step 3.

The factors 2 in Step 3 can be discarded since they cancel out in the calculation of $\mathbf{d}_k$ (see Eq. (5.20)). In Step 4, $\mathbf{H}_k$ is forced to become positive definite, if it is not positive definite, and, further, the Newton direction $\mathbf{d}_k$ is calculated without the direct inversion of $\mathbf{H}_k$.

Example 5.4 Solve the system of nonlinear equations
$$x_1^2 - x_2^2 + x_1 - 3x_2 = 2$$
$$x_1^3 - x_2^4 = -2$$
$$x_1^2 + x_2^3 + 2x_1 - x_2 = -1.1$$
by converting it into an unconstrained optimization problem and then applying the Gauss–Newton method.

Solution If we let
$$f_1(\mathbf{x}) = x_1^2 - x_2^2 + x_1 - 3x_2 - 2$$
$$f_2(\mathbf{x}) = x_1^3 - x_2^4 + 2$$
$$f_3(\mathbf{x}) = x_1^2 + x_2^3 + 2x_1 - x_2 + 1.1$$
the system of equations becomes $f_i(\mathbf{x}) = 0$ for $i = 1, 2, 3$. If we define the objective function
$$F(\mathbf{x}) = f_1^2(\mathbf{x}) + f_2^2(\mathbf{x}) + f_3^2(\mathbf{x})$$
then we can solve the system of equations by solving the unconstrained problem
$$\text{minimize } F(\mathbf{x}) = \sum_{i=1}^{n} f_i^2(\mathbf{x})$$

Since the objective function is in the form of a sum of squares, the Gauss–Newton method is applicable. To solve the problem, we define

$$\mathbf{f}(\mathbf{x}) = [f_1(\mathbf{x})\ f_2(\mathbf{x})\ f_3(\mathbf{x})]^T$$

and evaluate the Jacobian as

$$J(\mathbf{x}) = \begin{bmatrix} \nabla f_1(\mathbf{x})^T \\ \nabla f_2(\mathbf{x})^T \\ \nabla f_3(\mathbf{x})^T \end{bmatrix} = \begin{bmatrix} 2x_1 + 1 & -2x_2 - 3 \\ 3x_1^2 & -4x_2^3 \\ 2x_1 + 2 & 3x_2^2 - 1 \end{bmatrix}$$

The gradient and approximate Hessian of $F(\mathbf{x})$ required by the Gauss–Newton method can be evaluated by using Eqs. (5.33) and (5.34), respectively.

It turns out that precise solutions for the system of equations under consideration do not exist but the Gauss–Newton algorithm can be used to find two approximate solutions. The backtracking line search can be used to carry out the line search in Step 5 of Algorithm 5.6. With $\mathbf{x}_0 = [1.5\ -1.75]^T$ and $\varepsilon = 10^{-6}$, it took Algorithm 5.6 six iterations to converge to a solution point

$$\mathbf{x}^* = \begin{bmatrix} -0.366584 \\ -1.184677 \end{bmatrix}$$

at which the objective function $F(\mathbf{x})$ assumes the value of 7.5632×10^{-3}. On the other hand, with $\mathbf{x}_0 = [1.5\ 1.5]^T$ and $\varepsilon = 10^{-6}$, the algorithm converged after seven iterations to the solution point

$$\mathbf{x}^{**} = \begin{bmatrix} -0.931770 \\ -1.046161 \end{bmatrix}$$

at which $F(\mathbf{x})$ assumes the value of 4.6216×10^{-4}. Evidently, two approximate solutions have been obtained for the system of equations under consideration, $\mathbf{x}^*$ and $\mathbf{x}^{**}$, with the latter being a much better solution. ∎

Problems

5.1 The steepest-descent method is applied to solve the problem

$$\text{minimize}\ \ f(\mathbf{x}) = 2x_1^2 - 2x_1x_2 + x_2^2 + 2x_1 - 2x_2$$

and a sequence $\{\mathbf{x}_k\}$ is generated.

(a) Assuming that

$$\mathbf{x}_{2k+1} = \left[0\ \left(1 - \frac{1}{5^k} \right) \right]^T$$

show that

$$\mathbf{x}_{2k+3} = \left[0\ \left(1 - \frac{1}{5^{k+1}} \right) \right]^T$$

(b) Find the minimizer of $f(\mathbf{x})$ using the result in part (a).

5.2 The problem

$$\text{minimize } f(\mathbf{x}) = x_1^2 + 2x_2^2 + 4x_1 + 4x_2$$

is to be solved by using the steepest-descent method with an initial point $\mathbf{x}_0 = [0\ 0]^T$.

(a) By means of induction, show that

$$\mathbf{x}_k = \left[\frac{2}{3^k} - 2 \quad \left(-\frac{1}{3}\right)^k - 1 \right]^T$$

(b) Deduce the minimizer of $f(\mathbf{x})$.

5.3 Consider the minimization problem

$$\text{minimize } f(\mathbf{x}) = x_1^2 + x_2^2 - 0.2x_1x_2 - 2.2x_1 + 2.2x_2 + 2.2$$

(a) Find a point satisfying the first-order necessary conditions for a minimizer.
(b) Show that this point is the global minimizer.
(c) What is the rate of convergence of the steepest-descent method for this problem?
(d) Starting at $\mathbf{x} = [0\ 0]^T$, how many steepest-descent iterations would it take (at most) to reduce the function value to 10^{-10}?

5.4 (a) Solve the problem

$$\text{minimize } f(\mathbf{x}) = 5x_1^2 - 9x_1x_2 + 4.075x_2^2 + x_1$$

by applying the steepest-descent method with $\mathbf{x}_0 = [1\ 1]^T$ and $\varepsilon = 3 \times 10^{-6}$.
(b) Give a convergence analysis on the above problem to explain why the steepest-decent method requires a large number of iterations to reach the solution.

5.5 (a) Solve the unconstrained problem

$$\text{minimize } f(\mathbf{x}) = 5x_1^2 - 9x_1x_2 + 4.075x_2^2 + x_1$$

by using Algorithm 5.1 with the line search step performed employing the formula in Eq. (5.5)
(b) Repeat part (a) employing the Barzilai–Borwein formula in Eq. (5.15).
(c) Compare the algorithms in parts (a) and (b) in terms of the average CPU time required by the two algorithms.

5.6 Solve the problem

$$\text{minimize } f(\mathbf{x}) = (x_1 + 5)^2 + (x_2 + 8)^2 + (x_3 + 7)^2$$
$$+ 2x_1^2x_2^2 + 4x_1^2x_3^2$$

by applying Algorithm 5.1.

(a) Start with $\mathbf{x}_0 = [1\ 1\ 1]^T$ and $\varepsilon = 10^{-6}$. Verify the solution point using the second-order sufficient conditions.

(b) Repeat (a) using $\mathbf{x}_0 = [-2.3\ 0\ 0]^T$.

(c) Repeat (a) using $\mathbf{x}_0 = [0\ 2\ -12]^T$.

5.7 Solve the problem in Problem 5.6 by applying Algorithm 5.2. Try the same initial points as in Problem 5.6 (a)–(c). Compare the solutions obtained and the amount of computation required with that of Algorithm 5.1.

5.8 Solve the problem

$$\text{minimize}\ \ f(\mathbf{x}) = (x_1^2 + x_2^2 - 1)^2 + (x_1 + x_2 - 1)^2$$

by applying Algorithm 5.1. Use $\varepsilon = 10^{-6}$ and try the following initial points: $[4\ 4]^T, [4\ -4]^T, [-4\ 4]^T, [-4\ -4]^T$. Examine the solution points obtained.

5.9 Solve the problem in Problem 5.8 by applying Algorithm 5.2. Compare the computational efficiency of Algorithm 5.2 with that of Algorithm 5.1.

5.10 Solve the problem

$$\text{minimize}\ \ f(\mathbf{x}) = -x_2^2 e^{1-x_1^2-20(x_1-x_2)^2}$$

by applying Algorithm 5.1.

(a) Start with $\mathbf{x}_0 = [0.1\ 0.1]^T$ and $\varepsilon = 10^{-6}$. Examine the solution obtained.

(b) Start with $\mathbf{x}_0 = [0.8\ 0.1]^T$ and $\varepsilon = 10^{-6}$. Examine the solution obtained.

(c) Start with $\mathbf{x}_0 = [1.1\ 0.1]^T$ and $\varepsilon = 10^{-6}$. Examine the solution obtained.

5.11 Solve the problem in Problem 5.10 by applying Algorithm 5.2. Try the 3 initial points specified in Problem 5.10 (a)–(c) and examine the solutions obtained.

5.12 Solve the problem

$$\text{minimize}\ \ f(\mathbf{x}) = x_1^3 e^{x_2-x_1^2-10(x_1-x_2)^2}$$

by applying Algorithm 5.1. Use $\varepsilon = 10^{-6}$ and try the following initial points: $[-3\ -3]^T, [3\ 3]^T, [3\ -3]^T$, and $[-3\ 3]^T$. Examine the solution points obtained.

5.13 Solve Problem 5.12 by applying Algorithm 5.2. Examine the solution points obtained.

5.14 Solve the minimization problem in Problem 5.1 with $\mathbf{x}_0 = [0\ 0]^T$ by using the Newton method.

5.15 Solve the minimization problem in Problem 5.2 with $\mathbf{x}_0 = [0\ 0]^T$ by using the Newton method.

5.16 Modify the Newton algorithm described in Algorithm 5.3 by incorporating Eq. (5.21) into the algorithm. Give a step-by-step description of the modified algorithm.

5.17 Solve Problem 5.6 by applying the algorithm in Problem 5.16. Examine the solution points obtained and compare the algorithm's computational complexity with that of Algorithm 5.1.

5.18 Solve Problem 5.8 by applying the algorithm in Problem 5.16. Examine the solution points obtained and compare the amount of computation required with that of Algorithm 5.1.

5.19 Solve Problem 5.10 by applying the algorithm in Problem 5.16. Examine the solutions obtained and compare the algorithm's computational complexity with that of Algorithm 5.1.

5.20 Solve Problem 5.12 by applying the algorithm in Problem 5.16. Examine the solutions obtained and compare the amount of computation required with that of Algorithm 5.1.

5.21 (a) Find the global minimizer of the objective function

$$f(\mathbf{x}) = (x_1 + 10x_2)^2 + 5(x_3 - x_4)^2 + (x_2 - 2x_3)^4$$
$$+ 100(x_1 - x_4)^4$$

by using the fact that each term in the objective function is nonnegative.

(b) Solve the problem in part (a) using the steepest-descent method with $\varepsilon = 10^{-6}$ and try the initial points $[-2 \ -1 \ 1 \ 2]^T$ and $[200 \ -200 \ 100 \ -100]^T$.

(c) Solve the problem in part (a) using the modified Newton method in Problem 5.16 with the same termination tolerance and initial points as in (b).

(d) Solve the problem in part (a) using the Gauss–Newton method with the same termination tolerance and initial points as in (b).

(e) Based on the results of (b)–(d), compare the computational efficiency and solution accuracy of the three methods.

5.22 Solve Problem 5.6 by applying the Gauss–Newton method. Examine the solutions obtained and compare the results with those obtained first by using Algorithm 5.1 and then by using the algorithm in Problem 5.16.

5.23 Solve Problem 5.8 by applying the Gauss–Newton method. Examine the solutions obtained and compare the results with those obtained first by using Algorithm 5.1 and then by using the algorithm in Problem 5.16.

5.24 By proving that

$$f(\mathbf{x}) - \inf_{\delta} \hat{f}(\delta) = \frac{\lambda(\mathbf{x})^2}{2}$$

where $\hat{f}(\delta)$ is the second-order approximation of $f(\mathbf{x})$ at $\mathbf{x}$ and

$$\lambda(\mathbf{x}) = [\mathbf{g}(\mathbf{x})^T \mathbf{H}(\mathbf{x})^{-1}\mathbf{g}(\mathbf{x})]^{1/2}$$

is the Newton decrement (see Sect. 5.3.3), show that

$$f(\mathbf{x}) - f^* \approx \lambda(\mathbf{x})^2/2$$

where $f^* = \min f(\mathbf{x})$.

References

1. J. Barzilai and J. M. Borwein, "Two-point step size gradient methods," *IMA J. Numerical Analysis*, vol. 8, pp. 141–148, 1988.
2. Y.-H. Dai and L.-Z. Liao, "R-linear convergence of Barzilai and Borwein gradient method," *IMA J. Numerical Analysis*, vol. 22, pp. 1–10, 2002.
3. Y.-H. Dai and R. Fletcher, "Projected Barzilai-Borwein methods for large-scale box-constrained quadratic programming," *Numer. Math.*, vol. 100, pp. 21–47, 2005.
4. D. G. Luenberger and Y. Ye, *Linear and Nonlinear Programming*, 4th ed. New York: Springer, 2008.
5. S. M. Goldfeld, R. E. Quandt, and H. F. Trotter, "Maximization by quadratic hill-climbing," *Econometrica*, vol. 34, pp. 541–551, 1966.
6. P. B. Zwart, "Nonlinear programming: A quadratic analysis of ridge paralysis," Washington University, St. Louis, MO, Tech. Rep. COO-1493-21, Jan. 1969.
7. A. Matthews and D. Davies, "A comparison of modified Newton methods for unconstrained optimization," *Computer J.*, vol. 14, pp. 293–294, 1971.
8. S. Boyd and L. Vandenberghe, *Convex Optimization*. Cambridge, UK: Cambridge University Press, 2004.
9. J. E. Dennis, Jr. and R. B. Schnabel, *Numerical Methods for Unconstrained Optimization and Nonlinear Equations*. Englewood Cliffs, NJ: Prentice-Hall, 1983.

Conjugate-Direction Methods

6

6.1 Introduction

In the multidimensional optimization methods described so far, the direction of search in each iteration depends on the local properties of the objective function. Although a relation may exist between successive search directions, such a relation is incidental. In this chapter, methods are described in which the optimization is performed by using sequential search directions that bear a strict mathematical relationship to one another. An important class of methods of this type is a class based on a set of search directions known as *conjugate directions*.

Like the Newton method, conjugate-direction methods are developed for the quadratic optimization problem and are then extended to the general optimization problem. For a quadratic problem, convergence is achieved in a finite number of iterations.

Conjugate-direction methods have been found to be very effective in many types of problems and have been used extensively in the past. The four most important methods of this class are as follows:

1. Conjugate-gradient method
2. Fletcher–Reeves method
1. Powell's method
2. Partan method

The principles involved and specific algorithms based on these methods form the subject matter of this chapter.

© Springer Science+Business Media, LLC, part of Springer Nature 2021
A. Antoniou and W.-S. Lu, *Practical Optimization*, Texts in Computer Science,
https://doi.org/10.1007/978-1-0716-0843-2_6

6.2 Conjugate Directions

If $f(\mathbf{x}) \in C^1$ where $\mathbf{x} = [x_1 \ x_2 \ \cdots \ x_n]^T$, the problem

$$\underset{\mathbf{x}}{\text{minimize}} \ \ F = f(\mathbf{x})$$

can be solved by using the following algorithm:

Algorithm 6.1 Coordinate-descent algorithm
Step 1
Input $\mathbf{x}_1$ and initialize the tolerance ε.
Set $k = 1$.
Step 2
Set $\mathbf{d}_k = [0 \ 0 \ \cdots \ 0 \ d_k \ 0 \ \cdots \ 0]^T$.
Step 3
Find α_k, the value of α that minimizes $f(\mathbf{x}_k + \alpha \mathbf{d}_k)$, using a line search.
Set $\mathbf{x}_{k+1} = \mathbf{x}_k + \alpha_k \mathbf{d}_k$
Calculate $f_{k+1} = f(\mathbf{x}_{k+1})$.
Step 4
If $\|\alpha_k \mathbf{d}_k\|_2 < \varepsilon$ then output $\mathbf{x}^* = \mathbf{x}_{k+1}$ and $f(\mathbf{x}^*) = f_{k+1}$, and stop.
Step 5
If $k = n$, set $\mathbf{x}_1 = \mathbf{x}_{k+1}$, $k = 1$ and repeat from Step 2;
otherwise, set $k = k + 1$ and repeat from Step 2.

In this algorithm, an initial point $\mathbf{x}_1$ is assumed, and $f(\mathbf{x})$ is minimized in direction $\mathbf{d}_1$ to obtain a new point $\mathbf{x}_2$. The procedure is repeated for points $\mathbf{x}_2$, $\mathbf{x}_3$, ... and when $k = n$, the algorithm is reinitialized and the procedure is repeated until convergence is achieved. Evidently, this algorithm differs from those in Chap. 5 in that $f(\mathbf{x})$ is minimized repeatedly using a set of directions which bear a strict relationship to one another. The relationship among the various directions is that they form a set of *coordinate directions* since only one element of $\mathbf{x}_k$ is allowed to vary in each line search.

Algorithm 6.1, which is often referred to as a coordinate-descent algorithm, is not very effective or reliable in practice, since an oscillatory behavior can sometimes occur. However, by using another class of interrelated directions known as *conjugate directions*, some quite effective algorithms can be developed.

Definition 6.1 Conjugate directions

(a) Two distinct nonzero vectors $\mathbf{d}_1$ and $\mathbf{d}_2$ are said to be conjugate with respect to a real symmetric matrix $\mathbf{H}$, if
$$\mathbf{d}_1^T \mathbf{H} \mathbf{d}_2 = 0$$

(b) A finite set of distinct nonzero vectors $\{\mathbf{d}_0, \ \mathbf{d}_1, \ \ldots, \ \mathbf{d}_k\}$ is said to be *conjugate* with respect to a real symmetric matrix $\mathbf{H}$, if
$$\mathbf{d}_i^T \mathbf{H} \mathbf{d}_j = 0 \qquad \text{for all } i \neq j \qquad\qquad (6.1)$$

∎

If $\mathbf{H} = \mathbf{I}_n$, where $\mathbf{I}_n$ is the $n \times n$ identity matrix, then Eq. (6.1) can be expressed as

$$\mathbf{d}_i^T \mathbf{H} \mathbf{d}_j = \mathbf{d}_i^T \mathbf{I}_n \mathbf{d}_j = \mathbf{d}_i^T \mathbf{d}_j = 0 \quad \text{for } i \neq j$$

This is the well-known condition for *orthogonality* between vectors $\mathbf{d}_i$ and $\mathbf{d}_j$ and, in effect, conjugacy is a generalization of orthogonality.

If $\mathbf{d}_j$ for $j = 0, 1, \ldots, k$ are *eigenvectors* of $\mathbf{H}$ then

$$\mathbf{H} \mathbf{d}_j = \lambda_j \mathbf{d}_j$$

where the λ_j are the eigenvalues of $\mathbf{H}$. Hence, we have

$$\mathbf{d}_i^T \mathbf{H} \mathbf{d}_j = \lambda_j \mathbf{d}_i^T \mathbf{d}_j = 0 \quad \text{for } i \neq j$$

since $\mathbf{d}_i$ and $\mathbf{d}_j$ for $i \neq j$ are orthogonal [1]. In effect, the set of eigenvectors $\mathbf{d}_j$ constitutes a set of conjugate directions with respect to $\mathbf{H}$.

Theorem 6.1 *Linear independence of conjugate vectors If nonzero vectors $\mathbf{d}_0$, $\mathbf{d}_1$, $\ldots$, $\mathbf{d}_k$ form a conjugate set with respect to a positive definite matrix $\mathbf{H}$, then they are linearly independent.*

Proof Consider the system

$$\sum_{j=0}^{k} \alpha_j \mathbf{d}_j = \mathbf{0}$$

On premultiplying by $\mathbf{d}_i^T \mathbf{H}$, where $0 \leq i \leq k$, and then using Definition 6.1, we obtain

$$\sum_{j=0}^{k} \alpha_j \mathbf{d}_i^T \mathbf{H} \mathbf{d}_j = \alpha_i \mathbf{d}_i^T \mathbf{H} \mathbf{d}_i = 0$$

Since $\mathbf{H}$ is positive definite, we have $\mathbf{d}_i^T \mathbf{H} \mathbf{d}_i > 0$. Therefore, the above system has a solution if and only if $\alpha_j = 0$ for $j = 0, 1, \ldots, k$, that is, vectors $\mathbf{d}_i$ are linearly independent. ∎

The use of conjugate directions in the process of optimization can be demonstrated by considering the quadratic problem

$$\underset{\mathbf{x}}{\text{minimize}} \quad f(\mathbf{x}) = a + \mathbf{x}^T \mathbf{b} + \tfrac{1}{2}\mathbf{x}^T \mathbf{H} \mathbf{x} \tag{6.2}$$

where $a = f(\mathbf{0})$, $\mathbf{b}$ is the gradient of $f(\mathbf{x})$ at $\mathbf{x} = \mathbf{0}$, and $\mathbf{H}$ is the Hessian. The gradient of $f(\mathbf{x})$ at any point can be deduced as

$$\mathbf{g} = \mathbf{b} + \mathbf{H}\mathbf{x}$$

At the minimizer $\mathbf{x}^*$ of $f(\mathbf{x})$, $\mathbf{g} = \mathbf{0}$ and thus

$$\mathbf{H}\mathbf{x}^* = -\mathbf{b} \tag{6.3}$$

If $\mathbf{d}_0, \mathbf{d}_1, \ldots, \mathbf{d}_{n-1}$ are distinct conjugate directions in E^n, then they form a basis of E^n since they are linearly independent and span the E^n space. This means that all possible vectors in E^n can be expressed as linear combinations of directions $\mathbf{d}_0, \mathbf{d}_1, \ldots, \mathbf{d}_{n-1}$. Hence $\mathbf{x}^*$ can be expressed as

$$\mathbf{x}^* = \sum_{i=0}^{n-1} \alpha_i \mathbf{d}_i \tag{6.4}$$

where α_i for $i = 0, 1, \ldots, n-1$ are constants. If $\mathbf{H}$ is positive definite, then from Definition 6.1 we can write

$$\mathbf{d}_k^T \mathbf{H} \mathbf{x}^* = \sum_{i=0}^{n-1} \alpha_i \mathbf{d}_k^T \mathbf{H} \mathbf{d}_i = \alpha_k \mathbf{d}_k^T \mathbf{H} \mathbf{d}_k$$

and thus

$$\alpha_k = \frac{\mathbf{d}_k^T \mathbf{H} \mathbf{x}^*}{\mathbf{d}_k^T \mathbf{H} \mathbf{d}_k} \tag{6.5}$$

Now from Eq. (6.3)

$$\alpha_k = -\frac{\mathbf{d}_k^T \mathbf{b}}{\mathbf{d}_k^T \mathbf{H} \mathbf{d}_k} = -\frac{\mathbf{b}^T \mathbf{d}_k}{\mathbf{d}_k^T \mathbf{H} \mathbf{d}_k}$$

Therefore, Eq. (6.4) gives the minimizer as

$$\mathbf{x}^* = -\sum_{k=0}^{n-1} \frac{\mathbf{d}_k^T \mathbf{b}}{\mathbf{d}_k^T \mathbf{H} \mathbf{d}_k} \mathbf{d}_k \tag{6.6}$$

In effect, if n conjugate directions are known, an explicit expression for $\mathbf{x}^*$ can be obtained.

The significance of conjugate directions can be demonstrated by attempting to obtain $\mathbf{x}^*$ using a set of n nonzero orthogonal directions $\mathbf{p}_0, \mathbf{p}_1, \ldots, \mathbf{p}_{n-1}$. Proceeding as above, we can show that

$$\mathbf{x}^* = \sum_{k=0}^{n-1} \frac{\mathbf{p}_k^T \mathbf{x}^*}{\|\mathbf{p}_k\|_2^2} \mathbf{p}_k$$

Evidently, in this case, $\mathbf{x}^*$ depends on itself and, in effect, there is a distinct advantage in using conjugate directions.

6.3 Basic Conjugate-Directions Method

The computation of $\mathbf{x}^*$ through the use of Eq. (6.6) can be regarded as an iterative computation whereby n successive adjustments $\alpha_k \mathbf{d}_k$ are made to an initial point $\mathbf{x}_0 = \mathbf{0}$. Alternatively, the sequence generated by the recursive relation

$$\mathbf{x}_{k+1} = \mathbf{x}_k + \alpha \mathbf{d}_k$$

where

$$\alpha_k = -\frac{\mathbf{b}^T \mathbf{d}_k}{\mathbf{d}_k^T \mathbf{H} \mathbf{d}_k}$$

and $\mathbf{x}_0 = \mathbf{0}$ converges when $k = n - 1$ to

$$\mathbf{x}_n = \mathbf{x}^*$$

A similar result can be obtained for an arbitrary initial point $\mathbf{x}_0$ as is demonstrated by the following theorem.

Theorem 6.2 *Convergence of conjugate-directions method* If $\{\mathbf{d}_0, \mathbf{d}_1, \ldots, \mathbf{d}_{n-1}\}$ *is a set of nonzero conjugate directions,* $\mathbf{H}$ *is an* $n \times n$ *positive definite matrix, and the problem*

$$\underset{\mathbf{x}}{minimize} \ \ f(\mathbf{x}) = a + \mathbf{x}^T \mathbf{b} + \tfrac{1}{2}\mathbf{x}^T \mathbf{H}\mathbf{x} \tag{6.7}$$

is quadratic, then for any initial point $\mathbf{x}_0$ *the sequence generated by the relation*

$$\mathbf{x}_{k+1} = \mathbf{x}_k + \alpha_k \mathbf{d}_k \ \ \ \ for \ \ k \geq 0 \tag{6.8}$$

where

$$\alpha_k = -\frac{\mathbf{g}_k^T \mathbf{d}_k}{\mathbf{d}_k^T \mathbf{H}\mathbf{d}_k}$$

and

$$\mathbf{g}_k = \mathbf{b} + \mathbf{H}\mathbf{x}_k \tag{6.9}$$

converges to the unique solution $\mathbf{x}^*$ *of the quadratic problem in* n *iterations, i.e.,* $\mathbf{x}_n = \mathbf{x}^*.$

Proof Vector $\mathbf{x}^* - \mathbf{x}_0$ can be expressed as a linear combination of conjugate directions as

$$\mathbf{x}^* - \mathbf{x}_0 = \sum_{i=0}^{n-1} \alpha_i \mathbf{d}_i \tag{6.10}$$

Hence as in Eq. (6.5)

$$\alpha_k = \frac{\mathbf{d}_k^T \mathbf{H}(\mathbf{x}^* - \mathbf{x}_0)}{\mathbf{d}_k^T \mathbf{H}\mathbf{d}_k} \tag{6.11}$$

The iterative procedure in Eq. (6.8) will yield

$$\mathbf{x}_k - \mathbf{x}_0 = \sum_{i=0}^{k-1} \alpha_i \mathbf{d}_i \tag{6.12}$$

and so

$$\mathbf{d}_k^T \mathbf{H}(\mathbf{x}_k - \mathbf{x}_0) = \sum_{i=0}^{k-1} \alpha_i \mathbf{d}_k^T \mathbf{H}\mathbf{d}_i = 0$$

since $i \neq k$. Evidently,

$$\mathbf{d}_k^T \mathbf{H}\mathbf{x}_k = \mathbf{d}_k^T \mathbf{H}\mathbf{x}_0 \tag{6.13}$$

and thus Eqs. (6.11) and (6.13) give

$$\alpha_k = \frac{\mathbf{d}_k^T (\mathbf{H}\mathbf{x}^* - \mathbf{H}\mathbf{x}_k)}{\mathbf{d}_k^T \mathbf{H}\mathbf{d}_k} \tag{6.14}$$

From Eq. (6.9)

$$\mathbf{H}\mathbf{x}_k = \mathbf{g}_k - \mathbf{b} \tag{6.15}$$

and since $\mathbf{g}_k = \mathbf{0}$ at minimizer $\mathbf{x}_k$, we have

$$\mathbf{H}\mathbf{x}^* = -\mathbf{b} \tag{6.16}$$

Therefore, Eqs. (6.14)–(6.16) yield

$$\alpha_k = -\frac{\mathbf{d}_k^T \mathbf{g}_k}{\mathbf{d}_k^T \mathbf{H}\mathbf{d}_k} = -\frac{\mathbf{g}_k^T \mathbf{d}_k}{\mathbf{d}_k^T \mathbf{H}\mathbf{d}_k} \tag{6.17}$$

Now for $k = n$ Eqs. (6.12) and (6.10) yield

$$\mathbf{x}_n = \mathbf{x}_0 + \sum_{i=0}^{n-1} \alpha_i \mathbf{d}_i = \mathbf{x}^*$$

and, therefore, the iterative relation in Eq. (6.8) converges to $\mathbf{x}^*$ in n iterations.   ∎

By using Theorem 6.2 in conjunction with various techniques for the generation of conjugate directions, a number of distinct conjugate-direction methods can be developed.

Methods based on Theorem 6.2 have certain common properties. Two of these properties are given in the following theorem.

Theorem 6.3 *Orthogonality of gradient to a set of conjugate directions*

(a) *The gradient $\mathbf{g}_k$ is orthogonal to directions $\mathbf{d}_i$ for $0 \leq i < k$, that is,*

$$\mathbf{g}_k^T \mathbf{d}_i = \mathbf{d}_i^T \mathbf{g}_k = 0 \quad \text{for} \quad 0 \leq i < k$$

(b) *The assignment $\alpha = \alpha_k$ in Theorem 6.2 minimizes $f(\mathbf{x})$ on each line*

$$\mathbf{x} = \mathbf{x}_{k-1} + \alpha \mathbf{d}_i \quad \text{for} \quad 0 \leq i < k$$

Proof (a) We assume that

$$\mathbf{g}_k^T \mathbf{d}_i = 0 \quad \text{for} \quad 0 \leq i < k \tag{6.18}$$

and show that

$$\mathbf{g}_{k+1}^T \mathbf{d}_i = 0 \quad \text{for} \quad 0 \leq i < k + 1$$

From Eq. (6.9)

$$\mathbf{g}_{k+1} - \mathbf{g}_k = \mathbf{H}(\mathbf{x}_{k+1} - \mathbf{x}_k)$$

and from Eq. (6.8)

$$\mathbf{g}_{k+1} = \mathbf{g}_k + \alpha_k \mathbf{H}\mathbf{d}_k \tag{6.19}$$

Hence

$$\mathbf{g}_{k+1}^T \mathbf{d}_i = \mathbf{g}_k^T \mathbf{d}_i + \alpha_k \mathbf{d}_k^T \mathbf{H} \mathbf{d}_i \tag{6.20}$$

For $i = k$, Eqs. (6.20) and (6.17) give

$$\mathbf{g}_{k+1}^T \mathbf{d}_k = \mathbf{g}_k^T \mathbf{d}_k + \alpha_k \mathbf{d}_k^T \mathbf{H} \mathbf{d}_k = 0 \tag{6.21}$$

For $0 \le i < k$, Eq. (6.18) gives

$$\mathbf{g}_k^T \mathbf{d}_i = 0$$

and since $\mathbf{d}_i$ and $\mathbf{d}_k$ are conjugate

$$\mathbf{d}_k^T \mathbf{H} \mathbf{d}_i = 0$$

Hence Eq. (6.20) gives

$$\mathbf{g}_{k+1}^T \mathbf{d}_i = 0 \quad \text{for } 0 \le i < k \tag{6.22}$$

By combining Eqs. (6.21) and (6.22), we have

$$\mathbf{g}_{k+1}^T \mathbf{d}_i = 0 \quad \text{for } 0 \le i < k+1 \tag{6.23}$$

Now if $k = 0$, Eq. (6.23) gives $\mathbf{g}_1^T \mathbf{d}_i = 0$ for $0 \le i < 1$ and from Eqs. (6.18) and (6.23), we obtain

$$\begin{aligned}
\mathbf{g}_2^T \mathbf{d}_i &= 0 \quad \text{for } 0 \le i < 2 \\
\mathbf{g}_3^T \mathbf{d}_i &= 0 \quad \text{for } 0 \le i < 3
\end{aligned}$$

$$\vdots \qquad\qquad \vdots$$

$$\mathbf{g}_k^T \mathbf{d}_i = 0 \quad \text{for } 0 \le i < k$$

(b) Since

$$\begin{aligned}
\mathbf{g}_k^T \mathbf{d}_i \equiv \mathbf{g}^T(\mathbf{x}_k)\mathbf{d}_i &= \mathbf{g}(\mathbf{x}_{k-1} + \alpha \mathbf{d}_i)^T \mathbf{d}_i \\
&= \frac{df(\mathbf{x}_{k-1} + \alpha \mathbf{d}_i)}{d\alpha} = 0
\end{aligned}$$

$f(\mathbf{x})$ is minimized on each line

$$\mathbf{x} = \mathbf{x}_{k-1} + \alpha \mathbf{d}_i \quad \text{for } 0 \le i < k$$

∎

The implication of the second part of the above theorem is that $\mathbf{x}_k$ minimizes $f(\mathbf{x})$ with respect to the subspace spanned by the set of vectors $\{\mathbf{d}_0, \mathbf{d}_1, \ldots, \mathbf{d}_{k-1}\}$. Therefore, $\mathbf{x}_n$ minimizes $f(\mathbf{x})$ with respect to the space spanned by the set of vectors $\{\mathbf{d}_0, \mathbf{d}_1, \ldots, \mathbf{d}_{n-1}\}$, namely, E^n. This is another way of stating that $\mathbf{x}_n = \mathbf{x}^*$.

6.4 Conjugate-Gradient Method

An effective method for the generation of conjugate directions proposed by Hestenes and Stiefel [2] is the so-called conjugate-gradient method. In this method, directions are generated sequentially, one per iteration. For iteration $k + 1$, a new point $\mathbf{x}_{k+1}$ is generated by using the previous direction $\mathbf{d}_k$. Then a new direction $\mathbf{d}_{k+1}$ is generated by adding a vector $\beta_k \mathbf{d}_k$ to $-\mathbf{g}_{k+1}$, the negative of the gradient at the new point.

The conjugate-gradient method is based on the following theorem. This is essentially the same as Theorem 6.2 except that the method of generating conjugate directions is now defined.

Theorem 6.4 *Convergence of conjugate-gradient method*

(a) *If* $\mathbf{H}$ *is a positive definite matrix, then for any initial point* $\mathbf{x}_0$ *and an initial direction*

$$\mathbf{d}_0 = -\mathbf{g}_0 = -(\mathbf{b} + \mathbf{H}\mathbf{x}_0)$$

the sequence generated by the recursive relation

$$\mathbf{x}_{k+1} = \mathbf{x}_k + \alpha_k \mathbf{d}_k \tag{6.24}$$

where

$$\alpha_k = -\frac{\mathbf{g}_k^T \mathbf{d}_k}{\mathbf{d}_k^T \mathbf{H} \mathbf{d}_k} \tag{6.25}$$

$$\mathbf{g}_k = \mathbf{b} + \mathbf{H}\mathbf{x}_k \tag{6.26}$$

$$\mathbf{d}_{k+1} = -\mathbf{g}_{k+1} + \beta_k \mathbf{d}_k \tag{6.27}$$

$$\beta_k = \frac{\mathbf{g}_{k+1}^T \mathbf{H} \mathbf{d}_k}{\mathbf{d}_k^T \mathbf{H} \mathbf{d}_k} \tag{6.28}$$

converges to the unique solution $\mathbf{x}^*$ *of the problem given in Eq. (6.2).*
(b) *The gradient* $\mathbf{g}_k$ *is orthogonal to* $\{\mathbf{g}_0,\ \mathbf{g}_1,\ \ldots,\ \mathbf{g}_{k-1}\}$, *i.e.,*

$$\mathbf{g}_k^T \mathbf{g}_i = 0 \quad \text{for } 0 \le i < k$$

Proof (a) The proof of convergence is the same as in Theorem 6.2. What remains to prove is that directions $\mathbf{d}_0,\ \mathbf{d}_1,\ \ldots,\ \mathbf{d}_{n-1}$ form a conjugate set, that is,

$$\mathbf{d}_k^T \mathbf{H} \mathbf{d}_i = 0 \quad \text{for } 0 \le i < k \text{ and } 1 \le k \le n$$

The proof is by induction. We assume that

$$\mathbf{d}_k^T \mathbf{H} \mathbf{d}_i = 0 \quad \text{for } 0 \le i < k \tag{6.29}$$

and show that

$$\mathbf{d}_{k+1}^T \mathbf{H} \mathbf{d}_i = 0 \quad \text{for } 0 \le i < k + 1$$

Let $S(\mathbf{v}_0,\ \mathbf{v}_1,\ \ldots,\ \mathbf{v}_k)$ be the subspace spanned by vectors $\mathbf{v}_0,\ \mathbf{v}_1,\ \ldots,\ \mathbf{v}_k$. From Eq. (6.19)

$$\mathbf{g}_{k+1} = \mathbf{g}_k + \alpha_k \mathbf{H} \mathbf{d}_k \tag{6.30}$$

and hence for $k = 0$, we have

$$\mathbf{g}_1 = \mathbf{g}_0 + \alpha_0 \mathbf{H} \mathbf{d}_0 = \mathbf{g}_0 - \alpha_0 \mathbf{H} \mathbf{g}_0$$

since $\mathbf{d}_0 = -\mathbf{g}_0$. In addition, Eq. (6.27) yields

$$\mathbf{d}_1 = -\mathbf{g}_1 + \beta_0 \mathbf{d}_0 = -(1 + \beta_0)\mathbf{g}_0 + \alpha_0 \mathbf{H} \mathbf{g}_0$$

that is, $\mathbf{g}_1$ and $\mathbf{d}_1$ are linear combinations of $\mathbf{g}_0$ and $\mathbf{H}\mathbf{g}_0$, and so

$$S(\mathbf{g}_0, \mathbf{g}_1) = S(\mathbf{d}_0, \mathbf{d}_1) = S(\mathbf{g}_0, \mathbf{H}\mathbf{g}_0)$$

Similarly, for $k = 2$, we get

$$\mathbf{g}_2 = \mathbf{g}_0 - [\alpha_0 + \alpha_1(1 + \beta_0)]\mathbf{H}\mathbf{g}_0 + \alpha_0\alpha_1 \mathbf{H}^2 \mathbf{g}_0$$
$$\mathbf{d}_2 = -[1 + (1 + \beta_0)\beta_1]\mathbf{g}_0 + [\alpha_0 + \alpha_1(1 + \beta_0) + \alpha_0\beta_1]\mathbf{H}\mathbf{g}_0$$
$$-\alpha_0\alpha_1 \mathbf{H}^2 \mathbf{g}_0$$

and hence

$$S(\mathbf{g}_0, \mathbf{g}_1, \mathbf{g}_2) = S(\mathbf{g}_0, \mathbf{H}\mathbf{g}_0, \mathbf{H}^2\mathbf{g}_0)$$
$$S(\mathbf{d}_0, \mathbf{d}_1, \mathbf{d}_2) = S(\mathbf{g}_0, \mathbf{H}\mathbf{g}_0, \mathbf{H}^2\mathbf{g}_0)$$

By continuing the induction, we can show that

$$S(\mathbf{g}_0, \mathbf{g}_1, \ldots, \mathbf{g}_k) = S(\mathbf{g}_0, \mathbf{H}\mathbf{g}_0, \ldots, \mathbf{H}^k\mathbf{g}_0) \tag{6.31}$$
$$S(\mathbf{d}_0, \mathbf{d}_1, \ldots, \mathbf{d}_k) = S(\mathbf{g}_0, \mathbf{H}\mathbf{g}_0, \ldots, \mathbf{H}^k\mathbf{g}_0) \tag{6.32}$$

Now from Eq. (6.27)

$$\mathbf{d}_{k+1}^T \mathbf{H} \mathbf{d}_i = -\mathbf{g}_{k+1}^T \mathbf{H} \mathbf{d}_i + \beta_k \mathbf{d}_k^T \mathbf{H} \mathbf{d}_i \tag{6.33}$$

For $i = k$, the use of Eq. (6.28) gives

$$\mathbf{d}_{k+1}^T \mathbf{H} \mathbf{d}_k = -\mathbf{g}_{k+1}^T \mathbf{H} \mathbf{d}_k + \beta_k \mathbf{d}_k^T \mathbf{H} \mathbf{d}_k = 0 \tag{6.34}$$

For $i < k$, Eq. (6.32) shows that

$$\mathbf{H}\mathbf{d}_i \in S(\mathbf{d}_0, \mathbf{d}_1, \ldots, \mathbf{d}_k)$$

and thus $\mathbf{H}\mathbf{d}_i$ can be represented by the linear combination

$$\mathbf{H}\mathbf{d}_i = \sum_{i=0}^{k} a_i \mathbf{d}_i \tag{6.35}$$

where a_i for $i = 0, 1, \ldots, k$ are constants. Now from Eqs. (6.33) and (6.35)

$$\mathbf{d}_{k+1}^T \mathbf{H} \mathbf{d}_i = -\sum_{i=0}^{k} a_i \mathbf{g}_{k+1}^T \mathbf{d}_i + \beta_k \mathbf{d}_k^T \mathbf{H} \mathbf{d}_i$$
$$= 0 \quad \text{for} \quad i < k \tag{6.36}$$

The first term is zero from the orthogonality property of Theorem 6.3(a) whereas the second term is zero from the assumption in Eq. (6.29). By combining Eqs. (6.34) and (6.36), we have

$$\mathbf{d}_{k+1}^T \mathbf{H} \mathbf{d}_i = 0 \quad \text{for} \quad 0 \le i < k + 1 \tag{6.37}$$

For $k = 0$, Eq. (6.37) gives

$$\mathbf{d}_1^T \mathbf{H} \mathbf{d}_i = 0 \quad \text{for } 0 \le i < 1$$

and, therefore, from Eqs. (6.29) and (6.37), we have

$$\mathbf{d}_2^T \mathbf{H} \mathbf{d}_i = 0 \quad \text{for } 0 \le i < 2$$
$$\mathbf{d}_3^T \mathbf{H} \mathbf{d}_i = 0 \quad \text{for } 0 \le i < 3$$
$$\vdots \qquad\qquad \vdots$$
$$\mathbf{d}_k^T \mathbf{H} \mathbf{d}_i = 0 \quad \text{for } 0 \le i < k$$

(b) From Eqs. (6.31) and (6.32), $\mathbf{g}_0$, $\mathbf{g}_1$, ..., $\mathbf{g}_k$ span the same subspace as $\mathbf{d}_0$, $\mathbf{d}_1$, ..., $\mathbf{d}_k$ and, consequently, they are linearly independent. We can write

$$\mathbf{g}_i = \sum_{j=0}^{i} a_j \mathbf{d}_j$$

where a_j for $j = 0, 1, \ldots, i$ are constants. Therefore, from Theorem 6.3

$$\mathbf{g}_k^T \mathbf{g}_i = \sum_{j=0}^{i} a_j \mathbf{g}_k^T \mathbf{d}_j = 0 \quad \text{for } 0 \le i < k$$

∎

The expressions for α_k and β_k in the above theorem can be simplified somewhat. From Eq. (6.27)

$$-\mathbf{g}_k^T \mathbf{d}_k = \mathbf{g}_k^T \mathbf{g}_k - \beta_{k-1} \mathbf{g}_k^T \mathbf{d}_{k-1}$$

where

$$\mathbf{g}_k^T \mathbf{d}_{k-1} = 0$$

according to Theorem 6.3(a). Hence

$$-\mathbf{g}_k^T \mathbf{d}_k = \mathbf{g}_k^T \mathbf{g}_k$$

and, therefore, the expression for α_k in Eq. (6.25) is modified as

$$\alpha_k = \frac{\mathbf{g}_k^T \mathbf{g}_k}{\mathbf{d}_k^T \mathbf{H} \mathbf{d}_k} \tag{6.38}$$

On the other hand, from Eq. (6.19)

$$\mathbf{H} \mathbf{d}_k = \frac{1}{\alpha_k} (\mathbf{g}_{k+1} - \mathbf{g}_k)$$

and so

$$\mathbf{g}_{k+1}^T \mathbf{H} \mathbf{d}_k = \frac{1}{\alpha_k} (\mathbf{g}_{k+1}^T \mathbf{g}_{k+1} - \mathbf{g}_{k+1}^T \mathbf{g}_k) \tag{6.39}$$

Now from Eqs. (6.31) and (6.32)

$$\mathbf{g}_k \in S(\mathbf{d}_1, \ \mathbf{d}_2, \ \ldots, \ \mathbf{d}_k)$$

or

$$\mathbf{g}_k = \sum_{i=0}^{k} a_i \mathbf{d}_i$$

and as a result

$$\mathbf{g}_{k+1}^T \mathbf{g}_k = \sum_{i=0}^{k} a_i \mathbf{g}_{k+1}^T \mathbf{d}_i = 0 \tag{6.40}$$

by virtue of Theorem 6.3(a). Therefore, Eq. (6.28) and Eqs. (6.38)–(6.40) yield

$$\beta_k = \frac{\mathbf{g}_{k+1}^T \mathbf{g}_{k+1}}{\mathbf{g}_k^T \mathbf{g}_k}$$

The above principles and theorems lead to the following algorithm:

Algorithm 6.2 Conjugate-gradient algorithm
Step 1
Input $\mathbf{x}_0$ and initialize the tolerance ε.
Step 2
Compute $\mathbf{g}_0$ and set $\mathbf{d}_0 = -\mathbf{g}_0$, $k = 0$.
Step 3
Input $\mathbf{H}_k$, i.e., the Hessian at $\mathbf{x}_k$.
Compute

$$\alpha_k = \frac{\mathbf{g}_k^T \mathbf{g}_k}{\mathbf{d}_k^T \mathbf{H}_k \mathbf{d}_k}$$

Set $\mathbf{x}_{k+1} = \mathbf{x}_k + \alpha_k \mathbf{d}_k$ and calculate $f_{k+1} = f(\mathbf{x}_{k+1})$.
Step 4
If $\|\alpha_k \mathbf{d}_k\|_2 < \varepsilon$, output $\mathbf{x}^* = \mathbf{x}_{k+1}$ and $f(\mathbf{x}^*) = f_{k+1}$, and stop.
Step 5
 Compute $\mathbf{g}_{k+1}$.
Compute

$$\beta_k = \frac{\mathbf{g}_{k+1}^T \mathbf{g}_{k+1}}{\mathbf{g}_k^T \mathbf{g}_k}$$

Generate new direction

$$\mathbf{d}_{k+1} = -\mathbf{g}_{k+1} + \beta_k \mathbf{d}_k$$

Set $k = k + 1$, and repeat from Step 3.

A typical solution trajectory for the above algorithm for a 2-dimensional convex quadratic problem is illustrated in Fig. 6.1. Note that $\mathbf{x}_1 = \mathbf{x}_0 - \alpha_0 \mathbf{g}_0$, where α_0 is the value of α that minimizes $f(\mathbf{x}_0 - \alpha \mathbf{g}_0)$, as in the steepest-descent algorithm.

The main advantages of the conjugate-gradient algorithm are as follows:

1. The gradient is always finite and linearly independent of all previous direction vectors, except when the solution is reached.
2. The computations are relatively simple and only slightly more complicated by comparison to the computations in the steepest-descent method.
3. No line searches are required.
4. For convex quadratic problems, the algorithm converges in n iterations.

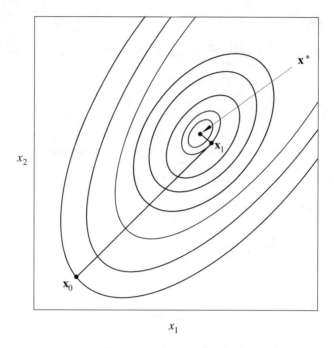

Fig. 6.1 Typical solution trajectory in conjugate-gradient algorithm for a quadratic problem

5. The first direction is a steepest-descent direction and it thus leads to a good reduction in $f(\mathbf{x})$ during the first iteration.
6. The algorithm has good convergence properties when applied for the solution of nonquadratic problems since the directions are based on gradient information.
7. Problems associated with the inversion of the Hessian are absent.

The disadvantages of the algorithm are:

1. The Hessian must be supplied, stored, and manipulated.
2. For nonquadratic problems convergence may not be achieved in rare occasions.

Example 6.1 (a) Applying the conjugate-gradient algorithm, i.e., Algorithm 6.2, minimize the objective function in Example 5.1. Use the same initial point as in Example 5.1, namely, $\mathbf{x}_0 = [0\ 0\ 0\ 0]^T$.

(b) Compare the computational efficiency of the conjugate-gradient algorithm with that of the steepest-descent algorithm, i.e., Algorithm 5.1, as described in Example 5.1(b) in terms of the average CPU time required by each algorithm.

Solution (a) The objective function, $f(\mathbf{x})$, in Example 5.1 is a convex quadratic function whose Hessian is given by

$$\mathbf{H} = \begin{bmatrix} 30 & 0 & 0 & 0 \\ 0 & 10 & 0 & 0 \\ 0 & 0 & 3 & 0 \\ 0 & 0 & 0 & 1 \end{bmatrix}$$

Hence the value of α_k that minimizes the value of $f(\mathbf{x}_k + \alpha_k \mathbf{d}_k)$ in Step 3 of Algorithm 6.2 is given by

$$\alpha_k = \frac{\mathbf{g}_k^T \mathbf{g},}{\mathbf{d}_k^T \mathbf{H} \mathbf{d}_k} = \frac{\mathbf{g}_k^T \mathbf{g}_k}{30 d_k(1)^2 + 10 d_k(2)^2 + 3 d_k(3)^2 + d_k(4)^2}$$

With $\mathbf{x}_0 = \mathbf{0}$ and $\varepsilon = 10^{-6}$, it took Algorithm 6.2 four iterations to converge to the solution

$$\mathbf{x}^* = \begin{bmatrix} 0.033333 \\ 0.100000 \\ 0.333333 \\ 1.000000 \end{bmatrix}$$

which is as expected since for convex quadratic problems the conjugate-gradient algorithm is known to converge in n iterations.

(b) The average CPU times required by Algorithm 5.2 first using the step-estimation formula in Eq. (5.5) and then using the Barzilai–Borwein step-size estimation formula in Eq. (5.13) were obtained by using the pseudocode described in the solution to Example 5.1(d). The normalized CPU times with respect to the CPU time required by Algorithm 5.2 using the step-size estimation in Eq. (5.5) were found to be 1 and 0.1786, respectively. This result demonstrates that the conjugate-gradient algorithm is more efficient. ∎

6.5 Minimization of Nonquadratic Functions

Like the Newton method, conjugate-direction methods are developed for the convex quadratic problem but are then applied for the solution of quadratic as well as non-quadratic problems. The fundamental assumption is made that if a steady reduction is achieved in the objective function in successive iterations, the neighborhood of the solution will eventually be reached. If $\mathbf{H}$ is positive definite near the solution, then convergence will, in principle, follow in at most n iterations. For this reason, conjugate-direction methods, like the Newton method, are said to have *quadratic termination*. In addition, the rate of convergence is quadratic, that is, the order of convergence is two.

The use of conjugate-direction methods for the solution of nonquadratic problems may sometimes be relatively inefficient in reducing the objective function, in particular if the initial point is far from the solution. In such a case, unreliable previous data are likely to accumulate in the current direction vector, since they are calculated on the basis of past directions. Under these circumstances, the solution trajectory may wander through suboptimal areas of the parameter space, and progress will be slow. This problem can be overcome by re-initializing these algorithms periodically, say, every n iterations, in order to obliterate previous unreliable information, and in order to provide new vigor to the algorithm through the use of a steepest-descent step. Most of the time, the information accumulated in the current direction is quite reliable and throwing it away is likely to increase the amount of computation. Nevertheless, this seems to be a fair price to pay if the robustness of the algorithm is increased.

6.6 Fletcher–Reeves Method

The Fletcher–Reeves method [3] is a variation of the conjugate-gradient method. Its main feature is that parameters α_k for $k = 0, 1, 2, \ldots$ are determined by minimizing $f(\mathbf{x} + \alpha \mathbf{d}_k)$ with respect to α using a line search as in the case of the steepest-descent or the Newton method. The difference between this method and the steepest-descent or the Newton method is that $\mathbf{d}_k$ is a conjugate direction with respect to $\mathbf{d}_{k-1}$, $\mathbf{d}_{k-2}$, $\ldots$, $\mathbf{d}_0$ rather than the steepest-descent or Newton direction.

If the problem to be solved is convex and quadratic and the directions are selected as in Eq. (6.27) with β_k given by Eq. (6.28), then

$$\frac{df(\mathbf{x}_k + \alpha \mathbf{d}_k)}{d\alpha} = \mathbf{g}_{k+1}^T \mathbf{d}_k = 0$$

for $k = 0, 1, 2, \ldots$. Further, the conjugacy of the set of directions assures that

$$\frac{df(\mathbf{x}_k + \alpha \mathbf{d}_i)}{d\alpha} = \mathbf{g}_{k+1}^T \mathbf{d}_i = 0 \qquad \text{for} \ \ 0 \le i \le k$$

or

$$\mathbf{g}_k^T \mathbf{d}_i = 0 \qquad \text{for} \ \ 0 \le i < k$$

as in Theorem 6.3. Consequently, the determination of α_k through a line search is equivalent to using Eq. (6.25). Since a line search entails more computation than Eq. (6.25), the Fletcher–Reeves modification would appear to be a retrograde step. Nevertheless, two significant advantages are gained as follows:

1. The modification renders the method more amenable to the minimization of nonquadratic problems since a larger reduction can be achieved in $f(\mathbf{x})$ along $\mathbf{d}_k$ at points outside the neighborhood of the solution. This is due to the fact that Eq. (6.25) will not yield the minimum along $\mathbf{d}_k$ in the case of a nonquadratic problem.
2. The modification obviates the derivation and calculation of the Hessian.

The Fletcher–Reeves algorithm can be shown to converge subject to the condition that the algorithm is re-initialized every n iterations. An implementation of the algorithm is as follows:

Algorithm 6.3 Fletcher-Reeves algorithm
Step 1
Input $\mathbf{x}_0$ and initialize the tolerance ε.
Step 2
Set $k = 0$.
Compute $\mathbf{g}_0$ and set $\mathbf{d}_0 = -\mathbf{g}_0$.
Step 3
Find α_k, the value of α that minimizes $f(\mathbf{x}_k + \alpha \mathbf{d}_k)$.
Set $\mathbf{x}_{k+1} = \mathbf{x}_k + \alpha_k \mathbf{d}_k$.
Step 4
If $\|\alpha_k \mathbf{d}_k\|_2 < \varepsilon$, output $\mathbf{x}^* = \mathbf{x}_{k+1}$ and $f(\mathbf{x}^*) = f_{k+1}$, and stop.
Step 5
If $k = n - 1$, set $\mathbf{x}_0 = \mathbf{x}_{k+1}$ and go to Step 2.
Step 6
Compute $\mathbf{g}_{k+1}$.
Compute

$$\beta_k = \frac{\mathbf{g}_{k+1}^T \mathbf{g}_{k+1}}{\mathbf{g}_k^T \mathbf{g}_k}$$

Set $\mathbf{d}_{k+1} = -\mathbf{g}_{k+1} + \beta_k \mathbf{d}_k$.
Set $k = k + 1$ and repeat from Step 3.

Example 6.2 (a) Applying the Fletcher–Reeves algorithm (Algorithm 6.3), minimize the Himmelblau function [4, p. 428]

$$f(\mathbf{x}) = (x_1^2 + x_2 - 11)^2 + (x_1 + x_2^2 - 7)^2$$

Try four initial points $\mathbf{x}_0 = [6 \ 6]^T$, $\mathbf{x}_0 = [6 \ -6]^T$, $\mathbf{x}_0 = [-6 \ 6]^T$, and $\mathbf{x}_0 = [-6 \ -6]^T$. and assume a termination tolerance $\varepsilon = 10^{-12}$.
  (b) Apply the steepest-descent algorithm (Algorithm 5.1) to the same problem and
compare the results with those obtained from part (a) using the same initial points
and termination tolerance.
  (c) Compare the computational efficiency of the Fletcher–Reeves algorithm with
the steepest-descent algorithm in terms of the average CPU time required by each
algorithm.

Solution (a) The objective function is a sum of two squared terms and hence it is
always nonnegative and assumes the value of zero at point $\mathbf{x} = [3 \ 2]^T$. Therefore, at
a global minimizer, $\mathbf{x}^*$, the equations

$$x_1^2 + x_2 - 11 = 0 \quad \text{and} \quad x_1 + x_2^2 - 7 = 0$$

are satisfied simultaneously. From the first equation we have

$$x_2 = 11 - x_1^2$$

which can be used in the second equation to eliminate variable x_2. This leads to

$$x_1 + (11 - x_1^2)^2 - 7 = 0$$

i.e.,

$$x_1^4 - 22x_1^2 + x_1 + 114 = 0$$

By solving this 4th-order equation and using $x_2 = 11 - x_1^2$, four global minimizers of $f(\mathbf{x})$ can be obtained as

$$\mathbf{x}_1^* = \begin{bmatrix} 3 \\ 2 \end{bmatrix}, \quad \mathbf{x}_2^* = \begin{bmatrix} 3.584428 \\ -1.848127 \end{bmatrix}$$

$$\mathbf{x}_3^* = \begin{bmatrix} -2.805118 \\ 3.131313 \end{bmatrix}, \quad \mathbf{x}_4^* = \begin{bmatrix} -3.779310 \\ -3.283186 \end{bmatrix}$$

We note that these global minimizers are widely separated from each other and one is located in each of the four quadrants.

The gradient of $f(\mathbf{x})$ can be obtained as

$$\nabla f(\mathbf{x}) = \begin{bmatrix} 4(x_1^2 + x_2 - 11)x_1 + 2(x_1 + x_2^2 - 7) \\ 2(x_1^2 + x_2 - 11) + 4(x_1 + x_2^2 - 7)x_2 \end{bmatrix}$$

With $\mathbf{x}_0 = [6\ 6]^T$, it took Algorithm 6.3 eleven iterations to converge to the solution

$$\mathbf{x}^* = \begin{bmatrix} 3.000000 \\ 2.000000 \end{bmatrix}$$

With $\mathbf{x}_0 = [6\ -6]^T$, Algorithm 6.3 converged after fifteen iterations to the solution

$$\mathbf{x}^* = \begin{bmatrix} 3.584428 \\ -1.848127 \end{bmatrix}$$

With $\mathbf{x}_0 = [-6\ 6]^T$, it took Algorithm 6.3 five iterations to converge to the solution

$$\mathbf{x}^* = \begin{bmatrix} -2.805118 \\ 3.131313 \end{bmatrix}$$

With $\mathbf{x}_0 = [-6\ -6]^T$, Algorithm 6.3 converged after six iterations to the solution

$$\mathbf{x}^* = \begin{bmatrix} -3.779310 \\ -3.283186 \end{bmatrix}$$

The Himmelblau function is a 4th-order polynomial in two variables and hence its Hessian is not a constant matrix but depends on $\mathbf{x}$. Furthermore, it can be verified that the function is not always convex and, as a result, different initial points can lead to different solutions as has been found out.

(b) With $\mathbf{x}_0 = [6\ 6]^T$, Algorithm 5.1 converged after fourteen iterations to the solution

$$\mathbf{x}^* = \begin{bmatrix} 3.000000 \\ 2.000000 \end{bmatrix}$$

With $\mathbf{x}_0 = [6\ -6]^T$, Algorithm 5.1 converged after fifteen iterations to the solution

$$\mathbf{x}^* = \begin{bmatrix} 3.584428 \\ -1.848127 \end{bmatrix}$$

With $\mathbf{x}_0 = [-6\ 6]^T$, Algorithm 5.1 converged after seven iterations to the solution

$$\mathbf{x}^* = \begin{bmatrix} -2.805118 \\ 3.131313 \end{bmatrix}$$

With $\mathbf{x}_0 = [-6\ -6]^T$, it took Algorithm 5.1 seven iterations to converge to the solution

$$\mathbf{x}^* = \begin{bmatrix} -3.779310 \\ -3.283186 \end{bmatrix}$$

(c) The average CPU times required by Algorithms 6.3 and 5.1 were obtained by using the pseudocode described in the solution to Example 5.1(d). The CPU times required by Algorithm 6.3 for the four cases normalized with respect to the CPU times required by Algorithm 5.1 were obtained as 0.86, 0.65, 0.81, and 0.89, respectively. Algorithm 6.3 converged to the same solution points but with reduced computation effort relative to that required by Algorithm 5.1.   ∎

6.7 Powell's Method

A conjugate-direction method which has been used extensively in the past is one due to Powell [5]. This method, like the conjugate-gradient method, is developed for the convex quadratic problem but it can be applied successfully to nonquadratic problems.

The distinctive feature of Powell's method is that conjugate directions are generated through a series of line searches. The technique used is based on the following theorem:

Theorem 6.5 *Generation of conjugate directions in Powell's method* *Let $\mathbf{x}_a^*$ and $\mathbf{x}_b^*$ be the minimizers obtained if the convex quadratic function*

$$f(\mathbf{x}) = a + \mathbf{x}^T \mathbf{b} + \tfrac{1}{2}\mathbf{x}^T \mathbf{H}\mathbf{x}$$

is minimized with respect to α on lines

$$\mathbf{x} = \mathbf{x}_a + \alpha \mathbf{d}_a$$

and

$$\mathbf{x} = \mathbf{x}_b + \alpha \mathbf{d}_b$$

respectively, as illustrated in Fig. 6.2.

If $\mathbf{d}_b = \mathbf{d}_a$, then vector $\mathbf{x}_b^* - \mathbf{x}_a^*$ is conjugate with respect to $\mathbf{d}_a$ (or $\mathbf{d}_b$).

Proof If $f(\mathbf{x}_a + \alpha \mathbf{d}_a)$ and $f(\mathbf{x}_b + \alpha \mathbf{d}_b)$ are minimized with respect to α, then

$$\frac{df(\mathbf{x}_a + \alpha \mathbf{d}_a)}{d\alpha} = \mathbf{d}_a^T \mathbf{g}(\mathbf{x}_a^*) = 0 \tag{6.41a}$$

$$\frac{df(\mathbf{x}_b + \alpha \mathbf{d}_b)}{d\alpha} = \mathbf{d}_b^T \mathbf{g}(\mathbf{x}_b^*) = 0 \tag{6.41b}$$

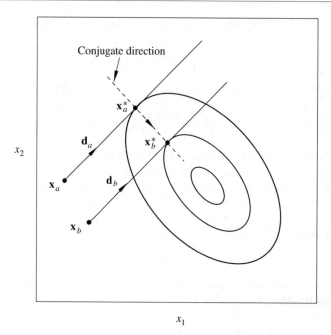

Fig. 6.2 Generation of a conjugate direction

as in the case of a steepest-descent step (see Sect. 5.2). Since

$$\mathbf{g}(\mathbf{x}_a^*) = \mathbf{b} + \mathbf{H}\mathbf{x}_a^* \qquad (6.42a)$$

$$\mathbf{g}(\mathbf{x}_b^*) = \mathbf{b} + \mathbf{H}\mathbf{x}_b^* \qquad (6.42b)$$

then for $\mathbf{d}_b = \mathbf{d}_a$, Eqs. (6.41a)–(6.42b) yield

$$\mathbf{d}_a^T \mathbf{H}(\mathbf{x}_b^* - \mathbf{x}_a^*) = 0$$

and, therefore, vector $\mathbf{x}_b^* - \mathbf{x}_a^*$ is conjugate with respect to direction $\mathbf{d}_a$ (or $\mathbf{d}_b$). ∎

In Powell's algorithm, an initial point $\mathbf{x}_{00}$ as well as n linearly independent directions $\mathbf{d}_{01}$, $\mathbf{d}_{02}$, ..., $\mathbf{d}_{0n}$ are assumed and a series of line searches are performed in each iteration. Although any set of initial linearly independent directions can be used, it is convenient to use a set of coordinate directions.

In the first iteration, $f(\mathbf{x})$ is minimized sequentially in directions $\mathbf{d}_{01}$, $\mathbf{d}_{02}$, ..., $\mathbf{d}_{0n}$ starting from point $\mathbf{x}_{00}$ to yield points $\mathbf{x}_{01}$, $\mathbf{x}_{02}$, ..., $\mathbf{x}_{0n}$, respectively, as depicted in Fig. 6.3a. Then a new direction $\mathbf{d}_{0(n+1)}$ is generated as

$$\mathbf{d}_{0(n+1)} = \mathbf{x}_{0n} - \mathbf{x}_{00}$$

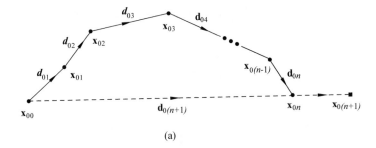

(a)

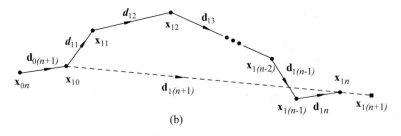

(b)

Fig. 6.3 First and second iterations in Powell's algorithm

and $f(\mathbf{x})$ is minimized in this direction to yield a new point $\mathbf{x}_{0(n+1)}$. The set of directions is then updated by letting

$$\mathbf{d}_{11} = \mathbf{d}_{02}$$
$$\mathbf{d}_{12} = \mathbf{d}_{03}$$
$$\vdots$$
$$\mathbf{d}_{1(n-1)} = \mathbf{d}_{0n}$$
$$\mathbf{d}_{1n} = \mathbf{d}_{0(n+1)} \tag{6.43}$$

The effect of the first iteration is to reduce $f(\mathbf{x})$ by an amount $\Delta f = f(\mathbf{x}_{00}) - f(\mathbf{x}_{0(n+1)})$ and simultaneously to delete $\mathbf{d}_{01}$ from and add $\mathbf{d}_{0(n+1)}$ to the set of directions.

The same procedure is repeated in the second iteration. Starting with point

$$\mathbf{x}_{10} = \mathbf{x}_{0(n+1)}$$

$f(\mathbf{x})$ is minimized sequentially in directions $\mathbf{d}_{11}$, $\mathbf{d}_{12}$, $\ldots$, $\mathbf{d}_{1n}$ to yield points $\mathbf{x}_{11}$, $\mathbf{x}_{12}$, $\ldots$, $\mathbf{x}_{1n}$, as depicted in Fig. 6.3b. Then a new direction $\mathbf{d}_{1(n+1)}$ is generated as

$$\mathbf{d}_{1(n+1)} = \mathbf{x}_{1n} - \mathbf{x}_{10}$$

and $f(\mathbf{x})$ is minimized in direction $\mathbf{d}_{1(n+1)}$ to yield point $\mathbf{x}_{1(n+1)}$. Since

$$\mathbf{d}_{1n} = \mathbf{d}_{0(n+1)}$$

by assignment (see Eq. (6.43)), $\mathbf{d}_{1(n+1)}$ is conjugate to $\mathbf{d}_{1n}$, according to Theorem 6.5. Therefore, if we let

$$\mathbf{d}_{21} = \mathbf{d}_{12}$$
$$\mathbf{d}_{22} = \mathbf{d}_{13}$$
$$\vdots$$
$$\mathbf{d}_{2(n-1)} = \mathbf{d}_{1n}$$
$$\mathbf{d}_{2n} = \mathbf{d}_{1(n+1)}$$

the new set of directions will include a pair of conjugate directions, namely, $\mathbf{d}_{2(n-1)}$ and $\mathbf{d}_{2n}$.

Proceeding in the same way, each new iteration will increase the number of conjugate directions by one, and since the first two iterations yield two conjugate directions, n iterations will yield n conjugate directions. Powell's method will thus require $n(n + 1)$ line searches since each iteration entails $(n + 1)$ line searches. An implementation of Powell's algorithm is as follows:

Algorithm 6.4 Powell's algorithm
Step 1
Input $\mathbf{x}_{00} = [x_{01} \ x_{02} \ \cdots \ x_{0n}]^T$ and initialize the tolerance ε.
Set

$$\mathbf{d}_{01} = \begin{bmatrix} x_{01} \ 0 \ \cdots \ 0 \end{bmatrix}^T$$
$$\mathbf{d}_{02} = \begin{bmatrix} 0 \ x_{02} \ \cdots \ 0 \end{bmatrix}^T$$
$$\vdots$$
$$\mathbf{d}_{0n} = \begin{bmatrix} 0 \ 0 \ \cdots \ x_{0n} \end{bmatrix}^T$$

Set $k = 0$.
Step 2
For $i = 1$ to n do:
Find α_{ki}, the value of α that minimizes $f(\mathbf{x}_{k(i-1)} + \alpha \mathbf{d}_{ki})$.
Set $\mathbf{x}_{ki} = \mathbf{x}_{k(i-1)} + \alpha_{ki}\mathbf{d}_{ki}$.
Step 3
Generate a new direction

$$\mathbf{d}_{k(n+1)} = \mathbf{x}_{kn} - \mathbf{x}_{k0}$$

Find $\alpha_{k(n+1)}$, the value of α that minimizes $f(\mathbf{x}_{kn} + \alpha \mathbf{d}_{k(n+1)})$.
Set

$$\mathbf{x}_{k(n+1)} = \mathbf{x}_{kn} + \alpha_{k(n+1)}\mathbf{d}_{k(n+1)}$$

Calculate $f_{k(n+1)} = f(\mathbf{x}_{k(n+1)})$.

Step 4

If $\|\alpha_{k(n+1)}\mathbf{d}_{k(n+1)}\|_2 < \varepsilon$, output $\mathbf{x}^* = \mathbf{x}_{k(n+1)}$ and $f(\mathbf{x}^*) = f_{k(n+1)}$, and stop.

Step 5

Update directions by setting

$$\mathbf{d}_{(k+1)1} = \mathbf{d}_{k2}$$
$$\mathbf{d}_{(k+1)2} = \mathbf{d}_{k3}$$
$$\vdots$$
$$\mathbf{d}_{(k+1)n} = \mathbf{d}_{k(n+1)}$$

Set $\mathbf{x}_{(k+1)0} = \mathbf{x}_{k(n+1)}$, $k = k + 1$, and repeat from Step 2.

In Step 1, $\mathbf{d}_{01}$, $\mathbf{d}_{02}$, ..., $\mathbf{d}_{0n}$ are assumed to be a set of coordinate directions. In Step 2, $f(\mathbf{x})$ is minimized along the path $\mathbf{x}_{k0}$, $\mathbf{x}_{k1}$, ..., $\mathbf{x}_{kn}$. In Step 3, $f(\mathbf{x})$ is minimized in the new conjugate direction. The resulting search pattern for the case of a quadratic 2-dimensional problem is illustrated in Fig. 6.4.

The major advantage of Powell's algorithm is that the Hessian need not be supplied, stored, or manipulated. Furthermore, by using a 1-D algorithm that is based on function evaluations for line searches, the need for the gradient can also be eliminated.

A difficulty associated with Powell's method is that linear dependence can sometimes arise, and the method may fail to generate a complete set of linearly independent directions that span E^n, even in the case of a convex quadratic problem. This may happen if the minimization of $f(\mathbf{x}_{k(j-1)} + \alpha\mathbf{d}_{kj})$ with respect to α in Step 2 of the

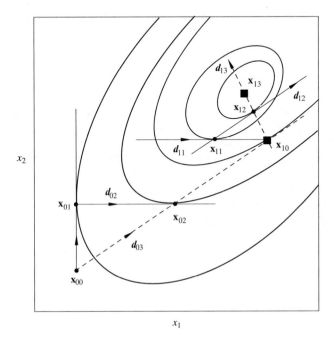

Fig. 6.4 Solution trajectory in Powell's algorithm for a quadratic problem

algorithm yields $\alpha_{kj} = 0$ for some j. In such a case, Step 3 will yield

$$\mathbf{d}_{k(n+1)} = \sum_{\substack{i=1 \\ i \neq j}}^{n} \alpha_{ki} \mathbf{d}_{ki}$$

that is, the new direction generated will not have a component in direction $\mathbf{d}_{kj}$, and since $\mathbf{d}_{kj}$ will eventually be dropped, a set of n directions will result that does not span E^n. This means that at least two directions will be linearly dependent and the algorithm will not converge to the solution.

The above problem can be avoided by discarding $\mathbf{d}_{kn}$ if linear dependence is detected in the hope that the use of the same set of directions in the next iteration will be successful in generating a new conjugate direction. This is likely to happen since the next iteration will start with a new point $\mathbf{x}_k$.

In principle, linear dependence would occur if at least one α_{ki} becomes zero, as was demonstrated above. Unfortunately, however, owing to the finite precision of computers, zero is an improbable value for α_{ki} and, therefore, checking the value of α_{ki} is an unreliable test for linear dependence. An alternative approach due to Powell is as follows.

If the direction vectors $\mathbf{d}_{ki}$ for $i = 1,\ 2,\ \dots,\ n$ are normalized such that

$$\mathbf{d}_{ki}^T \mathbf{H} \mathbf{d}_{ki} = 1 \qquad \text{for } i = 1,\ 2,\ \dots,\ n$$

then the determinant of matrix

$$\mathbf{D} = [\mathbf{d}_{k1}\ \mathbf{d}_{k2}\ \cdots\ \mathbf{d}_{kn}]$$

assumes a maximum value if and only if the directions $\mathbf{d}_{ki}$ belong to a conjugate set. Thus if a nonconjugate direction $\mathbf{d}_{1k}$ is dropped and conjugate direction $\mathbf{d}_{k(n+1)}$ is added to $\mathbf{D}$, the determinant of $\mathbf{D}$ will increase. On the other hand, if the addition of $\mathbf{d}_{k(n+1)}$ results in linear dependence in $\mathbf{D}$, the determinant of $\mathbf{D}$ will decrease. On the basis of these principles, Powell developed a modified algorithm in which a test is used to determine whether the new direction generated should or should not be used in the next iteration. The test also identifies which one of the n old directions should be replaced by the new direction so as to achieve the maximum increase in the determinant, and thus reduce the risk of linear dependence.

An alternative but very similar technique for the elimination of linear dependence in the set of directions was proposed by Zangwill [6]. This technique is more effective and more economical in terms of computation than Powell's modification and, therefore, it deserves to be considered in detail.

Zangwill's technique can be implemented by applying the following modifications to Powell's algorithm.

1. The initial directions in Step 1 are chosen to be the coordinate set of vectors of unit length such that

$$\mathbf{D}_0 = \begin{bmatrix} \mathbf{d}_{01} & \mathbf{d}_{02} & \cdots & \mathbf{d}_{0n} \end{bmatrix}$$

$$= \begin{bmatrix} 1 & 0 & \cdots & 0 \\ 0 & 1 & \cdots & 0 \\ \vdots & \vdots & & \vdots \\ 0 & 0 & \cdots & 1 \end{bmatrix}$$

and the determinant of $\mathbf{D}_0$, designated as Δ_0, is set to unity.

2. In Step 2, constants α_{ki} for $i = 1, 2, \ldots, n$ are determined as before, and the largest α_{ki} is then selected, i.e.,

$$\alpha_{km} = \max\{\alpha_{k1}, \alpha_{k2}, \ldots, \alpha_{kn}\}$$

3. In Step 3, a new direction is generated as before, and is then normalized to unit length so that

$$\mathbf{d}_{k(n+1)} = \frac{1}{\lambda_k}(\mathbf{x}_{kn} - \mathbf{x}_{k0})$$

where

$$\lambda_k = \|\mathbf{x}_{kn} - \mathbf{x}_{k0}\|_2$$

4. Step 4 is carried out as before. In Step 5, the new direction in item (3) is used to replace direction $\mathbf{d}_{km}$ provided that this substitution will maintain the determinant of

$$\mathbf{D}_k = \begin{bmatrix} \mathbf{d}_{k1} & \mathbf{d}_{k2} & \cdots & \mathbf{d}_{kn} \end{bmatrix}$$

finite and larger than a constant ε_1 in the range $0 < \varepsilon_1 \leq 1$, namely,

$$0 < \varepsilon_1 < \Delta_k = \det \mathbf{D}_k \leq 1$$

Otherwise, the most recent set of directions is used in the next iteration. Since

$$\Delta_k = \det[\mathbf{d}_{k1} \cdots \mathbf{d}_{k(m-1)} \mathbf{d}_{km} \mathbf{d}_{k(m+1)} \cdots \mathbf{d}_{kn}]$$

and

$$\mathbf{d}_{k(n+1)} = \frac{1}{\lambda_k} \sum_{i-1}^{n} \alpha_{ki} \mathbf{d}_{ki}$$

replacing $\mathbf{d}_{km}$ by $\mathbf{d}_{k(n+1)}$ yields

$$\Delta_k' = \frac{\alpha_{km}}{\lambda_k} \Delta_k$$

This result follows readily by noting that

(a) if a constant multiple of a column is added to another column, the determinant remains unchanged, and
(b) if a column is multiplied by a constant, the determinant is multiplied by the same constant.

From (a), the summation in Δ_k' can be eliminated and from (b) constant α_{km}/λ_k can be factored out. In this way, the effect of the substitution of $\mathbf{d}_{km}$ on the determinant of $\mathbf{D}_k$ is known. If

$$\frac{\alpha_{km}}{\lambda_k} \Delta_k > \varepsilon_1$$

we let

$$\mathbf{d}_{(k+1)m} = \mathbf{d}_{k(n+1)}$$

and

$$\mathbf{d}_{(k+1)i} = \mathbf{d}_{ki}$$

for $i = 1, 2, \ldots, m - 1, m + 1, \ldots, n$. Otherwise, we let

$$\mathbf{d}_{(k+1)i} = \mathbf{d}_{ki}$$

for $i = 1, 2, \ldots, n$. Simultaneously, the determinant Δ_k can be updated as

$$\delta_{k+1} = \begin{cases} \dfrac{\alpha_{km}}{\lambda_k} \Delta_k & \text{if } \dfrac{\alpha_{km}}{\lambda_k} \Delta_k > \varepsilon_1 \\ \Delta_k & \text{otherwise} \end{cases}$$

The net result of the above modifications is that the determinant of the direction matrix will always be finite and positive, which implies that the directions will always be linearly independent. The strategy in item (2) above of replacing the direction $\mathbf{d}_{ki}$ that yields the maximum α_{ki} ensures that the value of the determinant Δ_k is kept as large as possible so as to prevent linear dependence from arising in subsequent iterations.

The modified algorithm, which is often referred to as Zangwill's algorithm, can be shown to converge in the case of a convex quadratic problem. Its implementation is as follows:

Algorithm 6.5 Zangwill's algorithm
Step 1
Input $\mathbf{x}_{00}$ and initialize the tolerances ε and ε_1.
    Set
$$\mathbf{d}_{01} = \begin{bmatrix} 1 & 0 & \cdots & 0 \end{bmatrix}^T$$
$$\mathbf{d}_{02} = \begin{bmatrix} 0 & 1 & \cdots & 0 \end{bmatrix}^T$$
$$\vdots$$
$$\mathbf{d}_{0n} = \begin{bmatrix} 0 & 0 & \cdots & 1 \end{bmatrix}^T$$
Set $k = 0$, $\Delta_0 = 1$.
Step 2
For $i = 1$ to n do:
Find α_{ki}, the value of α that minimizes $f(\mathbf{x}_{k(i-1)} + \alpha \mathbf{d}_{ki})$.
Set $\mathbf{x}_{ki} = \mathbf{x}_{k(i-1)} + \alpha_{ki} \mathbf{d}_{ki}$.
Determine
$$\alpha_{km} = \max\{\alpha_{k1}, \alpha_{k2}, \ldots, \alpha_{kn}\}$$

Step 3
Generate a new direction
$$\mathbf{d}_{k(n+1)} = \mathbf{x}_{kn} - \mathbf{x}_{k0}$$
Find $\alpha_{k(n+1)}$, the value of α that minimizes $f(\mathbf{x}_{kn} + \alpha \mathbf{d}_{k(n+1)})$.
Set
$$\mathbf{x}_{k(n+1)} = \mathbf{x}_{kn} + \alpha_{k(n+1)} \mathbf{d}_{k(n+1)}$$
Calculate $f_{k(n+1)} = f(\mathbf{x}_{k(n+1)})$.
Calculate $\lambda_k = \|\mathbf{x}_{kn} - \mathbf{x}_{k0}\|_2$.

Step 4
If $\|\alpha_{k(n+1)} \mathbf{d}_{k(n+1)}\|_2 < \varepsilon$, output $\mathbf{x}^* = \mathbf{x}_{k(n+1)}$ and $f(\mathbf{x}^*) = f_{k(n+1)}$, and stop.
Step 5
If $\alpha_{km} \Delta_k / \lambda_k > \varepsilon_1$, then do:
    Set $\mathbf{d}_{(k+1)m} = \mathbf{d}_{k(n+1)}$ and $\mathbf{d}_{(k+1)i} = \mathbf{d}_{ki}$ for $i = 1, 2, \ldots, m-1$,
    $m+1, \ldots, n$.
    Set $\Delta_{k+1} = \dfrac{\alpha_{km}}{\lambda_k} \Delta_k$.
Otherwise, set
$\mathbf{d}_{(k+1)i} = \mathbf{d}_{ki}$ for $i = 1, 2, \ldots, n$, and $\Delta_{k+1} = \Delta_k$.
Set $\mathbf{x}_{(k+1)0} = \mathbf{x}_{k(n+1)}$, $k = k+1$, and repeat from Step 2.

6.8 Partan Method

In the early days of optimization, experimentation with two-variable functions re-
vealed the characteristic zigzag pattern in the solution trajectory in the steepest-
descent method. It was noted that in well-behaved functions, successive solution

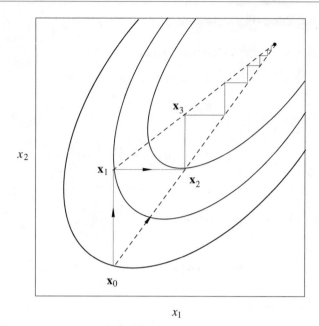

Fig. 6.5 Zigzag pattern of steepest-descent algorithm

points tend to coincide on two lines which intersect in the neighborhood of the minimizer, as depicted in Fig. 6.5. Therefore, an obvious strategy to attempt was to perform two steps of steepest descent followed by a search along the line connecting the initial point to the second solution point, as shown in Fig. 6.5. An iterative version of this approach was tried and found to converge to the solution. Indeed, for convex quadratic functions, convergence could be achieved in n iterations. The method has come to be known as the *parallel tangent method*, or *partan* for short, because of a special geometric property of the tangents to the contours in the case of quadratic functions.

The partan algorithm is illustrated in Fig. 6.6. An initial point $\mathbf{x}_0$ is assumed and two successive steepest-descent steps are taken to yield points $\mathbf{x}_1$ and $\mathbf{y}_1$. Then a line search is performed in the direction $\mathbf{y}_1 - \mathbf{x}_0$ to yield a point $\mathbf{x}_2$. This completes the first iteration. In the second iteration, a steepest-descent step is taken from point $\mathbf{x}_2$ to yield point $\mathbf{y}_2$, and a line search is performed along direction $\mathbf{y}_2 - \mathbf{x}_1$ to yield point $\mathbf{x}_3$, and so on. In effect, points $\mathbf{y}_1, \mathbf{y}_2, \ldots,$ in Fig. 6.6 are obtained by steepest-descent steps and points $\mathbf{x}_2, \mathbf{x}_3, \ldots$ are obtained by line searches along the directions $\mathbf{y}_2 - \mathbf{x}_1$, $\mathbf{y}_3 - \mathbf{x}_2, \ldots$.

In the case of a convex quadratic problem, the lines connecting $\mathbf{x}_1, \mathbf{x}_2, \ldots, \mathbf{x}_k$, which are not part of the algorithm, form a set of conjugate-gradient directions. This property can be demonstrated by assuming that $\mathbf{d}_0, \mathbf{d}_1, \ldots, \mathbf{d}_{k-1}$ form a set of conjugate-gradient directions and then showing that $\mathbf{d}_k$ is a conjugate-gradient direction with respect to $\mathbf{d}_0, \mathbf{d}_1, \ldots, \mathbf{d}_{k-1}$.

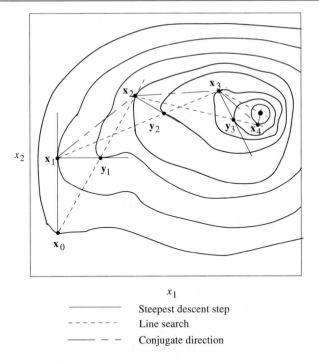

Steepest descent step

------------ Line search

------ — — Conjugate direction

Fig. 6.6 Solution trajectory for partan method for a nonquadratic problem

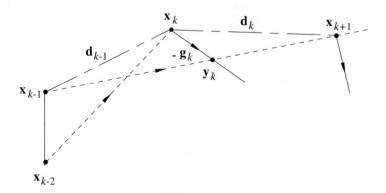

Fig. 6.7 Trajectory for kth iteration in partan method

Consider the steps illustrated in Fig. 6.7 and note that

$$\mathbf{g}_k^T \mathbf{d}_i = 0 \quad \text{for } 0 \leq i < k \tag{6.44}$$

on the basis of the above assumption and Theorem 6.3. From Eqs. (6.31) and (6.32), the gradient at point $\mathbf{x}_{k-1}$ can be expressed as

$$\mathbf{g}_{k-1} = \sum_{i=0}^{k-1} a_i \mathbf{d}_i$$

where a_i for $i = 0, 1, \ldots, k - 1$ are constants, and hence

$$\mathbf{g}_k^T \mathbf{g}_{k-1} = \mathbf{g}_k^T (\mathbf{b} + \mathbf{H}\mathbf{x}_{k-1}) = \sum_{i=0}^{k-1} a_i \mathbf{g}_k^T \mathbf{d}_i = 0 \tag{6.45}$$

or

$$\mathbf{g}_k^T \mathbf{b} = -\mathbf{g}_k^T \mathbf{H}\mathbf{x}_{k-1} \tag{6.46}$$

Since $\mathbf{y}_k$ is obtained by a steepest-descent step at point $\mathbf{x}_k$, we have

$$\mathbf{y}_k - \mathbf{x}_k = -\mathbf{g}_k$$

and

$$-\mathbf{g}(\mathbf{y}_k)^T \mathbf{g}_k = \mathbf{g}_k^T (\mathbf{b} + \mathbf{H}\mathbf{y}_k) = 0$$

or

$$\mathbf{g}_k^T \mathbf{b} = -\mathbf{g}_k^T \mathbf{H}\mathbf{y}_k \tag{6.47}$$

Hence Eqs. (6.46) and (6.47) yield

$$\mathbf{g}_k^T \mathbf{H}(\mathbf{y}_k - \mathbf{x}_{k-1}) = 0 \tag{6.48}$$

Since

$$\mathbf{y}_k - \mathbf{x}_{k-1} = \beta(\mathbf{x}_{k+1} - \mathbf{x}_{k-1})$$

where β is a constant, Eq. (6.48) can be expressed as

$$\mathbf{g}_k^T \mathbf{H}(\mathbf{x}_{k+1} - \mathbf{x}_{k-1}) = 0$$

or

$$\mathbf{g}_k^T \mathbf{H}\mathbf{x}_{k+1} = \mathbf{g}_k^T \mathbf{H}\mathbf{x}_{k-1} \tag{6.49}$$

We can now write

$$\mathbf{g}_k^T \mathbf{g}_{k+1} = \mathbf{g}_k^T (\mathbf{b} + \mathbf{H}\mathbf{x}_{k+1}) \tag{6.50}$$

and from Eqs. (6.45), (6.49), and (6.50), we have

$$\mathbf{g}_k^T \mathbf{g}_{k+1} = \mathbf{g}_k^T (\mathbf{b} + \mathbf{H}\mathbf{x}_{k-1})$$

$$= \mathbf{g}_k^T \mathbf{g}_{k-1} = 0 \tag{6.51}$$

Point $\mathbf{x}_{k+1}$ is obtained by performing a line search in direction $\mathbf{x}_{k+1} - \mathbf{y}_k$, and hence

$$\mathbf{g}_{k+1}^T (\mathbf{x}_{k+1} - \mathbf{y}_k) = 0 \tag{6.52}$$

From Fig. 6.7

$$\mathbf{x}_{k+1} = \mathbf{x}_k + \mathbf{d}_k \tag{6.53}$$

and

$$\mathbf{y}_k = \mathbf{x}_k - \alpha_k \mathbf{g}_k \tag{6.54}$$

where α_k is the value of α that minimizes $f(\mathbf{x}_k - \alpha \mathbf{g}_k)$. Thus Eqs. (6.52)–(6.54) yield

$$\mathbf{g}_{k+1}^T (\mathbf{d}_k + \alpha_k \mathbf{g}_k) = 0$$

or

$$\mathbf{g}_{k+1}^T \mathbf{d}_k + \alpha_k \mathbf{g}_k^T \mathbf{g}_{k+1} = 0 \tag{6.55}$$

Now from Eqs. (6.51) and (6.55)

$$\mathbf{g}_{k+1}^T \mathbf{d}_k = 0 \tag{6.56}$$

and on combining Eqs. (6.44) and (6.56), we obtain

$$\mathbf{g}_{k+1}^T \mathbf{d}_i = 0 \quad \text{for} \ \ 0 \le i < k + 1$$

that is, $\mathbf{x}_k$ satisfies Theorem 6.3.

6.9 Solution of Systems of Linear Equations

An important application of the conjugate-gradient method is to solve systems of linear equations. Consider the system of linear equations

$$\mathbf{Ax} = \mathbf{b} \tag{6.57}$$

where matrix $\mathbf{A} \in R^{n \times n}$ is symmetric and positive definite. Equation (6.57) is related to the convex quadratic problem

$$\text{minimize } f(\mathbf{x}) = \frac{1}{2}\mathbf{x}^T \mathbf{A} \mathbf{x} - \mathbf{x}^T \mathbf{b} \tag{6.58}$$

in the sense that the unique solution of Eq. (6.57) is the unique and global minimizer of $f(\mathbf{x})$ in Eq. (6.58). Therefore, a system of equations with the stated properties can be solved by minimizing a strictly convex quadratic function. An iterative procedure for the solution of the system of equations in Eq. (6.57) based on the conjugate-gradient method can be constructed as detailed below.

Let $\mathbf{x}_0$ be the initial point. The next point at iteration k can be expressed as

$$\mathbf{x}_{k+1} = \mathbf{x}_k + \alpha_k \mathbf{d}_k$$

where $\mathbf{d}_k$ is the current conjugate direction and α_k is the current step size. The *residue* of Eq. (6.57) at $\mathbf{x}_k$ can be defined as

$$\mathbf{r}_k = \mathbf{b} - \mathbf{Ax}_k \tag{6.59}$$

The residue $\mathbf{r}_k$ can be evaluated iteratively as

$$\mathbf{r}_{k+1} = \mathbf{b} - \mathbf{Ax}_{k+1} = \mathbf{b} - \mathbf{A}(\mathbf{x}_k + \alpha_k \mathbf{d}_k) = \mathbf{r}_k - \alpha_k \mathbf{A}\mathbf{d}_k$$

From Eq. (6.59), the gradient of $f(\mathbf{x})$ at $\mathbf{x}_k$ is related to residue $\mathbf{r}_k$ as

$$\mathbf{g}_k = \mathbf{Ax}_k - \mathbf{b} = -\mathbf{r}_k$$

Therefore, step size α_k in Step 3 of Algorithm 6.2 can be expressed as

$$\alpha_k = \frac{\mathbf{r}_k^T \mathbf{r}_k}{\mathbf{d}_k^T \mathbf{A}\mathbf{d}_k}$$

and, similarly, ratio β_k in Step 5 of Algorithm 6.2 can be expressed as

$$\beta_k = \frac{\mathbf{r}_{k+1}^T \mathbf{r}_{k+1}}{\mathbf{r}_k^T \mathbf{r}_k}$$

From Algorithm 6.2, it is known that the solution of the system of equations in Eq. (6.57) can be obtained in n iterations. However, there are cases where such a solution may not be sufficiently accurate due to numerical roundoff errors. A convenient way to terminate the procedure is to check whether the magnitude of residue $\mathbf{r}_k$ is less than a prescribed tolerance. An algorithm for solving the system of equations in Eq. (6.57) is as follows.

Algorithm 6.6 Conjugate-gradient algorithm for the solution of the system of linear equations in Eq. (6.57)
Step 1
Input $\mathbf{A}$, $\mathbf{b}$, $\mathbf{x}_0$ and initialize the termination tolerance ε.
Set $\mathbf{r}_0 = \mathbf{b} - \mathbf{A}\mathbf{x}_0$, $\mathbf{d}_0 = \mathbf{r}_0$, and $k = 0$.
Step 2
Compute

$$\alpha_k = \frac{\mathbf{r}_k^T \mathbf{r}_k}{\mathbf{d}_k^T \mathbf{A} \mathbf{d}_k}$$

and set

$$\mathbf{x}_{k+1} = \mathbf{x}_k + \alpha_k \mathbf{d}_k$$

Step 3
If $\|\mathbf{r}_k\|_2 < \varepsilon$, output $\mathbf{x}^* = \mathbf{x}_{k+1}$ and stop.
Step 4
Compute

$$\mathbf{r}_{k+1} = \mathbf{r}_k - \alpha_k \mathbf{A} \mathbf{d}_k$$

$$\beta_k = \frac{\mathbf{r}_{k+1}^T \mathbf{r}_{k+1}}{\mathbf{r}_k^T \mathbf{r}_k}$$

$$\mathbf{d}_{k+1} = \mathbf{r}_{k+1} + \beta_k \mathbf{d}_k$$

Set $k = k + 1$ and repeat from Step 2.

Example 6.3 Using Algorithm 6.6, solve the system of linear equations $\mathbf{A}\mathbf{x} = \mathbf{b}$ where

$$\mathbf{A} = \begin{bmatrix} 1 & \frac{1}{2} & \frac{1}{3} & \frac{1}{4} \\ \frac{1}{2} & \frac{1}{3} & \frac{1}{4} & \frac{1}{5} \\ \frac{1}{3} & \frac{1}{4} & \frac{1}{5} & \frac{1}{6} \\ \frac{1}{4} & \frac{1}{5} & \frac{1}{6} & \frac{1}{7} \end{bmatrix} \quad \text{and} \quad \mathbf{b} = \begin{bmatrix} 1 & 1 & 1 & 1 \end{bmatrix}^T$$

$\mathbf{A}$ is a positive-definite matrix known as the 4×4 *Hilbert* matrix [7].

Solution Hilbert matrices are known to have large condition numbers[1] and are, as a consequence, subject to numerical ill-conditioning. The condition number of an $n \times n$ Hilbert matrix is of order $(1 + \sqrt{2})^{4n}/\sqrt{n}$ and it grows very rapidly as n is increased. For example, the condition number of a 4×4 Hilbert matrix is 15,514.

---

[1]The condition number of a matrix is defined as the ratio of its largest to its smallest eigenvalue.

The exact inverse of a Hilbert matrix can be obtained by using the closed-form formula

$$\left(A^{-1}\right)_{i,j} = (-1)^{i+j}(i+j-1)\binom{n+i-1}{n-j}\binom{n+j-1}{n-i}\binom{i+j-2}{i-1}^2$$

In view of the availability of the above formula, Hilbert matrices are often used as test matrices for evaluating the robustness of numerical algorithms for the solution of systems of linear equations.

The exact inverse of the 4×4 Hilbert matrix can be obtained as

$$A^{-1} = \begin{bmatrix} 16 & -120 & 240 & -140 \\ -120 & 1200 & -2700 & 1680 \\ 240 & -2700 & 6480 & -4200 \\ -140 & 1680 & -4200 & 2800 \end{bmatrix}$$

Therefore, the exact solution of the system of linear equations $Ax = b$ is given by

$$x^{**} = A^{-1}b = \begin{bmatrix} -4 \\ 60 \\ -180 \\ 140 \end{bmatrix}$$

With the termination tolerance for the equation residue set to $\varepsilon = 10^{-6}$ and initial point $x_0 = 0$, it took Algorithm 6.6 four iterations to converge to the solution

$$x^* = \begin{bmatrix} -3.99999999837 \\ 60.00000000085 \\ -179.99999999932 \\ 140.00000000043 \end{bmatrix}$$

With the termination tolerance reduced to $\varepsilon = 10^{-9}$ and $x_0 = 0$, the algorithm converged after five iterations to the solution

$$x^* = \begin{bmatrix} -3.99999999996 \\ 59.99999999995 \\ -179.99999999997 \\ 139.99999999992 \end{bmatrix}$$

Evidently, the second solution is much closer to the exact solution than the first solution.   ■

Problems

6.1 Use the conjugate-gradient method to solve the optimization problem

$$\text{minimize } f(x) = \tfrac{1}{2}x^T Q x + b^T x$$

where $\mathbf{Q}$ is given by

$$\mathbf{Q} = \begin{bmatrix} \mathbf{Q}_1 & \mathbf{Q}_2 & \mathbf{Q}_3 & \mathbf{Q}_4 \\ \mathbf{Q}_2 & \mathbf{Q}_1 & \mathbf{Q}_2 & \mathbf{Q}_3 \\ \mathbf{Q}_3 & \mathbf{Q}_2 & \mathbf{Q}_1 & \mathbf{Q}_2 \\ \mathbf{Q}_4 & \mathbf{Q}_3 & \mathbf{Q}_2 & \mathbf{Q}_1 \end{bmatrix} \quad \text{with } \mathbf{Q}_1 = \begin{bmatrix} 12 & 8 & 7 & 6 \\ 8 & 12 & 8 & 7 \\ 7 & 8 & 12 & 8 \\ 6 & 7 & 8 & 12 \end{bmatrix}$$

$$\mathbf{Q}_2 = \begin{bmatrix} 3 & 2 & 1 & 0 \\ 2 & 3 & 2 & 1 \\ 1 & 2 & 3 & 2 \\ 0 & 1 & 2 & 3 \end{bmatrix}, \quad \mathbf{Q}_3 = \begin{bmatrix} 2 & 1 & 0 & 0 \\ 1 & 2 & 1 & 0 \\ 0 & 1 & 2 & 1 \\ 0 & 0 & 1 & 2 \end{bmatrix}, \quad \mathbf{Q}_4 = \mathbf{I}_4$$

and $\mathbf{b} = -[1\,1\,1\,1\,0\,0\,0\,0\,1\,1\,1\,1\,0\,0\,0\,0]^T$.

6.2 Use the Fletcher–Reeves algorithm to find the minimizer of the Rosenbrock function [8]
$$f(\mathbf{x}) = 100(x_2 - x_1^2)^2 + (1 - x_1)^2$$
Use $\varepsilon = 10^{-6}$ and try three initial points $\mathbf{x}_0 = [-2\ 2]^T$, $\mathbf{x}_0 = [2\ -2]^T$, and $\mathbf{x}_0 = [-2\ -2]^T$ and observe the results.

6.3 Solve Problem 5.4 by applying the conjugate-gradient algorithm (Algorithm 6.2).

(a) With $\varepsilon = 3 \times 10^{-7}$ and $\mathbf{x}_0 = [1\ 1]^T$, perform two iterations by following the steps described in Algorithm 6.2.
(b) Compare the results of the first iteration obtained by using the conjugate-gradient algorithm with those obtained by using the steepest-descent method.
(c) Compare the results of the second iteration obtained by using the conjugate-gradient algorithm with those obtained by using the steepest-descent method.

6.4 Solve Problem 5.6 by applying the Fletcher–Reeves algorithm.

(a) Examine the solution obtained and the amount of computation required.
(b) Compare the results obtained in part (a) with those of Problems 5.6, 5.17, and 5.22.

6.5 Solve Problem 5.8 by applying the Fletcher–Reeves algorithm.

(a) Examine the solution obtained and the amount of computation required.
(b) Compare the results obtained in part (a) with those of Problems 5.8, 5.18, and 5.23.

6.6 Solve Problem 5.10 by applying the Fletcher–Reeves algorithm.

   (*a*) Examine the solution obtained and the amount of computation required.
   (*b*) Compare the results obtained in part (*a*) with those of Problems 5.10 and
        5.19.

6.7 Solve Problem 5.4 by applying Powell's algorithm (Algorithm 6.4) and compare
     the results with those obtained in Problems 5.4 and 6.3.

6.8 Solve Problem 5.6 by applying Powell's algorithm and compare the results with
     those obtained in Problems 5.6, 5.17, 5.22 and 6.4.

6.9 Solve Problem 5.8 by applying Powell's algorithm and compare the results with
     those obtained in Problems 5.8, 5.18, 5.23, and 6.5.

6.10 Solve Problem 5.4 by applying Zangwill's algorithm and compare the results
      with those obtained in Problems 5.4, 6.3, and 6.7.

6.11 The conjugate-gradient method can also be used to solve systems of linear equa-
      tions whose coefficient matrix is not symmetric, not positive definite, and also
      not square [9]. Consider a system of linear equations $\mathbf{Ax} = \mathbf{b}$ where $\mathbf{A} \in R^{m \times n}$,
      $\mathbf{x} \in R^{n \times 1}$, and $\mathbf{b} \in R^{m \times 1}$.

   (*a*) Show that the solution of the so-called normal equation

   $$\mathbf{A}^T \mathbf{Ax} = \mathbf{A}^T \mathbf{b}$$

   always exists regardless of the values of m and n and the rank of $\mathbf{A}$ (Hint:
   Use the singular-value decomposition of $\mathbf{A}$).
   (*b*) Show that a minimizer of the least-squares problem

   $$\text{minimize} \ \|\mathbf{Ax} - \mathbf{b}\|_2^2$$

   is a solution of the normal equation in part (a), and vice versa.
   (*c*) Based on the results of parts (a) and (b), generalize Algorithm 6.6 to enable
   it to solve a linear equation of the type $\mathbf{Ax} = \mathbf{b}$ where $\mathbf{A}^T \mathbf{A}$ is nonsingular.

6.12 Solve the system of linear equations $\mathbf{Ax} = \mathbf{b}$ where

$$\mathbf{A} = \begin{bmatrix} 1 & 1 & 1 & 1 \\ 1 & 2 & 3 & 4 \\ 1 & 4 & 9 & 16 \\ 1 & 8 & 27 & 64 \end{bmatrix} \quad \text{and} \quad \mathbf{b} = \begin{bmatrix} 0 \\ -1 \\ 3 \\ 35 \end{bmatrix}$$

by applying Algorithm 6.6. Note that matrix $\mathbf{A}$ is a Vandermonde matrix [10] of
the form

$$
\mathbf{A} = \begin{bmatrix} v_1^0 & v_2^0 & v_3^0 & v_4^0 \\ v_1^1 & v_2^1 & v_3^1 & v_4^1 \\ v_1^2 & v_2^2 & v_3^2 & v_4^2 \\ v_1^3 & v_2^3 & v_3^3 & v_4^3 \end{bmatrix}
$$

with $[v_1 \; v_2 \; v_3 \; v_4] = [1 \; 2 \; 3 \; 4]$. Also note that matrix $\mathbf{A}$ is not a symmetric matrix. (Hint: Use the results obtained in the solution of Problem 6.11.)

References

1. R. A. Horn and C. R. Johnson, *Matrix Analysis*, 2nd ed. New York: Cambridge University Press, 2013.
2. M. R. Hestenes and E. L. Stiefel, "Methods of conjugate gradients for solving linear systems," *J. Res. Natl. Bureau Standards*, vol. 49, pp. 409–436, 1952.
3. R. Fletcher and C. M. Reeves, "Function minimization by conjugate gradients," *Computer J.*, vol. 7, pp. 149–154, 1964.
4. D. M. Himmelblau, *Applied Nonlinear Programming*. New York: McGraw-Hill, 1972.
5. M. J. D. Powell, "An efficient method for finding the minimum of a function of several variables without calculating derivatives," *Computer J.*, vol. 7, pp. 155–162, 1964.
6. W. I. Zangwill, "Minimizing a function without calculating derivatives," *Computer J.*, vol. 10, pp. 293–296, 1968.
7. M.-D. Choi, "Tricks or treats with the Hilbert matrix," *American Mathematical Monthly*, vol. 90, no. 5, pp. 301–312, 1983.
8. H. H. Rosenbrock, "An automatic method for finding the greatest or least value of a function," *Computer J.*, vol. 3, pp. 175–184, 1960.
9. J. R. Shewchuk, "An introduction to the conjugate gradient method without the agonizing pain," Research Report, School of Computer Science, Carnegie Mellon University, Pittsburgh, PA, 1994.
10. G. H. Golub and C. F. Van Loan, *Matrix Computations*, 3rd ed. Baltimore, MD: Johns Hopkins University Press, 1996.

Quasi-Newton Methods

7

7.1 Introduction

In Chap. 6, multidimensional optimization methods were considered in which the search for the minimizer is carried out by using a set of conjugate directions. An important feature of some of these methods (e.g., the Fletcher–Reeves and Powell's methods) is that explicit expressions for the second derivatives of $f(\mathbf{x})$ are not required. Another class of methods that do not require explicit expressions for the second derivatives is the class of *quasi-Newton methods*. These are sometimes referred to as *variable metric methods*.

As the name implies, the foundation of these methods is the classical Newton method described in Sect. 5.3. The basic principle in quasi-Newton methods is that the direction of search is based on an $n \times n$ direction matrix $\mathbf{S}$ which serves the same purpose as the inverse Hessian in the Newton method. This matrix is generated from available data and is contrived to be an approximation of $\mathbf{H}^{-1}$. Furthermore, as the number of iterations is increased, $\mathbf{S}$ becomes progressively a more accurate representation of $\mathbf{H}^{-1}$, and for convex quadratic objective functions it becomes identical to $\mathbf{H}^{-1}$ in $n + 1$ iterations.

Quasi-Newton methods, like most other methods, are developed for the convex quadratic problem and are then extended to the general problem. They rank among the most efficient methods available and are, therefore, used very extensively in numerous applications.

Several distinct quasi-Newton methods have evolved in recent years. In this chapter, we discuss in detail the four most important methods of this class, which are:

1. Rank-one method
2. Davidon–Fletcher–Powell method
3. Broyden–Fletcher–Goldfarb–Shanno method
4. Hoshino method
5. The Broyden family of methods

© Springer Science+Business Media, LLC, part of Springer Nature 2021
A. Antoniou and W.-S. Lu, *Practical Optimization*, Texts in Computer Science,
https://doi.org/10.1007/978-1-0716-0843-2_7

We then discuss briefly a number of alternative approaches and describe two interesting generalizations, one due to Broyden and the other due to Huang.

7.2 The Basic Quasi-Newton Approach

In the methods of Chap. 5, the point generated in the kth iteration is given by

$$\mathbf{x}_{k+1} = \mathbf{x}_k - \alpha_k \mathbf{S}_k \mathbf{g}_k \tag{7.1}$$

where

$$\mathbf{S}_k = \begin{cases} \mathbf{I}_n & \text{for the steepest-descent method} \\[2mm] \mathbf{H}_k^{-1} & \text{for the Newton method} \end{cases}$$

Let us examine the possibility of using some arbitrary $n \times n$ positive definite matrix $\mathbf{S}_k$ for the solution of the quadratic problem

$$\text{minimize } f(\mathbf{x}) = a + \mathbf{b}^T \mathbf{x} + \tfrac{1}{2} \mathbf{x}^T \mathbf{H} \mathbf{x}$$

By differentiating $f(\mathbf{x}_k - \alpha \mathbf{S}_k \mathbf{g}_k)$ with respect to α and then setting the result to zero, the value of α that minimizes $f(\mathbf{x}_k - \alpha \mathbf{S}_k \mathbf{g}_k)$ can be deduced as

$$\alpha_k = \frac{\mathbf{g}_k^T \mathbf{S}_k \mathbf{g}_k}{\mathbf{g}_k^T \mathbf{S}_k \mathbf{H} \mathbf{S}_k \mathbf{g}_k} \tag{7.2}$$

where

$$\mathbf{g}_k = \mathbf{b} + \mathbf{H} \mathbf{x}_k$$

is the gradient of $f(\mathbf{x})$ at $\mathbf{x} = \mathbf{x}_k$.

It can be shown that

$$f(\mathbf{x}_{k+1}) - f(\mathbf{x}^*) \le \left(\frac{1-r}{1+r} \right)^2 [f(\mathbf{x}_k) - f(\mathbf{x}^*)]$$

where r is the ratio of the smallest to the largest eigenvalue of $\mathbf{S}_k \mathbf{H}$ (see [1, Chap. 8] for proof). In effect, an algorithm based on Eqs. (7.1) and (7.2) would converge linearly with a convergence ratio

$$\beta = \left(\frac{1-r}{1+r} \right)^2$$

for any positive definite $\mathbf{S}_k$ (see Sect. 3.7). Convergence is fastest if $r = 1$, that is, if the eigenvalues of $\mathbf{S}_k \mathbf{H}$ are all equal. This means that the best results can be achieved by choosing

$$\mathbf{S}_k \mathbf{H} = \mathbf{I}_n$$

or

$$\mathbf{S}_k = \mathbf{H}^{-1}$$

Similarly, for the general optimization problem, we should choose some positive definite $\mathbf{S}_k$ which is equal to or, at least, approximately equal to $\mathbf{H}_k^{-1}$.

Quasi-Newton methods are methods that are motivated by the preceding observation. The direction of search is based on a positive definite matrix $\mathbf{S}_k$ which is generated from available data, and which is contrived to be an approximation for $\mathbf{H}_k^{-1}$. Several approximations are possible for $\mathbf{H}_k^{-1}$ and, consequently, a number of different quasi-Newton methods can be developed.

7.3   Generation of Matrix $\mathbf{S}_k$

Let $f(\mathbf{x}) \in C^2$ be a function in E^n and assume that the gradients of $f(\mathbf{x})$ at points $\mathbf{x}_k$ and $\mathbf{x}_{k+1}$ are designated as $\mathbf{g}_k$ and $\mathbf{g}_{k+1}$, respectively. If

$$\mathbf{x}_{k+1} = \mathbf{x}_k + \boldsymbol{\delta}_k \tag{7.3}$$

then the Taylor series gives the elements of $\mathbf{g}_{k+1}$ as

$$g_{(k+1)m} = g_{km} + \sum_{i=1}^{n} \frac{\partial g_{km}}{\partial x_{ki}} \delta_{ki} + \frac{1}{2} \sum_{i=1}^{n} \sum_{j=1}^{n} \frac{\partial^2 g_{km}}{\partial x_{ki} \partial x_{kj}} \delta_{ki} \delta_{kj} + \cdots$$

for $m = 1, 2, \ldots, n$. Now if $f(\mathbf{x})$ is quadratic, the second derivatives of $f(\mathbf{x})$ are constant and, in turn, the second derivatives of g_{km} are zero. Thus

$$g_{(k+1)m} = g_{km} + \sum_{i=1}^{n} \frac{\partial g_{km}}{\partial x_{ki}} \delta_{ki}$$

and since

$$g_{km} = \frac{\partial f_k}{\partial x_{km}}$$

we have

$$g_{(k+1)m} = g_{km} + \sum_{i=1}^{n} \frac{\partial^2 f_k}{\partial x_{ki} \partial x_{km}} \delta_{ki}$$

for $m = 1, 2, \ldots, n$. Therefore, $\mathbf{g}_{k+1}$ is given by

$$\mathbf{g}_{k+1} = \mathbf{g}_k + \mathbf{H} \boldsymbol{\delta}_k$$

where $\mathbf{H}$ is the Hessian of $f(\mathbf{x})$. Alternatively, we can write

$$\boldsymbol{\gamma}_k = \mathbf{H} \boldsymbol{\delta}_k \tag{7.4}$$

where

$$\boldsymbol{\delta}_k = \mathbf{x}_{k+1} - \mathbf{x}_k \tag{7.5}$$

$$\boldsymbol{\gamma}_k = \mathbf{g}_{k+1} - \mathbf{g}_k \tag{7.6}$$

The above analysis has shown that if the gradient of $f(\mathbf{x})$ is known at two points $\mathbf{x}_k$ and $\mathbf{x}_{k+1}$, a relation can be deduced that provides a certain amount of information about $\mathbf{H}$. Since there are n^2 unknowns in $\mathbf{H}$ (or $n(n+1)/2$ unknowns if $\mathbf{H}$ is assumed to be a real symmetric matrix) and Eq. (7.4) provides only n equations, $\mathbf{H}$ cannot be determined uniquely. However, if the gradient is evaluated sequentially at $n + 1$ points, say, $\mathbf{x}_0$, $\mathbf{x}_1$, ..., $\mathbf{x}_n$ such that the changes in $\mathbf{x}$, namely,

$$\delta_0 = \mathbf{x}_1 - \mathbf{x}_0$$
$$\delta_1 = \mathbf{x}_2 - \mathbf{x}_1$$
$$\vdots$$
$$\delta_{n-1} = \mathbf{x}_n - \mathbf{x}_{n-1}$$

form a set of linearly independent vectors, then sufficient information is obtained to determine $\mathbf{H}$ uniquely. To demonstrate this fact, n equations of the type given by Eq. (7.4) can be re-arranged as

$$[\boldsymbol{\gamma}_0 \; \boldsymbol{\gamma}_1 \; \cdots \; \boldsymbol{\gamma}_{n-1}] = \mathbf{H}[\delta_0 \; \delta_1 \; \cdots \; \delta_{n-1}] \qquad (7.7)$$

and, therefore,

$$\mathbf{H} = [\boldsymbol{\gamma}_0 \; \boldsymbol{\gamma}_1 \; \cdots \; \boldsymbol{\gamma}_{n-1}][\delta_0 \; \delta_1 \; \cdots \; \delta_{n-1}]^{-1}$$

The solution exists if δ_0, δ_1, ..., δ_{n-1} form a set of linearly independent vectors.

The above principles can be used to construct the following algorithm:

Algorithm 7.1 Alternative Newton algorithm
Step 1
Input $\mathbf{x}_{00}$ and initialize the tolerance ε.
Set $k = 0$.
Input a set of linearly independent vectors δ_0, δ_1, ..., δ_{n-1}.
Step 2
Compute $\mathbf{g}_{00}$.
Step 3
For $i = 0$ to $n - 1$ do:
   Set $\mathbf{x}_{k(i+1)} = \mathbf{x}_{ki} + \delta_i$.
   Compute $\mathbf{g}_{k(i+1)}$.
   Set $\boldsymbol{\gamma}_{ki} = \mathbf{g}_{k(i+1)} - \mathbf{g}_{ki}$.
Step 4
Compute $\mathbf{H} = [\boldsymbol{\gamma}_0 \; \boldsymbol{\gamma}_1 \; \cdots \; \boldsymbol{\gamma}_{n-1}][\delta_0 \; \delta_1 \; \cdots \; \delta_{n-1}]^{-1}$.
Compute $\mathbf{S}_k = \mathbf{H}_k^{-1}$.
Step 5
Set $\mathbf{d}_k = -\mathbf{S}_k \mathbf{g}_{k0}$.
Find α_k, the value of α that minimizes $f(\mathbf{x}_{k0} + \alpha \mathbf{d}_k)$.
Set $\mathbf{x}_{(k+1)0} = \mathbf{x}_{k0} + \alpha_k \mathbf{d}_k$.
Step 6
If $\|\alpha_k \mathbf{d}_k\|_2 < \varepsilon$, output $\mathbf{x}_k^* = \mathbf{x}_{(k+1)0}$ and $f(\mathbf{x}^*) = f(\mathbf{x}_{(k+1)0})$, and stop.
Step 7
Set $k = k + 1$ and repeat from Step 3.

The above algorithm is essentially an implementation of the Newton method except that a mechanism is incorporated for the generation of $\mathbf{H}^{-1}$ using computed data. For a convex quadratic problem, the algorithm will yield the solution in one iteration and it will thus be quite effective. For a nonquadratic problem, however, the algorithm has the same disadvantages as any other algorithm based on the Newton method (e.g., Algorithm 5.3). First, matrix inversion is required, which is undesirable; second, matrix H_k must be checked for positive definiteness and rendered positive definite, if necessary, in every iteration.

A strategy that leads to the elimination of matrix inversion is as follows. We assume that a positive definite real symmetric matrix S_k is available, which is an approximation of H^{-1}, and compute a quasi-Newton direction

$$d_k = -S_k g_k \qquad (7.8)$$

We then find α_k, the value of α that minimizes $f(x_k + \alpha d_k)$, as in the Newton method. For a convex quadratic problem, Eq. (7.2) gives

$$\alpha_k = \frac{g_k^T S_k g_k}{(S_k g_k)^T H(S_k g_k)} \qquad (7.9)$$

where S_k and H are positive definite. Evidently, α_k is greater than zero provided that x_k is not the solution point x^*. We then determine a change in x as

$$\delta_k = \alpha_k d_k \qquad (7.10)$$

and deduce a new point x_{k+1} using Eq. (7.3). By computing the gradient at points x_k and x_{k+1}, the change in the gradient, γ_k, can be determined using Eq. (7.6). We then apply a correction to S_k and generate

$$S_{k+1} = S_k + C_k \qquad (7.11)$$

where C_k is an $n \times n$ *correction matrix* which can be computed from available data. On applying the above procedure iteratively starting with an initial point x_0 and an initial positive definite matrix S_0, say, $S_0 = I_n$, the sequences $\delta_0, \delta_1, \ldots, \delta_k$, $\gamma_0, \gamma_1, \ldots, \gamma_k$, and $S_1, S_2, \ldots, S_{k+1}$ can be generated. If

$$S_{k+1}\gamma_i = \delta_i \qquad \text{for } 0 \le i \le k \qquad (7.12)$$

then for $k = n - 1$, we can write

$$S[\gamma_0 \; \gamma_1 \; \cdots \; \gamma_{n-1}] = [\delta_0 \; \delta_1 \; \cdots \; \delta_{n-1}]$$

or

$$S = [\delta_0 \; \delta_1 \; \cdots \; \delta_{n-1}][\gamma_0 \; \gamma_1 \; \cdots \; \gamma_{n-1}]^{-1} \qquad (7.13)$$

and from Eqs. (7.7) and (7.13), we have

$$S_n = H^{-1}$$

Now if $k = n$, Eqs. (7.8)–(7.10) yield

$$d_n = -H^{-1}g_n$$
$$\alpha_n = 1$$
$$\delta_n = -H^{-1}g_n$$

respectively, and, therefore, from Eq. (7.3)

$$\mathbf{x}_{n+1} = \mathbf{x}_n - \mathbf{H}^{-1}\mathbf{g}_n = \mathbf{x}^*$$

as in the Newton method.

The above procedure leads to a family of quasi-Newton algorithms which have the fundamental property that they terminate in $n + 1$ iterations ($k = 0, 1, \ldots, n$) in the case of a convex quadratic problem. The various algorithms of this class differ from one another in the formula used for the derivation of the correction matrix $\mathbf{C}_n$.

In any derivation of $\mathbf{C}_n$, $\mathbf{S}_{k+1}$ must satisfy Eq. (7.12) and the following properties are highly desirable:

1. Vectors $\boldsymbol{\delta}_0, \boldsymbol{\delta}_1, \ldots, \boldsymbol{\delta}_k$ should form a set of conjugate directions (see Chap. 6).
2. A positive definite $\mathbf{S}_k$ should give rise to a positive definite $\mathbf{S}_{k+1}$.

The first property will ensure that the excellent properties of conjugate-direction methods apply to the quasi-Newton method as well. The second property will ensure that $\mathbf{d}_k$ is a descent direction in every iteration, i.e., for $k = 0, 1, \ldots$. To demonstrate this fact, consider the point $\mathbf{x}_k + \boldsymbol{\delta}_k$, and let

$$\boldsymbol{\delta}_k = \alpha \mathbf{d}_k$$

where

$$\mathbf{d}_k = -\mathbf{S}_k \mathbf{g}_k$$

For $\alpha > 0$, the Taylor series in Eq. (2.4h) gives

$$f(\mathbf{x}_k + \boldsymbol{\delta}_k) = f(\mathbf{x}_k) + \mathbf{g}_k^T \boldsymbol{\delta}_k + \tfrac{1}{2}\boldsymbol{\delta}_k^T \mathbf{H}(\mathbf{x}_k + c\boldsymbol{\delta}_k)\boldsymbol{\delta}_k$$

where c is a constant in the range $0 \le c < 1$. On eliminating $\boldsymbol{\delta}_k$, we obtain

$$\begin{aligned} f(\mathbf{x}_k + \boldsymbol{\delta}_k) &= f(\mathbf{x}_k) - \alpha \mathbf{g}_k^T \mathbf{S}_k \mathbf{g}_k + o(\alpha \|\mathbf{d}_k\|_2) \\ &= f(\mathbf{x}_k) - [\alpha \mathbf{g}_k^T \mathbf{S}_k \mathbf{g}_k - o(\alpha \|\mathbf{d}_k\|_2)] \end{aligned}$$

where $o(\alpha \|\mathbf{d}_k\|_2)$ is the remainder, which approaches zero faster than $\alpha \|\mathbf{d}_k\|_2$. Now if $\mathbf{S}_k$ is positive definite, then for a sufficiently small $\alpha > 0$, we have

$$\alpha \mathbf{g}_k^T \mathbf{S}_k \mathbf{g}_k - o(\alpha \|\mathbf{d}_k\|_2) > 0$$

since $\alpha > 0$, $\mathbf{g}_k^T \mathbf{S}_k \mathbf{g}_k > 0$, and $o(\alpha \|\mathbf{d}_k\|_2) \to 0$. Therefore,

$$f(\mathbf{x}_k + \boldsymbol{\delta}_k) < f(\mathbf{x}_k) \tag{7.14}$$

that is, *if $\mathbf{S}_k$ is positive definite, then $\mathbf{d}_k$ is a descent direction.*

The importance of property (2) should, at this point, be evident. A positive definite $\mathbf{S}_0$ will give a positive definite $\mathbf{S}_1$ which will give a positive definite $\mathbf{S}_2$, and so on. Consequently, directions $\mathbf{d}_0, \mathbf{d}_1, \mathbf{d}_2, \ldots$ will all be descent directions, and this will assure the convergence of the algorithm.

7.4 Rank-One Method

The rank-one method owes its name to the fact that correction matrix $\mathbf{C}_k$ in Eq. (7.11) has a rank of unity. This correction was proposed independently by Broyden [2], Davidon [3], Fiacco and McCormick [4], Murtagh and Sargent [5], and Wolfe [6]. The derivation of the rank-one formula is as follows.

Assume that

$$\mathbf{S}_{k+1}\boldsymbol{\gamma}_k = \boldsymbol{\delta}_k \tag{7.15}$$

and let

$$\mathbf{S}_{k+1} = \mathbf{S}_k + \beta_k \boldsymbol{\xi}_k \boldsymbol{\xi}_k^T \tag{7.16}$$

where $\boldsymbol{\xi}_k$ is a column vector and β_k is a constant. The correction matrix $\beta_k \boldsymbol{\xi}_k \boldsymbol{\xi}_k^T$ is symmetric and has a rank of unity as can be demonstrated (see Problem 7.1). From Eqs. (7.15) and (7.16)

$$\boldsymbol{\delta}_k = \mathbf{S}_k \boldsymbol{\gamma}_k + \beta_k \boldsymbol{\xi}_k \boldsymbol{\xi}_k^T \boldsymbol{\gamma}_k \tag{7.17}$$

and hence

$$\begin{aligned}
\boldsymbol{\gamma}_k^T(\boldsymbol{\delta}_k - \mathbf{S}_k \boldsymbol{\gamma}_k) &= \beta_k \boldsymbol{\gamma}_k^T \boldsymbol{\xi}_k \boldsymbol{\xi}_k^T \boldsymbol{\gamma}_k \\
&= \beta_k (\boldsymbol{\xi}_k^T \boldsymbol{\gamma}_k)^2
\end{aligned} \tag{7.18}$$

Alternatively, from Eq. (7.17)

$$\begin{aligned}
(\boldsymbol{\delta}_k - \mathbf{S}_k \boldsymbol{\gamma}_k) &= \beta_k \boldsymbol{\xi}_k \boldsymbol{\xi}_k^T \boldsymbol{\gamma}_k = \beta_k (\boldsymbol{\xi}_k^T \boldsymbol{\gamma}_k) \boldsymbol{\xi}_k \\
(\boldsymbol{\delta}_k - \mathbf{S}_k \boldsymbol{\gamma}_k)^T &= \beta_k \boldsymbol{\gamma}_k^T \boldsymbol{\xi}_k \boldsymbol{\xi}_k^T = \beta_k (\boldsymbol{\xi}_k^T \boldsymbol{\gamma}_k) \boldsymbol{\xi}_k^T
\end{aligned}$$

since $\boldsymbol{\xi}_k^T \boldsymbol{\gamma}_k$ is a scalar. Hence

$$(\boldsymbol{\delta}_k - \mathbf{S}_k \boldsymbol{\gamma}_k)(\boldsymbol{\delta}_k - \mathbf{S}_k \boldsymbol{\gamma}_k)^T = \beta_k (\boldsymbol{\xi}_k^T \boldsymbol{\gamma}_k)^2 \beta_k \boldsymbol{\xi}_k \boldsymbol{\xi}_k^T \tag{7.19}$$

and from Eqs. (7.18) and (7.19), we have

$$\begin{aligned}
\beta_k \boldsymbol{\xi}_k \boldsymbol{\xi}_k^T &= \frac{(\boldsymbol{\delta}_k - \mathbf{S}_k \boldsymbol{\gamma}_k)(\boldsymbol{\delta}_k - \mathbf{S}_k \boldsymbol{\gamma}_k)^T}{\beta_k (\boldsymbol{\xi}_k^T \boldsymbol{\gamma}_k)^2} \\
&= \frac{(\boldsymbol{\delta}_k - \mathbf{S}_k \boldsymbol{\gamma}_k)(\boldsymbol{\delta}_k - \mathbf{S}_k \boldsymbol{\gamma}_k)^T}{\boldsymbol{\gamma}_k^T(\boldsymbol{\delta}_k - \mathbf{S}_k \boldsymbol{\gamma}_k)}
\end{aligned}$$

With the correction matrix known, $\mathbf{S}_{k+1}$ can be deduced from Eq. (7.16) as

$$\mathbf{S}_{k+1} = \mathbf{S}_k + \frac{(\boldsymbol{\delta}_k - \mathbf{S}_k \boldsymbol{\gamma}_k)(\boldsymbol{\delta}_k - \mathbf{S}_k \boldsymbol{\gamma}_k)^T}{\boldsymbol{\gamma}_k^T(\boldsymbol{\delta}_k - \mathbf{S}_k \boldsymbol{\gamma}_k)} \tag{7.20}$$

For a convex quadratic problem, this formula will generate $\mathbf{H}^{-1}$ on iteration $n-1$ provided that Eq. (7.12) holds. This indeed is the case as will be demonstrated by the following theorem.

Theorem 7.1 *Generation of inverse Hessian*   *If* $\mathbf{H}$ *is the Hessian of a convex quadratic problem and*

$$\boldsymbol{\gamma}_i = \mathbf{H}\boldsymbol{\delta}_i \quad \text{for } 0 \le i \le k \tag{7.21}$$

where $\boldsymbol{\delta}_i$ *for* $i = 0, \ldots, k$ *are given linearly independent vectors, then for any initial symmetric matrix* $\mathbf{S}_0$

$$\boldsymbol{\delta}_i = \mathbf{S}_{k+1}\boldsymbol{\gamma}_i \quad \text{for } 0 \le i \le k \tag{7.22}$$

where

$$\mathbf{S}_{i+1} = \mathbf{S}_i + \frac{(\boldsymbol{\delta}_i - \mathbf{S}_i\boldsymbol{\gamma}_i)(\boldsymbol{\delta}_i - \mathbf{S}_i\boldsymbol{\gamma}_i)^T}{\boldsymbol{\gamma}_i^T(\boldsymbol{\delta}_i - \mathbf{S}_i\boldsymbol{\gamma}_i)} \tag{7.23}$$

Proof We assume that

$$\boldsymbol{\delta}_i = \mathbf{S}_k\boldsymbol{\gamma}_i \quad \text{for } 0 \le i \le k - 1 \tag{7.24}$$

and show that

$$\boldsymbol{\delta}_i = \mathbf{S}_{k+1}\boldsymbol{\gamma}_i \quad \text{for } 0 \le i \le k$$

If $0 \le i \le k - 1$, Eq. (7.20) yields

$$\mathbf{S}_{k+1}\boldsymbol{\gamma}_i = \mathbf{S}_k\boldsymbol{\gamma}_i + \boldsymbol{\zeta}_k(\boldsymbol{\delta}_k - \mathbf{S}_k\boldsymbol{\gamma}_k)^T\boldsymbol{\gamma}_i$$

where

$$\boldsymbol{\zeta}_k = \frac{\boldsymbol{\delta}_k - \mathbf{S}_k\boldsymbol{\gamma}_k}{\boldsymbol{\gamma}_k^T(\boldsymbol{\delta}_k - \mathbf{S}_k\boldsymbol{\gamma}_k)}$$

Since $\mathbf{S}_k$ is symmetric, we can write

$$\mathbf{S}_{k+1}\boldsymbol{\gamma}_i = \mathbf{S}_k\boldsymbol{\gamma}_i + \boldsymbol{\zeta}_k(\boldsymbol{\delta}_k^T\boldsymbol{\gamma}_i - \boldsymbol{\gamma}_k^T\mathbf{S}_k\boldsymbol{\gamma}_i)$$

and if Eq. (7.24) holds, then

$$\mathbf{S}_{k+1}\boldsymbol{\gamma}_i = \boldsymbol{\delta}_i + \boldsymbol{\zeta}_k(\boldsymbol{\delta}_k^T\boldsymbol{\gamma}_i - \boldsymbol{\gamma}_k^T\boldsymbol{\delta}_i) \tag{7.25}$$

For $0 \le i \le k$

$$\boldsymbol{\gamma}_i = \mathbf{H}\boldsymbol{\delta}_i$$

and

$$\boldsymbol{\gamma}_k^T = \boldsymbol{\delta}_k^T\mathbf{H}$$

Hence for $0 \le i \le k - 1$, we have

$$\boldsymbol{\delta}_k^T\boldsymbol{\gamma}_i - \boldsymbol{\gamma}_k^T\boldsymbol{\delta}_i = \boldsymbol{\delta}_k^T\mathbf{H}\boldsymbol{\delta}_i - \boldsymbol{\delta}_k^T\mathbf{H}\boldsymbol{\delta}_i = 0$$

and from Eq. (7.25)

$$\boldsymbol{\delta}_i = \mathbf{S}_{k+1}\boldsymbol{\gamma}_i \quad \text{for } 0 \le i \le k - 1 \tag{7.26}$$

By assignment (see Eq. (7.15))

$$\delta_k = S_{k+1} \gamma_k \tag{7.27}$$

and on combining Eqs. (7.26) and (7.27), we obtain

$$\delta_i = S_{k+1} \gamma_i \quad \text{for } 0 \le i \le k \tag{7.28}$$

To complete the induction, we note that

$$\delta_i = S_1 \gamma_i \quad \text{for } 0 \le i \le 0$$

by assignment, and since Eq. (7.28) holds if Eq. (7.24) holds, we can write

$$\delta_i = S_2 \gamma_i \quad \text{for } 0 \le i \le 1$$
$$\delta_i = S_3 \gamma_i \quad \text{for } 0 \le i \le 2$$
$$\vdots \qquad\qquad \vdots$$
$$\delta_i = S_{k+1} \gamma_i \quad \text{for } 0 \le i \le k$$

∎

These principles lead to the following algorithm:

Algorithm 7.2   Basic quasi-Newton algorithm
Step 1
Input x_0 and initialize the tolerance ε.
Set $k = 0$ and $S_0 = I_n$.
Compute g_0.
Step 2
Set $d_k = -S_k g_k$.
Find α_k, the value of α that minimizes $f(x_k + \alpha d_k)$, using a line search.
Set $\delta_k = \alpha_k d_k$ and $x_{k+1} = x_k + \delta_k$.
Step 3
If $\|\delta_k\|_2 < \varepsilon$, output $x^* = x_{k+1}$ and $f(x^*) = f(x_{k+1})$, and stop.
Step 4
Compute g_{k+1} and set
$$\gamma_k = g_{k+1} - g_k$$
Compute S_{k+1} using Eq. (7.20).
Set $k = k + 1$ and repeat from Step 2.

In Step 2, the value of α_k is obtained by using a line search in order to render the algorithm more amenable to nonquadratic problems. However, for convex quadratic problems, α_k should be calculated by using Eq. (7.2) which should involve a lot less computation than a line search.

There are two serious problems associated with the rank-one method. First, a positive definite S_k may not yield a positive definite S_{k+1}, even for a convex quadratic problem, and in such a case the next direction will not be a descent direction. Second, the denominator in the correction formula may approach zero and may even become zero. If it approaches zero, numerical ill-conditioning will occur, and if it becomes zero the method will break down since S_{k+1} will become undefined.

From Eq. (7.20), we can write

$$
\begin{aligned}
\boldsymbol{\gamma}_i^T \mathbf{S}_{k+1} \boldsymbol{\gamma}_i &= \boldsymbol{\gamma}_i^T \mathbf{S}_k \boldsymbol{\gamma}_i + \frac{\boldsymbol{\gamma}_i^T (\boldsymbol{\delta}_k - \mathbf{S}_k \boldsymbol{\gamma}_k)(\boldsymbol{\delta}_k^T - \boldsymbol{\gamma}_k^T \mathbf{S}_k) \boldsymbol{\gamma}_i}{\boldsymbol{\gamma}_k^T (\boldsymbol{\delta}_k - \mathbf{S}_k \boldsymbol{\gamma}_k)} \\
&= \boldsymbol{\gamma}_i^T \mathbf{S}_k \boldsymbol{\gamma}_i + \frac{(\boldsymbol{\gamma}_i^T \boldsymbol{\delta}_k - \boldsymbol{\gamma}_i^T \mathbf{S}_k \boldsymbol{\gamma}_k)(\boldsymbol{\delta}_k^T \boldsymbol{\gamma}_i - \boldsymbol{\gamma}_k^T \mathbf{S}_k \boldsymbol{\gamma}_i)}{\boldsymbol{\gamma}_k^T (\boldsymbol{\delta}_k - \mathbf{S}_k \boldsymbol{\gamma}_k)} \\
&= \boldsymbol{\gamma}_i^T \mathbf{S}_k \boldsymbol{\gamma}_i + \frac{(\boldsymbol{\gamma}_i^T \boldsymbol{\delta}_k - \boldsymbol{\gamma}_i^T \mathbf{S}_k \boldsymbol{\gamma}_k)^2}{\boldsymbol{\gamma}_k^T (\boldsymbol{\delta}_k - \mathbf{S}_k \boldsymbol{\gamma}_k)}
\end{aligned}
$$

Therefore, if $\mathbf{S}_k$ is positive definite, a sufficient condition for $\mathbf{S}_{k+1}$ to be positive definite is

$$
\boldsymbol{\gamma}_k^T (\boldsymbol{\delta}_k - \mathbf{S}_k \boldsymbol{\gamma}_k) > 0
$$

The problems associated with the rank-one method can be overcome by checking the denominator of the correction formula in Step 4 of the algorithm. If it becomes zero or negative, $\mathbf{S}_{k+1}$ can be discarded and $\mathbf{S}_k$ can be used for the subsequent iteration. However, if this problem occurs frequently the possibility exists that $\mathbf{S}_{k+1}$ may not converge to $\mathbf{H}^{-1}$. Then the expected rapid convergence may not materialize.

Example 7.1 Applying the basic quasi-Newton algorithm (Algorithm 7.2), minimize the quadratic function

$$
f(\mathbf{x}) = 5x_1^2 - 9x_1 x_2 + 4.075 x_2^2 + x_1
$$

Start with the initial point $\mathbf{x}_0 = [0 \ 0]^T$ and assume a termination tolerance of $\varepsilon = 10^{-6}$.

Solution

The Hessian of $f(\mathbf{x})$ is given by

$$
\mathbf{H} = \begin{bmatrix} 10 & -9 \\ -9 & 8.15 \end{bmatrix}
$$

Since the values of the leading principal minors, namely, 10 and 0.5, are strictly positive, matrix $\mathbf{H}$ is positive definite and hence $f(\mathbf{x})$ is strictly convex. By setting the gradient of $f(\mathbf{x})$ to zero, i.e.,

$$
\nabla f(\mathbf{x}) = \begin{bmatrix} 10x_1 - 9x_2 + 1 \\ -9x_1 + 8.15x_2 \end{bmatrix} = 0
$$

the global minimizer of $f(\mathbf{x})$ can be obtained as

$$
\mathbf{x}^* = \begin{bmatrix} -16.3 \\ -18 \end{bmatrix}
$$

With $\mathbf{x}_0 = 0$ and $\mathbf{S}_0 = \mathbf{I}_2$, using the value of α_k given by the formula in Eq. (7.9), the first iteration of Algorithm 7.2 yields

$$
\mathbf{d}_0 = \begin{bmatrix} -1 \\ 0 \end{bmatrix}, \quad \alpha_0 = 0.1, \quad \text{and} \quad \mathbf{x}_1 = \begin{bmatrix} -0.1 \\ 0 \end{bmatrix}.
$$

Since $\|\boldsymbol{\delta}_0\|_2 = \|\mathbf{x}_1 - \mathbf{x}_0\|_2 = 0.1 > \varepsilon$ in Step 3 of Algorithm 7.2, the termination condition is not satisfied. Hence $\mathbf{S}_{k+1}$ is evaluated in Step 4 using the formula in Eq. (7.20) and the algorithm continues with the second iteration, which yields

$$\mathbf{S}_1 = \begin{bmatrix} 0.526316 & 0.473684 \\ 0.473684 & 0.526316 \end{bmatrix}$$

$$\mathbf{d}_1 = \begin{bmatrix} -0.426316 \\ -0.473684 \end{bmatrix}, \quad \alpha_1 = 38, \quad \text{and} \quad \mathbf{x}_2 = \begin{bmatrix} -16.3 \\ -18 \end{bmatrix}$$

Evidently, the value of $\mathbf{x}_2$ obtained by using Algorithm 7.2 is identical with the actual global solution of the problem, which was obtained analytically. ∎

Example 7.2 (*a*) Applying the basic quasi-Newton algorithm (Algorithm 7.2), minimize the Rosenbrock function

$$f(\mathbf{x}) = 100(x_1^2 - x_2)^2 + (x_1 - 1)^2$$

Try the four initial points $\mathbf{x}_0 = [3 \ 3]^T$, $\mathbf{x}_0 = [3 \ -1]^T$, $\mathbf{x}_0 = [-1 \ 3]^T$, and $\mathbf{x}_0 = [-1 \ -1]^T$. For a fair comparison with the Fletcher–Reeves algorithm, iterations should be terminated if $f(\mathbf{x}_k) < 10^{-8}$.

   (*b*) Apply the Fletcher–Reeves algorithm (Algorithm 6.3) to the same problem and compare the results obtained with those obtained in part (*a*) using the same initial points and termination condition.

   (*c*) Compare the computational efficiency of Algorithms 7.2 and 6.3 in terms of average CPU time required for each algorithm.

Solution (*a*) It is known that the Rosenbrock function has a unique global minimizer at $\mathbf{x}^* = [1 \ 1]^T$. The gradient of the Rosenbrock function is given by

$$\nabla f(\mathbf{x}) = \begin{bmatrix} 400(x_1^2 - x_2)x_1 + 2(x_1 - 1) \\ 200(x_2 - x_1^2) \end{bmatrix}$$

With $\mathbf{x}_0 = [3 \ 3]^T$, it took Algorithm 7.2 35 iterations to converge to the solution

$$\mathbf{x}^* = \begin{bmatrix} 0.999999 \\ 0.999997 \end{bmatrix}$$

With $\mathbf{x}_0 = [3 \ -1]^T$, Algorithm 7.2 converged after 7461 iterations to the solution

$$\mathbf{x}^* = \begin{bmatrix} 1 \\ 0.999999 \end{bmatrix}$$

With $\mathbf{x}_0 = [-1 \ 3]^T$, Algorithm 7.2 converged after 46 iterations to the solution

$$\mathbf{x}^* = \begin{bmatrix} 0.999986 \\ 0.999967 \end{bmatrix}$$

With $\mathbf{x}_0 = [-1 \ -1]^T$, Algorithm 7.2 converged after 37 iterations to the solution

$$\mathbf{x}^* = \begin{bmatrix} 0.999996 \\ 0.999992 \end{bmatrix}$$

(b) With $\mathbf{x}_0 = [3\, 3]^T$, it took Algorithm 6.3 32 iterations to converge to the solution

$$\mathbf{x}^* = \begin{bmatrix} 0.999996 \\ 0.999989 \end{bmatrix}$$

With $\mathbf{x}_0 = [3\ -1]^T$, Algorithm 6.3 converged after 30 iterations to the solution

$$\mathbf{x}^* = \begin{bmatrix} 0.999975 \\ 0.999949 \end{bmatrix}$$

With $\mathbf{x}_0 = [-1\, 3]^T$, Algorithm 6.3 converged after 137 iterations to the solution

$$\mathbf{x}^* = \begin{bmatrix} 0.999989 \\ 0.999972 \end{bmatrix}$$

With $\mathbf{x}_0 = [-1\ -1]^T$, Algorithm 6.3 converged after 77 iterations to the solution

$$\mathbf{x}^* = \begin{bmatrix} 1.000097 \\ 1.000192 \end{bmatrix}$$

The above results show that for the same termination criterion, Algorithms 6.3 and 7.2 achieve similar accuracies.

(c) The average CPU times required by Algorithms 7.2 and 6.3 were obtained by using the pseudocode described in the solution to Example 5.1. The CPU times required by Algorithm 7.2 for the four cases normalized with respect to the CPU times required by Algorithm 6.3 were obtained as 1.03, 7030.6, 0.36, and 0.58, respectively. Algorithm 7.2 converged to the same solution point with comparable or reduced computational effort relative to that required by Algorithm 6.3 in three cases but required a large number of iterations as well as a large CPU time to converge for the case where the initial point $\mathbf{x}_0 = [3\ -1]^T$ was used.   ∎

7.5 Davidon–Fletcher–Powell Method

An alternative quasi-Newton method is one proposed by Davidon [3] and later developed by Fletcher and Powell [7]. Although similar to the rank-one method, the Davidon–Fletcher–Powell (DFP) method has an important advantage. If the initial matrix $\mathbf{S}_0$ is positive definite, the updating formula for $\mathbf{S}_{k+1}$ will yield a sequence of positive definite matrices $\mathbf{S}_1, \mathbf{S}_2, \ldots, \mathbf{S}_n$. Consequently, the difficulty associated with the second term of the rank-one formula given by Eq. (7.20) will not arise. As a result every new direction will be a descent direction.

The updating formula for the DFP method is

$$\mathbf{S}_{k+1} = \mathbf{S}_k + \frac{\delta_k \delta_k^T}{\delta_k^T \gamma_k} - \frac{\mathbf{S}_k \gamma_k \gamma_k^T \mathbf{S}_k}{\gamma_k^T \mathbf{S}_k \gamma_k} \tag{7.29}$$

where the correction is an $n \times n$ symmetric matrix of *rank two*. The validity of this formula can be demonstrated by post-multiplying both sides by $\boldsymbol{\gamma}_k$, that is,

$$\mathbf{S}_{k+1}\boldsymbol{\gamma}_k = \mathbf{S}_k\boldsymbol{\gamma}_k + \frac{\boldsymbol{\delta}_k\boldsymbol{\delta}_k^T\boldsymbol{\gamma}_k}{\boldsymbol{\delta}_k^T\boldsymbol{\gamma}_k} - \frac{\mathbf{S}_k\boldsymbol{\gamma}_k\boldsymbol{\gamma}_k^T\mathbf{S}_k\boldsymbol{\gamma}_k}{\boldsymbol{\gamma}_k^T\mathbf{S}_k\boldsymbol{\gamma}_k}$$

Since $\boldsymbol{\delta}_k^T\boldsymbol{\gamma}_k$ and $\boldsymbol{\gamma}_k^T\mathbf{S}_k\boldsymbol{\gamma}_k$ are scalars, they can be cancelled out and so we have

$$\mathbf{S}_{k+1}\boldsymbol{\gamma}_k = \boldsymbol{\delta}_k \qquad (7.30)$$

as required.

The implementation of the DFP method is the same as in Algorithm 7.2 except that the rank-two formula of Eq. (7.29) is used in Step 4.

The properties of the DFP method are summarized by the following theorems.

Theorem 7.2 *Positive definiteness of* S *matrix* *If $\mathbf{S}_k$ is positive definite, then the matrix $\mathbf{S}_{k+1}$ generated by the DFP method is also positive definite.*

Proof For any nonzero vector $\mathbf{x} \in E^n$, Eq. (7.29) yields

$$\mathbf{x}^T\mathbf{S}_{k+1}\mathbf{x} = \mathbf{x}^T\mathbf{S}_k\mathbf{x} + \frac{\mathbf{x}^T\boldsymbol{\delta}_k\boldsymbol{\delta}_k^T\mathbf{x}}{\boldsymbol{\delta}_k^T\boldsymbol{\gamma}_k} - \frac{\mathbf{x}^T\mathbf{S}_k\boldsymbol{\gamma}_k\boldsymbol{\gamma}_k^T\mathbf{S}_k\mathbf{x}}{\boldsymbol{\gamma}_k^T\mathbf{S}_k\boldsymbol{\gamma}_k} \qquad (7.31)$$

For a real symmetric matrix $\mathbf{S}_k$, we can write

$$\mathbf{U}^T\mathbf{S}_k\mathbf{U} = \boldsymbol{\Lambda}$$

where $\mathbf{U}$ is a unitary matrix such that

$$\mathbf{U}^T\mathbf{U} = \mathbf{U}\mathbf{U}^T = \mathbf{I}_n$$

and $\boldsymbol{\Lambda}$ is a diagonal matrix whose diagonal elements are the eigenvalues of $\mathbf{S}_k$ (see Theorem 2.8). We can thus write

$$\begin{aligned}
\mathbf{S}_k &= \mathbf{U}\boldsymbol{\Lambda}\mathbf{U}^T = \mathbf{U}\boldsymbol{\Lambda}^{1/2}\boldsymbol{\Lambda}^{1/2}\mathbf{U}^T \\
&= (\mathbf{U}\boldsymbol{\Lambda}^{1/2}\mathbf{U}^T)(\mathbf{U}\boldsymbol{\Lambda}^{1/2}\mathbf{U}^T) \\
&= \mathbf{S}_k^{1/2}\mathbf{S}_k^{1/2}
\end{aligned}$$

If we let

$$\mathbf{u} = \mathbf{S}_k^{1/2}\mathbf{x} \quad \text{and} \quad \mathbf{v} = \mathbf{S}_k^{1/2}\boldsymbol{\gamma}_k$$

then Eq. (7.31) can be expressed as

$$\mathbf{x}^T\mathbf{S}_{k+1}\mathbf{x} = \frac{(\mathbf{u}^T\mathbf{u})(\mathbf{v}^T\mathbf{v}) - (\mathbf{u}^T\mathbf{v})^2}{\mathbf{v}^T\mathbf{v}} + \frac{(\mathbf{x}^T\boldsymbol{\delta}_k)^2}{\boldsymbol{\delta}_k^T\boldsymbol{\gamma}_k} \qquad (7.32)$$

From Step 2 of Algorithm 7.2, we have

$$\boldsymbol{\delta}_k = \alpha_k\mathbf{d}_k = -\alpha_k\mathbf{S}_k\mathbf{g}_k \qquad (7.33)$$

where α_k is the value of α that minimizes $f(\mathbf{x}_k + \alpha \mathbf{d}_k)$ at point $\mathbf{x} = \mathbf{x}_{k+1}$. Since $\mathbf{d}_k = -\mathbf{S}_k \mathbf{g}_k$ is a descent direction (see Eq. (7.14)), we have $\alpha_k > 0$. Furthermore,

$$\frac{df(\mathbf{x}_k + \alpha \mathbf{d}_k)}{d\alpha}\bigg|_{\alpha=\alpha_k} = \mathbf{g}(\mathbf{x}_k + \alpha_k \mathbf{d}_k)^T \mathbf{d}_k = \mathbf{g}_{k+1}^T \mathbf{d}_k = 0$$

(see Sect. 5.2.3) and thus

$$\alpha_k \mathbf{g}_{k+1}^T \mathbf{d}_k = \mathbf{g}_{k+1}^T \alpha_k \mathbf{d}_k = \mathbf{g}_{k+1}^T \boldsymbol{\delta}_k = \boldsymbol{\delta}_k^T \mathbf{g}_{k+1} = 0$$

Hence from Eq. (7.6), we can write

$$\boldsymbol{\delta}_k^T \boldsymbol{\gamma}_k = \boldsymbol{\delta}_k^T \mathbf{g}_{k+1} - \boldsymbol{\delta}_k^T \mathbf{g}_k = -\boldsymbol{\delta}_k^T \mathbf{g}_k$$

Now from Eq. (7.33), we get

$$\boldsymbol{\delta}_k^T \boldsymbol{\gamma}_k = -\boldsymbol{\delta}_k^T \mathbf{g}_k = -[-\alpha_k \mathbf{S}_k \mathbf{g}_k]^T \mathbf{g}_k = \alpha_k \mathbf{g}_k^T \mathbf{S}_k \mathbf{g}_k \tag{7.34}$$

and hence Eq. (7.32) can be expressed as

$$\mathbf{x}^T \mathbf{S}_{k+1} \mathbf{x} = \frac{(\mathbf{u}^T \mathbf{u})(\mathbf{v}^T \mathbf{v}) - (\mathbf{u}^T \mathbf{v})^2}{\mathbf{v}^T \mathbf{v}} + \frac{(\mathbf{x}^T \boldsymbol{\delta}_k)^2}{\alpha_k \mathbf{g}_k^T \mathbf{S}_k \mathbf{g}_k} \tag{7.35}$$

Since

$$\mathbf{u}^T \mathbf{u} = \|\mathbf{u}\|_2^2, \quad \mathbf{v}^T \mathbf{v} = \|\mathbf{v}\|_2^2, \quad \mathbf{u}^T \mathbf{v} = \|\mathbf{u}\|_2 \|\mathbf{v}\|_2 \cos \theta$$

where θ is the angle between vectors $\mathbf{u}$ and $\mathbf{v}$, Eq. (7.35) gives

$$\mathbf{x}^T \mathbf{S}_{k+1} \mathbf{x} = \frac{\|\mathbf{u}\|_2^2 \|\mathbf{v}\|_2^2 - (\|\mathbf{u}\|_2 \|\mathbf{v}\|_2 \cos \theta)^2}{\|\mathbf{v}\|_2^2} + \frac{(\mathbf{x}^T \boldsymbol{\delta}_k)^2}{\alpha_k \mathbf{g}_k^T \mathbf{S}_k \mathbf{g}_k}$$

The minimum value of the right-hand side of the above equation occurs when $\theta = 0$. In such a case, we have

$$\mathbf{x}^T \mathbf{S}_{k+1} \mathbf{x} = \frac{(\mathbf{x}^T \boldsymbol{\delta}_k)^2}{\alpha_k \mathbf{g}_k^T \mathbf{S}_k \mathbf{g}_k} \tag{7.36}$$

Since vectors $\mathbf{u}$ and $\mathbf{v}$ point in the same direction, we can write

$$\mathbf{u} = \mathbf{S}_k^{1/2} \mathbf{x} = \beta \mathbf{v} = \beta \mathbf{S}_k^{1/2} \boldsymbol{\gamma}_k = \mathbf{S}_k^{1/2} \beta \boldsymbol{\gamma}_k$$

and thus

$$\mathbf{x} = \beta \boldsymbol{\gamma}_k$$

where β is a positive constant. On eliminating $\mathbf{x}$ in Eq. (7.36) and then eliminating $\boldsymbol{\gamma}_k^T \boldsymbol{\delta}_k = \boldsymbol{\delta}_k^T \boldsymbol{\gamma}_k$ using Eq. (7.34), we get

$$\mathbf{x}^T \mathbf{S}_{k+1} \mathbf{x} = \frac{(\beta \boldsymbol{\gamma}_k^T \boldsymbol{\delta}_k)^2}{\alpha_k \mathbf{g}_k^T \mathbf{S}_k \mathbf{g}_k} = \alpha_k \beta^2 \mathbf{g}_k^T \mathbf{S}_k \mathbf{g}_k$$

Now for any $\theta \geq 0$, we have

$$\mathbf{x}^T \mathbf{S}_{k+1} \mathbf{x} \geq \alpha_k \beta^2 \mathbf{g}_k^T \mathbf{S}_k \mathbf{g}_k \tag{7.37}$$

Therefore, if $\mathbf{x} = \mathbf{x}_k$ is not the minimizer $\mathbf{x}^*$ (i.e., $\mathbf{g}_k \neq \mathbf{0}$), we have

$$\mathbf{x}^T \mathbf{S}_{k+1} \mathbf{x} > 0 \qquad \text{for } \mathbf{x} \neq \mathbf{0}$$

since $\alpha_k > 0$ and $\mathbf{S}_k$ is positive definite. In effect, *a positive definite* $\mathbf{S}_k$ *will yield a positive definite* $\mathbf{S}_{k+1}$. ∎

It is important to note that the above result holds for any $\alpha_k > 0$ for which

$$\boldsymbol{\delta}_k^T \boldsymbol{\gamma}_k = \boldsymbol{\delta}_k^T \mathbf{g}_{k+1} - \boldsymbol{\delta}_k^T \mathbf{g}_k > 0 \qquad (7.38)$$

even if $f(\mathbf{x})$ is not minimized at point $\mathbf{x}_{k+1}$, as can be verified by eliminating $\mathbf{x}$ in Eq. (7.32) and then using the inequality in Eq. (7.38) (see Problem 7.2). Consequently, if $\boldsymbol{\delta}_k^T \mathbf{g}_{k+1} > \boldsymbol{\delta}_k^T \mathbf{g}_k$, the positive definiteness of $\mathbf{S}_{k+1}$ can be assured even in the case where the minimization of $f(\mathbf{x}_k + \alpha \mathbf{d}_k)$ is inexact. The inequality in Eq. (7.38) will be put to good use later in the construction of a practical quasi-Newton algorithm (see Algorithm 7.3).

Theorem 7.3 *Conjugate directions in DFP method*

(a) If the line searches in Step 2 of the DFP algorithm are exact and $f(\mathbf{x})$ is a convex quadratic function, then the directions generated $\boldsymbol{\delta}_0, \boldsymbol{\delta}_1, \ldots, \boldsymbol{\delta}_k$ form a conjugate set, i.e.,

$$\boldsymbol{\delta}_i^T \mathbf{H} \boldsymbol{\delta}_j = 0 \qquad for \ 0 \leq i < j \leq k \qquad (7.39)$$

(b) If

$$\boldsymbol{\gamma}_i = \mathbf{H} \boldsymbol{\delta}_i \qquad for \ 0 \leq i \leq k \qquad (7.40)$$

then

$$\boldsymbol{\delta}_i = \mathbf{S}_{k+1} \boldsymbol{\gamma}_i \qquad for \ 0 \leq i \leq k \qquad (7.41)$$

Proof As for Theorem 7.1, the proof is by induction. We assume that

$$\boldsymbol{\delta}_i^T \mathbf{H} \boldsymbol{\delta}_j = 0 \qquad for \ 0 \leq i < j \leq k - 1 \qquad (7.42)$$

$$\boldsymbol{\delta}_i = \mathbf{S}_k \boldsymbol{\gamma}_i \qquad for \ 0 \leq i \leq k - 1 \qquad (7.43)$$

and show that Eqs. (7.39) and (7.41) hold.

(*a*) From Eqs. (7.4) and (7.6), we can write

$$\begin{aligned}
\mathbf{g}_k &= \mathbf{g}_{k-1} + \mathbf{H} \boldsymbol{\delta}_{k-1} \\
&= \mathbf{g}_{k-2} + \mathbf{H} \boldsymbol{\delta}_{k-2} + \mathbf{H} \boldsymbol{\delta}_{k-1} \\
&= \mathbf{g}_{k-3} + \mathbf{H} \boldsymbol{\delta}_{k-3} + \mathbf{H} \boldsymbol{\delta}_{k-2} + \mathbf{H} \boldsymbol{\delta}_{k-1} \\
&\vdots \\
&= \mathbf{g}_{i+1} + \mathbf{H}(\boldsymbol{\delta}_{i+1} + \boldsymbol{\delta}_{i+2} + \cdots + \boldsymbol{\delta}_{k-1})
\end{aligned}$$

Thus for $0 \leq i \leq k - 1$, we have

$$\boldsymbol{\delta}_i^T \mathbf{g}_k = \boldsymbol{\delta}_i^T \mathbf{g}_{i+1} + \boldsymbol{\delta}_i^T \mathbf{H}(\boldsymbol{\delta}_{i+1} + \boldsymbol{\delta}_{i+2} + \cdots + \boldsymbol{\delta}_{k-1}) \qquad (7.44)$$

If an exact line search is used in Step 2 of Algorithm 7.2, then $f(\mathbf{x})$ is minimized exactly at point $\mathbf{x}_{i+1}$, and hence

$$\boldsymbol{\delta}_i^T \mathbf{g}_{i+1} = 0 \qquad (7.45)$$

(see proof of Theorem 7.2). Now for $0 \leq i \leq k - 1$, Eq. (7.42) gives

$$\boldsymbol{\delta}_i^T \mathbf{H}(\boldsymbol{\delta}_{i+1} + \boldsymbol{\delta}_{i+2} + \cdots + \boldsymbol{\delta}_{k-1}) = 0 \qquad (7.46)$$

and from Eqs. (7.44)–(7.46), we get

$$\boldsymbol{\delta}_i^T \mathbf{g}_k = 0$$

Alternatively, from Eqs. (7.43) and (7.40) we can write

$$\boldsymbol{\delta}_i^T \mathbf{g}_k = (\mathbf{S}_k \boldsymbol{\gamma}_i)^T \mathbf{g}_k = (\mathbf{S}_k \mathbf{H} \boldsymbol{\delta}_i)^T \mathbf{g}_k$$
$$= \boldsymbol{\delta}_i^T \mathbf{H} \mathbf{S}_k \mathbf{g}_k = 0$$

Further, on eliminating $\mathbf{S}_k \mathbf{g}_k$ using Eq. (7.33)

$$\boldsymbol{\delta}_i^T \mathbf{g}_k = -\frac{1}{\alpha_k} \boldsymbol{\delta}_i^T \mathbf{H} \boldsymbol{\delta}_k = 0$$

and since $\alpha_k > 0$, we have

$$\boldsymbol{\delta}_i^T \mathbf{H} \boldsymbol{\delta}_k = 0 \qquad \text{for } 0 \leq i \leq k - 1 \qquad (7.47)$$

Now on combining Eqs. (7.42) and (7.47)

$$\boldsymbol{\delta}_i^T \mathbf{H} \boldsymbol{\delta}_j = 0 \qquad \text{for } 0 \leq i < j \leq k \qquad (7.48)$$

To complete the induction, we can write

$$\boldsymbol{\delta}_0^T \mathbf{g}_1 = (\mathbf{S}_1 \boldsymbol{\gamma}_0)^T \mathbf{g}_1 = (\mathbf{S}_1 \mathbf{H} \boldsymbol{\delta}_0)^T \mathbf{g}_1$$
$$= \boldsymbol{\delta}_0^T \mathbf{H} \mathbf{S}_1 \mathbf{g}_1$$
$$= -\frac{1}{\alpha_1} \boldsymbol{\delta}_0^T \mathbf{H} \boldsymbol{\delta}_1$$

and since $f(\mathbf{x})$ is minimized exactly at point $\mathbf{x}_1$, we have $\boldsymbol{\delta}_0^T \mathbf{g}_1 = 0$ and

$$\boldsymbol{\delta}_i^T \mathbf{H} \boldsymbol{\delta}_j = 0 \qquad \text{for } 0 \leq i < j \leq 1$$

Since Eq. (7.48) holds if Eq. (7.42) holds, we can write

$$\boldsymbol{\delta}_i^T \mathbf{H} \boldsymbol{\delta}_j = 0 \qquad \text{for } 0 \leq i < j \leq 2$$
$$\boldsymbol{\delta}_i^T \mathbf{H} \boldsymbol{\delta}_j = 0 \qquad \text{for } 0 \leq i < j \leq 3$$

$$\vdots \qquad\qquad \vdots$$

$$\boldsymbol{\delta}_i^T \mathbf{H} \boldsymbol{\delta}_j = 0 \qquad \text{for } 0 \leq i < j \leq k$$

that is, *the directions $\boldsymbol{\delta}_i$ for $= 1, \ldots, k$ form a conjugate set.*

(b) From Eq. (7.43)

$$\boldsymbol{\gamma}_k^T \boldsymbol{\delta}_i = \boldsymbol{\gamma}_k^T \mathbf{S}_k \boldsymbol{\gamma}_i \qquad \text{for } 0 \leq i \leq k - 1 \qquad (7.49)$$

On the other hand, Eq. (7.40) yields

$$\boldsymbol{\gamma}_k^T \boldsymbol{\delta}_i = \boldsymbol{\delta}_k^T \mathbf{H} \boldsymbol{\delta}_i \qquad \text{for } 0 \leq i \leq k - 1 \qquad (7.50)$$

and since $\boldsymbol{\delta}_0, \boldsymbol{\delta}_1, \ldots, \boldsymbol{\delta}_k$ form a set of conjugate vectors from part (a), Eqs. (7.49) and (7.50) yield

$$\boldsymbol{\gamma}_k^T \boldsymbol{\delta}_i = \boldsymbol{\gamma}_k^T \mathbf{S}_k \boldsymbol{\gamma}_i = \boldsymbol{\delta}_k^T \mathbf{H} \boldsymbol{\delta}_i = 0 \qquad \text{for } 0 \leq i \leq k - 1 \qquad (7.51)$$

By noting that

$$\delta_k^T = \gamma_k^T S_{k+1} \quad \text{and} \quad H\delta_i = \gamma_i$$

can be expressed as

$$\gamma_k^T \delta_i = \gamma_k^T S_k \gamma_i = \gamma_k^T S_{k+1} \gamma_i = 0 \qquad \text{for } 0 \le i \le k-1$$

therefore,

$$\delta_i = S_k \gamma_i = S_{k+1} \gamma_i \qquad \text{for } 0 \le i \le k-1 \tag{7.52}$$

Now from Eq. (7.30)

$$\delta_k = S_{k+1} \gamma_k \tag{7.53}$$

and on combining Eqs. (7.52) and (7.53), we obtain

$$\delta_i = S_{k+1} \gamma_i \qquad \text{for } 0 \le i \le k$$

The induction can be completed as in Theorem 7.1.   ∎

For $k = n - 1$, Eqs. (7.40) and (7.41) can be expressed as

$$[S_n H - \lambda I]\delta_i = 0 \qquad \text{for } 0 \le i \le n-1$$

with $\lambda = 1$. In effect, vectors δ_i are eigenvectors that correspond to the unity eigenvalue for matrix $S_n H$. Since they are linearly independent, we have

$$S_n = H^{-1}$$

that is, *in a quadratic problem S_{k+1} becomes the Hessian on iteration $n - 1$.*

Example 7.3 The DFP algorithm is essentially Algorithm 7.2 using the formula in Eq. (7.29) for the evaluation of matrix S_{k+1} in Step 4. Using this algorithm, minimize the quadratic function in Example 7.1, namely,

$$f(\mathbf{x}) = 5x_1^2 - 9x_1 x_2 + 4.075x_2^2 + x_1$$

employing the same initial point and termination tolerance as in Example 7.1.

Solution With $\mathbf{x}_0 = [0\ 0]^T$ and $\delta_0 = I_2$, using Eq. (7.9) the first iteration of the DFP algorithm yields

$$\mathbf{d}_0 = \begin{bmatrix} -1 \\ 0 \end{bmatrix}, \quad \alpha_0 = 0.1, \quad \text{and} \quad \mathbf{x}_1 = \begin{bmatrix} -0.1 \\ 0 \end{bmatrix}$$

These intermediate results are identical with those obtained in Example 7.1 as the first iteration of the rank-one and DFP algorithms involves the same steepest-descent search. Since $\|\delta_0\|_2 = \|\mathbf{x}_1 - \mathbf{x}_0\|_2 = 0.1 > \varepsilon$, the termination criterion in Step 3 of the DFP algorithm is not satisfied and S_{k+1} is evaluated using the formula in Eq. (7.29). The algorithm then continues with the second iteration, which yields

$$S_1 = \begin{bmatrix} 0.547514 & 0.497238 \\ 0.497238 & 0.552486 \end{bmatrix}$$

$$\mathbf{d}_1 = \begin{bmatrix} -0.447514 \\ -0.497238 \end{bmatrix}, \quad \alpha_1 = 36.2, \quad \text{and} \quad \mathbf{x}_2 = \begin{bmatrix} -16.3 \\ -18 \end{bmatrix}$$

We note that $\mathbf{x}_2$ is identical with the second iterate obtained in Example 7.1 and, as shown in Example 7.1, this is the actual global minimizer of $f(\mathbf{x})$.   ∎

Example 7.4 (*a*) Applying the DFP algorithm, minimize the Rosenbrock function

$$f(\mathbf{x}) = 100(x_1^2 - x_2)^2 + (x_1 - 1)^2$$

For the sake of consistency, use the same initial point and termination criterion as in Example 7.2.

(*b*) Compare the computational efficiency of the DFP algorithm with the rank-one algorithm (Algorithm 7.2) in terms of average CPU time required to converge.

Solution (*a*) With $\mathbf{x}_0 = [3\ 3]^T$, it took the DFP algorithm 46 iterations to converge to the solution

$$\mathbf{x}^* = \begin{bmatrix} 0.999998 \\ 0.999993 \end{bmatrix}$$

With $\mathbf{x}_0 = [3\ -1]^T$, the DFP algorithm converged after 27 iterations to the solution

$$\mathbf{x}^* = \begin{bmatrix} 0.999985 \\ 0.999973 \end{bmatrix}$$

With $\mathbf{x}_0 = [-1\ 3]^T$, the DFP algorithm converged after 103 iterations to the solution

$$\mathbf{x}^* = \begin{bmatrix} 0.999996 \\ 0.999992 \end{bmatrix}$$

With $\mathbf{x}_0 = [-1\ -1]^T$, the DFP algorithm converged after 27 iterations to the solution

$$\mathbf{x}^* = \begin{bmatrix} 0.999982 \\ 0.999963 \end{bmatrix}$$

(*b*) The average CPU times required by the DFP algorithm and the rank-one algorithm (Algorithm 7.2) were obtained as in the solution to Example 5.1. The CPU times required by the DFP algorithm for the four cases normalized with respect to the CPU times required by Algorithm 6.3 were obtained as 1.34, 1.34×10^{-4}, 1.56, and 0.56, respectively. Although these results do not identify clearly the more efficient of the two algorithms, the DFP algorithm tends to be more robust with respect to the use of different initial points. ∎

7.5.1 Alternative Form of DFP Formula

An alternative form of the DFP formula can be generated by using the Sherman–Morrison formula (see [8,9] and Sect. A.4), which states that an $n \times n$ matrix

$$\hat{\mathbf{U}} = \mathbf{U} + \mathbf{V}\mathbf{W}\mathbf{X}^T$$

where $\mathbf{V}$ and $\mathbf{X}$ are $n \times m$ matrices, $\mathbf{W}$ is an $m \times m$ matrix, and $m \leq n$, has an inverse

$$\hat{\mathbf{U}}^{-1} = \mathbf{U}^{-1} - \mathbf{U}^{-1}\mathbf{V}\mathbf{Y}^{-1}\mathbf{X}^T\mathbf{U}^{-1} \qquad (7.54)$$

where

$$\mathbf{Y} = \mathbf{W}^{-1} + \mathbf{X}^T\mathbf{U}^{-1}\mathbf{V}$$

The DFP formula can be written as

$$\mathbf{S}_{k+1} = \mathbf{S}_k + \mathbf{X}\mathbf{W}\mathbf{X}^T$$

where

$$\mathbf{X} = \left[\frac{\delta_k}{\sqrt{(\delta_k^T \gamma_k)}} \quad \frac{\mathbf{S}_k \gamma_k}{\sqrt{(\gamma_k^T \mathbf{S}_k \gamma_k)}} \right] \quad \text{and} \quad \mathbf{W} = \begin{bmatrix} 1 & 0 \\ 0 & -1 \end{bmatrix}$$

and hence Eq. (7.54) yields

$$\mathbf{S}_{k+1}^{-1} = \mathbf{S}_k^{-1} - \mathbf{S}_k^{-1} \mathbf{X} \mathbf{Y}^{-1} \mathbf{X}^T \mathbf{S}_k^{-1} \tag{7.55}$$

where

$$\mathbf{Y} = \mathbf{W}^{-1} + \mathbf{X}^T \mathbf{S}_k^{-1} \mathbf{X}$$

By letting

$$\mathbf{S}_{k+1}^{-1} = \mathbf{P}_{k+1}, \quad \mathbf{S}_k^{-1} = \mathbf{P}_k$$

and then deducing $\mathbf{Y}^{-1}$, Eq. (7.55) yields

$$\mathbf{P}_{k+1} = \mathbf{P}_k + \left(1 + \frac{\delta_k^T \mathbf{P}_k \delta_k}{\delta_k^T \gamma_k}\right) \frac{\gamma_k \gamma_k^T}{\delta_k^T \gamma_k} - \frac{(\gamma_k \delta_k^T \mathbf{P}_k + \mathbf{P}_k \delta_k \gamma_k^T)}{\delta_k^T \gamma_k} \tag{7.56}$$

This formula can be used to generate a sequence of approximations for the Hessian $\mathbf{H}$.

7.6   Broyden–Fletcher–Goldfarb–Shanno Method

Another recursive formula for generating a sequence of approximations for $\mathbf{H}^{-1}$ is one proposed by Broyden [2], Fletcher [10], Goldfarb [11], and Shanno [12] at about the same time. This is referred to as the *BFGS updating formula* [13,14] and is given by

$$\mathbf{S}_{k+1} = \mathbf{S}_k + \left(1 + \frac{\gamma_k^T \mathbf{S}_k \gamma_k}{\gamma_k^T \delta_k}\right) \frac{\delta_k \delta_k^T}{\gamma_k^T \delta_k} - \frac{(\delta_k \gamma_k^T \mathbf{S}_k + \mathbf{S}_k \gamma_k \delta_k^T)}{\gamma_k^T \delta_k} \tag{7.57}$$

This formula is said to be *the dual of the DFP formula* given in Eq. (7.29) and it can be obtained by letting

$$\mathbf{P}_{k+1} = \mathbf{S}_{k+1}, \quad \mathbf{P}_k = \mathbf{S}_k$$
$$\gamma_k = \delta_k, \quad \delta_k = \gamma_k$$

in Eq. (7.56). As may be expected, for convex quadratic functions, the BFGS formula has the following properties:

1. $\mathbf{S}_{k+1}$ becomes identical to $\mathbf{H}^{-1}$ for $k = n - 1$.
2. Directions $\delta_0, \delta_1, \ldots, \delta_k$ form a conjugate set.
3. $\mathbf{S}_{k+1}$ is positive definite if $\mathbf{S}_k$ is positive definite.
4. The inequality in Eq. (7.38) applies.

An alternative form of the BFGS formula can be obtained as

$$\mathbf{P}_{k+1} = \mathbf{P}_k + \frac{\boldsymbol{\gamma}_k \boldsymbol{\gamma}_k^T}{\boldsymbol{\gamma}_k^T \boldsymbol{\delta}_k} - \frac{\mathbf{P}_k \boldsymbol{\delta}_k \boldsymbol{\delta}_k^T \mathbf{P}_k}{\boldsymbol{\delta}_k^T \mathbf{P}_k \boldsymbol{\delta}_k}$$

by letting

$$\mathbf{S}_{k+1} = \mathbf{P}_{k+1}, \quad \mathbf{S}_k = \mathbf{P}_k$$
$$\boldsymbol{\delta}_k = \boldsymbol{\gamma}_k, \quad \boldsymbol{\gamma}_k = \boldsymbol{\delta}_k$$

in Eq. (7.29) or by applying the Sherman–Morrison formula to Eq. (7.57). This is the dual of Eq. (7.56).

Example 7.5 The BFGS algorithm is essentially Algorithm 7.2 using the formula in Eq. (7.57) for the evaluation of matrix S_{k+1} in Step 4. Using this algorithm, minimize the quadratic function in Example 7.1, namely,

$$f(\mathbf{x}) = 5x_1^2 - 9x_1x_2 + 4.075x_2^2 + x_1$$

employing the same initial point $\mathbf{x}_0 = [0 \ 0]^T$ and termination tolerance as in Example 7.1.

Solution With $\mathbf{x}_0 = [0 \ 0]^T$ and $\mathbf{S}_0 = \mathbf{I}_2$ using Eq. (7.9), the first iteration of the BFGS algorithm yields

$$\mathbf{d}_0 = \begin{bmatrix} -1 \\ 0 \end{bmatrix}, \quad \alpha_0 = 0.1, \quad \text{and} \quad \mathbf{x}_1 = \begin{bmatrix} -0.1 \\ 0 \end{bmatrix}$$

which is identical with the solution obtained in Examples 7.1 and 7.3, because, as before, the first iteration of the BFGS algorithm involves a steepest-descent search. Since $\|\boldsymbol{\delta}_0\|_2 = \|\mathbf{x}_1 - \mathbf{x}_0\|_2 = 0.1 > \varepsilon$, the termination criterion in Step 3 of the BFGS algorithm is not satisfied and hence it evaluates matrix S_{k+1} using the formula in Eq. (7.57). It then continues with the second iteration, which yields

$$\mathbf{S}_1 = \begin{bmatrix} 0.91 & 0.9 \\ 0.9 & 1 \end{bmatrix}$$

$$\mathbf{d}_1 = \begin{bmatrix} -0.81 \\ -0.9 \end{bmatrix}, \quad \alpha_1 = 20, \quad \text{and} \quad \mathbf{x}_2 = \begin{bmatrix} -16.3 \\ -18 \end{bmatrix}$$

Evidently, $\mathbf{x}_2$ is identical with the second iterate obtained in Examples 7.1 and 7.3, which is the actual global minimizer of $f(\mathbf{x})$. ∎

Example 7.6 (*a*) Applying the BFGS algorithm, minimize the Rosenbrock function

$$f(\mathbf{x}) = 100(x_1^2 - x_2)^2 + (x_1 - 1)^2$$

For the sake of consistency, use the same initial point and termination criterion as in Example 7.4.

(*b*) Compare the computational efficiency of the DFP and BFGS algorithms in terms of average CPU time required.

Solution (*a*) With $\mathbf{x}_0 = [3\ 3]^T$, it took the BFGS algorithm 26 iterations to converge to the solution

$$\mathbf{x}^* = \begin{bmatrix} 1.000025 \\ 1.000044 \end{bmatrix}$$

With $\mathbf{x}_0 = [3\ -1]^T$, the BFGS algorithm converged after 25 iterations to the solution

$$\mathbf{x}^* = \begin{bmatrix} 1 \\ 0.999999 \end{bmatrix}$$

With $\mathbf{x}_0 = [-1\ 3]^T$, the BFGS algorithm converged after 34 iterations to the solution

$$\mathbf{x}^* = \begin{bmatrix} 0.999993 \\ 0.999986 \end{bmatrix}$$

With $\mathbf{x}_0 = [-1\ -1]^T$, it took the BFGS algorithm 29 iterations to converge to the solution

$$\mathbf{x}^* = \begin{bmatrix} 0.999983 \\ 0.999965 \end{bmatrix}$$

(*b*) The average CPU times required by the BFGS and DFP algorithms were obtained as in the solution to Example 5.1. The CPU times required by the BFGS algorithm for the four cases normalized with respect to the CPU times required by the DFP were obtained as 0.51, 0.86, 0.34, and 0.98, respectively. The numerical results in Examples 7.2–7.6 suggest that the BFGS algorithm is a robust and efficient algorithm for use in many unconstrained optimization problems. ■

7.7   Hoshino Method

The application of the principle of duality (i.e., the application of the Sherman–Morrison formula followed by the replacement of $\mathbf{P}_k$, $\mathbf{P}_{k+1}$, $\boldsymbol{\gamma}_k$, and $\boldsymbol{\delta}_k$ by $\mathbf{S}_k$, $\mathbf{S}_{k+1}$, $\boldsymbol{\delta}_k$, and $\boldsymbol{\gamma}_k$) to the rank-one formula results in one and the same formula. For this reason, the rank-one formula is said to be *self-dual*. Another self-dual formula which was found to give good results is one due to Hoshino [15]. Like the DFP and BFGS formulas, the Hoshino formula is of *rank two*. It is given by

$$\mathbf{S}_{k+1} = \mathbf{S}_k + \theta_k \boldsymbol{\delta}_k \boldsymbol{\delta}_k^T - \psi_k (\boldsymbol{\delta}_k \boldsymbol{\gamma}_k^T \mathbf{S}_k + \mathbf{S}_k \boldsymbol{\gamma}_k \boldsymbol{\delta}_k^T + \mathbf{S}_k \boldsymbol{\gamma}_k \boldsymbol{\gamma}_k^T \mathbf{S}_k)$$

where

$$\theta_k = \frac{\boldsymbol{\gamma}_k^T \boldsymbol{\delta}_k + 2\boldsymbol{\gamma}_k^T \mathbf{S}_k \boldsymbol{\gamma}_k}{\boldsymbol{\gamma}_k^T \boldsymbol{\delta}_k (\boldsymbol{\gamma}_k^T \boldsymbol{\delta}_k + \boldsymbol{\gamma}_k^T \mathbf{S}_k \boldsymbol{\gamma}_k)} \quad \text{and} \quad \psi_k = \frac{1}{(\boldsymbol{\gamma}_k^T \boldsymbol{\delta}_k + \boldsymbol{\gamma}_k^T \mathbf{S}_k \boldsymbol{\gamma}_k)}$$

The inverse of $\mathbf{S}_{k+1}$, designated as $\mathbf{P}_{k+1}$, can be obtained by applying the Sherman–Morrison formula.

7.8 The Broyden Family

An updating formula which is of significant theoretical as well as practical interest
is one due to Broyden [2]. This formula entails an independent parameter ϕ_k and is
given by

$$S_{k+1} = (1 - \phi_k)S_{k+1}^{DFP} + \phi_k S_{k+1}^{BFGS} \qquad (7.58)$$

Evidently, if $\phi_k = 1$ or 0 the Broyden formula reduces to the BFGS or DFP formula,
and if

$$\phi_k = \frac{\delta_k^T \gamma_k}{\delta_k^T \gamma_k \pm \gamma_k^T S_k \gamma_k}$$

the rank-one or Hoshino formula is obtained.

   If the formula of Eq. (7.58) is used in Step 4 of Algorithm 7.2, a Broyden method
is obtained which has the properties summarized in Theorems 7.4A–7.4C below.
These are generic properties that apply to all the methods described so far.

Theorem 7.4A *Properties of Broyden method*   *If a Broyden method is applied to
a convex quadratic function and exact line searches are used, it will terminate after
$m \leq n$ iterations. The following properties apply for all $k = 0, 1, \ldots, m$:*

(a) $\delta_i = S_{k+1}\gamma_i \ for \ 0 \leq i \leq k$
(b) $\delta_i^T H\delta_j = 0 \ for \ 0 \leq i < j \leq k$
(c)   *If $m = n - 1$, then $S_m = H^{-1}$*

Theorem 7.4B *If $S_0 = I_n$, then a Broyden method with exact line searches is equivalent to the Fletcher–Reeves conjugate gradient method (see Sect. 6.6) provided that
$f(\mathbf{x})$ is a convex quadratic function. Integer m in Theorem 7.4A is the least number
of independent vectors in the sequence*

$$\mathbf{g}_0, \ H\mathbf{g}_0, \ \ldots, \ H^{n-1}\mathbf{g}_0$$

Theorem 7.4C *If $f(\mathbf{x}) \in C^1$, a Broyden method with exact line searches has the
property that $\mathbf{x}_{k+1}$ and the BFGS component of the Broyden formula are independent
of $\phi_0, \ \phi_1, \ \ldots, \ \phi_{k-1}$ for all $k \geq 1$.*

The proofs of these theorems are given by Fletcher [14].

7.8.1 Fletcher Switch Method

A particularly successful method of the Broyden family is one proposed by Fletcher
[13]. In this method, parameter ϕ_k in Eq. (7.58) is switched between zero and unity
throughout the optimization. The choice of ϕ_k in any iteration is based on the rule

$$\phi_k = \begin{cases} 0 & \text{if } \delta_k^T H\delta_k > \delta_k^T P_{k+1}\delta_k \\ 1 & \text{otherwise} \end{cases}$$

where $\mathbf{H}$ is the Hessian of $f(\mathbf{x})$, and $\mathbf{P}_{k+1}$ is the approximation of $\mathbf{H}$ generated by the updating formula. In effect, Fletcher's method compares $\mathbf{P}_{k+1}$ with $\mathbf{H}$ in direction δ_k, and if the above condition is satisfied then the DFP formula is used. Alternatively, the BFGS formula is used. The Hessian is not available in quasi-Newton methods but on assuming a convex quadratic problem, it can be eliminated. From Eq. (7.4)

$$\mathbf{H}\delta_k = \gamma_k$$

and, therefore, the above test becomes

$$\delta_k^T \gamma_k > \delta_k^T \mathbf{P}_{k+1} \delta_k \tag{7.59}$$

This test is convenient to use when an approximation for $\mathbf{H}$ is to be used in the implementation of the algorithm. An alternative, but equivalent, test which is applicable to the case where approximation for $\mathbf{H}^{-1}$ is to be used can be readily obtained from Eq. (7.59). We can write

$$\delta_k^T \gamma_k > \delta_k^T \mathbf{S}_{k+1}^{-1} \delta_k$$

and since

$$\delta_k = \mathbf{S}_{k+1} \gamma_k$$

according to Eq. (7.12), we have

$$\delta_k^T \gamma_k > \gamma_k^T \mathbf{S}_{k+1} \mathbf{S}_{k+1}^{-1} \mathbf{S}_{k+1} \gamma_k$$
$$> \gamma_k^T \mathbf{S}_{k+1} \gamma_k$$

since $\mathbf{S}_{k+1}$ is symmetric.

7.9 The Huang Family

Another family of updating formulas is one due to Huang [16]. This is a more general family which encompasses the rank-one, DFP, BFGS as well as some other formulas. It is of the form

$$\mathbf{S}_{k+1} = \mathbf{S}_k + \frac{\delta_k(\theta\delta_k + \phi\mathbf{S}_k^T\gamma_k)^T}{(\theta\delta_k + \phi\mathbf{S}_k^T\gamma_k)^T\gamma_k} - \frac{\mathbf{S}_k\gamma_k(\psi\delta_k + \omega\mathbf{S}_k^T\gamma_k)^T}{(\psi\delta_k + \omega\mathbf{S}_k^T\gamma_k)^T\gamma_k}$$

where θ, ϕ, ψ, and ω are independent parameters. The formulas that can be generated from the Huang formula are given in Table 7.1. The McCormick formula [17] is

$$\mathbf{S}_{k+1} = \mathbf{S}_k + \frac{(\delta_k - \mathbf{S}_k\gamma_k)\delta_k^T}{\delta_k^T\gamma_k}$$

Table 7.1 The Huang family

Formula	Parameters
Rank-one	$\theta = 1,\ \phi = -1,\ \psi = 1,\ \omega = -1$
DFP	$\theta = 1,\ \phi = 0,\ \psi = 0,\ \omega = 1$
BFGS	$\begin{cases} \dfrac{\phi}{\theta} = \dfrac{-\boldsymbol{\delta}_k^T \boldsymbol{\gamma}_k}{\boldsymbol{\delta}_k^T \boldsymbol{\gamma}_k + \boldsymbol{\gamma}_k^T \mathbf{S}_k \boldsymbol{\gamma}_k}, \\ \psi = 1,\ \omega = 0 \end{cases}$
McCormick	$\theta = 1,\ \phi = 0,\ \psi = 1,\ \omega = 0$
Pearson	$\theta = 0,\ \phi = 1,\ \psi = 0,\ \omega = 1$

whereas that of Pearson [18] is given by

$$\mathbf{S}_{k+1} = \mathbf{S}_k + \frac{(\boldsymbol{\delta}_k - \mathbf{S}_k \boldsymbol{\gamma}_k)\boldsymbol{\gamma}_k^T \mathbf{S}_k}{\boldsymbol{\gamma}_k^T \mathbf{S}_k \boldsymbol{\gamma}_k}$$

7.10 Practical Quasi-Newton Algorithm

A practical quasi-Newton algorithm that eliminates the problems associated with Algorithms 7.1 and 7.2 is detailed below. This is based on Algorithm 7.2 and uses a slightly modified version of Fletcher's inexact line search (Algorithm 4.6). The algorithm is flexible, efficient, and very reliable, and has been found to be very effective for the design of digital filters and equalizers (see [19, Chap. 16]).

Algorithm 7.3 Practical quasi-Newton algorithm
Step 1 (Initialize algorithm)
a. Input $\mathbf{x}_0$ and ε_1.
b. Set $k = m = 0$.
c. Set $\rho = 0.1,\ \sigma = 0.7,\ \tau = 0.1,\ \chi = 0.75,\ \hat{M} = 600$, and $\varepsilon_2 = 10^{-10}$.
d. Set $\mathbf{S}_0 = \mathbf{I}_n$.
e. Compute f_0 and $\mathbf{g}_0$, and set $m = m + 2$. Set $f_{00} = f_0$ and $\Delta f_0 = f_0$.
Step 2 (Initialize line search)
a. Set $\mathbf{d}_k = -\mathbf{S}_k \mathbf{g}_k$.
b. Set $\alpha_L = 0$ and $\alpha_U = 10^{99}$.
c. Set $f_L = f_0$ and compute $f_L' = \mathbf{g}(\mathbf{x}_k + \alpha_L \mathbf{d}_k)^T \mathbf{d}_k$.
d. (Estimate α_0)
    If $|f_L'| > \varepsilon_2$, then compute $\alpha_0 = -2\Delta f_0/f_L'$; otherwise, set $\alpha_0 = 1$.
    If $\alpha_0 \leq 0$ or $\alpha_0 > 1$, then set $\alpha_0 = 1$.
Step 3
a. Set $\boldsymbol{\delta}_k = \alpha_0 \mathbf{d}_k$ and compute $f_0 = f(\mathbf{x}_k + \boldsymbol{\delta}_k)$.
b. Set $m = m + 1$.
Step 4 (Interpolation)
If $f_0 > f_L + \rho (\alpha_0 - \alpha_L) f_L'$ and $|f_L - f_0| > \varepsilon_2$ and $m < \hat{M}$, then do:
a. If $\alpha_0 < \alpha_U$, then set $\alpha_U = \alpha_0$.

b. Compute $\breve{\alpha}_0$ using Eq. (4.57).

c. Compute $\breve{\alpha}_{0L} = \alpha_L + \tau(\alpha_U - \alpha_L)$; if $\breve{\alpha}_0 < \breve{\alpha}_{0L}$, then set $\breve{\alpha}_0 = \breve{\alpha}_{0L}$.

d. Compute $\breve{\alpha}_{0U} = \alpha_U - \tau(\alpha_U - \alpha_L)$; if $\breve{\alpha}_0 > \breve{\alpha}_{0U}$, then set $\breve{\alpha}_0 = \breve{\alpha}_{0U}$.

e. Set $\alpha_0 = \breve{\alpha}_0$ and go to Step 3.

Step 5

Compute $f_0' = \mathbf{g}(\mathbf{x}_k + \alpha_0 \mathbf{d}_k)^T \mathbf{d}_k$ and set $m = m + 1$.

Step 6 (Extrapolation)

If $f_0' < \sigma f_L'$ and $|f_L - f_0| > \varepsilon_2$ and $m < \hat{M}$, then do:

a. Compute $\Delta \alpha_0 = (\alpha_0 - \alpha_L) f_0' / (f_L' - f_0')$ (see Eq. (4.58)).

b. If $\Delta \alpha_0 \leq 0$, then set $\breve{\alpha}_0 = 2\alpha_0$; otherwise, set $\breve{\alpha}_0 = \alpha_0 + \Delta \alpha_0$.

c. Compute $\breve{\alpha}_{0U} = \alpha_0 + \chi(\alpha_U - \alpha_0)$; if $\breve{\alpha}_0 > \breve{\alpha}_{0U}$, then set $\breve{\alpha}_0 = \breve{\alpha}_{0U}$.

d. Set $\alpha_L = \alpha_0, \alpha_0 = \breve{\alpha}_0, f_L = f_0, f_L' = f_0'$ and go to Step 3.

Step 7 (Check termination criteria and output results)

a. Set $\mathbf{x}_{k+1} = \mathbf{x}_k + \boldsymbol{\delta}_k$.

b. Set $\Delta f_0 = f_{00} - f_0$.

c. If ($\|\boldsymbol{\delta}_k\|_2 < \varepsilon_1$ and $|\Delta f_0| < \varepsilon_1$) or $m \geq \hat{M}$, then output $\breve{\mathbf{x}} = \mathbf{x}_{k+1}$, $f(\breve{\mathbf{x}}) = f_{k+1}$, and stop.

d. Set $f_{00} = f_0$.

Step 8 (Prepare for the next iteration)

a. Compute $\mathbf{g}_{k+1}$ and set $\boldsymbol{\gamma}_k = \mathbf{g}_{k+1} - \mathbf{g}_k$.

b. Compute $D = \boldsymbol{\delta}_k^T \boldsymbol{\gamma}_k$; if $D \leq 0$, then set $\mathbf{S}_{k+1} = \mathbf{I}_n$; otherwise, compute $\mathbf{S}_{k+1}$ using Eq. (7.29) for the DFP method or Eq. (7.57) for the BFGS method.

c. Set $k = k + 1$ and go to Step 2.

The computational complexity of an algorithm can be determined by estimating the amount of computation required, which is not always an easy task. In optimization algorithms of the type described in Chaps. 5–7, most of the computational effort is associated with function and gradient evaluations and by counting the function and gradient evaluations, a measure of the computational complexity of the algorithm can be obtained. In Algorithm 7.3, this is done through index m which is increased by one for each evaluation of $f_0, \mathbf{g}_0$, or f_0' in Steps 1, 3, and 5. Evidently, we assume here that a function evaluation requires the same computational effort as a gradient evaluation, which may not be valid, since each gradient evaluation involves the evaluation of n first derivatives. A more precise measure of computational complexity could be obtained by finding the number of additions, multiplications, and divisions associated with each function and each gradient evaluation and then modifying Steps 1, 3, and 5 accordingly.

Counting the number of function evaluations can serve another useful purpose. An additional termination mechanism can be incorporated in the algorithm that can be used to abort the search for a minimum if the number of function evaluations becomes unreasonably large and exceeds some upper limit, say, $\hat{M}$. In Algorithm 7.3, interpolation is performed in Step 4 and extrapolation is performed in Step 6 only if $m < \hat{M}$, and if $m \geq \hat{M}$ the algorithm is terminated in Step 7c. This additional termination mechanism is useful when the problem being solved does not have a well-defined local minimum.

Although a positive definite matrix $\mathbf{S}_k$ will ensure that $\mathbf{d}_k$ is a descent direction for function $f(\mathbf{x})$ at point $\mathbf{x}_k$, sometimes the function $f(\mathbf{x}_k + \alpha \mathbf{d}_k)$ may have a very shallow minimum with respect to α and finding such a minimum can waste a large amount of computation. The same problem can sometimes arise if $f(\mathbf{x}_k + \alpha \mathbf{d}_k)$ does not have a well-defined minimizer or in cases where $|f_L - f_0|$ is very small and of the same order of magnitude as the roundoff errors. To avoid these problems, interpolation or extrapolation is carried out only if the expected reduction in the function $f(\mathbf{x}_k + \alpha \mathbf{d}_k)$ is larger than ε_2. In such a case, the algorithm continues with the next iteration unless the termination criteria in Step 7c are satisfied.

The estimate of α_0 in Step 2d can be obtained by assuming that the function $f(\mathbf{x}_k + \alpha \mathbf{d}_k)$ can be represented by a quadratic polynomial of α and that the reduction achieved in $f(\mathbf{x}_k + \alpha \mathbf{d}_k)$ by changing α from 0 to α_0 is equal to Δf_0, the total reduction achieved in the previous iteration. Under these assumptions, we can write

$$f_L - f_0 = \Delta f_0 \tag{7.60}$$

and from Eq. (4.57)

$$\alpha_0 = \breve{\alpha} \approx \alpha_L + \frac{(\alpha_0 - \alpha_L)^2 f_L'}{2[f_L - f_0 + (\alpha_0 - \alpha_L)f_L']} \tag{7.61}$$

Since $\alpha_L = 0$, Eqs. (7.60) and (7.61) give

$$\alpha_0 \approx \frac{\alpha_0^2 f_L'}{2[\Delta f_0 + \alpha_0 f_L']}$$

Now solving for α_0, we get

$$\alpha_0 \approx -\frac{2\Delta f_0}{f_L'}$$

This estimate of α is reasonable for points far away from the solution but can become quite inaccurate as the minimizer is approached and could even become negative due to numerical ill-conditioning. For these reasons, if the estimate is equal to or less than zero or greater than unity, it is replaced by unity in Step 2d, which is the value of α that would minimize $f(\mathbf{x}_k + \alpha \mathbf{d}_k)$ in the case of a convex quadratic problem. Recall that practical problems tend to become convex and quadratic in the neighborhood of a local minimizer.

The most important difference between the inexact line search in Algorithm 4.6 and that used in Algorithm 7.3 is related to a very real problem that can arise in practice. The first derivatives f_0' and f_L' may on occasion satisfy the inequalities

$$\alpha_0 f_L' < \alpha_L f_0' \quad \text{and} \quad f_L' > f_0'$$

and the quadratic extrapolation in Step 6 would yield

$$\breve{\alpha} = \alpha_0 + \frac{(\alpha_0 - \alpha_L)f_0'}{f_L' - f_0'} = \frac{\alpha_0 f_L' - \alpha_L f_0'}{f_L' - f_0'} < 0$$

that is, it will predict a negative α. This would correspond to a maximum of $f(\mathbf{x}_k + \alpha \mathbf{d}_k)$ since $\alpha \mathbf{d}_k$ is a descent direction only if α is positive. In such a case, $\Delta \alpha_0 = \breve{\alpha} - \alpha_0$

would assume a negative value in Step 6*b* and to ensure that α is changed in the descent direction, the value $2\alpha_0$ is assigned to $\breve{\alpha}_0$. This new value could turn out to be unreasonably large and could exceed the most recent upper bound α_U. Although this is not catastrophic, unnecessary computations would need to be performed to return to the neighborhood of the solution, and to avoid the problem a new and more reasonable value of $\breve{\alpha}_0$ in the current bracket is used in Step 6*c*. A value of $\chi = 0.75$ will ensure that $\breve{\alpha}_0$ is no closer to α_U than 25% of the permissible range. Note that under the above circumstances, the inexact search of Algorithm 4.6 may fail to exit Step 7.

If the DFP or BFGS updating formula is used in Step 8*b* and the condition in Eq. (7.38) is satisfied, then a positive definite matrix $\mathbf{S}_k$ will result in a positive definite $\mathbf{S}_{k+1}$, as was discussed just after the proof of Theorem 7.2. We will now demonstrate that if the Fletcher inexact line search is used and the search is not terminated until the inequality in Eq. (4.59) is satisfied, then Eq. (7.38) is, indeed, satisfied. When the search is terminated in the *k*th iteration, we have $\alpha_0 \equiv \alpha_k$ and from Step 3 of the algorithm $\boldsymbol{\delta}_k = \alpha_k \mathbf{d}_k$. Now from Eqs. (7.38) and (4.59), we obtain

$$\boldsymbol{\delta}_k^T \boldsymbol{\gamma}_k = \boldsymbol{\delta}_k^T \mathbf{g}_{k+1} - \boldsymbol{\delta}_k^T \mathbf{g}_k$$
$$= \alpha_k (\mathbf{g}_{k+1}^T \mathbf{d}_k - \mathbf{g}_k^T \mathbf{d}_k)$$
$$\geq \alpha_k (\sigma - 1) \mathbf{g}_k^T \mathbf{d}_k$$

If $\mathbf{d}_k$ is a descent direction, then $\mathbf{g}_k^T \mathbf{d}_k < 0$ and $\alpha_k > 0$. Since $\sigma < 1$ in Fletcher's inexact line search, we conclude that

$$\boldsymbol{\delta}_k^T \boldsymbol{\gamma}_k > 0$$

and, in effect, the positive definiteness of $\mathbf{S}_k$ is assured. In exceptional circumstances, the inexact line search may not force the condition in Eq. (4.59), for example, when interpolation or extrapolation is aborted, if $|f_L - f_0| < \varepsilon_2$, and a nonpositive definite $\mathbf{S}_{k+1}$ matrix may occur. To safeguard against this possibility and ensure that a descent direction is achieved in every iteration, the quantity $\boldsymbol{\delta}_k^T \boldsymbol{\gamma}_k$ is checked in Step 8*b* and if it is found to be negative or zero, the identity matrix $\mathbf{I}_n$ is assigned to $\mathbf{S}_{k+1}$. This is not catastrophic and it may actually be beneficial since the next change in $\mathbf{x}$ will be in the steepest-descent direction.

The algorithm will be terminated in Step 7*c* if the distance between two successive points and the reduction in the objective function $f(\mathbf{x})$ are less than ε_1. One could, of course, use different tolerances for $\mathbf{x}$ and $f(\mathbf{x})$ and, depending on the problem, one of the two conditions may not even be required.

As may be recalled, the DFP and BFGS updating formulas are closely interrelated through the principle of *duality* and one can be obtained from the other and vice versa through the use of the Sherman–Morrison formula (see Sect. 7.6). Consequently, there are no clear theoretical advantages that apply to the one and not the other formula. Nevertheless, extensive experimental results reported by Fletcher [13] show that the use of the BFGS formula tends to yield algorithms that are somewhat more efficient in a number of different problems. This is consistent with the experience of the authors.

Problems

7.1 Let $\boldsymbol{\xi}$ be a nonzero column vector. Show that matrix $\mathbf{M} = \boldsymbol{\xi}\boldsymbol{\xi}^T$ has a rank of one and is symmetric and positive semidefinite.

7.2 In a quasi-Newton algorithm, $\mathbf{S}_{k+1}$ is obtained from a positive definite matrix $\mathbf{S}_k$ by using the DFP updating formula. Show that the condition

$$\boldsymbol{\delta}_k^T \boldsymbol{\gamma}_k > 0$$

will ensure that $\mathbf{S}_{k+1}$ is positive definite.

7.3 Minimize the objective function in Problem 5.4 by applying the DFP algorithm (e.g., Algorithm 7.3 with the DFP updating formula) using $\mathbf{x}_0 = [0\ 0]^T$ and $\varepsilon = 3 \times 10^{-7}$. Compare the results with those obtained in Problem 5.4.

7.4 Minimize the objective function in Problem 5.6 by applying the DFP algorithm using $\mathbf{x}_0 = [1\ 1\ 1]^T$ and $\varepsilon = 10^{-6}$. Compare the results with those obtained in Problems 5.6 and 6.4.

7.5 Minimize the objective function in Problem 5.8 by applying the DFP algorithm using $\varepsilon = 10^{-6}$, $\mathbf{x}_0 = [4\ -4]^T$, and $\mathbf{x}_0 = [-4\ 4]^T$. Compare the results with those obtained in Problems 5.8 and 6.5.

7.6 Minimize the objective function in Problem 5.10 by applying the DFP algorithm using $\mathbf{x}_0 = [0.1\ 0.1]^T$ and $\varepsilon = 10^{-6}$. Compare the results with those obtained in Problems 5.10 and 6.6.

7.7 It is known that the function

$$f(\mathbf{x}) = 100(x_1^3 - x_2)^2 + (x_1 - 1)^2$$

has a unique global minimizer $\mathbf{x}^* = [1\ 1]^T$.
Applying the DFP algorithm, minimize the above objective function. Try three initial points $\mathbf{x}_0 = [4\ 2]^T$, $\mathbf{x}_0 = [2\ -4]^T$ and $\mathbf{x}_0 = [-2\ 9]^T$ and assume a termination tolerance $\varepsilon = 10^{-6}$.

7.8 Minimize the objective function in Problem 5.4 by applying the BFGS algorithm (e.g., Algorithm 7.3 with the BFGS updating formula) using $\mathbf{x}_0 = [0\ 0]^T$ and $\varepsilon = 3 \times 10^{-7}$. Compare the results with those obtained in Problems 5.4 and 7.3.

7.9 Minimize the objective function in Problem 5.6 by applying the BFGS algorithm using $\mathbf{x}_0 = [1\ 1\ 1]^T$ and $\varepsilon = 10^{-6}$. Compare the results with those obtained in Problems 5.6, 6.4, and 7.4.

7.10 Minimize the objective function in Problem 5.8 by applying the BFGS algorithm using $\varepsilon = 10^{-6}$, $\mathbf{x}_0 = [4 \ -4]^T$, and $\mathbf{x}_0 = [-4 \ 4]^T$. Compare the results with those obtained in Problems 5.8, 6.5, and 7.5.

7.11 Minimize the objective function in Problem 5.10 by applying the BFGS algorithm using $\mathbf{x}_0 = [0.1 \ 0.1]^T$ and $\varepsilon = 10^{-6}$. Compare the results with those obtained in Problems 5.10, 6.6, and 7.6.

7.12 (a) Applying the BFGS algorithm, minimize the objective function in Problem 7.7. Try the same initial points and assume the same termination tolerance as in Problem 7.7.

(b) Compare the computational efficiencies of the DFP and BFGS algorithms in terms of the average CPU time required.

7.13 Using the program constructed in Problem 7.7, minimize the function

$$f(\mathbf{x}) = 100[(x_3 - 10\theta)^2 + (r - 1)^2] + x_3^2$$

where

$$\theta = \begin{cases} \dfrac{1}{2\pi} \tan^{-1}\left(\dfrac{x_2}{x_1}\right) & \text{for } x_1 > 0 \\ 0.25 & \text{for } x_1 = 0 \\ 0.5 + \dfrac{1}{2\pi} \tan^{-1}\left(\dfrac{x_2}{x_1}\right) & \text{for } x_1 < 0 \end{cases}$$

and

$$r = \sqrt{(x_1^2 + x_2^2)}$$

Repeat with the program constructed in Problem 7.12 and compare the results obtained.

7.14 Using the program constructed in Problem 7.7, minimize the function

$$f(\mathbf{x}) = (x_1 + 10x_2)^2 + 5(x_3 - x_4)^2 + (x_2 - 2x_3)^4 + 100(x_1 - x_4)^4$$

Repeat with the program constructed in Problem 7.12 and compare the results obtained.

7.15 Using the program constructed in Problem 7.7, minimize the function

$$f(\mathbf{x}) = \sum_{i=2}^{5} [100(x_i - x_{i-1}^2)^2 + (1 - x_i)^2]$$

Repeat with the program constructed in Problem 7.12 and compare the results obtained.

7.16 An interesting variant of the BFGS method is to modify the formula in Eq. (7.57) by replacing $\mathbf{S}_k$ by the identity matrix, which gives

$$\mathbf{S}_{k+1} = \mathbf{I} + \left(1 + \frac{\boldsymbol{\gamma}_k^T \boldsymbol{\gamma}_k}{\boldsymbol{\gamma}_k^T \boldsymbol{\delta}_k}\right) \frac{\boldsymbol{\delta}_k \boldsymbol{\delta}_k^T}{\boldsymbol{\gamma}_k^T \boldsymbol{\delta}_k} - \frac{\boldsymbol{\delta}_k \boldsymbol{\gamma}_k^T + \boldsymbol{\gamma}_k \boldsymbol{\delta}_k^T}{\boldsymbol{\gamma}_k^T \boldsymbol{\delta}_k}$$

Since $\mathbf{S}_{k+1}$ is now determined without reference to $\mathbf{S}_k$, the above updating formula is known as a *memoryless* BFGS formula [1, Chap. 10].

Verify that the memoryless BFGS method can be implemented without explicitly updating matrix $\mathbf{S}_k$. Instead, point $\mathbf{x}_k$ is updated as

$$\mathbf{x}_{k+1} = \mathbf{x}_k + \alpha_k \mathbf{d}_k$$

where α_k is determined by using a line search, and $\mathbf{d}_k$ is updated using the formula

$$\mathbf{d}_{k+1} = -\mathbf{g}_{k+1} + \eta_1 \boldsymbol{\gamma}_k + (\eta_2 - \eta_3)\boldsymbol{\delta}_k$$

where

$$\eta_1 = \boldsymbol{\delta}_k^T \mathbf{g}_{k+1}/\eta_4, \quad \eta_2 = \boldsymbol{\gamma}_k^T \mathbf{g}_{k+1}/\eta_4$$

$$\eta_3 = \left(1 + \frac{\boldsymbol{\gamma}_k^T \boldsymbol{\gamma}_k}{\eta_4}\right)\eta_1, \quad \eta_4 = \boldsymbol{\gamma}_k^T \boldsymbol{\delta}_k$$

7.17 Minimize the objective function in Problem 5.8 by applying the memoryless BFGS method using $\varepsilon = 10^{-6}$, $\mathbf{x}_0 = [4 \ -4]^T$, and $\mathbf{x}_0 = [-4 \ 4]^T$. Compare the results with those obtained in Problems 5.8, 6.5, 7.5, and 7.10.

References

1. D. G. Luenberger and Y. Ye, *Linear and Nonlinear Programming*, 4th ed. New York: Springer, 2008.
2. C. G. Broyden, "Quasi-Newton methods and their application to function minimization," *Maths. Comput.*, vol. 21, pp. 368–381, 1965.
3. W. C. Davidon, "Variable metric method for minimization," AEC Res. and Dev. Report ANL-5990, 1959.
4. A. V. Fiacco and G. P. McCormick, *Nonlinear Programming: Sequential Unconstrained Minimization Techniques*. New York: Wiley, 1968, (republished by SIAM in 1990).
5. B. A. Murtagh and R. W. H. Sargent, "A constrained minimization method with quadratic convergence," in *Optimization*, R. Fletcher, Ed. London: Academic Press, 1969, pp. 215–246.
6. P. Wolfe, "Methods of nonlinear programming," in *Nonlinear Programming*, J. Abadie, Ed. New York: Interscience, Wiley, 1967, pp. 97–131.
7. R. Fletcher and M. J. D. Powell, "A rapidly convergent descent method for minimization," *Computer J.*, vol. 6, pp. 163–168, 1963.
8. T. Kailath, *Linear Systems*. Englewood Cliffs, NJ: Prentice-Hall, 1980.

9. P. E. Gill, W. Murray, and M. H. Wright, *Numerical Linear Algebra and Optimization, Vol. I.* Reading: Addison-Wesley, 1991.

10. R. Fletcher, "A new approach to variable metric algorithms," *Computer J.*, vol. 13, pp. 317–322, 1970.

11. D. Goldfarb, "A family of variable metric methods derived by variational means," *Maths. Comput.*, vol. 24, pp. 23–26, 1970.

12. D. F. Shanno, "Conditioning of quasi-Newton methods for function minimization," *Maths. Comput.*, vol. 24, pp. 647–656, 1970.

13. R. Fletcher, *Practical Methods of Optimization,* vol. 1. New York: Wiley, 1980.

14. R. Fletcher, *Practical Methods of Optimization*, 2nd ed. New York: Wiley, 1987.

15. S. Hoshino, "A formulation of variable metric methods," *J. Inst. Maths. Applns.*, vol. 10, pp. 394–403, 1972.

16. H. Y. Huang, "Unified approach to quadratically convergent algorithms for function minimization," *J. Opt. Theo. Applns.*, vol. 5, pp. 405–423, 1970.

17. G. P. McCormick and J. D. Pearson, "Variable metric methods and unconstrained optimization," in *Optimization*, R. Fletcher, Ed. London: Academic Press, 1969.

18. J. D. Pearson, "Variable metric methods of minimization," *Computer J.*, vol. 12, pp. 171–178, 1969.

19. A. Antoniou, *Digital Signal Processing: Signals, Systems, and Filters.* New York: McGraw-Hill, 2005.

Minimax Methods

8

8.1 Introduction

In many scientific and engineering applications it is often necessary to minimize the maximum of some quantity with respect to one or more independent variables. Algorithms that can be used to solve problems of this type are said to be *minimax algorithms*. In the case where the quantity of interest depends on a real-valued parameter w that belongs to a set $\mathcal{S}$, the objective function can be represented by $f(\mathbf{x}, w)$ and the solution of the minimax problem pertaining to $f(\mathbf{x}, w)$ amounts to finding a vector variable $\mathbf{x}$ that minimizes the maximum of $f(\mathbf{x}, w)$ over $w \in \mathcal{S}$. There is also a discrete version of this problem in which the continuous parameter w is sampled to obtain discrete values $\mathcal{S}_d = \{w_i : i = 1, \ldots, L\} \subset \mathcal{S}$ and the corresponding minimax optimization problem is to find a vector $\mathbf{x}$ that minimizes the maximum of $f(\mathbf{x}, w_i)$ over $w_i \in \mathcal{S}_d$.

This chapter is concerned with efficient minimax algorithms. In Sect. 8.2, we illustrate minimax optimization using an example from digital signal processing. Two minimax algorithms due to Charalambous [1,2] are studied in Sect. 8.3 and improved versions of these algorithms using a technique of nonuniform variable sampling [3] are presented in Sect. 8.4.

8.2 Problem Formulation

A minimax problem pertaining to objective function $f(\mathbf{x}, w)$ can be formally stated as

$$\underset{\mathbf{x}}{\text{minimize}} \ \max_{w \in \mathcal{S}} f(\mathbf{x}, w) \tag{8.1a}$$

© Springer Science+Business Media, LLC, part of Springer Nature 2021
A. Antoniou and W.-S. Lu, *Practical Optimization*, Texts in Computer Science,
https://doi.org/10.1007/978-1-0716-0843-2_8

where S is a compact set on the w axis, and if $f(\mathbf{x}, w)$ is sampled with respect to w we have

$$\underset{\mathbf{x}}{\text{minimize}} \quad \underset{w_i \in \mathcal{S}_d}{\max} f(\mathbf{x}, w_i) \qquad (8.1\text{b})$$

where $\mathcal{S}_d = \{w_i : i = 1, 2, \ldots, L\}$ is a discrete version of set S. Obviously, the problems in Eqs. (8.1a) and (8.1b) are closely interrelated, and subject to the condition that the sampling of S is sufficiently dense, an approximate solution of the problem in Eq. (8.1a) can be obtained by solving the discrete problem in Eq. (8.1b).

As an illustrative example, let us consider a problem encountered in the field of digital signal processing whereby a digital filter needs to be designed [4, Chap. 16].[1] In this design problem, we require a transfer function of the form

$$H(z) = \frac{\displaystyle\sum_{i=0}^{N} a_i z^{-i}}{1 + \displaystyle\sum_{i=1}^{N} b_i z^{-i}} \qquad (8.2)$$

where z is a complex variable and a_i, b_i are real coefficients (see Sect. B.5.1) such that the amplitude response of the filter

$$M(\mathbf{x}, \omega) = |H(e^{j\omega T})| \qquad (8.3)$$

approximates a specified amplitude response $M_0(\omega)$. Vector $\mathbf{x}$ in Eq. (8.3) is defined as

$$\mathbf{x} = [a_0\ a_1\ \cdots\ a_N\ b_1\ \cdots\ b_N]^T$$

and ω denotes the frequency, which can assume values in the range of interest Ω. In the case of a lowpass digital filter, the desired amplitude response, $M_0(\omega)$, is assumed to be a piecewise constant function, as illustrated in Fig. 8.1 (see Sect. B.9.1). The difference between $M(\mathbf{x}, \omega)$ and $M_0(\omega)$, which is, in effect, the approximation error, can be expressed as

$$e(\mathbf{x}, \omega) = M(\mathbf{x}, \omega) - M_0(\omega) \qquad (8.4)$$

(see Sect. B.9.3).

The design of a digital filter can be accomplished by minimizing one of the norms described in Sect. A.8.1. If the L_1 or L_2 norm is minimized, then the sum of the magnitudes or the sum of the squares of the elemental errors is minimized. The minimum error thus achieved usually turns out to be unevenly distributed with respect to frequency and may exhibit large peaks which are often objectionable. If prescribed amplitude response specifications are to be met, the magnitude of the largest elemental error should be minimized and, therefore, the L_∞ norm of the error function should be used. Since the L_∞ norm of the error function $e(\mathbf{x}, \omega)$ in

---

[1]See Appendix B for a brief summary of the basics of digital filters.

Fig. 8.1 Formulation of objective function

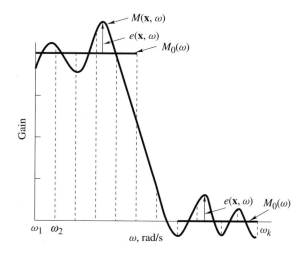

Eq. (8.4) is numerically equal to $\max_{\omega \in \Omega} |e(\mathbf{x}, \omega)|$, the minimization of the L_∞ norm can be expressed as

$$\underset{\mathbf{x}}{\text{minimize}} \ \max_{\omega \in \Omega} |e(\mathbf{x}, \omega)| \tag{8.5}$$

This is a minimax problem of the type stated in Eq. (8.1a) where the objective function is the magnitude of the approximation error, i.e., $f(\mathbf{x}, \omega) = |e(\mathbf{x}, \omega)|$.

The application of minimax algorithms for the design of digital filters usually yields designs in which the error is uniformly distributed with respect to frequency.

8.3 Minimax Algorithms

The most fundamental algorithm for the minimax optimization problem in Eq. (8.5) is the so-called *least-pth* algorithm, which involves minimizing an objective function in the form of a sum of elemental error functions, each raised to the pth power, for increasing values of p, say, $p = 2, 4, 8, \ldots$, etc.

Let $\omega_1, \omega_2, \ldots, \omega_K$ be K frequencies in Ω and define vector

$$\mathbf{e}(\mathbf{x}) = [e_1(\mathbf{x}) \ e_2(\mathbf{x}) \ \cdots \ e_n(\mathbf{x})]^T$$

where $e_i(\mathbf{x}) \equiv e(\mathbf{x}, \omega_i)$ is evaluated using Eq. (8.4). If we denote the L_p norm of vector $\mathbf{e}(\mathbf{x})$ at $\mathbf{x} = \mathbf{x}_k$ as $\Psi_k(\mathbf{x})$, i.e.,

$$\Psi_k(\mathbf{x}) = \|\mathbf{e}(\mathbf{x})\|_p = \left[\sum_{i=1}^{K} |e_i(\mathbf{x})|^p \right]^{1/p}$$

then we have

$$\lim_{p \to \infty} \Psi_k(\mathbf{x}) = \lim_{p \to \infty} \|\mathbf{e}(\mathbf{x})\|_p = \|\mathbf{e}(\mathbf{x})\|_\infty = \max_{1 \le i \le K} |e_i(\mathbf{x})| \equiv \widehat{E}(\mathbf{x})$$

In other words, by minimizing function $\Psi_k(\mathbf{x})$ for increasing power of p, the minimization of the L_∞ norm of $\mathbf{e}(\mathbf{x})$ can be achieved.

In a practical design, the approximation error $\|\mathbf{e}(\mathbf{x})\|_\infty$ is always strictly greater than zero and thus function $\Psi_k(\mathbf{x})$ can be expressed as

$$\Psi_k(\mathbf{x}) = \widehat{E}(\mathbf{x}) \left\{ \sum_{i=1}^{K} \left[\frac{|e_i(\mathbf{x})|}{\widehat{E}(\mathbf{x})} \right]^p \right\}^{1/p} \tag{8.6a}$$

where

$$e_i(\mathbf{x}) \equiv e(\mathbf{x}, \, \omega_i) \tag{8.6b}$$

$$\widehat{E}(\mathbf{x}) = \max_{1 \le i \le K} |e_i(\mathbf{x})| \tag{8.6c}$$

These principles lead readily to the so-called *least-pth minimax algorithm*, which is as follows [1]:

Algorithm 8.1 Least-pth minimax algorithm
Step 1
Input $\breve{\mathbf{x}}_0$ and ε_1. Set $k = 1$, $p = 2$, $\mu = 2$, and $\widehat{E}_0 = 10^{99}$.
Step 2
Initialize frequencies $\omega_1, \, \omega_2, \, \ldots, \, \omega_K$.
Step 3
Using $\breve{\mathbf{x}}_{k-1}$ as initial point, minimize $\Psi_k(\mathbf{x})$ in Eq. (8.6a) with respect to $\mathbf{x}$, to obtain $\breve{\mathbf{x}}_k$. Set $\widehat{E}_k = \widehat{E}(\breve{\mathbf{x}}_k)$.
Step 4
If $|\widehat{E}_{k-1} - \widehat{E}_k| < \varepsilon_1$, then output $\breve{\mathbf{x}}_k$ and $\widehat{E}_k$, and stop. Otherwise, set $p = \mu p$ and $k = k + 1$, and go to step 3.

The underlying principle for the above algorithm is that the minimax problem is solved by solving a sequence of closely related problems whereby the solution of one problem renders the solution of the next one more tractable. Parameter μ in step 1, which must obviously be an integer, should not be too large in order to avoid numerical ill-conditioning. A value of 2 gives good results.

The minimization in step 3 can be carried out by using any unconstrained optimization algorithm, for example, Algorithm 7.3 described in Sect. 7.10. The gradient of $\Psi_k(\mathbf{x})$ is given by [1]

$$\nabla \Psi_k(\mathbf{x}) = \left\{ \sum_{i=1}^{K} \left[\frac{|e_i(\mathbf{x})|}{\widehat{E}(\mathbf{x})} \right]^p \right\}^{(1/p)-1} \sum_{i=1}^{K} \left[\frac{|e_i(\mathbf{x})|}{\widehat{E}(\mathbf{x})} \right]^{p-1} \nabla |e_i(\mathbf{x})| \tag{8.7}$$

The preceding algorithm works very well, except that it requires a considerable amount of computation. An alternative and much more efficient minimax algorithm is one described in [5,6]. This algorithm is based on principles developed by Charalambous [2] and involves the minimization of the objective function

$$\Psi(\mathbf{x}, \lambda, \xi) = \sum_{i \in I_1} \tfrac{1}{2} \lambda_i [\phi_i(\mathbf{x}, \, \xi)]^2 + \sum_{i \in I_2} \tfrac{1}{2} [\phi_i(\mathbf{x}, \, \xi)]^2 \tag{8.8}$$

where ξ and λ_i for $i = 1, 2, \ldots, K$ are constants and

$$\phi_i(\mathbf{x}, \xi) = |e_i(\mathbf{x})| - \xi$$

$$I_1 = \{i : \phi_i(\mathbf{x}, \xi) > 0 \text{ and } \lambda_i > 0\} \tag{8.9a}$$

$$I_2 = \{i : \phi_i(\mathbf{x}, \xi) > 0 \text{ and } \lambda_i = 0\} \tag{8.9b}$$

The halves in Eq. (8.8) are included for the purpose of simplifying the expression for the gradient (see Eq. (8.11)).

If

(a) the second-order sufficiency conditions for a minimum of $\widehat{E}(\mathbf{x})$ hold at $\breve{\mathbf{x}}$,

(b) $\lambda_i = \breve{\lambda}_i$ for $i = 1, 2, \ldots, K$ where $\breve{\lambda}_i$ are the minimax multipliers corresponding to the minimum point $\breve{\mathbf{x}}$ of $\widehat{E}(\mathbf{x})$, and

(c) $\widehat{E}(\breve{\mathbf{x}} - \xi)$ is sufficiently small

then it can be proved that $\breve{\mathbf{x}}$ is a *strong* local minimum point of function $\Psi(\mathbf{x}, \boldsymbol{\lambda}, \xi)$ given by Eq. (8.8) (see [2] for details). In practice, the conditions in (a) are satisfied for most practical problems. Consequently, if multipliers λ_i are forced to approach the minimax multipliers $\breve{\lambda}_i$ and ξ is forced to approach $\widehat{E}(\breve{\mathbf{x}})$, then the minimization of $\widehat{E}(\mathbf{x})$ can be accomplished by minimizing $\Psi(\mathbf{x}, \boldsymbol{\lambda}, \xi)$ with respect to $\mathbf{x}$. A minimax algorithm based on these principles is as follows:

Algorithm 8.2 Charalambous minimax algorithm
Step 1
Input $\breve{\mathbf{x}}_0$ and ε_1. Set $k = 1$, $\xi_1 = 0$, $\lambda_{11} = \lambda_{12} = \cdots = \lambda_{1K} = 1$, and $\widehat{E}_0 = 10^{99}$.
Step 2
Initialize frequencies $\omega_1, \omega_2, \ldots, \omega_K$.
Step 3
Using $\breve{\mathbf{x}}_{k-1}$ as initial point, minimize $\Psi(\mathbf{x}, \boldsymbol{\lambda}_k, \xi_k)$ with respect to $\mathbf{x}$ to obtain $\breve{\mathbf{x}}_k$. Set

$$\widehat{E}_k = \widehat{E}(\breve{\mathbf{x}}_k) = \max_{1 \leq i \leq K} |e_i(\breve{\mathbf{x}}_k)| \tag{8.10}$$

Step 4
Compute

$$\Phi_k = \sum_{i \in I_1} \lambda_{ki} \phi_i(\breve{\mathbf{x}}_k, \xi_k) + \sum_{i \in I_2} \phi_i(\breve{\mathbf{x}}_k, \xi_k)$$

and update

$$
\lambda_{(k+1)i} = \begin{cases} \lambda_{ki}\phi_i(\breve{\mathbf{x}}_k, \xi_k)/\Phi_k & \text{for } i \in I_1 \\ \phi_i(\breve{\mathbf{x}}_k, \xi_k)/\Phi_k & \text{for } i \in I_2 \\ 0 & \text{for } i \in I_3 \end{cases}
$$

for $i = 1, 2, \ldots, K$ where

$$
I_1 = \{i : \ \phi_i(\breve{\mathbf{x}}_k, \xi_k) > 0 \text{ and } \lambda_{ki} > 0\}
$$

$$
I_2 = \{i : \ \phi_i(\breve{\mathbf{x}}_k, \xi_k) > 0 \text{ and } \lambda_{ki} = 0\}
$$

and

$$
I_3 = \{i : \ \phi_i(\breve{\mathbf{x}}_k, \xi_k) \le 0\}
$$

Step 5
Compute

$$
\xi_{k+1} = \sum_{i=1}^{K} \lambda_{(k+1)i} |e_i(\breve{\mathbf{x}})|
$$

Step 6
If $|\widehat{E}_{k-1} - \widehat{E}_k| < \varepsilon_1$, then output $\breve{\mathbf{x}}_k$ and $\widehat{E}_k$, and stop. Otherwise, set $k = k+1$ and go to step 3.

The gradient of $\Psi(\mathbf{x}, \lambda_k, \xi_k)$, which is required in step 3 of the algorithm, is given by

$$
\nabla\Psi(\mathbf{x}, \lambda_k, \xi_k) = \sum_{i\in I_1} \lambda_{ki}\phi_i(\mathbf{x}, \xi_k)\nabla|e_i(\mathbf{x})| + \sum_{i\in I_2} \phi_i(\mathbf{x}, \xi_k)\nabla|e_i(\mathbf{x})| \qquad (8.11)
$$

Constant ξ is a lower bound of the minimum of $\widehat{E}(\mathbf{x})$ and as the algorithm progresses, it approaches $\widehat{E}(\breve{\mathbf{x}})$ from below. Consequently, the number of functions $\phi_i(\mathbf{x}, \xi)$ that do not satisfy either Eq. (8.9a) or Eq. (8.9b) increases rapidly with the number of iterations. Since the derivatives of these functions are unnecessary in the minimization of $\Psi(\mathbf{x}, \lambda, \xi)$, they need not be evaluated. This increases the efficiency of the algorithm quite significantly.

As in Algorithm 8.1, the minimization in step 3 of Algorithm 8.2 can be carried out by using Algorithm 7.3.

Example 8.1 Consider the overdetermined system of linear equations

$$
\begin{aligned}
3x_1 - 4x_2 + 2x_3 - x_4 &= -17.4 \\
-2x_1 + 3x_2 + 6x_3 - 2x_4 &= -1.2 \\
x_1 + 2x_2 + 5x_3 + x_4 &= 7.35 \\
-3x_1 + x_2 - 2x_3 + 2x_4 &= 9.41 \\
7x_1 - 2x_2 + 4x_3 + 3x_4 &= 4.1 \\
10x_1 - x_2 + 8x_3 + 5x_4 &= 12.3
\end{aligned}
$$

which can be expressed as

$$\mathbf{Ax} = \mathbf{b} \tag{8.12a}$$

where

$$\mathbf{A} = \begin{bmatrix} 3 & -4 & 2 & -1 \\ -2 & 3 & 6 & -2 \\ 1 & 2 & 5 & 1 \\ -3 & 1 & -2 & 2 \\ 7 & -2 & 4 & 3 \\ 10 & -1 & 8 & 5 \end{bmatrix}, \quad \mathbf{b} = \begin{bmatrix} -17.4 \\ -1.2 \\ 7.35 \\ 9.41 \\ 4.1 \\ 12.3 \end{bmatrix} \tag{8.12b}$$

(a) Find the least-squares solution of Eq. (8.12a), $\mathbf{x}_{ls}$, by solving the minimization problem

$$\text{minimize } \|\mathbf{Ax} - \mathbf{b}\|_2 \tag{8.13}$$

(b) Find the minimax solution of Eq. (8.12a), $\mathbf{x}_{minimax}$, by applying Algorithm 8.2 to solve the minimization problem

$$\text{minimize } \|\mathbf{Ax} - \mathbf{b}\|_\infty \tag{8.14}$$

(c) Compare the magnitudes of the equation errors for the solutions $\mathbf{x}_{ls}$ and $\mathbf{x}_{minimax}$.

Solution (a) The square of the L_2 norm $\|\mathbf{Ax} - \mathbf{b}\|_2$ is found to be

$$\|\mathbf{Ax} - \mathbf{b}\|_2^2 = \mathbf{x}^T \mathbf{A}^T \mathbf{A} \mathbf{x} - 2\mathbf{x}^T \mathbf{A}^T \mathbf{b} + \mathbf{b}^T \mathbf{b}$$

It is easy to verify that matrix $\mathbf{A}^T \mathbf{A}$ is positive definite; hence $\|\mathbf{Ax} - \mathbf{b}\|_2^2$ is a strictly globally convex function whose unique minimizer is given by

$$\mathbf{x}_{ls} = (\mathbf{A}^T \mathbf{A})^{-1} \mathbf{A}^T \mathbf{b} = \begin{bmatrix} 0.6902 \\ 3.6824 \\ -0.7793 \\ 3.1150 \end{bmatrix} \tag{8.15}$$

(b) By denoting

$$\mathbf{A} = \begin{bmatrix} \mathbf{a}_1^T \\ \mathbf{a}_2^T \\ \vdots \\ \mathbf{a}_6^T \end{bmatrix}, \quad \mathbf{b} = \begin{bmatrix} b_1 \\ b_2 \\ \vdots \\ b_6 \end{bmatrix}$$

we can write

$$\mathbf{Ax} - \mathbf{b} = \begin{bmatrix} \mathbf{a}_1^T \mathbf{x} - b_1 \\ \mathbf{a}_2^T \mathbf{x} - b_2 \\ \vdots \\ \mathbf{a}_6^T \mathbf{x} - b_6 \end{bmatrix}$$

and the L_∞ norm $\|\mathbf{Ax} - \mathbf{b}\|_\infty$ can be expressed as

$$\|\mathbf{Ax} - \mathbf{b}\|_\infty = \max_{1 \le i \le 6} |\mathbf{a}_i^T \mathbf{x} - b_i|$$

Hence the problem in Eq. (8.14) becomes

$$\underset{\mathbf{x}}{\text{minimize}} \quad \max_{1 \le i \le 6} |e_i(\mathbf{x})| \tag{8.16}$$

where

$$e_i(\mathbf{x}) = \mathbf{a}_i^T \mathbf{x} - b_i$$

which is obviously a minimax problem. The gradient of $e_i(\mathbf{x})$ is simply given by

$$\nabla e_i(\mathbf{x}) = \mathbf{a}_i$$

By using the least-squares solution $\mathbf{x}_{ls}$ obtained in part (a) as the initial point and $\varepsilon_1 = 4 \times 10^{-6}$, it took Algorithm 8.2 four iterations to converge to the solution

$$\mathbf{x}_{minimax} = \begin{bmatrix} 0.7592 \\ 3.6780 \\ -0.8187 \\ 3.0439 \end{bmatrix} \tag{8.17}$$

In this example as well as Examples 8.2 and 8.3, the unconstrained optimization required in Step 3 was carried out using a quasi-Newton BFGS algorithm which was essentially Algorithm 7.3 with a slightly modified version of Step 8b as follows:

Step 8b′
Compute $D = \boldsymbol{\delta}_k^T \boldsymbol{\gamma}_k$. If $D \le 0$, then set $\mathbf{S}_{k+1} = \mathbf{I}_n$, otherwise, compute $\mathbf{S}_{k+1}$ using Eq. (7.57).

(c) Using Eqs. (8.15) and (8.17), the magnitudes of the equation errors for solutions $\mathbf{x}_{ls}$ and $\mathbf{x}_{minimax}$ were found to be

$$|\mathbf{Ax}_{ls} - \mathbf{b}| = \begin{bmatrix} 0.0677 \\ 0.0390 \\ 0.0765 \\ 0.4054 \\ 0.2604 \end{bmatrix} \quad \text{and} \quad |\mathbf{Ax}_{minimax} - \mathbf{b}| = \begin{bmatrix} 0.2844 \\ 0.2844 \\ 0.2843 \\ 0.2844 \\ 0.2844 \\ 0.2844 \end{bmatrix}$$

As can be seen, the minimax algorithm tends to equalize the equation errors.   ∎

8.4   Improved Minimax Algorithms

To achieve good results with the above minimax algorithms, the sampling of the objective function $f(\mathbf{x}, w)$ with respect to w must be dense; otherwise, the error in the objective function may develop spikes in the intervals between sampling points during the minimization. This problem is usually overcome by using a fairly large value of K of the order of 20 to 30 times the number of variables, depending on the type of optimization problem. For example, if a 10th-order digital filter is to be designed, i.e., $N = 10$ in Eq. (8.2), the objective function depends on 21 variables and a value of K as high as 630 may be required. In such a case, each function evaluation in the minimization of the objective function would involve computing the gain of the filter as many as 630 times. A single optimization may sometimes necessitate 300 to 600 function evaluations, and a minimax algorithm like Algorithm 8.1 or 8.2 may require 5 to 10 unconstrained optimizations to converge. Consequently, up to 3.8 million function evaluations may be required to complete a design.

A technique will now be described that can be used to suppress spikes in the error function without using a large value of K [3]. The technique entails the application of *nonuniform variable sampling* and it is described in terms of the filter-design problem considered earlier. The steps involved are as follows:

1. Evaluate the error function in Eq. (8.4) with respect to a dense set of uniformly-spaced frequencies that span the frequency band of interest, say, $\bar{w}_1, \bar{w}_2, \ldots, \bar{w}_L$, where L is fairly large of the order of $10 \times K$.
2. Segment the frequency band of interest into K intervals.
3. For each of the K intervals, find the frequency that yields maximum error. Let these frequencies be $\widehat{w}_i$ for $i = 1, 2, \ldots, K$.
4. Use frequencies $\widehat{w}_i$ as sample frequencies in the evaluation of the objective function, i.e., set $w_i = \widehat{w}_i$ for $i = 1, 2, \ldots, K$.

By applying the above nonuniform sampling technique before the start of the second and subsequent optimizations, *frequency points at which spikes are beginning to form are located and are used as sample points in the next optimization*. In this way, the error at these frequencies is reduced and the formation of spikes is prevented.

Assume that a digital filter is required to have a specified amplitude response with respect to a frequency band B which extends from $\bar{w}_1$ to $\bar{w}_L$, and let $\bar{w}_1, \bar{w}_2, \ldots, \bar{w}_L$ be uniformly spaced frequencies such that

$$\bar{w}_i = \bar{w}_{i-1} + \Delta w$$

for $i = 2, 3, \ldots, L$ where

$$\Delta w = \frac{\bar{w}_L - \bar{w}_1}{L - 1} \tag{8.18}$$

These frequency points may be referred to as *virtual sample points*. Band B can be segmented into K intervals, say, Ω_1 to Ω_K such that Ω_1 and Ω_K are of width $\Delta w/2$, Ω_2 and Ω_{K-1} are of width $l\Delta w$, and Ω_i for $i = 3, 4, \ldots, K - 2$ are of width $2l\Delta w$ where l is an integer. These requirements can be satisfied by letting

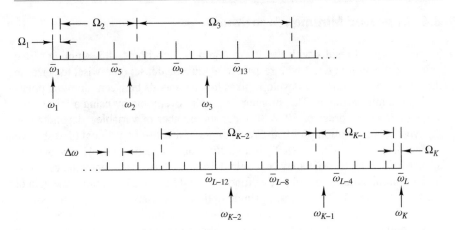

Fig. 8.2 Segmentation of frequency axis.

$$\Omega_1 = \left\{\omega : \bar{\omega}_1 \leq \omega < \bar{\omega}_1 + \tfrac{1}{2}\Delta\omega\right\}$$
$$\Omega_2 = \left\{\omega : \bar{\omega}_1 + \tfrac{1}{2}\Delta\omega \leq \omega < \bar{\omega}_1 + (l + \tfrac{1}{2})\Delta\omega\right\}$$
$$\Omega_i = \left\{\omega : \bar{\omega}_1 + \left[(2i-5)l + \tfrac{1}{2}\right]\Delta\omega \leq \omega < \bar{\omega}_1 + \left[(2i-3)l + \tfrac{1}{2}\right]\Delta\omega\right\}$$

for $i = 3, 4, \ldots, K - 2$

$$\Omega_{K-1} =$$
$$\left\{\omega : \bar{\omega}_1 + \left[(2K-7)l + \tfrac{1}{2}\right]\Delta\omega \leq \omega < \bar{\omega}_1 + \left[(2K-6)l + \tfrac{1}{2}\right]\Delta\omega\right\}$$

and

$$\Omega_K = \left\{\omega : \bar{\omega}_1 + \left[(2K-6)l + \tfrac{1}{2}\right]\Delta\omega \leq \omega \leq \bar{\omega}_L\right\}$$

where

$$\bar{\omega}_L = \bar{\omega}_1 + [(2K-6)l + 1]\Delta\omega. \tag{8.19}$$

The scheme is feasible if

$$L = (2K - 6)l + 2 \tag{8.20}$$

according to Eqs. (8.18) and (8.19), and is illustrated in Fig. 8.2 for the case where $K = 8$ and $l = 5$.

In the above segmentation scheme, there is only one sample in each of intervals Ω_1 and Ω_K, l samples in each of intervals Ω_2 and Ω_{K-1}, and $2l$ samples in each of intervals $\Omega_3, \Omega_4, \ldots, \Omega_{K-2}$, as can be seen in Fig. 8.2. Thus step 3 of the technique will yield $\hat{\omega}_1 = \bar{\omega}_1$ and $\hat{\omega}_K = \bar{\omega}_L$, i.e., the lower and upper band edges are forced to remain sample frequencies throughout the optimization. This strategy leads to two advantages: (*a*) the error at the band edges is always minimized, and (*b*) a somewhat higher sampling density is maintained near the band edges, where spikes are more likely to occur.

In the above technique, the required amplitude response, $M_0(\omega)$, needs to be specified with respect to a *dense* set of frequency points. If $M_0(\omega)$ is piecewise constant as in Fig. 8.1, then the required values of $M_0(\omega)$ can be easily obtained. If, on the other hand, $M_0(\omega)$ is specified by an array of numbers, the problem can be overcome through the use of interpolation. Let us assume that the amplitude response is specified at frequencies $\tilde{\omega}_1$ to $\tilde{\omega}_S$, where $\tilde{\omega}_1 = \bar{\omega}_1$ and $\tilde{\omega}_S = \bar{\omega}_L$. The required amplitude response for any frequency interval spanned by four successive specification points, say, $\tilde{\omega}_j \leq \omega \leq \tilde{\omega}_{j+3}$, can be represented by a third-order polynomial of ω of the form

$$M_0(\omega) = a_{0j} + a_{1j}\omega + a_{2j}\omega^2 + a_{3j}\omega^3 \tag{8.21}$$

and by varying j from 1 to $S - 3$, a set of $S - 3$ third-order polynomials can be obtained which can be used to interpolate the amplitude response to any desired degree of resolution. To achieve maximum interpolation accuracy, each of these polynomials should as far as possible be used at the center of the frequency range of its validity. Hence the first and last polynomials should be used for the frequency ranges $\tilde{\omega}_1 \leq \omega < \tilde{\omega}_3$ and $\tilde{\omega}_{S-2} \leq \omega \leq \tilde{\omega}_S$, respectively, and the jth polynomial for $2 \leq j \leq S - 4$ should be used for the frequency range $\tilde{\omega}_{j+1} \leq \omega < \tilde{\omega}_{j+2}$.

Coefficients a_{ij} for $i = 0, 1, \ldots, 3$ and $j = 1$ to $S - 3$ can be determined by computing $\tilde{\omega}_m$, $(\tilde{\omega}_m)^2$, and $(\tilde{\omega}_m)^3$ for $m = j, j + 1, \ldots, j + 3$, and then constructing the system of simultaneous equations

$$\tilde{\boldsymbol{\Omega}}_j \mathbf{a}_j = \mathbf{M}_{0j} \tag{8.22}$$

where

$$\mathbf{a}_j = \begin{bmatrix} a_{0j} & \cdots & a_{3j} \end{bmatrix}^T \quad \text{and} \quad \mathbf{M}_{0j} = \begin{bmatrix} M_0(\tilde{\omega}_j) & \cdots & M_0(\tilde{\omega}_{j+3}) \end{bmatrix}^T$$

are column vectors and $\tilde{\boldsymbol{\Omega}}_j$ is the 4×4 matrix given by

$$\tilde{\boldsymbol{\Omega}}_j = \begin{bmatrix} 1 & \tilde{\omega}_j & (\tilde{\omega}_j)^2 & (\tilde{\omega}_j)^3 \\ 1 & \tilde{\omega}_{j+1} & (\tilde{\omega}_{j+1})^2 & (\tilde{\omega}_{j+1})^3 \\ 1 & \tilde{\omega}_{j+2} & (\tilde{\omega}_{j+2})^2 & (\tilde{\omega}_{j+2})^3 \\ 1 & \tilde{\omega}_{j+3} & (\tilde{\omega}_{j+3})^2 & (\tilde{\omega}_{j+3})^3 \end{bmatrix}$$

Therefore, from Eq. (8.22) we have

$$\mathbf{a}_j = \tilde{\boldsymbol{\Omega}}_j^{-1} \mathbf{M}_{0j}. \tag{8.23}$$

The above nonuniform sampling technique can be incorporated in Algorithm 8.1 by replacing steps 1, 2, and 4 as shown below. The filter to be designed is assumed to be a single-band filter, for the sake of simplicity, although the technique is applicable to filters with an arbitrary number of bands.

Algorithm 8.3 Modified version of Algorithm 8.1
Step 1

(a) Input $\breve{\mathbf{x}}_0$ and ε_1. Set $k = 1, p = 2, \mu = 2$, and $\widehat{E}_0 = 10^{99}$. Initialize K.

(b) Input the required amplitude response $M_0(\tilde{\omega}_m)$ for $m = 1, 2, \ldots, S$.

(c) Compute L and $\Delta\omega$ using Eqs. (8.20) and (8.18), respectively.

(d) Compute coefficients a_{ij} for $i = 0, 1, \ldots, 3$ and $j = 1$ to $S - 3$ using Eq. (8.23).

(e) Compute the required ideal amplitude response for $\bar{\omega}_1, \bar{\omega}_2, \ldots, \bar{\omega}_L$ using Eq. (8.21).

Step 2
Set $\omega_1 = \bar{\omega}_1, \omega_2 = \bar{\omega}_{1+l}, \omega_i = \bar{\omega}_{2(i-2)l+1}$ for $i = 3, 4, \ldots, K - 2, \omega_{K-1} = \bar{\omega}_{L-l}$, and $\omega_K = \bar{\omega}_L$.
Step 3
Using $\breve{\mathbf{x}}_{k-1}$ as initial value, minimize $\Psi_k(\mathbf{x})$ in Eq. (8.6a) with respect to $\mathbf{x}$, to obtain $\breve{\mathbf{x}}_k$. Set $\widehat{E}_k = \widehat{E}(\breve{\mathbf{x}}_k)$.
Step 4

(a) Compute $|e_i(\breve{\mathbf{x}}_k)|$ for $i = 1, 2, \ldots, L$ using Eqs. (8.4) and (8.6b).

(b) Determine frequencies $\widehat{\omega}_i$ for $i = 1, 2, \ldots, K$ and

$$\widehat{P}_k = \widehat{P}(\breve{\mathbf{x}}_k) = \max_{1 \leq i \leq L} |e_i(\breve{\mathbf{x}}_k)| \tag{8.24}$$

(c) Set $\widehat{\omega}_i$ for $i = 1, 2, \ldots, K$.

(d) If $|\widehat{E}_{k-1} - \widehat{E}_k| < \varepsilon_1$ and $|\widehat{P}_k - \widehat{E}_k| < \varepsilon_1$, then output $\breve{\mathbf{x}}_k$ and $\widehat{E}_k$, and stop. Otherwise, set $p = \mu p, k = k + 1$ and go to step 3.

The above nonuniform variable sampling technique can be applied to Algorithm 8.2 by replacing steps 1, 2, and 6 as follows:

Algorithm 8.4 Modified version of Algorithm 8.2
Step 1

(a) Input $\breve{\mathbf{x}}_0$ and ε_1. Set $k = 1, \xi_1 = 0, \lambda_{11} = \lambda_{12} = \cdots = \lambda_{1K} = 1$, and $\widehat{E}_0 = 10^{99}$. Initialize K.

(b) Input the required amplitude response $M_0(\tilde{\omega}_m)$ for $m = 1, 2, \ldots, S$.

(c) Compute L and $\Delta\omega$ using Eqs. (8.20) and (8.18), respectively.

(d) Compute coefficients a_{ij} for $i = 0, 1, \ldots, 3$ and $j = 1$ to $S - 3$ using Eq. (8.23).

(e) Compute the required ideal amplitude response for $\bar{\omega}_1, \bar{\omega}_2, \ldots, \bar{\omega}_L$ using Eq. (8.21).

Step 2
Set $\omega_1 = \bar{\omega}_1$, $\omega_2 = \bar{\omega}_{1+l}$, $\omega_i = \bar{\omega}_{2(i-2)l+1}$ for $i = 3, 4, \ldots, K-2$, $\omega_{K-1} = \bar{\omega}_{L-l}$, and $\omega_K = \bar{\omega}_L$.

Step 3
Using $\breve{\mathbf{x}}_{k-1}$ as initial value, minimize $\Psi(\mathbf{x}, \lambda_k, \xi_k)$ with respect to $\mathbf{x}$ to obtain $\breve{\mathbf{x}}_k$. Set

$$\widehat{E}_k = \widehat{E}(\breve{\mathbf{x}}_k) = \max_{1 \le i \le K} |e_i(\breve{\mathbf{x}}_k)|$$

Step 4
Compute

$$\Phi_k = \sum_{i \in I_1} \lambda_{ki} \phi_i(\breve{\mathbf{x}}_k, \xi_k) + \sum_{i \in I_2} \phi_i(\breve{\mathbf{x}}_k, \xi_k)$$

and update

$$\lambda_{(k+1)i} = \begin{cases} \lambda_{ki} \phi_i(\breve{\mathbf{x}}_k, \xi_k)/\Phi_k & \text{for } i \in I_1 \\ \phi_i(\breve{\mathbf{x}}_k, \xi_k)/\Phi_k & \text{for } i \in I_2 \\ 0 & \text{for } i \in I_3 \end{cases}$$

for $i = 1, 2, \ldots, K$ where

$$I_1 = \{i : \phi_i(\breve{\mathbf{x}}_k, \xi_k) > 0 \text{ and } \lambda_{ki} > 0\}$$

$$I_2 = \{i : \phi_i(\breve{\mathbf{x}}_k, \xi_k) > 0 \text{ and } \lambda_{ki} = 0\}$$

and

$$I_3 = \{i : \phi_i(\breve{\mathbf{x}}_k, \xi_k) \le 0\}$$

Step 5
Compute

$$\xi_{k+1} = \sum_{i=1}^{K} \lambda_{(k+1)i} |e_i(\breve{\mathbf{x}})|$$

Step 6

(a) Compute $|e_i(\breve{\mathbf{x}}_k)|$ for $i = 1, 2, \ldots, L$ using Eqs. (8.4) and (8.6b).

(b) Determine frequencies $\widehat{\omega}_i$ for $i = 1, 2, \ldots, K$ and

$$\widehat{P}_k = \widehat{P}(\breve{\mathbf{x}}_k) = \max_{1 \le i \le L} |e_i(\breve{\mathbf{x}}_k)|$$

(c) Set $\omega_i = \widehat{\omega}_i$ for $i = 1, 2, \ldots, K$.

(d) If $|\widehat{E}_{k-1} - \widehat{E}_k| < \varepsilon_1$ and $|\widehat{P}_k - \widehat{E}_k| < \varepsilon_1$, then output $\breve{\mathbf{x}}_k$ and $\widehat{E}_k$, and stop. Otherwise, set $k = k + 1$ and go to step 3.

In step 2, the initial sample frequencies ω_1 and ω_K are assumed to be at the left-hand and right-hand band edges, respectively; ω_2 and ω_{K-1} are taken to be the last and first frequencies in intervals Ω_2 and Ω_{K-1}, respectively; and each of frequencies $\omega_3, \omega_4, \ldots, \omega_{K-2}$ is set near the center of the respective interval $\Omega_3, \Omega_4, \ldots, \Omega_{K-2}$. This assignment is illustrated in Fig. 8.2.

Without the nonuniform sampling technique, the number of samples K should be chosen to be of the order of 20 to 30 times the number of variables, depending on the selectivity of the filter, as was mentioned in the first paragraph of Sect. 8.4. If the above technique is used, the number of virtual sample points is approximately equal to $2l \times K$, according to Eq. (8.20). As l is increased above unity, the frequencies of maximum error, $\widehat{\omega}_i$, become progressively more precise, owing to the increased resolution; however, the amount of computation required in step 4 of Algorithm 8.3 or step 6 of Algorithm 8.4 is proportionally increased. Eventually, a situation of diminishing returns is reached whereby further increases in l bring about only slight improvements in the precision of the $\widehat{\omega}_i$'s. With $l = 5$, a value of K in the range of 2 to 6 times the number of variables was found to give good results for a diverse range of designs. In effect, the use of the nonuniform sampling technique in the minimax algorithms described would lead to a reduction in the amount of computation of the order of 75 percent.

Example 8.2 (a) Applying Algorithm 8.1, design a 10th-order lowpass IIR digital filter assuming a transfer function of the form given in Eq. (8.2). The desired amplitude response is

$$M_0(\omega) = \begin{cases} 1 & \text{for } 0 \le \omega \le \omega_p \text{ rad/s} \\ 0 & \text{for } \omega_a \le \omega \le \pi \text{ rad/s} \end{cases} \tag{8.25}$$

where $\omega_p = 0.4\pi$, $\omega_a = 0.5\pi$, and the sampling frequency is 2π.

(b) Applying Algorithm 8.3, design the digital filter specified in part (a).

Solution (a) Using Eqs. (8.2) and (8.3), the amplitude response of the filter is obtained as

$$M(\mathbf{x}, \omega) = \left| \frac{a_0 + a_1 e^{-j\omega} + \cdots + a_N e^{-jN\omega}}{1 + b_1 e^{-j\omega} + \cdots + b_N e^{-jN\omega}} \right| \tag{8.26}$$

If we denote

$$\mathbf{a} = \begin{bmatrix} a_0 \\ a_1 \\ \vdots \\ a_N \end{bmatrix}, \quad \mathbf{b} = \begin{bmatrix} b_1 \\ b_2 \\ \vdots \\ b_N \end{bmatrix}, \quad \mathbf{c}(\omega) = \begin{bmatrix} 1 \\ \cos\omega \\ \vdots \\ \cos N\omega \end{bmatrix}, \quad \text{and} \quad \mathbf{s}(\omega) = \begin{bmatrix} 0 \\ \sin\omega \\ \vdots \\ \sin N\omega \end{bmatrix}$$

then $\mathbf{x} = [\mathbf{a}^T \ \mathbf{b}^T]^T$. Thus the error function in Eq. (8.4) can be expressed as

$$\begin{aligned} e_i(\mathbf{x}) &= M(\mathbf{x}, \omega_i) - M_0(\omega_i) \\ &= \frac{\{[\mathbf{a}^T \mathbf{c}(\omega_i)]^2 + [\mathbf{a}^T \mathbf{s}(\omega_i)]^2\}^{1/2}}{\{[1 + \mathbf{b}^T \hat{\mathbf{c}}(\omega_i)]^2 + [\mathbf{b}^T \hat{\mathbf{s}}(\omega_i)]^2\}^{1/2}} - M_0(\omega_i) \end{aligned} \tag{8.27}$$

where

$$\hat{\mathbf{c}}(\omega) = \begin{bmatrix} \cos \omega \\ \vdots \\ \cos N\omega \end{bmatrix} \quad \text{and} \quad \hat{\mathbf{s}}(\omega) = \begin{bmatrix} \sin \omega \\ \vdots \\ \sin N\omega \end{bmatrix}$$

The gradient of the objective function $\Psi_k(\mathbf{x})$ can be obtained as

$$\nabla |e_i(\mathbf{x})| = \text{sgn}\,[e_i(\mathbf{x})]\nabla e_i(\mathbf{x}) \tag{8.28a}$$

by using Eqs. (8.7) and (8.27), where

$$\nabla e_i(\mathbf{x}) = \begin{bmatrix} \dfrac{\partial e_i(\mathbf{x})}{\partial \mathbf{a}} \\[2mm] \dfrac{\partial e_i(\mathbf{x})}{\partial \mathbf{b}} \end{bmatrix} \tag{8.28b}$$

$$\frac{\partial e_i(\mathbf{x})}{\partial \mathbf{a}} = \frac{M(\mathbf{x},\,\omega_i)\{[\mathbf{a}^T\mathbf{c}(\omega_i)]\mathbf{c}(\omega_i) + [\mathbf{a}^T\mathbf{s}(\omega_i)]\mathbf{s}(\omega_i)\}}{[\mathbf{a}^T\mathbf{c}(\omega_i)]^2 + [\mathbf{a}^T\mathbf{s}(\omega_i)]^2} \tag{8.28c}$$

$$\frac{\partial e_i(\mathbf{x})}{\partial \mathbf{b}} = -\frac{M(\mathbf{x},\,\omega_i)\{[1 + \mathbf{b}^T\hat{\mathbf{c}}(\omega_i)]\hat{\mathbf{c}}(\omega_i) + [\mathbf{b}^T\hat{\mathbf{s}}(\omega_i)]\hat{\mathbf{s}}(\omega_i)\}}{[1 + \mathbf{b}^T\hat{\mathbf{c}}(\omega_i)]^2 + [\mathbf{b}^T\hat{\mathbf{s}}(\omega_i)]^2} \tag{8.28d}$$

The above minimax problem was solved by using a MATLAB program that implements Algorithm 8.1. The program accepts the parameters ω_p, ω_a, K, and ε_1 as inputs and produces the filter coefficient vectors $\mathbf{a}$ and $\mathbf{b}$ as output. Step 3 of the algorithm was implemented using the quasi-Newton BFGS algorithm alluded to in Example 8.1 with a termination tolerance ε_2. The program also generates plots for the approximation error $|e(\mathbf{x},\,\omega)|$ and the amplitude response of the filter designed. The initial point was taken to be $\mathbf{x}_0 = [\mathbf{a}_0^T \ \mathbf{b}_0^T]^T$ where $\mathbf{a}_0 = [1 \ 1 \ \cdots \ 1]^T$ and $\mathbf{b}_0 = [0 \ 0 \ \cdots \ 0]^T$. The number of actual sample points, K, was set to 600, i.e., 267 and 333 in the frequency ranges $0 \le \omega \le 0.4\pi$ and $0.5\pi \le \omega \le \pi$, respectively, and ε_1 and ε_2 were set to 10^{-6} and 10^{-9}, respectively. The algorithm required seven iterations and 198.20 s of CPU time on a 3.1 GHz Pentium 4 PC to converge to the solution point $\mathbf{x} = [\mathbf{a}^T \ \mathbf{b}^T]^T$ where

$$\mathbf{a} = \begin{bmatrix} 0.00735344 \\ 0.02709762 \\ 0.06800724 \\ 0.12072224 \\ 0.16823049 \\ 0.18671705 \\ 0.16748698 \\ 0.11966157 \\ 0.06704789 \\ 0.02659087 \\ 0.00713664 \end{bmatrix}, \quad \mathbf{b} = \begin{bmatrix} -3.35819120 \\ 8.39305902 \\ -13.19675182 \\ 16.35127992 \\ -14.94617828 \\ 10.68550651 \\ -5.65665532 \\ 2.15596724 \\ -0.52454530 \\ 0.06260344 \end{bmatrix}$$

Note that design problems of this type have multiple possible solutions and the designer would often need to experiment with different initial points as well as different values of K, ε_1, and ε_2, in order to achieve a good design.

The transfer function of a digital filter must have poles inside the unit circle of the z plane to assure the stability of the filter (see Sect. B.7). Since the minimax algorithms of this chapter are unconstrained, no control can be exercised on the pole positions and, therefore, a transfer function may be obtained that represents an unstable filter. Fortunately, the problem can be eliminated through a well-known stabilization technique. In this technique, all the poles of the transfer function that are located outside the unit circle are replaced by their reciprocals and the transfer function is then multiplied by an appropriate multiplier constant which is equal to the reciprocal of the product of these poles (see [4, p. 535]). For example, if

$$H(z) = \frac{N(z)}{D(z)} = \frac{N(z)}{D'(z) \prod_{i=1}^{k}(z - p_{u_i})} \tag{8.29}$$

is a transfer function with k poles $p_{u_1}, p_{u_2}, \ldots, p_{u_k}$ that lie outside the unit circle, then a stable transfer function that yields the same amplitude response can be obtained as

$$H'(z) = H_0 \frac{N(z)}{D'(z) \prod_{i=1}^{k}(z - 1/p_{u_i})} = \frac{\sum_{i=0}^{N} a_i' z^i}{1 + \sum_{i=1}^{N} b_i' z^i} \tag{8.30a}$$

where

$$H_0 = \frac{1}{\prod_{i=1}^{k} p_{u_i}} \tag{8.30b}$$

In the design problem considered above, the poles of $H(z)$ were obtained as shown in column 2 of Table 8.1 by using command $\texttt{roots}$ of MATLAB. Since $|p_i| > 1$ for $i = 1$ and 2, a complex-conjugate pair of poles are located outside the unit circle, which render the filter unstable. By applying the above stabilization technique, the poles in column 3 of Table 8.1 were obtained and multiplier constant H_0 was calculated as $H_0 = 0.54163196$.

Table 8.1 Poles of the IIR filters for Example 8.2(a)

i	Poles of the unstable filter	Poles of the stabilized filter
1	$0.51495917 + 1.25741370j$	$0.27891834 + 0.68105544j$
2	$0.51495917 - 1.25741370j$	$0.27891834 - 0.68105544j$
3	$0.23514844 + 0.92879138j$	$0.23514844 + 0.92879138j$
4	$0.23514844 - 0.92879138j$	$0.23514844 - 0.92879138j$
5	$0.24539982 + 0.82867789j$	$0.24539982 + 0.82867789j$
6	$0.24539982 - 0.82867789j$	$0.24539982 - 0.82867789j$
7	$0.32452615 + 0.46022220j$	$0.32452615 + 0.46022220j$
8	$0.32452615 - 0.46022220j$	$0.32452615 - 0.46022220j$
9	$0.35906202 + 0.16438481j$	$0.35906202 + 0.16438481j$
10	$0.35906202 - 0.16438481j$	$0.35906202 - 0.16438481j$

By using Eq. (8.30a), coefficients $\mathbf{a}'$ and $\mathbf{b}'$ were obtained as

$$\mathbf{a}' = \begin{bmatrix} 0.00398286 \\ 0.01467694 \\ 0.03683489 \\ 0.06538702 \\ 0.09111901 \\ 0.10113192 \\ 0.09071630 \\ 0.06481253 \\ 0.03631528 \\ 0.01440247 \\ 0.00386543 \end{bmatrix}, \quad \mathbf{b}' = \begin{bmatrix} -2.88610955 \\ 5.98928394 \\ -8.20059471 \\ 8.75507027 \\ -7.05776764 \\ 4.44624218 \\ -2.10292453 \\ 0.72425530 \\ -0.16255342 \\ 0.01836567 \end{bmatrix}$$

The largest magnitude of the poles of the modified transfer function is 0.9581, and thus the filter is stable.

The approximation error $|e(\mathbf{x}, \omega)|$ over the passband and stopband is plotted in Fig. 8.3 and the amplitude response of the filter is shown in Fig. 8.4.

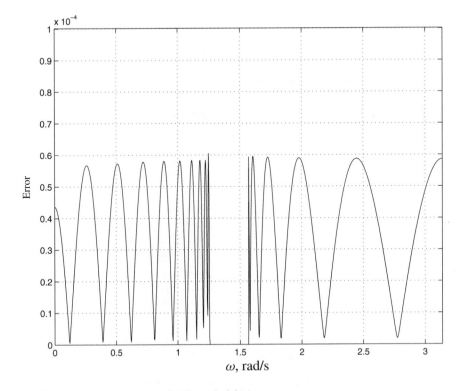

Fig. 8.3 Error $|e(\mathbf{x}, \omega)|$ versus ω for Example 8.2(a)

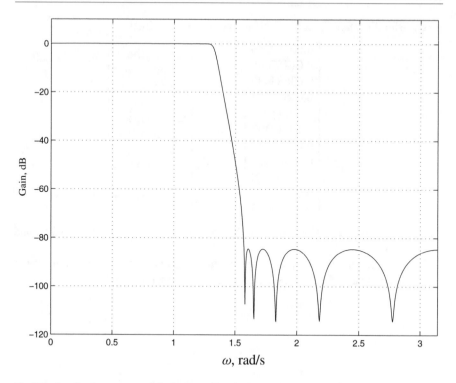

Fig. 8.4   Amplitude response of the lowpass filter for Example 8.2(*a*)

(*b*) For part (*b*), the number of sampling points was set to 65, i.e., 29 and 36 in the frequency ranges $0 \leq \omega \leq \omega_p$ and $\omega_a \leq \omega \leq \pi$, respectively. The initial point and parameters ε_1 and ε_2 were the same as in part (*a*), and parameter l was set to 5. It took Algorithm 8.3 six iterations and 18.73 s of CPU time to converge to the solution point $\mathbf{x} = [\mathbf{a}^T \ \mathbf{b}^T]$ where

$$
\mathbf{a} = \begin{bmatrix} 0.00815296 \\ 0.03509437 \\ 0.09115541 \\ 0.16919427 \\ 0.24129855 \\ 0.27357739 \\ 0.24813555 \\ 0.17915173 \\ 0.09963780 \\ 0.03973358 \\ 0.00981327 \end{bmatrix}, \quad
\mathbf{b} = \begin{bmatrix} -2.02896582 \\ 3.98574025 \\ -3.65125139 \\ 2.56127374 \\ -0.11412527 \\ -1.16704564 \\ 1.36351210 \\ -0.77298905 \\ 0.25851314 \\ -0.03992105 \end{bmatrix}
$$

As can be verified, a complex-conjugate pair of poles of the transfer function obtained are located outside the unit circle. By applying the stabilization technique described in part (*a*), the coefficients of the modified transfer function were obtained as

$$\mathbf{a}' = \begin{bmatrix} 0.00584840 \\ 0.02517440 \\ 0.06538893 \\ 0.12136889 \\ 0.17309179 \\ 0.19624651 \\ 0.17799620 \\ 0.12851172 \\ 0.07147364 \\ 0.02850227 \\ 0.00703940 \end{bmatrix}, \quad \mathbf{b}' = \begin{bmatrix} -1.83238201 \\ 3.04688036 \\ -2.42167890 \\ 1.31022752 \\ 0.27609329 \\ -0.90732976 \\ 0.84795926 \\ -0.43579279 \\ 0.13706106 \\ -0.02054213 \end{bmatrix}$$

The largest magnitude of the poles of the modified transfer function is 0.9537 and thus the filter is stable.

The approximation error $|e(\mathbf{x}, \omega)|$ over the passband and stopband is plotted in Fig. 8.5 and the amplitude response of the filter is depicted in Fig. 8.6.   ∎

The next example illustrates the application of Algorithms 8.2 and 8.4.

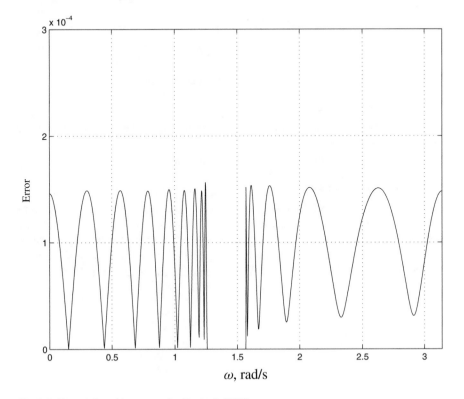

Fig. 8.5 Error $|e(\mathbf{x}, \omega)|$ versus ω for Example 8.2(*b*)

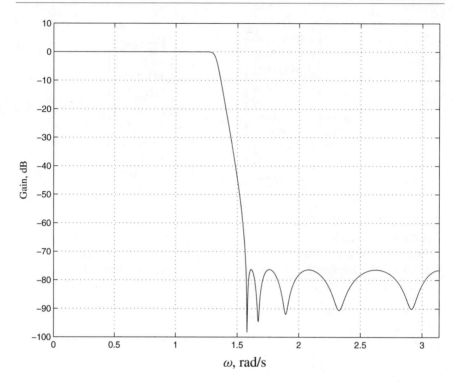

Fig. 8.6 Amplitude response of the lowpass filter for Example 8.2(*b*)

Example 8.3 (*a*) Applying Algorithm 8.2, design the 10th-order lowpass digital filter specified in Example 8.2(a).

(*b*) Applying Algorithm 8.4, carry out the same design.

Solution (*a*) The required design was obtained by using a MATLAB program that implements Algorithm 8.2 following the approach outlined in the solution of Example 8.2. The number of actual sample points, K, was set to 650, i.e., 289 and 361 in the frequency ranges $0 \leq \omega \leq \omega_p$ and $\omega_a \leq \omega \leq \pi$, respectively, and ε_1 and ε_2 were set to 3×10^{-9} and 10^{-15}, respectively. The initial point $\mathbf{x}_0$ was the same as in part (*a*) of Example 8.2. Algorithm 8.2 required eight iterations and 213.70 s of CPU time to converge to the solution point $\mathbf{x} = [\mathbf{a}^T \ \mathbf{b}^T]^T$ where

$$
\mathbf{a} = \begin{bmatrix} 0.05487520 \\ 0.23393481 \\ 0.59719051 \\ 1.09174124 \\ 1.53685612 \\ 1.71358243 \\ 1.53374494 \\ 1.08715408 \\ 0.59319673 \\ 0.23174666 \\ 0.05398863 \end{bmatrix}, \quad
\mathbf{b} = \begin{bmatrix} -5.21138732 \\ 18.28000994 \\ -39.14255091 \\ 66.45234153 \\ -78.76751214 \\ 76.41046395 \\ -50.05505315 \\ 25.84116347 \\ -6.76718946 \\ 0.68877840 \end{bmatrix}
$$

As can be shown, the transfer function obtained has three complex-conjugate pairs of poles that are located outside the unit circle. By applying the stabilization technique described in part (a) of Example 8.2, the coefficients of the modified transfer function were obtained as

$$
\mathbf{a'} = \begin{bmatrix} 0.00421864 \\ 0.01798421 \\ 0.04591022 \\ 0.08392980 \\ 0.11814890 \\ 0.13173509 \\ 0.11790972 \\ 0.08357715 \\ 0.04560319 \\ 0.01781599 \\ 0.00415048 \end{bmatrix}, \quad
\mathbf{b'} = \begin{bmatrix} -2.49921097 \\ 4.87575840 \\ -6.01897510 \\ 5.92269310 \\ -4.27567184 \\ 2.41390695 \\ -0.98863984 \\ 0.28816806 \\ -0.05103514 \\ 0.00407073 \end{bmatrix}
$$

The largest magnitude of the modified transfer function is 0.9532 and thus the filter is stable.

The approximation error $|e(\mathbf{x}, \omega)|$ over the passband and stopband is plotted in Fig. 8.7 and the amplitude response of the filter is depicted in Fig. 8.8.

(b) As in Example 8.2(b), the number of sampling points was set to 65, i.e., 29 and 36 in the frequency ranges $0 \leq \omega \leq \omega_p$ and $\omega_a \leq \omega \leq \pi$, respectively, and ε_1 and ε_2 were set to $\varepsilon = 10^{-9}$ and $\varepsilon_2 = 10^{-15}$, respectively. The initial point $\mathbf{x}_0$ was the same as in part (a) and parameter l was set to 4. Algorithm 8.4 required sixteen iterations and 48.38 s of CPU time to converge to a solution point $\mathbf{x} = [\mathbf{a}^T \ \mathbf{b}^T]^T$ where

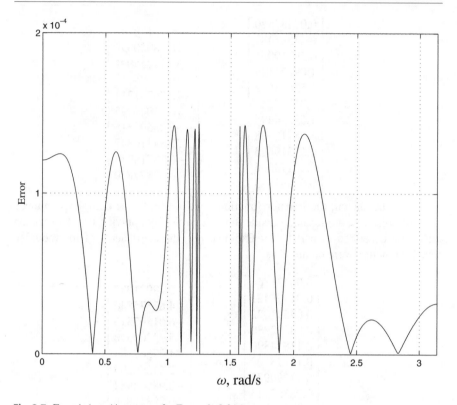

Fig. 8.7 Error $|e(\mathbf{x}, \omega)|$ versus ω for Example 8.3(a)

$$\mathbf{a} = \begin{bmatrix} 0.01307687 \\ 0.05061800 \\ 0.12781582 \\ 0.22960471 \\ 0.32150671 \\ 0.35814899 \\ 0.32167525 \\ 0.22984873 \\ 0.12803465 \\ 0.05073663 \end{bmatrix}, \quad \mathbf{b} = \begin{bmatrix} -4.25811576 \\ 11.94976697 \\ -20.27972610 \\ 27.10889061 \\ -26.10756891 \\ 20.09430301 \\ -11.29104740 \\ 4.74405652 \\ -1.28479278 \\ 0.16834783 \end{bmatrix}$$

These coefficients correspond to an unstable transfer function with one pair of poles outside the unit circle. By applying the stabilization technique, the coefficients of the modified transfer function were obtained as

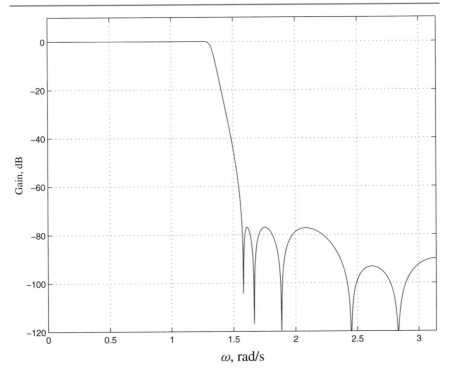

Fig. 8.8 Amplitude response of the lowpass filter for Example 8.3(a)

$$
\mathbf{a}' = \begin{bmatrix} 0.00392417 \\ 0.01518969 \\ 0.03835557 \\ 0.06890085 \\ 0.09647924 \\ 0.10747502 \\ 0.09652981 \\ 0.06897408 \\ 0.03842123 \\ 0.01522528 \\ 0.00393926 \end{bmatrix}, \quad
\mathbf{b}' = \begin{bmatrix} -2.80254807 \\ 5.74653112 \\ -7.72509562 \\ 8.13565547 \\ -6.44870979 \\ 3.99996323 \\ -1.85761204 \\ 0.62780900 \\ -0.13776283 \\ 0.01515986 \end{bmatrix}
$$

The largest magnitude of the poles of the modified transfer function is 0.9566 and thus the filter is stable.

The approximation error $|e(\mathbf{x}, \omega)|$ over the passband and stopband is plotted in Fig. 8.9 and the amplitude response of the filter is depicted in Fig. 8.10. ∎

From the designs carried out in Examples 8.2 and 8.3, we note that the use of the least-pth method with uniform sampling in Example 8.2(a) resulted in the lowest minimax error but a very large density of sample points was required to achieve a good design, which translates into a large amount of computation. Through the use

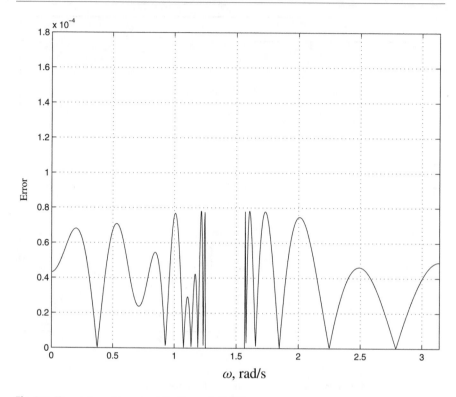

Fig. 8.9 Error $|e(\mathbf{x}, \omega)|$ versus ω for Example 8.3(b)

of nonuniform variable sampling in Example 8.2(b), a design of practically the same quality was achieved with much less computation.

It should be mentioned that in the Charalambous algorithm, the value of ξ becomes progressively larger and approaches the minimum value of the objective function from below as the optimization progresses. As a result, the number of sample points that remain active is progressively reduced, i.e., the sizes of index sets I_1 and I_2 become progressively smaller. Consequently, by avoiding the computation of the partial derivatives of $e_i(\mathbf{x})$ for $i \in I_3$ through careful programming, the evaluation of gradient $\nabla \Psi$ (see Eq. (8.11)) can be carried out much more efficiently. In the above examples, we have not taken advantage of the above technique but our past experience has shown that when it is fully implemented, the Charalambous algorithm usually requires from 10 to 40% of the computation required by the least-pth method, depending on the application.

Finally, it should be mentioned that with optimization there is always an element of chance in obtaining a good design and, therefore, one would need to carry out a large number of different designs using a large set of randomly chosen initial points to be able to compare two alternative design algorithms such as Algorithms 8.1 and 8.2.

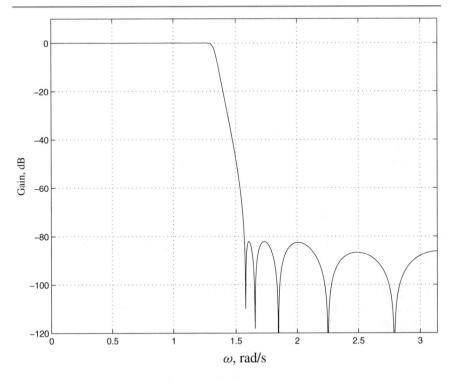

Fig. 8.10 Amplitude response of the lowpass filter for Example 8.3(*b*)

Problems

8.1 Consider the overdetermined system of nonlinear equations

$$x_1^2 - x_2^2 - x_1 - 3x_2 = 2$$
$$x_1^3 - x_2^4 = -2$$
$$x_1^2 + x_2^3 + 2x_1 - x_2 = -1.1$$

(*a*) Using the Gauss-Newton method, find a solution for the above equations, $\mathbf{x}_{gn}$, by minimizing

$$F(\mathbf{x}) = \sum_{i=1}^{3} f_i^2(\mathbf{x})$$

where

$$f_1(\mathbf{x}) = x_1^2 - x_2^2 - x_1 - 3x_2 - 2$$
$$f_2(\mathbf{x}) = x_1^3 - x_2^4 + 2$$
$$f_3(\mathbf{x}) = x_1^2 + x_2^3 + 2x_1 - x_2 + 1.1$$

(b) Applying Algorithm 8.2, find a minimax solution, $\mathbf{x}_{minimax}$, by solving the minimax problem

$$\underset{\mathbf{x}}{\text{minimize}} \quad \underset{1 \leq i \leq 3}{\max} |f_i(\mathbf{x})|$$

(c) Evaluate and compare the equation errors for the solutions $\mathbf{x}_{gn}$ and $\mathbf{x}_{minimax}$.

8.2 Verify the expression for the gradient $\nabla e_i(\mathbf{x})$ given in Eqs. (8.28a), (8.28c), and (8.28d).

8.3 Applying Algorithm 8.1, design a 12th-order highpass digital filter, assuming a desired amplitude response

$$M_0(\omega) = \begin{cases} 0 & \text{for } 0 \leq \omega \leq 0.45\pi \text{ rad/s} \\ 1 & \text{for } 0.5\pi \leq \omega \leq \pi \text{ rad/s} \end{cases}$$

The transfer function of the filter is of the form given by Eq. (8.2).

8.4 Repeat Problem 8.3 by applying Algorithm 8.2.

8.5 Repeat Problem 8.3 by applying Algorithm 8.3.

8.6 Repeat Problem 8.3 by applying Algorithm 8.4.

8.7 Applying Algorithm 8.1, design a 12th-order bandpass filter, assuming a desired amplitude response

$$M_0(\omega) = \begin{cases} 0 & \text{for } 0 \leq \omega \leq 0.3\pi \text{ rad/s} \\ 1 & \text{for } 0.375\pi \leq \omega \leq 0.625\pi \text{ rad/s} \\ 0 & \text{for } 0.7\pi \leq \omega \leq \pi \text{ rad/s} \end{cases}$$

The transfer function of the filter is of the form given by Eq. (8.2).

8.8 Repeat Problem 8.7 by applying Algorithm 8.2.

8.9 Repeat Problem 8.7 by applying Algorithm 8.3.

8.10 Repeat Problem 8.7 by applying Algorithm 8.4.

8.11 Applying Algorithm 8.1, design a 12th-order bandstop filter, assuming a desired amplitude response

$$M_0(\omega) = \begin{cases} 1 & \text{for } 0 \leq \omega \leq 0.35\pi \text{ rad/s} \\ 0 & \text{for } 0.425\pi \leq \omega \leq 0.575\pi \text{ rad/s} \\ 1 & \text{for } 0.65\pi \leq \omega \leq \pi \text{ rad/s} \end{cases}$$

The transfer function of the filter is of the form given by Eq. (8.2).

8.12 Repeat Problem 8.11 by applying Algorithm 8.2.

8.13 Repeat Problem 8.11 by applying Algorithm 8.3.

8.14 Repeat Problem 8.11 by applying Algorithm 8.4.

References

1. C. Charalambous, "A unified review of optimization," *IEEE Trans. Microwave Theory and Techniques*, vol. MTT-22, pp. 289–300, Mar. 1974.

2. C. Charalambous, "Acceleration of the least-*p*th algorithm for minimax optimization with engineering applications," *Mathematical Programming*, vol. 17, pp. 270–297, 1979.

3. A. Antoniou, "Improved minimax optimisation algorithms and their application in the design of recursive digital filters," *Proc. Inst. Elect. Eng., part G*, vol. 138, pp. 724–730, Dec. 1991.

4. A. Antoniou, *Digital Signal Processing: Signals, Systems, and Filters*. New York: McGraw-Hill, 2005.

5. C. Charalambous and A. Antoniou, "Equalisation of recursive digital filters," *Proc. Inst. Elect. Eng., part G*, vol. 127, pp. 219–225, Oct. 1980.

6. C. Charalambous, "Design of 2-dimensional circularly-symmetric digital filters," *Proc. Inst. Elect. Eng., part G*, vol. 129, pp. 47–54, Apr. 1982.

Applications of Unconstrained Optimization

9

9.1 Introduction

Optimization problems occur in many disciplines, for example, in engineering, physical sciences, social sciences, and commerce. In this chapter, we demonstrate the usefulness of the unconstrained optimization algorithms studied in this book by applying them to a number of problems in engineering. Applications of various constrained optimization algorithms will be presented in Chap. 16.

Optimization is particularly useful in the various branches of engineering like electrical, mechanical, chemical, and aeronautical engineering. The applications we consider here and in Chap. 16 are in the areas of digital signal processing, pattern recognition, automatic control, robotics, and telecommunications. For each selected application, sufficient background material is provided to assist the reader to understand the application. The steps involved are the problem formulation phase, which converts the problem at hand into an unconstrained optimization problem, and the solution phase, which involves selecting and applying an appropriate optimization algorithm.

In Sect. 9.2, we introduce a machine-learning technique for the classification of handwritten digits known as *softmax regression*. In Sect. 9.3, we consider a problem known as *inverse kinematics of robotic manipulators* which entails a system of nonlinear equations. The problem is first converted into an unconstrained minimization problem and then various methods studied earlier are applied and the results obtained are compared in terms of solution accuracy and computational efficiency. Throughout the discussion, the advantages of using an optimization-based solution method relative to a conventional closed-form method are stressed. In Sect. 9.4, we obtain weighted least-squares and minimax designs of finite-duration impulse-response (FIR) digital filters using unconstrained optimization.

© Springer Science+Business Media, LLC, part of Springer Nature 2021   239
A. Antoniou and W.-S. Lu, *Practical Optimization*, Texts in Computer Science,
https://doi.org/10.1007/978-1-0716-0843-2_9

Fig. 9.1 Sample
handwritten-digits from
MNIST

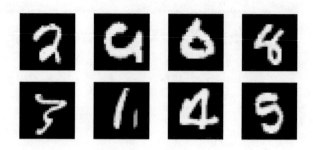

9.2 Classification of Handwritten Digits

In this section, we introduce a machine-learning technique known as *softmax regression* and then apply it to the classification of handwritten-digits. The parameters involved are optimized using a gradient descent algorithm.

9.2.1 Handwritten-Digit Recognition Problem

In a handwritten-digit recognition (HWDR) problem, we are given a data set, known as *training data*, to tune the parameters of a recognition algorithm so as to minimize the recognition error. The training data consist of ten classes of handwritten-digits $\{\mathcal{D}_j, \ j = 0, \ 1, \ \ldots, \ 9\}$ where class $\mathcal{D}_j$ contains digits that are labeled as $j = 0, \ 1, \ \ldots, \ 9$. We seek an approach that utilizes the available training data to optimally tune a multiclass classifier to classify handwritten-digits that are not contained in the training data. Research for HWDR using machine-learning techniques has been active in the past several decades [1–4] due primarily to its broad applications in postal mail sorting and routing, automatic address reading, and bank check processing, etc. The challenge arising from HWDR lies in the fact that handwritten-digits within a given digit class vary widely in shape, line width, and style, even when they are normalized in size and properly centralized. Reference [5] describes a database, known as the Modified National Institute of Standards and Technology (MNIST) database (see Sect. 9.2.4.1 for details), of handwritten-digits, that has been widely used as a source of training and testing data for HWDR. Figure 9.1 displays a small sample of digits from the MNIST database where each digit is represented by a gray-scale digital image of size 28×28 pixels.

9.2.2 Histogram of Oriented Gradients

It is well known that appropriate pre-processing of the raw input data to extract more effective features is often useful [6]. In an HWDR problem, raw input data assume the form $\mathcal{D}_j = \{(\mathbf{D}_i^{(j)}, l_j) \text{ for } i = 1, \ 2, \ \ldots, \ n_j\}$ where $\mathbf{D}_i^{(j)}$ is a digital image representing the ith digit in class $\mathcal{D}_j$ and l_j is the label of the digits in class $\mathcal{D}_j$. For

Fig. 9.2 **a** Digit 5 as an image I, **b** $\nabla_x \mathbf{D}$ as an image, **c** $\nabla_y \mathbf{D}$ as an image, and **d** norm $\|\nabla \mathbf{D}\|_2$ as an image

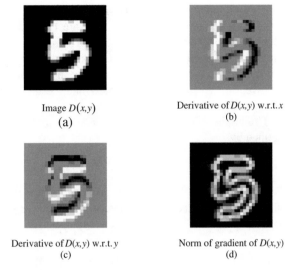

Image $D(x,y)$
(a)

Derivative of $D(x,y)$ w.r.t. x
(b)

Derivative of $D(x,y)$ w.r.t. y
(c)

Norm of gradient of $D(x,y)$
(d)

HWDR, we always have $l_j = j$. Images $\{\mathbf{D}_k^{(j)}\}$ are not very effective for the solution of the HWDR problem because they do not provide directional information of the digits explicitly. A feature descriptor known as *the histogram of oriented gradients* (HOG) [7] has been found to be effective for digital images.

For the sake of convenience, each of the images used in our illustration of the HOG is assumed to be of dimension 28×28 pixels to ensure that the construction of the HOG of a given image is directly applicable to the MNIST data sets. Consider an image $\mathbf{D}$ with normalized light intensity such as that in Fig. 9.2a. The HOG of the image can be constructed by dividing $\mathbf{D}$ into a total of 196 non-overlapping sub-images, each of size 2×2 pixels, which can be referred to as the *cells* of the image. In effect, there are 14 rows and 14 columns of cells. We denote the cell in the ith row and jth column as $C(i, j)$.

Next, we form image *blocks* where each block is a 2×2 cell sub-image; in effect, an image block is a 4×4 sub-matrix. Unlike cells, image blocks overlap and are constructed as follows. Starting from the 4 cells in the upper left corner of the image, the blocks are formed row by row and the last block is constructed using the 4 cells at the bottom right corner. The first block, denoted as $B(1, 1)$, is a square of size 4×4 which consists of the 4 cells from the upper left corner of the image, namely, the 4 cells $\{C(1, 1), C(1, 2), C(2, 1), C(2, 2)\}$. The next block, denoted as $B(1, 2)$, consists of the 4 cells $\{C(1, 2), C(1, 3), C(2, 2), C(2, 3)\}$ and, as can be seen, it has 50% overlap with block $B(1, 1)$. The rest of the image blocks involving the first and second rows of cells can be similarly defined with the last block being $B(1, 13)$ which consists of the 4 cells $\{C(1, 13), C(1, 14), C(2, 13), C(2, 14)\}$ and has 50% overlap with block $B(1, 12)$. The second row of blocks involves the second and third rows of cells and it starts from block $B(2, 1)$ which has 50% overlap with block $B(1, 1)$ and consists of the 4 cells $\{C(2, 1), C(2, 2), C(3, 1), C(3, 2)\}$. The remaining blocks $B(2, 2), \ldots, B(2, 13)$ are defined in a similar manner. The formation process continues until block $B(13, 13)$, consisting of cells $\{C(13, 13), C(13, 14), C(14, 13), C(14, 14)\}$, is formed. Evidently, there are 13 rows and 13 columns of overlapping blocks and, in effect, there is a total of 169 blocks. The manner in which the image blocks are formed immediately implies that each inner cell (i.e., a cell that does not contain boundary pixels) is involved in four blocks and each boundary cell is involved in two blocks. On the other hand, each of the cells in the corners of the image is involved in only one block.

The gradient $\nabla \mathbf{D}$ of image $\mathbf{D}$ is a vector field over the image's domain and is defined by

$$\nabla \mathbf{D} = \begin{bmatrix} \partial \mathbf{D}/\partial x \\ \partial \mathbf{D}/\partial y \end{bmatrix}$$

For digital images, however, the partial derivatives must be approximated by partial differences along the x and y directions, respectively, as

$$\frac{\partial \mathbf{D}}{\partial x} \approx \frac{\mathbf{D}(x + \delta, y) - \mathbf{D}(x - \delta, y)}{2\delta} \tag{9.1a}$$

$$\frac{\partial \mathbf{D}}{\partial y} \approx \frac{\mathbf{D}(x, y + \delta) - \mathbf{D}(x, y - \delta)}{2\delta} \tag{9.1b}$$

In practice, the approximated gradient in Eqs. (9.1a) and (9.1b) over the entire image domain is evaluated by using two-dimensional discrete convolution as

$$\frac{\partial \mathbf{D}}{\partial x} \approx \mathbf{D} \otimes h_x, \quad \frac{\partial \mathbf{D}}{\partial y} \approx \mathbf{D} \otimes h_y,$$

where h_x and h_y are given by

$$h_x = [-1 \ 0 \ 1], \ h_y = \begin{bmatrix} 1 \\ 0 \\ -1 \end{bmatrix}$$

In what follows, the discrete gradient of image $\mathbf{D}$ is denoted by $\{\nabla_x \mathbf{D}, \nabla_y \mathbf{D}\}$. As an example, the components of the discrete gradient, $\nabla_x \mathbf{D}$ and $\nabla_y \mathbf{D}$ for the image in Fig. 9.2a are shown in Fig. 9.2b and c, respectively. The gradient provides two important pieces of information about the image's local structure, namely, the magnitude $\|\nabla \mathbf{D}\|_2$ and angle $\Theta(x, y)$ of the gradient, which can be computed by using the formulas

$$\|\nabla \mathbf{D}\|_2 = \sqrt{(\nabla_x \mathbf{D})^2 + (\nabla_y \mathbf{D})^2} \tag{9.2a}$$

$$\Theta = \tan^{-1}\left(\frac{\nabla_y \mathbf{D}}{\nabla_x \mathbf{D}}\right) \tag{9.2b}$$

Figure 9.2d shows the magnitude of $\nabla \mathbf{D}$ defined in Eq. (9.2a) for the image in Fig. 9.2a. The arctangent in Eq. (9.2b) takes the *sign* of the gradient into account, i.e., it is the four-quadrant arctangent in the range between $-\pi$ and π.

To facilitate the definition of the HOG of an image, we begin by dividing the angle range $(-\pi, \pi)$ evenly into β bins with the first and second bins associated with the ranges $-\pi$ to $-\pi + \Delta$ and $-\pi + \Delta$ to $-\pi + 2\Delta$ where $\Delta = 2\pi/\beta$, and so on. The 169 image blocks defined earlier can be arranged as a sequence

$\{B(1, 1), \ldots, B(1, 13), B(2, 1), \ldots, B(2, 13), \ldots, B(13, 1), \ldots, B(13, 13)\}$.
We can now associate each block $B(i, j)$ with a column vector $\mathbf{h}_{i,j}$ of length β,
which is initially set to $\mathbf{h}_{i,j} = \mathbf{0}$, and then associate each component of $\mathbf{h}_{i,j}$ with a
bin defined above. Since $B(i, j)$ is of size 4×4, it produces 16 pairs $\{\Theta, \|\nabla \mathbf{D}\|_2\}$.
For each pair $\{\Theta, \|\nabla \mathbf{D}\|_2\}$, the magnitude $\|\nabla \mathbf{D}\|_2$ is added to the corresponding
component of $\mathbf{h}_{i,j}$. Vector $\mathbf{h}_{i,j}$ so obtained is called the HOG of block $B(i, j)$ and
the HOG of image D is defined by stacking all $\{\mathbf{h}_{i,j}\}$ in an orderly fashion as

$$
\mathbf{x} = \begin{bmatrix} \mathbf{h}_{1,1} \\ \vdots \\ \mathbf{h}_{1,13} \\ \mathbf{h}_{2,1} \\ \vdots \\ \mathbf{h}_{13,13} \end{bmatrix} \tag{9.3}
$$

For an image of size 28×28, the length of $\mathbf{x}$ in Eq. (9.3) is equal to $169 \times \beta$. Typically
the value of β (i.e., the number of bins per histogram) is in the range 7 to 9. With
$\beta = 7$, for example, the vector $\mathbf{x}$ in Eq. (9.3) is of length 1183. In a supervised
learning problem, for a class of training data that contains N images where each
image $\mathbf{D}_i$ is associated with a HOG vector $\mathbf{x}_i$, the HOG feature of the data set is a
matrix of size 1183 by N in the form

$$
\mathbf{X} = [\mathbf{x}_1 \ \ \mathbf{x}_2 \ \ \cdots \ \ \mathbf{x}_N]
$$

As an example, Fig. 9.3a and b depict the HOG matrices of 1,600 randomly selected
digits "1" and digits "5" from MNIST, respectively. It is observed that in each HOG
matrix there exists a great deal of similarity between its columns. Since each column
represents the HOG features of a digit in the same class, the similarity demonstrates
robustness of the HOG features against deformation among the digits. On the other
hand, the two HOG matrices from Fig. 9.3a and b look quite different from each
other, and this demonstrates HOG's potential for handwritten-digit recognition.

The fact that a given cell is present in multiple image blocks explains why the
HOG is insensitive to slight deformation of position and angle of the contents in the
image.

9.2.3   Softmax Regression for Use in Multiclass Classification

To better understand softmax regression, we begin with a brief review of the classic
logistic regression for binary classification.

9.2.3.1   Use of Logistic Regression for Binary Classification

In a binary classification problem, the training data assume the form $\mathcal{D} = \{(\mathbf{x}_i, l_i),$
$i = 1, 2, \ldots, N\}$ with feature vector $\mathbf{x}_i \in R^n$ and label $l_i \in \{0, 1\}$. The training

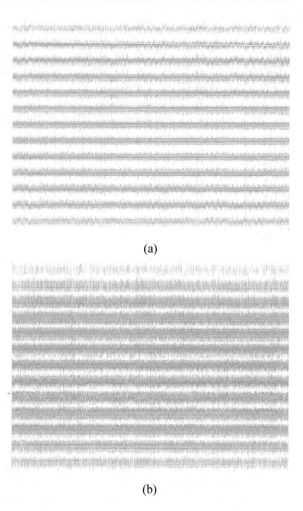

(a)

(b)

data are used to optimize a parameter vector

$$\boldsymbol{\theta} = \begin{bmatrix} \mathbf{w} \\ b \end{bmatrix}$$

such that a *logistic sigmoid* function of the form

$$h_\theta = \frac{1}{1 + e^{-(\mathbf{w}^T \mathbf{x} + b)}} = \frac{1}{1 + e^{-\boldsymbol{\theta}^T \hat{\mathbf{x}}}} \tag{9.4}$$

classifies a *test input* **x** as belonging to class C_0 if $h_\theta(\mathbf{x}) > 0.5$ or class C_1 if
$h_\theta(\mathbf{x}) < 0.5$ where

$$\hat{\mathbf{x}} = \begin{bmatrix} \mathbf{x}_i \\ 1 \end{bmatrix} \tag{9.5}$$

As the value of $h_\theta(\mathbf{x})$ always falls between 0 and 1, binary classification based
on logistic function $h_\theta(\mathbf{x})$ in Eq. (9.4) has a probabilistic interpretation if $h_\theta(\mathbf{x})$ is

regarded as the probability of **x** belonging to class C_0. In statistics, there is a method of estimating model parameters such that the model fits given observations. The method is known as *maximum-likelihood estimation* (MLE) and it entails optimizing the parameters in the likelihood (probability) function to achieve maximum probability. In probability theory, a sequence of random variables is said to be independent and identically distributed (i.i.d.) if all the random variables are mutually independent and have the same probability distribution. In the context of the binary classification problem, the components of vector $\boldsymbol{\theta}$ are the parameters to be estimated and if we regard $\{\mathbf{x}_i, i = 1, 2, \ldots, N\}$ in the training data as i.i.d. random variables, then the likelihood is the probability of observing this data set, which is given by

$$\prod_{i=1}^{N} P(l_i | \mathbf{x}_i) \tag{9.6}$$

where $P(l_i | \mathbf{x}_i)$ denotes the conditional probability of label l_i given sample $\mathbf{x}_i$. With the above probabilistic interpretation of function $h_\theta(\mathbf{x})$ in mind, $P(l_i | \mathbf{x}_i)$ is related to $h_\theta(\mathbf{x})$ as

$$P(l_i | \mathbf{x}_i, \boldsymbol{\theta}) = \begin{cases} h_\theta(\mathbf{x}_i) & \text{for } l_i = 0 \\ 1 - h_\theta(\mathbf{x}_i) & \text{for } l_i = 1 \end{cases} \tag{9.7}$$

From Eqs. (9.4) and (9.7), conditional probability $P(l_i | \mathbf{x}_i)$ in Eq. (9.7) can be expressed as

$$P(l_i | \mathbf{x}_i, \boldsymbol{\theta}) = \frac{1}{1 + e^{-(1-2l_i)(\mathbf{w}^T \mathbf{x}_i + b)}} = \frac{1}{1 + e^{-(1-2l_i)\boldsymbol{\theta}^T \hat{\mathbf{x}}_i}}$$

which establishes an explicit connection of the conditional probability to parameter $\boldsymbol{\theta}$. The optimal $\boldsymbol{\theta}$ can be obtained by maximizing the conditional probability in Eq. (9.6) with respect to $\boldsymbol{\theta}$. Since maximizing the conditional probability in Eq. (9.6) is equivalent to minimizing negative logarithm of the probability, we formulate the MLE problem as the unconstrained problem

$$\underset{\boldsymbol{\theta}}{\text{minimize}} \quad f(\boldsymbol{\theta}) = \frac{1}{N} \sum_{i=1}^{N} \ln(1 + e^{-(1-2l_i)\boldsymbol{\theta}^T \hat{\mathbf{x}}_i})$$

The gradient and Hessian of $f(\boldsymbol{\theta})$ are given by

$$\nabla f(\boldsymbol{\theta}) = -\frac{1}{N} \sum_{i=1}^{N} \frac{(1 - 2l_i)\hat{\mathbf{x}}_i}{1 + e^{(1-2l_i)\boldsymbol{\theta}^T \hat{\mathbf{x}}}}$$

and

$$\nabla^2 f(\boldsymbol{\theta}) = -\frac{1}{N} \sum_{i=1}^{N} \frac{(1 - 2l_i)^2 e^{(1-2l_i)\boldsymbol{\theta}^T \hat{\mathbf{x}}} \hat{\mathbf{x}}_i \hat{\mathbf{x}}_i^T}{\left(1 + e^{(1-2l_i)\boldsymbol{\theta}^T \hat{\mathbf{x}}}\right)^2}$$

respectively. Evidently, $\nabla^2 f(\boldsymbol{\theta})$ is positive semidefinite and hence $f(\boldsymbol{\theta})$ is convex.

9.2.3.2   Use of Softmax Regression for Multiclass Classification

For binary classification problems for classes C_0 and C_1, the usefulness of logistic regression lies in the fact that the posterior probability of class C_0 is related to a logistic sigmoid of a linear function of input data [6]. *Softmax* regression is a generalization of logistic regression to deal with classification problems where the number of classes involved, K, is more than two. The training data in this case are given by $\mathcal{D} = \{(\mathbf{x}_i, l_i), i = 1, 2, \ldots, N\}$ where $\mathbf{x}_i \in R^n$ and $l_i \in \{0, 1, \ldots, K - 1\}$, and the probabilities of an observed data vector $\mathbf{x}$ belonging to class C_i for $i = 0, 1, \ldots, K - 1$ for a given parameter matrix $\boldsymbol{\Theta}$ are given by

$$
\begin{bmatrix}
P(l = 0|\mathbf{x}, \boldsymbol{\Theta}) \\
P(l = 1|\mathbf{x}, \boldsymbol{\Theta}) \\
\vdots \\
P(l = K - 1|\mathbf{x}, \boldsymbol{\Theta})
\end{bmatrix}
= \frac{1}{\sum_{j=0}^{K-1} e^{\boldsymbol{\theta}_j^T \hat{\mathbf{x}}}}
\begin{bmatrix}
e^{\boldsymbol{\theta}_0^T \hat{\mathbf{x}}} \\
e^{\boldsymbol{\theta}_1^T \hat{\mathbf{x}}} \\
\vdots \\
e^{\boldsymbol{\theta}_{K-1}^T \hat{\mathbf{x}}}
\end{bmatrix}
\tag{9.8}
$$

where $\boldsymbol{\Theta} = [\boldsymbol{\theta}_0\ \boldsymbol{\theta}_1\ \cdots\ \boldsymbol{\theta}_{K-1}]$. Using Eq. (9.8), a test vector $\mathbf{x}$ is classified to be in class C_{i*} if $i^* = \arg\left(\max_{0 \le i \le K-1} e^{\boldsymbol{\theta}_i^T \hat{\mathbf{x}}}\right)$ which gives

$$
i^* = \arg\left(\max_{0 \le i \le K-1} \boldsymbol{\theta}_i^T \hat{\mathbf{x}}\right)
\tag{9.9}
$$

Note that Eq. (9.8) also covers the binary classification case $K = 2$. With $K = 2$, Eq. (9.8) becomes

$$
\begin{bmatrix}
P(l = 0 \mid \mathbf{x}, \boldsymbol{\Theta}) \\
P(l = 1 \mid \mathbf{x}, \boldsymbol{\Theta})
\end{bmatrix}
= \frac{1}{e^{\boldsymbol{\theta}_0^T \hat{\mathbf{x}}} + e^{\boldsymbol{\theta}_1^T \hat{\mathbf{x}}}}
\begin{bmatrix}
e^{\boldsymbol{\theta}_0^T \hat{\mathbf{x}}} \\
e^{\boldsymbol{\theta}_1^T \hat{\mathbf{x}}}
\end{bmatrix}
$$

where

$$
\frac{1}{e^{\boldsymbol{\theta}_0^T \hat{\mathbf{x}}} + e^{\boldsymbol{\theta}_1^T \hat{\mathbf{x}}}}
\begin{bmatrix}
e^{\boldsymbol{\theta}_0^T \hat{\mathbf{x}}} \\
e^{\boldsymbol{\theta}_1^T \hat{\mathbf{x}}}
\end{bmatrix}
=
\begin{bmatrix}
\frac{1}{1 + e^{-\tilde{\boldsymbol{\theta}}^T \hat{\mathbf{x}}}} \\
1 - \frac{1}{1 + e^{-\tilde{\boldsymbol{\theta}}^T \hat{\mathbf{x}}}}
\end{bmatrix}
=
\begin{bmatrix}
h_{\tilde{\boldsymbol{\theta}}}(\mathbf{x}) \\
1 - h_{\tilde{\boldsymbol{\theta}}}(\mathbf{x})
\end{bmatrix}
$$

with $\tilde{\boldsymbol{\theta}} = \boldsymbol{\theta}_0 - \boldsymbol{\theta}_1$, and $h_{\tilde{\boldsymbol{\theta}}}(\mathbf{x})$ is defined by Eq. (9.4). If we rename $\tilde{\boldsymbol{\theta}}$ as $\boldsymbol{\theta}$ for the sake of simplicity, then we obtain

$$
\begin{bmatrix}
P(l = 0 \mid \mathbf{x}, \boldsymbol{\theta}) \\
P(l = 1 \mid \mathbf{x}, \boldsymbol{\theta})
\end{bmatrix}
=
\begin{bmatrix}
h_{\boldsymbol{\theta}}(\mathbf{x}) \\
1 - h_{\boldsymbol{\theta}}(\mathbf{x})
\end{bmatrix}
$$

which coincides with Eq. (9.7).

For a classification problem with training data $\mathcal{D} = \{(\mathbf{x}_i, l_i), i = 1, 2, \ldots, N\}$ where $\{\mathbf{x}_i\}$ can be treated as a set of i.i.d. random variables, given model parameter $\boldsymbol{\Theta}$ the probability of observing data set $\mathcal{D}$ is equal to $\prod_{i=1}^{N} P(l_i \mid \mathbf{x}_i, \boldsymbol{\Theta})$. From Eq. (9.8), we can write

$$
P(l_i \mid \mathbf{x}_i, \boldsymbol{\Theta}) = \frac{e^{\boldsymbol{\theta}_{l_i}^T \hat{\mathbf{x}}_i}}{\sum_{j=0}^{K-1} e^{\boldsymbol{\theta}_j^T \hat{\mathbf{x}}_i}}
$$

and hence

$$\prod_{i=1}^{N} P(l_i \mid \mathbf{x}_i, \boldsymbol{\Theta}) = \prod_{i=1}^{N} \frac{e^{\boldsymbol{\theta}_{l_i}^T \hat{\mathbf{x}}_i}}{\sum_{j=0}^{K-1} e^{\boldsymbol{\theta}_j^T \hat{\mathbf{x}}_i}} \tag{9.10}$$

This expression connects explicitly the conditional probability of observing data set $\mathcal{D}$ to parameter matrix $\boldsymbol{\Theta}$ and hence allows one to optimize $\boldsymbol{\Theta}$ in the classifier in Eq. (9.9) by maximizing the conditional probability in Eq. (9.10) with respect to $\boldsymbol{\Theta}$. Since maximizing the conditional probability is equivalent to minimizing the negative of the logarithm of the probability, the problem at hand can be formulated as the unconstrained problem

$$\underset{\boldsymbol{\Theta}}{\text{minimize}} \ \ f(\boldsymbol{\Theta}) = -\frac{1}{N} \sum_{i=1}^{N} \ln \left(\frac{e^{\boldsymbol{\theta}_{l_i}^T \hat{\mathbf{x}}_i}}{\sum_{j=0}^{K-1} e^{\boldsymbol{\theta}_{l_j}^T \hat{\mathbf{x}}_i}} \right) \tag{9.11}$$

The objective function in Eq. (9.11) and its gradient can be evaluated efficiently if we organize the training data $\mathcal{D} = \{(\mathbf{x}_i, l_i), i = 1, 2, \ldots, N\}$ as follows. We divide the training data into K subsets, denoted as $\mathcal{D}_0, \mathcal{D}_1, \ldots, \mathcal{D}_{K-1}$, such that all $\mathbf{x}_i$ in subset S_j have the same label j, namely, $\mathcal{D}_j = \{\mathbf{x}_i : l_i = j\}$. If we evaluate function $f(\boldsymbol{\Theta})$ in Eq. (9.11) according to the K subsets, then parameter $\boldsymbol{\theta}_{l_i}$ involved in the jth subset is always equal to $\boldsymbol{\theta}_j$; hence we have

$$
\begin{aligned}
f(\boldsymbol{\Theta}) &= -\frac{1}{N} \sum_{i=1}^{N} \ln \left(\frac{e^{\boldsymbol{\theta}_{l_i}^T \hat{\mathbf{x}}_i}}{\sum_{j=0}^{K-1} e^{\boldsymbol{\theta}_{l_j}^T \hat{\mathbf{x}}_i}} \right) = -\frac{1}{N} \sum_{i=1}^{N} \boldsymbol{\theta}_{l_i}^T \hat{\mathbf{x}}_i + \frac{1}{N} \sum_{i=1}^{N} \ln \left(\sum_{j=0}^{K-1} e^{\boldsymbol{\theta}_j^T \hat{\mathbf{x}}_i} \right) \\
&= -\frac{1}{N} \sum_{j=0}^{K-1} \sum_{\hat{\mathbf{x}}_i \in S_j} \boldsymbol{\theta}_j^T \hat{\mathbf{x}}_i + \frac{1}{N} \sum_{i=1}^{N} \ln \left(\sum_{j=0}^{K-1} e^{\boldsymbol{\theta}_j^T \hat{\mathbf{x}}_i} \right) \\
&= -\sum_{j=0}^{K-1} \left(\boldsymbol{\theta}_j^T \cdot \frac{1}{N} \sum_{\hat{\mathbf{x}}_i \in S_j} \hat{\mathbf{x}}_i \right) + \frac{1}{N} \sum_{i=1}^{N} \ln \left(\sum_{j=0}^{K-1} e^{\boldsymbol{\theta}_j^T \hat{\mathbf{x}}_i} \right) \\
&= -\sum_{j=0}^{K-1} (\boldsymbol{\theta}_j^T \cdot \tilde{\mathbf{x}}_j) + \frac{1}{N} \sum_{i=1}^{N} \ln \left(\sum_{j=0}^{K-1} e^{\boldsymbol{\theta}_j^T \hat{\mathbf{x}}_i} \right)
\end{aligned} \tag{9.12}
$$

where

$$\tilde{\mathbf{x}}_j = \frac{1}{N} \sum_{\hat{\mathbf{x}}_i \in S_j} \hat{\mathbf{x}}_i$$

Following Eq. (9.12), the gradient of $f(\boldsymbol{\Theta})$ can be evaluated as

$$\nabla f(\boldsymbol{\Theta}) = \begin{bmatrix} \nabla_{\boldsymbol{\theta}_0} f \\ \nabla_{\boldsymbol{\theta}_1} f \\ \vdots \\ \nabla_{\boldsymbol{\theta}_{K-1}} f \end{bmatrix}$$

where $\nabla_{\theta_k} f$ is given by

$$\nabla_{\theta_k} f = -\tilde{\mathbf{x}}_k + \frac{1}{N} \sum_{i=1}^{N} \frac{e^{\theta_k^T \hat{\mathbf{x}}_i} \hat{\mathbf{x}}_i}{\sum_{j=0}^{K-1} e^{\theta_j^T \hat{\mathbf{x}}_i}} = -\tilde{\mathbf{x}}_k + \frac{1}{N} \sum_{i=1}^{N} P(l = k | \mathbf{x}_i, \boldsymbol{\Theta}) \quad (9.13)$$

for $k = 0, 1, \ldots, K - 1$. The Hessian of $f(\boldsymbol{\Theta})$ is a block matrix of the form $\nabla^2 f(\boldsymbol{\Theta}) = \{\nabla_{\theta_k, \theta_l}^2 f \text{ for } k, l = 0, 1, \ldots, K - 1\}$ where

$$\nabla_{\theta_k, \theta_l}^2 f = \begin{cases} \dfrac{1}{N} \displaystyle\sum_{i=1}^{N} \left(\dfrac{\sum_{j=0, j\neq k}^{K-1} e^{(\theta_k + \theta_j)^T \hat{\mathbf{x}}_i}}{\left(\sum_{j=0}^{K-1} e^{\theta_j^T \hat{\mathbf{x}}_i}\right)^2} \right) \hat{\mathbf{x}}_i \hat{\mathbf{x}}_i^T & \text{if } l = k \\[3em] -\dfrac{1}{N} \displaystyle\sum_{i=1}^{N} \left(\dfrac{e^{(\theta_k + \theta_l)^T \hat{\mathbf{x}}_i}}{\left(\sum_{j=0}^{K-1} e^{\theta_j^T \hat{\mathbf{x}}_i}\right)^2} \right) \hat{\mathbf{x}}_i \hat{\mathbf{x}}_i^T & \text{if } l \neq k \end{cases} \quad (9.14)$$

The Hessian matrix turns out to be positive semidefinite and hence the objective function $f(\boldsymbol{\Theta})$ is convex. We can prove this by showing that $\mathbf{v}^T \nabla^2 f(\boldsymbol{\Theta}) \mathbf{v} \geq 0$ for any $\mathbf{v} \in R^{K(n+1)}$. To this end, we divide vector $\mathbf{v}$ into a set of K vectors $\{\mathbf{v}_i, i = 0, 1, \ldots, K - 1\}$, each of length $n + 1$, and use Eq. (9.14) to express $\mathbf{v}^T \nabla^2 f(\boldsymbol{\Theta}) \mathbf{v}$ as a

$$\mathbf{v}^T \nabla^2 f(\boldsymbol{\Theta}) \mathbf{v} = -\frac{1}{N} \sum_{i=1}^{N} \frac{1}{\left(\sum_{j=0}^{K-1} e^{\theta_j^T \hat{\mathbf{x}}_i}\right)^2} \left(\sum_{k=0}^{K-1} \sum_{l=0}^{K-1} e^{(\theta_k + \theta_l)^T \hat{\mathbf{x}}_i} (\tau_{k,i}^2 - \tau_{k,i} \tau_{l,i}) \right)$$

$$\tag{9.15}$$

where $\tau_{k,i} = \mathbf{v}_k^T \hat{\mathbf{x}}_i$. If we let $e_{k,i} = e^{\theta_k^T \hat{\mathbf{x}}_i}$, then $e^{(\theta_k + \theta_l)^T \hat{\mathbf{x}}_i} = e_{k,i} \cdot e_{l,i}$ and thus we can estimate

$$\sum_{k=0}^{K-1} \sum_{l=0}^{K-1} e^{(\theta_k + \theta_l)^T \hat{\mathbf{x}}_i} (\tau_{k,i}^2 - \tau_{k,i} \tau_{l,i}) = \sum_{k=0}^{K-1} \sum_{l=0}^{K-1} e_{k,i} e_{l,i} (\tau_{k,i}^2 - \tau_{k,i} \tau_{l,i})$$

$$= \sum_{k=0}^{K-1} \sum_{l=0}^{K-1} e_{k,i} e_{l,i} \tau_{k,i}^2 - \sum_{k=0}^{K-1} \sum_{l=0}^{K-1} e_{k,i} e_{l,i} \tau_{k,i} \tau_{l,i}$$

$$\geq \sum_{k=0}^{K-1} \sum_{l=0}^{K-1} e_{k,i} e_{l,i} \tau_{k,i}^2 - \frac{1}{2} \sum_{k=0}^{K-1} \sum_{l=0}^{K-1} e_{k,i} e_{l,i} (\tau_{k,i}^2 + \tau_{l,i}^2)$$

$$= 0$$

Therefore, $\mathbf{v}^T \nabla^2 f(\boldsymbol{\Theta}) \mathbf{v} \geq 0$.

It is important to note that objective function $f(\boldsymbol{\Theta})$ is invariant under a constant shift of parameter $\boldsymbol{\Theta}$, that is, $f(\boldsymbol{\Theta}) = f(\boldsymbol{\Theta}_v)$ where $\boldsymbol{\Theta}_\psi = [\theta_0 - \boldsymbol{\Psi} \ \theta_1 - \boldsymbol{\Psi} \cdots \theta_{K-1} - \boldsymbol{\Psi}]$ and $\boldsymbol{\Psi}$ is a constant vector. This is because

$$f(\boldsymbol{\Theta}_\psi) = -\frac{1}{N} \sum_{i=1}^{N} \ln \left(\frac{e^{(\theta_j - \boldsymbol{\Psi})^T \hat{\mathbf{x}}_i}}{\sum_{j=0}^{K-1} e^{(\theta_j - \boldsymbol{\Psi})^T \hat{\mathbf{x}}_i}} \right) = -\frac{1}{N} \sum_{i=1}^{N} \ln \left(\frac{e^{-\boldsymbol{\Psi}^T \hat{\mathbf{x}}_i} e^{\theta_l^T \hat{\mathbf{x}}_i}}{\sum_{j=0}^{K-1} e^{-\boldsymbol{\Psi}^T \hat{\mathbf{x}}_i} e^{\theta_j^T \hat{\mathbf{x}}_i}} \right)$$

$$= -\frac{1}{N} \sum_{i=1}^{N} \ln \left(\frac{e^{\theta_l^T \hat{\mathbf{x}}_i}}{\sum_{j=0}^{K-1} e^{\theta_j^T \hat{\mathbf{x}}_i}} \right) = f(\boldsymbol{\Theta})$$

Consequently, the minimizer of the problem in Eq. (9.11) is not unique. This problem can be fixed by adding an additional term, known as *weight decay* term [6], to the objective function in Eq. (9.12), in which case the problem at hand becomes

$$\underset{\boldsymbol{\Theta}}{\text{minimize}} \quad f_\mu(\boldsymbol{\Theta}) = -\sum_{j=0}^{K-1} (\boldsymbol{\theta}_j^T \cdot \tilde{\mathbf{x}}_j) + \frac{1}{N} \sum_{i=1}^{N} \ln \left(\sum_{j=0}^{K-1} e^{\boldsymbol{\theta}_j^T \hat{\mathbf{x}}} \right) + \frac{\mu}{2} \|\boldsymbol{\Theta}\|_F^2 \quad (9.16)$$

where $\mu > 0$ is a small constant and $\|\boldsymbol{\Theta}\|_F^2$ is the squared Frobenius norm of $\boldsymbol{\Theta}$ given by $\|\boldsymbol{\Theta}\|_F^2 = \sum_{j=0}^{K-1} \|\boldsymbol{\theta}_j\|_2^2$.

The gradient of $f_\mu(\boldsymbol{\Theta})$ is given by

$$\nabla f_\mu(\boldsymbol{\Theta}) = \begin{bmatrix} \nabla_{\theta_0} f_\mu \\ \nabla_{\theta_1} f_\mu \\ \vdots \\ \nabla_{\theta_{K-1}} f_\mu \end{bmatrix}$$

where

$$\nabla_{\boldsymbol{\theta}_k} f_\mu = \nabla_{\boldsymbol{\theta}_k} f + \mu \boldsymbol{\theta}_k$$

for $k = 0, 1, \ldots, K-1$ and $\nabla_{\boldsymbol{\theta}_k} f$ is given by Eq. (9.13).

The Hessian of $f_\mu(\boldsymbol{\Theta})$ is equal to $\nabla^2 f_\mu(\boldsymbol{\Theta}) = \nabla^2 f(\boldsymbol{\Theta}) + \mu \mathbf{I}$ which is strictly convex and, therefore, the problem in Eq. (9.16) has a unique minimizer. It should be mentioned that including the weight decay also tends to prevent the magnitudes of the components of $\boldsymbol{\Theta}$ from becoming too large, which in turn prevents the model in Eq. (9.8) from becoming overly complex for a given training data set [6].

9.2.4   Use of Softmax Regression for the Classification of Handwritten Digits

We can at this point use the softmax regression technique described in Sect. 9.2.3 to solve the classification problem of handwritten digits which are taken from the MNIST database. We begin with some technical details about the MNIST database, then present an accelerated gradient-descent algorithm that can be used to solve the problem in Eq. (9.16), and after that we evaluate the performance of the classification technique.

9.2.4.1   The MNIST Database

Database MNIST contains a *training* data set $\mathcal{D}$ and a *testing* data set $\mathcal{T}$. The training data set assumes the form $\mathcal{D} = \{(\mathbf{D}_i, l_i), i = 1, 2, \ldots, N\}$ with $N = 60,000$ where each $\mathbf{D}_i$ is a gray-scale image of a handwritten digit of size 28×28 pixels which is labeled as digit $l_i \in \{0, 1, \ldots, 9\}$. Data set $\mathcal{D}$ can be subdivided into ten classes, denoted as $\mathcal{D}_j$ for $j = 0, 1, \ldots, 9$, according to the ten possible label values; hence we have $\mathcal{D} = \bigcup_{j=0}^{9} \mathcal{D}_j$ where $\mathcal{D}_j = \{(\mathbf{D}_i^{(j)}, j), i = 1, 2, \ldots, n_j\}$ contains all the digits that are associated with the same label j. The sizes of the ten data classes vary from the smallest $n_5 = 5,421$ to the largest $n_1 = 6,742$. The testing data set $\mathcal{T} = \{(\mathbf{D}_i^{(t)}, l_i^{(t)}), i = 1, 2, \ldots, 10,000\}$ is a separate set of 10,000 handwritten digits of size 28×28 that is used to evaluate the performance of a classifier.

9.2.4.2 Classification of Handwritten Digits Using the MNIST Data Set

From Eq. (9.9), the optimal classifier is defined by

$$i^* = \arg \left(\max_{0 \leq i \leq K-1} \boldsymbol{\theta}_i^{*T} \hat{\mathbf{x}} \right) \tag{9.17}$$

where $K = 10$ and $\boldsymbol{\Theta}^* = [\boldsymbol{\theta}_0^* \ \boldsymbol{\theta}_1^* \ \cdots \ \boldsymbol{\theta}_{K-1}^*]$ is the global minimizer of the problem in Eq. (9.16). To specify the objective function, we randomly select 1500 samples from each data class $\mathcal{D}_j = \{(\mathbf{D}_i^{(j)}, j), i = 1, 2, \ldots, n_j\}$ described in Sect. 9.2.4.1, then compute the HOG with $b = 7$ for each sample (see Sect. 9.2.2) to generate feature vectors for digit j as $\mathcal{X}_j = \{(\mathbf{x}_i^{(j)}, j), i = 1, 2, \ldots, 1500\}$ with $\mathbf{x}_i^{(j)} \in R^{1183 \times 1}$. Then the data set required to define the objective function in Eq. (9.16) is constructed as $\mathcal{X} = \bigcup_{j=0}^{9} \mathcal{X}_j$, and the vectors $\{\hat{\mathbf{x}}_i, i = 1, 2, \ldots, N\}$ in Eq. (9.16) are related to data set $\mathcal{X}$ as $\hat{\mathbf{x}}_i = \begin{bmatrix} \mathbf{x}_i \\ 1 \end{bmatrix}$ for $i = 1, 2, \ldots, N$ where $\mathbf{x}_i \in \mathcal{X}$ (see Eq. (9.5)). Under these circumstances, the total number of variables to be optimized, $\boldsymbol{\Theta} = [\boldsymbol{\theta}_0 \ \boldsymbol{\theta}_2 \ \cdots \ \boldsymbol{\theta}_9]$, is equal to $10 \times 1184 = 11840$.

Considering the complexity of our problem in terms of the number of variables and number of terms in evaluating the objective function and its gradient, which exceeds 15000 terms, the optimization algorithms suitable for the problem at hand are limited to those that use only gradient information. The steepest-descent algorithm would work but it tends to be rather inefficient. An improved version of the steepest-descent algorithm for convex functions, known as Nesterov's accelerated gradient (NAG) descent algorithm [8], has been found to be suitable for the problem in Eq. (9.16). Below we outline the NAG algorithm.

The NAG algorithm can be applied to unconstrained problems in which the objective function $f(\mathbf{x})$ is convex and satisfies the condition

$$\|\nabla f(\mathbf{x}) - \nabla f(\mathbf{y})\|_2 \leq \frac{1}{\alpha} \|\mathbf{x} - \mathbf{y}\|_2 \tag{9.18}$$

for some constant $\alpha > 0$. The steps involved are detailed below.

Algorithm 9.1 Nesterov's accelerated gradient-descent algorithm
Step 1
Input initial point $\mathbf{x}_0$, constants μ and α, and tolerance ε. Set $\mathbf{y}_1 = \mathbf{x}_0$, $t_1 = 1$, and $k = 1$.
Step 2
Compute $\mathbf{x}_k = \mathbf{y}_k - \alpha \nabla f(\mathbf{y}_k)$
Step 3
Compute $t_{k+1} = \dfrac{1 + \sqrt{1 + 4t_k^2}}{2}$
Step 4
Update $\mathbf{y}_{k+1} = \mathbf{x}_k + \left(\dfrac{t_k - 1}{t_{k+1}}\right)(\mathbf{x}_k - \mathbf{x}_{k-1})$
Step 5
If $f(\mathbf{x}_{k-1}) - f(\mathbf{x}_k) < \varepsilon$, output $\mathbf{x}_k$ as the solution and stop; otherwise set $k = k + 1$ and repeat from Step 2.

It can be shown that the NAG algorithm converges to the minimizer $\mathbf{x}^*$ with rate $O(1/k^2)$ in the sense that

$$f(\mathbf{x}_k) - f(\mathbf{x}^*) \le \frac{2\|\mathbf{x}_0 - \mathbf{x}^*\|_2^2}{\alpha k^2}$$

(see [8] for details). The algorithmic steps described above show that the NAG algorithm is essentially a variant of the conventional steepest-descent algorithm (see Algorithm 5.1) where the gradient descent step (i.e., Step 2) is taken at a point obtained as the weighted sum of the last two iterates (see Step 4). It turns out that with appropriate weights calculated in Step 3, the weighted sum of the last two iterates is a smooth direction that helps to avoid the typical zigzag profile of steepest-descent iterates and hence improves the convergence rate. Also note that the algorithm uses a constant step size α. As a result, a NAG iteration is even more economical relative to a conventional steepest-descent iteration, which uses a line search to determine the value of α in each iteration.

It is important to stress that in order to apply Algorithm 9.1 the objective function must be convex and satisfy Eq. (9.18). Concerning the problem in Eq. (9.16), its objective function has been shown to be convex. To show that the objective function satisfies Eq. (9.18), we use a result from convex analysis that the convex function $f_\mu(\boldsymbol{\Theta})$ in Eq. (9.16) satisfies Eq. (9.18) if and only if there exists a constant $L > 0$ such that $L\mathbf{I} - \nabla^2 f_\mu(\boldsymbol{\Theta})$ is positive semidefinite [9]. We prove this by showing that $\mathbf{v}^T(L\mathbf{I} - \nabla^2 f_\mu(\boldsymbol{\Theta}))\mathbf{v} \ge 0$ for any $\mathbf{v}$. To this end, we write $\nabla^2 f_\mu(\boldsymbol{\Theta}) = \nabla^2 f(\boldsymbol{\Theta}) + \mu\mathbf{I}$ and use Eq. (9.15) to estimate $\mathbf{v}^T \nabla^2 f(\boldsymbol{\Theta})\mathbf{v}$. By definition, $\tau_{k,i} = \mathbf{v}_k^T \hat{\mathbf{x}}_i$ and hence on using the Cauchy-Schwartz inequality we have

$$|\tau_{k,i}^2 - \tau_{k,i}\tau_{l,i}| \le \tau_{k,i}^2 + |\tau_{k,i}| \cdot |\tau_{l,i}| \le (\|\mathbf{v}_k\|_2^2 + \|\mathbf{v}_k\|_2 \|\mathbf{v}_l\|_2)\|\hat{\mathbf{x}}_i\|_2^2 \le 2\|\mathbf{v}\|_2^2 \cdot \|\hat{\mathbf{x}}_i\|_2^2$$

Therefore, from Eq. (9.15) we obtain

$$\mathbf{v}^T \nabla^2 f(\boldsymbol{\Theta}) \mathbf{v} = \frac{1}{N} \sum_{i=1}^{N} \frac{1}{\left(\sum_{j=0}^{K-1} e^{\boldsymbol{\theta}_j^T \hat{\mathbf{x}}_i}\right)^2} \left(\sum_{k=0}^{K-1} \sum_{l=0}^{K-1} e^{(\theta_k + \theta_l)^T \hat{\mathbf{x}}_i} (\tau_{k,i}^2 - \tau_{k,i} \tau_{l,i}) \right)$$

$$\leq \frac{2\|\mathbf{v}\|_2^2}{N} \sum_{i=1}^{N} \frac{\|\hat{\mathbf{x}}_i\|_2^2}{\left(\sum_{j=0}^{K-1} e^{\boldsymbol{\theta}_j^T \hat{\mathbf{x}}_i}\right)^2} \left(\sum_{k=0}^{K-1} \sum_{l=0}^{K-1} e^{(\theta_k + \theta_l)^T \hat{\mathbf{x}}_i} \right)$$

$$= \frac{2\|\mathbf{v}\|_2^2}{N} \sum_{i=1}^{N} \|\hat{\mathbf{x}}_i\|_2^2 = L_1 \|\mathbf{v}\|_2^2$$

where $L_1 = \frac{2}{N} \sum_{i=1}^{N} \|\hat{\mathbf{x}}_i\|_2^2$. It follows that

$$\mathbf{v}^T \nabla^2 f_\mu(\boldsymbol{\Theta}) \mathbf{v} = \mathbf{v}^T \nabla^2 f(\boldsymbol{\Theta}) \mathbf{v} + \mu \|\mathbf{v}\|_2^2 \leq L_1 \|\mathbf{v}\|_2^2 + \mu \|\mathbf{v}\|_2^2 = (L_1 + \mu)\|\mathbf{v}\|_2^2$$
$$= L\|\mathbf{v}\|_2^2$$

where $L = L_1 + \mu$. This shows that $\mathbf{v}^T (L\mathbf{I} - \nabla^2 f_\mu(\boldsymbol{\Theta}))\mathbf{v} \geq 0$. Consequently, $f_\mu(\boldsymbol{\Theta})$ in Eq. (9.16) satisfies the condition in Eq. (9.18) and Algorithm 9.1 is applicable to the problem in Eq. (9.16). At this point, it should be mentioned that once the value of L is known, $1/L$ can be used as constant α in Eq. (9.18).

Algorithm 9.1 was applied to the problem in Eq. (9.16) with $N = 15000$, $K = 10$, $\boldsymbol{\Theta}_0 = \mathbf{0} \in R^{1184 \times 10}$, $\mu = 10^{-3}$, $\alpha = 0.7$, and $\varepsilon = 10^{-7}$. It took the algorithm 309 iterations to converge to a solution $\boldsymbol{\Theta}^*$ at which $f_\mu(\boldsymbol{\Theta}^*) = 1.335170 \times 10^{-2}$. The profiles of $f_\mu(\boldsymbol{\Theta}_k)$ versus the number of iterations k is shown in Fig. 9.4.

To evaluate the performance of the optimal solution $\boldsymbol{\Theta}^* = [\boldsymbol{\theta}_0^* \ \boldsymbol{\theta}_1^* \ \cdots \ \boldsymbol{\theta}_{K-1}^*]$, the softmax-based classifier in Eq. (9.17), namely,

$$i^* = \arg \left(\max_{0 \leq i \leq K-1} \boldsymbol{\theta}_i^{*T} \hat{\mathbf{x}} \right)$$

Fig. 9.4 Objective function $f_\mu(\boldsymbol{\Theta})$ versus number of iterations k

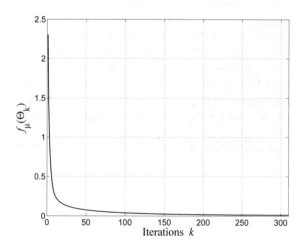

was applied to classify the MNIST testing data set $\mathcal{T} = \{(\mathbf{D}_i^{(t)}, l_i^{(t)}), i = 1, 2, \ldots,$ 10000\}$ where each $\hat{\mathbf{x}}$ assumes the form $\hat{\mathbf{x}} = \begin{bmatrix} \mathbf{x} \\ 1 \end{bmatrix}$, and $\mathbf{x}$ is the HOG of a digit $\mathbf{D}_i^{(t)}$ from $\mathcal{T}$. The classifier recognized 9820 digits correctly and hence it has achieved an error rate of 1.8%.

9.3   Inverse Kinematics for Robotic Manipulators

9.3.1   Position and Orientation of a Manipulator

Typically an industrial robot, also known as a robotic manipulator, comprises a chain of mechanical links with one end fixed relative to the ground and the other end, known as the *end-effector*, free to move. Motion is made possible in a manipulator by moving the joint of each link about its axis with an electric or hydraulic actuator.

One of the basic problems in robotics is the description of the position and orientation of the end-effector in terms of the joint variables. There are two types of joints: rotational joints for rotating the associated robot link, and translational joints for pushing and pulling the associated robot link along a straight line. However, joints in industrial robots are almost always rotational. Fig. 9.5 shows a three-joint industrial robot, where the three joints can be used to rotate links 1, 2, and 3. In this case, the end-effector is located at the end of link 3, whose position and orientation can be conveniently described relative to a fixed coordinate system which is often referred to as a *frame* in robotics. As shown in Fig. 9.5, frame {0} is attached to the robot base and is fixed relative to the ground. Next, frames {1}, {2}, and {3} are considered to be attached to joint axes 1, 2, and 3, respectively, and are subject to the following rules:

- The z axis of frame $\{i\}$ is along the joint axis i for $i = 1, 2, 3$.
- The x axis of frame $\{i\}$ is perpendicular to the z axes of frames $\{i\}$ and $\{i + 1\}$ for $i = 1, 2, 3$.
- The y axis of frame $\{i\}$ is determined such that frame $\{i\}$ is a standard right-hand coordinate system.
- Frame {4} is considered to be attached to the end of link 3 in such a way that the axes of frames {3} and {4} are in parallel and the distance between the z axes of these two frames is zero.

Having assigned the frames, the relation between two consecutive frames can be characterized by the so-called *Denavit-Hartenberg (DH) parameters*[10], which are defined as follows:

As can be observed in Fig. 9.5, parameters d_1, a_2, and d_4 represent the lengths of links 1, 2, and 3, respectively, d_3 represents the offset between link 1 and link 2, and a_3 represents the offset between link 2 and link 3. The above frame assignment also

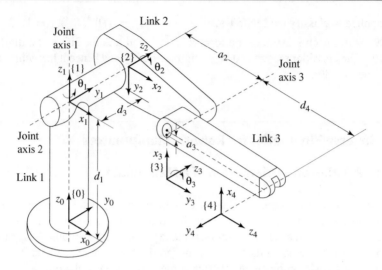

Fig. 9.5 A three-link robotic manipulator

a_i: distance from the z_i axis to the z_{i+1} axis measured along the x_i axis
α_i: angle between the z_i axis and the z_{i+1} axis measured about the x_i axis
d_i: distance from the x_{i-1} axis to the x_i axis measured along the z_i axis
θ_i: angle between the x_{i-1} axis and the x_i axis measured about the z_i axis

Table 9.1 DH parameters of 3-link robot

i	α_{i-1}	a_{i-1}	d_i	θ_i
1	$0°$	0	d_1	θ_1
2	$-90°$	0	0	θ_2
3	$0°$	a_2	d_3	θ_3
4	$-90°$	a_3	d_4	$0°$

fixes angles α_0, α_1, α_2, and α_3 as $\alpha_0 = 0°$, $\alpha_1 = -90°$, $\alpha_2 = 0°$, and $\alpha_3 = -90°$. The DH parameters of the three-joint robot in Fig. 9.5 are summarized in Table 9.1. The only variable parameters are θ_1, θ_2, and θ_3, which represent the rotation angles of joints 1, 2, and 3, respectively.

Since the DH parameters a_{i-1}, α_{i-1}, d_i, and θ_i characterize the relation between frames $\{i-1\}$ and $\{i\}$, they can be used to describe the position and orientation of frame $\{i\}$ in relation to those of frame $\{i-1\}$. To this end, we define the so-called *homogeneous transformation* in terms of the 4×4 matrix

$$
{}^{i-1}_{i}\mathbf{T} = \left[\begin{array}{c|c} {}^{i-1}_{i}\mathbf{R} & {}^{i-1}\mathbf{p}_{i\,\mathrm{ORG}} \\ \hline 0\ 0\ 0 & 1 \end{array} \right]_{4\times4}
\tag{9.19}
$$

where vector ${}^{i-1}\mathbf{p}_{i\,\mathrm{ORG}}$ denotes the position of the origin of frame $\{i\}$ in frame $\{i-1\}$, and matrix ${}^{i-1}_{i}\mathbf{R}$ is an orthogonal matrix whose columns denote the x-, y-, and z-coordinate vectors of frame $\{i\}$ with respect to frame $\{i-1\}$. With the DH parameters a_{i-1}, α_{i-1}, d_i, and θ_i known, the homogeneous transformation in Eq. (9.19) can be expressed as [10]

$$
{}^{i-1}_{i}\mathbf{T} =
\begin{bmatrix}
c\theta_i & -s\theta_i & 0 & a_{i-1} \\
s\theta_i c\alpha_{i-1} & c\theta_i c\alpha_{i-1} & -\alpha_{i-1} & -s\alpha_{i-1}d_i \\
s\theta_i s\alpha_{i-1} & c\theta_i s\alpha_{i-1} & \alpha_{i-1} & \alpha_{i-1}d_i \\
0 & 0 & 0 & 1
\end{bmatrix}
\tag{9.20}
$$

where $s\theta$ and $c\theta$ denote $\sin\theta$ and $\cos\theta$, respectively. The significance of the above formula is that it can be used to evaluate the position and orientation of the end-effector as

$$
{}^{0}_{N}\mathbf{T} = {}^{0}_{1}\mathbf{T}\,{}^{1}_{2}\mathbf{T}\cdots{}^{N-1}_{N}\mathbf{T}
\tag{9.21}
$$

where each ${}^{i-1}_{i}\mathbf{T}$ on the right-hand side can be obtained using Eq. (9.20). The formula in Eq. (9.21) is often referred to as the *equation of forward kinematics*.

Example 9.1 Derive closed-form formulas for the position and orientation of the robot tip in Fig. 9.5 in terms of joint angles θ_1, θ_2, and θ_3.

Solution Using Table 9.1 and Eq. (9.20), the homogeneous transformations ${}^{i-1}_{i}\mathbf{T}$ for $i = 1$, 2, 3, and 4 are obtained as

$$
{}^{0}_{1}\mathbf{T} =
\begin{bmatrix}
c_1 & -s_1 & 0 & 0 \\
s_1 & c_1 & 0 & 0 \\
0 & 0 & 1 & d_1 \\
0 & 0 & 0 & 1
\end{bmatrix}, \quad
{}^{1}_{2}\mathbf{T} =
\begin{bmatrix}
c_2 & -s_2 & 0 & 0 \\
0 & 0 & 1 & 0 \\
-s_2 & -c_2 & 0 & 0 \\
0 & 0 & 0 & 1
\end{bmatrix}
$$

$$
{}^{2}_{3}\mathbf{T} =
\begin{bmatrix}
c_3 & -s_3 & 0 & a_2 \\
s_3 & c_3 & 0 & 0 \\
0 & 0 & 1 & d_3 \\
0 & 0 & 0 & 1
\end{bmatrix}, \quad
{}^{3}_{4}\mathbf{T} =
\begin{bmatrix}
1 & 0 & 0 & a_3 \\
0 & 0 & 1 & d_4 \\
0 & -1 & 0 & 0 \\
0 & 0 & 0 & 1
\end{bmatrix}
$$

With $N = 4$, Eq. (9.21) gives

$$
{}^{0}_{4}\mathbf{T} = {}^{0}_{1}\mathbf{T}\,{}^{1}_{2}\mathbf{T}\,{}^{2}_{3}\mathbf{T}\,{}^{3}_{4}\mathbf{T}
$$

$$
=
\begin{bmatrix}
c_1c_{23} & s_1 & -c_1s_{23} & c_1(a_2c_2 + a_3c_{23} - d_4s_{23}) - d_3s_1 \\
s_1c_{23} & -c_1 & -s_1s_{23} & s_1(a_2c_2 + a_3c_{23} - d_4s_{23}) + d_3c_1 \\
-s_{23} & 0 & -c_{23} & d_1 - a_2s_2 - a_3s_{23} - d_4c_{23} \\
0 & 0 & 0 & 1
\end{bmatrix}
$$

where $c_1 = \cos\theta_1$, $s_1 = \sin\theta_1$, $c_{23} = \cos(\theta_2 + \theta_3)$, and $s_{23} = \sin(\theta_2 + \theta_3)$. Therefore, the position and orientation of the robot tip in frame $\{0\}$ are given by

$$^0\mathbf{p}_{4ORG} = \begin{bmatrix} c_1(a_2c_2 + a_3c_{23} - d_4s_{23}) - d_3s_1 \\ s_1(a_2c_2 + a_3c_{23} - d_4s_{23}) + d_3c_1 \\ d_1 - a_2s_2 - a_3s_{23} - d_4c_{23} \end{bmatrix} \tag{9.22}$$

and

$$^0_4\mathbf{R} = \begin{bmatrix} c_1c_{23} & s_1 & -c_1s_{23} \\ s_1c_{23} & -c_1 & -s_1s_{23} \\ -s_{23} & 0 & -c_{23} \end{bmatrix}$$

respectively.   ■

9.3.2 Inverse Kinematics Problem

The joint angles of manipulator links are usually measured using sensors such as optical encoders that are attached to the link actuators. As discussed in Sect. 9.3.1, when the joint angles θ_1, θ_2, $\ldots$, θ_n are known, the position and orientation of the end-effector can be evaluated using Eq. (9.21). A related and often more important problem is the *inverse kinematics problem* which is as follows: find the joint angles θ_i for $1 \leq i \leq n$ with which the manipulator's end-effector would achieve a *prescribed* position and orientation. The significance of the inverse kinematics lies in the fact that the tasks to be accomplished by a robot are usually in terms of trajectories in the Cartesian space that the robot's end-effector must follow. Under these circumstances, the position and orientation for the end-effector are known and the problem is to find the correct values of the joint angles that would move the robot's end-effector to the desired position and orientation.

Mathematically, the inverse kinematics problem can be described as the problem of finding the values θ_i for $1 \leq i \leq n$ that would satisfy Eq. (9.21) for a given $^0_N\mathbf{T}$. Since Eq. (9.21) is highly nonlinear, the problem of finding its solutions is not a trivial one [10]. For example, if a prescribed position of the end-effector for the three-link manipulator in Fig. 9.5 is given by $^0\mathbf{p}_{4ORG} = [p_x \; p_y \; p_z]^T$, then Eq. (9.22) gives

$$\begin{aligned} c_1(a_2c_2 + a_3c_{23} - d_4s_{23}) - d_3s_1 &= p_x \\ s_1(a_2c_2 + a_3c_{23} - d_4s_{23}) + d_3c_1 &= p_y \\ d_1 - a_2s_2 - a_3s_{23} - d_4c_{23} &= p_z \end{aligned} \tag{9.23}$$

In the next section, we describe an optimization approach for the solution of the inverse kinematics problem on the basis of Eq. (9.23).

9.3.3 Solution of Inverse Kinematics Problem

If we let

$$\mathbf{x} = [\theta_1 \ \theta_2 \ \theta_3]^T \tag{9.24a}$$

$$f_1(\mathbf{x}) = c_1(a_2c_2 + a_3c_{23} - d_4s_{23}) - d_3s_1 - p_x \tag{9.24b}$$

$$f_2(\mathbf{x}) = s_1(a_2c_2 + a_3c_{23} - d_4s_{23}) + d_3c_1 - p_y \tag{9.24c}$$

$$f_3(\mathbf{x}) = d_1 - a_2s_2 - a_3s_{23} - d_4c_{23} - p_z \tag{9.24d}$$

then Eq. (9.23) is equivalent to

$$f_1(\mathbf{x}) = 0 \tag{9.25a}$$

$$f_2(\mathbf{x}) = 0 \tag{9.25b}$$

$$f_3(\mathbf{x}) = 0 \tag{9.25c}$$

To solve this system of nonlinear equations, we construct the objective function

$$F(\mathbf{x}) = f_1^2(\mathbf{x}) + f_2^2(\mathbf{x}) + f_3^2(\mathbf{x})$$

and notice that vector $\mathbf{x}^*$ solves Eqs. (9.25a)–(9.25c) if and only if $F(\mathbf{x}^*) = 0$. Since function $F(\mathbf{x})$ is nonnegative, finding a solution point $\mathbf{x}$ for Eqs. (9.25a)–(9.25c) amounts to finding a minimizer $\mathbf{x}^*$ at which $F(\mathbf{x}^*) = 0$. In other words, we can convert the inverse kinematics problem at hand into the unconstrained minimization problem

$$\text{minimize} \quad F(\mathbf{x}) = \sum_{k=1}^{3} f_k^2(\mathbf{x}) \tag{9.26}$$

An advantage of this approach over conventional methods for inverse kinematics problems [10] is that when the desired position $[p_x \ p_y \ p_z]^T$ is *not* within the manipulator's reach, the conventional methods will fail to work and a conclusion that no solution exists will be drawn. With the optimization approach, however, minimizing function $F(\mathbf{x})$ will still yield a minimizer, say, $\mathbf{x}^* = [\theta_1^* \ \theta_2^* \ \theta_3^*]^T$, although the objective function $F(\mathbf{x})$ would not become zero at $\mathbf{x}^*$. In effect, an *approximate* solution of the problem would be obtained, which could be entirely satisfactory in many engineering applications.

To apply the minimization algorithms studied earlier, we let

$$\mathbf{f}(\mathbf{x}) = \begin{bmatrix} f_1(\mathbf{x}) \\ f_2(\mathbf{x}) \\ f_3(\mathbf{x}) \end{bmatrix}$$

and evaluate the gradient of $F(\mathbf{x})$ as

$$\mathbf{g}(\mathbf{x}) = 2\mathbf{J}^T(\mathbf{x})\mathbf{f}(\mathbf{x}) \tag{9.27}$$

where the Jacobian matrix $\mathbf{J}(\mathbf{x})$ is given by

$$\mathbf{J}(\mathbf{x}) = [\nabla f_1(\mathbf{x}) \ \nabla f_2(\mathbf{x}) \ \nabla f_3(\mathbf{x})]^T$$

$$= \begin{bmatrix} -q_3s_1 - d_3c_1 & q_4c_1 & q_2c_1 \\ q_3c_1 - d_3s_1 & q_4s_1 & q_2s_1 \\ 0 & -q_3 & -q_1 \end{bmatrix}$$

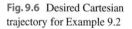

Fig. 9.6 Desired Cartesian
trajectory for Example 9.2

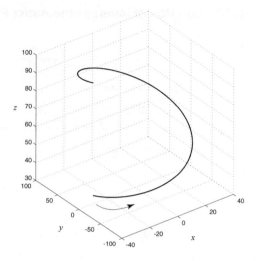

with $q_1 = a_3 c_{23} - d_4 s_{23}, q_2 = -a_3 s_{23} - d_4 c_{23}, q_3 = a_2 c_2 + q_1$, and $q_4 = -a_2 s_2 + q_2$.
The Hessian of $F(\mathbf{x})$ is given by

$$\mathbf{H}(\mathbf{x}) = 2\mathbf{J}^T(\mathbf{x})\mathbf{J}(\mathbf{x}) + 2\sum_{k=1}^{3} f_k(\mathbf{x})\nabla^2 f_k(\mathbf{x}) \tag{9.28}$$

where $\nabla^2 f_k(\mathbf{x})$ is the Hessian of $f_k(\mathbf{x})$ (see Prob. 9.4).

   The application of unconstrained optimization to the inverse kinematic problem
is illustrated by the following example.

Example 9.2 In the three-link manipulator depicted in Fig. 9.5, $d_1 = 66.04$ cm,
$d_3 = 14.91$ cm, $d_4 = 43.31$ cm, $a_2 = 43.18$ cm, and $a_3 = 2.03$ cm. By applying
a steepest-descent (SD), Newton (N), Gauss-Newton (GN), Fletcher-Reeves (FR)
algorithm and then a quasi-Newton (QN) algorithm based on the Broyden-Fletcher-
Goldfarb-Shanno updating formula in Eq. (7.57), determine the joint angles $\theta_i(t)$ for
$i = 1, 2, 3$ and $-\pi \le t \le \pi$ such that the manipulator's end-effector tracks the
desired trajectory $\mathbf{p}_d(t) = [p_x(t)\ p_y(t)\ p_z(t)]^T$ where

$$p_x(t) = 30\cos t, \quad p_y(t) = 100\sin t, \quad p_z(t) = 10t + 66.04$$

for $-\pi \le t \le \pi$ as illustrated in Fig. 9.6.

Solution The problem was solved by applying Algorithms 5.1, 5.6, and 6.3 as the
steepest-descent, Gauss-Newton, and Fletcher-Reeves algorithm, respectively, using
the inexact line search in Steps 1 to 6 of Algorithm 7.3 in each case. The Newton algorithm used was essentially Algorithm 5.3 incorporating the Hessian-matrix
modification in Eq. (5.21) as detailed below:

Algorithm 9.2   Newton algorithm
Step 1
Input $\mathbf{x}_0$ and initialize the tolerance ε.
Set $k = 0$.
Step 2
Compute $\mathbf{g}_k$ and $\mathbf{H}_k$.
Step 3
Compute the eigenvalues of $\mathbf{H}_k$ (see Sec. A.5).
Determine the smallest eigenvalue of $\mathbf{H}_k$, λ_{min}.
Modify matrix $\mathbf{H}_k$ to

$$\hat{\mathbf{H}}_k = \begin{cases} \mathbf{H}_k & \text{if } \lambda_{min} > 0 \\ \mathbf{H}_k + \gamma \mathbf{I}_n & \text{if } \lambda_{min} \leq 0 \end{cases}$$

where

$$\gamma = -1.05\lambda_{min} + 0.1$$

Step 4
Compute $\hat{\mathbf{H}}_k^{-1}$ and $\mathbf{d}_k = -\hat{\mathbf{H}}_k^{-1}\mathbf{g}_k$

Step 5
Find α_k, the value of α that minimizes $f(\mathbf{x}_k + \alpha\mathbf{d}_k)$, using the inexact line search in Steps 1 to 6 of Algorithm 7.3.
Step 6
Set $\mathbf{x}_{k+1} = \mathbf{x}_k + \alpha_k\mathbf{d}_k$.
Compute $f_{k+1} = f(\mathbf{x}_{k+1})$.
Step 7
If $\|\alpha_k\mathbf{d}_k\|_2 < \varepsilon$, then output $\mathbf{x}^* = \mathbf{x}_{k+1}$ and $f(\mathbf{x}^*) = f_{k+1}$, and stop; otherwise, set $k = k + 1$ and repeat from Step 2.

The quasi-Newton algorithm used was essentially Algorithm 7.3 with a slightly modified version of Step 8b as follows:

Step 8b′
Compute $D = \boldsymbol{\delta}_k^T \boldsymbol{\gamma}_k$. If $D \leq 0$, then set $\mathbf{S}_{k+1} = \mathbf{I}_n$, otherwise, compute $\mathbf{S}_{k+1}$ using Eq. (7.57).

At $t = t_k$, the desired trajectory can be described in terms of its Cartesian coordinates as

$$\mathbf{p}_d(t_k) = \begin{bmatrix} p_x(t_k) \\ p_y(t_k) \\ p_z(t_k) \end{bmatrix} = \begin{bmatrix} 30\cos t_k \\ 100\sin t_k \\ 10t_k + 66.04 \end{bmatrix}$$

where $-\pi \leq t_k \leq \pi$. Assuming 100 uniformly spaced sample points, the solution of the system of equations in Eqs. (9.25a)–(9.25c) can be obtained by solving the minimization problem in Eq. (9.26) for $k = 1, 2, \ldots, 100$, i.e., for $t_k = -\pi, \ldots, \pi$, using the specified DH parameters. Since the gradient and Hessian of $F(\mathbf{x})$ are available (see Eqs. (9.27) and (9.28)), the problem can be solved using each of the

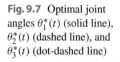

Fig. 9.7 Optimal joint angles $\theta_1^*(t)$ (solid line), $\theta_2^*(t)$ (dashed line), and $\theta_3^*(t)$ (dot-dashed line)

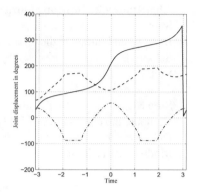

five optimization algorithms specified in the description of the problem to obtain a minimizer $\mathbf{x}^*(t_k)$ in each case. If the objective function $F(\mathbf{x})$ turns out to be zero at $\mathbf{x}^*(t_k)$, then $\mathbf{x}^*(t_k)$ satisfies Eqs. (9.25a)–(9.25c), and the joint angles specified by $\mathbf{x}^*(t_k)$ lead the manipulator's end-effector to the desired position precisely. On the other hand, if $F[\mathbf{x}^*(t_k)]$ is nonzero, then $\mathbf{x}^*(t_k)$ is taken as an approximate solution of the inverse kinematics problem at instant t_k.

Once the minimizer $\mathbf{x}^*(t_k)$ is obtained, the above steps can be repeated at $t = t_{k+1}$ to obtain solution point $\mathbf{x}^*(t_{k+1})$. Since t_{k+1} differs from t_k only by a small amount and the profile of optimal joint angles is presumably continuous, $\mathbf{x}^*(t_{k+1})$ is expected to be in the vicinity of $\mathbf{x}^*(t_k)$. Therefore, the previous solution $\mathbf{x}^*(t_k)$ can be used as a reasonable initial point for the next optimization.[1]

The five optimization algorithms were applied to the problem at hand and were all found to work although with different performance in terms of solution accuracy and computational complexity. The solution obtained using the BFGS algorithm, $\mathbf{x}^*(t_k) = [\theta_1^*(t_k) \ \theta_2^*(t_k) \ \theta_3^*(t_k)]^T$ for $1 \leq k \leq 100$, is plotted in Fig. 9.7; the tracking profile of the end-effector is plotted as the dotted curve in Fig. 9.8 and is compared with the desired trajectory, which is plotted as the solid curve. It turns out that the desired positions $\mathbf{p}_d(t_k)$ for $20 \leq k \leq 31$ and $70 \leq k \leq 81$ are beyond the manipulator's reach. As a result, we see in Fig. 9.8 that there are two small portions of the tracking profile that deviate from the desired trajectory, but even in this case, the corresponding $\mathbf{x}^*(t_k)$ still offers a reasonable approximate solution. The remaining part of the tracking profile coincides with the desired trajectory almost perfectly which simply means that for the desired positions within the manipulator's work space, $\mathbf{x}^*(t_k)$ offers a nearly exact solution.

The performance of the five algorithms in terms of the normalized CPU time and iterations per sample point and the error at sample points within and outside the work space is summarized in Table 9.2. The data supplied are in the form of

---

[1]Choosing the initial point on the basis of *any* knowledge about the solution instead of a random initial point can lead to a large reduction in the amount of computation in most optimization problems.

Fig. 9.8 End-effector's
profile (dotted line) and the
desired trajectory (solid line)

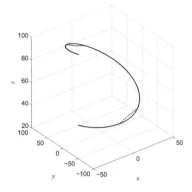

Table 9.2 Performance comparisons for Example 9.2

Algorithm	Normalized CPU time	Average number of iterations per sample point	Average error within work space	Average error outside work space
SD	1.0	47	0.0488	4.3750
N	0.0553	4	4.0161×10^{-8}	4.3737
GN	0.1060	5	1.5578×10^{-4}	4.3739
FR	0.4516	14	0.1778	4.3743
BFGS	0.1244	6	2.8421×10^{-5}	4.3737

averages with respect to 100 runs of the algorithms using random initializations. As can be seen, the average errors within the manipulator's work space for the solutions $\mathbf{x}^*(t_k)$ obtained using the steepest-descent and Fletcher-Reeves algorithms are much larger than those obtained using the Newton, Gauss-Newton, and BFGS algorithms, although the solutions obtained are still acceptable considering the relatively large size of the desired trajectory. The best results in terms of efficiency as well as accuracy are obtained by using the Newton and BFGS algorithms.

∎

9.4   Design of Digital Filters

In this section, we will apply unconstrained optimization to the design of FIR digital filters. Different designs are possible depending on the type of FIR filter required and the formulation of the objective function. The theory and design principles of digital filters are quite extensive [11] and are beyond the scope of this book. To facilitate the understanding of the application of unconstrained optimization to the design of digital filters, we present a brief review of the highlights of the theory, properties, and

characterization of digital filters in Appendix B, which should prove quite adequate
in the present context.

The one design aspect of digital filters that can be handled quite efficiently with
optimization is the approximation problem whereby the parameters of the filter have
to be chosen to achieve a specified type of frequency response. Below, we examine
two different designs (see Sect. B.9). In one design, we formulate a weighted least-
squares objective function, i.e., one based on the square of the L_2 norm, for the design
of linear-phase FIR filters, and in other we obtain a minimax objective function, i.e.,
one based on the L_∞ norm.

The L_p norm of a vector where $p \geq 1$ is defined in Sect. A.8.1. Similarly, the L_p
norm of a function $F(\omega)$ of a continuous variable ω can be defined with respect to
the interval $[a, b]$ as

$$\|F(\omega)\|_p = \left(\int_a^b |F(\omega)|^p \, d\omega \right)^{1/p}$$

where $p \geq 1$ and if

$$\int_a^b |F(\omega)|^p \, d\omega \leq K < \infty$$

the L_p norm of $F(\omega)$ exists. If $F(\omega)$ is bounded with respect to the interval $[a, b]$,
i.e., $|F(\omega)| \leq M$ for $\omega \in [a, b]$ where M is finite, then the L_∞ norm of $F(\omega)$ is
defined as

$$\|F(\omega)\|_\infty = \max_{a \leq \omega \leq b} |F(\omega)|$$

and as in the case of the L_∞ norm of a vector, it can be verified that

$$\lim_{p \to \infty} \|F(\omega)\|_p = \|F(\omega)\|_\infty$$

(see Sect. A.8.1).

9.4.1 Weighted Least-Squares Design of FIR Filters

As shown in Sect. B.5.1, an FIR filter is completely specified by its transfer function
which assumes the form

$$H(z) = \sum_{n=0}^{N} h_n z^{-n}$$

where the coefficients h_n for $n = 0, 1, \ldots, n$ represent the impulse response of the
filter.

9.4.1.1   Specified Frequency Response

Assuming a normalized sampling frequency of 2π, which corresponds to a normalized sampling period $T = 1$ s, the frequency response of an FIR filter is obtained as $H(e^{j\omega})$ by letting $z = e^{j\omega}$ in the transfer function (see Sect. B.8). In practice, the frequency response is required to approach some desired frequency response, $H_d(\omega)$, to within a specified error. Hence an FIR filter can be designed by formulating an objective function based on the difference between the actual and desired frequency responses (see Sect. B.9.3). Except in some highly specialized applications, the transfer function coefficients (or impulse response values) of a digital filter are real and, consequently, knowledge of the frequency response of the filter with respect to the positive half of the baseband fully characterizes the filter (see Sect. B.8). Under these circumstances, a weighted least-squares objective function that can be used to design FIR filters can be constructed as

$$e(\mathbf{x}) = \int_0^\pi W(\omega)|H(e^{j\omega}) - H_d(\omega)|^2 \, d\omega \tag{9.29}$$

where $\mathbf{x} = [h_0 \, h_1 \, \cdots \, h_N]^T$ is an $N+1$-dimensional variable vector representing the transfer function coefficients, ω is a normalized frequency variable which is assumed to be in the range 0 to π rad/s, and $W(\omega)$ is a predefined weighting function. The design is accomplished by finding the vector $\mathbf{x}^*$ that minimizes $e(\mathbf{x})$, and this can be efficiently done by means of unconstrained optimization.

Weighting is used to emphasize or deemphasize the objective function with respect to one or more ranges of ω. Without weighting, an optimization algorithm would tend to minimize the objective function uniformly with respect to ω. Thus if the objective function is multiplied by a weighting constant larger than unity for values of ω in a certain critical range but is left unchanged for all other frequencies, a reduced value of the objective function will be achieved with respect to the critical frequency range. This is due to the fact that the weighted objective function will tend to be minimized uniformly and thus the actual unweighted objective function will tend to be scaled down in proportion to the inverse of the weighting constant in the critical range of ω relative to its value at other frequencies. Similarly, if a weighting constant of value less than unity is used for a certain uncritical frequency range, an increased value of the objective will be the outcome with respect to the uncritical frequency range. Weighting is very important in practice because through the use of suitable scaling, the designer is often able to design a more economical filter for the required specifications. In the above example, the independent variable is frequency. In other applications, it could be time or some other independent parameter.

An important step in an optimization-based design is to express the objective function in terms of variable vector $\mathbf{x}$ *explicitly*. This facilitates the evaluation of the gradient and Hessian of the objective function. To this end, if we let

$$\mathbf{c}(\omega) = [1 \, \cos\omega \, \cdots \, \cos N\omega]^T$$
$$\mathbf{s}(\omega) = [0 \, \sin\omega \, \cdots \, \sin N\omega]^T$$

the frequency response of the filter can be expressed as

$$H(e^{j\omega}) = \sum_{n=0}^{N} h_n \cos n\omega - j \sum_{n=0}^{N} h_n \sin n\omega = \mathbf{x}^T \mathbf{c}(\omega) - j\mathbf{x}^T \mathbf{s}(\omega) \tag{9.30}$$

If we let

$$H_d(\omega) = H_r(\omega) - jH_i(\omega) \tag{9.31}$$

where $H_r(\omega)$ and $-H_i(\omega)$ are the real and imaginary parts of $H_d(\omega)$, respectively, then Eqs. (9.30) and (9.31) give

$$
\begin{aligned}
|H(e^{j\omega}) - H_d(\omega)|^2 &= [\mathbf{x}^T \mathbf{c}(\omega) - H_r(\omega)]^2 + [\mathbf{x}^T \mathbf{s}(\omega) - H_i(\omega)]^2 \\
&= \mathbf{x}^T [\mathbf{c}(\omega)\mathbf{c}^T(\omega) + \mathbf{s}(\omega)\mathbf{s}^T(\omega)]\mathbf{x} \\
&\quad -2\mathbf{x}^T [\mathbf{c}(\omega)H_r(\omega) + \mathbf{s}(\omega)H_i(\omega)] + |H_d(\omega)|^2
\end{aligned}
$$

Therefore, the objective function in Eq. (9.29) can be expressed as a quadratic function with respect to $\mathbf{x}$ of the form

$$e(\mathbf{x}) = \mathbf{x}^T \mathbf{Q}\mathbf{x} - 2\mathbf{x}^T \mathbf{b} + \kappa \tag{9.32}$$

where κ is a constant[2] and

$$\mathbf{Q} = \int_0^{\pi} W(\omega)[\mathbf{c}(\omega)\mathbf{c}^T(\omega) + \mathbf{s}(\omega)\mathbf{s}^T(\omega)]\, d\omega \tag{9.33}$$

$$\mathbf{b} = \int_0^{\pi} W(\omega)[H_r(\omega)\mathbf{c}(\omega) + H_i(\omega)\mathbf{s}(\omega)]\, d\omega$$

Matrix $\mathbf{Q}$ in Eq. (9.33) is positive definite (see Prob. 9.5). Hence the objective function $e(\mathbf{x})$ in Eq. (9.32) is globally strictly convex and has a unique global minimizer $\mathbf{x}^*$ given by

$$\mathbf{x}^* = \mathbf{Q}^{-1}\mathbf{b}$$

For the design of high-order FIR filters, matrix $\mathbf{Q}$ in Eq. (9.33) is of a large size and the methods described in Sect. 6.4 can be used to find the minimizer without obtaining the inverse of matrix $\mathbf{Q}$.

9.4.1.2 Linear Phase Response

The frequency response of an FIR digital filter of order N (or length $N + 1$) with linear phase response is given by

$$H(e^{j\omega}) = e^{-j\omega N/2} A(\omega) \tag{9.34}$$

Assuming an even-order filter, function $A(\omega)$ in Eq. (9.34) can be expressed as

$$A(\omega) = \sum_{n=0}^{N/2} a_n \cos n\omega$$

---

[2]Symbol κ will be used to represent a constant throughout this chapter.

$$a_n = \begin{cases} h_{N/2} & \text{for } n = 0 \\ 2h_{N/2-n} & \text{for } n \neq 0 \end{cases} \tag{9.35}$$

(see Sect. B.9.2) and if the desired frequency response is assumed to be of the form

$$H_d(\omega) = e^{-j\omega N/2} A_d(\omega)$$

then the least-squares objective function

$$e_l(\mathbf{x}) = \int_0^\pi W(\omega)[A(\omega) - A_d(\omega)]^2 \, d\omega \tag{9.36}$$

can be constructed where the variable vector is given by

$$\mathbf{x} = [a_0 \ a_1 \ \cdots \ a_{N/2}]^T$$

If we now let

$$\mathbf{c}_l(\omega) = [1 \ \cos \omega \ \cdots \ \cos N\omega/2]^T$$

$A(\omega)$ can be written in terms of the inner product $\mathbf{x}^T \mathbf{c}_l(\omega)$ and the objective function $e_l(\mathbf{x})$ in Eq. (9.36) can be expressed as

$$e_l(\mathbf{x}) = \mathbf{x}^T \mathbf{Q}_l \mathbf{x} - 2\mathbf{x}^T \mathbf{b}_l + \kappa \tag{9.37}$$

where κ is a constant, as before, with

$$\mathbf{Q}_l = \int_0^\pi W(\omega) \mathbf{c}_l(\omega) \mathbf{c}_l^T(\omega) \, d\omega \tag{9.38a}$$

$$\mathbf{b}_l = \int_0^\pi W(\omega) A_d(\omega) \mathbf{c}_l(\omega) \, d\omega \tag{9.38b}$$

Like matrix $\mathbf{Q}$ in Eq. (9.33), matrix $\mathbf{Q}_l$ in Eq. (9.38a) is positive definite; hence, like the objective function $e(\mathbf{x})$ in Eq. (9.32), $e_l(\mathbf{x})$ in Eq. (9.37) is globally strictly convex and its unique global minimizer is given in closed form by

$$\mathbf{x}_l^* = \mathbf{Q}_l^{-1} \mathbf{b}_l \tag{9.39}$$

For filters of order less than 200, matrix $\mathbf{Q}_l$ in Eq. (9.39) is of size less than 100, and the formula in Eq. (9.39) requires a moderate amount of computation. For higher-order filters, the closed-form solution given in Eq. (9.39) becomes computationally very demanding and methods that do not require the computation of the inverse of matrix $\mathbf{Q}_l$ such as those studied in Sect. 6.4 would be preferred.

Example 9.3 (a) Applying the above method, formulate the design of an even-order linear-phase lowpass FIR filter assuming the desired amplitude response

$$A_d(\omega) = \begin{cases} 1 & \text{for } 0 \leq \omega \leq \omega_p \\ 0 & \text{for } \omega_a \leq \omega \leq \pi \end{cases}$$

where ω_p and ω_a are the passband and stopband edges, respectively (see Sect. A.8.1). Assume a normalized sampling frequency of 2π rad/s.

(b) Using the formulation in part (a), design FIR filters with $\omega_p = 0.45\pi$ and $\omega_a = 0.5\pi$ for filter orders of 20, 40, 60, and 80.

Solution (a) A suitable weighting function $W(\omega)$ for this problem is

$$W(\omega) = \begin{cases} 1 & \text{for } 0 \leq \omega \leq \omega_p \\ \gamma & \text{for } \omega_a \leq \omega \leq \pi \\ 0 & \text{elsewhere} \end{cases}$$

The value of γ can be chosen to emphasize or deemphasize the error function in the stopband relative to that in the passband. Since $W(\omega)$ is piecewise constant, the matrix $\mathbf{Q}_l$ in Eq. (9.38a) can be written as

$$\mathbf{Q}_l = \mathbf{Q}_{l1} + \mathbf{Q}_{l2}$$

where

$$\mathbf{Q}_{l1} = \int_0^{\omega_p} \mathbf{c}_l(\omega)\mathbf{c}_l^T(\omega)\, d\omega = \{q_{ij}^{(1)}\} \qquad \text{for } 1 \leq i, \ j \leq \frac{N+2}{2} \qquad (9.40a)$$

and

$$\mathbf{Q}_{l2} = \gamma \int_{\omega_a}^{\pi} \mathbf{c}_l(\omega)\mathbf{c}_l^T(\omega)\, d\omega = \{q_{ij}^{(2)}\} \qquad \text{for } 1 \leq i, \ j \leq \frac{N+2}{2} \qquad (9.40b)$$

with

$$q_{ij}^{(1)} = \begin{cases} \dfrac{\omega_p}{2} + \dfrac{\sin[2(i-1)\omega_p]}{4(i-1)} & \text{for } i = j \\[3mm] \dfrac{\sin[(i-j)\omega_p]}{2(i-j)} + \dfrac{\sin[(i+j-2)\omega_p]}{2(i+j-2)} & \text{for } i \neq j \end{cases} \qquad (9.41a)$$

and

$$q_{ij}^{(2)} = \begin{cases} \gamma\left[\dfrac{(\pi-\omega_a)}{2} - \dfrac{\sin[2(i-1)\omega_a]}{4(i-1)}\right] & \text{for } i = j \\[3mm] -\dfrac{\gamma}{2}\left[\dfrac{\sin[(i-j)\omega_a]}{(i-j)} + \dfrac{\sin[(i+j-2)\omega_a]}{(i+j-2)}\right] & \text{for } i \neq j \end{cases} \qquad (9.41b)$$

Note that for $i = j = 1$, the expressions in Eqs. (9.41a) and (9.41b) are evaluated by taking the limit as $i \to 1$, which implies that

$$q_{11}^{(1)} = \omega_p \quad \text{and} \quad q_{11}^{(2)} = \gamma(\pi - \omega_a) \qquad (9.41c)$$

Vector $\mathbf{b}_l$ in Eq. (9.38b) is given by

$$\mathbf{b}_l = \int_0^{\omega_p} \mathbf{c}_l(\omega)\, d\omega = \{b_n\} \qquad (9.42a)$$

where

$$b_n = \frac{\sin[(n-1)\omega_p]}{(n-1)} \qquad \text{for } 1 \leq n \leq \frac{N+2}{2} \qquad (9.42b)$$

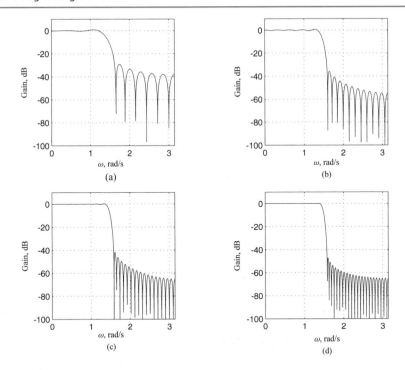

Fig. 9.9 Amplitude responses of the filters in Example 9.3: **a** $N = 20$, **b** $N = 40$, **c** $N = 60$, **d** $N = 80$

As before, for $n = 1$, the expression in Eq. (9.42b) is evaluated by taking the limit as $n \to 1$, which gives

$$b_1 = \omega_p \qquad\qquad (9.42c)$$

(*b*) Optimal weighted least-squares designs for the various values of N were obtained by computing the minimizer $\mathbf{x}_l^*$ given by Eq. (9.39) and then evaluating the filter coefficients $\{h_i\}$ using Eq. (9.35). The weighting constant γ was assumed to be 25. The amplitude responses of the FIR filters obtained are plotted in Fig. 9.9. ∎

9.4.2 Minimax Design of FIR Filters

The Parks-McClellan algorithm and its variants have been the most efficient tools for the minimax design of FIR digital filters [11–13]. However, these algorithms apply only to the class of linear-phase FIR filters. The group delay introduced by these filters is constant and independent of frequency in the entire baseband (see Sect. B.8) but it can be quite large. In practice, a variable group delay in stopbands is of little concern and by allowing the phase response to be nonlinear in stopbands, FIR filters can be designed with approximately constant group delay with respect to the passband(s), which is significantly reduced relative to that achieved with filters that have a constant group delay throughout the entire baseband.

This section presents a least-pth approach to the design of low-delay FIR filters. For FIR filters, the weighted L_p error function with an even integer p can be shown to be globally convex.[3] This property, in conjunction with the availability of the gradient and Hessian of the objective function in closed form, enables us to develop an unconstrained optimization method for the design problem at hand.

9.4.2.1   Objective Function

Given a desired frequency response $H_d(\omega)$ for an FIR filter, we want to determine the coefficients $\{h_n\}$ in the transfer function

$$H(z) = \sum_{n=0}^{N} h_n z^{-n}$$

such that the weighted L_{2p} approximation error

$$f(\mathbf{h}) = \left[\int_0^\pi W(\omega)|H(e^{j\omega}) - H_d(\omega)|^{2p} \, d\omega \right]^{1/2p} \tag{9.43}$$

is minimized, where $W(\omega) \geq 0$ is a weighting function, p is a positive integer, and $\mathbf{h} = [h_0 \ h_1 \ \cdots \ h_N]^T$.

If we let

$$H_d(\omega) = H_{dr}(\omega) - j H_{di}(\omega)$$
$$\mathbf{c}(\omega) = [1 \ \cos\omega \ \cdots \ \cos N\omega]^T$$
$$\mathbf{s}(\omega) = [0 \ \sin\omega \ \cdots \ \sin N\omega]^T$$

then Eq. (9.43) becomes

$$f(\mathbf{h}) = \left\{ \int_0^\pi W[(\mathbf{h}^T\mathbf{c} - H_{dr})^2 + (\mathbf{h}^T\mathbf{s} - H_{di})^2]^p \, d\omega \right\}^{1/2p}$$

where for simplicity the frequency dependence of W, $\mathbf{c}$, $\mathbf{s}$, H_{dr}, and H_{di} has been omitted. Now if we let

$$e_2(\omega) = [\mathbf{h}^T\mathbf{c}(\omega) - H_{dr}(\omega)]^2 + [\mathbf{h}^T\mathbf{s}(\omega) - H_{di}(\omega)]^2$$

then the objective function can be expressed as

$$f(\mathbf{h}) = \left[\int_0^\pi W(\omega)e_2^p(\omega) \, d\omega \right]^{1/2p} \tag{9.44}$$

---

[3]Note that this property does not apply to infinite-duration impulse response (IIR) filters [11].

9.4.2.2   Gradient and Hessian of $f(\mathbf{h})$

Using Eq. (9.44), the gradient and Hessian of objective function $f(\mathbf{h})$ can be readily obtained as

$$\nabla f(\mathbf{h}) = f^{1-2p}(\mathbf{h}) \int_0^\pi W(\omega) e_2^{p-1}(\omega) \mathbf{q}(\omega) \, d\omega \qquad (9.45a)$$

where

$$\mathbf{q}(\omega) = [\mathbf{h}^T \mathbf{c}(\omega) - H_{dr}(\omega)]\mathbf{c}(\omega) + [\mathbf{h}^T \mathbf{s}(\omega) - H_{di}(\omega)]\mathbf{s}(\omega) \qquad (9.45b)$$

and

$$\nabla^2 f(\mathbf{h}) = \mathbf{H}_1 + \mathbf{H}_2 - \mathbf{H}_3 \qquad (9.45c)$$

where

$$\mathbf{H}_1 = 2(p-1)f^{1-2p}(\mathbf{h}) \int_0^\pi W(\omega) e_2^{p-2}(\omega) \mathbf{q}(\omega)\mathbf{q}^T(\omega) \, d\omega \qquad (9.45d)$$

$$\mathbf{H}_2 = f^{1-2p}(\mathbf{h}) \int_0^\pi W(\omega) e_2^{p-1}(\omega) [\mathbf{c}(\omega)\mathbf{c}^T(\omega) + \mathbf{s}(\omega)\mathbf{s}^T(\omega)] \, d\omega \qquad (9.45e)$$

$$\mathbf{H}_3 = (2p-1)f^{-1}(\mathbf{h}) \nabla f(\mathbf{h}) \nabla^T f(\mathbf{h}) \qquad (9.45f)$$

respectively.

Of central importance to the present algorithm is the property that for each and every positive integer p, the weighted L_{2p} objective function defined in Eq. (9.43) is convex in the entire parameter space $\mathcal{R}^{N+1}$. This property can be proved by showing that the Hessian $\nabla^2 f(\mathbf{h})$ is positive semidefinite for all $\mathbf{h} \in R^{N+1}$ (see Prob. 9.9).

9.4.2.3   Design Algorithm

It is now quite clear that an FIR filter whose frequency response approximates a rather arbitrary frequency response $H_d(\omega)$ to within a given tolerance in the minimax sense can be obtained by minimizing $f(\mathbf{h})$ in Eq. (9.43) with a sufficiently large p. It follows from the above discussion that for a given p, $f(\mathbf{h})$ has a unique global minimizer. Therefore, any descent minimization algorithm, e.g., the steepest-descent, Newton, and quasi-Newton methods studied in previous chapters, can, in principle, be used to obtain the minimax design regardless of the initial design chosen. The amount of computation required to obtain the design is largely determined by the choice of optimization method as well as the initial point assumed.

A reasonable initial point can be deduced by using the L_2-optimal design obtained by minimizing $f(\mathbf{h})$ in Eq. (9.43) with $p = 1$. We can write

$$f(\mathbf{h}) = (\mathbf{h}^T \mathbf{Q}\mathbf{h} - 2\mathbf{h}^T \mathbf{p} + \kappa)^{1/2} \qquad (9.46a)$$

where

$$\mathbf{Q} = \int_0^\pi W(\omega)[\mathbf{c}(\omega)\mathbf{c}^T(\omega) + \mathbf{s}(\omega)\mathbf{s}^T(\omega)] \, d\omega \qquad (9.46b)$$

$$\mathbf{p} = \int_0^\pi W(\omega)[H_{dr}(\omega)\mathbf{c}(\omega) + H_{di}(\omega)\mathbf{s}(\omega)] \, d\omega \qquad (9.46c)$$

Since $\mathbf{Q}$ is positive definite; the global minimizer of $f(\mathbf{h})$ in Eq. (9.46a) can be obtained as the solution of the linear equation

$$\mathbf{Qh} = \mathbf{p} \qquad\qquad (9.47)$$

We note that $\mathbf{Q}$ in Eq. (9.47) is a symmetric Toeplitz matrix[4] for which fast algorithms are available to compute the solution of Eq. (9.47) [14].

The minimization of convex objective function $f(\mathbf{h})$ can be accomplished in a number of ways. Since the gradient and Hessian of $f(\mathbf{h})$ are available in closed form and $\nabla^2 f(\mathbf{h})$ is positive semidefinite, the Newton method and the family of quasi-Newton methods are among the most appropriate.

From Eqs. (9.44) and (9.45a)–(9.45f), we note that $f(\mathbf{h})$, $\nabla f(\mathbf{h})$, and $\nabla^2 f(\mathbf{h})$ all involve integration which can be carried out using numerical methods. In computing $\nabla^2 f(\mathbf{h})$, the error introduced in the numerical integration can cause the Hessian to lose its positive definiteness but the problem can be easily fixed by modifying $\nabla^2 f(\mathbf{h})$ to $\nabla^2 f(\mathbf{h}) + \varepsilon \mathbf{I}$ where ε is a small positive scalar.

9.4.2.4 Direct and Sequential Optimizations

With a power p, weighting function $W(\omega)$, and an initial $\mathbf{h}$, say, $\mathbf{h}_0$, chosen, the design can be obtained directly or indirectly.

In a direct optimization, one of the unconstrained optimization methods is applied to minimize the L_{2p} objective function in Eq. (9.44) directly. Based on rather extensive trials, it was found that to achieve a near-minimax design, the value of p should be set to 20 for order $N \leq 10$ and a p in the range $2N$ to $2.5N$ should be used for filters of order higher than 10.

In sequential optimization, an L_{2p} optimization is first carried out with $p = 1$. The minimizer thus obtained, $\mathbf{h}^*$, is then used as the initial point in another optimization with $p = 2$. The same procedure is repeated for $p = 4, 8, 16, \ldots$ until the reduction in the objective function between two successive optimizations is less than a prescribed tolerance.

Example 9.4 Using the above direct and sequential approaches first with a Newton and then with a quasi-Newton algorithm, design a lowpass FIR filter of order $N = 54$ that would have approximately constant passband group delay of 23 s. Assume idealized passband and stopband gains of 1 and 0, respectively, passband edge $\omega_p = 0.45\pi$, stopband edge $\omega_a = 0.55\pi$, and a normalized sampling frequency $\omega_s = 2\pi$. Also let $W(\omega) = 1$ in both the passband and stopband and $W(\omega) = 0$ elsewhere.

Solution The design was carried out using the direct approach with $p = 128$ and the sequential approach with $p = 2, 4, 8, \ldots, 128$ by minimizing the objective function in Eq. (9.44) with the Newton algorithm and a quasi-Newton algorithm

---

[4]A Toeplitz matrix is a matrix whose entries along each diagonal are constant [14].

with the BFGS updating formula in Eq. (7.57). The Newton algorithm used was essentially the same as Algorithm 9.2 (see solution of Example 9.2) except that Step 3 was replaced by the following modified Step 3:

Step 3′
Modify matrix $\mathbf{H}_k$ to $\hat{\mathbf{H}}_k = \mathbf{H}_k + 0.1\mathbf{I}_n$

The quasi-Newton algorithm used was Algorithm 7.3 with the modifications described in the solution of Example 9.2.

A lowpass FIR filter that would satisfy the required specifications can be obtained by assuming a complex-valued idealized frequency response of the form

$$H_d(\omega) = \begin{cases} e^{-j23\omega} & \text{for } \omega \in [0, \omega_p] \\ 0 & \text{for } \omega \in [\omega_a, \omega_s/2] \end{cases}$$
$$= \begin{cases} e^{-j23\omega} & \text{for } \omega \in [0, 0.45\pi] \\ 0 & \text{for } \omega \in [0.55\pi, \pi] \end{cases}$$

(see Sect. B.9.2). The integrations in Eqs. (9.44), (9.45a), and (9.45c) can be carried out by using one of several available numerical methods for integration. A fairly simple and economical approach which works well in optimization is as follows: Given a continuous function $f(\omega)$ of ω, an approximate value of its integral over the interval $[a, b]$ can be obtained as

$$\int_a^b f(\omega)d\omega \approx \delta \sum_{i=1}^K f(\omega_i)$$

where $\delta = (b-a)/K$ and $\omega_1 = a+\delta/2, \omega_2 = a+3\delta/2, \ldots, \omega_K = a+(2K-1)\delta/2$. That is, we divide interval $[a, b]$ into K subintervals, add the values of the function at the midpoints of the K subintervals, and then multiply the sum obtained by δ.

The objective function in Eq. (9.44) was expressed as

$$f(\mathbf{h}) = \left[\int_0^{0.45\pi} e_2^p(\omega)d\omega\right]^{1/2p} + \left[\int_{0.55\pi}^{\pi} e_2^p(\omega)d\omega\right]^{1/2p}$$

and each integral was evaluated using the above approach with $K = 500$. The integrals in Eqs. (9.45a) and (9.45c) were evaluated in the same way.

The initial $\mathbf{h}$ was obtained by applying L_2 optimization to Eqs. (9.46a)–(9.46c). All trials converged to the same near-minimax design, and the sequential approach turned out to be more efficient than the direct approach. The Newton and quasi-Newton algorithms required 21.1 and 40.7 s of CPU time, respectively, on a PC with a Pentium 4, 3.2 GHz CPU. The amplitude response, passband error, and group delay characteristic of the filter obtained are plotted in Fig. 9.10a, b, and c, respectively. We note that an equiripple amplitude response was achieved in both the passband and stopband. The passband group delay varies between 22.9 and 23.1 s but it is not equiripple. This is because the minimax optimization was carried out for the *complex-valued* frequency response $H_d(\omega)$, not the phase response alone (see Eq. (9.43)). ∎

Fig. 9.10 Minimax design
of a lowpass filter with low
passband group delay for
Example 9.4: **a** Frequency
response, **b** magnitude of the
passband error, and **c**
passband group delay

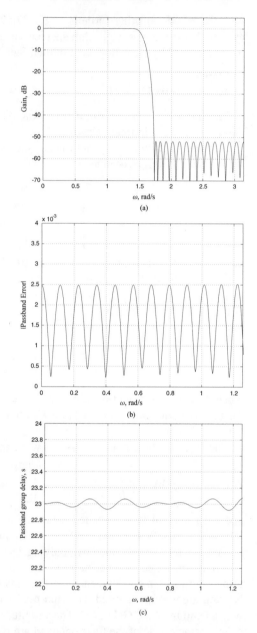

Example 9.5 Using the above direct and sequential approaches first with a Newton
and then with a quasi-Newton algorithm, design a bandpass FIR filter of order $N =$
160 that would have approximately constant passband group delay of 65 s. Assume
idealized passband and stopband gains of 1 and 0, respectively, lower and upper
passband edges $\omega_{p1} = 0.4\pi$ and $\omega_{p2} = 0.6\pi$, respectively, lower and upper stopband
edges $\omega_{a1} = 0.375\pi$ and $\omega_{a2} = 0.625\pi$, respectively, and normalized sampling

frequency$= 2\pi$. Also let $W(\omega) = 1$ in the passband, $W(\omega) = 50$ in the stopbands, and $W(\omega) = 0$ elsewhere.

Solution The required design was carried out using the direct approach with $p = 128$ and the sequential approach with $p = 2, 4, 8, \ldots, 128$ by minimizing the objective function in Eq. (9.44) with the Newton and quasi-Newton algorithms described in Example 9.4.

A bandpass FIR filter that would satisfy the required specifications was obtained by assuming a complex-valued idealized frequency response of the form

$$H_d(\omega) = \begin{cases} e^{-j65\omega} & \text{for } \omega \in [\omega_{p1}, \omega_{p2}] \\ 0 & \text{for } \omega \in [0, \omega_{a1}] \bigcup [\omega_{a2}, \omega_s/2] \end{cases}$$

$$= \begin{cases} e^{-j65\omega} & \text{for } \omega \in [0.4\pi, 0.6\pi] \\ 0 & \text{for } \omega \in [0, 0.375\pi] \bigcup [0.625\pi, \pi] \end{cases}$$

(see Sect. B.9.2). The objective function in Eq. (9.44) was expressed as

$$f(\mathbf{h}) = \left[\int_0^{0.375\pi} 50 e_2^p(\omega) d\omega \right]^{1/2p} + \left[\int_{0.4\pi}^{0.6\pi} e_2^p(\omega) d\omega \right]^{1/2p}$$

$$+ \left[\int_{0.625\pi}^{\pi} 50 e_2^p(\omega) d\omega \right]^{1/2p}$$

and the integrals at the right-hand side were evaluated using the numerical method in the solution of Example 9.4 with $K = 382, 236, 382$ respectively. The integrals in Eqs. (9.45a) and (9.45c) were similarly evaluated in order to obtain the gradient and Hessian of the problem.

As in Example 9.4, the sequential approach was more efficient. The Newton and quasi-Newton algorithms required 173.5 and 201.8 s, respectively, on a Pentium 4 PC.

The amplitude response, passband error, and group delay characteristic are plotted in Fig. 9.11a, b, and c, respectively. We note that an equiripple amplitude response has been achieved in both the passband and stopband. ∎

We conclude this chapter with some remarks on the numerical results of Examples 9.2, 9.4, and 9.5. Quasi-Newton algorithms, in particular algorithms using an inexact line-search along with the BFGS updating formula (e.g., Algorithm 7.3), are known to be very robust and efficient relative to other gradient-based algorithms [15, 16]. However, the basic Newton algorithm used for these problems, namely, Algorithm 9.2, turned out to be more efficient than the quasi-Newton algorithm. This is largely due to certain unique features of the problems considered, which favor the basic Newton algorithm. The problem in Example 9.2 is a simple problem with only three independent variables and a well-defined gradient and Hessian that can be easily computed through closed form formulas. Furthermore, the inversion of the Hessian is almost a trivial task. The problems in Examples 9.4 and 9.5 are significantly more complex than that in Example 9.2; however, their gradients and Hessians are fairly easy to compute accurately and efficiently through closed-form formulas as in

Fig. 9.11 Minimax design
of a bandpass filter with low
passband group delay for
Example 9.5: **a** Frequency
response, **b** magnitude of
passband error, **c** passband
group delay

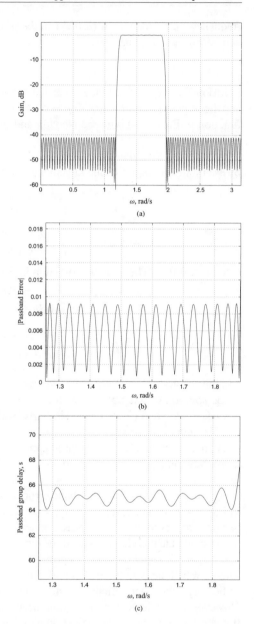

Example 9.2. In addition, these problems are convex with unique global minimums which are easy to locate. On the other hand, a large number of variables in the problem tends to be an impediment in quasi-Newton algorithms because, as was shown in Chap. 7, these algorithms would, in theory, require n iterations in an n-variable problem to compute the inverse-Hessian in a well-defined convex quadratic problem (see proof of Theorem 7.3), more in nonconvex nonquadratic problems. However, in highly nonlinear multimodal problems[5] with a moderate number of independent variables, quasi-Newton algorithms are usually the most efficient.

9.5   Source Localization

Source localization is a class of optimization problems whereby the distances

$$d_i = \|\mathbf{x} - \mathbf{s}_i\|_2 \quad \text{for } i = 1, \, 2, \, \dots, \, m$$

between a radiating source located at a point $\mathbf{x}$ and several sensors located at $\{\mathbf{s}_i\}$ are available and the location of point $\mathbf{x}$ needs to be determined. Distances d_i are often referred to as *range measurements*.

A variant of the above problem is one whereby (1) an additional sensor located at point $\mathbf{s}_0$ is placed at the origin of a co-ordinate system, i.e., at $\mathbf{s}_0 = \mathbf{0}$, (2) the range-difference measurements

$$\hat{d}_i = \|\mathbf{x} - \mathbf{s}_i\|_2 - \|\mathbf{x}\|_2 \quad \text{for } i = 1, \, 2, \, \dots, \, m$$

are available, and (3) the location of point $\mathbf{x}$ needs to be determined. Range-difference measurements can, in practice, be obtained from the time differences of arrival as measured by an array of passive sensors.

Source localization problems are very common in wireless communications, navigation, geophysics, and many other areas [17–20]. In this section, we formulate optimization problems of this type as unconstrained problems and apply Newton and quasi-Newton algorithms for their solution.

9.5.1   Source Localization Based on Range Measurements

For the sake of simplicity, we consider source localization problems on a plane, i.e., $\mathbf{s}_i \in R^2$ and $\mathbf{x} \in R^2$. In practice, measurement noise is always present and hence the range measurements are given by

$$r_i = \|\mathbf{x} - \mathbf{s}_i\|_2 + n_i \quad \text{for } i = 1, \, 2, \, \dots, \, m$$

where n_i represents the unknown value of the noise component for the ith measurement. The source localization problem entails estimating the value of vector $\mathbf{x}$ given

---

[5]Problems with multiple minima.

the locations of the m sensors, $\{s_i, \ i = 1, 2, \ldots, m\}$, and range measurements $\{r_i, \ i = 1, 2, \ldots, m\}$. A natural way to address the problem is to formulate the localization problem as the unconstrained minimization problem

$$\underset{\mathbf{x}}{\text{minimize}} \ \ f(\mathbf{x}) = \frac{1}{2} \sum_{i=1}^{m} (\|\mathbf{x} - \mathbf{s}_i\|_2 - r_i)^2 \tag{9.48}$$

For such a problem, it is reasonable to assume that $\mathbf{x} \neq \mathbf{s}_i$ for $1 \leq i \leq m$ as the sensors are, in practice, located at some distance from the radiating source. In this situation, the objective function in Eq. (9.48) is differentiable and its gradient and Hessian are given by

$$\mathbf{g}(\mathbf{x}) = \sum_{i=1}^{m} \left(1 - \frac{r_i}{\|\mathbf{x} - \mathbf{s}_i\|_2} \right) (\mathbf{x} - \mathbf{s}_i) \tag{9.49}$$

and

$$\mathbf{H}(\mathbf{x}) = \sum_{i=1}^{m} \frac{r_i}{\|\mathbf{x} - \mathbf{s}_i\|_2^3} (\mathbf{x} - \mathbf{s}_i)(\mathbf{x} - \mathbf{s}_i)^T + \tau \mathbf{I} \tag{9.50a}$$

respectively, where $\mathbf{I}$ represents the 2×2 identity matrix and

$$\tau = m - \sum_{i=1}^{m} \frac{r_i}{\|\mathbf{x} - \mathbf{s}_i\|_2} \tag{9.50b}$$

From Eqs. (9.50a) and (9.50b), we note that $f(\mathbf{x})$ becomes nonconvex when τ is negative and of a large magnitude.

A localization problem like the above can be solved by using any one of the unconstrained optimization algorithms described in the previous chapters. To illustrate the formulation of the problem for the case of the Newton algorithm (Algorithm 5.3 in Sect. 5.3), let us consider a localization system with five sensors located at [20]

$$\mathbf{s}_1 = \begin{bmatrix} 6 \\ 4 \end{bmatrix}, \quad \mathbf{s}_2 = \begin{bmatrix} 0 \\ -10 \end{bmatrix}, \quad \mathbf{s}_3 = \begin{bmatrix} 5 \\ -3 \end{bmatrix}, \quad \mathbf{s}_4 = \begin{bmatrix} 1 \\ -4 \end{bmatrix}, \quad \text{and} \ \ \mathbf{s}_5 = \begin{bmatrix} 3 \\ -3 \end{bmatrix} \tag{9.51}$$

The source is located at $\mathbf{x}_s = [-2 \ 3]^T$ and the exact and noisy range measurements are given by

$$\{d_i\} = \{8.0622, \ 13.1529, \ 9.2195, \ 7.6157, \ 7.8102\}$$

and

$$\{r_i\} = \{8.0051, \ 13.0112, \ 9.1138, \ 7.7924, \ 8.0210\} \tag{9.52}$$

respectively.

A contour plot of $f(\mathbf{x})$ over the region

$$\mathcal{R} = \{x : -7 \leq x_1 \leq 15, \ -15 \leq x_2 \leq 7\}$$

is shown in Fig. 9.12. As can be seen, there are two minimizers, three maximizers, and four saddle points over the region.

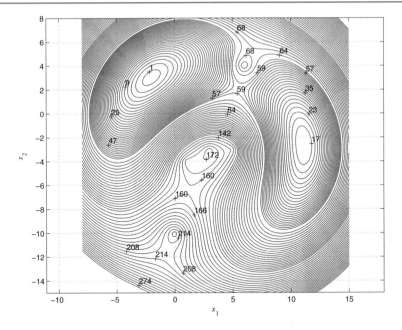

Fig. 9.12 Contours of $f(\mathbf{x})$ over $\mathcal{R}$

In Step 2 of the Newton algorithm it is necessary to force the Hessian to become positive definite if it is not positive definite already. To this end, we can write the Hessian in Eq. (9.50a) as

$$\mathbf{H}(\mathbf{x}) = \mathbf{H}_1(\mathbf{x}) + \tau \mathbf{I}$$

where τ is a small positive constant and

$$\mathbf{H}(\mathbf{x}) = \sum_{i=1}^{m} \frac{r_i}{\|\mathbf{x} - \mathbf{s}_i\|_2^3} (\mathbf{x} - \mathbf{s}_i)(\mathbf{x} - \mathbf{s}_i)^T$$

An eigen-decomposition of $\mathbf{H}_1(\mathbf{x})$ is given by

$$\mathbf{H}_1(\mathbf{x}) = \mathbf{U} \boldsymbol{\Lambda} \mathbf{U}^T$$

where $\mathbf{U}$ is orthogonal and $\boldsymbol{\Lambda} = \text{diag} \{\lambda_1, \lambda_2\}$. Hence the Hessian can be expressed as

$$\mathbf{H}(\mathbf{x}) = \mathbf{U} \hat{\boldsymbol{\Lambda}} \mathbf{U}^T$$

where $\hat{\boldsymbol{\Lambda}} = \text{diag} \{\lambda_1 + \tau, \lambda_2 + \tau\}$. To assure a positive definite Hessian, which will, in turn, assure a descent step in the Newton algorithm, the Hessian can be modified as

$$\tilde{\mathbf{H}}(\mathbf{x}) = \mathbf{U} \tilde{\boldsymbol{\Lambda}} \mathbf{U}^T$$

where

$$\tilde{\boldsymbol{\Lambda}} = \text{diag} \{\max(\lambda_1 + \tau, \ 0.05), \ \max(\lambda_2 + \tau, \ 0.05)\}$$

Under these circumstances, the search direction in the kth iteration of the Newton algorithm is given by

$$\mathbf{d}_k = -\mathbf{U}\tilde{\mathbf{\Lambda}}^{-1}\mathbf{U}^T g(\mathbf{x}_k)$$

where $g(\mathbf{x})$ is given by Eq. (9.49).

Using the Newton algorithm, the global and local minimizers of $f(\mathbf{x})$ were obtained as

$$\mathbf{x}^* = \begin{bmatrix} -1.990678 \\ 3.047388 \end{bmatrix} \quad \text{and} \quad \mathbf{x}_* = \begin{bmatrix} 11.115212 \\ -2.678506 \end{bmatrix} \tag{9.53}$$

respectively. At these points, the objective function assumes the values

$$f(\mathbf{x}^*) = 0.1048 \quad \text{and} \quad f(\mathbf{x}_*) = 15.0083$$

The global minimizer is very close to the exact location of the source, namely, $\mathbf{x}_s = [-2\ 3]^T$. Due to the presence of noise, the exact location of $\mathbf{x}_s$ would not normally be obtained in practice.

Example 9.6 Using the Newton algorithm described in Sect. 9.5.1 and the quasi-Newton algorithm incorporating the BFGS updating formula as described in Sect. 7.6, solve the five-sensor source location problem described in Sect. 9.5.1 assuming that the sensor locations and noisy range measurements are given by Eqs. (9.51) and (9.52), respectively.

Solution The performance of the two algorithms was evaluated by applying each algorithm using a total of 98×98 initial points that are uniformly distributed over the region $R = \{\mathbf{x} : -8 \le x_1 \le 8,\ -8 \le x_2 \le 8\}$. From Fig. 9.12, it is observed that the region includes the global minimizer of the objective function $f(\mathbf{x})$ in Eq. (9.48) as well as several maximizers and saddle points.

Figure 9.13 shows a profile of the values of the minimized objective function obtained using the Newton algorithm versus the initial points $\mathbf{x}_0$ from region $\mathcal{R}$. Among the 9604 initial points, there are 7476 $\mathbf{x}_0$'s for which the Newton algorithm converged to the global minimizer $\mathbf{x}^*$ given in Eq. (9.53). For each of the remaining 2128 initial points, the algorithm converged to the local minimizer $\mathbf{x}_*$. The profile of the values of the minimized objective function obtained with the quasi-Newton algorithm versus the initial points $\mathbf{x}_0$ in $\mathcal{R}$ is shown in Fig. 9.14. The quasi-Newton algorithm converged to the global minimizer starting with 7453 initial points in region $\mathcal{R}$ and for the remaining 2151 initial points it converged to $\mathbf{x}_*$.

Normalizing the average CPU time required by the Newton algorithm per run to unity, the average normalized CPU time required by the quasi-Newton algorithm per run is obtained as 1.2545.

In summary, the Newton and quasi-Newton algorithms converged to the global minimizer $\mathbf{x}^*$ starting from initial points distributed over a fairly large area in the vicinity of $\mathbf{x}^*$ while requiring similar computational effort. ∎

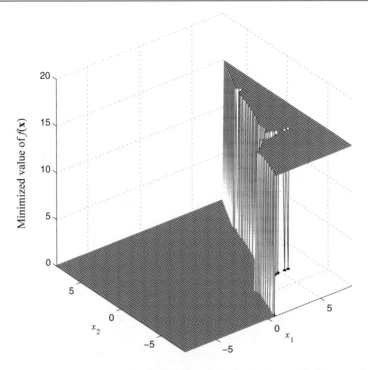

Fig. 9.13 Values of the minimized objective function obtained using the modified Newton algorithm versus the initial points $\mathbf{x}_0$ over the region $\mathcal{R}$

9.5.2   Source Localization Based on Range-Difference Measurements

In source location problems based on range-difference measurements, it is assumed that there exists an additional sensor, $\mathbf{s}_0$, at the origin, namely, at $\mathbf{s}_0 = \mathbf{0}$. The range-difference d_i is defined as the difference between the distance from sensor $\mathbf{s}_i$ to source $\mathbf{x}$ and the distance from sensor $\mathbf{s}_0$ to source $\mathbf{x}$, i.e.,

$$d_i = \|\mathbf{x} - \mathbf{s}_i\|_2 - \|\mathbf{x}\|_2 \quad \text{for } i = 1, 2, \ldots, m$$

Localization problems of this type entail estimating the components of $\mathbf{x}$, given the locations of the $m + 1$ sensors, $\{\mathbf{s}_i, \ i = 0, 1, \ldots, m\}$, and noise-contaminated range-difference measurements $\{r_i, \ i = 1, 2, \ldots, m\}$ where

$$r_i = d_i + \varepsilon_i \quad \text{for } i = 1, 2, \ldots, m \tag{9.54}$$

and ε_i is the value of the noise for the ith measurement. The localization problem in such applications can be formulated as the nonlinear least-squares problem

$$\text{minimize } f(\mathbf{x}) = \frac{1}{2} \sum_{i=1}^{m} (\|\mathbf{x} - \mathbf{s}_i\|_2 - \|\mathbf{x}\|_2 - r_i)^2 \tag{9.55}$$

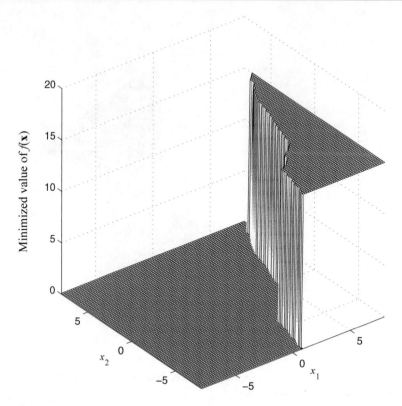

Fig. 9.14 Values of the minimized objective function obtained using the BFGS algorithm versus the initial points $\mathbf{x}_0$ over the region $\mathcal{R}$

If $\mathbf{x} \neq \mathbf{s}_i$ for $i = 0, 1, \ldots, m$, then $f(\mathbf{x})$ is differentiable. Its gradient and Hessian can be evaluated as

$$\mathbf{g}(\mathbf{x}) = \sum_{i=1}^{m} c_i (\mathbf{q}_i - \tilde{\mathbf{x}}) \tag{9.56a}$$

and

$$\mathbf{H}(\mathbf{x}) = \sum_{i=1}^{m} [(\mathbf{q}_i - \tilde{\mathbf{x}})(\mathbf{q}_i - \tilde{\mathbf{x}})^T + c_i(\mathbf{Q}_{1i} - \mathbf{Q}_2)] \tag{9.56b}$$

respectively, where

$$c_i = \|\mathbf{x} - \mathbf{s}_i\|_2 - \|\mathbf{x}\|_2 - r_i, \quad \mathbf{q}_i = \frac{\mathbf{x} - \mathbf{s}_i}{\|\mathbf{x} - \mathbf{s}_i\|_2}, \quad \tilde{\mathbf{x}} = \frac{\mathbf{x}}{\|\mathbf{x}\|_2}$$

$$\mathbf{Q}_{1i} = \frac{1}{\|\mathbf{x} - \mathbf{s}_i\|_2}(\mathbf{I} - \mathbf{q}_i \mathbf{q}_i^T), \quad \text{and} \quad \mathbf{Q}_2 = \frac{1}{\|\mathbf{x}\|_2}(\mathbf{I} - \tilde{\mathbf{x}}\tilde{\mathbf{x}}^T)$$

As in the Newton algorithm used for the solution of the source localization problem based on range measurements, the Hessian $\mathbf{H}(\mathbf{x})$ in Eq. (9.56b) may or may not be positive definite. Let the eigendecomposition of $\mathbf{H}(\mathbf{x})$ be

$$\mathbf{H}(\mathbf{x}) = \mathbf{U}\boldsymbol{\Lambda}\mathbf{U}^T$$

where $\boldsymbol{\Lambda}$ is the diagonal matrix with the eigenvalue values λ_1 and λ_2 of $\mathbf{H}(\mathbf{x})$ on its diagonal. The Hessian can be modified to

$$\tilde{\mathbf{H}}(\mathbf{x}) = \mathbf{U}\tilde{\boldsymbol{\Lambda}}\mathbf{U}^T \tag{9.57}$$

where

$$\tilde{\boldsymbol{\Lambda}} = \text{diag}\{\max(\lambda_1, \; 0.05), \; \max(\lambda_2, \; 0.05)\}$$

to ensure that $\tilde{\mathbf{H}}(\mathbf{x})$ is positive definite. Since $\mathbf{H}(\mathbf{x})$ is of very small size, computing $\tilde{\mathbf{H}}(\mathbf{x})$ in Eq. (9.57) is economical and hence the Newton algorithm in conjunction with $\tilde{\mathbf{H}}(\mathbf{x})$ in Eq. (9.57) is a good choice for solving the problem in Eq. (9.54). This is especially the case when the number of sensors, m, is small. If m is large, the size of the Hessian continues to be 2×2. However, as can be seen from Eq. (9.56b), the evaluation of $\mathbf{H}(\mathbf{x})$ becomes expensive for a large m and in such a case a quasi-Newton algorithm would be more attractive as it requires only gradient information which can be obtained efficiently using Eq. (9.56a).

Example 9.7 In a six-sensor source-location system, sensor $\mathbf{s}_0$ is located at the origin and the other five sensors are located at

$$\mathbf{s}_1 = \begin{bmatrix} -5 \\ -13 \end{bmatrix}, \quad \mathbf{s}_2 = \begin{bmatrix} -12 \\ 1 \end{bmatrix}, \quad \mathbf{s}_3 = \begin{bmatrix} -1 \\ -5 \end{bmatrix}, \quad \mathbf{s}_4 = \begin{bmatrix} -9 \\ -12 \end{bmatrix}, \quad \text{and} \quad \mathbf{s}_5 = \begin{bmatrix} -3 \\ -12 \end{bmatrix}$$

The radiating source is located at $\mathbf{x}_s = [-5 \; 11]^T$ [20]. The noisy range-difference measurements in Eq. (9.54) are given by

$$\{r_i\} = \{11.8829, \; 0.1803, \; 4.6399, \; 11.2402, \; 10.8183\}$$

Using the Newton algorithm described in Sect. 9.5.1 and the quasi-Newton algorithm incorporating the BFGS updating formula as described in Sect. 7.6, estimate the location of the radiating source, $\mathbf{x}_s$, by solving the unconstrained problem in Eq. (9.55).

Solution The Newton and quasi-Newton algorithms were applied to the source local-ization problem with a total of 98×98 initial points that are uniformly placed over the region $\mathcal{R} = \{x : -13 \le x_1 \le 3, \; 0 \le x_2 \le 16\}$. The Newton algorithm converged from every initial point to the minimizer

$$\mathbf{x}^* = \begin{bmatrix} -4.988478 \\ 10.637469 \end{bmatrix}$$

which is very close to the true location of the radiating source, $\mathbf{x}_s$. The quasi-Newton algorithm converged to the same solution $\mathbf{x}^*$ from 9592 initial points but it failed to converge with the other 12 initial points, most of which are in the area $\{x : 0.43 \le x_1 \le 0.53, \; 1.7 \le x_2 \le 2.3\}$. Normalizing the average CPU time required by the Newton algorithm per run to unity, the average CPU time required by the quasi-Newton algorithm was found to be 0.8640. Evidently, the Newton and quasi-Newton algorithms converged to the global minimizer $\mathbf{x}^*$ starting from initial points distributed over a fairly large area in the vicinity of $\mathbf{x}^*$ while requiring similar computational effort.

Problems

9.1 (a) Show that the gradient $\nabla\mathbf{D}$ defined by Eqs. (9.1a) and (9.1b) is invariant to a constant shift of its pixel values, where each pixel is modified by adding the same constant s.

(b) Show that the normalized gradient $\nabla\mathbf{D}/\|\nabla\mathbf{D}\|_2$ is invariant to constant multiplication of its pixel values, where each pixel is modified by multiplying by the same constant c.

9.2 Consider an image $\mathbf{D}$ whose HOG is given by $\mathbf{x}$ in Eq. (9.3). If image $\mathbf{D}$ is modified to $\hat{\mathbf{D}}$ by an illumination change, i.e., $\hat{\mathbf{D}} = c\mathbf{D}$ for some constant $c > 0$, show that the HOG of $\hat{\mathbf{D}}$ is given by $\hat{\mathbf{x}} = c\mathbf{x}$.

9.3 Show that the solution index i^* obtained from Eq. (9.9) assures that $P(l = i^*|\mathbf{x}, \boldsymbol{\Theta})$ is the largest probability among $\{P(l = i|\mathbf{x}, \boldsymbol{\Theta})$ for $i = 0,\ 1,\ \ldots,\ K-1\}$.

9.4 Derive formulas for the evaluation of $\nabla^2 f_k(\mathbf{x})$ for $k = 1,\ 2$, and 3 for the set of functions $f_k(\mathbf{x})$ given by Eqs. (9.24a)–(9.24d).

9.5 Show that for a nontrivial weighting function $W(\omega) \geq 0$, the matrix $\mathbf{Q}$ given by Eq. (9.33) is positive definite.

9.6 Derive the expressions of $\mathbf{Q}_l$ and $\mathbf{b}_l$ given in Eqs. (9.40a)–(9.42c).

9.7 Write a MATLAB program to implement the unconstrained optimization algorithm for the weighted least-squares design of linear-phase lowpass FIR digital filters studied in Sect. 9.4.1.2.

9.8 Develop an unconstrained optimization algorithm for the weighted least-squares design of linear-phase highpass digital filters.

9.9 Prove that the objective function given in Eq. (9.43) is globally convex. Hint: Show that for any $\mathbf{y} \in R^{N+1}$, $\mathbf{y}^T \nabla^2 f(\mathbf{h})\mathbf{y} \geq 0$.

9.10 Develop a method based on unconstrained optimization for the design of FIR filters with low passband group delay allowing coefficients with complex values.

9.11 Consider the double inverted pendulum control system described in Example 1.2, where $\alpha = 16$, $\beta = 8$, $T_0 = 0.8$, $\Delta t = 0.02$, and $K = 40$. The initial state is set to $\mathbf{x}(0) = [\pi/6 \ 1 \ \pi/6 \ 1]^T$ and the constraints on the magnitude of control actions are $|u(i)| \le m$ for $i = 0, 1, \ldots, K - 1$ with $m = 112$.

(a) Use the singular-value decomposition technique (see Sect. A.9, especially Eqs. (A.43), (A.44a), and (A.44b)) to eliminate the equality constraint $\mathbf{a}(\mathbf{u}) = \mathbf{0}$ in Eq. (1.9b).

(b) Convert the constrained problem obtained from part (a) to an unconstrained problem of the augmented objective function

$$F_\tau(\mathbf{u}) = \mathbf{u}^T\mathbf{u} - \tau \sum_{i=0}^{K-1} \ln[m - u(i)] - \tau \sum_{i=0}^{K-1} \ln[m + u(i)]$$

where the barrier parameter τ is fixed to a positive value in each round of minimization, which is then reduced to a smaller value at a fixed rate in the next round of minimization.

Note that in each round of minimization, a line search step should be carefully executed where the step size α is limited to a *finite* interval $[0, \bar{\alpha}]$ that is determined by the constraints $|u(i)| \le m$ for $0 \le i \le K - 1$.

9.12 (a) Verify the gradient given by Eq. (9.49).

(b) Verify the Hessian given by Eqs. (9.50a) and (9.50b).

(c) Verify the gradient given by Eq. (9.56a).

(d) Verify the Hessian given by Eq. (9.56b).

References

1. Y. LeCun, B. Boser, J. S. Denker, D. Henderson, R. E. Howard, W. Hubbardand, and L. D. Jackel, "Backpropagation applied to handwritten zip code recognition," *Neural Computation*, vol. 1, pp. 541–551, 1989.

2. Y. LeCun, L. D. Jackel, L. Bottou, A. Brunot, C. Cortes, J. S. Denker, H. Drucker, I. Guyon, U. A. Muller, E. Sackinger, P. Simard, and V. Vapnik, "Set-membership affine projection algorithm with variable data-reuse factor," in *Int. Conf. on Artificial Neural Networks*, May 1995, pp. 53–60.

3. G. E. Hinton, P. Dayan, and M. Revow, "Modeling the manifolds of images of handwritten digits," *IEEE Trans. Neural Networks*, vol. 8, no. 1, pp. 65–74, 1997.

4. Y. LeCun, L. Bottou, Y. Bengio, and P. Haffner, "Gradient-based learning applied to document recognition," *Proc. IEEE*, vol. 86, pp. 2278–2324, 1998.

5. Y. LeCun, C. Cortes, and C. J. C. Burges, "The MNIST database of handwritten digits," Available online: http://yann.lecun.com/exdb/mnist.

6. C. M. Bishop, *Pattern Recognition and Machine Learning*. New York: Springer, 2006.

7. N. D. Dalal and B. Triggs, "Histogram of oriented gradients for human detection," in *Proc. Computer Vision and Pattern Recognition*, San Diego, June 2005, pp. 886–893.

8. Y. Nesterov, "A method for solving a convex programming problem with rate of convergence $O(1/k^2)$," *Soviet Math. Doklady*, vol. 269, no. 3, pp. 543–547, 1983, in Russian.

9. Y. E. Nesterov, *Introductory Lectures on Convex Optimization — A Basic Course*.   Norwell, MA: Kluwer Academic Publishers, 2004.

10. J. J. Craig, *Introduction to Robotics*, 2nd ed.   Reading, MA: Addison-Wesley, 1989.

11. A. Antoniou, *Digital Filters: Analysis, Design, and Signal Processing Applications*.   New York: McGraw-Hill, 2018.

12. T. W. Parks and J. H. McClellan, "Chebyshev approximation for nonrecursive digital filters with linear phase," *IEEE Trans. Circuit Theory*, vol. 19, no. 2, pp. 189–194, 1972.

13. T. W. Parks and C. S. Burrus, *Digital Filter Design*.   New York: Wiley, 1987.

14. G. H. Golub and C. F. Van Loan, *Matrix Computations*, 4th ed.   Baltimore, MD: Johns Hopkins University Press, 2013.

15. R. Fletcher, *Practical Methods of Optimization,* vol. 1.   New York: Wiley, 1980.

16. R. Fletcher, *Practical Methods of Optimization*, 2nd ed.   New York: Wiley, 1987.

17. J. O. Smith and J. S. Abel, "Closed-form least-squares source location estimation from range-difference measurements," *IEEE Trans. Signal Process.*, vol. 12, pp. 1661–1669, 1987.

18. K. W. Cheung, W. K. Ma, and H. C. So, "Accurate approximation algorithm for TOA-based maximum likelihood mobile location using semidefinite programming," in *Proc. IEEE Intern. Conf. on Acoust., Speech, and Signal Process.*, Montreal, Canada, May 2004, pp. 145–148.

19. P. Stoica and J. Li, "Source localization from range-difference measurements," *IEEE Signal Process. Mag.*, vol. 23, pp. 63–66, 2006.

20. A. Beck, P. Stoica, and J. Li, "Exact and approximate solutions of source localization problems," *IEEE Trans. Signal Process.*, vol. 56, no. 5, pp. 1770–1777, 2008.

Fundamentals of Constrained Optimization

10

10.1 Introduction

The material presented so far dealt largely with principles, methods, and algorithms for unconstrained optimization. In this and the next five chapters, we build on the introductory principles of constrained optimization discussed in Sects. 1.4–1.6 and proceed to examine the underlying theory and structure of some very sophisticated and efficient constrained optimization algorithms.

The presence of constraints gives rise to a number of technical issues that are not encountered in unconstrained problems. For example, a search along the direction of the negative of the gradient of the objective function is a well-justified technique for unconstrained minimization. However, in a constrained optimization problem points along such a direction may not satisfy the constraints and in such a case the search will not yield a solution of the problem. Consequently, new methods for determining feasible search directions have to be sought.

Many powerful techniques developed for constrained optimization problems are based on unconstrained optimization methods. If the constraints are simply given in terms of lower and/or upper limits on the parameters, the problem can be readily converted into an unconstrained problem. Furthermore, methods of transforming a constrained minimization problem into a sequence of unconstrained minimizations of an appropriate auxiliary function exist.

The purpose of this chapter is to lay a theoretical foundation for the development of various algorithms for constrained optimization. Equality and inequality constraints are discussed in general terms in Sect. 10.2. After a brief discussion on the classification of constrained optimization problems in Sect. 10.3, several variable transformation techniques for converting optimization problems with simple constraints into unconstrained problems are studied in Sect. 10.4. One of the most important concepts in constrained optimization, the concept of *Lagrange multipliers*, is introduced and a geometric interpretation of Lagrange multipliers is given in Sect. 10.5. The first-order necessary conditions for a point $\mathbf{x}^*$ to be a solution of a constrained problem, known as the *Karush-Kuhn-Tucker conditions*, are studied

© Springer Science+Business Media, LLC, part of Springer Nature 2021
A. Antoniou and W.-S. Lu, *Practical Optimization*, Texts in Computer Science,
https://doi.org/10.1007/978-1-0716-0843-2_10

in Sect. 10.6 and the second-order conditions are discussed in Sect. 10.7. As in the unconstrained case, the concept of convexity plays an important role in the study of constrained optimization and it is discussed in Sect. 10.8. Finally, the concept of duality, which is of significant importance in the development and unification of optimization theory, is addressed in Sect. 10.9.

10.2   Constraints

10.2.1   Notation and Basic Assumptions

In its most general form, a constrained optimization problem is to find a vector $\mathbf{x}^*$ that solves the problem

$$\text{minimize}\ \ f(\mathbf{x}) \tag{10.1a}$$

$$\text{subject to:}\ \ a_i(\mathbf{x}) = 0 \quad \text{for } i = 1,\ 2,\ \ldots,\ p \tag{10.1b}$$

$$c_j(\mathbf{x}) \leq 0 \quad \text{for } j = 1,\ 2,\ \ldots,\ q \tag{10.1c}$$

Throughout the chapter, we assume that the objective function $f(\mathbf{x})$ as well as the functions involved in the constraints in Eqs. (10.1b) and (10.1c), namely, $\{a_i(\mathbf{x})$ for $i = 1,\ 2,\ \ldots,\ p\}$ and $\{c_j(\mathbf{x})$ for $j = 1,\ 2,\ \ldots,\ q\}$, are continuous and have continuous second partial derivatives, i.e., $a_i(\mathbf{x})$, $c_j(\mathbf{x}) \in C^2$. Let $\mathcal{R}$ denote the feasible region for the problem in Eqs. (10.1a)–(10.1c) which was defined in Sect. 1.5 as the set of points satisfying Eqs. (10.1b) and (10.1c), i.e.,

$$\mathcal{R} = \{\mathbf{x} : a_i(\mathbf{x}) = 0 \text{ for } i = 1,\ 2,\ \ldots,\ p,\ c_j(\mathbf{x}) \leq 0 \text{ for } j = 1,\ 2,\ \ldots,\ q\}$$

In this chapter as well as the rest of the book, we often need to compare two vectors or matrices entry by entry. For two matrices $\mathbf{A} = \{a_{ij}\}$ and $\mathbf{B} = \{b_{ij}\}$ of the same dimension, we use $\mathbf{A} \geq \mathbf{B}$ to denote $a_{ij} \geq b_{ij}$ for all i, j. Consequently, $\mathbf{A} \geq \mathbf{0}$ means $a_{ij} \geq 0$ for all i, j. We write $\mathbf{A} \succ \mathbf{0}$, $\mathbf{A} \succeq \mathbf{0}$, $\mathbf{A} \prec \mathbf{0}$, and $\mathbf{A} \preceq \mathbf{0}$ to denote that matrix $\mathbf{A}$ is positive definite, positive semidefinite, negative definite, and negative semidefinite, respectively.

10.2.2   Equality Constraints

The set of equality constraints

$$a_1(\mathbf{x}) = 0$$

$$\vdots \tag{10.2}$$

$$a_p(\mathbf{x}) = 0$$

defines a hypersurface in R^n. Using vector notation, we can write

$$\mathbf{a}(\mathbf{x}) = [a_1(\mathbf{x})\ a_2(\mathbf{x})\ \cdots\ a_p(\mathbf{x})]^T$$

and from Eq. (10.2), we have

$$\mathbf{a}(\mathbf{x}) = \mathbf{0} \tag{10.3}$$

Definition 10.1 (*Regular point*) A point $\mathbf{x}$ is called a *regular point* of the constraints in Eq. (10.2) if $\mathbf{x}$ satisfies Eq. (10.2) and column vectors $\nabla a_1(\mathbf{x})$, $\nabla a_2(\mathbf{x})$, ..., $\nabla a_p(\mathbf{x})$ are linearly independent.   ∎

The definition states, in effect, that $\mathbf{x}$ is a regular point of the constraints if it is a solution of Eq. (10.2) and the Jacobian $\mathbf{J}_e = [\nabla a_1(\mathbf{x}) \ \nabla a_2(\mathbf{x}) \ \cdots \ \nabla a_p(\mathbf{x})]^T$ has full row rank. The importance of a point $\mathbf{x}$ being regular for a given set of equality constraints lies in the fact that a tangent plane of the hypersurface determined by the constraints at a regular point $\mathbf{x}$ is well defined. Later in this chapter, the term 'tangent plane' will be used to express and describe important necessary as well as sufficient conditions for constrained optimization problems. Since $\mathbf{J}_e$ is a $p \times n$ matrix, it would not be possible for $\mathbf{x}$ to be a regular point of the constraints if $p > n$. This leads to an upper bound for the number of independent equality constraints, i.e., $p \le n$. Furthermore, if $p = n$, in many cases the number of vectors $\mathbf{x}$ that satisfy Eq. (10.2) is finite and the optimization problem becomes a trivial one. For these reasons, we shall assume that $p < n$ throughout the rest of the book.

Example 10.1 Discuss and sketch the feasible region described by the equality constraints

$$- x_1 + x_3 - 1 = 0 \tag{10.4a}$$
$$x_1^2 + x_2^2 - 2x_1 = 0 \tag{10.4b}$$

Solution The Jacobian of the constraints is given by

$$\mathbf{J}_e(\mathbf{x}) = \begin{bmatrix} -1 & 0 & 1 \\ 2x_1 - 2 & 2x_2 & 0 \end{bmatrix}$$

which has rank 2 except at $\mathbf{x} = [1 \ 0 \ x_3]^T$. Since $\mathbf{x} = [1 \ 0 \ x_3]^T$ does not satisfy the constraint in Eq. (10.4b), any point $\mathbf{x}$ satisfying Eqs. (10.4a) and (10.4b) is regular. The constraints in Eqs. (10.4a) and (10.4b) describe a curve which is the intersection between the cylinder in Eq. (10.4b) and the plane in Eq. (10.4a). To display the curve, we derive a parametric representation for the curve as follows. Eq. (10.4b) can be written as

$$(x_1 - 1)^2 + x_2^2 = 1$$

which suggests the parametric expressions

$$x_1 = 1 + \cos t \tag{10.5a}$$
$$x_2 = \sin t \tag{10.5b}$$

for x_1 and x_2. Now Eqs. (10.5a) and (10.5b) in conjunction with Eq. (10.4a) give

$$x_3 = 2 + \cos t \tag{10.5c}$$

With parameter t varying from 0 to 2π, Eqs. (10.5a)–(10.5c) describe the curve shown in Fig. 10.1.   ∎

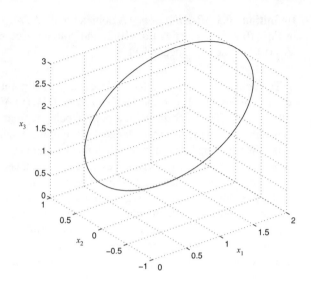

A particularly important class of equality constraints is the class of linear constraints where functions $a_i(\mathbf{x})$ are all linear. In this case, Eq. (10.2) becomes a system of linear equations which can be represented by the equation

$$\mathbf{Ax} = \mathbf{b} \tag{10.6}$$

where $\mathbf{A} \in R^{p \times n}$ is numerically equal to the Jacobian, i.e., $\mathbf{A} = \mathbf{J}_e$, and $\mathbf{b} \in R^{p \times 1}$. Since the Jacobian is a constant matrix, any solution point of Eq. (10.6) is a regular point if $\text{rank}(\mathbf{A}) = p$. If $\text{rank}(\mathbf{A}) = p' < p$, then there are two possibilities: either

$$\text{rank}([\mathbf{A} \; \mathbf{b}]) > \text{rank}(\mathbf{A}) \tag{10.7}$$

or

$$\text{rank}([\mathbf{A} \; \mathbf{b}]) = \text{rank}(\mathbf{A}) \tag{10.8}$$

If Eq. (10.7) is satisfied, then we conclude that contradictions exist in Eq. (10.6), and a careful examination of Eq. (10.6) is necessary to eliminate such contradictions. If Eq. (10.8) holds with $\text{rank}(\mathbf{A}) = p' < p$, then simple algebraic manipulations can be used to reduce Eq. (10.6) to an equivalent set of p' equality constraints

$$\hat{\mathbf{A}}\mathbf{x} = \hat{\mathbf{b}} \tag{10.9}$$

where $\hat{\mathbf{A}} \in R^{p' \times n}$ has rank p' and $\hat{\mathbf{b}} \in R^{p' \times 1}$. Further, linear equality constraints in the form of Eq. (10.9) with a full row rank $\hat{\mathbf{A}}$ can be eliminated so as to convert the problem to an unconstrained problem or to reduce the number of parameters involved. The reader is referred to Sect. 10.4.1.1 for the details.

When $\text{rank}(\mathbf{A}) = p' < p$, a numerically reliable way to reduce Eqs. (10.6)–(10.9) is to apply the singular-value decomposition (SVD) to matrix $\mathbf{A}$. The basic theory pertaining to the SVD can be found in Sect. A.9. Applying the SVD to $\mathbf{A}$, we obtain

$$\mathbf{A} = \mathbf{U\Sigma V}^T \tag{10.10}$$

where $\mathbf{U} \in R^{p \times p}$ and $\mathbf{V} \in R^{n \times n}$ are orthogonal matrices and

$$\Sigma = \begin{bmatrix} \mathbf{S} & \mathbf{0} \\ \mathbf{0} & \mathbf{0} \end{bmatrix}_{p \times n}$$

with $\mathbf{S} = \text{diag}\{\sigma_1, \sigma_2, \ldots, \sigma_{p'}\}$, and $\sigma_1 \geq \sigma_2 \geq \cdots \geq \sigma_{p'} > 0$. It follows that

$$\mathbf{A} = \mathbf{U} \begin{bmatrix} \hat{\mathbf{A}} \\ \mathbf{0} \end{bmatrix}$$

with $\hat{\mathbf{A}} = \mathbf{S}[\mathbf{v}_1 \ \mathbf{v}_2 \ \cdots \ \mathbf{v}_{p'}]^T \in R^{p' \times n}$ where $\mathbf{v}_i$ denotes the ith column of $\mathbf{V}$, and Eq. (10.6) becomes

$$\begin{bmatrix} \hat{\mathbf{A}} \\ \mathbf{0} \end{bmatrix} \mathbf{x} = \begin{bmatrix} \hat{\mathbf{b}} \\ \mathbf{0} \end{bmatrix}$$

This leads to Eq. (10.9) where $\hat{\mathbf{b}}$ is formed by using the first p' entries of $\mathbf{U}^T \mathbf{b}$. Evidently, any solution point of Eq. (10.9) is a regular point.

In MATLAB, the SVD of a matrix $\mathbf{A}$ is performed by using command svd. The decomposition in Eq. (10.10) can be obtained by using

```
[U, SIGMA, V]=svd(A);
```

The command svd can also be used to compute the rank of a matrix. We use svd(A) to compute the singular values of $\mathbf{A}$, and the number of the nonzero singular values of $\mathbf{A}$ is the rank of $\mathbf{A}$.[1]

Example 10.2 Simplify the linear equality constraints

$$\begin{aligned} x_1 - 2x_2 + 3x_3 + 2x_4 &= 4 \\ 2x_2 - x_3 &= 1 \\ 2x_1 - 10x_2 + 9x_3 + 4x_4 &= 5 \end{aligned} \tag{10.11}$$

Solution It can be readily verified that rank$(\mathbf{A}) = $ rank$([\mathbf{A} \ \mathbf{b}]) = 2$. Hence the constraints in Eq. (10.11) can be reduced to a set of two equality constraints. The SVD of $\mathbf{A}$ yields

$$\mathbf{U} = \begin{bmatrix} 0.2717 & -0.8003 & -0.5345 \\ -0.1365 & -0.5818 & 0.8018 \\ 0.9527 & 0.1449 & 0.2673 \end{bmatrix}$$

$$\Sigma = \begin{bmatrix} 14.8798 & 0 & 0 & 0 \\ 0 & 1.6101 & 0 & 0 \\ 0 & 0 & 0 & 0 \end{bmatrix}$$

$$\mathbf{V} = \begin{bmatrix} 0.1463 & -0.3171 & 0.6331 & -0.6908 \\ -0.6951 & -0.6284 & -0.3161 & -0.1485 \\ 0.6402 & -0.3200 & -0.6322 & -0.2969 \\ 0.2926 & -0.6342 & 0.3156 & 0.6423 \end{bmatrix}$$

---

[1]The rank of a matrix can also be found by using MATLAB command rank.

Therefore, the reduced set of equality constraints is given by

$$2.1770x_1 - 10.3429x_2 + 9.5255x_3 + 4.3540x_4 = 5.7135 \quad (10.12a)$$
$$a - 0.5106x_1 - 1.0118x_2 - 0.5152x_3 - 1.0211x_4 = -3.0587 \quad (10.12b)$$

∎

10.2.3   Inequality Constraints

In this section, we discuss the class of inequality constraints. The discussion will be focused on their difference from as well as their relation to equality constraints. In addition, the convexity of a feasible region defined by linear inequalities will be addressed.

Consider the constraints

$$c_1(\mathbf{x}) \leq 0$$
$$c_2(\mathbf{x}) \leq 0$$
$$\vdots \qquad\qquad\qquad\qquad (10.13)$$
$$c_q(\mathbf{x}) \leq 0$$

Unlike the number of equality constraints, the number of inequality constraints, q, is not required to be less than n. For example, if we consider the case where all $c_j(\mathbf{x})$ for $1 \leq j \leq q$ are linear functions, then the constraints in Eq. (10.13) represent a polyhedron with q facets, and the number of facets in such a polyhedron is obviously unlimited.

The next two issues are concerned with the inequalities in Eq. (10.13). For a feasible point $\mathbf{x}$, these inequalities can be divided into two classes, the set of constraints with $c_i(\mathbf{x}) = 0$, which are called *active constraints*, and the set of constraints with $c_i(\mathbf{x}) < 0$, which are called *inactive constraints*. Since $c_i(\mathbf{x})$ are continuous functions, the constraints that are inactive at $\mathbf{x}$ will remain so in a sufficiently small neighborhood of $\mathbf{x}$. This means that the local properties of $\mathbf{x}$ will not be affected by the inactive constraints. On the other hand, when $c_i(\mathbf{x}) = 0$ the point $\mathbf{x}$ is on the boundary determined by the active constraints. Hence directions exist that would violate some of these constraints. In other words, active constraints restrict the feasible region of the neighborhoods of $\mathbf{x}$. For example, consider a constrained problem with the feasible region shown as the shaded area in Fig. 10.2. The problem involves three inequality constraints $c_1(\mathbf{x}) \leq 0$ and $c_2(\mathbf{x}) \leq 0$ are inactive while $c_3(\mathbf{x}) \leq 0$ is active at point $\mathbf{x} = \bar{\mathbf{x}}$ since $\bar{\mathbf{x}}$ is on the boundary characterized by $c_3(\mathbf{x}) = 0$. It can be observed that local searches in a neighborhood of $\bar{\mathbf{x}}$ will not be affected by the first two constraints but will be restricted to one side of the tangent line to the curve $c_3(\mathbf{x}) = 0$ at $\bar{\mathbf{x}}$. The concept of active constraints is an important one as it can be used to reduce the number of constraints that must be taken into account in a particular iteration and, therefore, often leads to improved computational efficiency.

Another approach to deal with inequality constraints is to convert them into equality constraints. For the sake of simplicity, we consider the problem

$$\text{minimize } f(\mathbf{x}) \qquad \mathbf{x} \in R^n \qquad (10.14a)$$

Fig. 10.2 Active and
inactive constraints

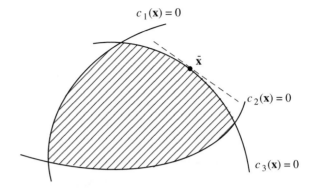

$$\text{subject to:} \quad c_i(\mathbf{x}) \le 0 \quad \text{for } i = 1, \, 2, \, \ldots, \, q \tag{10.14b}$$

which involves only inequality constraints. The constraints in Eq. (10.14b) are equivalent to

$$\hat{c}_1 = c_1(\mathbf{x}) + y_1 = 0$$
$$\hat{c}_2 = c_2(\mathbf{x}) + y_2 = 0$$
$$\vdots \tag{10.15a}$$
$$\hat{c}_q = c_q(\mathbf{x}) + y_q = 0$$
$$y_i \ge 0 \quad \text{for } 1 \le i \le q \tag{10.15b}$$

where $y_1, \, y_2, \, \ldots, \, y_q$ are called *slack variables*. The constraints in Eq. (10.15b) can be eliminated by using the simple variable substitutions

$$y_i = \hat{y}_i^2 \quad \text{for } 1 \le i \le q$$

If we let

$$\hat{\mathbf{x}} = [x_1 \; \cdots \; x_n \; \hat{y}_1 \; \cdots \; \hat{y}_q]^T$$

then the problem in Eqs. (10.14a) and (10.14b) can be formulated as

$$\text{minimize } f(\hat{\mathbf{x}}) \quad \hat{\mathbf{x}} \in E^{n+q} \tag{10.16a}$$
$$\text{subject to:} \quad \hat{c}_i(\hat{\mathbf{x}}) = 0 \quad \text{for } i = 1, \, 2, \, \ldots, \, q \tag{10.16b}$$

The idea of introducing slack variables to reformulate an optimization problem has been used successfully in the past, especially in linear programming, to transform a nonstandard problem into a standard problem (see Chap. 11 for the details).

We conclude this section by showing that there is a close relation between the linearity of inequality constraints and the convexity of the feasible region defined by the constraints. Although determining whether or not the region characterized by the inequality constraints in Eq. (10.13) is convex is not always easy, it can be readily shown that a feasible region defined by Eq. (10.13) with linear $c_i(\mathbf{x})$ is a *convex polyhedron*.

To demonstrate that this indeed is the case, we can write the linear inequality constraints as

$$\mathbf{Cx} \leq \mathbf{d} \tag{10.17}$$

with $\mathbf{C} \in R^{q \times n}$, $\mathbf{d} \in R^{q \times 1}$. Let $\mathcal{R} = \{\mathbf{x} : \mathbf{Cx} \leq \mathbf{d}\}$ and assume that $\mathbf{x}_1, \mathbf{x}_2 \in \mathcal{R}$. For $\lambda \in [0, 1]$, the point $\mathbf{x} = \lambda \mathbf{x}_1 + (1 - \lambda)\mathbf{x}_2$ satisfies Eq. (10.17) because

$$\mathbf{Cx} = \lambda \mathbf{Cx}_1 + (1 - \lambda)\mathbf{Cx}_2$$
$$\leq \lambda \mathbf{d} + (1 - \lambda)\mathbf{d} = \mathbf{d}$$

Therefore, $\mathbf{Cx} \leq \mathbf{d}$ defines a convex set (see Sect. 2.7). In the literature, inequality constraints are sometimes given in the form

$$c_1(\mathbf{x}) \geq 0$$
$$\vdots \tag{10.18}$$
$$c_q(\mathbf{x}) \geq 0$$

A similar argument can be used to show that if $c_i(\mathbf{x})$ for $1 \leq i \leq q$ in Eq. (10.18) are all linear functions, then the feasible region defined by Eq. (10.18) is convex.

10.3   Classification of Constrained Optimization Problems

In Sect. 1.6, we provided an introductory discussion on the various branches of mathematical programming. Here, we re-examine the classification issue paying particular attention to the structure of constrained optimization problems.

Constrained optimization problems can be classified according to the nature of the objective function and the constraints. For specific classes of problems, there often exist methods that are particularly suitable for obtaining solutions quickly and reliably. For example, for linear programming problems, the *simplex method* of Dantzig [1] and the *primal-dual interior-point methods* [2] have proven very efficient. For general convex programming problems, several *interior-point methods* that are particularly efficient have recently been developed [3,4].

Before discussing the classification, we formally describe the different types of minimizers of a general constrained optimization problem. In the following definitions, $\mathcal{R}$ denotes the feasible region of the problem in Eqs. (10.1a)–(10.1c) and the set of points $\{\mathbf{x} : ||\mathbf{x} - \mathbf{x}^*|| \leq \delta\}$ with $\delta > 0$ is said to be a *ball* centered at $\mathbf{x}^*$.

Definition 10.2 (*Local constrained minimizer*)   Point $\mathbf{x}^*$ is a local constrained minimizer of the problem in Eqs. (10.1a)–(10.1c) if there exists a ball $\mathcal{B}_{x^*} = \{\mathbf{x} : ||\mathbf{x} - \mathbf{x}^*|| \leq \delta\}$ with $\delta > 0$ such that $\mathcal{D}_{x^*} = \mathcal{B}_{x^*} \cap \mathcal{R}$ is nonempty and $f(\mathbf{x}^*) = \min\{f(\mathbf{x}) : \mathbf{x} \in \mathcal{D}_{x^*}\}$. ∎

Definition 10.3 (*Global constrained minimizer*)   Point $\mathbf{x}^*$ is a global constrained minimizer of the problem in Eqs. (10.1a)–(10.1c) if

$$\mathbf{x}^* \in \mathcal{R} \quad \text{and} \quad f(\mathbf{x}^*) = \min\{f(\mathbf{x}) : \mathbf{x} \in \mathcal{R}\} \quad ∎$$

Definition 10.4 (*Strong local constrained minimizer*)   A constrained minimizer $\mathbf{x}^*$ is called a strong local minimizer if there exists a ball $\mathcal{B}_{x^*}$ such that $\mathcal{D}_{x^*} = \mathcal{B}_{x^*} \cap \mathcal{R}$ is nonempty and $\mathbf{x}^*$ is the only constrained minimizer in $\mathcal{D}_{x^*}$.   ∎

10.3.1   Linear Programming

The standard form of a linear programming (LP) problem can be stated as

$$\text{minimize } f(\mathbf{x}) = \mathbf{c}^T \mathbf{x} \tag{10.19a}$$

$$\text{subject to: } \mathbf{Ax} = \mathbf{b} \tag{10.19b}$$
$$\mathbf{x} \geq \mathbf{0} \tag{10.19c}$$

where $\mathbf{c} \in R^{n \times 1}$, $\mathbf{A} \in R^{p \times n}$, and $\mathbf{b} \in R^{p \times 1}$ are given. In words, we need to find a vector $\mathbf{x}^*$ that minimizes a linear objective function subject to the linear equality constraints in Eq. (10.19b) and the nonnegativity bounds in Eq. (10.19c).

LP problems may also be encountered in the nonstandard form

$$\text{minimize } \mathbf{c}^T \mathbf{x} \tag{10.20a}$$

$$\text{subject to: } \mathbf{Ax} \leq \mathbf{b} \tag{10.20b}$$

By introducing slack variables in terms of vector $\mathbf{y}$ as

$$\mathbf{y} = \mathbf{b} - \mathbf{Ax}$$

Equation 10.20b can be expressed as

$$\mathbf{Ax} + \mathbf{y} = \mathbf{b} \tag{10.21a}$$

and

$$\mathbf{y} \geq \mathbf{0} \tag{10.21b}$$

If we express variable $\mathbf{x}$ as the difference of two nonnegative vectors $\mathbf{x}^+ \geq \mathbf{0}$ and $\mathbf{x}^- \geq \mathbf{0}$, i.e.,

$$\mathbf{x} = \mathbf{x}^+ - \mathbf{x}^-$$

and let

$$\hat{\mathbf{x}} = \begin{bmatrix} \mathbf{x}^+ \\ \mathbf{x}^- \\ \mathbf{y} \end{bmatrix}$$

then the objective function becomes

$$\hat{\mathbf{c}}^T \hat{\mathbf{x}} = [\mathbf{c}^T \ -\mathbf{c}^T \ \mathbf{0}]\hat{\mathbf{x}}$$

and the constraints in Eqs. (10.21a) and (10.21b) can be written as

$$[\mathbf{A} \ -\mathbf{A} \ \mathbf{I}] \hat{\mathbf{x}} = \mathbf{b}$$

and

$$\hat{\mathbf{x}} \geq \mathbf{0}$$

Therefore, the problem in Eqs. (10.20a) and (10.20b) can be stated as the standard LP problem

$$\text{minimize } \hat{\mathbf{c}}^T \hat{\mathbf{x}} \tag{10.22a}$$

$$\text{subject to: } \hat{\mathbf{A}}\hat{\mathbf{x}} = \mathbf{b} \tag{10.22b}$$

$$\hat{\mathbf{x}} \geq \mathbf{0} \tag{10.22c}$$

where

$$\hat{\mathbf{c}} = \begin{bmatrix} \mathbf{c} \\ -\mathbf{c} \\ \mathbf{0} \end{bmatrix} \quad \text{and} \quad \hat{\mathbf{A}} = [\mathbf{A} \ -\mathbf{A} \ \mathbf{I}]$$

The simplex and other methods that are very effective for LP problems will be studied in Chaps. 11 and 12.

10.3.2 Quadratic Programming

The simplest, yet the most frequently encountered class of constrained nonlinear optimization problems is the class of quadratic programming (QP) problems. In these problems, the objective function is quadratic and the constraints are linear, i.e.,

$$\text{minimize } f(\mathbf{x}) = \tfrac{1}{2}\mathbf{x}^T \mathbf{H} \mathbf{x} + \mathbf{x}^T \mathbf{p} + c \tag{10.23a}$$

$$\text{subject to: } \mathbf{A}\mathbf{x} = \mathbf{b} \tag{10.23b}$$

$$\mathbf{C}\mathbf{x} \leq \mathbf{d} \tag{10.23c}$$

In many applications, the Hessian of $f(\mathbf{x})$, $\mathbf{H}$, is positive semidefinite. This implies that $f(\mathbf{x})$ is a globally convex function. Since the feasible region defined by Eqs. (10.23b) and (10.23c) is always convex, QP problems with positive semidefinite $\mathbf{H}$ can be regarded as a special class of convex programming problems which will be further addressed in Sect. 10.3.3. Algorithms for solving QP problems will be studied in Chap. 13.

10.3.3 Convex Programming

In a convex programming (CP) problem, a parameter vector is sought that minimizes a *convex* objective function subject to a set of constraints that define a *convex* feasible region for the problem [3,4]. Evidently, LP and QP problems with positive semidefinite Hessian matrices can be viewed as CP problems.

There are other types of CP problems that are of practical importance in engineering and science. As an example, consider the problem

$$\text{minimize} \quad \ln(\det\mathbf{P}^{-1}) \tag{10.24a}$$
$$\text{subject to:} \quad \mathbf{P} \succ \mathbf{0} \tag{10.24b}$$
$$\mathbf{v}_i^T \mathbf{P} \mathbf{v}_i \le 1 \quad \text{for } i = 1, \ 2, \ \dots, \ L \tag{10.24c}$$

where vectors $\mathbf{v}_i$ for $1 \le i \le L$ are given and the elements of matrix $\mathbf{P} = \mathbf{P}^T$ are the variables. It can be shown that if $\mathbf{P} \succ \mathbf{0}$ (i.e., $\mathbf{P}$ is positive definite), then $\ln(\det\mathbf{P}^{-1})$ is a convex function of $\mathbf{P}$ (see Prob. 10.6). In addition, if $\mathbf{p} = \mathbf{P}(:)$ denotes the vector obtained by lexicographically ordering the elements of matrix $\mathbf{P}$, then the set of vectors $\mathbf{p}$ satisfying the constraints in Eqs. (10.24b) and (10.24c) is convex and, therefore, Eqs. (10.24a)–(10.24c) describe a CP problem. Algorithms for solving CP problems will be studied in Chap. 14.

10.3.4 General Constrained Optimization Problem

The problem in Eqs. (10.1a)–(10.1c) will be referred to as a *general constrained optimization* (GCO) *problem* if either $f(\mathbf{x})$ has a nonlinearity of higher order than second order and is not globally convex or at least one constraint is not convex.

Example 10.3 Classify the constrained problem (see [5]):

$$\text{minimize} \quad f(\mathbf{x}) = \tfrac{1}{27\sqrt{3}}[(x_1 - 3)^2 - 9]x_2^3$$
$$\text{subject to:} \quad -x_1/\sqrt{3} + x_2 \le 0$$
$$-x_1 - \sqrt{3}x_2 \le 0$$
$$x_1 + \sqrt{3}x_2 \le 6$$
$$-x_1 \le 0$$
$$-x_2 \le 0$$

Solution The Hessian of $f(\mathbf{x})$ is given by

$$\mathbf{H}(\mathbf{x}) = \frac{2}{27\sqrt{3}} \begin{bmatrix} x_2^3 & 3(x_1 - 3)x_2^2 \\ 3(x_1 - 3)x_2^2 & 3[(x_1 - 3)^2 - 9]x_2 \end{bmatrix}$$

Note that $\mathbf{x} = [3 \ 1]^T$ satisfies all the constraints but $\mathbf{H}(\mathbf{x})$ is indefinite at point $\mathbf{x}$; hence $f(\mathbf{x})$ is not convex in the feasible region and the problem is a GCO problem. ∎

Very often GCO problems have multiple solutions that correspond to a number of distinct local minimizers. An effective way to obtain a good local solution in such a problem, especially when a reasonable initial point, say, $\mathbf{x}_0$, can be identified, is to tackle the problem by using a *sequential QP method*. In these methods, the highly nonlinear objective function is approximated in the neighborhood of point $\mathbf{x}_0$ in terms of a convex quadratic function while the nonlinear constraints are approximated in terms of linear constraints. In this way, the QP problem can be solved efficiently to obtain a solution, say, $\mathbf{x}_1$. The GCO problem is then approximated in the neighborhood of point $\mathbf{x}_1$ to yield a new QP problem whose solution is $\mathbf{x}_2$. This process is continued until a certain convergence criterion, such as $||\mathbf{x}_k - \mathbf{x}_{k+1}||$ or $|f(\mathbf{x}_k) - f(\mathbf{x}_{k+1})| < \varepsilon$ where ε is a prescribed termination tolerance, is met. Sequential QP methods will be studied in detail in Chap. 15.

Another approach for the solution of a GCO problem is to reformulate the problem as a sequential unconstrained problem in which the objective function is modified taking the constraints into account. The *barrier function methods* are representatives of this class of approaches, and will be investigated in Chap. 15.

10.4   Simple Transformation Methods

A transformation method is a method that solves the problem in Eqs. (10.1a)–(10.1c) by transforming the constrained optimization problem into an unconstrained optimization problem [6,7].

In this section, we shall study several simple transformation methods that can be applied when the equality constraints are linear equations or simple nonlinear equations, and when the inequality constraints are lower and/or upper bounds.

10.4.1   Variable Elimination

10.4.1.1   Linear Equality Constraints
Consider the optimization problem

$$\text{minimize} \ \ f(\mathbf{x}) \tag{10.25a}$$

$$\text{subject to:} \ \ \ \mathbf{Ax} = \mathbf{b} \tag{10.25b}$$

$$c_i(\mathbf{x}) \leq 0 \quad \text{for } 1 \leq i \leq q \tag{10.25c}$$

where $\mathbf{A} \in R^{p \times n}$ has full row rank, i.e., rank$(\mathbf{A}) = p$ with $p < n$. It can be shown that all solutions of Eq. (10.25b) are characterized by

$$\mathbf{x} = \mathbf{A}^+ \mathbf{b} + [\mathbf{I}_n - \mathbf{A}^+ \mathbf{A}]\hat{\boldsymbol{\phi}} \tag{10.26}$$

where A^+ denotes the Moore-Penrose pseudo-inverse of A [8], I_n is the $n \times n$ identity matrix, and $\hat{\phi}$ is an arbitrary n-dimensional parameter vector (see Prob. 10.7). The solutions expressed in Eq. (10.26) can be simplified considerably by using the SVD. As A has full row rank, the SVD of A gives

$$A = U\Sigma V^T$$

where $U \in R^{p \times p}$ and $V \in R^{n \times n}$ are orthogonal and $\Sigma = [S\ 0] \in R^{p \times n}$, $S = \text{diag}\{\sigma_1, \sigma_2, \ldots, \sigma_p\}$ where $\sigma_1 \geq \cdots \geq \sigma_p > 0$. Hence we have

$$A^+ = A^T(AA^T)^{-1} = V \begin{bmatrix} S^{-1} \\ 0 \end{bmatrix} U^T$$

and

$$I_n - A^+A = V \begin{bmatrix} 0 & 0 \\ 0 & I_{n-p} \end{bmatrix} V^T = V_r V_r^T$$

where $V_r = [v_{p+1}\ v_{p+2}\ \cdots\ v_n]$ contains the last $r = n - p$ columns of V. Therefore, Eq. (10.26) becomes

$$x = V_r\phi + A^+b \tag{10.27}$$

where $\phi \in R^{r \times 1}$ is an arbitrary r-dimensional vector. In words, Eq. (10.27) is a complete characterization of all solutions that satisfy Eq. (10.25b). Substituting Eq. (10.27) into Eqs. (10.25a) and (10.25c), we obtain the equivalent optimization problem

$$\underset{\phi}{\text{minimize}}\quad f(V_r\phi + A^+b) \tag{10.28}$$
$$\text{subject to: } c_i(V_r\phi + A^+b) \leq 0 \quad \text{for } 1 \leq i \leq q$$

in which the linear equality constraints are eliminated and the number of parameters is reduced from $n = \dim(x)$ to $r = \dim(\phi)$.

We note two features of the problem in Eq. (10.28). First, the size of the problem as compared with that of the problem in Eqs. (10.25a)–(10.25c) is reduced from n to $r = n - p$ and if ϕ^* is a solution of the problem in Eqs. (10.27), (10.28) implies that x^* given by

$$x^* = V_r\phi^* + A^+b \tag{10.29}$$

is a solution of the problem in Eqs. (10.25a)–(10.25c). Second, the linear relationship between x and ϕ as shown in Eq. (10.27) means that the degree of nonlinearity of the objective function $f(x)$ is preserved in the constrained problem of Eq. (10.28). If, for example, Eqs. (10.25a)–(10.25c) represent an LP or QP problem, then the problem in Eq. (10.28) is also an LP or QP problem. Moreover, it can be shown that if the problem in Eqs. (10.25a)–(10.25c) is a CP problem, then the reduced problem in Eq. (10.28) is also a CP problem.

A weak point of the above method is that performing the SVD of matrix A is computationally demanding, especially when the size of A is large. An alternative method that does not require the SVD is as follows. Assume that A has full row rank and let $\mathbf{P} \in R^{n \times n}$ be a permutation matrix that would permute the columns of $\mathbf{A}$ such that

$$\mathbf{A}\mathbf{x} = \mathbf{A}\mathbf{P}\mathbf{P}^T\mathbf{x} = [\mathbf{A}_1 \ \mathbf{A}_2]\hat{\mathbf{x}}$$

where $\mathbf{A}_1 \in R^{p \times p}$ consists of p linearly independent columns of $\mathbf{A}$ and $\hat{\mathbf{x}} = \mathbf{P}^T\mathbf{x}$ is simply a vector obtained by re-ordering the components of $\mathbf{x}$ accordingly. If we denote

$$\hat{\mathbf{x}} = \begin{bmatrix} \tilde{\mathbf{x}} \\ \boldsymbol{\psi} \end{bmatrix} \tag{10.30}$$

with $\tilde{\mathbf{x}} \in R^{p \times 1}$, $\boldsymbol{\psi} \in R^{r \times 1}$, then Eq. (10.25b) becomes

$$\mathbf{A}_1\tilde{\mathbf{x}} + \mathbf{A}_2\boldsymbol{\psi} = \mathbf{b}$$

i.e.,

$$\tilde{\mathbf{x}} = \mathbf{A}_1^{-1}\mathbf{b} - \mathbf{A}_1^{-1}\mathbf{A}_2\boldsymbol{\psi}$$

It follows that

$$\begin{aligned} \mathbf{x} = \mathbf{P}\hat{\mathbf{x}} = \mathbf{P}\begin{bmatrix} \tilde{\mathbf{x}} \\ \boldsymbol{\psi} \end{bmatrix} &= \mathbf{P}\begin{bmatrix} \mathbf{A}_1^{-1}\mathbf{b} - \mathbf{A}_1^{-1}\mathbf{A}_2\boldsymbol{\psi} \\ \boldsymbol{\psi} \end{bmatrix} \\ &\equiv \mathbf{W}\boldsymbol{\psi} + \tilde{\mathbf{b}} \end{aligned} \tag{10.31}$$

where

$$\mathbf{W} = \mathbf{P}\begin{bmatrix} -\mathbf{A}_1^{-1}\mathbf{A}_2 \\ \mathbf{I}_r \end{bmatrix} \in R^{n \times r}$$

$$\tilde{\mathbf{b}} = \mathbf{P}\begin{bmatrix} \mathbf{A}_1^{-1}\mathbf{b} \\ \mathbf{0} \end{bmatrix} \in R^{n \times 1}$$

The optimization problem in Eqs. (10.25a)–(10.25c) is now reduced to

$$\underset{\boldsymbol{\psi}}{\text{minimize}} \ f(\mathbf{W}\boldsymbol{\psi} + \tilde{\mathbf{b}}) \tag{10.32a}$$

$$\text{subject to:} \ c_i(\mathbf{W}\boldsymbol{\psi} + \tilde{\mathbf{b}}) \leq 0 \quad \text{for } 1 \leq i \leq q \tag{10.32b}$$

Note that the new parameter vector $\boldsymbol{\psi}$ is actually a collection of r components from $\mathbf{x}$.

Example 10.4 Apply the above variable elimination method to minimize

$$f(\mathbf{x}) = \tfrac{1}{2}\mathbf{x}^T\mathbf{H}\mathbf{x} + \mathbf{x}^T\mathbf{p} + c \tag{10.33}$$

subject to the constraints in Eq. (10.11), where $\mathbf{x} = [x_1 \ x_2 \ x_3 \ x_4]^T$.

Solution Since rank$(\mathbf{A}) = $ rank$([\mathbf{A} \ \mathbf{b}]) = 2$, the three constraints in Eq. (10.11) are consistent but redundant. It can be easily verified that the first two constraints in Eq. (10.11) are linearly independent; hence if we let

$$\mathbf{x} = \begin{bmatrix} \tilde{\mathbf{x}} \\ \boldsymbol{\psi} \end{bmatrix} \quad \text{with} \quad \tilde{\mathbf{x}} = \begin{bmatrix} x_1 \\ x_2 \end{bmatrix} \quad \text{and} \quad \boldsymbol{\psi} = \begin{bmatrix} x_3 \\ x_4 \end{bmatrix}$$

then Eq. (10.11) is equivalent to

$$\begin{bmatrix} 1 & -2 \\ 0 & 2 \end{bmatrix} \tilde{\mathbf{x}} + \begin{bmatrix} 3 & 2 \\ -1 & 0 \end{bmatrix} \boldsymbol{\psi} = \begin{bmatrix} 4 \\ 1 \end{bmatrix}$$

i.e.,

$$\tilde{\mathbf{x}} = \begin{bmatrix} -2 & -2 \\ \frac{1}{2} & 0 \end{bmatrix} \boldsymbol{\psi} + \begin{bmatrix} 5 \\ \frac{1}{2} \end{bmatrix} \equiv \mathbf{W}\boldsymbol{\psi} + \tilde{\mathbf{b}} \tag{10.34}$$

It follows that if we partition $\mathbf{H}$ and $\mathbf{p}$ in Eq. (10.33) as

$$\mathbf{H} = \begin{bmatrix} \mathbf{H}_{11} & \mathbf{H}_{12} \\ \mathbf{H}_{12}^T & \mathbf{H}_{22} \end{bmatrix} \quad \text{and} \quad \mathbf{p} = \begin{bmatrix} \mathbf{p}_1 \\ \mathbf{p}_2 \end{bmatrix}$$

with $\mathbf{H}_{11} \in R^{2\times2}$, $\mathbf{H}_{22} \in R^{2\times2}$, $\mathbf{p}_1 \in R^{2\times1}$, $\mathbf{p}_2 \in R^{2\times1}$, then Eq. (10.33) becomes

$$f(\boldsymbol{\psi}) = \tfrac{1}{2}\boldsymbol{\psi}^T \hat{\mathbf{H}} \boldsymbol{\psi} + \boldsymbol{\psi}^T \hat{\mathbf{p}} + \hat{c} \tag{10.35}$$

where

$$\hat{\mathbf{H}} = \mathbf{W}^T \mathbf{H}_{11} \mathbf{W} + \mathbf{H}_{12}^T \mathbf{W} + \mathbf{W}^T \mathbf{H}_{12} + \mathbf{H}_{22}$$
$$\hat{\mathbf{p}} = \mathbf{H}_{12}^T \tilde{\mathbf{b}} + \mathbf{W}^T \mathbf{H}_{11} \tilde{\mathbf{b}} + \mathbf{p}_2 + \mathbf{W}^T \mathbf{p}_1$$
$$\hat{c} = \tfrac{1}{2}\tilde{\mathbf{b}}^T \mathbf{H}_{11} \tilde{\mathbf{b}} + \tilde{\mathbf{b}}^T \mathbf{p}_2 + c$$

The problem now reduces to minimizing $f(\boldsymbol{\psi})$ without constraints. By writing

$$\hat{\mathbf{H}} = [\mathbf{W}^T \ \mathbf{I}] \mathbf{H} \begin{bmatrix} \mathbf{W} \\ \mathbf{I} \end{bmatrix}$$

we note that $\hat{\mathbf{H}}$ is positive definite if $\mathbf{H}$ is positive definite. In such a case, the unique minimizer of the problem is given by

$$\mathbf{x}^* = \begin{bmatrix} \tilde{\mathbf{x}}^* \\ \boldsymbol{\psi}^* \end{bmatrix}$$

with

$$\boldsymbol{\psi}^* = -\hat{\mathbf{H}}^{-1}\hat{\mathbf{p}} \quad \text{and} \quad \tilde{\mathbf{x}}^* = \mathbf{W}\boldsymbol{\psi}^* + \tilde{\mathbf{b}} \qquad \blacksquare$$

10.4.1.2 Nonlinear Equality Constraints

When the equality constraints are nonlinear, no general methods are available for variable elimination since solving a system of nonlinear equations is far more involved than solving a system of linear equations, if not impossible. However, in many cases the constraints can be appropriately manipulated to yield an equivalent constraint set in which some variables are expressed in terms of the rest of the variables so that the constraints can be partially or completely eliminated.

Example 10.5 Use nonlinear variable substitution to simplify the constrained problem

$$\text{minimize } f(\mathbf{x}) = -x_1^4 - 2x_2^4 - x_3^4 - x_1^2 x_2^2 - x_1^2 x_3^2 \tag{10.36}$$

$$\text{subject to: } a_1(\mathbf{x}) = x_1^4 + x_2^4 + x_3^4 - 25 = 0 \tag{10.37a}$$

$$a_2(\mathbf{x}) = 8x_1^2 + 14x_2^2 + 7x_3^2 - 56 = 0 \tag{10.37b}$$

Solution By writing Eq. (10.37b) as

$$x_3^2 = -\tfrac{8}{7}x_1^2 - 2x_2^2 + 8$$

the constraint in Eq. (10.37b) as well as variable x_3 in Eqs. (10.36) and (10.37a) can be eliminated, and an equivalent minimization problem can be formulated as

$$\text{minimize } f(\mathbf{x}) = -\tfrac{57}{49}x_1^4 - 6x_2^4 - \tfrac{25}{7}x_1^2 x_2^2 + \tfrac{72}{7}x_1^2 + 32x_2^2 \tag{10.38}$$

$$\text{subject to: } a_1(\mathbf{x}) = \tfrac{113}{49}x_1^4 + 5x_2^4 + \tfrac{32}{7}x_1^2 x_2^2 - \tfrac{128}{7}x_1^2 - 32x_2^2 + 39 = 0 \tag{10.39}$$

To eliminate Eq. (10.39), we write the equation as

$$5x_2^4 + (\tfrac{32}{7}x_1^2 - 32)x_2^2 + (\tfrac{113}{49}x_1^4 - \tfrac{128}{7}x_1^2 + 39) = 0$$

and treat it as a quadratic equation of x_2^2. In this way

$$x_2^2 = -(\tfrac{16}{35}x_1^2 - \tfrac{16}{5}) \pm \frac{1}{10}\sqrt{\left(-\tfrac{212}{49}x_1^4 + \tfrac{512}{7}x_1^2 + 244\right)} \tag{10.40}$$

By substituting Eq. (10.40) into Eq. (10.38), we obtain a minimization problem with only one variable.

The plus and minus signs in Eq. (10.40) mean that we have to deal with two separate cases, and the minimizer can be determined by comparing the results for the two cases. It should be noted that the polynomial under the square root in Eq. (10.40) assumes a negative value for large x_1; therefore, the one-dimensional minimization problem must be solved on an interval where the square root yields real values. ∎

10.4.2   Variable Transformations

10.4.2.1   Nonnegativity Bounds
The nonnegativity bound

$$x_i \geq 0$$

can be eliminated by using the variable transformation [7]

$$x_i = y_i^2 \tag{10.41}$$

Similarly, the constraint $x_i \geq d$ can be eliminated by using the transformation

$$x_i = d + y_i^2 \tag{10.42}$$

and one can readily verify that $x_i \leq d$ can be eliminated by using the transformation

$$x_i = d - y_i^2 \tag{10.43}$$

Although these transformations are simple and easy to use, these bounds are eliminated at the cost of increasing the degree of nonlinearity of the objective function as well as the remaining constraints, which may, in turn, reduce the efficiency of the optimization process.

Example 10.6 Apply a variable transformation to simplify the constrained problem

$$\text{minimize } f(\mathbf{x}) = -x_1^2 - 2x_2^2 - x_3^2 - x_1x_2 - x_1x_3 \tag{10.44}$$

$$\text{subject to: } b_1(\mathbf{x}) = x_1^2 + x_2^2 + x_3^2 - 25 = 0 \tag{10.45a}$$

$$b_2(\mathbf{x}) = 8x_1 + 14x_2 + 7x_3 - 56 = 0 \tag{10.45b}$$

$$x_i \geq 0 \quad i = 1,\ 2,\ 3 \tag{10.45c}$$

Solution The nonnegativity bounds in the problem can be eliminated by using the transformation in Eq. (10.41). While eliminating Eq. (10.45c), the transformation changes Eqs. (10.44), (10.45a), and (10.45b) to Eqs. (10.36), (10.37a), and (10.37b), respectively, where the y_i's have been renamed as x_i's.   ∎

10.4.2.2   Interval-Type Constraints
The hyperbolic tangent function defined by

$$y = \tanh(z) = \frac{e^z - e^{-z}}{e^z + e^{-z}} \tag{10.46}$$

is a differentiable monotonically increasing function that maps the entire 1-D space $-\infty < z < \infty$ onto the interval $-1 < y < 1$ as can be seen in Fig. 10.3. This in conjunction with the linear transformation

$$x = \frac{(b-a)}{2}y + \frac{b+a}{2} \tag{10.47}$$

transforms the infinite interval $(-\infty,\ \infty)$ into the open interval $(a,\ b)$. By writing $\tanh(z)$ as

$$\tanh(z) = \frac{e^{2z} - 1}{e^{2z} + 1}$$

we note that evaluating $\tanh(z)$ has about the same numerical complexity as the exponential function.

Fig. 10.3 The hyperbolic
tangent function

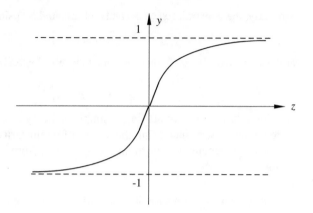

An alternative transformation for Eq. (10.46) is one that uses the inverse tangent
function

$$y = \frac{2}{\pi} \tan^{-1} z \tag{10.48}$$

which is also differentiable and monotonically increasing. As the transformations
in Eqs. (10.46) and (10.48) are nonlinear, applying them to eliminate interval-type
constraints will in general increase the nonlinearity of the objective function as well
as the remaining constraints.

Example 10.7 In certain engineering problems, an nth-order polynomial

$$p(z) = z^n + d_{n-1} z^{n-1} + \cdots + d_1 z + d_0$$

is required to have zeros inside the unit circle of the z plane, for example, the denominator of the transfer function in discrete-time systems and digital filters [9]. Such
polynomials are sometimes called *Schur polynomials*.

Find a suitable transformation for coefficients d_0 and d_1 which would ensure that
the second-order polynomial

$$p(z) = z^2 + d_1 z + d_0$$

is always a Schur polynomial.

Solution The zeros of $p(z)$ are located inside the unit circle if and only if [9]

$$d_0 < 1$$
$$d_1 - d_0 < 1 \tag{10.49}$$
$$d_1 + d_0 > -1$$

The region described by the constraints in Eq. (10.49) is the triangle shown in
Fig. 10.4. For a fixed $d_0 \in (-1, \ 1)$, the line segment inside the triangle shown as a
dashed line is characterized by d_1 varying from $-(1+d_0)$ to $1+d_0$. As d_0 varies from

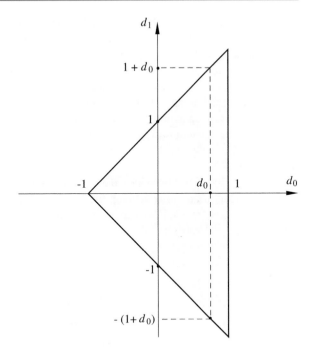

Fig. 10.4 Region of the d_1 versus d_0 plane for which $p(z)$ is a Schur polynomial

-1 to 1, the line segment will cover the entire triangle. This observation suggests the transformation

$$d_0 = \tanh(b_0) \qquad (10.50)$$
$$d_1 = [1 + \tanh(b_0)] \tanh(b_1)$$

which provides a one-to-one correspondence between points in the triangle in the $(d_0,\ d_1)$ space and points in the entire $(b_0,\ b_1)$ space. In other words, $p(z)$ is transformed into the polynomial

$$p(z) = z^2 + [1 + \tanh(b_0)] \tanh(b_1)z + \tanh(b_0) \qquad (10.51)$$

which is always a Schur polynomial for *any* finite values of b_0 and b_1.

This characterization of second-order Schur polynomials has been found to be useful in the design of stable recursive digital filters [10].   ■

10.5 Lagrange Multipliers

Lagrange multipliers play a crucial role in the study of constrained optimization. On the one hand, the conditions imposed on the Lagrange multipliers are always an integral part of various necessary and sufficient conditions and, on the other, they provide a natural connection between constrained and corresponding unconstrained optimization problems; each individual Lagrange multiplier can be interpreted as the rate of change in the objective function with respect to changes in the associated constraint function [7]. In simple terms, if $\mathbf{x}^*$ is a local minimizer of a constrained minimization problem, then in addition to $\mathbf{x}^*$ being a feasible point, the gradient of the objective function at $\mathbf{x}^*$ has to be a linear combination of the gradients of the constraint functions, and the Lagrange multipliers are the coefficients in that linear combination. Moreover, the Lagrange multipliers associated with inequality constraints have to be nonnegative and the multipliers associated with inactive inequality constraints have to be zero. Collectively, these conditions are known as the *Karush-Kuhn-Tucker conditions* (KKT).

In what follows, we introduce the concept of Lagrange multipliers through a simple application and then develop the KKT conditions for an arbitrary problem with equality constraints. Let us consider the minimization of the objective function $f(x_1, x_2, x_3, x_4)$ subject to the equality constraints

$$a_1(x_1, x_2, x_3, x_4) = 0 \qquad (10.52a)$$

$$a_2(x_1, x_2, x_3, x_4) = 0 \qquad (10.52b)$$

If these constraints can be expressed as

$$x_3 = h_1(x_1, x_2) \qquad (10.53a)$$

$$x_4 = h_2(x_1, x_2) \qquad (10.53b)$$

then they can be eliminated by substituting Eqs. (10.53a) and (10.53b) into the objective function, which will assume the form $f[x_1, x_2, h_1(x_1, x_2), h_2(x_1, x_2)]$. If $\mathbf{x}^* = [x_1^* \ x_2^* \ x_3^* \ x_4^*]^T$ is a local minimizer of the original constrained optimization problem, then $\hat{\mathbf{x}}^* = [x_1^* \ x_2^*]^T$ is a local minimizer of the problem

$$\text{minimize} \ \ f[x_1, x_2, h_1(x_1, x_2), h_2(x_1, x_2)]$$

Therefore, it follows that at $\hat{\mathbf{x}}^*$ we have

$$\nabla f = \begin{bmatrix} \dfrac{\partial f}{\partial x_1} \\[2mm] \dfrac{\partial f}{\partial x_2} \end{bmatrix} = \mathbf{0}$$

Since variables x_3 and x_4 in the constraints of Eqs. (10.53a) and (10.53b) are related to variables x_1 and x_2, the use of the chain rule for the partial derivatives in ∇f gives

$$\frac{\partial f}{\partial x_1} + \frac{\partial f}{\partial x_3}\frac{\partial h_1}{\partial x_1} + \frac{\partial f}{\partial x_4}\frac{\partial h_2}{\partial x_1} = 0$$

$$\frac{\partial f}{\partial x_2} + \frac{\partial f}{\partial x_3}\frac{\partial h_1}{\partial x_2} + \frac{\partial f}{\partial x_4}\frac{\partial h_2}{\partial x_2} = 0$$

From Eqs. (10.52a)–(10.53b), we have

$$\frac{\partial a_1}{\partial x_1} + \frac{\partial a_1}{\partial x_3}\frac{\partial h_1}{\partial x_1} + \frac{\partial a_1}{\partial x_4}\frac{\partial h_2}{\partial x_1} = 0$$

$$\frac{\partial a_1}{\partial x_2} + \frac{\partial a_1}{\partial x_3}\frac{\partial h_1}{\partial x_2} + \frac{\partial a_1}{\partial x_4}\frac{\partial h_2}{\partial x_2} = 0$$

$$\frac{\partial a_2}{\partial x_1} + \frac{\partial a_2}{\partial x_3}\frac{\partial h_1}{\partial x_1} + \frac{\partial a_2}{\partial x_4}\frac{\partial h_2}{\partial x_1} = 0$$

$$\frac{\partial a_2}{\partial x_2} + \frac{\partial a_2}{\partial x_3}\frac{\partial h_1}{\partial x_2} + \frac{\partial a_2}{\partial x_4}\frac{\partial h_2}{\partial x_2} = 0$$

The above six equations can now be expressed as

$$\begin{bmatrix} \nabla^T f(\mathbf{x}^*) \\ \nabla^T a_1(\mathbf{x}^*) \\ \nabla^T a_2(\mathbf{x}^*) \end{bmatrix} \begin{bmatrix} 1 & 0 \\ 0 & 1 \\ \frac{\partial h_1}{\partial x_1} & \frac{\partial h_1}{\partial x_2} \\ \frac{\partial h_2}{\partial x_1} & \frac{\partial h_2}{\partial x_2} \end{bmatrix} = \mathbf{0} \tag{10.54}$$

This equation implies that $\nabla f(\mathbf{x}^*)$, $\nabla a_1(\mathbf{x}^*)$, and $\nabla a_2(\mathbf{x}^*)$ are linearly dependent (see Prob. 10.9). Hence there exist constants α, β, γ which are not all zero such that

$$\alpha \nabla f(\mathbf{x}^*) + \beta \nabla a_1(\mathbf{x}^*) + \gamma \nabla a_2(\mathbf{x}^*) = \mathbf{0} \tag{10.55}$$

If we assume that $\mathbf{x}^*$ is a regular point of the constraints, then α in Eq. (10.55) cannot be zero and Eq. (10.55) can be simplified to

$$\nabla f(\mathbf{x}^*) + \lambda_1^* \nabla a_1(\mathbf{x}^*) + \lambda_2^* \nabla a_2(\mathbf{x}^*) = \mathbf{0} \tag{10.56}$$

and, therefore

$$\nabla f(\mathbf{x}^*) = -\lambda_1^* \nabla a_1(\mathbf{x}^*) - \lambda_2^* \nabla a_2(\mathbf{x}^*)$$

where $\lambda_1^* = \beta/\alpha$ and $\lambda_2^* = \gamma/\alpha$. In words, we conclude that *at a local minimizer of the constrained optimization problem, the gradient of the objective function is a linear combination of the gradients of the constraints.* Constants λ_1^* and λ_2^* in Eq. (10.56) are called the *Lagrange multipliers* of the constrained problem. In the rest of this section, we examine the concept of Lagrange multipliers from a different perspective.

10.5.1 Equality Constraints

We now consider the constrained optimization problem

$$\text{minimize } f(\mathbf{x}) \tag{10.57a}$$
$$\text{subject to: } a_i(\mathbf{x}) = 0 \quad \text{for } i = 1, 2, \ldots, p \tag{10.57b}$$

following an approach used by Fletcher in [7, Chap.9]. Let $\mathbf{x}^*$ be a local minimizer of the problem in Eqs. (10.57a) and (10.57b). By using the Taylor series of constraint function $a_i(\mathbf{x})$ at $\mathbf{x}^*$, we can write

$$a_i(\mathbf{x}^* + \mathbf{s}) = a_i(\mathbf{x}^*) + \mathbf{s}^T \nabla a_i(\mathbf{x}^*) + o(\|\mathbf{s}\|)$$
$$= \mathbf{s}^T \nabla a_i(\mathbf{x}^*) + o(\|\mathbf{s}\|) \tag{10.58}$$

since $a_i(\mathbf{x}^*) = 0$. It follows that $a_i(\mathbf{x}^* + \mathbf{s}) = 0$ is equivalent to

$$\mathbf{s}^T \nabla a_i(\mathbf{x}^*) = 0 \quad \text{for } i = 1, 2, \ldots, p \tag{10.59}$$

In other words, $\mathbf{s}$ is feasible if and only if it is orthogonal to the gradients of the constraint functions. Now we project the gradient $\nabla f(\mathbf{x}^*)$ orthogonally onto the space spanned by $\{\nabla a_1(\mathbf{x}^*), \nabla a_2(\mathbf{x}^*), \ldots, \nabla a_p(\mathbf{x}^*)\}$. If we denote the projection as

$$-\sum_{i=1}^{p} \lambda_i^* \nabla a_i(\mathbf{x}^*)$$

then $\nabla f(\mathbf{x}^*)$ can be expressed as

$$\nabla f(\mathbf{x}^*) = -\sum_{i=1}^{p} \lambda_i^* \nabla a_i(\mathbf{x}^*) + \mathbf{d} \tag{10.60}$$

where $\mathbf{d}$ is orthogonal to $\nabla a_i(\mathbf{x}^*)$ for $i = 1, 2, \ldots, p$.

In what follows, we show that if $\mathbf{x}^*$ is a local minimizer then $\mathbf{d}$ must be zero. The proof is accomplished by contradiction. Assume that $\mathbf{d} \neq \mathbf{0}$ and let $\mathbf{s} = -\mathbf{d}$. Since $\mathbf{s}$ is orthogonal to $\nabla a_i(\mathbf{x}^*)$ by virtue of Eq. (10.59), $\mathbf{s}$ is feasible at $\mathbf{x}^*$. Now we use Eq. (10.60) to obtain

$$\mathbf{s}^T \nabla f(\mathbf{x}^*) = \mathbf{s}^T \left(-\sum_{i=1}^{p} \lambda_i^* \nabla a_i(\mathbf{x}^*) + \mathbf{d} \right) = -||\mathbf{d}||^2 < 0$$

This means that $\mathbf{s}$ is a descent direction at $\mathbf{x}^*$ which contradicts the fact that $\mathbf{x}^*$ is a minimizer. Therefore, $\mathbf{d} = \mathbf{0}$ and Eq. (10.60) becomes

$$\nabla f(\mathbf{x}^*) = -\sum_{i=1}^{p} \lambda_i^* \nabla a_i(\mathbf{x}^*) \tag{10.61}$$

In effect, for an arbitrary constrained problem with equality constraints, the gradient of the objective function at a local minimizer *is equal to the linear combination of the gradients of the equality constraint functions with the Lagrange multipliers as the coefficients.*

For the problem in Eqs. (10.1a)–(10.1c) with equality as well as inequality constraints, Eq. (10.61) needs to be modified to include those inequality constraints that are *active* at $\mathbf{x}^*$. This more general case is treated in Sect. 10.6.

Example 10.8 Determine the Lagrange multipliers for the optimization problem

$$\text{minimize } f(\mathbf{x})$$
$$\text{subject to: } \mathbf{A}\mathbf{x} = \mathbf{b}$$

where $\mathbf{A} \in R^{p \times n}$ is assumed to have full row rank. Also discuss the case where the constraints are nonlinear.

Solution Eq. (10.61) in this case becomes

$$\mathbf{g}^* = -\mathbf{A}^T \boldsymbol{\lambda}^* \tag{10.62}$$

where $\boldsymbol{\lambda}^* = [\lambda_1^* \ \lambda_2^* \ \cdots \ \lambda_p^*]^T$ and $\mathbf{g}^* = \nabla f(\mathbf{x}^*)$. By virtue of Eq. (10.62), the Lagrange multipliers are uniquely determined as

$$\boldsymbol{\lambda}^* = -(\mathbf{AA}^T)^{-1}\mathbf{Ag}^* = -(\mathbf{A}^T)^+\mathbf{g}^* \tag{10.63}$$

where $(\mathbf{A}^T)^+$ denotes the Moore-Penrose pseudo-inverse of $\mathbf{A}^T$.

For the case of nonlinear equality constraints, a similar conclusion can be reached in terms of the Jacobian of the constraints in Eq. (10.57b). If we let

$$\mathbf{J}_e(\mathbf{x}) = [\nabla a_1(\mathbf{x}) \ \nabla a_2(\mathbf{x}) \ \cdots \ \nabla a_p(\mathbf{x})]^T \tag{10.64}$$

then the Lagrange multipliers λ_i^* for $1 \le i \le p$ in Eq. (10.61) are uniquely determined as

$$\boldsymbol{\lambda}^* = -[\mathbf{J}_e^T(\mathbf{x}^*)]^+\mathbf{g}^* \tag{10.65}$$

provided that $\mathbf{J}_e(\mathbf{x})$ has full row rank at $\mathbf{x}^*$.   ∎

The concept of Lagrange multipliers can also be explained from a different perspective. If we introduce the function

$$L(\mathbf{x}, \boldsymbol{\lambda}) = f(\mathbf{x}) + \sum_{i=1}^{p} \lambda_i a_i(\mathbf{x}) \tag{10.66}$$

as the *Lagrangian* of the optimization problem, then the condition in Eq. (10.61) and the constraints in Eq. (10.57b) can be written as

$$\nabla_x L(\mathbf{x}, \boldsymbol{\lambda}) = \mathbf{0} \quad \text{for } \{\mathbf{x}, \boldsymbol{\lambda}\} = \{\mathbf{x}^*, \boldsymbol{\lambda}^*\} \tag{10.67a}$$

and

$$\nabla_\lambda L(\mathbf{x}, \boldsymbol{\lambda}) = \mathbf{0} \quad \text{for } \{\mathbf{x}, \boldsymbol{\lambda}\} = \{\mathbf{x}^*, \boldsymbol{\lambda}^*\} \tag{10.67b}$$

respectively. The numbers of equations in Eqs. (10.67a) and (10.67b) are n and p, respectively, and the total number of equations is consistent with the number of parameters in $\mathbf{x}$ and $\boldsymbol{\lambda}$, i.e., $n + p$. Now if we define the gradient operator ∇ as

$$\nabla = \begin{bmatrix} \nabla_x \\ \nabla_\lambda \end{bmatrix}$$

then Eqs. (10.67a) and (10.67b) can be expressed as

$$\nabla L(\mathbf{x}, \boldsymbol{\lambda}) = \mathbf{0} \quad \text{for } \{\mathbf{x}, \boldsymbol{\lambda}\} = \{\mathbf{x}^*, \boldsymbol{\lambda}^*\} \tag{10.68}$$

From the above analysis, we see that the Lagrangian incorporates the constraints into a modified objective function in such a way that a constrained minimizer $\mathbf{x}^*$ is connected to an *unconstrained* minimizer $\{\mathbf{x}^*, \boldsymbol{\lambda}^*\}$ for the augmented objective function $L(\mathbf{x}, \boldsymbol{\lambda})$ where the augmentation is achieved with the p Lagrange multipliers.

Example 10.9 Solve the problem

$$\text{minimize } f(\mathbf{x}) = \tfrac{1}{2}\mathbf{x}^T\mathbf{Hx} + \mathbf{x}^T\mathbf{p}$$

$$\text{subject to: } \mathbf{Ax} = \mathbf{b}$$

where $\mathbf{H} \succ \mathbf{0}$ and $\mathbf{A} \in R^{p \times n}$ has full row rank.

Solution In Example 10.4 we solved a similar problem by eliminating the equality constraints. Here, we define the Lagrangian

$$L(\mathbf{x},\ \boldsymbol{\lambda}) = \tfrac{1}{2}\mathbf{x}^T\mathbf{H}\mathbf{x} + \mathbf{x}^T\mathbf{p} + \boldsymbol{\lambda}^T(\mathbf{A}\mathbf{x} - \mathbf{b})$$

and apply the condition in Eq. (10.68) to obtain

$$\nabla L(\mathbf{x},\ \boldsymbol{\lambda}) = \begin{bmatrix} \mathbf{H}\mathbf{x} + \mathbf{p} + \mathbf{A}^T\boldsymbol{\lambda} \\ \\ \mathbf{A}\mathbf{x} - \mathbf{b} \end{bmatrix}$$

$$= \begin{bmatrix} \mathbf{H} & \mathbf{A}^T \\ \mathbf{A} & \mathbf{0} \end{bmatrix}\begin{bmatrix} \mathbf{x} \\ \boldsymbol{\lambda} \end{bmatrix} + \begin{bmatrix} \mathbf{p} \\ -\mathbf{b} \end{bmatrix} = \mathbf{0} \qquad (10.69)$$

Since $\mathbf{H} \succ \mathbf{0}$ and $\text{rank}(\mathbf{A}) = p$, we can show that the matrix

$$\begin{bmatrix} \mathbf{H} & \mathbf{A}^T \\ \mathbf{A} & \mathbf{0} \end{bmatrix}$$

is nonsingular (see[11, Chap. 11]) and, therefore, Eq. (10.69) has the unique solution

$$\begin{bmatrix} \mathbf{x}^* \\ \boldsymbol{\lambda}^* \end{bmatrix} = \begin{bmatrix} \mathbf{H} & \mathbf{A}^T \\ \mathbf{A} & \mathbf{0} \end{bmatrix}^{-1}\begin{bmatrix} -\mathbf{p} \\ \mathbf{b} \end{bmatrix}$$

It follows that

$$\mathbf{x}^* = -\mathbf{H}^{-1}(\mathbf{A}^T\boldsymbol{\lambda}^* - \mathbf{p}) \qquad (10.70a)$$

where

$$\boldsymbol{\lambda}^* = -(\mathbf{A}\mathbf{H}^{-1}\mathbf{A}^T)^{-1}(\mathbf{A}\mathbf{H}^{-1}\mathbf{p} + \mathbf{b}) \qquad (10.70b)$$

   In Sect. 10.8, it will be shown that $\mathbf{x}^*$ given by Eq. (10.70a) with $\boldsymbol{\lambda}^*$ determined using Eq. (10.70b) is the unique, global minimizer of the constrained minimization problem.   ∎

10.5.2 Tangent Plane and Normal Plane

The first derivative of a smooth function of one variable indicates the direction along which the function increases. Similarly, the gradient of a smooth multivariable function indicates the direction along which the function increases at the greatest rate. This fact can be verified by using the first-order approximation of the Taylor series of the function, namely,

$$f(\mathbf{x}^* + \boldsymbol{\delta}) = f(\mathbf{x}^*) + \boldsymbol{\delta}^T\nabla f(\mathbf{x}^*) + o(||\boldsymbol{\delta}||)$$

If $||\boldsymbol{\delta}||$ is small, then the value of the function increases by $\boldsymbol{\delta}^T\nabla f(\mathbf{x}^*)$ which reaches the maximum when the direction of $\boldsymbol{\delta}$ coincides with that of $\nabla f(\mathbf{x}^*)$.

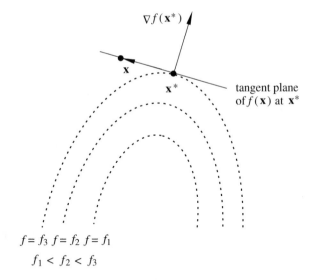

Fig. 10.5 Relation of tangent plane to contours and gradient

$$f = f_3 \; f = f_2 \; f = f_1$$
$$f_1 < f_2 < f_3$$

Two interrelated concepts that are closely related to the gradients of the objective function and the constraints of the optimization problem in Eqs. (10.57a) and (10.57b) are the *tangent plane* and *normal plane*.

The tangent plane of a smooth function $f(\mathbf{x})$ at a given point $\mathbf{x}^*$ can be defined in two ways as follows. If $C_{\mathbf{x}^*}$ is the contour surface of $f(\mathbf{x})$ that passes through point $\mathbf{x}^*$, then we can think of the tangent plane as a hyperplane in R^n that touches $C_{\mathbf{x}^*}$ at and only at point $\mathbf{x}^*$. Alternatively, the tangent plane can be defined as a hyperplane that passes through point $\mathbf{x}^*$ with $\nabla f(\mathbf{x}^*)$ as the normal. For example, for $n = 2$ the contours, tangent plane, and gradient of a smooth function are related to each other as illustrated in Fig. 10.5.

Following the above discussion, the tangent plane at point $\mathbf{x}^*$ can be defined analytically as the set

$$\mathcal{T}_{\mathbf{x}^*} = \{\mathbf{x} : \; \nabla f(\mathbf{x}^*)^T (\mathbf{x} - \mathbf{x}^*) = 0\}$$

In other words, *a point $\mathbf{x}$ lies on the tangent plane if the vector that connects $\mathbf{x}^*$ to $\mathbf{x}$ is orthogonal to the gradient $\nabla f(\mathbf{x}^*)$*, as can be seen in Fig. 10.5.

Proceeding in the same way, a tangent plane can be defined for a surface that is characterized by *several* equations. Let S be the surface defined by the equations

$$a_i(\mathbf{x}) = 0 \quad \text{for } i = 1, \, 2, \, \ldots, \, p$$

and assume that $\mathbf{x}^*$ is a point satisfying these constraints, i.e., $\mathbf{x}^* \in S$. The tangent plane of S at $\mathbf{x}^*$ is given by

$$\mathcal{T}_{\mathbf{x}^*} = \{\mathbf{x} : \; \mathbf{J}_e(\mathbf{x}^*)(\mathbf{x} - \mathbf{x}^*) = \mathbf{0}\} \tag{10.71}$$

where $\mathbf{J}_e$ is the Jacobian defined by Eq. (10.64). From Eq. (10.71), we conclude that the tangent plane of S is actually an $(n - p)$-dimensional hyperplane in space R^n.

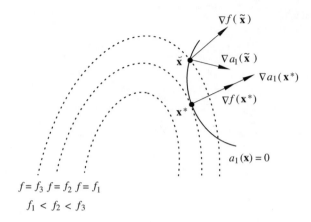

Fig. 10.6 Geometrical interpretation of Eq. (10.61): $\nabla f(\tilde{\mathbf{x}})$ lies in $N_{\mathbf{x}^*}$ if $\tilde{\mathbf{x}} = \mathbf{x}^*$ where $\mathbf{x}^*$ is a minimizer

For example, in the case of Fig. 10.5 we have $n = 2$ and $p = 1$ and hence the tangent plane degenerates into a straight line.

The normal plane can similarly be defined. Given a set of equations $a_i(\mathbf{x}) = 0$ for $1 \le i \le p$ and a point $\mathbf{x}^* \in S$, the normal plane at $\mathbf{x}^*$ is given by

$$N_{\mathbf{x}^*} = \{\mathbf{x} : \mathbf{x} - \mathbf{x}^* = \sum_{i=1}^{p} \alpha_i \nabla a_i(\mathbf{x}^*) \text{ for } \alpha_i \in R\} \tag{10.72}$$

It follows that $\{N_{\mathbf{x}^*} - \mathbf{x}^*\}$ is the range of matrix $\mathbf{J}_e^T(\mathbf{x}^*)$, and hence it is a p-dimensional subspace in R^n. More importantly, $\mathcal{T}_{\mathbf{x}^*}$ and $N_{\mathbf{x}^*}$ are orthogonal to each other.

10.5.3   Geometrical Interpretation

On the basis of the preceding definitions, a geometrical interpretation of the necessary condition in Eq. (10.61) is possible [7,12] as follows: *If $\mathbf{x}^*$ is a constrained local minimizer, then the vector $\nabla f(\mathbf{x}^*)$ must lie in the normal plane $N_{\mathbf{x}^*}$.*

A two-variable example is illustrated in Fig. 10.6 where several contours of the objective function $f(x_1, x_2)$ and the only equality constraint $a_1(x_1, x_2) = 0$ are depicted. Note that at feasible point $\tilde{\mathbf{x}}$, $\nabla f(\tilde{\mathbf{x}})$ lies exactly in the normal plane generated by $\nabla a_1(\tilde{\mathbf{x}})$ only when $\tilde{\mathbf{x}}$ coincides with $\mathbf{x}^*$, a minimizer of $f(\mathbf{x})$ subject to constraint $a_1(\mathbf{x}) = 0$.

Equation 10.61 may also hold when $\mathbf{x}^*$ is a minimizer as illustrated in Fig. 10.7a and b, or a maximizer as shown in Fig. 10.7c or $\mathbf{x}^*$ is neither a minimizer nor a maximizer. In addition, for a local minimizer, the Lagrange multipliers can be either positive as in Fig. 10.7a or negative as in Fig. 10.7b.

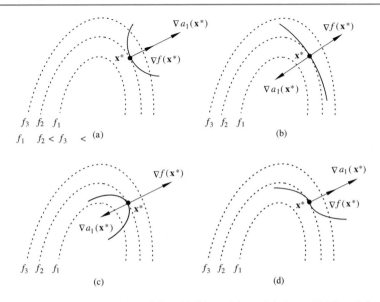

Fig. 10.7 Geometrical interpretation of Eq. (10.61): **a** $\mathbf{x}^*$ is a minimizer with $\lambda^* < 0$; **b** $\mathbf{x}^*$ is a minimizer with $\lambda^* > 0$; **c** $\mathbf{x}^*$ is a maximizer; **d** $\mathbf{x}^*$ is neither a minimizer nor a maximizer

Example 10.10 Construct the geometrical interpretation of Eq. (10.61) for the three-variable problem

$$\text{minimize } f(\mathbf{x}) = x_1^2 + x_2^2 + \tfrac{1}{4}x_3^2$$
$$\text{subject to: } a_1(\mathbf{x}) = -x_1 + x_3 - 1 = 0$$
$$a_2(\mathbf{x}) = x_1^2 + x_2^2 - 2x_1 = 0$$

Solution As was discussed in Example 10.1, the above constraints describe the curve obtained as the intersection of the cylinder $a_2(\mathbf{x}) = 0$ with the plane $a_1(\mathbf{x}) = 0$. Figure 10.8 shows that the constrained problem has a global minimizer $\mathbf{x}^* = [0\ 0\ 1]^T$. At $\mathbf{x}^*$, the tangent plane in Eq. (10.71) becomes a line that passes through $\mathbf{x}^*$ and is parallel with the x_2 axis while the normal plane $\mathcal{N}_{\mathbf{x}^*}$ is the plane spanned by

$$\nabla a_1(\mathbf{x}^*) = \begin{bmatrix} -1 \\ 0 \\ 1 \end{bmatrix} \quad \text{and} \quad \nabla a_2(\mathbf{x}^*) = \begin{bmatrix} -2 \\ 0 \\ 0 \end{bmatrix}$$

which is identical to plane $x_2 = 0$. Note that at $\mathbf{x}^*$

$$\nabla f(\mathbf{x}^*) = \begin{bmatrix} 0 \\ 0 \\ \tfrac{1}{2} \end{bmatrix}$$

As is expected, $\nabla f(\mathbf{x}^*)$ lies in the normal plane $\mathcal{N}_{\mathbf{x}^*}$ (see Fig. 10.8) and can be expressed as

$$\nabla f(\mathbf{x}^*) = -\lambda_1^* \nabla a_1(\mathbf{x}^*) - \lambda_2^* \nabla a_2(\mathbf{x}^*)$$

where $\lambda_1^* = -\tfrac{1}{2}$ and $\lambda_2^* = \tfrac{1}{4}$ are the Lagrange multipliers.   ∎

Fig. 10.8 An interpretation
of Eq. (10.61) for Example
10.10

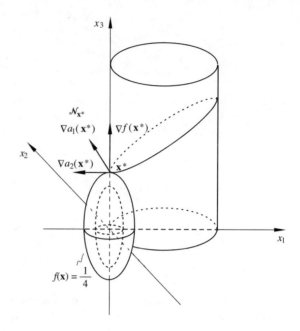

10.6 First-Order Necessary Conditions

The necessary conditions for a point $\mathbf{x}^*$ to be a local minimizer are useful in two
situations: (*a*) They can be used to exclude those points that do not satisfy the
necessary conditions from the candidate points; (*b*) they become sufficient conditions
when the objective function in question is convex (see Sect. 10.8 for details).

10.6.1 Equality Constraints

Based on the discussion in Sect. 10.5, the first-order necessary conditions for a
minimum for the problem in Eqs. (10.57a) and (10.57b) can be summarized in terms
of the following theorem.

Theorem 10.1 (First-order necessary conditions for a minimum, equality constraints) *If* $\mathbf{x}^*$ *is a constrained local minimizer of the problem in Eqs. (10.57a) and
(10.57b) and is a regular point of the constraints in Eq. (10.57b), then*

(*a*)

$$a_i(\mathbf{x}^*) = 0 \ \ for \ \ i = 1, \ 2, \ \dots, \ p \tag{10.73}$$

(b) and there exist Lagrange multipliers λ_i^* for $i = 1, 2, \ldots, p$ such that

$$\nabla f(\mathbf{x}^*) + \sum_{i=1}^{p} \lambda_i^* \nabla a_i(\mathbf{x}^*) = 0 \quad \blacksquare \tag{10.74}$$

Equation 10.74 can be expressed in terms of the Jacobian $\mathbf{J}_e(\mathbf{x})$ (see Eq. (10.64)) as

$$\mathbf{g}(\mathbf{x}^*) + \mathbf{J}_e^T(\mathbf{x}^*)\boldsymbol{\lambda}^* = 0$$

where $\mathbf{g}(\mathbf{x}) = \nabla f(\mathbf{x})$. In other words, *if $\mathbf{x}^*$ is a local minimizer* of the problem in Eqs. (10.57a) and (10.57b), *then there exists a vector $\boldsymbol{\lambda}^* \in R^p$ such that the $(n + p)$-dimensional vector $[\mathbf{x}^{*T} \ \boldsymbol{\lambda}^{*T}]^T$ satisfies the $n + p$ nonlinear equations*

$$\begin{bmatrix} \mathbf{g}(\mathbf{x}^*) + \mathbf{J}_e^T(\mathbf{x}^*)\boldsymbol{\lambda}^* \\ \mathbf{a}(\mathbf{x}^*) \end{bmatrix} = 0 \tag{10.75}$$

Theorem 10.1 can be related to the first-order necessary conditions for a minimum for the case of unconstrained minimization in Theorem 2.1 (see Sect. 2.5) as follows. If function $f(\mathbf{x})$ is minimized without constraints, we can consider the problem as the special case of the problem in Eqs. (10.57a) and (10.57b) where the number of constraints is reduced to zero. In such a case, condition (*a*) of Theorem 10.1 is satisfied automatically and condition (*a*) of Theorem 2.1 must hold. On the other hand, condition (*b*) becomes

$$\nabla f(\mathbf{x}^*) = 0$$

which is condition (*b*) of Theorem 2.1.

If $\mathbf{x}^*$ is a local minimizer and $\boldsymbol{\lambda}^*$ is the associated vector of Lagrange multipliers, the set $\{\mathbf{x}^*, \ \boldsymbol{\lambda}^*\}$ may be referred to as the *minimizer set* or *minimizer* for short.

Example 10.11 Find the points that satisfy the necessary conditions for a minimum for the problem in Example 10.10.

Solution We have

$$\mathbf{g}(\mathbf{x}) = \begin{bmatrix} 2x_1 \\ 2x_2 \\ \frac{1}{2}x_3 \end{bmatrix}, \quad \mathbf{J}_e^T(\mathbf{x}) = \begin{bmatrix} -1 & 2x_1 - 2 \\ 0 & 2x_2 \\ 1 & 0 \end{bmatrix}$$

Hence Eq. (10.75) becomes

$$2x_1 - \lambda_1 + \lambda_2(2x_1 - 2) = 0$$
$$2x_2 + 2\lambda_2 x_2 = 0$$
$$x_3 + 2\lambda_1 = 0$$
$$-x_1 + x_3 - 1 = 0$$
$$x_1^2 + x_2^2 - 2x_1 = 0$$

Solving the above system of equations, we obtain two solutions, i.e.,

$$\mathbf{x}_1^* = \begin{bmatrix} 0 \\ 0 \\ 1 \end{bmatrix} \quad \text{and} \quad \boldsymbol{\lambda}_1^* = \begin{bmatrix} -\frac{1}{2} \\ \frac{1}{4} \end{bmatrix}$$

and

$$\mathbf{x}_2^* = \begin{bmatrix} 2 \\ 0 \\ 3 \end{bmatrix} \quad \text{and} \quad \boldsymbol{\lambda}_2^* = \begin{bmatrix} -\frac{3}{2} \\ -\frac{11}{4} \end{bmatrix}$$

The first solution, $\{\mathbf{x}_1^*, \boldsymbol{\lambda}_1^*\}$, is the global minimizer set as can be observed in Fig. 10.8. Later on in Sect. 10.7, we will show that $\{\mathbf{x}_2^*, \boldsymbol{\lambda}_2^*\}$ is not a minimizer set. ∎

10.6.2   Inequality Constraints

Consider now the general constrained optimization problem in Eqs. (10.1a)–(10.1c) and let $\mathbf{x}^*$ be a local minimizer. The set $\mathcal{J}(\mathbf{x}^*) \subseteq \{1, 2, \ldots, q\}$ is the set of indices j for which the constraints $c_j(\mathbf{x}) \le 0$ are *active* at $\mathbf{x}^*$, i.e., $c_j(\mathbf{x}^*) = 0$. At point $\mathbf{x}^*$, the feasible directions are characterized only by the equality constraints *and* those inequality constraints $c_j(\mathbf{x})$ with $j \in \mathcal{J}(\mathbf{x}^*)$, and are not influenced by the inequality constraints that are *inactive*. As a matter of fact, for an inactive constraint $c_j(\mathbf{x}) \le 0$, the feasibility of $\mathbf{x}^*$ implies that

$$c_j(\mathbf{x}^*) < 0$$

This leads to

$$c_j(\mathbf{x}^* + \boldsymbol{\delta}) < 0$$

for any $\boldsymbol{\delta}$ with a sufficiently small $||\boldsymbol{\delta}||$.

If there are K active inequality constraints at $\mathbf{x}^*$ and

$$\mathcal{J}(\mathbf{x}^*) = \{j_1, j_2, \ldots, j_K\} \tag{10.76}$$

then Eq. (10.61) needs to be modified to

$$\nabla f(\mathbf{x}^*) + \sum_{i=1}^{p} \lambda_i^* \nabla a_i(\mathbf{x}^*) + \sum_{k=1}^{K} \mu_{j_k}^* \nabla c_{j_k}(\mathbf{x}^*) = \mathbf{0} \tag{10.77}$$

In words, Eq. (10.77) states that *the gradient at $\mathbf{x}^*$, $\nabla f(\mathbf{x}^*)$, is a linear combination of the gradients of all the constraint functions that are active at $\mathbf{x}^*$.*

An argument similar to that used in Sect. 10.5.2 to explain why Eq. (10.77) must hold for a local minimum of the problem in Eqs. (10.1a)–(10.1c) is as follows [7]. We start by assuming that $\mathbf{x}^*$ is a regular point for the constraints that are active at $\mathbf{x}^*$. Let j_k be one of the indices from $\mathcal{J}(\mathbf{x}^*)$ and assume that $\mathbf{s}$ is a feasible vector at $\mathbf{x}^*$. Using the Taylor series of $c_{j_k}(\mathbf{x})$, we can write

$$c_{j_k}(\mathbf{x}^* + \mathbf{s}) = c_{j_k}(\mathbf{x}^*) + \mathbf{s}^T \nabla c_{j_k}(\mathbf{x}^*) + o(||\mathbf{s}||)$$
$$= \mathbf{s}^T \nabla c_{j_k}(\mathbf{x}^*) + o(||\mathbf{s}||)$$

Since $\mathbf{s}$ is feasible, $c_{j_k}(\mathbf{x}^* + \mathbf{s}) \leq 0$ which leads to

$$\mathbf{s}^T \nabla c_{j_k}(\mathbf{x}^*) \leq 0 \tag{10.78}$$

Now we orthogonally project $\nabla f(\mathbf{x}^*)$ onto the space spanned by $\mathcal{S} = \{\nabla a_i(\mathbf{x}^*)$ for $1 \leq i \leq p$ and $\nabla c_{j_k}(\mathbf{x}^*)$ for $1 \leq k \leq K\}$. Since the projection is on $\mathcal{S}$, it can be expressed as a linear combination of vectors $\{\nabla a_i(\mathbf{x}^*)$ for $1 \leq i \leq p$ and $\nabla c_{j_k}(\mathbf{x}^*)$ for $1 \leq k \leq K\}$, i.e.,

$$-\sum_{i=1}^{p} \lambda_i^* \nabla a_i(\mathbf{x}^*) - \sum_{k=1}^{K} \mu_{j_k}^* \nabla c_{j_k}(\mathbf{x}^*)$$

for some λ_i^*'s and $\mu_{j_k}^*$'s. If we denote the difference between $\nabla f(\mathbf{x}^*)$ and this projection by $\mathbf{d}$, then we can write

$$\nabla f(\mathbf{x}^*) = -\sum_{i=1}^{p} \lambda_i^* \nabla a_i(\mathbf{x}^*) - \sum_{k=1}^{K} \mu_{j_k}^* \nabla c_{j_k}(\mathbf{x}^*) + \mathbf{d} \tag{10.79}$$

Since $\mathbf{d}$ is orthogonal to $\mathcal{S}$, $\mathbf{d}$ is orthogonal to $\nabla a_i(\mathbf{x}^*)$ and $\nabla c_{j_k}(\mathbf{x}^*)$; hence $\mathbf{s} = -\mathbf{d}$ is a feasible direction (see Eqs. (10.59) and (10.78)); however, Eq. (10.79) gives

$$\mathbf{s}^T \nabla f(\mathbf{x}^*) = -||\mathbf{d}||^2 < 0$$

meaning that $\mathbf{s}$ would be a descent direction at $\mathbf{x}^*$. This contradicts the fact that $\mathbf{x}^*$ is a local minimizer. Therefore, $\mathbf{d} = \mathbf{0}$ and Eq. (10.77) holds. Constants λ_i^* and $\mu_{j_k}^*$ in Eq. (10.77) are the Lagrange multipliers for equality and inequality constraints, respectively. Unlike the Lagrange multipliers associated with equality constraints, which can be either positive or negative, those associated with *active* inequality constraints must be nonnegative, i.e.,

$$\mu_{j_k}^* \geq 0 \quad \text{for } 1 \leq k \leq K \tag{10.80}$$

We demonstrate the validity of Eq. (10.80) by contradiction. Suppose that $\mu_{j_{k*}}^* < 0$ for some j_{k*}. Since the gradients in $\mathcal{S}$ are linearly independent, the system

$$\begin{bmatrix} \nabla^T a_1(\mathbf{x}^*) \\ \vdots \\ \nabla^T a_p(\mathbf{x}^*) \\ \nabla^T c_{j_1}(\mathbf{x}^*) \\ \vdots \\ \nabla^T c_{j_K}(\mathbf{x}^*) \end{bmatrix} \mathbf{s} = \begin{bmatrix} 0 \\ \vdots \\ 0 \\ 1 \\ 0 \\ \vdots \\ 0 \end{bmatrix}$$

has a solution for $\mathbf{s}$, where the vector on the right-hand side of the above equation has only one nonzero entry corresponding to $\nabla^T c_{j_{k*}}(\mathbf{x}^*)$. Hence we have a vector $\mathbf{s}$ satisfying the equations

$$\mathbf{s}^T \nabla a_i(\mathbf{x}^*) = 0 \quad \text{for } 1 \leq i \leq p$$

$$\mathbf{s}^T \nabla c_{j_k}(\mathbf{x}^*) = \begin{cases} 1 & \text{for } k = k^* \\ 0 & \text{otherwise} \end{cases}$$

It follows from Eqs. (10.59) and (10.78) that $\mathbf{s}$ is feasible. By virtue of Eq. (10.77), we obtain

$$\mathbf{s}^T \nabla f(\mathbf{x}^*) = \mu_{j_{k*}}^* < 0$$

Hence $\mathbf{s}$ is a descent direction at $\mathbf{x}^*$, which contradicts the fact that $\mathbf{x}^*$ is a local minimizer. This demonstrates the validity of Eq. (10.80). The following theorem, known as the *KKT conditions* theorem [13,14], summarizes the above discussion.

Theorem 10.2 (Karush-Kuhn-Tucker conditions) *If $\mathbf{x}^*$ is a local minimizer of the problem in Eqs. (10.1a)–(10.1c) and is regular for the constraints that are active at $\mathbf{x}^*$, then*

(a)
$$a_i(\mathbf{x}^*) = 0 \ \ for \ \ 1 \le i \le p,$$

(b)
$$c_j(\mathbf{x}^*) \le 0 \ \ for \ \ 1 \le j \le q,$$

(c) there exist Lagrange multipliers λ_i^* for $1 \le i \le p$ and μ_j^* for $1 \le j \le q$ such that

$$\nabla f(\mathbf{x}^*) + \sum_{i=1}^{p} \lambda_i^* \nabla a_i(\mathbf{x}^*) + \sum_{j=1}^{q} \mu_j^* \nabla c_j(\mathbf{x}^*) = \mathbf{0} \qquad (10.81)$$

(d)
$$\lambda_i^* a_i(\mathbf{x}^*) = 0 \ \ for \ \ 1 \le i \le p, \qquad (10.82a)$$
$$\mu_j^* c_j(\mathbf{x}^*) = 0 \ \ for \ \ 1 \le j \le q, \qquad (10.82b)$$

(e) and
$$\mu_j^* \ge 0 \ \ for \ \ 1 \le j \le q. \qquad (10.83)$$

∎

Some remarks on the KKT conditions stated in Theorem 10.2 are in order. Conditions (a) and (b) simply mean that $\mathbf{x}^*$ must be a feasible point. The $p+q$ equations in Eqs. (10.82a) and (10.82b) are often referred to as the *complementarity KKT conditions*. They state that λ_i^* and $a_i(\mathbf{x}^*)$ cannot be nonzero simultaneously, and μ_j^* and $c_j(\mathbf{x}^*)$ cannot be nonzero simultaneously. Note that condition (a) implies the condition in Eq. (10.82a) regardless of whether λ_i^* is zero or not. For the equality conditions in Eq. (10.82b), we need to distinguish those constraints that are active at $\mathbf{x}^*$, i.e.,

$$c_j(\mathbf{x}^*) = 0 \quad \text{for } j \in \mathcal{J}(\mathbf{x}^*) = \{j_1, \ j_2, \ \dots, \ j_K\}$$

from those that are inactive at $\mathbf{x}^*$, i.e.,

$$c_j(\mathbf{x}^*) < 0 \quad \text{for } j \in \{1, \ 2, \ \dots, \ q\} \backslash \mathcal{J}(\mathbf{x}^*)$$

where $\mathcal{I} \backslash \mathcal{J}$ denotes the system indices in $\mathcal{I}$, that are not in $\mathcal{J}$. From Eq. (10.82b),

$$\mu_j^* = 0 \quad \text{for } j \in \{1, \ 2, \ \dots, \ q\} \backslash \mathcal{J}(\mathbf{x}^*)$$

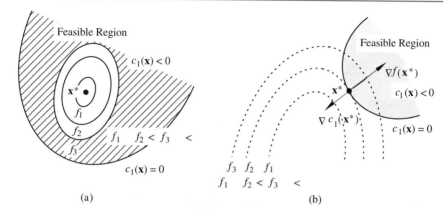

Fig. 10.9 Nonnegativity of Lagrange multipliers: **a** $\mathbf{x}^*$ is a minimizer in the interior of the feasible region; **b** $\mathbf{x}^*$ is a minimizer on the boundary of the feasible region

which reduces Eq. (10.81) to Eq. (10.77); however, μ_j may be nonzero for $j \in \mathcal{J}(\mathbf{x}^*)$. Condition ($e$) states that

$$\mu_j^* \geq 0 \quad \text{for } j \in \mathcal{J}(\mathbf{x}^*) \tag{10.84}$$

The nonnegativity of the Lagrange multipliers associated with inequality constraints can be explained using Fig. 10.9. For the sake of simplicity, let us assume that $p = 0$ and $q = 1$ in which case the optimization problem would involve only one inequality constraint, namely,

$$c_1(\mathbf{x}) \leq 0 \tag{10.85}$$

If the minimizer $\mathbf{x}^*$ happens to be inside the feasible region $\mathcal{R}$ defined by the constraint in Eq. (10.85) (see Fig. 10.9a), then $\nabla f(\mathbf{x}^*) = \mathbf{0}$ and $\mu_1^* = 0$. If $\mathbf{x}^*$ is on the boundary of $\mathcal{R}$ (see Fig. 10.9b), then Eq. (10.81) implies that

$$\nabla f(\mathbf{x}^*) + \mu_1^* \nabla c_1(\mathbf{x}^*) = \mathbf{0}$$

As can be seen in Fig. 10.9b, $\nabla c_1(\mathbf{x}^*)$ is a vector pointing towards the interior of the feasible region, since $c_1(\mathbf{x}^*) = 0$ and $c(\mathbf{x}) < 0$ inside $\mathcal{R}$, and similarly $\nabla f(\mathbf{x}^*)$ is a vector pointing towards the interior of $\mathcal{R}$. This in conjunction with the above equation implies that $\nabla f(\mathbf{x}^*)$ and $\nabla c_1(\mathbf{x}^*)$ must be in the same direction and hence $\mu_1^* > 0$. It should be stressed that the nonnegativity of the Lagrange multipliers holds only for those multipliers associated with *inequality* constraints. As was illustrated in Fig. 10.7a and b, nonzero Lagrange multipliers associated with equality constraints can be either positive or negative.

There are a total of p (equality) $+K$ (inequality) Lagrange multipliers that may be nonzero, and there are n entries in parameter vector $\mathbf{x}$. It is interesting to note that the KKT conditions involve the same number of equations, i.e.,

$$\mathbf{g}(\mathbf{x}^*) - \mathbf{J}_e^T(\mathbf{x}^*)\boldsymbol{\lambda}^* - \hat{\mathbf{J}}_{ie}^T(\mathbf{x}^*)\hat{\boldsymbol{\mu}}^* = \mathbf{0} \tag{10.86a}$$

$$\mathbf{a}(\mathbf{x}^*) = \mathbf{0} \tag{10.86b}$$

$$\hat{\mathbf{c}}(\mathbf{x}^*) = \mathbf{0} \tag{10.86c}$$

where

$$\hat{\boldsymbol{\mu}}^* = [\mu_{j1}^* \ \mu_{j2}^* \ \cdots \ \mu_{jK}^*]^T \tag{10.87a}$$

$$\hat{\mathbf{J}}_{ie}(\mathbf{x}^*) = [\nabla c_{j1}(\mathbf{x}^*) \ \nabla c_{j2}(\mathbf{x}^*) \ \cdots \ \nabla c_{jK}(\mathbf{x}^*)]^T \tag{10.87b}$$

$$\hat{\mathbf{c}}(\mathbf{x}^*) = [c_{j1}(\mathbf{x}^*) \ c_{j2}(\mathbf{x}^*) \ \cdots \ c_{jK}(\mathbf{x}^*)]^T \tag{10.87c}$$

Example 10.12 Solve the constrained minimization problem

$$\text{minimize } f(\mathbf{x}) = x_1^2 + x_2^2 - 14x_1 - 6x_2$$

$$\text{subject to: } c_1(\mathbf{x}) = -2 + x_1 + x_2 \le 0$$
$$c_2(\mathbf{x}) = -3 + x_1 + 2x_2 \le 0$$

by applying the KKT conditions.

Solution The KKT conditions imply that

$$2x_1 - 14 + \mu_1 + \mu_2 = 0$$
$$2x_2 - 6 + \mu_1 + 2\mu_2 = 0$$
$$\mu_1(-2 + x_1 + x_2) = 0$$
$$\mu_2(-3 + x_1 + 2x_2) = 0$$
$$\mu_1 \ge 0$$
$$\mu_2 \ge 0$$

One way to find the solution in this simple case is to consider all possible cases with regard to active constraints and verify the nonnegativity of the μ_i's obtained [11, Chap. 11].

Case 1 No active constraints

If there are no active constraints, we have $\mu_1^* = \mu_2^* = 0$, which leads to

$$\mathbf{x}^* = \begin{bmatrix} 7 \\ 3 \end{bmatrix}$$

Obviously, this $\mathbf{x}^*$ violates both constraints and it is not a solution.

Case 2 One constraint active

If only the first constraint is active, then we have $\mu_2^* = 0$, and

$$2x_1 - 14 + \mu_1 = 0$$
$$2x_2 - 6 + \mu_1 = 0$$
$$-2 + x_1 + x_2 = 0$$

Solving this system of equations, we obtain

$$\mathbf{x}^* = \begin{bmatrix} 3 \\ -1 \end{bmatrix} \quad \text{and} \quad \mu_1^* = 8$$

Since $\mathbf{x}^*$ also satisfies the second constraint, $\mathbf{x}^* = [3 \; -1]^T$ and $\boldsymbol{\mu}^* = [8 \; 0]^T$ satisfy the KKT conditions.

If only the second constraint is active, then $\mu_1^* = 0$ and the KKT conditions become

$$2x_1 - 14 + \mu_2 = 0$$
$$2x_2 - 6 + 2\mu_2 = 0$$
$$-3 + x_1 + 2x_2 = 0$$

The solution of this system of equations is given by

$$\mathbf{x}^* = \begin{bmatrix} 5 \\ -1 \end{bmatrix} \quad \text{and} \quad \mu_2^* = 4$$

As $\mathbf{x}^*$ violates the first constraint, the above $\mathbf{x}^*$ and $\boldsymbol{\mu}^*$ do not satisfy the KKT conditions.

Case 3 Both constraints active

If both constraints are active, we have

$$2x_1 - 14 + \mu_1 + \mu_2 = 0$$
$$2x_2 - 6 + \mu_1 + 2\mu_2 = 0$$
$$-2 + x_1 + x_2 = 0$$
$$-3 + x_1 + 2x_2 = 0$$

The solution to this system of equations is given by

$$\mathbf{x}^* = \begin{bmatrix} 1 \\ 1 \end{bmatrix} \quad \text{and} \quad \boldsymbol{\mu}^* = \begin{bmatrix} 20 \\ -8 \end{bmatrix}$$

Since $\mu_2^* < 0$, this is not a solution of the optimization problem.

Therefore, the only candidate for a minimizer of the problem is

$$\mathbf{x}^* = \begin{bmatrix} 3 \\ -1 \end{bmatrix}, \quad \boldsymbol{\mu}^* = \begin{bmatrix} 8 \\ 0 \end{bmatrix}$$

As can be observed in Fig. 10.10, the above point is actually the global minimizer. ∎

10.7 Second-Order Conditions

As in the unconstrained case, there are second-order conditions

(a) that must be satisfied for a point to be a local minimizer (i.e., necessary conditions), and
(b) that will assure that a point is a local minimizer (i.e., sufficient conditions).

The conditions in the constrained case are more complicated than their unconstrained counterparts due to the involvement of the various constraints, as may be expected.

Fig. 10.10 Contours of
$f(\mathbf{x})$ and the two constraints
for Example 10.12

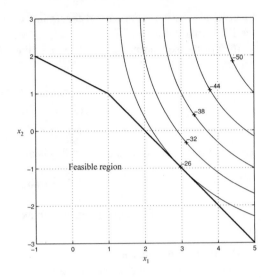

10.7.1 Second-Order Necessary Conditions

Suppose $\mathbf{x}^*$ is a local minimizer for the equality-constrained problem in Eqs. (10.57a)
and (10.57b) and is a regular point of the constraints in Eq. (10.57b). A second-order
condition can be derived by examining the behavior of $f(\mathbf{x})$ in a neighborhood of
$\mathbf{x}^*$. If $\mathbf{s}$ is a feasible direction at $\mathbf{x}^*$, then $a_i(\mathbf{x}^* + \mathbf{s}) = 0$ for $1 \leq i \leq p$, which in
conjunction with Eq. (10.66) implies that

$$f(\mathbf{x}^* + \mathbf{s}) = L(\mathbf{x}^* + \mathbf{s}, \ \boldsymbol{\lambda}^*) \tag{10.88}$$

where $\boldsymbol{\lambda}^*$ satisfies Eq. (10.74). By using the Taylor expansion of $L(\mathbf{x}^* + \mathbf{s}, \ \boldsymbol{\lambda}^*)$ at
$\{\mathbf{x}^*, \ \boldsymbol{\lambda}^*\}$ and Theorem 10.1, we obtain

$$f(\mathbf{x}^* + \mathbf{s}) = L(\mathbf{x}^*, \ \boldsymbol{\lambda}^*) + \mathbf{s}^T \nabla_x L(\mathbf{x}^*, \ \boldsymbol{\lambda}^*) + \tfrac{1}{2}\mathbf{s}^T \nabla_x^2 L(\mathbf{x}^*, \ \boldsymbol{\lambda}^*)\mathbf{s} + o(||\mathbf{s}||^2)$$
$$= f(\mathbf{x}^*) + \tfrac{1}{2}\mathbf{s}^T \nabla_x^2 L(\mathbf{x}^*, \ \boldsymbol{\lambda}^*)\mathbf{s} + o(||\mathbf{s}||^2) \tag{10.89}$$

Using an argument similar to that used in the proof of Theorem 2.2, it can be shown
that by virtue of $\mathbf{x}^*$ being a local minimizer, we have

$$\mathbf{s}^T \nabla_x^2 L(\mathbf{x}^*, \ \boldsymbol{\lambda}^*)\mathbf{s} \geq 0 \tag{10.90}$$

From Eqs. (10.59) and (10.64), it is clear that $\mathbf{s}$ is feasible at $\mathbf{x}^*$ if

$$\mathbf{J}_e(\mathbf{x}^*)\mathbf{s} = \mathbf{0}$$

i.e., $\mathbf{s} \in \mathcal{N}[\mathbf{J}_e(\mathbf{x}^*)]$, which is the null space of $\mathbf{J}_e(\mathbf{x}^*)$. Since this null space can be
characterized by a basis of the space, Eq. (10.90) is equivalent to the positive semidef-
initeness of $\mathbf{N}^T(\mathbf{x}^*)\nabla_x^2 L(\mathbf{x}^*, \boldsymbol{\lambda}^*)\mathbf{N}(\mathbf{x}^*)$ where $\mathbf{N}(\mathbf{x}^*)$ is a matrix whose columns form
a basis of $\mathcal{N}[\mathbf{J}_e(\mathbf{x}^*)]$. These results can be summarized in terms of Theorem 10.3.

Theorem 10.3 (Second-order necessary conditions for a minimum, equality constraints)   *If* $\mathbf{x}^*$ *is a constrained local minimizer of the problem in Eqs. (10.57a) and (10.57b) and is a regular point of the constraints in Eq. (10.57b), then*

(a)
$$a_i(\mathbf{x}^*) = 0 \quad for \ i = 1, \ 2, \ \ldots, \ p,$$

(b) there exist λ_i^* *for* $i = 1, \ 2, \ \ldots, \ p$ *such that*

$$\nabla f(\mathbf{x}^*) + \sum_{i=1}^{p} \lambda_i^* \nabla a_i(\mathbf{x}^*) = \mathbf{0},$$

(c) and
$$\mathbf{N}^T(\mathbf{x}^*)\nabla_x^2 L(\mathbf{x}^*, \ \boldsymbol{\lambda}^*)\mathbf{N}(\mathbf{x}^*) \succeq 0. \quad \blacksquare \tag{10.91}$$

Example 10.13 In Example 10.11 it was found that

$$\mathbf{x}_2^* = \begin{bmatrix} 2 \\ 0 \\ 3 \end{bmatrix} \quad and \quad \boldsymbol{\lambda}_2^* = \begin{bmatrix} -\frac{3}{2} \\ -\frac{11}{4} \end{bmatrix}$$

satisfy the first-order necessary conditions for a minimum for the problem of Example 10.10. Check whether the second-order necessary conditions for a minimum are satisfied.

Solution We can write

$$\nabla_x^2 L(\mathbf{x}_2^*, \ \boldsymbol{\lambda}_2^*) = \begin{bmatrix} -\frac{7}{2} & 0 & 0 \\ 0 & -\frac{7}{2} & 0 \\ 0 & 0 & \frac{1}{2} \end{bmatrix}$$

and

$$\mathbf{J}_e(\mathbf{x}_2^*) = \begin{bmatrix} -1 & 0 & 1 \\ 2 & 0 & 0 \end{bmatrix}$$

It can be readily verified that the null space of $\mathbf{J}_e(\mathbf{x}_2^*)$ is the one-dimensional space spanned by $\mathbf{N}(x_2^*) = [0 \ 1 \ 0]^T$. Since

$$\mathbf{N}^T(\mathbf{x}_2^*)\nabla_x^2 L(\mathbf{x}_2^*, \ \boldsymbol{\lambda}_2^*)\mathbf{N}(\mathbf{x}_2^*) = -\frac{7}{2} < 0 \tag{10.92}$$

we conclude that $\{\mathbf{x}_2^*, \ \boldsymbol{\lambda}_2^*\}$ does not satisfy the second-order necessary conditions. $\blacksquare$

For the general constrained optimization problem in Eqs. (10.1a)–(10.1c) a second-order condition similar to Eq. (10.91) can be derived as follows. Let $\mathbf{x}^*$ be a local minimizer of the problem in Eqs. (10.1a)–(10.1c) and $\mathcal{J}(\mathbf{x}^*)$ be the index set for the inequality constraints that are active at $\mathbf{x}^*$ (see Eq. (10.76)). A direction $\mathbf{s}$ is said to be feasible at $\mathbf{x}^*$ if

$$a_i(\mathbf{x}^* + \mathbf{s}) = 0 \quad \text{for } 1 \le i \le p \tag{10.93a}$$
$$c_j(\mathbf{x}^* + \mathbf{s}) = 0 \quad \text{for } j \in \mathcal{J}(\mathbf{x}^*) \tag{10.93b}$$

Recall that the Lagrangian for the problem in Eqs. (10.1a)–(10.1c) is defined by

$$L(\mathbf{x}, \boldsymbol{\lambda}, \boldsymbol{\mu}) = f(\mathbf{x}) + \sum_{i=1}^{p} \lambda_i a_i(\mathbf{x}) + \sum_{j=1}^{q} \mu_j c_j(\mathbf{x}) \tag{10.94}$$

If $\boldsymbol{\lambda}^*$ and $\boldsymbol{\mu}^*$ are the Lagrange multipliers described in Theorem 10.2, then the constraints in Eqs. (10.1b) and (10.1c) and the complementarity condition in Eqs. (10.82a) and (10.82b) imply that

$$f(\mathbf{x}^*) = L(\mathbf{x}^*, \boldsymbol{\lambda}^*, \boldsymbol{\mu}^*) \tag{10.95}$$

From Eqs. (10.81), (10.93a), (10.93b), and (10.95), we have

$$\begin{aligned}
f(\mathbf{x}^* + \mathbf{s}) &= L(\mathbf{x}^* + \mathbf{s}, \boldsymbol{\lambda}^*, \boldsymbol{\mu}^*) \\
&= L(\mathbf{x}^*, \boldsymbol{\lambda}^*, \boldsymbol{\mu}^*) + \mathbf{s}^T \nabla_x L(\mathbf{x}^*, \boldsymbol{\lambda}^*, \boldsymbol{\mu}^*) \\
&\quad + \tfrac{1}{2} \mathbf{s}^T \nabla_x^2 L(\mathbf{x}^*, \boldsymbol{\lambda}^*, \boldsymbol{\mu}^*) \mathbf{s} + o(||\mathbf{s}||^2) \\
&= f(\mathbf{x}^*) + \tfrac{1}{2} \mathbf{s}^T \nabla_x^2 L(\mathbf{x}^*, \boldsymbol{\lambda}^*, \boldsymbol{\mu}^*) \mathbf{s} + o(||\mathbf{s}||^2)
\end{aligned}$$

This in conjunction with the fact that $f(\mathbf{x}^*) \leq f(\mathbf{x}^* + \mathbf{s})$ implies that

$$\mathbf{s}^T \nabla_x^2 L(\mathbf{x}^*, \boldsymbol{\lambda}^*, \boldsymbol{\mu}^*) \mathbf{s} \geq 0 \tag{10.96}$$

for any $\mathbf{s}$ feasible at $\mathbf{x}^*$.

From Eqs. (10.93a) and (10.93b), the feasible directions at $\mathbf{x}^*$ are those directions that are orthogonal to the gradients of the constraints that are active at $\mathbf{x}^*$, namely,

$$\mathbf{J}(\mathbf{x}^*)\mathbf{s} = \begin{bmatrix} \mathbf{J}_e(\mathbf{x}^*) \\ \hat{\mathbf{J}}_{ie}(\mathbf{x}^*) \end{bmatrix} \mathbf{s} = \mathbf{0} \tag{10.97}$$

where $\hat{\mathbf{J}}_{ie}(\mathbf{x})$ is given by Eq. (10.87b). Hence the feasible directions at $\mathbf{x}^*$ are characterized by the null space of $\mathbf{J}(\mathbf{x}^*)$, denoted as $\mathcal{N}[\mathbf{J}(\mathbf{x}^*)]$, and the condition in Eq. (10.96) assures the positive semidefiniteness of $\mathbf{N}^T(\mathbf{x}^*)\nabla_x^2 L(\mathbf{x}^*, \boldsymbol{\lambda}^*, \boldsymbol{\mu}^*)\mathbf{N}(\mathbf{x}^*)$ where $\mathbf{N}(\mathbf{x}^*)$ is a matrix whose columns form a basis of $\mathcal{N}[\mathbf{J}(\mathbf{x}^*)]$. A set of necessary conditions for the general constrained optimization problem in Eqs. (10.1a)–(10.1c) can now be summarized in terms of the following theorem.

Theorem 10.4 (Second-order necessary conditions for a minimum, general constrained problem) *If $\mathbf{x}^*$ is a constrained local minimizer of the problem in Eqs. (10.1a)–(10.1c) and is a regular point of the constraints in Eqs. (10.1b) and (10.1c), then*

(a)

$$a_i(\mathbf{x}^*) = 0 \quad for \ 1 \leq i \leq p,$$

(b)

$$c_j(\mathbf{x}^*) \leq 0 \quad for \ 1 \leq j \leq q,$$

(c) *there exist Lagrange multipliers λ_i^*'s and μ_j^*'s such that*

$$\nabla f(\mathbf{x}^*) + \sum_{i=1}^{p} \lambda_i^* \nabla a_i(\mathbf{x}^*) + \sum_{j=1}^{q} \mu_j^* \nabla c_j(\mathbf{x}^*) = \mathbf{0}$$

(d)

$$\lambda_i^* a_i(\mathbf{x}^*) = 0 \quad for \;\; 1 \le i \le p \; and \; \mu_j^* c_j(\mathbf{x}^*) = 0 \quad for \;\; 1 \le j \le q,$$

(e)

$$\mu_j^* \ge 0 \quad for \;\; 1 \le j \le q,$$

(f) *and*

$$\mathbf{N}^T(\mathbf{x}^*)\nabla_x^2 L(\mathbf{x}^*, \; \boldsymbol{\lambda}^*, \; \boldsymbol{\mu}^*)\mathbf{N}(\mathbf{x}^*) \succeq \mathbf{0}. \quad \blacksquare \qquad (10.98)$$

10.7.2 Second-Order Sufficient Conditions

For the constrained problem in Eqs. (10.57a) and (10.57b), second-order sufficient conditions for a point $\mathbf{x}^*$ to be a local minimizer can be readily obtained from Eq. (10.89), where $\{\mathbf{x}^*, \boldsymbol{\lambda}^*\}$ is assumed to satisfy the first-order necessary conditions described in Theorem 10.1. Using an argument similar to that used in the proof of Theorem 2.4, we can show that a point $\mathbf{x}^*$ that satisfies the conditions in Theorem 10.1 is a local minimizer if the matrix

$$\mathbf{N}^T(\mathbf{x}^*)\nabla_x^2 L(\mathbf{x}^*, \; \boldsymbol{\lambda}^*)\mathbf{N}(\mathbf{x}^*)$$

is positive definite.

Theorem 10.5 (Second-order sufficient conditions for a minimum, equality con-straints) *If $\mathbf{x}^*$ is a regular point of the constraints in Eq. (10.57b), then it is a strong local minimizer of Eqs. (10.57a) and (10.57b) if*

(a)

$$a_i(\mathbf{x}^*) = 0 \quad for \;\; 1 \le i \le p,$$

(b) *there exist Lagrange multipliers λ_i^* for $i = 1, \; 2, \; \ldots, \; p$ such that*

$$\nabla f(\mathbf{x}^*) + \sum_{i=1}^{p} \lambda_i^* \nabla a_i(\mathbf{x}^*) = \mathbf{0},$$

(c) *and*

$$\mathbf{N}^T(\mathbf{x}^*)\nabla_x^2 L(\mathbf{x}^*, \; \boldsymbol{\lambda}^*)\mathbf{N}(\mathbf{x}) \succ \mathbf{0}, \qquad (10.99)$$

i.e., $\nabla_x^2 L(\mathbf{x}^*, \; \boldsymbol{\lambda}^*)$ is positive definite in the null space $\mathcal{N}[\mathbf{J}(\mathbf{x}^*)]$. $\blacksquare$

Example 10.14 Check whether the second-order sufficient conditions for a mini-mum are satisfied in the minimization problem of Example 10.10.

Solution We compute

$$\nabla_x^2 L(\mathbf{x}_1^*, \boldsymbol{\lambda}_1^*) = \begin{bmatrix} \frac{5}{2} & 0 & 0 \\ 0 & \frac{5}{2} & 0 \\ 0 & 0 & \frac{1}{2} \end{bmatrix}$$

which is positive definite in the entire E^3. Hence Theorem 10.5 implies that $\mathbf{x}_1^*$ is a strong local minimizer.   ■

A set of sufficient conditions for point $\mathbf{x}^*$ to be a local minimizer for the general constrained problem in Eqs. (10.1a)–(10.1c) is given by the following theorem.

Theorem 10.6 (Second-order sufficient conditions for a minimum, general constrained problem) *A point $\mathbf{x}^* \in R^n$ is a strong local minimizer of the problem in Eqs. (10.1a)–(10.1c) if*

(a)

$$a_i(\mathbf{x}^*) = 0 \ \ for \ 1 \le i \le p,$$

(b)

$$c_j(\mathbf{x}^*) \le 0 \ \ for \ 1 \le j \le q,$$

(c) $\mathbf{x}^*$ is a regular point of the constraints that are active at $\mathbf{x}^*$,
(d) there exist λ_i^*'s and μ_j^*'s such that

$$\nabla f(\mathbf{x}^*) + \sum_{i=1}^{p} \lambda_i^* \nabla a_i(\mathbf{x}^*) + \sum_{j=1}^{q} \mu_j^* \nabla c_j(\mathbf{x}^*) = \mathbf{0},$$

(e)

$$\lambda_i^* a_i(\mathbf{x}^*) = 0 \ \ for \ 1 \le i \le p \ and \ \mu_j^* c_j(\mathbf{x}^*) = 0 \ \ for \ 1 \le j \le q,$$

(f)

$$\mu_j^* \ge 0 \ \ for \ 1 \le j \le q,$$

(g) and

$$\mathbf{N}^T(\mathbf{x}^*) \nabla_x^2 L(\mathbf{x}^*, \boldsymbol{\lambda}^*, \boldsymbol{\mu}^*) \mathbf{N}(\mathbf{x}^*) \succ \mathbf{0} \tag{10.100}$$

where $\mathbf{N}(\mathbf{x}^*)$ is a matrix whose columns form a basis of the null space of $\tilde{\mathbf{J}}(\mathbf{x}^*)$ defined by

$$\tilde{\mathbf{J}}(\mathbf{x}^*) = \begin{bmatrix} \mathbf{J}_e(\mathbf{x}^*) \\ \tilde{\mathbf{J}}_{ie}(\mathbf{x}^*) \end{bmatrix} \tag{10.101}$$

The Jacobian $\tilde{\mathbf{J}}_{ie}(\mathbf{x}^*)$ is the matrix whose rows are composed of those gradients of inequality constraints that are active at $\mathbf{x}^*$, i.e., $\nabla^T c_j(\mathbf{x}^*)$, with $c_j(\mathbf{x}^*) = 0$ and $\mu_j^* > 0$.

Proof Let us suppose that $\mathbf{x}^*$ satisfies conditions (a) to (g) but is not a strong local minimizer. Under these circumstances there would exist a sequence of feasible points $\mathbf{x}_k \to \mathbf{x}^*$ such that $f(\mathbf{x}_k) \leq f(\mathbf{x}^*)$. If we write $\mathbf{x}_k = \mathbf{x}^* + \delta_k \mathbf{s}_k$ with $||\mathbf{s}_k|| = 1$ for all k, then we may assume that $\delta_k > 0$, and $\mathbf{s}_k \to \mathbf{s}^*$ for some vector $\mathbf{s}^*$ with $||\mathbf{s}^*|| = 1$ and

$$f(\mathbf{x}^* + \delta_k \mathbf{s}_k) - f(\mathbf{x}^*) \leq 0$$

which leads to

$$\nabla^T f(\mathbf{x}^*)\mathbf{s}^* \leq 0 \tag{10.102}$$

Since $\mathbf{s}^*$ is feasible at $\mathbf{x}^*$, we have

$$\nabla^T a_i(\mathbf{x}^*)\mathbf{s}^* = 0 \tag{10.103}$$

If $\mathcal{J}_p(\mathbf{x}^*)$ is the index set for inequality constraints that are active at $\mathbf{x}^*$ and are associated with strictly positive Lagrange multipliers, then

$$c_j(\mathbf{x}_k) - c_j(\mathbf{x}^*) = c_j(\mathbf{x}_k) \leq 0 \qquad \text{for } j \in \mathcal{J}_p(\mathbf{x}^*)$$

i.e.,

$$c_j(\mathbf{x}^* + \delta_k \mathbf{s}_k) - c_j(\mathbf{x}^*) \leq 0$$

which leads to

$$\nabla^T c_j(\mathbf{x}^*)\mathbf{s}^* \leq 0 \qquad \text{for } j \in \mathcal{J}_p(\mathbf{x}^*) \tag{10.104}$$

Now the inequality in Eq. (10.104) cannot occur since conditions (d), (e), (f) in conjunction with Eqs. (10.102) and (10.103) would imply that

$$0 \geq \nabla^T f(\mathbf{x}^*)\mathbf{s}^* = -\sum_{i=1}^{p} \lambda_i^* \nabla^T a_i(\mathbf{x}^*)\mathbf{s}^* - \sum_{j=1}^{q} \mu_j^* \nabla^T c_j(\mathbf{x}^*)\mathbf{s}^* > 0$$

i.e., $0 > 0$ which is a contradiction. Hence

$$\nabla^T c_j(\mathbf{x}^*)\mathbf{s}^* = 0 \qquad \text{for } j \in \mathcal{J}_p(\mathbf{x}^*) \tag{10.105}$$

From Eqs. (10.103) and (10.104), it follows that $\mathbf{s}^*$ belongs to the null space of $\tilde{\mathbf{J}}(\mathbf{x}^*)$ and so condition (g) implies that $\mathbf{s}^{*T} \nabla_x^2 L(\mathbf{x}^*, \boldsymbol{\lambda}^*, \boldsymbol{\mu}^*)\mathbf{s}^* > 0$. Since $\mathbf{x}_k \to \mathbf{x}^*$, we have $\mathbf{s}_k^T \nabla_x^2 L(\mathbf{x}^*, \boldsymbol{\lambda}^*, \boldsymbol{\mu}^*)\mathbf{s}_k > 0$ for a sufficiently large k. Using the condition in (d), the Taylor expansion of $L(\mathbf{x}_k, \boldsymbol{\lambda}^*, \boldsymbol{\mu}^*)$ at $\mathbf{x}^*$ gives

$$\begin{aligned} L(\mathbf{x}_k^*, \boldsymbol{\lambda}^*, \boldsymbol{\mu}^*) &= L(\mathbf{x}^*, \boldsymbol{\lambda}^*, \boldsymbol{\mu}^*) + \delta_k \mathbf{s}_k^T \nabla_x L(\mathbf{x}^*, \boldsymbol{\lambda}^*, \boldsymbol{\mu}^*) \\ &\quad + \tfrac{1}{2}\delta_k^2 \mathbf{s}_k^T \nabla_x^2 L(\mathbf{x}^*, \boldsymbol{\lambda}^*, \boldsymbol{\mu}^*)\mathbf{s}_k + o(\delta_k^2) \\ &= f(\mathbf{x}^*) + \tfrac{1}{2}\delta_k^2 \mathbf{s}_k^T \nabla_x^2 L(\mathbf{x}^*, \boldsymbol{\lambda}^*, \boldsymbol{\mu}^*)\mathbf{s}_k + o(\delta_k^2) \end{aligned}$$

This in conjunction with the inequalities $f(\mathbf{x}_k) \geq L(\mathbf{x}_k, \boldsymbol{\lambda}^*, \boldsymbol{\mu}^*)$ and $f(\mathbf{x}_k) \leq f(\mathbf{x}^*)$ leads to

$$0 \geq f(\mathbf{x}_k) - f(\mathbf{x}^*) \geq \tfrac{1}{2}\delta_k^2 \mathbf{s}_k^T \nabla_x^2 L(\mathbf{x}^*, \boldsymbol{\lambda}^*, \boldsymbol{\mu}^*)\mathbf{s}_k + o(\delta_k^2) \tag{10.106}$$

So, for a sufficiently large k the right-hand side of Eq. (10.106) becomes strictly positive, which leads to the contradiction $0 > 0$. This completes the proof.   ∎

Example 10.15 Use Theorem 10.6 to check the solution of the minimization problem discussed in Example 10.12.

Solution The candidate for a local minimizer was found to be

$$\mathbf{x}^* = \begin{bmatrix} 3 \\ -1 \end{bmatrix}, \quad \boldsymbol{\mu}^* = \begin{bmatrix} 8 \\ 0 \end{bmatrix}$$

Since the constraints are linear,

$$\nabla_x^2 L(\mathbf{x}^*, \boldsymbol{\lambda}^*, \boldsymbol{\mu}^*) = \nabla^2 f(\mathbf{x}^*) = \begin{bmatrix} 2 & 0 \\ 0 & 2 \end{bmatrix}$$

which is positive definite in the entire E^2. Therefore, $\{\mathbf{x}^*, \boldsymbol{\mu}^*\}$ satisfies all the conditions of Theorem 10.6 and hence $\mathbf{x}^*$ is a strong local minimizer. As was observed in Fig. 10.10, $\mathbf{x}^*$ is actually the global minimizer of the problem.   ∎

10.8 Convexity

Convex functions and their basic properties were studied in Sect. 2.7 and the unconstrained optimization of convex functions was discussed in Sect. 2.8. The concept of convexity is also important in constrained optimization. In unconstrained optimization, the properties of convex functions are of interest when these functions are defined over a convex set. In a constrained optimization, the objective function is minimized with respect to the feasible region, which is characterized by the constraints imposed. As may be expected, the concept of convexity can be fully used to achieve useful optimization results when both the objective function and the feasible region are convex. In Sect. 10.3, these problems were referred to as CP problems. A typical problem of this class can be formulated as

$$\begin{aligned} &\text{minimize } f(\mathbf{x}) \\ \text{subject to: } &a_i(\mathbf{x}) = \mathbf{a}_i^T \mathbf{x} - b_i = 0 \quad \text{for } 1 \le i \le p \\ &c_j(\mathbf{x}) \le 0 \quad \text{for } 1 \le j \le q \end{aligned} \tag{10.107}$$

where $f(\mathbf{x})$ and $c_j(\mathbf{x})$ for $1 \le j \le q$ are convex functions. The main results, which are analogous to those in Sect. 2.8, are described by the next two theorems.

Theorem 10.7 (Globalness and convexity of minimizers in CP problems)

(a) If $\mathbf{x}^*$ is a local minimizer of a CP problem, then $\mathbf{x}^*$ is also a global minimizer.
(b) The set of minimizers of a CP problem, denoted as S, is convex.
(c) If the objective function $f(\mathbf{x})$ is strictly convex in the feasible region $\mathcal{R}$, then the global minimizer is unique.

Proof (a) If $\mathbf{x}^*$ is a local minimizer that is not a global minimizer, then there is a feasible $\hat{\mathbf{x}}$ such that $f(\hat{\mathbf{x}}) < f(\mathbf{x}^*)$. If we let $\mathbf{x}_\tau = \tau\hat{\mathbf{x}} + (1 - \tau)\mathbf{x}^*$ for $0 < \tau < 1$, then the convexity of $f(\mathbf{x})$ implies that

$$f(\mathbf{x}_\tau) \le \tau f(\hat{\mathbf{x}}) + (1 - \tau)f(\mathbf{x}^*) < f(\mathbf{x}^*)$$

no matter how close $\mathbf{x}_\tau$ is to $\mathbf{x}^*$. This contradicts the assumption that $\mathbf{x}^*$ is a local minimizer since $f(\mathbf{x}^*)$ is supposed to assume the smallest value in a sufficiently small neighborhood of $\mathbf{x}^*$. Hence $\mathbf{x}^*$ is a global minimizer.

(b) Let $\mathbf{x}_a, \mathbf{x}_b \in S$. From part (a), it follows that $\mathbf{x}_a$ and $\mathbf{x}_b$ are global minimizers. If $\mathbf{x}_\tau = \tau\mathbf{x}_a + (1 - \tau)\mathbf{x}_b$ for $0 \le \tau \le 1$, then the convexity of $f(\mathbf{x})$ leads to

$$f(\mathbf{x}_\tau) \le \tau f(\mathbf{x}_a) + (1 - \tau)f(\mathbf{x}_b) = f(\mathbf{x}_a)$$

Since $\mathbf{x}_a$ is a global minimizer, $f(\mathbf{x}_\tau) \ge f(\mathbf{x}_a)$. Hence $f(\mathbf{x}_\tau) = f(\mathbf{x}_a)$, i.e., $\mathbf{x}_\tau \in S$ for each τ, thus S is convex.

(c) Suppose that the solution set S contains two distinct points $\mathbf{x}_a$ and $\mathbf{x}_b$ and $\mathbf{x}_\tau$ is defined as in part (b) with $0 < \tau < 1$. Since $\mathbf{x}_a \ne \mathbf{x}_b$ and $\tau \in (0, 1)$, we have $\mathbf{x}_\tau \ne \mathbf{x}_a$. By using the strict convexity of $f(\mathbf{x})$, we would conclude that $f(\mathbf{x}_\tau) < f(\mathbf{x}_a)$ which contradicts the assumption that $\mathbf{x}_a \in S$. Therefore, the global minimizer is unique. ■

It turns out that in a CP problem, the KKT conditions become sufficient for $\mathbf{x}^*$ to be a global minimizer as stated in the following theorem.

Theorem 10.8 (Sufficiency of KKT conditions in CP problems) *If $\mathbf{x}^*$ is a regular point of the constraints in Eq. (10.107) and satisfies the KKT conditions stated in Theorem 10.2 where $f(\mathbf{x})$ and $c_j(\mathbf{x})$ for $1 \le j \le q$ are convex, then $\mathbf{x}^*$ is a global minimizer.*

Proof For a feasible point $\hat{\mathbf{x}}$ with $\hat{\mathbf{x}} \ne \mathbf{x}^*$, we have $a_i(\hat{\mathbf{x}}) = 0$ for $1 \le i \le p$ and $c_j(\hat{\mathbf{x}}) \le 0$ for $1 \le j \le q$. In terms of the notation used in Theorem 10.2, we can write

$$f(\hat{\mathbf{x}}) \ge f(\hat{\mathbf{x}}) + \sum_{j=1}^{q} \mu_j^* c_j(\hat{\mathbf{x}})$$

Since $f(\mathbf{x})$ and $c_j(\mathbf{x})$ are convex, then from Theorem 2.12, we have

$$f(\hat{\mathbf{x}}) \ge f(\mathbf{x}^*) + \nabla^T f(\mathbf{x}^*)(\hat{\mathbf{x}} - \mathbf{x}^*)$$

and

$$c_j(\hat{\mathbf{x}}) \ge c_j(\mathbf{x}^*) + \nabla^T c_j(\mathbf{x}^*)(\hat{\mathbf{x}} - \mathbf{x}^*)$$

It follows that

$$f(\hat{\mathbf{x}}) \ge f(\mathbf{x}^*) + \nabla^T f(\mathbf{x}^*)(\hat{\mathbf{x}} - \mathbf{x}^*) + \sum_{j=1}^{q} \mu_j^* \nabla^T c_j(\mathbf{x}^*)(\hat{\mathbf{x}} - \mathbf{x}^*) + \sum_{j=1}^{q} \mu_j^* c_j(\mathbf{x}^*)$$

In light of the complementarity conditions in Eq. (10.82b), the last term in the above inequality is zero and hence we have

$$f(\hat{\mathbf{x}}) \geq f(\mathbf{x}^*) + [\nabla f(\mathbf{x}^*) + \sum_{j=1}^{q} \mu_j^* \nabla c_j(\mathbf{x}^*)]^T (\hat{\mathbf{x}} - \mathbf{x}^*) \qquad (10.108)$$

Since $a_i(\hat{\mathbf{x}}) = a_i(\mathbf{x}^*) = 0$, we get

$$0 = a_i(\hat{\mathbf{x}}) - a_i(\mathbf{x}^*) = \mathbf{a}_i^T(\hat{\mathbf{x}} - \mathbf{x}^*) = \nabla^T a_i(\mathbf{x}^*)(\hat{\mathbf{x}} - \mathbf{x}^*)$$

Multiplying the above equality by $-\lambda_i^*$ and then adding it to the inequality in Eq. (10.108) for $1 \leq i \leq p$, we obtain

$$f(\hat{\mathbf{x}}) \geq f(\mathbf{x}^*) + [\nabla f(\mathbf{x}^*) + \sum_{i=1}^{p} \lambda_i^* \nabla a_i(\mathbf{x}^*) + \sum_{j=1}^{q} \mu_j^* \nabla c_j(\mathbf{x}^*)]^T (\hat{\mathbf{x}} - \mathbf{x}^*)$$

From Eq. (10.81), the last term in the above inequality is zero, which leads to $f(\hat{\mathbf{x}}) \geq f(\mathbf{x}^*)$. This shows that $\mathbf{x}^*$ is a global minimizer. ∎

10.9 Duality

Duality as applied to optimization is essentially a similarity property between two functions which enables a problem transformation that leads to an indirect but sometimes more efficient solution method. In a duality-based method, the original problem, which is referred to as the *primal* problem, is transformed into a corresponding problem whose parameters are the Lagrange multipliers of the primal. The transformed problem is called the *dual* problem. In the case where the number of inequality constraints is much greater than the dimension of $\mathbf{x}$, solving the dual problem for the Lagrange multipliers and then finding $\mathbf{x}^*$ for the primal problem is an attractive alternative. Below we provide a brief review of this approach. An in-depth exposition of the subject can be found in [15].

We begin by introducing a function called the *Lagrange dual function*. Consider the general convex programming (CP) problem in Eq. (10.107) whose Lagrangian is given by

$$L(\mathbf{x}, \boldsymbol{\lambda}, \boldsymbol{\mu}) = f(\mathbf{x}) + \sum_{i=1}^{p} \lambda_i(\mathbf{a}_i^T - b_i) + \sum_{j=1}^{q} \mu_j c_j(\mathbf{x}) \qquad (10.109)$$

Definition 10.5 (*Lagrange dual function*) The Lagrange dual function of the problem in Eq. (10.107) is defined as

$$q(\boldsymbol{\lambda}, \boldsymbol{\mu}) = \inf_{\mathbf{x}} L(\mathbf{x}, \boldsymbol{\lambda}, \boldsymbol{\mu}) \qquad (10.110)$$

for $\boldsymbol{\lambda} \in R^p$ and $\boldsymbol{\mu} \in R^q$ with $\boldsymbol{\mu} \geq \mathbf{0}$ where $\inf_{\mathbf{x}} L(\mathbf{x}, \boldsymbol{\lambda}, \boldsymbol{\mu})$ is the *infimum* of the objective function, which is defined as the function's *largest lower bound*. Note that the Lagrangian in Eq. (10.109) is *convex* with respect to $\mathbf{x}$. If $L(\mathbf{x}, \boldsymbol{\lambda}, \boldsymbol{\mu})$ is unbounded from below for some $\{\boldsymbol{\lambda}, \boldsymbol{\mu}\}$, then the value of $q(\boldsymbol{\lambda}, \boldsymbol{\mu})$ is assumed to be $-\infty$.   ∎

The dual function $q(\boldsymbol{\lambda}, \boldsymbol{\mu})$ in Eq. (10.110) is concave with respect to $\{\boldsymbol{\lambda}, \boldsymbol{\mu}\}$. This property follows from the fact that for $\boldsymbol{\lambda}_1, \boldsymbol{\lambda}_2 \in R^p$ and $\boldsymbol{\mu}_1, \boldsymbol{\mu}_2 \in R^q$ with $\boldsymbol{\mu}_1, \boldsymbol{\mu}_2 \geq \mathbf{0}$, we have

$$
q(t\boldsymbol{\lambda}_1 + (1-t)\boldsymbol{\lambda}_2, t\boldsymbol{\mu}_1 + (1-t)\boldsymbol{\mu}_2)
$$

$$
= \inf_{\mathbf{x}} L(\mathbf{x}, t\boldsymbol{\lambda}_1 + (1-t)\boldsymbol{\lambda}_2, t\boldsymbol{\mu}_1 + (1-t)\boldsymbol{\mu}_2)
$$

$$
= \inf_{\mathbf{x}} \left[(t+1-t)f(\mathbf{x}) + \sum_{i=1}^{p}(t\lambda_{1,i} + (1-t)\lambda_{2,i})(a_i^T\mathbf{x} - b_i) \right.
$$

$$
\left. + \sum_{j=1}^{q}(t\mu_{1,j} + (1-t)\mu_{2,j})c_j(\mathbf{x}) \right]
$$

$$
\geq t \inf_{\mathbf{x}} \left[f(\mathbf{x}) + \sum_{i=1}^{p}\lambda_{1,i}(a_i^T\mathbf{x} - b_i) + \sum_{j=1}^{q}\mu_{1,j}c_j(\mathbf{x}) \right]
$$

$$
+ (1-t)\inf_{\mathbf{x}} \left[f(\mathbf{x}) + \sum_{i=1}^{p}\lambda_{2,i}(a_i^T\mathbf{x} - b_i) + \sum_{j=1}^{q}\mu_{2,j}c_j(\mathbf{x}) \right]
$$

$$
= tq(\boldsymbol{\lambda}_1, \boldsymbol{\mu}_1) + (1-t)q(\boldsymbol{\lambda}_2, \boldsymbol{\mu}_2)
$$

where $t \in (0, 1)$.

Definition 10.6 (*Lagrange dual problem*) The Lagrange dual problem with respect to the CP problem in Eq. (10.107) is defined as

$$
\begin{array}{cc}
\text{maximize} & q(\boldsymbol{\lambda}, \boldsymbol{\mu}) \\
\boldsymbol{\lambda}, \boldsymbol{\mu} & \\
\text{subject to:} & \boldsymbol{\mu} \geq \mathbf{0}
\end{array}
\qquad (10.111)
$$
∎

For any feasible $\mathbf{x}$ for the problem in Eq. (10.107) and any feasible $\{\boldsymbol{\lambda}, \boldsymbol{\mu}\}$ for the problem in Eq. (10.111), we have

$$
f(\mathbf{x}) \geq q(\boldsymbol{\lambda}, \boldsymbol{\mu})
\qquad (10.112)
$$

This is because

$$
L(\mathbf{x}, \boldsymbol{\lambda}, \boldsymbol{\mu}) = f(\mathbf{x}) + \sum_{i=1}^{p}\lambda_i(a_i^T\mathbf{x} - b_i) + \sum_{j=1}^{q}\mu_j c_j(\mathbf{x})
$$

$$
= f(\mathbf{x}) + \sum_{j=1}^{q}\mu_j c_j(\mathbf{x}) \leq f(\mathbf{x})
$$

and thus

$$
q(\boldsymbol{\lambda}, \boldsymbol{\mu}) = \inf_{\mathbf{x}} L(\mathbf{x}, \boldsymbol{\lambda}, \boldsymbol{\mu}) \leq L(\mathbf{x}, \boldsymbol{\lambda}, \boldsymbol{\mu}) \leq f(\mathbf{x})
$$

The problems in Eqs. (10.107) and (10.111) are commonly referred to as the *primal* and *dual* problems, respectively.   The property in Eq. (10.112) leads naturally to the concept of *duality gap* between the primal and dual objective functions, which is given by

$$\delta(\mathbf{x},\ \boldsymbol{\lambda},\ \boldsymbol{\mu}) = f(\mathbf{x}) - q(\boldsymbol{\lambda},\ \boldsymbol{\mu}) \tag{10.113}$$

For a feasible $(\mathbf{x},\ \boldsymbol{\lambda},\ \boldsymbol{\mu})$, it follows from Eq. (10.112) that the duality gap is always nonnegative.

Now if $\mathbf{x}^*$ is a solution of the primal problem in Eq. (10.107), then from Eq. (10.112) we obtain

$$f(\mathbf{x}^*) \geq q(\boldsymbol{\lambda},\ \boldsymbol{\mu}) \tag{10.114}$$

In other words, the dual function at any feasible $\{\boldsymbol{\lambda},\ \boldsymbol{\mu}\}$ is a lower bound of the optimal value of the primal objective function, $f(\mathbf{x}^*)$. Furthermore, the tightest lower bound that can be achieved with the dual function $q(\boldsymbol{\lambda},\ \boldsymbol{\mu})$ can be found by maximizing the dual function $q(\boldsymbol{\lambda},\ \boldsymbol{\mu})$ subject to the constraint $\boldsymbol{\mu} \geq \mathbf{0}$. This maximization problem is the Lagrange dual problem as formulated in Eq. (10.111). If we now denote the solution of Eq. (10.111) as $(\boldsymbol{\lambda}^*,\ \boldsymbol{\mu}^*)$, then we have

$$f(\mathbf{x}^*) \geq q(\boldsymbol{\lambda}^*,\ \boldsymbol{\mu}^*) \tag{10.115}$$

Based on Eq. (10.115), the concept of *strong* and *weak* duality can now be introduced as follows.

Definition 10.7 (*Strong and weak duality*) Let $\mathbf{x}^*$ and $(\boldsymbol{\lambda}^*,\ \boldsymbol{\mu}^*)$ be the solutions of the primal problem in Eq. (10.107) and dual problem in Eq. (10.111), respectively. Strong duality holds if $f(\mathbf{x}^*) = q(\boldsymbol{\lambda}^*,\ \boldsymbol{\mu}^*)$, i.e., the optimal duality gap is zero; on the other hand, weak duality holds if $f(\mathbf{x}^*) > q(\boldsymbol{\lambda}^*,\ \boldsymbol{\mu}^*)$.   ■

It can be shown that if the interior of the feasible region of the primal problem is nonempty, then strong duality holds [15].

Example 10.16 Find the Lagrange dual of the LP problem

$$\begin{aligned} &\text{minimize}\ \ \mathbf{c}^T\mathbf{x} \\ &\text{subject to: } \mathbf{Ax} = \mathbf{b} \quad\ \ \mathbf{x} \geq \mathbf{0} \end{aligned} \tag{10.116}$$

Solution If we express $\mathbf{x} \geq \mathbf{0}$ as $-\mathbf{x} \leq \mathbf{0}$, then the Lagrangian of the LP problem can be expressed as

$$L(\mathbf{x},\ \boldsymbol{\lambda},\ \boldsymbol{\mu}) = \mathbf{c}^T\mathbf{x} + (\mathbf{Ax} - \mathbf{b})^T\boldsymbol{\lambda} - \mathbf{x}^T\boldsymbol{\mu}$$

and thus

$$q(\boldsymbol{\lambda},\ \boldsymbol{\mu}) = \inf_{\mathbf{x}}\{\mathbf{c}^T\mathbf{x} + (\mathbf{Ax} - \mathbf{b})^T\boldsymbol{\lambda} - \mathbf{x}^T\boldsymbol{\mu}\} = \inf_{\mathbf{x}}\{(\mathbf{c} + \mathbf{A}^T\boldsymbol{\lambda} - \boldsymbol{\mu})^T\mathbf{x} - \mathbf{b}^T\boldsymbol{\lambda}\} \tag{10.117}$$

For a given $\{\boldsymbol{\lambda}, \boldsymbol{\mu}\}$ such that $\mathbf{c} + \mathbf{A}^T\boldsymbol{\lambda} - \boldsymbol{\mu} \neq \mathbf{0}$, it follows from Eq. (10.117) that $q(\boldsymbol{\lambda}, \boldsymbol{\mu}) = -\infty$. To assure a well-defined dual function $q(\boldsymbol{\lambda}, \boldsymbol{\mu})$, we assume that $\mathbf{c} + \mathbf{A}^T\boldsymbol{\lambda} - \boldsymbol{\mu} = \mathbf{0}$. This leads to

$$q(\boldsymbol{\lambda}, \boldsymbol{\mu}) = \inf_{\mathbf{x}}(-\mathbf{b}^T\boldsymbol{\lambda}) = -\mathbf{b}^T\boldsymbol{\lambda}$$

Hence the Lagrange dual of the problem in Eq. (10.116) is given by

$$\text{maximize} \quad -\mathbf{b}^T\boldsymbol{\lambda}$$
$$\text{subject to: } \boldsymbol{\mu} \geq \mathbf{0}$$

Since $\mathbf{c} + \mathbf{A}^T\boldsymbol{\lambda} - \boldsymbol{\mu} = \mathbf{0}$ and $\boldsymbol{\mu} = \mathbf{c} + \mathbf{A}^T\boldsymbol{\lambda}$, the above problem becomes

$$\text{maximize} \quad -\mathbf{b}^T\boldsymbol{\lambda}$$
$$\text{subject to: } -\mathbf{c} - \mathbf{A}^T\boldsymbol{\lambda} \leq \mathbf{0}$$

which is equivalent to

$$\text{minimize} \quad \mathbf{b}^T\boldsymbol{\lambda}$$
$$\text{subject to: } (-\mathbf{A}^T)\boldsymbol{\lambda} \leq \mathbf{c} \quad \blacksquare$$

Example 10.17 Find the Lagrange dual of the QP problem

$$\text{minimize} \quad \tfrac{1}{2}\mathbf{x}^T\mathbf{H}\mathbf{x} + \mathbf{p}^T\mathbf{x} \tag{10.118}$$
$$\text{subject to: } \mathbf{A}\mathbf{x} \leq \mathbf{b}$$

where $\mathbf{H}$ is positive definite.

Solution The Lagrangian of the QP problem is given by

$$L(\mathbf{x}, \boldsymbol{\mu}) = \tfrac{1}{2}\mathbf{x}^T\mathbf{H}\mathbf{x} + \mathbf{p}^T\mathbf{x} + \boldsymbol{\mu}^T(\mathbf{A}\mathbf{x} - \mathbf{b})$$

Hence

$$q(\boldsymbol{\mu}) = \inf_{\mathbf{x}} \left\{ \tfrac{1}{2}\mathbf{x}^T\mathbf{H}\mathbf{x} + \mathbf{p}^T\mathbf{x} + \boldsymbol{\mu}^T(\mathbf{A}\mathbf{x} - \mathbf{b}) \right\} \tag{10.119}$$

where the infimum is attained at $\mathbf{x} = -\mathbf{H}^{-1}(\mathbf{p} + \mathbf{A}^T\boldsymbol{\mu})$. By substituting this solution into Eq. (10.119), we obtain

$$q(\boldsymbol{\mu}) = -\tfrac{1}{2}\boldsymbol{\mu}^T\mathbf{A}\mathbf{H}^{-1}\mathbf{A}^T\boldsymbol{\mu} - \boldsymbol{\mu}^T(\mathbf{A}\mathbf{H}^{-1}\mathbf{p} + \mathbf{b}) - \tfrac{1}{2}\mathbf{p}^T\mathbf{H}^{-1}\mathbf{p} \tag{10.120}$$

If we let $\mathbf{P} = \mathbf{A}\mathbf{H}^{-1}\mathbf{A}^T$, $\mathbf{t} = \mathbf{A}\mathbf{H}^{-1}\mathbf{p} + \mathbf{b}$, and neglect the constant term in Eq. (10.120), the dual function becomes $q(\boldsymbol{\mu}) = -\tfrac{1}{2}\boldsymbol{\mu}^T\mathbf{P}\boldsymbol{\mu} - \boldsymbol{\mu}^T\mathbf{t}$ and hence the Lagrange dual of Eq. (10.118) assumes the form

$$\text{minimize} \quad \tfrac{1}{2}\boldsymbol{\mu}^T\mathbf{P}\boldsymbol{\mu} + \boldsymbol{\mu}^T\mathbf{t} \tag{10.121}$$
$$\text{subject to: } \boldsymbol{\mu} \geq \mathbf{0}$$

Note that by definition, matrix $\mathbf{P}$ in Eq. (10.121) is positive definite. Consequently, like the primal problem, the dual problem is a convex QP problem. Note also that the primal problem in Eq. (10.118) involves n variables and q constraints while the dual problem in Eq. (10.121) involves q variables as well as q constraints. Consequently, if $q < n$ it would be preferable to solve the dual problem rather than the primal problem because the dual problem would entail reduced computational effort. $\quad \blacksquare$

Problems

10.1 A trigonometric polynomial is given by

$$A(\omega) = \sum_{k=0}^{n} a_k \cos k\omega \qquad\qquad \text{(P10.1)}$$

and Ω_p, Ω_a are sets given by

$$\Omega_p = \{\omega_{p0}, \omega_{p1}, \ldots, \omega_{pN}\} \subseteq [0, \omega_p]$$
$$\Omega_a = \{\omega_{a0}, \omega_{a1}, \ldots, \omega_{aM}\} \subseteq [\omega_a, \pi]$$

with $\omega_p \leq \omega_a$. Coefficients a_k for $k = 0, 1, \ldots, n$ are required in Eq. (P10.1) such that the upper bound δ in

$$|A(\omega) - 1| \leq \delta \qquad \text{for } \omega \in \Omega_p \qquad\qquad \text{(P10.2)}$$

and

$$|A(\omega)| \leq \delta \qquad \text{for } \omega \in \Omega_a \qquad\qquad \text{(P10.3)}$$

is minimized. Formulate the above problem as a constrained minimization problem.

10.2 Consider the trigonometric polynomial $A(\omega)$ given in Prob. P10.1. Suppose we need to find a_k, for $k = 0, 1, \ldots, n$ such that

$$J = \int_0^{\omega_p} [A(\omega) - 1]^2 d\omega + \int_{\omega_a}^{\pi} W(\omega) A^2(\omega) d\omega \qquad\qquad \text{(P10.4)}$$

is minimized subject to constraints in Eqs. (P10.2) and (P10.3), where $W(\omega) \geq 0$ is a weighting function, and δ is treated as a known positive scalar. Formulate the above problem as a constrained optimization.

10.3 (a) Write a MATLAB function to examine whether the equality constraints in $\mathbf{Ax} = \mathbf{b}$ are (i) inconsistent, or (ii) consistent but redundant, or (iii) consistent without redundancy.
   (b) Modify the MATLAB function obtained from part (a) so that if $\mathbf{Ax} = \mathbf{b}$ is found to be consistent but redundant, the constraints are reduced to $\hat{\mathbf{A}}\mathbf{x} = \hat{\mathbf{b}}$ such that (i) $\hat{\mathbf{A}}\mathbf{x} = \hat{\mathbf{b}}$ describes the same feasible region and (ii) the constraints in $\hat{\mathbf{A}}\mathbf{x} = \hat{\mathbf{b}}$ are not redundant.

10.4 In Sect. 10.3.1, it was shown that the LP problem in Eqs. (10.20a) and (10.20b) can be converted into the standard-form LP problem of Eqs. (10.19a)–(10.19c). Show that the standard-form LP problem in Eqs. (10.19a)–(10.19c) can be converted into the problem in Eqs. (10.20a) and (10.20b). Hint: Use Eq. (10.27).

10.5 (a) Apply the result of Prob. 10.4 to convert the LP problem

$$\text{minimize } f(\mathbf{x}) = x_1 + 2x_2 + 11x_3 + 2x_4$$
$$\text{subject to: } a_1(\mathbf{x}) = x_1 + x_2 + x_3 + 2x_4 = 3$$
$$a_2(\mathbf{x}) = x_2 + 2x_3 + 4x_4 = 3$$
$$a_3(\mathbf{x}) = 2x_3 + x_4 = 2$$
$$c_i(\mathbf{x}) = -x_i \le 0 \quad \text{for } i = 1, 2, 3, 4$$

into the problem in Eqs. (10.20a) and (10.20b).

(b) Solve the LP problem obtained in part (a).

(c) Use the result of part (b) to solve the standard-form LP problem in part (a).

10.6 (a) Prove that if $\mathbf{P}$ is positive definite, then $\ln(\det \mathbf{P}^{-1})$ is a convex function of $\mathbf{P}$.

(b) Prove that if $\mathbf{p} = \mathbf{P}(:)$ denotes the vector obtained by lexicographically ordering matrix $\mathbf{P}$, then the set of vectors satisfying the constraints in Eqs. (10.24b) and (10.24c) is convex.

10.7 Prove that all solutions of $\mathbf{Ax} = \mathbf{b}$ are characterized by Eq. (10.26). To simplify the proof, assume that $\mathbf{A} \in R^{p \times n}$ has full row rank. In this case the pseudo-inverse of $\mathbf{A}^+$ is given by

$$\mathbf{A}^+ = \mathbf{A}^T (\mathbf{A}\mathbf{A}^T)^{-1}$$

Fig. P10.8 Feasible region for Prob. 10.8

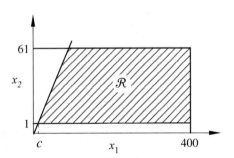

10.8 The feasible region shown in Fig. P10.8 can be described by

$$\mathcal{R}: \begin{cases} c < x_1 < 400 \\ 1 < x_2 < 61 \\ \phantom{1 < } x_2 < x_1/c \end{cases}$$

where $c > 0$ is a constant such that $400/c > 61$. Find variable transformations $x_1 = T_1(t_1, t_2)$ and $x_2 = T_2(t_1, t_2)$ such that $-\infty < t_1, t_2 < \infty$ describe the same feasible region.

10.9 Show that $\nabla f(\mathbf{x})$, $\nabla a_1(\mathbf{x})$, and $\nabla a_2(\mathbf{x})$ that satisfy Eq. (10.54) are linearly dependent.

Hint: Apply the singular-value decomposition to

$$
\begin{bmatrix}
1 & 0 \\
0 & 1 \\
\frac{\partial h_1}{\partial x_1} & \frac{\partial h_1}{\partial x_2} \\
\frac{\partial h_2}{\partial x_1} & \frac{\partial h_2}{\partial x_2}
\end{bmatrix}
$$

10.10 (a) Provide an example to demonstrate that $\mathbf{Ax} \le \mathbf{b}$ does not imply $\mathbf{MAx} \le \mathbf{Mb}$ in general, even if $\mathbf{M}$ is positive definite.

(b) Which condition on $\mathbf{M}$ will ensure that $\mathbf{Ax} \le \mathbf{b}$ implies $\mathbf{MAx} \le \mathbf{Mb}$?

10.11 Use two methods, namely, Eq. (10.27) and the Lagrange multiplier method, to solve the problem

$$\text{minimize } f(\mathbf{x}) = \tfrac{1}{2}\mathbf{x}^T \mathbf{Hx} + \mathbf{x}^T \mathbf{p}$$

$$\text{subject to: } \mathbf{Ax} = \mathbf{b}$$

where

$$
\mathbf{H} =
\begin{bmatrix}
\mathbf{H}_1 & \mathbf{H}_2 & \mathbf{H}_3 & \mathbf{H}_4 \\
\mathbf{H}_2 & \mathbf{H}_1 & \mathbf{H}_2 & \mathbf{H}_3 \\
\mathbf{H}_3 & \mathbf{H}_2 & \mathbf{H}_1 & \mathbf{H}_2 \\
\mathbf{H}_4 & \mathbf{H}_3 & \mathbf{H}_2 & \mathbf{H}_1
\end{bmatrix}
$$

with

$$
\mathbf{H}_1 =
\begin{bmatrix}
10 & 8 & 7 & 6 \\
8 & 10 & 8 & 7 \\
7 & 8 & 10 & 8 \\
6 & 7 & 8 & 10
\end{bmatrix}
$$

$$
\mathbf{H}_2 =
\begin{bmatrix}
3 & 2 & 1 & 0 \\
2 & 3 & 2 & 1 \\
1 & 2 & 3 & 2 \\
0 & 1 & 2 & 3
\end{bmatrix},
\quad
\mathbf{H}_3 =
\begin{bmatrix}
2 & 1 & 0 & 0 \\
1 & 2 & 1 & 0 \\
0 & 1 & 2 & 1 \\
0 & 0 & 1 & 2
\end{bmatrix},
\quad \mathbf{H}_4 = \mathbf{I}_4
$$

$$
\mathbf{p} =
\begin{bmatrix}
\mathbf{p}_1 \\
\mathbf{p}_2 \\
\mathbf{p}_3 \\
\mathbf{p}_4
\end{bmatrix},
\quad
\mathbf{p}_1 =
\begin{bmatrix}
1 \\ -1 \\ 2 \\ -2
\end{bmatrix},
\quad
\mathbf{p}_2 =
\begin{bmatrix}
0 \\ 0 \\ 0 \\ 0
\end{bmatrix},
\quad
\mathbf{p}_3 =
\begin{bmatrix}
2 \\ 2 \\ -4 \\ 4
\end{bmatrix},
\quad
\mathbf{p}_4 =
\begin{bmatrix}
0 \\ 0 \\ 0 \\ 0
\end{bmatrix}
$$

$$\mathbf{A} = [\mathbf{H}_3 \ \mathbf{H}_2 \ \mathbf{H}_1 \ \mathbf{H}_4], \quad \mathbf{b} = [1 \ 14 \ 3 \ -4]^T$$

10.12 Consider the feasible region $\mathcal{R}$ defined by

$$\mathcal{R}: \ a_i(\mathbf{x}) = 0 \qquad \text{for } i = 1, \ 2, \ \ldots, \ p$$
$$c_j(\mathbf{x}) \leq 0 \qquad \text{for } j = 1, \ 2, \ \ldots, \ q$$

At a feasible point $\mathbf{x}$, let $\mathcal{J}(\mathbf{x})$ be the active index set for the inequality constraints at $\mathbf{x}$, and define the sets $\mathcal{F}(\mathbf{x})$ and $F(\mathbf{x})$ as

$$\mathcal{F}(\mathbf{x}) = \{\mathbf{s} : \ \mathbf{s} \text{ is feasible at } \mathbf{x}\}$$

and

$$F(\mathbf{x}) = \{\mathbf{s} : \ \mathbf{s}^T \nabla a_i(\mathbf{x}) = 0 \qquad \text{for } i = 1, \ 2, \ \ldots, \ p$$
$$\text{and } \mathbf{s}^T \nabla c_j(\mathbf{x}) \leq 0 \qquad \text{for } j \in \mathcal{J}(\mathbf{x})\}$$

respectively. Prove that $\mathcal{F}(\mathbf{x}) \subseteq F(\mathbf{x})$, i.e., set $F(\mathbf{x})$ contains set $\mathcal{F}(\mathbf{x})$.

10.13 Prove that if at a feasible $\mathbf{x}$ one of the following conditions is satisfied, then $\mathcal{F}(\mathbf{x}) = F(\mathbf{x})$:

(i) The constraints that are active at $\mathbf{x}$ are all linear.
(ii) Vectors $\nabla a_i(\mathbf{x})$ for $i = 1, \ 2, \ \ldots, \ p$ and $\nabla c_j(\mathbf{x})$ for those $c_j(\mathbf{x})$ that are active at $\mathbf{x}$ are linearly independent.

10.14 In the literature, the assumption that $\mathcal{F}(\mathbf{x}) = F(\mathbf{x})$ is known as the *constraint qualification* of $\mathbf{x}$. Verify that the constraint qualification assumption does not hold at $\mathbf{x} = \mathbf{0}$ when the constraints are given by

$$c_1(\mathbf{x}) = -x_1^3 + x_2$$
$$c_2(\mathbf{x}) = -x_2$$

Hint: Check the vector $\mathbf{s} = [-1 \ 0]^T$.

10.15 Consider the constrained minimization problem (see [13])

$$\text{minimize} \ f(\mathbf{x}) = (x_1 - 2)^2 + x_2^2$$
$$\text{subject to:} \ c_1(\mathbf{x}) = -x_1 \leq 0$$
$$c_2(\mathbf{x}) = -x_2 \leq 0$$
$$c_3(\mathbf{x}) = (x_1 - 1)^3 + x_2 \leq 0$$

(a) Using a graphical solution, show that $\mathbf{x}^* = [1 \ 0]^T$ is the global minimizer.
(b) Verify that $\mathbf{x}^*$ is not a regular point.
(c) Show that there exist no $\mu_2 \geq 0$ and $\mu_3 \geq 0$ such that

$$\nabla f(\mathbf{x}^*) + \mu_2 \nabla c_2(\mathbf{x}^*) + \mu_3 \nabla c_3(\mathbf{x}^*) = \mathbf{0}$$

10.16 Given column vectors $v_1,\ v_2,\ \ldots,\ v_q$, define the polyhedral cone C as

$$C = \{v : \ v = \sum_{i=1}^{q} \mu_i\, v_i,\ \mu_i \geq 0\}$$

Prove that C is closed and convex.

10.17 Let $\mathbf{g}$ be a vector that does not belong to set C in Prob. 10.16. Prove that there exists a hyperplane $s^T \mathbf{x} = 0$ that separates C and $\mathbf{g}$.

10.18 Given column vectors $v_1,\ v_2,\ \ldots,\ v_q$ and $\mathbf{g}$, show that the set

$$S = \{\mathbf{s} : \ \mathbf{s}^T \mathbf{g} < 0 \text{ and } \mathbf{s}^T v_i \geq 0,\ \text{for } i = 1,\ 2,\ \ldots,\ q\}$$

is empty if and only if there exist $\mu_i \geq 0$ such that

$$\mathbf{g} = \sum_{i=1}^{q} \mu_i\, v_i$$

(This is known as Farkas' lemma.)
Hint: Use the results of Probs. 10.16 and 10.17.

10.19 Let $\mathcal{J}(\mathbf{x}^*) = \{j_1,\ j_2,\ \ldots,\ j_K\}$ be the active index set at $\mathbf{x}^*$ for the constraints in Eq. (10.1c). Show that the set

$$S = \{\mathbf{s} : \ \mathbf{s}^T \nabla f(\mathbf{x}^*) < 0,\ \mathbf{s}^T \nabla a_i(\mathbf{x}^*) = 0 \ \text{for } i = 1,\ 2,\ \ldots,\ p,$$
$$\text{and } \mathbf{s}^T \nabla c_j(\mathbf{x}^*) \leq 0 \ \text{for } j \in \mathcal{J}(\mathbf{x}^*)\}$$

is empty if and only if there exist multipliers λ_i^* for $1 \leq i \leq p$ and $\mu_j^* \geq 0$, such that

$$\nabla f(\mathbf{x}^*) + \sum_{i=1}^{p} \lambda_i^* \nabla a_i(\mathbf{x}^*) + \sum_{j \in \mathcal{J}(\mathbf{x}^*)} \mu_j^* \nabla c_j(\mathbf{x}^*) = \mathbf{0}$$

(This is known as the Extension of Farkas' lemma.)

10.20 Using the KKT conditions, find solution candidates for the following CP problem

$$\text{minimize } x_1^2 + x_2^2 - 2x_1 - 4x_2 + 9$$
$$\text{subject to: } -x_1 + 1 \leq 0$$
$$-x_2 \leq 0$$
$$\tfrac{1}{2}x_1 + x_2 - \tfrac{3}{2} \leq 0$$

10.21 (*a*) Obtain the KKT conditions for the solution points of the problem

$$\text{minimize } f(\mathbf{x}) = -x_1^3 + x_2^3 - 2x_1 x_3^2$$
$$\text{subject to: } 2x_1 + x_2^2 + x_3 - 5 = 0$$
$$-5x_1^2 + x_2^2 + x_3 \leq -2$$
$$-x_1 \leq 0, \; -x_2 \leq 0, \; -x_3 \leq 0$$

(*b*) Vector $\mathbf{x}^* = [1 \; 0 \; 3]^T$ is known to be a local minimizer of the problem in part (*a*). Find λ_i^* and μ_i^* for $1 \leq i \leq 4$ at $\mathbf{x}^*$ and verify that $\mu_i^* \geq 0$ for $1 \leq i \leq 4$.

(*c*) Examine the second-order conditions for set $(\mathbf{x}^*, \boldsymbol{\lambda}^*, \boldsymbol{\mu}^*)$.

10.22 Consider the minimization problem

$$\text{minimize } f(\mathbf{x}) = \mathbf{c}^T \mathbf{x}$$
$$\text{subject to: } \mathbf{A}\mathbf{x} = \mathbf{0}$$
$$\|\mathbf{x}\| \leq 1$$

where $\|\mathbf{x}\|$ denotes the Euclidean norm of $\mathbf{x}$.

(*a*) Show that this is a CP problem.
(*b*) Derive the KKT conditions for the solution points of the problem.
(*c*) Show that if $\mathbf{c} - \mathbf{A}^T \boldsymbol{\lambda} \neq \mathbf{0}$ where $\boldsymbol{\lambda}$ satisfies

$$\mathbf{A}\mathbf{A}^T \boldsymbol{\lambda} = \mathbf{A}\mathbf{c}$$

then the minimizer is given by

$$\mathbf{x}^* = -\frac{\mathbf{c} - \mathbf{A}^T \boldsymbol{\lambda}}{\|\mathbf{c} - \mathbf{A}^T \boldsymbol{\lambda}\|}$$

Otherwise, any feasible $\mathbf{x}$ is a solution.

10.23 Consider the minimization problem

$$\text{minimize } f(\mathbf{x}) = \mathbf{c}^T \mathbf{x}$$
$$\text{subject to: } \|\mathbf{A}\mathbf{x}\| \leq 1$$

(*a*) Show that this is a CP problem.
(*b*) Derive the KKT conditions for the solution points of the problem.
(*c*) Show that if the solution of the equation

$$\mathbf{A}^T \mathbf{A}\mathbf{y} = \mathbf{c}$$

is nonzero, then the minimizer is given by

$$\mathbf{x}^* = -\frac{\mathbf{y}}{\|\mathbf{A}\mathbf{y}\|}$$

Otherwise, any feasible $\mathbf{x}$ is a solution.

10.24 (*a*) Obtain the Lagrange dual function of the following convex problem

$$\text{minimize } \mathbf{x}^T \mathbf{x}$$
$$\text{subject to: } \mathbf{A}\mathbf{x} = \mathbf{b}$$

(*b*) Repeat part (*a*) for the convex problem

$$\text{minimize } \tfrac{1}{2}\mathbf{x}^T \mathbf{H}\mathbf{x} + \mathbf{p}^T \mathbf{x} + c$$
$$\text{subject to: } \tfrac{1}{2}\mathbf{x}^T \mathbf{H}_i \mathbf{x} + \mathbf{p}_i^T \mathbf{x} + c_i \leq 0 \quad \text{for } i = 1,\ 2,\ \dots,\ q$$

where $\mathbf{H}$ and $\mathbf{H}_i$ for $1,\ 2,\ \dots,\ q$ are positive definite.

References

1. G. B. Dantzig, *Linear Programming and Extensions.* Princeton, NJ: Princeton University Press, 1963.
2. S. J. Wright, *Primal-Dual Interior-Point Methods.* Philadelphia, PA: SIAM, 1997.
3. S. Boyd, L. El Ghaoui, E. Feron, and V. Balakrishnan, *Linear Matrix Inequalities in System and Control Theory.* Philadelphia, PA: SIAM, 1994.
4. Y. Nesterov and A. Nemirovskii, *Interior-Point Polynomial Algorithms in Convex Programming.* Philadelphia, PA: SIAM, 1994.
5. J. T. Betts, "An accelerated multiplier method for nonlinear programming," *JOTA*, vol. 21, no. 2, pp. 137–174, 1977.
6. G. Van der Hoek, "Reduction methods in nonlinear programming," in *Mathematical Centre Tracts,* vol. 126. Amsterdam: Mathematisch Centrum, 1980.
7. R. Fletcher, *Practical Methods of Optimization*, 2nd ed. New York: Wiley, 1987.
8. G. H. Golub and C. F. Van Loan, *Matrix Computations*, 4th ed. Baltimore, MD: Johns Hopkins University Press, 2013.
9. A. Antoniou, *Digital Signal Processing: Signals, Systems, and Filters.* New York: McGraw-Hill, 2005.
10. W.-S. Lu, "A parameterization method for the design of IIR digital filters with prescribed stability margin," in *Proc. IEEE Int. Symp. Circuits and Systems*, Hong Kong, Jun. 1997, pp. 2381–2384.
11. D. G. Luenberger and Y. Ye, *Linear and Nonlinear Programming*, 4th ed. New York: Springer, 2008.
12. E. K. P. Chong and S. H. Żak, *An Introduction to Optimization.* New York: Wiley, 1996.
13. W. Karush, "Minima of functions of several variables with inequalities as side constraints," Master's thesis, University of Chicago, Chicago, 1939.
14. H. W. Kuhn and A. W. Tucker, "Nonlinear programming," in *Proc. 2nd Berkeley Symposium*, Berkeley, CA, 1951, pp. 481–492.
15. S. Boyd and L. Vandenberghe, *Convex Optimization.* Cambridge, UK: Cambridge University Press, 2004.

Linear Programming Part I: The Simplex Method

11.1 Introduction

Linear programming (LP) problems occur in a diverse range of real-life applications in economic analysis and planning, operations research, computer science, medicine, and engineering. In such problems, it is known that any minima occur at the vertices of the feasible region and can be determined through a 'brute-force' or exhaustive approach by evaluating the objective function at all the vertices of the feasible region. However, the number of variables involved in a practical LP problem is often very large and an exhaustive approach would entail a considerable amount of computation. In 1947, Dantzig developed a method for the solution of LP problems known as the *simplex method* [1,2]. Although in the worst case, the simplex method is known to require an exponential number of iterations, for typical standard-form problems the number of iterations required is just a small multiple of the problem dimension [3]. For this reason, the simplex method has been the primary method for solving LP problems since its introduction.

In Sect. 11.2, the general theory of constrained optimization developed in Chap. 10 is applied to derive optimality conditions for LP problems. The geometrical features of LP problems are discussed and connected to the several issues that are essential in the development of the simplex method. In Sect. 11.3, the simplex method is presented for alternative-form LP problems as well as for standard-form LP problems from a linear-algebraic perspective.

11.2 General Properties

11.2.1 Formulation of LP Problems

In Sect. 10.3.1, the standard-form LP problem was stated as

$$\text{minimize } f(\mathbf{x}) = \mathbf{c}^T \mathbf{x} \tag{11.1a}$$

© Springer Science+Business Media, LLC, part of Springer Nature 2021
A. Antoniou and W.-S. Lu, *Practical Optimization*, Texts in Computer Science,
https://doi.org/10.1007/978-1-0716-0843-2_11

$$\text{subject to:} \quad \mathbf{A}\mathbf{x} = \mathbf{b} \qquad\qquad (11.1b)$$

$$\mathbf{x} \geq \mathbf{0} \qquad\qquad (11.1c)$$

where $\mathbf{c} \in R^{n \times 1}$ with $\mathbf{c} \neq \mathbf{0}$, $\mathbf{A} \in R^{p \times n}$, and $\mathbf{b} \in R^{p \times 1}$ are given. Throughout this chapter, we assume that $\mathbf{A}$ is of full row rank, i.e., rank$(\mathbf{A}) = p$. For the standard-form LP problem in Eqs. (11.1a)–(11.1c) to be a meaningful LP problem, full row rank in $\mathbf{A}$ implies that $p < n$.

For a fixed scalar β, the equation $\mathbf{c}^T \mathbf{x} = \beta$ describes an affine manifold in the n-dimensional Euclidean space E^n (see Sect. A.15). For example, with $n = 2$, $\mathbf{c}^T \mathbf{x} = \beta$ represents a line and $\mathbf{c}^T \mathbf{x} = \beta$ for $\beta = \beta_1$, β_2, ... represents a family of parallel lines. The normal of these lines is $\mathbf{c}$, and for this reason vector $\mathbf{c}$ is often referred to as the *normal vector* of the objective function.

Another LP problem which is often encountered in practice involves minimizing a linear function subject to inequality constraints, i.e.,

$$\text{minimize} \ f(\mathbf{x}) = \mathbf{c}^T \mathbf{x} \qquad\qquad (11.2a)$$

$$\text{subject to:} \quad \mathbf{A}\mathbf{x} \leq \mathbf{b} \qquad\qquad (11.2b)$$

where $\mathbf{c} \in R^{n \times 1}$ with $\mathbf{c} \neq \mathbf{0}$, $\mathbf{A} \in R^{p \times n}$, and $\mathbf{b} \in R^{p \times 1}$ are given. This will be referred to as the *alternative-form* LP problem hereafter. If we let

$$\mathbf{A} = \begin{bmatrix} \mathbf{a}_1^T \\ \mathbf{a}_2^T \\ \vdots \\ \mathbf{a}_p^T \end{bmatrix}, \quad \mathbf{b} = \begin{bmatrix} b_1 \\ b_2 \\ \vdots \\ b_p \end{bmatrix}$$

then the p constraints in Eq. (11.2b) can be written as

$$\mathbf{a}_i^T \mathbf{x} \leq b_i \quad \text{for } i = 1, \ 2, \ \ldots, \ p$$

where vector $\mathbf{a}_i$ is the normal of the ith inequality constraint, and $\mathbf{A}$ is usually referred to as the *constraint matrix*.

By introducing a p-dimensional slack vector variable $\mathbf{y}$, Eq. (11.2b) can be reformulated as

$$\mathbf{A}\mathbf{x} + \mathbf{y} = \mathbf{b} \quad \text{for } \mathbf{y} \geq \mathbf{0}$$

Furthermore, vector variable $\mathbf{x}$ can be decomposed as

$$\mathbf{x} = \mathbf{x}^+ - \mathbf{x}^- \quad \text{with } \mathbf{x}^+ \geq \mathbf{0} \ \text{ and } \ \mathbf{x}^- \geq \mathbf{0}$$

Hence if we let

$$\hat{\mathbf{x}} = \begin{bmatrix} \mathbf{x}^+ \\ \mathbf{x}^- \\ \mathbf{y} \end{bmatrix}, \quad \hat{\mathbf{c}} = \begin{bmatrix} \mathbf{c} \\ -\mathbf{c} \\ \mathbf{0} \end{bmatrix}, \quad \hat{\mathbf{A}} = [\mathbf{A} \ -\mathbf{A} \ \mathbf{I}_p]$$

then Eqs. (11.2a) and (11.2b) can be expressed as a standard-form LP problem, i.e.,

$$\text{minimize} \ \ f(\mathbf{x}) = \hat{\mathbf{c}}^T \hat{\mathbf{x}} \tag{11.3a}$$

$$\text{subject to:} \ \ \hat{\mathbf{A}} \hat{\mathbf{x}} = \mathbf{b} \tag{11.3b}$$

$$\hat{\mathbf{x}} \geq \mathbf{0} \tag{11.3c}$$

Likewise, the most general LP problem with both equality and inequality constraints, i.e.,

$$\text{minimize} \ \ f(\mathbf{x}) = \mathbf{c}^T \mathbf{x} \tag{11.4}$$

$$\text{subject to:} \ \ \mathbf{A}\mathbf{x} = \mathbf{b}$$

$$\mathbf{C}\mathbf{x} \leq \mathbf{d}$$

can be expressed as a standard-form LP problem with respect to an augmented variable $\hat{\mathbf{x}}$. It is primarily for these reasons that the standard-form LP problem in Eqs. (11.1a)–(11.1c) has been employed most often as the prototype for the description and implementation of various LP algorithms. Nonstandard LP problems, particularly the problem in Eqs. (11.2a) and (11.2b), may be encountered directly in a variety of applications. Although the problem in Eqs. (11.2a) and (11.2b) can be reformulated as a standard-form LP problem, the increase in problem size leads to reduced computational efficiency which can sometimes be a serious problem particularly when the number of inequality constraints is large. In what follows, the underlying principles pertaining to the LP problems in Eqs. (11.1a)–(11.1c) and Eqs. (11.2a)–(11.2b) will be described separately to enable us to solve each of these problems directly without the need of converting the one form into the other.

11.2.2   Optimality Conditions

Since linear functions are convex (or concave), an LP problem can be viewed as a convex programming problem. By applying Theorems 10.8 and 10.2 to the problem in Eqs. (11.1a)–(11.1c), the following theorem can be deduced.

Theorem 11.1 (Karush-Kuhn-Tucker conditions for standard-form LP problem) *If* $\mathbf{x}^*$ *is regular for the constraints that are active at* $\mathbf{x}^*$, *then it is a global solution of the LP problem in Eqs. (11.1a)–(11.1c) if and only if*

(a)

$$\mathbf{A}\mathbf{x}^* = \mathbf{b} \tag{11.5a}$$

(b)

$$\mathbf{x}^* \geq \mathbf{0} \tag{11.5b}$$

(c) *there exist Lagrange multipliers* $\boldsymbol{\lambda}^* \in R^{p \times 1}$ *and* $\boldsymbol{\mu}^* \in R^{n \times 1}$ *such that* $\boldsymbol{\mu}^* \geq \mathbf{0}$
   and

$$\mathbf{c} + \mathbf{A}^T \boldsymbol{\lambda}^* - \boldsymbol{\mu}^* = \mathbf{0} \tag{11.5c}$$

(d)

$$\mu_i^* x_i^* = 0 \quad \text{for } 1 \leq i \leq n \tag{11.5d}$$

∎

   The first two conditions in Eqs. (11.5a) and (11.5b) simply say that solution $\mathbf{x}^*$ must be a feasible point. In Eq. (11.5c), constraint matrix $\mathbf{A}$ and vector $\mathbf{c}$ are related through the Lagrange multipliers $\boldsymbol{\lambda}^*$ and $\boldsymbol{\mu}^*$.
   An immediate observation on the basis of Eqs. (11.5a)–(11.5d) is that in most cases solution $\mathbf{x}^*$ cannot be strictly feasible. Here we take the term 'strictly feasible points' to mean those points that satisfy the equality constraints with $x_i^* > 0$ for $1 \leq i \leq n$. From Eq. (11.5d), $\boldsymbol{\mu}^*$ must be a zero vector for a strictly feasible point $\mathbf{x}^*$ to be a solution. Hence Eq. (11.5c) becomes

$$\mathbf{c} + \mathbf{A}^T \boldsymbol{\lambda}^* = \mathbf{0} \tag{11.6}$$

In other words, for a strictly feasible point to be a minimizer of the standard-form LP problem in Eqs. (11.1a)–(11.1c), the n-dimensional vector $\mathbf{c}$ must lie in the p-dimensional subspace spanned by the p columns of $\mathbf{A}^T$. Since $p < n$, the probability that Eq. (11.6) is satisfied is very small. Therefore, any solutions of the problem are very likely to be located on the *boundary* of the feasible region.

Example 11.1 Solve the LP problem

$$\text{minimize } f(\mathbf{x}) = x_1 + 4x_2 \tag{11.7a}$$

$$\text{subject to: } x_1 + x_2 = 1 \tag{11.7b}$$

$$\mathbf{x} \geq \mathbf{0} \tag{11.7c}$$

Solution As shown in Fig. 11.1, the feasible region of the above problem is the segment of the line $x_1 + x_2 = 1$ in the first quadrant, the dashed lines are contours of the form $f(\mathbf{x}) = \text{constant}$, and the arrow points to the steepest-descent direction of $f(\mathbf{x})$. We have

$$\mathbf{c} = \begin{bmatrix} 1 \\ 4 \end{bmatrix} \quad \text{and} \quad \mathbf{A}^T = \begin{bmatrix} 1 \\ 1 \end{bmatrix}$$

Fig. 11.1 LP problem in
Example 11.1

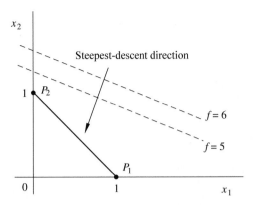

Since $\mathbf{c}$ and $\mathbf{A}^T$ are linearly independent, Eq. (11.6) cannot be satisfied and, therefore, no interior feasible point can be a solution. This leaves two end points to verify. From Fig. 11.1 it is evident that the unique minimizer is $\mathbf{x}^* = [1 \ 0]^T$.

At $\mathbf{x}^*$ the constraint in Eq. (11.7b) and the second constraint in Eq. (11.7c) are active, and since the Jacobian of these constraints, namely,

$$\begin{bmatrix} 1 & 1 \\ 0 & 1 \end{bmatrix}$$

is nonsingular, $\mathbf{x}^*$ is a regular point. Now Eq. (11.5d) gives $\mu_1^* = 0$, which leads to Eq. (11.5c) with

$$\lambda^* = -1 \quad \text{and} \quad \mu_2^* = 3$$

This confirms that $\mathbf{x}^* = [1 \ 0]^T$ is indeed a global solution.

Note that if the objective function is changed to

$$f(\mathbf{x}) = \mathbf{c}^T \mathbf{x} = 4x_1 + 4x_2$$

then Eq. (11.6) is satisfied with $\lambda^* = -4$ and any feasible point becomes a global solution. In fact, the objective function remains constant in the feasible region, i.e.,

$$f(\mathbf{x}) = 4(x_1 + x_2) = 4 \quad \text{for } \mathbf{x} \in \mathcal{R}$$

A graphical interpretation of this situation is shown in Fig. 11.2.   ∎

Note that the conditions in Theorems 10.2 and 10.8 are also applicable to the alternative-form LP problem in Eqs. (11.2a) and (11.2b) since the problem is, in effect, a convex programming (CP) problem. These conditions can be summarized in terms of the following theorem.

Fig. 11.2 LP problem in Example 11.1 with $f(\mathbf{x}) = 4x_1 + 4x_2$

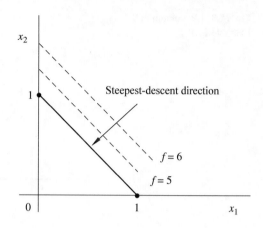

Theorem 11.2 (Necessary and sufficient conditions for a minimum in alternative-form LP problem) *If* $\mathbf{x}^*$ *is regular for the constraints in Eq. (11.2b) that are active at* $\mathbf{x}^*$, *then it is a global solution of the problem in Eqs. (11.2a) and (11.2b) if and only if*

(a)

$$\mathbf{A}\mathbf{x}^* \leq \mathbf{b} \tag{11.8a}$$

(b) *there exists a* $\boldsymbol{\mu}^* \in R^{p \times 1}$ *such that* $\boldsymbol{\mu}^* \geq \mathbf{0}$ *and*

$$\mathbf{c} + \mathbf{A}^T \boldsymbol{\mu}^* = \mathbf{0} \tag{11.8b}$$

(c)

$$\mu_i^* (\mathbf{a}_i^T \mathbf{x}^* - b_i) = 0 \quad \text{for } 1 \leq i \leq p \tag{11.8c}$$

where $\mathbf{a}_i^T$ *is the ith row of* $\mathbf{A}$. ∎

The observation made with regard to Theorem 11.1, namely, that the solutions of the problem are very likely to be located on the boundary of the feasible region, also applies to Theorem 11.2. As a matter of fact, if $\mathbf{x}^*$ is a strictly feasible point satisfying Eq. (11.8c), then $\mathbf{A}\mathbf{x}^* < \mathbf{b}$ and the complementarity condition in Eq. (11.8c) implies that $\boldsymbol{\mu}^* = \mathbf{0}$. Hence Eq. (11.8b) cannot be satisfied unless $\mathbf{c} = \mathbf{0}$, which would lead to a meaningless LP problem. In other words, any solutions of Eqs. (11.8a)–(11.8c) can only occur on the boundary of the feasible region defined by Eq. (11.2b).

Fig. 11.3 Feasible region in Example 11.2

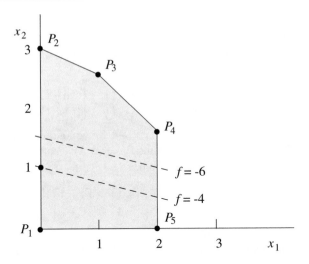

Example 11.2 Solve the LP problem

$$\text{minimize } f(\mathbf{x}) = -x_1 - 4x_2$$

subject to:
$$- x_1 \leq 0$$
$$x_1 \leq 2$$
$$-x_2 \leq 0$$
$$x_1 + x_2 - 3.5 \leq 0$$
$$x_1 + 2x_2 - 6 \leq 0$$

Solution The five constraints can be expressed as $\mathbf{A}\mathbf{x} \leq \mathbf{b}$ with

$$\mathbf{A} = \begin{bmatrix} -1 & 0 \\ 1 & 0 \\ 0 & -1 \\ 1 & 1 \\ 1 & 2 \end{bmatrix}, \quad \mathbf{b} = \begin{bmatrix} 0 \\ 2 \\ 0 \\ 3.5 \\ 6 \end{bmatrix}$$

The feasible region is the polygon shown in Fig. 11.3.

Since the solution cannot be inside the polygon, we consider the five edges of the polygon. We note that at any point $\mathbf{x}$ on an edge other than the five vertices P_i for $1 \leq i \leq 5$ only *one* constraint is active. This means that only one of the five μ_i's is nonzero. At such an $\mathbf{x}$, Eq. (11.8b) becomes

$$\mathbf{c} = \begin{bmatrix} -1 \\ -4 \end{bmatrix} = -\mu_i \mathbf{a}_i \tag{11.9}$$

where $\mathbf{a}_i$ is the transpose of the ith row in $\mathbf{A}$. Since each $\mathbf{a}_i$ is linearly independent of $\mathbf{c}$, no μ_i exists that satisfies Eq. (11.9). This then leaves the five vertices for verification. At point $P_1 = [0 \; 0]^T$, both the first and third constraints are active and Eq. (11.8b) becomes

$$\begin{bmatrix} -1 \\ -4 \end{bmatrix} = \begin{bmatrix} 1 & 0 \\ 0 & 1 \end{bmatrix} \begin{bmatrix} \mu_1 \\ \mu_3 \end{bmatrix}$$

which gives $\mu_1 = -1$ and $\mu_3 = -4$. Since condition (b) of Theorem 11.2 is violated, P_1 is not a solution. At point $P_2 = [0 \; 3]^T$, both the first and fifth constraints are active, and Eq. (11.8b) becomes

$$\begin{bmatrix} -1 \\ -4 \end{bmatrix} = \begin{bmatrix} 1 & -1 \\ 0 & -2 \end{bmatrix} \begin{bmatrix} \mu_1 \\ \mu_5 \end{bmatrix}$$

which gives $\mu_1 = 1$ and $\mu_5 = 2$. Since the rest of the μ_i's are all zero, conditions (a)–(c) of Theorem 11.2 are satisfied with $\boldsymbol{\mu} \equiv \boldsymbol{\mu}^* = [1 \; 0 \; 0 \; 0 \; 2]^T$ and $P_2 = [0 \; 3]^T$ is a minimizer, i.e., $\mathbf{x} \equiv \mathbf{x}^* = P_2$. One can go on to check the rest of the vertices to confirm that point P_2 is the unique solution to the problem. However, the uniqueness of the solution is obvious from Fig. 11.3.

We conclude the example with two remarks on the solution's uniqueness. Later on, we will see that the solution can also be verified by using the positivity of those μ_i's that are associated with active inequality constraints (see Theorem 11.7 in Sect. 11.2.4.2). If we consider minimizing the linear function $f(\mathbf{x}) = \mathbf{c}^T \mathbf{x}$ with $\mathbf{c} = [-1 \; -2]^T$ subject to the same constraints as above, then the contours defined by $f(\mathbf{x}) = $ constant are in parallel with edge $\overline{P_2 P_3}$. Hence any point on $\overline{P_2 P_3}$ is a solution and, therefore, we do not have a unique solution. ∎

11.2.3 Geometry of an LP Problem

11.2.3.1 Facets, Edges, and Vertices

The optimality conditions and the two examples discussed in Sect. 11.2.2 indicate that points on the boundary of the feasible region are of critical importance in LP. For the two-variable case, the feasible region $\mathcal{R}$ defined by Eq. (11.2b) is a polygon, and the facets and edges of $\mathcal{R}$ are the same. For problems with $n > 2$, they represent different geometrical structures which are increasingly difficult to visualize and formal definitions for these structures are, therefore, necessary.

In general, the feasible region defined by $\mathcal{R} = \{\mathbf{x} : \mathbf{A}\mathbf{x} \geq \mathbf{b}\}$ is a convex polyhedron. A set of points, $\mathcal{F}$, in the n-dimensional space E^n is said to be a *face* of a convex polyhedron $\mathcal{R}$ if the condition $\mathbf{p}_1, \mathbf{p}_2 \in \mathcal{F}$ implies that $(\mathbf{p}_1 + \mathbf{p}_2)/2 \in \mathcal{F}$. The dimension of a face is defined as the dimension of $\mathcal{F}$. Depending on its dimension, a face can be a facet, an edge, or a vertex. If l is the dimension of a face $\mathcal{F}$, then a *facet* of $\mathcal{F}$ is an $(l-1)$-dimensional face, an *edge* of $\mathcal{F}$ is a one-dimensional face, and a *vertex* of $\mathcal{F}$ is a zero-dimensional face [4]. As an example, Fig. 11.4 shows the convex polyhedron defined by the constraints

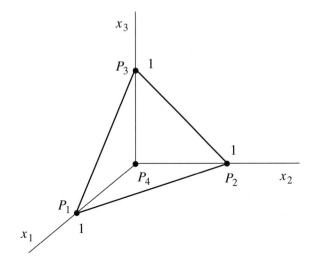

Fig. 11.4 Polyhedron defined by Eq. (11.10) and its facets, edges, and vertices

$$x_1 + x_2 + x_3 \leq 1$$

$$x_1 \geq 0, \quad x_2 \geq 0, \quad x_3 \geq 0$$

i.e.,

$$\mathbf{A}\mathbf{x} \leq \mathbf{b} \qquad (11.10)$$

with

$$\mathbf{A} = \begin{bmatrix} 1 & 1 & 1 \\ -1 & 0 & 0 \\ 0 & -1 & 0 \\ 0 & 0 & -1 \end{bmatrix}, \quad \mathbf{b} = \begin{bmatrix} 1 \\ 0 \\ 0 \\ 0 \end{bmatrix}$$

The polyhedron is a three-dimensional face which has four facets, six edges, and four vertices.

In the case where $n = 2$, a feasible region defined by $\mathbf{A}\mathbf{x} \leq \mathbf{b}$ becomes a polygon and facets become edges. As can be seen in Fig. 11.3, the vertices of a polygon are the points where *two* inequality constraints become active. In the case where $n = 3$, Fig. 11.4 suggests that vertices are the points where *three* inequality constraints become active. In general, we define a vertex point as follows [3].

Definition 11.1 A vertex is a feasible point P at which there exist at least n active constraints which contain n linearly independent constraints where n is the dimension of $\mathbf{x}$. Vertex P is said to be *nondegenerate* if exactly n constraints are active at P or *degenerate* if more than n constraints are active at P. ∎

Definition 11.1 covers the general case where both equality and inequality constraints are present. Linearly independent active constraints are the constraints that are active at P and the matrix whose rows are the vectors associated with the active constraints is of full row rank. At point P_1 in Fig. 11.1, for example, the equality constraint in Eq. (11.7b) and one of the inequality constraints, i.e., $x_2 \geq 0$, are active. This in conjunction with the nonsingularity of the associated matrix

$$\begin{bmatrix} 1 & 1 \\ 0 & 1 \end{bmatrix}$$

implies that P_1 is a nondegenerate vertex. It can be readily verified that point P_2 in Fig. 11.1, points P_i for $i = 1, 2, \ldots, 5$ in Fig. 11.3, and points P_i for $i = 1, 2, \ldots, 4$ in Fig. 11.4 are also nondegenerate vertices.

As another example, the feasible region characterized by the constraints

$$x_1 + x_2 + x_3 \leq 1$$
$$0.5x_1 + 2x_2 + x_3 \leq 1$$
$$x_1 \geq 0, \ x_2 \geq 0, \ x_3 \geq 0$$

i.e.,

$$\mathbf{Ax} \leq \mathbf{b} \tag{11.11}$$

with

$$\mathbf{A} = \begin{bmatrix} 1 & 1 & 1 \\ 0.5 & 2 & 1 \\ -1 & 0 & 0 \\ 0 & -1 & 0 \\ 0 & 0 & -1 \end{bmatrix}, \quad \mathbf{b} = \begin{bmatrix} 1 \\ 1 \\ 0 \\ 0 \\ 0 \end{bmatrix}$$

is illustrated in Fig. 11.5. The convex polyhedron has five facets, eight edges, and five vertices. At vertex P_5 four constraints are active but since $n = 3$, P_5 is degenerate. The other four vertices, namely, P_1, P_2, P_3 and P_4, are nondegenerate.

11.2.3.2 Feasible Descent Directions

A vector $\mathbf{d} \in R^{n \times 1}$ is said to be a *feasible descent direction* at a feasible point $\mathbf{x} \in R^{n \times 1}$ if $\mathbf{d}$ is a feasible direction as defined by Def. 2.4 and the linear objective function strictly decreases along $\mathbf{d}$, i.e., $f(\mathbf{x} + \alpha \mathbf{d}) < f(\mathbf{x})$ for $\alpha > 0$, where $f(\mathbf{x}) = \mathbf{c}^T \mathbf{x}$. Evidently, this implies that

$$\frac{1}{\alpha}[f(\mathbf{x} + \alpha \mathbf{d}) - f(\mathbf{x})] = \mathbf{c}^T \mathbf{d} < 0 \tag{11.12}$$

Fig. 11.5 A feasible region
with a degenerate vertex

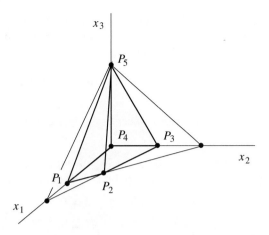

For the problem in Eqs. (11.2a) and (11.2b), we denote as $\mathbf{A}_a$ the matrix whose rows
are the rows of $\mathbf{A}$ that are associated with the constraints which are active at $\mathbf{x}$. We
call $\mathbf{A}_a$ the *active constraint matrix* at $\mathbf{x}$. If $\mathcal{J} = \{j_1, \; j_2, \; \ldots, \; j_K\}$ is the set of
indices that identify active constraints at $\mathbf{x}$, then

$$\mathbf{A}_a = \begin{bmatrix} \mathbf{a}_{j1}^T \\ \mathbf{a}_{j2}^T \\ \vdots \\ \mathbf{a}_{jK}^T \end{bmatrix} \tag{11.13}$$

satisfies the system of equations

$$\mathbf{a}_j^T \mathbf{x} = b_j \qquad \text{for } j \in \mathcal{J}$$

For $\mathbf{d}$ to be a feasible direction, we must have

$$\mathbf{A}_a(\mathbf{x} + \alpha \mathbf{d}) \leq \mathbf{b}_a$$

where $\mathbf{b}_a = [b_{j1} \; b_{j2} \; \cdots \; b_{jK}]^T$. It follows that

$$\mathbf{A}_a \mathbf{d} \leq \mathbf{0}$$

which in conjunction with Eq. (11.12) characterizes a feasible descent direction $\mathbf{d}$
such that

$$\mathbf{A}_a \mathbf{d} \leq \mathbf{0} \quad \text{and} \quad \mathbf{c}^T \mathbf{d} < 0 \tag{11.14}$$

Since $\mathbf{x}^*$ is a solution of the problem in Eqs. (11.2a) and (11.2b) if and only if no
feasible descent directions exist at $\mathbf{x}^*$, we can state the following theorem.

Theorem 11.3 (Necessary and sufficient conditions for a minimum in alternative-form LP problem) *Point* $\mathbf{x}^*$ *is a solution of the problem in Eqs. (11.2a) and (11.2b) if and only if it is feasible and*

$$\mathbf{c}^T \mathbf{d} \geq 0 \quad \text{for all } \mathbf{d} \text{ with } \mathbf{A}_{a*} \mathbf{d} \leq \mathbf{0} \tag{11.15}$$

where $\mathbf{A}_{a*}$ *is the active constraint matrix at* $\mathbf{x}^*$.   ∎

For the standard-form LP problem in Eqs. (11.1a)–(11.1c), a feasible descent direction $\mathbf{d}$ at a feasible point $\mathbf{x}^*$ satisfies the constraints

$$\mathbf{Ad} = \mathbf{0}$$
$$d_j \geq 0 \quad \text{for } j \in \mathcal{J}_*$$

and

$$\mathbf{c}^T \mathbf{d} \leq 0$$

where $\mathcal{J}_* = \{j_1, \ j_2, \ \ldots, \ j_K\}$ is the set of indices for the constraints in Eq. (11.1c) that are active at $\mathbf{x}^*$. This leads to the following theorem.

Theorem 11.4 (Necessary and sufficient conditions for a minimum in standard-form LP problem) *Point* $\mathbf{x}^*$ *is a solution of the LP problem in Eqs. (11.1a)–(11.1c) if and only if it is a feasible point and*

$$\mathbf{c}^T \mathbf{d} \geq 0 \text{ for all } \mathbf{d} \text{ with } \mathbf{d} \in \mathcal{N}(\mathbf{A}) \text{ and } d_j \geq 0 \text{ for } j \in \mathcal{J}_* \tag{11.16}$$

where $\mathcal{N}(\mathbf{A})$ *denotes the null space of* $\mathbf{A}$.   ∎

11.2.3.3 Finding a Vertex

Examples 11.1 and 11.2 discussed in Sect. 11.2.2 indicate that any solutions of the LP problems in Eqs. (11.1a)–(11.1c) and Eqs. (11.2a)–(11.2b) can occur at vertex points. In Sect. 11.2.3.4, it will be shown that under some reasonable conditions, a vertex minimizer always exists. In what follows, we describe an iterative strategy that can be used to find a minimizer vertex for the LP problem in Eqs. (11.2a) and (11.2b) starting with a feasible point $\mathbf{x}_0$.

In the kth iteration, if the active constraint matrix at $\mathbf{x}_k$, $\mathbf{A}_{a_k}$, has rank n, then $\mathbf{x}_k$ itself is already a vertex. So let us assume that rank$(\mathbf{A}_{a_k}) < n$. From a linear algebra perspective, the basic idea here is to generate a feasible point $\mathbf{x}_{k+1}$ such that the active constraint matrix at $\mathbf{x}_{k+1}$, $\mathbf{A}_{a_{k+1}}$, is an *augmented* version of $\mathbf{A}_{a_k}$ with rank$(\mathbf{A}_{a_{k+1}})$ increased by one. In other words, $\mathbf{x}_{k+1}$ is a point such that (a) it is feasible, (b) all the constraints that are active at $\mathbf{x}_k$ remain active at $\mathbf{x}_{k+1}$, and (c) there is a new active constraint at $\mathbf{x}_{k+1}$, which was inactive at $\mathbf{x}_k$. In this way, a vertex can be identified in a finite number of steps.

Let

$$\mathbf{x}_{k+1} = \mathbf{x}_k + \alpha_k \mathbf{d}_k \tag{11.17}$$

To assure that all active constraints at $\mathbf{x}_k$ remain active at $\mathbf{x}_{k+1}$, we must have

$$\mathbf{A}_{a_k} \mathbf{x}_{k+1} = \mathbf{b}_{a_k}$$

where $\mathbf{b}_{a_k}$ is composed of the entries of $\mathbf{b}$ that are associated with the constraints which are active at $\mathbf{x}_k$. Since $\mathbf{A}_{a_k} \mathbf{x}_k = \mathbf{b}_{a_k}$, it follows that

$$\mathbf{A}_{a_k} \mathbf{d}_k = \mathbf{0} \tag{11.18}$$

Since $\text{rank}(\mathbf{A}_{a_k}) < n$, the solutions of Eq. (11.18) form the null space of $\mathbf{A}_{a_k}$ of dimension $n - \text{rank}(\mathbf{A}_{a_k})$. Now for a fixed $\mathbf{x}_k$ and $\mathbf{d}_k \in \mathcal{N}(\mathbf{A}_{a_k})$, we call an inactive constraint $\mathbf{a}_i^T \mathbf{x}_k - b_i < 0$ *increasing* with respect to $\mathbf{d}_k$ if $\mathbf{a}_i^T \mathbf{d}_k > 0$. If the ith constraint is an increasing constraint with respect to $\mathbf{d}_k$, then moving from $\mathbf{x}_k$ to $\mathbf{x}_{k+1}$ along $\mathbf{d}_k$, the constraint becomes

$$\begin{aligned} \mathbf{a}_i^T \mathbf{x}_{k+1} - b_i &= \mathbf{a}_i^T (\mathbf{x}_k + \alpha_k \mathbf{d}_k) - b_i \\ &= (\mathbf{a}_i^T \mathbf{x}_k - b_i) + \alpha_k \mathbf{a}_i^T \mathbf{d}_k \end{aligned}$$

with $\mathbf{a}_i^T \mathbf{x}_k - b_i < 0$ and $\mathbf{a}_i^T \mathbf{d}_k > 0$. A positive α_k that makes the ith constraint active at point $\mathbf{x}_{k+1}$ can be identified as

$$\alpha_k = -\frac{(\mathbf{a}_i^T \mathbf{x}_k - b_i)}{\mathbf{a}_i^T \mathbf{d}_k} \tag{11.19}$$

It should be stressed, however, that moving the point along $\mathbf{d}_k$ also affects other inactive constraints and care must be taken to ensure that the value of α_k used does not lead to an infeasible $\mathbf{x}_{k+1}$. From the above discussion, we note two problems that need to be addressed, namely, how to find a direction $\mathbf{d}_k$ in the null space $\mathcal{N}(\mathbf{A}_{a_k})$ such that there is at least one decreasing constraint with respect to $\mathbf{d}_k$ and, if such a $\mathbf{d}_k$ is found, how to determine the step size α_k in Eq. (11.17).

Given $\mathbf{x}_k$ and $\mathbf{A}_{a_k}$, we can find an inactive constraint whose normal $\mathbf{a}_i^T$ is linearly independent of the rows of $\mathbf{A}_{a_k}$. It follows that the system of equations

$$\begin{bmatrix} \mathbf{A}_{a_k} \\ \mathbf{a}_i^T \end{bmatrix} \mathbf{d}_k = \begin{bmatrix} \mathbf{0} \\ 1 \end{bmatrix} \tag{11.20}$$

has a solution $\mathbf{d}_k$ with $\mathbf{d}_k \in \mathcal{N}(\mathbf{A}_{a_k})$ and $\mathbf{a}_i^T \mathbf{d}_k > 0$. Having determined $\mathbf{d}_k$, the set of indices corresponding to increasing constraints with respect to $\mathbf{d}_k$ can be defined as

$$\mathcal{I}_k = \{i : \mathbf{a}_i^T \mathbf{x}_k - b_i < 0, \ \mathbf{a}_i^T \mathbf{d}_k > 0\}$$

The value of α_k can be determined as the value for which $\mathbf{x}_k + \alpha_k \mathbf{d}_k$ intersects the nearest new constraint. Hence, α_k can be calculated as

$$\alpha_k = \min_{i \in \mathcal{I}_k} \left(-\frac{(\mathbf{a}_i^T \mathbf{x}_k - b_i)}{\mathbf{a}_i^T \mathbf{d}_k} \right) \tag{11.21}$$

If $i = i^*$ is an index in $\mathcal{I}_k$ that yields the α_k in Eq. (11.21), then it is quite clear that at point $\mathbf{x}_{k+1} = \mathbf{x}_k + \alpha_k \mathbf{d}_k$ the active constraint matrix becomes

$$\mathbf{A}_{a_{k+1}} = \begin{bmatrix} \mathbf{A}_{a_k} \\ \mathbf{a}_{i*}^T \end{bmatrix} \tag{11.22}$$

where $\text{rank}(\mathbf{A}_{a_{k+1}}) = \text{rank}(\mathbf{A}_{a_k}) + 1$. By repeating the above steps, a feasible point $\mathbf{x}_K$ with $\text{rank}(\mathbf{A}_{a_K}) = n$ will eventually be reached, and point $\mathbf{x}_K$ is then deemed to be a vertex.

Example 11.3 Starting from point $\mathbf{x}_0 = [1 \ 1]^T$, apply the iterative procedure described above to find a vertex for the LP problem in Example 11.2.

Solution Since the components of the residual vector at $\mathbf{x}_0$, namely,

$$\mathbf{r}_0 = \mathbf{A}\mathbf{x}_0 - \mathbf{b} = \begin{bmatrix} -1 \\ -1 \\ -1 \\ -1.5 \\ -3 \end{bmatrix}$$

are all negative, there are no active constraints at $\mathbf{x}_0$. If the first constraint (whose residual is the smallest) is chosen to form equation Eq. (11.20), we have

$$[-1 \ 0]\mathbf{d}_0 = 1$$

which has a (nonunique) solution $\mathbf{d}_0 = [-1 \ 0]^T$. The set $\mathcal{I}_0$ in this case contains only one index, i.e.,

$$\mathcal{I}_0 = \{1\}$$

Using Eq. (11.21), we obtain $\alpha_0 = 1$ with $i^* = 1$. Hence

$$\mathbf{x}_1 = \mathbf{x}_0 + \alpha_0 \mathbf{d}_0 = \begin{bmatrix} 1 \\ 1 \end{bmatrix} + \begin{bmatrix} -1 \\ 0 \end{bmatrix} = \begin{bmatrix} 0 \\ 1 \end{bmatrix}$$

with

$$\mathbf{A}_{a_1} = [-1 \ 0]$$

At point $\mathbf{x}_1$, the residual vector is given by

$$\mathbf{r}_1 = \mathbf{A}\mathbf{x}_1 - \mathbf{b} = \begin{bmatrix} 0 \\ -2 \\ -1 \\ -2.5 \\ -4 \end{bmatrix}$$

Now if the third constraint (whose residual is the smallest) is chosen to form

$$\begin{bmatrix} \mathbf{A}_{a_1} \\ \mathbf{a}_3^T \end{bmatrix} \mathbf{d}_1 = \begin{bmatrix} -1 & 0 \\ 0 & -1 \end{bmatrix} \mathbf{d}_1 = \begin{bmatrix} 0 \\ 1 \end{bmatrix}$$

we obtain $\mathbf{d}_1 = [0 \ -1]^T$. It follows that

$$\mathcal{I}_1 = \{3\}$$

From Eq. (11.21), $\alpha_1 = 1$ with $i^* = 3$ and, therefore,

$$\mathbf{x}_2 = \mathbf{x}_1 + \alpha_1 \mathbf{d}_1 = \begin{bmatrix} 0 \\ 0 \end{bmatrix}$$

with

$$\mathbf{A}_{a_2} = \begin{bmatrix} -1 & 0 \\ 0 & -1 \end{bmatrix}$$

Since $\text{rank}(\mathbf{A}_{a_2}) = 2 = n$, $\mathbf{x}_2$ is a vertex. A graphical illustration of this solution procedure is shown in Fig. 11.6. ∎

The iterative strategy described above can also be applied to the standard-form LP problem in Eqs. (11.1a)–(11.1c). Note that the presence of the equality constraints in Eq. (11.1b) means that at any feasible point $\mathbf{x}_k$, the active constraint matrix $\mathbf{A}_{a_k}$ always contains $\mathbf{A}$ as a submatrix.

Example 11.4 Find a vertex for the convex polygon

$$x_1 + x_2 + x_3 = 1$$

such that

$$\mathbf{x} \geq \mathbf{0}$$

starting with $\mathbf{x}_0 = [\frac{1}{3} \ \frac{1}{3} \ \frac{1}{3}]^T$.

Fig. 11.6 Search path for a
vertex starting from point $\mathbf{x}_0$

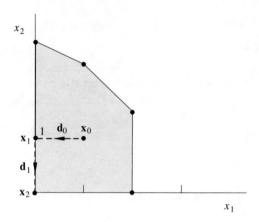

Solution At $\mathbf{x}_0$, matrix $\mathbf{A}_{a_0}$ is given by $\mathbf{A}_{a_0} = [1\ 1\ 1]$. Note that the residual vector at $\mathbf{x}_k$ for the standard-form LP problem is always given by $\mathbf{r}_k = \mathbf{x}_k$. Hence $\mathbf{r}_0 = \left[\frac{1}{3}\ \frac{1}{3}\ \frac{1}{3}\right]^T$. If the first inequality constraint is chosen to form Eq. (11.20), then we have

$$\begin{bmatrix} 1 & 1 & 1 \\ 1 & 0 & 0 \end{bmatrix} \mathbf{d}_0 = \begin{bmatrix} 0 \\ -1 \end{bmatrix}$$

which has a (nonunique) solution $\mathbf{d}_0 = [-1\ 1\ 0]^T$. It follows that

$$\mathcal{I}_0 = \{1\}$$

and

$$\alpha_0 = \tfrac{1}{3} \quad \text{with } i^* = 1$$

Hence

$$\mathbf{x}_1 = \mathbf{x}_0 + \alpha_0 \mathbf{d}_0 = \begin{bmatrix} 0 \\ \frac{2}{3} \\ \frac{1}{3} \end{bmatrix}$$

At $\mathbf{x}_1$,

$$\mathbf{A}_{a_1} = \begin{bmatrix} 1 & 1 & 1 \\ 1 & 0 & 0 \end{bmatrix}$$

and $\mathbf{r}_1 = \left[0\ \frac{2}{3}\ \frac{1}{3}\right]^T$. Choosing the third inequality constraint to form Eq. (11.20), we have

Fig. 11.7 Search path for a
vertex for Example 11.4

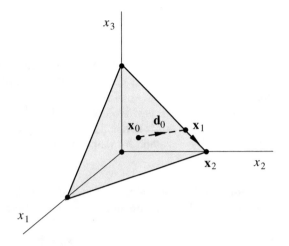

$$\begin{bmatrix} 1 & 1 & 1 \\ 1 & 0 & 0 \\ 0 & 0 & 1 \end{bmatrix} \mathbf{d}_1 = \begin{bmatrix} 0 \\ 0 \\ -1 \end{bmatrix}$$

which leads to $\mathbf{d}_1 = [0 \ 1 \ -1]^T$. Consequently,

$$\mathcal{I}_1 = \{3\}$$

and

$$\alpha_1 = \tfrac{1}{3} \quad \text{with } i^* = 3$$

Therefore,

$$\mathbf{x}_2 = \mathbf{x}_1 + \alpha_1 \mathbf{d}_1 = \begin{bmatrix} 0 \\ 1 \\ 0 \end{bmatrix}$$

At $\mathbf{x}_2$,

$$\mathbf{A}_{a_2} = \begin{bmatrix} 1 & 1 & 1 \\ 1 & 0 & 0 \\ 0 & 0 & 1 \end{bmatrix}$$

Hence rank($\mathbf{A}_{a_2}$) = 3, indicating that $\mathbf{x}_2$ is a vertex. The search path that leads to
vertex $\mathbf{x}_2$ is illustrated in Fig. 11.7.   ∎

11.2.3.4   Two Implementation Issues

There are two issues that have to be dealt with when using the method in Sect. 11.2.3.3 to find a vertex. First, as in the case of any iterative optimization method, we need to identify a feasible point. As will be shown shortly, this problem can itself be treated as an LP problem, which is often referred to as a *phase-1 LP problem*. Second, in order to move from point $\mathbf{x}_k$ to point $\mathbf{x}_{k+1}$, we need to identify a constraint, say, the ith constraint, which is inactive at $\mathbf{x}_k$ such that $\mathbf{a}_i^T$ is linearly independent of the rows of $\mathbf{A}_{a_k}$. Obviously, this is a rank determination problem. Later on in this subsection, we will describe a method for rank determination based on the QR decomposition of matrix $\mathbf{A}_{a_k}$.

I. Finding a feasible point.   Finding a feasible point for the LP problem in Eqs. (11.2a) and (11.2b) amounts to finding a vector $\mathbf{x}_0 \in R^{n \times 1}$ such that

$$\mathbf{A}\mathbf{x}_0 \leq \mathbf{b}$$

To this end, we consider the modified constraints

$$\mathbf{A}\mathbf{x} + \phi\mathbf{e} \leq \mathbf{b} \qquad (11.23)$$

where ϕ is an auxiliary scalar variable and $\mathbf{e} = [1 \; 1 \; \cdots \; 1]^T$. Evidently, if $\mathbf{x} = \mathbf{0}$ and $\phi = \phi_0 = \min(0, \; b_1, \; b_2, \; \ldots, \; b_p)$ in Eq. (11.23) where b_i is the ith component of $\mathbf{b}$ in Eq. (11.23), then $\phi \leq 0$ and Eq. (11.23) is satisfied because

$$\mathbf{A}\mathbf{0} + \phi_0\mathbf{e} \leq \mathbf{b}$$

In other words, if we define the augmented vector $\hat{\mathbf{x}}$ as

$$\hat{\mathbf{x}} = \begin{bmatrix} \mathbf{x} \\ \phi \end{bmatrix}$$

then the initial value

$$\hat{\mathbf{x}}_0 = \begin{bmatrix} \mathbf{0} \\ \phi_0 \end{bmatrix} \qquad (11.24)$$

satisfies the constraints in Eq. (11.23). This suggests that a phase-1 LP problem can be formulated as

$$\text{minimize} \; -\phi \qquad (11.25a)$$

$$\text{subject to:}\quad \mathbf{A}\mathbf{x} + \phi\mathbf{e} \le \mathbf{b} \tag{11.25b}$$

$$\phi \le 0 \tag{11.25c}$$

A feasible initial point for this problem is given by Eq. (11.24). If the solution is assumed to be

$$\hat{\mathbf{x}}^* = \begin{bmatrix} \mathbf{x}^* \\ \phi^* \end{bmatrix} = \begin{bmatrix} \mathbf{x}^* \\ 0 \end{bmatrix}$$

then at $\hat{\mathbf{x}}^*$ the constraints in Eq. (11.25b) become $\mathbf{A}\mathbf{x}^* \le \mathbf{b}$ and hence $\mathbf{x}^*$ is a feasible point for the original LP problem. If $\phi^* < 0$, we conclude that no feasible point exists for constraints $\mathbf{A}\mathbf{x} \le \mathbf{b}$ and ϕ^* then represents a single perturbation of the constraint in Eq. (11.2b) with minimum L_∞ norm to ensure feasibility. In effect, point $\mathbf{x}^*$ would become feasible if the constraints were modified to

$$\mathbf{A}\mathbf{x} \le \tilde{\mathbf{b}} \quad \text{with } \tilde{\mathbf{b}} = \mathbf{b} - \phi^*\mathbf{e} \tag{11.26}$$

II. Finding a linearly independent $\mathbf{a}_i^T$. Assume that at $\mathbf{x}_k$, rank$(\mathbf{A}_{a_k}) = r_k$ with $r_k < n$. Finding a normal vector $\mathbf{a}_i^T$ associated with an inactive constraint at $\mathbf{x}_k$ such that $\mathbf{a}_i^T$ is linearly independent of the rows of $\mathbf{A}_{a_k}$ is equivalent to finding an $\mathbf{a}_i^T$ such that rank$(\hat{\mathbf{A}}_{a_k}) = r_k + 1$ where

$$\hat{\mathbf{A}}_{a_k} = \begin{bmatrix} \mathbf{A}_{a_k} \\ \mathbf{a}_i^T \end{bmatrix} \tag{11.27}$$

An effective way of finding the rank of a matrix obtained through finite-precision computations is to perform QR decomposition with column pivoting, which can be done through the use of the Householder QR decomposition described in Sect. A.12.2 (see also [5, Chap. 5]). On applying this procedure to a matrix $\mathbf{M} \in R^{n \times m}$ with $m \le n$, after r steps of the procedure we obtain

$$\mathbf{M}\mathbf{P}^{(r)} = \mathbf{Q}^{(r)}\mathbf{R}^{(r)}$$

where $\mathbf{Q}^{(r)} \in R^{n \times n}$ is an orthogonal matrix, $\mathbf{P}^{(r)} \in R^{m \times m}$ is a permutation matrix, and

$$\mathbf{R}^{(r)} = \begin{bmatrix} \mathbf{R}_{11}^{(r)} & \mathbf{R}_{12}^{(r)} \\ \mathbf{0} & \mathbf{R}_{22}^{(r)} \end{bmatrix}$$

where $\mathbf{R}_{11}^{(r)} \in R^{r \times r}$ is nonsingular and upper triangular. If $||\mathbf{R}_{22}^{(r)}||_2$ is negligible, then the numerical rank of $\mathbf{M}$ is deemed to be r. A reasonable condition for terminating the QR decomposition is

$$||\mathbf{R}_{22}^{(r)}||_2 \leq \varepsilon ||\mathbf{M}||_2 \tag{11.28}$$

where ε is some small machine-dependent parameter. When Eq. (11.28) is satisfied, block $\mathbf{R}_{22}^{(r)}$ is set to zero and the QR decomposition of $\mathbf{M}$ becomes

$$\mathbf{MP} = \mathbf{QR}$$

where $\mathbf{P} = \mathbf{P}^{(r)}$, $\mathbf{Q} = \mathbf{Q}^{(r)}$, and

$$\mathbf{R} = \begin{bmatrix} \mathbf{R}_{11}^{(r)} & \mathbf{R}_{12}^{(r)} \\ \mathbf{0} & \mathbf{0} \end{bmatrix} \tag{11.29}$$

For matrix $\mathbf{A}_{a_k}$ in Eq. (11.27), the above QR decomposition can be applied to $\mathbf{A}_{a_k}^T \in R^{n \times r}$ with $r < n$, i.e.,

$$\mathbf{A}_{a_k}^T \mathbf{P} = \mathbf{QR} \tag{11.30}$$

where $\mathbf{R}$ has the form of Eq. (11.29), and the size of $\mathbf{R}_{11}^{(r)}$ gives the rank of $\mathbf{A}_{a_k}$. A nice feature of the QR decomposition method is that if matrix $\mathbf{A}_{a_k}$ is altered in some way, for example, by adding a rank-one matrix or appending a row (or column) to it or deleting a row (or column) from it, the QR decomposition of the altered matrix can be obtained based on the QR decomposition of matrix $\mathbf{A}_{a_k}$ with a computationally simple updating procedure (see Sect. A.12 and [5, Chap. 12]). In the present case, we are interested in the QR decomposition of $\hat{\mathbf{A}}_{a_k}^T$ in Eq. (11.27), which is obtained from $\mathbf{A}_{a_k}^T$ by appending $\mathbf{a}_i$ as the last column. If we let

$$\hat{\mathbf{P}} = \begin{bmatrix} \mathbf{P} & \mathbf{0} \\ \mathbf{0} & 1 \end{bmatrix}$$

it follows from Eq. (11.30) that

$$\begin{aligned} \mathbf{Q}^T \hat{\mathbf{A}}_{a_k}^T \hat{\mathbf{P}} &= \mathbf{Q}^T [\mathbf{A}_{a_k}^T \ \mathbf{a}_i] \hat{\mathbf{P}} \\ &= [\mathbf{Q}^T \mathbf{A}_{a_k}^T \mathbf{P} \ \ \mathbf{Q}^T \mathbf{a}_i] = [\mathbf{R} \ \mathbf{w}_i] \\ &= \begin{bmatrix} \mathbf{R}_{11}^{(r)} & \mathbf{R}_{12}^{(r)} & \vdots \\ - - - & - - - & \vdots \ \mathbf{w}_i \\ \mathbf{0} & \mathbf{0} & \vdots \end{bmatrix} \end{aligned} \tag{11.31}$$

where $\mathbf{w}_i = \mathbf{Q}^T \mathbf{a}_i$ is a column vector with n entries. Note that if we apply $n - r + 1$ Givens rotations (see Sect. A.11.2 and [5, Chap. 5]) $\mathbf{J}_l^T$ for $1 \leq l \leq n - r - 1$ to $\mathbf{w}_i$ successively so that

$$
\mathbf{J}_{n-r-1}^T \cdots \mathbf{J}_2^T \cdot \mathbf{J}_1^T \mathbf{w}_i = \begin{bmatrix} \psi_1 \\ \vdots \\ \psi_r \\ \psi_{r+1} \\ 0 \\ \vdots \\ 0 \end{bmatrix} \Big\} (n-r-1) \text{ zeros} \tag{11.32}
$$

then the structure of $\mathbf{R}$ is not changed. Now by defining

$$
\mathbf{J} = \mathbf{J}_1 \times \mathbf{J}_2 \times \cdots \times \mathbf{J}_{n-r-1} \quad \text{and} \quad \hat{\mathbf{Q}} = \mathbf{Q}\mathbf{J}
$$

$$
\hat{\mathbf{Q}}^T \hat{\mathbf{A}}_{a_k}^T \hat{\mathbf{P}} = \mathbf{J}[\mathbf{R}\ \mathbf{w}_i] = \begin{bmatrix} \hat{\mathbf{R}}_{11}^{(r)} & \hat{\mathbf{R}}_{12}^{(r)} & \begin{matrix} \psi_1 \\ \vdots \\ \psi_r \\ \psi_{r+1} \\ 0 \end{matrix} \\ \mathbf{0} & \mathbf{0} & \begin{matrix} \vdots \\ 0 \\ 0 \end{matrix} \end{bmatrix} \tag{11.33}
$$

where $\hat{\mathbf{R}}_{11}^{(r)}$ is an $r \times r$ nonsingular upper triangular matrix. If ψ_{r+1} is negligible, then rank$(\hat{\mathbf{A}}_{a_k})$ may be deemed to be r; hence $\mathbf{a}_i^T$ is not linearly independent of the rows of $\mathbf{A}_{a_k}$. However, if ψ_{r+1} is not negligible, then Eq. (11.33) shows that rank$(\hat{\mathbf{A}}_{a_k}) = r + 1$ so that $\mathbf{a}_i^T$ is a desirable vector for Eq. (11.20). By applying a permutation matrix $\mathbf{P}_a$ to Eq. (11.33) to interchange the $(r + 1)$th column with the last column of the matrix on the right-hand side, the updated QR decomposition of $\hat{\mathbf{A}}_{a_k}^T$ is obtained as

$$
\mathbf{A}_{a_k}^T \tilde{\mathbf{P}} = \hat{\mathbf{Q}}\hat{\mathbf{R}}
$$

where $\tilde{\mathbf{P}} = \hat{\mathbf{P}}\mathbf{P}_a$ is a permutation matrix and $\hat{\mathbf{R}}$ is given by

$$
\mathbf{R} = \begin{bmatrix} \hat{\mathbf{R}}_{11}^{(r+1)} & \tilde{\mathbf{R}}_{12}^{(r+1)} \\ \mathbf{0} & \mathbf{0} \end{bmatrix}
$$

with

$$
\hat{\mathbf{R}}_{11}^{(r+1)} = \begin{bmatrix} \mathbf{R} & \begin{matrix} \psi_1 \\ \vdots \\ \psi_r \end{matrix} \\ \mathbf{0} & \psi_{r+1} \end{bmatrix}
$$

11.2.4 Vertex Minimizers

11.2.4.1 Finding a Vertex Minimizer

The iterative method for finding a vertex described in Sect. 11.2.3.3 does not involve the objective function $f(\mathbf{x}) = \mathbf{c}^T\mathbf{x}$. Consequently, the vertex obtained may not be a minimizer (see Example 11.3). However, as will be shown in the next theorem, if we start the iterative method at a minimizer, a vertex would eventually be reached without increasing the objective function which would, therefore, be a vertex minimizer.

Theorem 11.5 (Existence of a vertex minimizer in alternative-form LP problem) *If the minimum of $f(\mathbf{x})$ in the alternative-form LP problem of Eqs. (11.2a) and (11.2b) is finite, then there is a vertex minimizer.*

Proof If $\mathbf{x}_0$ is a minimizer, then $\mathbf{x}_0$ is finite and satisfies the conditions stated in Theorem 11.2. Hence there exists a $\boldsymbol{\mu}^* \geq \mathbf{0}$ such that

$$\mathbf{c} + \mathbf{A}^T\boldsymbol{\mu}^* = \mathbf{0} \tag{11.34}$$

By virtue of the complementarity condition in Eqs. 11.8c, 11.34 can be written as

$$\mathbf{c} + \mathbf{A}_{a_0}^T\boldsymbol{\mu}_a^* = \mathbf{0} \tag{11.35}$$

where $\mathbf{A}_{a_0}$ is the active constraint matrix at $\mathbf{x}_0$ and $\boldsymbol{\mu}_a^*$ is composed of the entries of $\boldsymbol{\mu}^*$ that correspond to the active constraints. If $\mathbf{x}_0$ is not a vertex, the method described in Sect. 11.2.3.3 can be applied to yield a point $\mathbf{x}_1 = \mathbf{x}_0 + \alpha_0\mathbf{d}_0$ which is closer to a vertex, where $\mathbf{d}_0$ is a feasible direction that satisfies the condition $\mathbf{A}_{a_0}\mathbf{d}_0 = \mathbf{0}$ (see Eq. (11.18)). It follows that at $\mathbf{x}_1$ the objective function remains the same as at $\mathbf{x}_0$, i.e.,

$$f(\mathbf{x}_1) = \mathbf{c}^T\mathbf{x}_1 = \mathbf{c}^T\mathbf{x}_0 - \alpha_0\boldsymbol{\mu}_a^{*T}\mathbf{A}_{a_0}\mathbf{d}_0 = \mathbf{c}^T\mathbf{x}_0 = f(\mathbf{x}_0)$$

which means that $\mathbf{x}_1$ is a minimizer. If $\mathbf{x}_1$ is not yet a vertex, then the process is continued to generate minimizers $\mathbf{x}_2$, $\mathbf{x}_3$, ... until a vertex minimizer is reached. ∎

Theorem 11.5 also applies to the standard-form LP problem in Eqs. (11.1a)–(11.1c). To prove this, let $\mathbf{x}_0$ be a finite minimizer of Eqs. (11.1a)–(11.1c). It follows from Eq. (11.5c) that

$$\mathbf{c} = -\mathbf{A}^T\boldsymbol{\lambda}^* + \boldsymbol{\mu}^* \tag{11.36}$$

The complementarity condition implies that Eq. (11.36) can be written as

$$\mathbf{c} = -\mathbf{A}^T\boldsymbol{\lambda}^* + \mathbf{I}_0^T\boldsymbol{\mu}_a^* \tag{11.37}$$

where $\mathbf{I}_0$ consists of the rows of the $n \times n$ identity matrix that are associated with the inequality constraints in Eq. (11.3c) that are active at $\mathbf{x}_0$, and $\boldsymbol{\mu}_a^*$ is composed of the entries of $\boldsymbol{\mu}^*$ that correspond to the active (inequality) constraints. At $\mathbf{x}_0$, the active constraint matrix $\mathbf{A}_{a0}$ is given by

$$\mathbf{A}_{a0} = \begin{bmatrix} -\mathbf{A} \\ \mathbf{I}_0 \end{bmatrix} \qquad (11.38)$$

Hence Eq. (11.37) becomes

$$\mathbf{c} = \mathbf{A}_{a0}^T \boldsymbol{\eta}_a^* \quad \text{with} \quad \boldsymbol{\eta}_a^* = \begin{bmatrix} \boldsymbol{\lambda}^* \\ \boldsymbol{\mu}_a^* \end{bmatrix} \qquad (11.39)$$

which is the counterpart of Eq. (11.35) for the problem in Eqs. (11.1a)–(11.1c). The rest of the proof is identical with that of Theorem 11.5. We can, therefore, state the following theorem.

Theorem 11.6 (Existence of a vertex minimizer in standard-form LP problem)   *If the minimum of $f(\mathbf{x})$ in the LP problem of Eqs. (11.1a)–(11.1c) is finite, then a vertex minimizer exists.*   ∎

11.2.4.2   Uniqueness
A key feature in the proofs of Theorems 11.5 and 11.6 is the connection of vector $\mathbf{c}$ to the active constraints as described by Eqs. (11.35) and (11.39) through the Lagrange multipliers $\boldsymbol{\mu}^*$ and $\boldsymbol{\lambda}^*$. As will be shown in the next theorem, the Lagrange multipliers also play a critical role in the uniqueness of a vertex minimizer.

Theorem 11.7 (Uniqueness of minimizer of alternative-form LP problem)   *Let $\mathbf{x}^*$ be a vertex minimizer of the LP problem in Eqs. (11.2a) and (11.2b) at which*

$$\mathbf{c} + \mathbf{A}_{a*}^T \boldsymbol{\mu}_a^* = \mathbf{0}$$

where $\boldsymbol{\mu}_a^* \geq \mathbf{0}$ is defined in the proof of Theorem 11.5. If $\boldsymbol{\mu}_a^* > \mathbf{0}$, then $\mathbf{x}^*$ is the unique vertex minimizer of Eqs. (11.2a) and (11.2b).

Proof   Let us suppose that there is another vertex minimizer $\tilde{\mathbf{x}} \neq \mathbf{x}^*$. We can write

$$\tilde{\mathbf{x}} = \mathbf{x}^* + \mathbf{d}$$

with $\mathbf{d} = \tilde{\mathbf{x}} - \mathbf{x}^* \neq \mathbf{0}$. Since both $\mathbf{x}^*$ and $\tilde{\mathbf{x}}$ are feasible, $\mathbf{d}$ is a feasible direction, which implies that $\mathbf{A}_{a*}\mathbf{d} \leq \mathbf{0}$. Since $\mathbf{x}^*$ is a vertex, $\mathbf{A}_{a*}$ is nonsingular; hence $\mathbf{A}_{a*}\mathbf{d} \leq \mathbf{0}$

together with $\mathbf{d} \neq \mathbf{0}$ implies that at least one component of $\mathbf{A}_{a*}\mathbf{d}$, say, $(\mathbf{A}_{a*}\mathbf{d})_i$, is strictly negative. We then have

$$0 = f(\tilde{\mathbf{x}}) - f(\mathbf{x}^*) = \mathbf{c}^T\tilde{\mathbf{x}} - \mathbf{c}^T\mathbf{x}^* = \mathbf{c}^T\mathbf{d}$$
$$= -\boldsymbol{\mu}_a^{*T}\mathbf{A}_{a*}\mathbf{d} \geq -(\boldsymbol{\mu}_a^*)_i \cdot (\mathbf{A}_{a*}\mathbf{d})_i > 0$$

The above contradiction implies that another minimizer $\tilde{\mathbf{x}}$ cannot exist.   ∎

For the standard-form LP problem in Eqs. (11.1a)–(11.1c), the following theorem applies.

Theorem 11.8 (Uniqueness of minimizer of standard-form LP problem) *Consider the LP problem in Eqs. (11.1a)–(11.1c) and let* $\mathbf{x}^*$ *be a vertex minimizer at which*

$$\mathbf{c} = \mathbf{A}_{a*}^T\boldsymbol{\eta}_a^*$$

with

$$\mathbf{A}_{a*} = \begin{bmatrix} -\mathbf{A} \\ \mathbf{I}_* \end{bmatrix}, \quad \boldsymbol{\eta}_a^* = \begin{bmatrix} \boldsymbol{\lambda}^* \\ \boldsymbol{\mu}_a^* \end{bmatrix}$$

where $\mathbf{I}_*$ *consists of the rows of the* $n \times n$ *identity matrix that are associated with the inequality constraints in Eq. (11.1c) that are active at* $\mathbf{x}^*$, $\boldsymbol{\lambda}^*$ *and* $\boldsymbol{\mu}^*$ *are the Lagrange multipliers in Eq. (11.5c), and* $\boldsymbol{\mu}_a^*$ *consists of the entries of* $\boldsymbol{\mu}^*$ *associated with active (inequality) constraints. If* $\boldsymbol{\mu}_a^* > \mathbf{0}$, *then* $\mathbf{x}^*$ *is the unique vertex minimizer of the problem in Eqs. (11.1a)–(11.1c).*   ∎

Theorem 11.8 can be proved by assuming that there is another minimizer $\tilde{\mathbf{x}}$ and then using an argument similar to that in the proof of Theorem 11.7 with some minor modifications. Direction $\mathbf{c}$ being feasible implies that

$$\mathbf{A}_{a*}\mathbf{d} = \begin{bmatrix} -\mathbf{A}\mathbf{d} \\ \mathbf{I}_*\mathbf{d} \end{bmatrix} = \begin{bmatrix} \mathbf{0} \\ \mathbf{I}_*\mathbf{d} \end{bmatrix} \geq \mathbf{0} \tag{11.40}$$

where $\mathbf{I}_*\mathbf{d}$ consists of the components of $\mathbf{d}$ that are associated with the active (inequality) constraints at $\mathbf{x}^*$. Since $\mathbf{A}_{a*}$ is nonsingular, Eq. (11.40) in conjunction with $\mathbf{d} \neq \mathbf{0}$ implies that at least one component of $\mathbf{I}_*\mathbf{d}$, say, $(\mathbf{I}_*\mathbf{d})_i$, is strictly positive. This yields the contradiction

$$0 = f(\tilde{\mathbf{x}}) - f(\mathbf{x}^*) = \mathbf{c}^T\mathbf{d}$$
$$= [\boldsymbol{\lambda}^{*T} \ \boldsymbol{\mu}_a^{*T}]\mathbf{A}_a\mathbf{d} = \boldsymbol{\mu}_a^{*T}\mathbf{I}_*\mathbf{d} \geq (\boldsymbol{\mu}_a^*)_i \cdot (\mathbf{I}_*\mathbf{d})_i$$
$$> 0$$

The strict positiveness of the Lagrange multiplier $\boldsymbol{\mu}_a^*$ is critical for the uniqueness of the solution. As a matter of fact, if the vertex minimizer $\mathbf{x}^*$ is *nondegenerate* (see Definition 11.1 in Sect. 11.2.3.1), then any zero entries in $\boldsymbol{\mu}_a^*$ imply the nonuniqueness of the solution. The reader is referred to [3, Sect. 7.7] for the details.

11.3 Simplex Method

11.3.1 Simplex Method for Alternative-Form LP Problem

In this section, we consider a general method for the solution of the LP problem in Eqs. (11.2a) and (11.2b) known as the *simplex method*. It was shown in Sect. 11.2.4 that if the minimum value of the objective function in the feasible region is finite, then a vertex minimizer exists. Let $\mathbf{x}_0$ be a vertex and assume that it is not a minimizer. The simplex method generates an adjacent vertex $\mathbf{x}_1$ with $f(\mathbf{x}_1) < f(\mathbf{x}_0)$ and continues doing so until a vertex minimizer is reached.

11.3.1.1 Nondegenerate Case

To simplify our discussion, we assume that all vertices are nondegenerate, i.e., at a vertex there are exactly n active constraints. This assumption is often referred to as the *nondegeneracy assumption* [3] in the literature.

Given a vertex $\mathbf{x}_k$, a vertex $\mathbf{x}_{k+1}$ is said to be *adjacent* to $\mathbf{x}_k$ if $\mathbf{A}_{a_{k+1}}$ differs from $\mathbf{A}_{a_k}$ by *one* row. In terms of the notation used in Sect. 11.2.3.2, we denote $\mathbf{A}_{a_k}$ as

$$\mathbf{A}_{a_k} = \begin{bmatrix} \mathbf{a}_{j_1}^T \\ \mathbf{a}_{j_2}^T \\ \vdots \\ \mathbf{a}_{j_n}^T \end{bmatrix}$$

where $\mathbf{a}_{j_l}$ is the normal of the j_lth constraint in Eq. (11.2b). Associated with $\mathbf{A}_{a_k}$ is the index set

$$\mathcal{J}_k = \{j_1, \ j_2, \ \ldots, \ j_n\}$$

Obviously, if $\mathcal{J}_k$ and $\mathcal{J}_{k+1}$ have exactly $(n-1)$ members, vertices $\mathbf{x}_k$ and $\mathbf{x}_{k+1}$ are adjacent. At vertex $\mathbf{x}_k$, the simplex method verifies whether $\mathbf{x}_k$ is a vertex minimizer, and if it is not, it finds an adjacent vertex $\mathbf{x}_{k+1}$ that yields a *reduced* value of the objective function. Since a vertex minimizer exists and there is only a finite number of vertices, the simplex method will find a solution after a finite number of iterations.

Under the nondegeneracy assumption, $\mathbf{A}_{a_k}$ is square and nonsingular. Hence there exists a $\boldsymbol{\mu}_k \in R^{n \times 1}$ such that

$$\mathbf{c} + \mathbf{A}_{a_k}^T \boldsymbol{\mu}_k = \mathbf{0} \tag{11.41}$$

Since $\mathbf{x}_k$ is a feasible point, by virtue of Theorem 11.2 we conclude that $\mathbf{x}_k$ is a vertex minimizer if and only if

$$\boldsymbol{\mu}_k \geq \mathbf{0} \tag{11.42}$$

In other words, $\mathbf{x}_k$ is not a vertex minimizer if and only if at least one component of $\boldsymbol{\mu}_k$, say, $(\boldsymbol{\mu}_k)_l$, is negative.

Assume that $\mathbf{x}_k$ is not a vertex minimizer and let

$$(\boldsymbol{\mu}_k)_l < 0 \tag{11.43}$$

The simplex method finds an edge as a feasible descent direction $\mathbf{d}_k$ that points from $\mathbf{x}_k$ to an adjacent vertex $\mathbf{x}_{k+1}$ given by

$$\mathbf{x}_{k+1} = \mathbf{x}_k + \alpha_k \mathbf{d}_k \tag{11.44}$$

It was shown in Sect. 11.2.3.2 that a feasible descent direction $\mathbf{d}_k$ is characterized by

$$\mathbf{A}_{a_k} \mathbf{d}_k \leq \mathbf{0} \quad \text{and} \quad \mathbf{c}^T \mathbf{d}_k < 0 \tag{11.45}$$

To find an edge that satisfies Eq. (11.45), we denote the lth coordinate vector (i.e., the lth column of the $n \times n$ identity matrix) as $\mathbf{e}_l$ and examine vector $\mathbf{d}_k$ that solves the equation

$$\mathbf{A}_{a_k} \mathbf{d}_k = -\mathbf{e}_l \tag{11.46}$$

From Eq. (11.46), we note that $\mathbf{A}_{a_k} \mathbf{d}_k \leq \mathbf{0}$. From Eqs. (11.41), (11.43), and (11.46), we have

$$\mathbf{c}^T \mathbf{d}_k = -\boldsymbol{\mu}_k^T \mathbf{A}_{a_k} \mathbf{d}_k = \boldsymbol{\mu}_k^T \mathbf{e}_l = (\boldsymbol{\mu}_k)_i < 0$$

and hence $\mathbf{d}_k$ satisfies Eq. (11.45) and, therefore, it is a feasible descent direction. Moreover, for $i \neq l$ Eq. (11.46) implies that

$$\mathbf{a}_{j_i}^T (\mathbf{x}_k + \alpha \mathbf{d}_k) = \mathbf{a}_{j_i}^T \mathbf{x}_k + \alpha \mathbf{a}_{j_i}^T \mathbf{d}_k = b_{j_i}$$

Therefore, there are exactly $n - 1$ constraints that are active at $\mathbf{x}_k$ and remain active at $\mathbf{x}_k + \alpha \mathbf{d}_k$. This means that $\mathbf{x}_k + \alpha \mathbf{d}_k$ with $\alpha > 0$ is an edge that connects $\mathbf{x}_k$ to an adjacent vertex $\mathbf{x}_{k+1}$ with $f(\mathbf{x}_{k+1}) < f(\mathbf{x}_k)$. By using an argument similar to that in Sect. 11.2.3.3, the right step size α_k can be identified as

$$\alpha_k = \min_{i \in I_k} \left(\frac{-(\mathbf{a}_i^T \mathbf{x}_k - b_i)}{\mathbf{a}_i^T \mathbf{d}_k} \right) \tag{11.47}$$

where I_k contains the indices of the constraints that are inactive at $\mathbf{x}_k$ with $\mathbf{a}_i^T \mathbf{d}_k > 0$, i.e.,

$$I_k = \{i : \mathbf{a}_i^T \mathbf{x}_k - b_i < 0 \text{ and } \mathbf{a}_i^T \mathbf{d}_k > 0\} \tag{11.48}$$

Once α_k is calculated, the next vertex $\mathbf{x}_{k+1}$ is determined by using Eq. (11.44).
Now if $i^* \in \mathcal{I}_k$ is the index that achieves the minimum in Eq. (11.47), i.e.,

$$\alpha_k = \frac{-(\mathbf{a}_{i*}^T \mathbf{x}_k - b_{i*})}{\mathbf{a}_{i*}^T \mathbf{d}_k}$$

then at $\mathbf{x}_{k+1}$ the i^*th constraint becomes active. With the j_lth constraint leaving $\mathbf{A}_{a_k}$
and the i^*th constraint entering $\mathbf{A}_{a_{k+1}}$, there are exactly n active constraints at $\mathbf{x}_{k+1}$
and $\mathbf{A}_{a_{k+1}}$ given by

$$\mathbf{A}_{a_{k+1}} = \begin{bmatrix} \mathbf{a}_{j_1}^T \\ \vdots \\ \mathbf{a}_{j_{l-1}}^T \\ \mathbf{a}_{i*}^T \\ \mathbf{a}_{j_{l+1}}^T \\ \vdots \\ \mathbf{a}_{j_n}^T \end{bmatrix} \tag{11.49}$$

and the index set is given by

$$\mathcal{J}_{k+1} = \{j_1, \ldots, j_{l-1}, i^*, j_{l+1}, \ldots, j_n\} \tag{11.50}$$

A couple of remarks on the method described are in order. First, when the Lagrange
multiplier vector $\boldsymbol{\mu}_k$ determined by using Eq. (11.41) contains more than one negative
component, a 'textbook rule' is to select the index l in Eq. (11.46) that corresponds to
the most negative component in $\boldsymbol{\mu}_k$ [3]. Second, Eq. (11.47) can be modified to deal
with the LP problem in Eqs. (11.2a) and (11.2b) with an unbounded minimum. If
the LP problem at hand does not have a bounded minimum, then at some iteration k
the index set $\mathcal{I}_k$ will become empty which signifies an unbounded solution of the LP
problem. Below, we summarize an algorithm that implements the simplex method
and use two examples to illustrate its application.

Algorithm 11.1 Simplex algorithm for the alternative-form LP problem in Eqs. (11.2a) and (11.2b), nondegenerate vertices

Step 1
Input vertex $\mathbf{x}_0$, and form $\mathbf{A}_{a_0}$ and $\mathcal{J}_0$.
Set $k = 0$.
Step 2
Solve

$$\mathbf{A}_{a_k}^T \boldsymbol{\mu}_k = -\mathbf{c} \tag{11.51}$$

for $\boldsymbol{\mu}_k$.
If $\boldsymbol{\mu}_k \geq \mathbf{0}$, stop ($\mathbf{x}_k$ is a vertex minimizer); otherwise, select the index l that
corresponds to the most negative component in $\boldsymbol{\mu}_k$.

Step 3
Solve

$$\mathbf{A}_{a_k} \mathbf{d}_k = -\mathbf{e}_l \qquad (11.52)$$

for $\mathbf{d}_k$.

Step 4
Compute the residual vector

$$\mathbf{r}_k = \mathbf{A}\mathbf{x}_k - \mathbf{b} = (r_i)_{i=1}^p \qquad (11.53a)$$

If the index set

$$\mathcal{I}_k = \{i : r_i < 0 \text{ and } \mathbf{a}_i^T \mathbf{d}_k > 0\} \qquad (11.53b)$$

is empty, stop (the objective function tends to $-\infty$ in the feasible region); otherwise, compute

$$\alpha_k = \min_{i \in \mathcal{I}_k} \left(\frac{-r_i}{\mathbf{a}_i^T \mathbf{d}_k} \right) \qquad (11.53c)$$

and record the index i^* with $\alpha_k = -r_{i*}/(\mathbf{a}_{i*}^T \mathbf{d}_k)$.

Step 5
Set

$$\mathbf{x}_{k+1} = \mathbf{x}_k + \alpha_k \mathbf{d}_k \qquad (11.54)$$

Update $\mathbf{A}_{a_{k+1}}$ and $\mathcal{J}_{k+1}$ using Eqs. (11.49) and (11.50), respectively.
Set $k = k + 1$ and repeat from Step 2.

Example 11.5 Solve the LP problem in Example 11.2 with initial vertex $\mathbf{x}_0 = [2 \ 1.5]^T$ using the simplex method.

Solution From Example 11.2 and Fig. 11.3, the objective function is given by

$$f(\mathbf{x}) = \mathbf{c}^T \mathbf{x} = -x_1 - 4x_2$$

and the constraints are given by $\mathbf{A}\mathbf{x} \le \mathbf{b}$ with

$$\mathbf{A} = \begin{bmatrix} -1 & 0 \\ 1 & 0 \\ 0 & -1 \\ 1 & 1 \\ 1 & 2 \end{bmatrix} \quad \text{and} \quad \mathbf{b} = \begin{bmatrix} 0 \\ 2 \\ 0 \\ 3.5 \\ 6 \end{bmatrix}$$

We note that at vertex $\mathbf{x}_0$, the second and fourth constraints are active and hence

$$\mathbf{A}_{a_0} = \begin{bmatrix} 1 & 0 \\ 1 & 1 \end{bmatrix}, \quad \mathcal{J}_0 = \{2,\ 4\}$$

Solving $\mathbf{A}_{a_0}^T \boldsymbol{\mu}_0 = -\mathbf{c}$ for $\boldsymbol{\mu}_0$ where $\mathbf{c} = [-1\ \ -4]^T$, we obtain $\boldsymbol{\mu}_0 = [-3\ \ 4]^T$. This shows that $\mathbf{x}_0$ is not a minimizer and $l = 1$. Next we solve

$$\mathbf{A}_{a_0} \mathbf{d}_0 = -\mathbf{e}_1$$

for $\mathbf{d}_0$ to obtain $\mathbf{d}_0 = [-1\ \ 1]^T$. From Fig. 11.3, it is evident that $\mathbf{d}_0$ is a feasible descent direction at $\mathbf{x}_0$. The residual vector at $\mathbf{x}_0$ is given by

$$\mathbf{r}_0 = \mathbf{A}\mathbf{x}_0 - \mathbf{b} = \begin{bmatrix} -2 \\ 0 \\ -1.5 \\ 0 \\ -1 \end{bmatrix}$$

which shows that the first, third, and fifth constraints are inactive at $\mathbf{x}_0$. Furthermore,

$$\begin{bmatrix} \mathbf{a}_1^T \\ \mathbf{a}_3^T \\ \mathbf{a}_5^T \end{bmatrix} \mathbf{d}_0 = \begin{bmatrix} -1 & 0 \\ 0 & -1 \\ 1 & 2 \end{bmatrix} \begin{bmatrix} -1 \\ 1 \end{bmatrix} = \begin{bmatrix} 1 \\ -1 \\ 1 \end{bmatrix}$$

Hence

$$\mathcal{I}_0 = \{1,\ 5\}$$

and

$$\alpha_0 = \min \left(\frac{-r_1}{\mathbf{a}_1^T \mathbf{d}_0},\ \frac{-r_5}{\mathbf{a}_5^T \mathbf{d}_0} \right) = 1$$

The next vertex is obtained as

$$\mathbf{x}_1 = \mathbf{x}_0 + \alpha_0 \mathbf{d}_0 = \begin{bmatrix} 1 \\ 2.5 \end{bmatrix}$$

with

$$\mathbf{A}_{a_1} = \begin{bmatrix} 1 & 2 \\ 1 & 1 \end{bmatrix} \quad \text{and} \quad \mathcal{J}_1 = \{5,\ 4\}$$

This completes the first iteration.

The second iteration starts by solving $A_{a_1}^T \mu_1 = -c$ for μ_1. It is found that $\mu_1 = [3 \ -2]^T$. Hence x_1 is not a minimizer and $l = 2$. By solving

$$A_{a_1} d_1 = -e_2$$

we obtain the feasible descent direction $d_1 = [-2 \ 1]^T$. Next we compute the residual vector at x_1 as

$$r_1 = Ax_1 - b = \begin{bmatrix} -1 \\ -1 \\ -2.5 \\ 0 \\ 0 \end{bmatrix}$$

which indicates that the first three constraints are inactive at x_1. By evaluating

$$\begin{bmatrix} a_1^T \\ a_2^T \\ a_3^T \end{bmatrix} d_1 = \begin{bmatrix} -1 & 0 \\ 1 & 0 \\ 0 & -1 \end{bmatrix} \begin{bmatrix} -2 \\ 1 \end{bmatrix} = \begin{bmatrix} 2 \\ -2 \\ -1 \end{bmatrix}$$

we obtain

$$\mathcal{I}_1 = \{1\}$$

and

$$\alpha_1 = \frac{-r_1}{a_1^T d_1} = \frac{1}{2}$$

This leads to

$$x_2 = x_1 + \alpha_1 d_1 = \begin{bmatrix} 0 \\ 3 \end{bmatrix}$$

with

$$A_{a_2} = \begin{bmatrix} 1 & 2 \\ -1 & 0 \end{bmatrix} \quad \text{and} \quad \mathcal{J}_2 = \{5, \ 1\}$$

which completes the second iteration.

Vertex x_2 is confirmed to be a minimizer at the beginning of the third iteration since the equation

$$A_{a_2}^T \mu_2 = -c$$

yields nonnegative Lagrange multipliers $\mu_2 = [2 \ 3]^T$.   ∎

Fig. 11.8 Feasible region
for Example 11.6

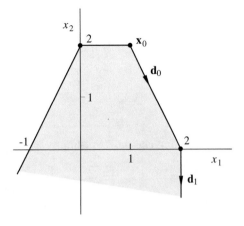

Example 11.6 Solve the LP problem

$$
\begin{aligned}
\text{minimize } f(\mathbf{x}) &= x_1 + x_2 \\
\text{subject to:} \qquad x_1 &\le 2 \\
x_2 &\le 2 \\
-2x_1 + x_2 &\le 2 \\
2x_1 + x_2 &\le 4
\end{aligned}
$$

Solution

The constraints can be written as $\mathbf{A}\mathbf{x} \le \mathbf{b}$ with

$$
\mathbf{A} = \begin{bmatrix} 1 & 0 \\ 0 & 1 \\ -2 & 1 \\ 2 & 1 \end{bmatrix} \quad \text{and} \quad \mathbf{b} = \begin{bmatrix} 2 \\ 2 \\ 2 \\ 4 \end{bmatrix}
$$

The feasible region defined by the constraints is illustrated in Fig. 11.8. Note that the feasible region is unbounded.

Assume that we are given vertex $\mathbf{x}_0 = [1 \ 2]^T$ to start Algorithm 11.1. At $\mathbf{x}_0$, the second and fourth constraints are active and so

$$
\mathbf{A}_{a0} = \begin{bmatrix} 0 & 1 \\ 2 & 1 \end{bmatrix} \quad \text{and} \quad \mathcal{J}_0 = \{2, 4\}
$$

Equation $\mathbf{A}_{a0}^T \boldsymbol{\mu}_0 = -\mathbf{c}$ yields $\boldsymbol{\mu}_0 = [-\frac{1}{2} \ -\frac{1}{2}]^T$ and hence $\mathbf{x}_0$ is not a minimizer.

Since both components of $\boldsymbol{\mu}_0$ are negative, we can choose index l to be either 1 or

2. Choosing $l = 1$, Eq. (11.46) becomes $\mathbf{A}_{a_0}\mathbf{d}_0 = -\mathbf{e}_1$ which gives $\mathbf{d}_0 = [\frac{1}{2} \ -1]^T$. The residual vector at $\mathbf{x}_0$ is given by

$$\mathbf{r}_0 = \mathbf{A}\mathbf{x}_0 - \mathbf{b} = \begin{bmatrix} -1 \\ 0 \\ -2 \\ 0 \end{bmatrix}$$

Hence the first and third constraints are inactive at $\mathbf{x}_0$. We now compute

$$\begin{bmatrix} \mathbf{a}_1^T \\ \mathbf{a}_3^T \end{bmatrix}\mathbf{d}_0 = \begin{bmatrix} 1 & 0 \\ -2 & 1 \end{bmatrix}\begin{bmatrix} \frac{1}{2} \\ -1 \end{bmatrix} = \begin{bmatrix} \frac{1}{2} \\ -2 \end{bmatrix}$$

to identify index set $\mathcal{I}_0 = \{1\}$. Hence

$$\alpha_0 = \frac{-r_1}{\mathbf{a}_1^T\mathbf{d}_0} = 2$$

and the next vertex is given by

$$\mathbf{x}_1 = \mathbf{x}_0 + \alpha_0\mathbf{d}_0 = \begin{bmatrix} 2 \\ 0 \end{bmatrix}$$

with

$$\mathbf{A}_{a_1} = \begin{bmatrix} 1 & 0 \\ 2 & 1 \end{bmatrix} \quad \text{and} \quad \mathcal{J}_1 = \{1, 4\}$$

Next we examine whether or not $\mathbf{x}_1$ is a minimizer by solving $\mathbf{A}_{a_1}^T\boldsymbol{\mu}_1 = -\mathbf{c}$. This gives $\boldsymbol{\mu}_1 = [3 \ -2]^T$ indicating that $\mathbf{x}_1$ is not a minimizer and $l = 2$. Solving $\mathbf{A}_{a_1}\mathbf{d}_1 = -\mathbf{e}_2$ for $\mathbf{d}_1$, we obtain $\mathbf{d}_1 = [0 \ -1]^T$. At $\mathbf{x}_1$ the residual vector is given by

$$\mathbf{r}_1 = \mathbf{A}\mathbf{x}_1 - \mathbf{b} = \begin{bmatrix} 0 \\ -2 \\ -6 \\ 0 \end{bmatrix}$$

Hence the second and third constraints are inactive. Next we evaluate

$$\begin{bmatrix} \mathbf{a}_2^T \\ \mathbf{a}_3^T \end{bmatrix}\mathbf{d}_1 = \begin{bmatrix} 0 & 1 \\ -2 & 1 \end{bmatrix}\begin{bmatrix} 0 \\ -1 \end{bmatrix} = \begin{bmatrix} -1 \\ -1 \end{bmatrix}$$

Since $\mathcal{I}_1$ is empty, we conclude that the solution of this LP problem is un-bounded.   ∎

11.3.1.2   Degenerate Case

When some of the vertices associated with the problem are degenerate, Algorithm 11.1 needs several minor modifications. At a degenerate vertex, say, $\mathbf{x}_k$, the number of active constraints is larger than n minus the dimension of variable vector $\mathbf{x}$. Consequently, the number of rows in matrix $\mathbf{A}_{a_k}$ is larger than n and matrix $\mathbf{A}_{a_k}$ should be replaced in Steps 2 and 3 of Algorithm 11.1 by a matrix $\hat{\mathbf{A}}_{a_k}$ that is composed of n *linearly independent* rows of $\mathbf{A}_{a_k}$. Likewise, $\mathbf{A}_{a_0}$ in Step 1 and $\mathbf{A}_{a_{k+1}}$ in Step 5 should be replaced by $\hat{\mathbf{A}}_{a_0}$ and $\hat{\mathbf{A}}_{a_{k+1}}$, respectively.

The set of constraints corresponding to the rows in $\hat{\mathbf{A}}_{a_k}$ is called a *working set* of active constraints and in the literature $\hat{\mathbf{A}}_{a_k}$ is often referred to as a *working-set matrix*.

Associated with $\hat{\mathbf{A}}_{a_k}$ is the *working index set* denoted as

$$\mathcal{W}_k = \{w_1, \ w_2, \ \ldots, \ w_n\}$$

which contains the indices of the rows of $\hat{\mathbf{A}}_{a_k}$ as they appear in matrix $\mathbf{A}$. Some additional modifications of the algorithm in terms of the notation just introduced are to replace $\mathcal{J}_0$ in Step 1 and $\mathcal{J}_{k+1}$ in Step 5 by $\mathcal{W}_0$ and $\mathcal{W}_{k+1}$, respectively, and to redefine the index set $\mathcal{I}_k$ in Eq. (11.48) as

$$\mathcal{I}_k = \{i \, : \, i \notin \mathcal{W}_k \text{ and } \mathbf{a}_i^T \mathbf{d}_k > 0\} \tag{11.55}$$

Relative to $\mathcal{I}_k$ in Eq. (11.48), the modified $\mathcal{I}_k$ in Eq. (11.55) also includes the indices of the constraints that are active at $\mathbf{x}_k$ but are excluded from $\hat{\mathbf{A}}_{a_k}$ and which satisfy the inequality $\mathbf{a}_i^T \mathbf{d}_k > 0$.

Obviously, for a nondegenerate vertex $\mathbf{x}_k$, $\hat{\mathbf{A}}_{a_k} = \mathbf{A}_{a_k}$ and there is only *one* working set of active constraints that includes all the active constraints at $\mathbf{x}_k$ and $\mathcal{I}_k$ does not contain indices of any active constraints. For a degenerate vertex $\mathbf{x}_k$, however, $\hat{\mathbf{A}}_{a_k}$ is not unique and, as Eq. (11.55) indicates, $\mathcal{I}_k$ may contain indices of active constraints. When $\mathcal{I}_k$ does include the index of an active constraint, the associated residual is zero. Consequently, the step size α_k computed using Eq. (11.53c) is also zero, which implies that $\mathbf{x}_{k+1} = \mathbf{x}_k$. Although under such circumstances the working index set $\mathcal{W}_{k+1}$ will differ from $\mathcal{W}_k$, the possibility of generating an infinite sequence of working index sets without moving from a given vertex does exist. For an example where such 'cycling' occurs, see [3, Sect. 8.3.2].

Cycling can be avoided by using an approach proposed by Bland [6]. The approach is known as *Bland's least-index rule* for deleting and adding constraints and is as follows:

1. In Step 2 of Algorithm 11.1, if the Lagrange multiplier $\boldsymbol{\mu}_k$ has more than one negative component, then index l is selected as the smallest index in the working index set $\mathcal{W}_k$ corresponding to a negative component of $\boldsymbol{\mu}_k$, i.e.,

$$l = \min_{w_i \in \mathcal{W}_k, \ (\boldsymbol{\mu}_k)_i < 0} (w_i) \tag{11.56}$$

2. In Step 4, if there is more than one index that yields the optimum α_k in Eq. (11.53c), then the associated constraints are called *blocking constraints*, and i^* is determined as the smallest index of a blocking constraint.

The steps of the modified simplex algorithm are as follows.

Algorithm 11.2 Simplex algorithm for the alternative-form LP problem in Eqs. (11.2a) and (11.2b), degenerate vertices

Step 1
Input vertex $\mathbf{x}_0$ and form a working-set matrix $\hat{\mathbf{A}}_{a_0}$ and a working-index set $\mathcal{W}_0$.
Set $k = 0$.

Step 2
Solve

$$\hat{\mathbf{A}}_{a_k}^T \boldsymbol{\mu}_k = -\mathbf{c} \tag{11.57}$$

for $\boldsymbol{\mu}_k$.
If $\boldsymbol{\mu}_k \geq \mathbf{0}$, stop (vertex $\mathbf{x}_k$ is a minimizer); otherwise, select index l using Eq. (11.56).

Step 3
Solve

$$\hat{\mathbf{A}}_{a_k} \mathbf{d}_k = -\mathbf{e}_l \tag{11.58}$$

for $\mathbf{d}_k$.

Step 4
Form index set $\mathcal{I}_k$ using Eq. (11.55).
If $\mathcal{I}_k$ is empty, stop (the objective function tends to $-\infty$ in the feasible region).

Step 5
Compute the residual vector

$$\mathbf{r}_k = \mathbf{A}\mathbf{x}_k - \mathbf{b} = (r_i)_{i=1}^p$$

parameter

$$\delta_i = \frac{-r_i}{\mathbf{a}_i^T \mathbf{d}_k} \quad \text{for } i \in \mathcal{I}_k \tag{11.59}$$

and

$$\alpha_k = \min_{i \in \mathcal{I}_k}(\delta_i)$$

Record index i^* as

$$i^* = \min_{\delta_i = \alpha_k}(i)$$

Step 6
Set $\mathbf{x}_{k+1} = \mathbf{x}_k + \alpha_k \mathbf{d}_k$.

Update $\hat{\mathbf{A}}_{a_{k+1}}$ by deleting row $\mathbf{a}_l^T$ and adding row $\mathbf{a}_{i*}^T$ and update index set $\mathcal{W}_{k+1}$ accordingly.
Set $k = k + 1$ and repeat from Step 2.

Example 11.7 Solve the LP problem

$$\text{minimize } f(\mathbf{x}) = -2x_1 - 3x_2 + x_3 + 12x_4$$
$$\text{subject to:} \quad -x_1 \le 0, \ -x_2 \le 0, \ -x_3 \le 0, \ -x_4 \le 0$$
$$-2x_1 - 9x_2 + x_3 + 9x_4 \le 0$$
$$\frac{1}{3}x_1 + x_2 - \frac{1}{3}x_3 - 2x_4 \le 0$$

(See [3, p. 351].)

Solution We start with $\mathbf{x}_0 = [0\ 0\ 0\ 0]^T$ which is obviously a degenerate vertex. Applying Algorithm 11.2, the first iteration results in the following computations:

$$\hat{\mathbf{A}}_{a0} = \begin{bmatrix} -1 & 0 & 0 & 0 \\ 0 & -1 & 0 & 0 \\ 0 & 0 & -1 & 0 \\ 0 & 0 & 0 & -1 \end{bmatrix}$$

$$\mathcal{W} = \{1, 2, 3, 4\}$$
$$\boldsymbol{\mu}_0 = [-2 \ -3 \ 1 \ 12]^T$$
$$l = 1$$
$$\mathbf{d}_0 = [1\ 0\ 0\ 0]^T$$
$$\mathbf{r}_0 = [0\ 0\ 0\ 0\ 0\ 0]^T$$
$$\mathcal{I}_0 = \{6\}$$
$$\alpha_0 = 0$$
$$i^* = 6$$
$$\mathbf{x}_1 = \mathbf{x}_0 = [0\ 0\ 0\ 0]^T$$

$$\hat{\mathbf{A}}_{a_1} = \begin{bmatrix} \frac{1}{3} & 1 & -\frac{1}{3} & -2 \\ 0 & -1 & 0 & 0 \\ 0 & 0 & -1 & 0 \\ 0 & 0 & 0 & -1 \end{bmatrix}$$

$$\mathcal{W}_1 = \{6, 2, 3, 4\}$$

Note that although $\mathbf{x}_1 = \mathbf{x}_0$, $\hat{\mathbf{A}}_{a_1}$ differs from $\hat{\mathbf{A}}_{a0}$. Repeating from Step 2, the second iteration ($k = 1$) gives

$$\boldsymbol{\mu}_1 = [6\ 3\ -1\ 0]^T$$
$$l = 3$$
$$\mathbf{d}_1 = [1\ 0\ 1\ 0]^T$$
$$\mathbf{r}_1 = [0\ 0\ 0\ 0]^T$$
$$\mathcal{I}_1 = \{\phi\}$$

where $\mathcal{I}_1$ is an empty set. Therefore, in the feasible region the objective function tends to $-\infty$.

As a matter of fact, all the points along the feasible descent direction $\mathbf{d}_1$ are feasible, i.e.,

$$\mathbf{x} = \mathbf{x}_1 + \alpha \mathbf{d}_1 = [\alpha \; 0 \; \alpha \; 0]^T \qquad \text{for } \alpha > 0$$

where $f(\mathbf{x}) = -\alpha$ approaches $-\infty$ as $\alpha \to +\infty$.   ∎

11.3.2   Simplex Method for Standard-Form LP Problems

11.3.2.1   Basic and Nonbasic Variables

For a standard-form LP problem of the type given in Eqs. (11.1a)–(11.1c) with a matrix $\mathbf{A}$ of full row rank, the p equality constraints in Eq. (11.1b) are always treated as active constraints. As was discussed in Sect. 10.4.1, these constraints reduce the number of 'free' variables from n to $n - p$. In other words, the p equality constraints can be used to express p dependent variables in terms of $n - p$ independent variables. Let $\mathbf{B}$ be a matrix that consists of p linearly independent columns of $\mathbf{A}$. If the variable vector $\mathbf{x}$ is partitioned accordingly, then we can write the equality constraint in Eq. (11.2b) as

$$\mathbf{A}\mathbf{x} = [\mathbf{B} \; \mathbf{N}] \begin{bmatrix} \mathbf{x}_B \\ \mathbf{x}_N \end{bmatrix} = \mathbf{B}\mathbf{x}_B + \mathbf{N}\mathbf{x}_N = \mathbf{b} \tag{11.60}$$

The variables contained in $\mathbf{x}_B$ and $\mathbf{x}_N$ are called *basic* and *nonbasic* variables, respectively. Since $\mathbf{B}$ is nonsingular, the basic variables can be expressed in terms of the nonbasic variables as

$$\mathbf{x}_B = \mathbf{B}^{-1}\mathbf{b} - \mathbf{B}^{-1}\mathbf{N}\mathbf{x}_N \tag{11.61}$$

At vertex $\mathbf{x}_k$, there are at least n active constraints. Hence in addition to the p equality constraints, there are at least $n - p$ inequality constraints that become active at $\mathbf{x}_k$. Therefore, for a standard-form LP problem a vertex contains at least $n - p$ zero components. The next theorem describes an interesting property of $\mathbf{A}$.

Theorem 11.9 (Linear independence of columns in matrix $\mathbf{A}$)   *The columns of $\mathbf{A}$ corresponding to strictly positive components of a vertex $\mathbf{x}_k$ are linearly independent.*   ∎

Proof We adopt the proof used in [3]. Let $\hat{\mathbf{B}}$ be formed by the columns of $\mathbf{A}$ that correspond to strictly positive components of $\mathbf{x}_k$, and let $\hat{\mathbf{x}}_k$ be the collection of the positive components of $\mathbf{x}_k$. If $\hat{\mathbf{B}}\hat{\mathbf{w}} = \mathbf{0}$ for some nonzero $\hat{\mathbf{w}}$, then it follows that

$$\mathbf{A}\mathbf{x}_k = \hat{\mathbf{B}}\hat{\mathbf{x}}_k = \hat{\mathbf{B}}(\hat{\mathbf{x}}_k + \alpha\hat{\mathbf{w}}) = \mathbf{b}$$

for any scalar α. Since $\hat{\mathbf{x}}_k > \mathbf{0}$, there exists a sufficiently small $\alpha_+ > 0$ such that

$$\hat{\mathbf{y}}_k = \hat{\mathbf{x}}_k + \alpha\hat{\mathbf{w}} > \mathbf{0} \qquad \text{for } -\alpha_+ \le \alpha \le \alpha_+$$

Now let $\mathbf{y}_k \in R^{n \times 1}$ be such that the components of $\mathbf{y}_k$ corresponding to $\hat{\mathbf{x}}_k$ are equal to the components of $\hat{\mathbf{y}}_k$ and the remaining components of $\mathbf{y}_k$ are zero. Evidently, we have

$$\mathbf{A}\mathbf{y}_k = \hat{\mathbf{B}}\hat{\mathbf{y}}_k = \mathbf{b}$$

and

$$\mathbf{y}_k \ge \mathbf{0}$$

Note that with $\alpha = 0$, $\mathbf{y}_k = \mathbf{x}_k$ is a vertex, and when α varies from $-\alpha_+$ to α_+, vertex $\mathbf{x}_k$ would lie between two feasible points on a straight line, which is a contradiction. Hence $\hat{\mathbf{w}}$ must be zero and the columns of $\hat{\mathbf{B}}$ are linearly independent. ∎

By virtue of Theorem 11.9, we can use the columns of $\hat{\mathbf{B}}$ as a set of core basis vectors to construct a nonsingular square matrix $\mathbf{B}$. If $\hat{\mathbf{B}}$ already contains p columns, we assume that $\mathbf{B} = \hat{\mathbf{B}}$; otherwise, we augment $\hat{\mathbf{B}}$ with additional columns of $\mathbf{A}$ to obtain a square nonsingular $\mathbf{B}$. Let the index set associated with $\mathbf{B}$ at $\mathbf{x}_k$ be denoted as $\mathcal{I}_\beta = \{\beta_1, \beta_2, \ldots, \beta_p\}$. With matrix $\mathbf{B}$ so formed, matrix $\mathbf{N}$ in Eq. (11.60) can be constructed with those $n - p$ columns of $\mathbf{A}$ that are not in $\mathbf{B}$. Let $\mathcal{I}_N = \{\nu_1, \nu_2, \ldots, \nu_{n-p}\}$ be the index set for the columns of $\mathbf{N}$ and let $\mathbf{I}_N$ be the $(n-p) \times n$ matrix composed of rows $\nu_1, \nu_2, \ldots, \nu_{n-p}$ of the $n \times n$ identity matrix. With this notation, it is clear that at vertex $\mathbf{x}_k$ the active constraint matrix $\mathbf{A}_{a_k}$ contains the working-set matrix

$$\hat{\mathbf{A}}_{a_k} = \begin{bmatrix} \mathbf{A} \\ \mathbf{I}_N \end{bmatrix} \tag{11.62}$$

as an $n \times n$ submatrix. It can be shown that matrix $\hat{\mathbf{A}}_{a_k}$ in Eq. (11.62) is nonsingular. In fact if $\hat{\mathbf{A}}_{a_k}\mathbf{x} = \mathbf{0}$ for some $\mathbf{x}$, then we have

$$\mathbf{B}\mathbf{x}_B + \mathbf{N}\mathbf{x}_N = \mathbf{0} \quad \text{and} \quad \mathbf{x}_N = \mathbf{0}$$

It follows that

$$\mathbf{x}_B = -\mathbf{B}^{-1}\mathbf{N}\mathbf{x}_N = \mathbf{0}$$

and hence $\mathbf{x} = \mathbf{0}$. Therefore, $\hat{\mathbf{A}}_{a_k}$ is nonsingular. In summary, at a vertex $\mathbf{x}_k$ a working set of active constraints for the application of the simplex method can be obtained with three simple steps as follows:

(a) Select the columns in matrix $\mathbf{A}$ that correspond to the strictly positive components of $\mathbf{x}_k$ to form matrix $\hat{\mathbf{B}}$.

(b) If the number of columns in $\hat{\mathbf{B}}$ is equal to p, take $\mathbf{B} = \hat{\mathbf{B}}$; otherwise, $\hat{\mathbf{B}}$ is augmented with additional columns of $\mathbf{A}$ to form a square nonsingular matrix $\mathbf{B}$.

(c) Determine the index set $\mathcal{I}_N$ and form matrix $\mathbf{I}_N$.

Example 11.8 Identify working sets of active constraints at vertex $\mathbf{x} = [3\ 0\ 0\ 0]^T$ for the LP problem

$$\text{minimize }\ f(\mathbf{x}) = x_1 - 2x_2 - x_4$$
$$\text{subject to:} \qquad 3x_1 + 4x_2 + x_3 = 9$$
$$2x_1 + x_2 + x_4 = 6$$
$$x_1 \geq 0,\ x_2 \geq 0,\ x_3 \geq 0,\ x_4 \geq 0$$

Solution It is easy to verify that point $\mathbf{x} = [3\ 0\ 0\ 0]^T$ is a degenerate vertex at which there are five active constraints. Since x_1 is the only strictly positive component, $\hat{\mathbf{B}}$ contains only the first column of $\mathbf{A}$, i.e.,

$$\hat{\mathbf{B}} = \begin{bmatrix} 3 \\ 2 \end{bmatrix}$$

Matrix $\hat{\mathbf{B}}$ can be augmented, for example, by using the second column of $\mathbf{A}$ to generate a nonsingular $\mathbf{B}$ as

$$\mathbf{B} = \begin{bmatrix} 3 & 4 \\ 2 & 1 \end{bmatrix}$$

This leads to

$$\mathcal{I}_N = \{3,\ 4\} \quad \text{and} \quad \hat{\mathbf{A}}_a = \begin{bmatrix} 3 & 4 & 1 & 0 \\ 2 & 1 & 0 & 1 \\ 0 & 0 & 1 & 0 \\ 0 & 0 & 0 & 1 \end{bmatrix}$$

Since vertex $\mathbf{x}$ is degenerate, matrix $\hat{\mathbf{A}}_a$ is not unique. As a reflection of this nonuniqueness, there are two possibilities for augmenting $\hat{\mathbf{B}}$. Using the third column of $\mathbf{A}$ for the augmentation, we have

$$\mathbf{B} = \begin{bmatrix} 3 & 1 \\ 2 & 0 \end{bmatrix}$$

which gives

$$\mathcal{I}_N = \{2,\ 4\}$$

and

$$\hat{\mathbf{A}}_a = \begin{bmatrix} 3 & 4 & 1 & 0 \\ 2 & 1 & 0 & 1 \\ 0 & 1 & 0 & 0 \\ 0 & 0 & 0 & 1 \end{bmatrix}$$

Alternatively, augmenting $\hat{\mathbf{B}}$ with the fourth column of $\mathbf{A}$ yields

$$\mathbf{B} = \begin{bmatrix} 3 & 0 \\ 2 & 1 \end{bmatrix}$$

which gives

$$\mathcal{I}_N = \{2, \ 3\}$$

and

$$\hat{\mathbf{A}}_a = \begin{bmatrix} 3 & 4 & 1 & 0 \\ 2 & 1 & 0 & 1 \\ 0 & 1 & 0 & 0 \\ 0 & 0 & 1 & 0 \end{bmatrix}$$

It can be easily verified that all three $\hat{\mathbf{A}}_a$'s are nonsingular.   ∎

11.3.2.2   Algorithm for Standard-Form LP Problem

Like Algorithms 11.1 and 11.2, an algorithm for the standard-form LP problem based on the simplex method can start with a vertex, and the steps of Algorithm 11.2 can serve as a framework for the implementation. A major difference from Algorithms 11.1 and 11.2 is that the special structure of the working-set matrix $\hat{\mathbf{A}}_{a_k}$ in Eq. (11.62) can be utilized in Steps 2 and 3, which would result in reduced computational complexity.

At a vertex $\mathbf{x}_k$, the nonsingularity of the working-set matrix $\hat{\mathbf{A}}_{a_k}$ given by Eq. (11.62) implies that there exist $\lambda_k \in R^{p \times 1}$ and $\hat{\boldsymbol{\mu}}_k \in R^{(n-p) \times 1}$ such that

$$\mathbf{c} = \hat{\mathbf{A}}_{a_k}^T \begin{bmatrix} -\lambda_k \\ \hat{\boldsymbol{\mu}}_k \end{bmatrix} = -\mathbf{A}^T \lambda_k + \mathbf{I}_N^T \hat{\boldsymbol{\mu}}_k \tag{11.63}$$

If $\boldsymbol{\mu}_k \in R^{n \times 1}$ is the vector with zero basic variables and the components of $\hat{\boldsymbol{\mu}}_k$ as its nonbasic variables, then Eq. (11.63) can be expressed as

$$\mathbf{c} = -\mathbf{A}^T \lambda_k + \boldsymbol{\mu}_k \tag{11.64}$$

By virtue of Theorem 11.1, vertex $\mathbf{x}_k$ is a minimizer if and only if $\hat{\boldsymbol{\mu}}_k \geq \mathbf{0}$. If we use a permutation matrix, $\mathbf{P}$, to rearrange the components of $\mathbf{c}$ in accordance with the partition of $\mathbf{x}_k$ into basic and nonbasic variables as in Eq. (11.60), then Eq. (11.63) gives

$$\mathbf{Pc} = \begin{bmatrix} \mathbf{c}_B \\ \mathbf{c}_N \end{bmatrix} = -\mathbf{PA}^T \lambda_k + \mathbf{PI}_N^T \hat{\boldsymbol{\mu}}_k$$

$$= -\begin{bmatrix} \mathbf{B}^T \\ \mathbf{N}^T \end{bmatrix} \lambda_k + \begin{bmatrix} \mathbf{0} \\ \hat{\boldsymbol{\mu}}_k \end{bmatrix}$$

It follows that

$$\mathbf{B}^T \lambda_k = -\mathbf{c}_B \tag{11.65}$$

and

$$\hat{\boldsymbol{\mu}}_k = \mathbf{c}_N + \mathbf{N}^T \lambda_k \tag{11.66}$$

Since $\mathbf{B}$ is nonsingular, λ_k and $\hat{\boldsymbol{\mu}}_k$ can be computed using Eqs. (11.65) and (11.66), respectively. Note that the system of equations that need to be solved is of size $p \times p$ rather than $n \times n$ as in Step 2 of Algorithms 11.1 and 11.2.

If some entry in $\hat{\boldsymbol{\mu}}_k$ is negative, then $\mathbf{x}_k$ is not a minimizer and a search direction $\mathbf{d}_k$ needs to be determined. Note that the Lagrange multipliers $\hat{\boldsymbol{\mu}}_k$ are *not* related to the equality constraints in Eq. (11.1b) but are related to those bound constraints in Eq. (11.1c) that are active and are associated with the nonbasic variables. If the search direction $\mathbf{d}_k$ is partitioned according to the basic and nonbasic variables, $\mathbf{x}_B$ and $\mathbf{x}_N$, into $\mathbf{d}_k^{(B)}$ and $\mathbf{d}_k^{(N)}$, respectively, and if $(\hat{\boldsymbol{\mu}}_k)_l < 0$, then assigning

$$\mathbf{d}_k^{(N)} = \mathbf{e}_l \tag{11.67}$$

where $\mathbf{e}_l$ is the lth column of the $(n-p) \times (n-p)$ identity matrix, yields a search direction $\mathbf{d}_k$ that makes the v_lth constraint inactive without affecting other bound constraints that are associated with the nonbasic variables. In order to assure the feasibility of $\mathbf{d}_k$, it is also required that $\mathbf{Ad}_k = \mathbf{0}$ (see Theorem 11.4). This requirement can be described as

$$\mathbf{Ad}_k = \mathbf{Bd}_k^{(B)} + \mathbf{Nd}_k^{(N)} = \mathbf{Bd}_k^{(B)} + \mathbf{Ne}_l = \mathbf{0} \tag{11.68}$$

where $\mathbf{Ne}_l$ is actually the v_lth column of $\mathbf{A}$. Hence $\mathbf{d}_k^{(B)}$ can be determined by solving the system of equations

$$\mathbf{Bd}_k^{(B)} = -\mathbf{a}_{v_l} \tag{11.69a}$$

where

$$\mathbf{a}_{v_l} = \mathbf{Ne}_l \tag{11.69b}$$

Together, Eqs. (11.67), (11.69a), and (11.69b) determine the search direction $\mathbf{d}_k$. From Eqs. (11.63), (11.67), and (11.68), it follows that

$$\mathbf{c}^T\mathbf{d}_k = -\lambda_k^T\mathbf{A}\mathbf{d}_k + \hat{\boldsymbol{\mu}}_k^T\mathbf{I}_N\mathbf{d}_k = \hat{\boldsymbol{\mu}}_k^T\mathbf{d}_k^{(N)} = \hat{\boldsymbol{\mu}}_k^T\mathbf{e}_l$$
$$= (\hat{\boldsymbol{\mu}}_k)_l < 0$$

Therefore, $\mathbf{d}_k$ is a feasible descent direction. From Eqs. (11.67), (11.69a), and (11.69b), it is observed that unlike the cases of Algorithms 11.1 and 11.2 where finding a feasible descent search direction requires the solution of a system of n equations (see Eqs. (11.52) and (11.58)), the present algorithm involves the solution of a system of p equations.

Considering the determination of step size α_k, we note that a point $\mathbf{x}_k + \alpha\mathbf{d}_k$ with any α satisfies the constraints in Eq. (11.1b), i.e.,

$$\mathbf{A}(\mathbf{x}_k + \alpha\mathbf{d}_k) = \mathbf{A}\mathbf{x}_k + \alpha\mathbf{A}\mathbf{d}_k = \mathbf{b}$$

Furthermore, Eq. (11.67) indicates that with any positive α, $\mathbf{x}_k + \alpha\mathbf{d}_k$ does not violate the constraints in Eq. (11.1c) that are associated with the nonbasic variables. Therefore, the only constraints that are sensitive to step size α_k are those that are associated with the basic variables and are decreasing along direction $\mathbf{d}_k$. When limited to the basic variables, $\mathbf{d}_k$ becomes $\mathbf{d}_k^{(B)}$. Since the normals of the constraints in Eq. (11.1c) are simply coordinate vectors, a bound constraint associated with a basic variable is decreasing along $\mathbf{d}_k$ if the associated component in $\mathbf{d}_k^{(B)}$ is negative. In addition, the special structure of the inequality constraints in Eq. (11.1c) also implies that the residual vector, when limited to basic variables in $\mathbf{x}_B$, is $\mathbf{x}_B$ itself.

The above analysis leads to a simple step that can be used to determine the index set

$$\mathcal{I}_k = \{i : (\mathbf{d}_k^{(B)})_i < 0\} \tag{11.70}$$

and, if $\mathcal{I}_k$ is not empty, to determine α_k as

$$\alpha_k = \min_{i \in \mathcal{I}_k} \left[\frac{(\mathbf{x}_k^{(B)})_i}{(-\mathbf{d}_k^{(B)})_i} \right] \tag{11.71}$$

where $\mathbf{x}_k^{(B)}$ denotes the vector for the basic variables of $\mathbf{x}_k$. If i^* is the index in $\mathcal{I}_k$ that achieves α_k, then the i^*th component of $\mathbf{x}_k^{(B)} + \alpha_k\mathbf{d}_k^{(B)}$ is zero. This zero component is then interchanged with the lth component of $\mathbf{x}_k^{(N)}$, which is now not zero but α_k. The vector $\mathbf{x}_k^{(B)} + \alpha\mathbf{d}_k^{(B)}$ after this updating becomes $\mathbf{x}_{k+1}^{(B)}$ and, of course, $\mathbf{x}_{k+1}^{(N)}$ remains a zero vector. Matrices $\mathbf{B}$ and $\mathbf{N}$ as well as the associated index sets $\mathcal{I}_B$ and $\mathcal{I}_N$ also need to be updated accordingly. An algorithm based on the above principles is as follows.

Algorithm 11.3 Simplex algorithm for the standard-form LP problem of Eqs. (11.1a)–(11.1c)

Step 1
Input vertex $\mathbf{x}_0$, set $k = 0$, and form $\mathbf{B}$, $\mathbf{N}$, $\mathbf{x}_0^{(B)}$, $\mathcal{I}_B = \{\beta_1^{(0)}, \beta_2^{(0)}, \ldots, \beta_p^{(0)}\}$, and $\mathcal{I}_N = \{v_1^{(0)}, v_2^{(0)}, \ldots, v_{n-p}^{(0)}\}$.

Step 2
Partition vector $\mathbf{c}$ into $\mathbf{c}_B$ and $\mathbf{c}_N$.
Solve Eq. (11.65) for λ_k and compute $\hat{\boldsymbol{\mu}}_k$ using Eq. (11.66).
If $\hat{\boldsymbol{\mu}}_k \geq \mathbf{0}$, stop ($\mathbf{x}_k$ is a vertex minimizer); otherwise, select the index l that corresponds to the most negative component in $\hat{\boldsymbol{\mu}}_k$.

Step 3
Solve Eq. (11.69a) for $\mathbf{d}_k^{(B)}$ where $\mathbf{a}_{v_l}$ is the $v_l^{(k)}$th column of $\mathbf{A}$.

Step 4
Form index set $\mathcal{I}_k$ in Eq. (11.70).
If $\mathcal{I}_k$ is empty then stop (the objective function tends to $-\infty$ in the feasible region); otherwise, compute α_k using Eq. (11.71) and record the index i^* with
$$\alpha_k = (\mathbf{x}_k^{(B)})_{i*}/(-\mathbf{d}^{(B)})_{i*}.$$

Step 5
Compute $\mathbf{x}_{k+1}^{(B)} = \mathbf{x}_k^{(B)} + \alpha_k \mathbf{d}_k^{(B)}$ and replace its i^*th zero component by α_k.
Set $\mathbf{x}_{k+1}^{(N)} = \mathbf{0}$.
Update $\mathbf{B}$ and $\mathbf{N}$ by interchanging the lth column of $\mathbf{N}$ with the i^*th column of $\mathbf{B}$.

Step 6
Update $\mathcal{I}_B$ and $\mathcal{I}_N$ by interchanging index $v_l^{(k)}$ of $\mathcal{I}_N$ with index $\beta_{i*}^{(B)}$ of $\mathcal{I}_B$.
Use the $\mathbf{x}_{k+1}^{(B)}$ and $\mathbf{x}_{k+1}^{(N)}$ obtained in Step 5 in conjunction with $\mathcal{I}_B$ and $\mathcal{I}_N$ to form $\mathbf{x}_{k+1}$.
Set $k = k + 1$ and repeat from Step 2.

Example 11.9 Solve the standard-form LP problem

$$\text{minimize } f(\mathbf{x}) = 2x_1 + 9x_2 + 3x_3$$

$$\text{subject to:} \qquad -2x_1 + 2x_2 + x_3 - x_4 = 1$$
$$x_1 + 4x_2 - x_3 - x_5 = 1$$
$$x_1 \geq 0, \ x_2 \geq 0, \ x_3 \geq 0, \ x_4 \geq 0, \ x_5 \geq 0$$

Solution From Eqs. (11.1a)–(11.1c)

$$\mathbf{A} = \begin{bmatrix} -2 & 2 & 1 & -1 & 0 \\ 1 & 4 & -1 & 0 & -1 \end{bmatrix}, \quad \mathbf{b} = \begin{bmatrix} 1 \\ 1 \end{bmatrix}$$

and

$$\mathbf{c} = [2\ 9\ 3\ 0\ 0]^T$$

To identify a vertex, we set $x_1 = x_3 = x_4 = 0$ and solve the system

$$\begin{bmatrix} 2 & 0 \\ 4 & -1 \end{bmatrix} \begin{bmatrix} x_2 \\ x_5 \end{bmatrix} = \begin{bmatrix} 1 \\ 1 \end{bmatrix}$$

for x_2 and x_5. This leads to $x_2 = 1/2$ and $x_5 = 1$; hence

$$\mathbf{x}_0 = [0\ \tfrac{1}{2}\ 0\ 0\ 1]^T$$

is a vertex. Associated with $\mathbf{x}_0$ are $\mathcal{I}_B = \{2,\ 5\}$, $\mathcal{I}_N = \{1,\ 3,\ 4\}$

$$\mathbf{B} = \begin{bmatrix} 2 & 0 \\ 4 & -1 \end{bmatrix}, \quad \mathbf{N} = \begin{bmatrix} -2 & 1 & -1 \\ 1 & -1 & 0 \end{bmatrix}, \quad \text{and} \quad \mathbf{x}_0^{(B)} = \begin{bmatrix} \tfrac{1}{2} \\ 1 \end{bmatrix}$$

Partitioning $\mathbf{c}$ into

$$\mathbf{c}_B = [9\ 0]^T \quad \text{and} \quad \mathbf{c}_N = [2\ 3\ 0]^T$$

and solving Eq. (11.65) for $\boldsymbol{\lambda}_0$, we obtain $\boldsymbol{\lambda}_0 = [-\tfrac{9}{2}\ 0]^T$. Hence Eq. (11.66) gives

$$\hat{\boldsymbol{\mu}}_0 = \begin{bmatrix} 2 \\ 3 \\ 0 \end{bmatrix} + \begin{bmatrix} -2 & 1 \\ 1 & -1 \\ -1 & 0 \end{bmatrix} \begin{bmatrix} -\tfrac{9}{2} \\ 0 \end{bmatrix} = \begin{bmatrix} 11 \\ -\tfrac{3}{2} \\ \tfrac{9}{2} \end{bmatrix}$$

Since $(\hat{\boldsymbol{\mu}}_0)_2 < 0$, $\mathbf{x}_0$ is not a minimizer, and $l = 2$. Next, we solve Eq. (11.69a) for $\mathbf{d}_0^{(B)}$ with $v_2^{(0)} = 3$ and $\mathbf{a}_3 = [1\ -1]^T$, which yields

$$\mathbf{d}_0^{(B)} = \begin{bmatrix} -\tfrac{1}{2} \\ -3 \end{bmatrix} \quad \text{and} \quad \mathcal{I}_0 = \{1,\ 2\}$$

Hence

$$\alpha_0 = \min\left(1, \tfrac{1}{3}\right) = \tfrac{1}{3} \quad \text{and} \quad i^* = 2$$

To find $\mathbf{x}_1^{(B)}$, we compute

$$\mathbf{x}_0^{(B)} + \alpha_0 \mathbf{d}_0^{(B)} = \begin{bmatrix} \tfrac{1}{3} \\ 0 \end{bmatrix}$$

and replace its i^*th component by α_0, i.e.,

$$\mathbf{x}_1^{(B)} = \begin{bmatrix} \frac{1}{3} \\ \frac{1}{3} \end{bmatrix} \quad \text{with} \quad \mathbf{x}_1^{(N)} = \begin{bmatrix} 0 \\ 0 \end{bmatrix}$$

Now we update $\mathbf{B}$ and $\mathbf{N}$ as

$$\mathbf{B} = \begin{bmatrix} 2 & 1 \\ 4 & -1 \end{bmatrix} \quad \text{and} \quad \mathbf{N} = \begin{bmatrix} -2 & 0 & -1 \\ 1 & -1 & 0 \end{bmatrix}$$

and update $\mathcal{I}_B$ and $\mathcal{I}_N$ as $\mathcal{I}_B = \{2,\ 3\}$ and $\mathcal{I}_N = \{1,\ 5,\ 4\}$. The vertex obtained is

$$\mathbf{x}_1 = \begin{bmatrix} 0 & \frac{1}{3} & \frac{1}{3} & 0 & 0 \end{bmatrix}^T$$

to complete the first iteration.

The second iteration starts with the partitioning of $\mathbf{c}$ into

$$\mathbf{c}_B = \begin{bmatrix} 9 \\ 3 \end{bmatrix} \quad \text{and} \quad \mathbf{c}_N = \begin{bmatrix} 2 \\ 0 \\ 0 \end{bmatrix}$$

Solving Eq. (11.65) for $\boldsymbol{\lambda}_1$, we obtain $\boldsymbol{\lambda}_1 = [-\frac{7}{2}\ -\frac{1}{2}]^T$ which leads to

$$\hat{\boldsymbol{\mu}}_1 = \begin{bmatrix} 2 \\ 0 \\ 0 \end{bmatrix} + \begin{bmatrix} -2 & 1 \\ 0 & -1 \\ -1 & 0 \end{bmatrix} \begin{bmatrix} -\frac{7}{2} \\ -\frac{1}{2} \end{bmatrix} = \begin{bmatrix} \frac{17}{2} \\ \frac{1}{2} \\ \frac{7}{2} \end{bmatrix}$$

Since $\hat{\boldsymbol{\mu}}_1 > \mathbf{0}$, $\mathbf{x}_1$ is the unique vertex minimizer.   ∎

We conclude this section with a remark on the degenerate case. For a standard-form LP problem, a vertex $\mathbf{x}_k$ is degenerate if it has more than $n - p$ zero components. With the notation used in Sect. 11.3.2.1, the matrix $\hat{\mathbf{B}}$ associated with a degenerate vertex contains fewer than p columns and hence the index set $\mathcal{I}_B$ contains at least one index that corresponds to a *zero* component of $\mathbf{x}_k$. Consequently, the index set $\mathcal{I}_k$ defined by Eq. (11.70) may contain an index corresponding to a zero component of $\mathbf{x}_k$. If this happens, then obviously the step size determined using Eq. (11.71) is $\alpha_k = 0$, which would lead to $\mathbf{x}_{k+1} = \mathbf{x}_k$ and from this point on, cycling would occur. In order to prevent cycling, modifications should be made in Steps 2 and 4 of Algorithm 11.3, for example, using Bland's least-index rule.

11.3.3 Tabular Form of the Simplex Method

For LP problems of very small size, the simplex method can be applied in terms of a *tabular form* in which the input data such as $\mathbf{A}$, $\mathbf{b}$, and $\mathbf{c}$ are used to form a table which evolves in a more explicit manner as simplex iterations proceed.

Consider the standard-form LP problem in Eqs. (11.1a)–(11.1c) and assume that at vertex $\mathbf{x}_k$ the equality constraints are expressed as

$$\mathbf{x}_k^{(B)} + \mathbf{B}^{-1}\mathbf{N}\mathbf{x}_k^{(N)} = \mathbf{B}^{-1}\mathbf{b} \tag{11.72}$$

From Eq. (11.64), the objective function is given by

$$\mathbf{c}^T\mathbf{x}_k = \boldsymbol{\mu}_k^T\mathbf{x}_k - \boldsymbol{\lambda}_k^T\mathbf{A}\mathbf{x}_k$$
$$= \mathbf{O}^T\mathbf{x}_k^{(B)} + \hat{\boldsymbol{\mu}}_k^T\mathbf{x}_k^{(N)} - \boldsymbol{\lambda}_k^T\mathbf{b} \tag{11.73}$$

So the important data at the kth iteration can be put together in a tabular form as shown in Table 11.1 from which we observe the following:

(a) If $\hat{\boldsymbol{\mu}}_k \geq 0$, $\mathbf{x}_k$ is a minimizer.
(b) Otherwise, an appropriate rule can be used to choose a negative component in $\hat{\boldsymbol{\mu}}_k$, say, $(\hat{\mu}_k)_l < 0$. As can be seen in Eqs. (11.69a) and (11.69b), the column in $\mathbf{B}^{-1}\mathbf{N}$ that is right above $(\hat{\mu}_k)_l$ gives $-\mathbf{d}_k^{(B)}$. In the discussion that follows, this column will be referred to as the pivot column. In addition, the variable in $\mathbf{x}_N^T$ that corresponds to $(\hat{\mu}_k)_l$ is the variable chosen as a *basic* variable.
(c) Since $\mathbf{x}_k^{(N)} = \mathbf{0}$, Eq. (11.72) implies that $\mathbf{x}_k^{(B)} = \mathbf{B}^{-1}\mathbf{b}$. Therefore, the far-right p-dimensional vector gives $\mathbf{x}_k^{(B)}$.
(d) Since $\mathbf{x}_k^{(N)} = \mathbf{0}$, Eq. (11.73) implies that the number in the lower-right corner of Table 11.1 is equal to $-f(\mathbf{x}_k)$.

Taking the LP problem discussed in Example 11.9 as an example, at $\mathbf{x}_0$ the table assumes the form shown in Table 11.2. Since $(\hat{\mu}_0)_2 < 0$, $\mathbf{x}_0$ is not a minimizer. As was shown above, $(\hat{\mu}_0)_2 < 0$ also suggests that x_3 is the variable in $\mathbf{x}_0^{(N)}$ that will become a basic variable, and the vector above $(\hat{\mu}_0)_2$, $\left[\frac{1}{2}\ 3\right]^T$, is the pivot column $-\mathbf{d}_0^{(B)}$. It follows from Eqs. (11.70) and (11.71) that only the *positive* components of the pivot column should be used to compute the ratio $(\mathbf{x}_0^{(B)})_i/(-\mathbf{d}_0^{(B)})_i$ where $\mathbf{x}_0^{(B)}$ is the far-right column in the table. The index that yields the minimum ratio is

Table 11.1 Simplex method, kth iteration

$\mathbf{x}_B^T$	$\mathbf{x}_N^T$	
$\mathbf{I}$	$\mathbf{B}^{-1}\mathbf{N}$	$\mathbf{B}^{-1}\mathbf{b}$
$\mathbf{O}^T$	$\hat{\boldsymbol{\mu}}_k^T$	$\boldsymbol{\lambda}_k^T\mathbf{b}$

Table 11.2 Simplex method, Example 11.9

Basic variables		Nonbasic variables			$\mathbf{B}^{-1}\mathbf{b}$	$\boldsymbol{\lambda}_k^T \mathbf{b}$
x_2	x_5	x_1	x_3	x_4		
1	0	-1	$\frac{1}{2}$	$-\frac{1}{2}$	$\frac{1}{2}$	
0	1	-5	3	-2	1	
0	0	11	$-\frac{3}{2}$	$\frac{9}{2}$		$-\frac{9}{2}$

Table 11.3 Simplex method, Example 11.9 cont'd

Basic variables		Nonbasic variables			$\mathbf{B}^{-1}\mathbf{b}$	$\boldsymbol{\lambda}_k^T \mathbf{b}$
x_2	x_5	x_1	x_3	x_4		
1	$-\frac{1}{6}$	$-\frac{1}{6}$	0	$-\frac{1}{6}$	$\frac{1}{3}$	
0	$\frac{1}{3}$	$-\frac{5}{3}$	1	$-\frac{2}{3}$	$\frac{1}{3}$	
0	0	11	$-\frac{3}{2}$	$\frac{9}{2}$		$-\frac{9}{2}$

Table 11.4 Simplex method, Example 11.9 cont'd

Basic variables		Nonbasic variables			$\mathbf{B}^{-1}\mathbf{b}$	$\boldsymbol{\lambda}_k^T \mathbf{b}$
x_2	x_3	x_1	x_5	x_4		
1	0	$-\frac{1}{6}$	$-\frac{1}{6}$	$-\frac{1}{6}$	$\frac{1}{3}$	
0	1	$-\frac{5}{3}$	$\frac{1}{3}$	$-\frac{2}{3}$	$\frac{1}{3}$	
0	0	$\frac{17}{2}$	$\frac{1}{2}$	$\frac{7}{2}$		-4

$i^* = 2$. This suggests that the second basic variable, x_5, should be exchanged with x_3 to become a nonbasic variable. To transform x_3 into the second basic variable, we use elementary row operations to transform the pivot column into the i^*th coordinate vector. In the present case, we add $-1/6$ times the second row to the first row, and then multiply the second row by $1/3$. The table assumes the form in Table 11.3.

Next we interchange the columns associated with variables x_3 and x_5 to form the updated basic and nonbasic variables, and then add $3/2$ times the second row to the last row to eliminate the nonzero Lagrange multiplier associated with variable x_3. This leads to the table shown as Table 11.4.

The Lagrange multipliers $\hat{\boldsymbol{\mu}}_1$ in the last row of Table 11.4 are all positive and hence $\mathbf{x}_1$ is the unique minimizer. Vector $\mathbf{x}_1$ is specified by $\mathbf{x}_1^{(B)} = \left[\frac{1}{3}\ \frac{1}{3}\right]^T$ in the far-right column and $\mathbf{x}_1^{(N)} = [0\ 0\ 0]^T$. In conjunction with the composition of the basic and nonbasic variables, $\mathbf{x}_1^{(B)}$ and $\mathbf{x}_1^{(N)}$ yield

$$\mathbf{x}_1 = \left[0\ \tfrac{1}{3}\ \tfrac{1}{3}\ 0\ 0\right]^T$$

At $\mathbf{x}_1$, the lower-right corner of Table 11.4 gives the minimum of the objective function as $f(\mathbf{x}_1) = 4$.

11.3.4 Computational Complexity

As in any iterative algorithm, the computational complexity of a simplex algorithm depends on both the number of iterations it requires to converge and the amount of computation in each iteration.

11.3.4.1 Computations Per Iteration

For an LP problem of the type given in Eqs. (11.2a) and (11.2b) with nondegenerate vertices, the major computational effort in each iteration is to solve two transposed $n \times n$ linear systems, i.e.,

$$\mathbf{A}_{a_k}^T \boldsymbol{\mu}_k = -\mathbf{c} \quad \text{and} \quad \mathbf{A}_{a_k} \mathbf{d}_k = -\mathbf{e}_l \tag{11.74}$$

(see Steps 2 and 3 of Algorithm 11.1). For the degenerate case, matrix $\mathbf{A}_{a_k}$ in Eq. (11.74) is replaced by working-set matrix $\hat{\mathbf{A}}_{a_k}$ which has the same size as $\mathbf{A}_{a_k}$. For the problem in Eqs. (11.1a)–(11.1c), the computational complexity in each iteration is largely related to solving two transposed $p \times p$ linear systems, namely,

$$\mathbf{B}^T \boldsymbol{\lambda}_k = -\mathbf{c}_B \quad \text{and} \quad \mathbf{B}\mathbf{d}_k^{(B)} = -\mathbf{a}_{v_l} \tag{11.75}$$

(see Steps 2 and 3 of Algorithm 11.3). Noticing the similarity between the systems in Eqs. (11.74) and (11.75), we conclude that the computational efficiency in each iteration depends critically on how efficiently two transposed linear systems of a given size are solved. A reliable and efficient approach to solve a linear system of equations in which the number of unknowns is equal to the number of equations (often called a *square system*) with a nonsingular asymmetric system matrix such as $\mathbf{A}_{a_k}$ in Eq. (11.74) and $\mathbf{B}$ in Eq. (11.75) is to use one of several matrix factorization-based methods. These include the lower-upper (LU) factorization with pivoting and the Householder orthogonalization-based QR factorization [3,5]. The number of floating-point operations (flops) required to solve an n-variable square system using the LU factorization and QR factorization methods are $2n^3/3$ and $4n^3/3$, respectively, (see Sect. A.12). It should be stressed that although the QR factorization requires more flops, it is comparable with the LU factorization in efficiency when memory traffic and vectorization overhead are taken into account [5, Chap. 5]. Another desirable feature of the QR factorization method is the guaranteed numerical stability, particularly when the system is ill-conditioned.

For the systems in Eqs. (11.74) and (11.75), there are two important features that can lead to further reduction in the amount of computation. First, each of the two systems involves a pair of matrices that are the *transposes* of each other. So when matrix factorization is performed for the first system, the transposed version of the factorization can be utilized to solve the second system. Second, in each iteration, the matrix is obtained from the matrix used in the preceding iteration through a *rank-one modification*. Specifically, Step 5 of Algorithm 11.1 updates $\mathbf{A}_{a_k}$ by replacing one of its rows with the normal vector of the constraint that just becomes active, while Step 5 of Algorithm 11.3 updates $\mathbf{B}$ by replacing one of its columns with the column in $\mathbf{N}$ that corresponds to the new basic variable. Let

$$
\mathbf{A}_{a_k} = \begin{bmatrix} \mathbf{a}_{j_1}^T \\ \mathbf{a}_{j_2}^T \\ \vdots \\ \mathbf{a}_{j_n}^T \end{bmatrix}
$$

and assume that $\mathbf{a}_{i*}^T$ is used to replace $\mathbf{a}_{j_l}^T$ in the updating of $\mathbf{A}_{a_k}$ to $\mathbf{A}_{a_{k+1}}$. Under these circumstances

$$
\mathbf{A}_{a_{k+1}} = \mathbf{A}_{a_k} + \mathbf{\Delta}_a \tag{11.76}
$$

where $\mathbf{\Delta}_a$ is the rank-one matrix

$$
\mathbf{\Delta}_a = \mathbf{e}_{j_l}(\mathbf{a}_{i*}^T - \mathbf{a}_{j_l}^T)
$$

with $\mathbf{e}_{j_l}$ being the j_lth coordinate vector. Similarly, if we denote matrix $\mathbf{B}$ in the kth and $(k+1)$th iterations as $\mathbf{B}_k$ and $\mathbf{B}_{k+1}$, respectively, then

$$
\mathbf{B}_{k+1} = \mathbf{B}_k + \mathbf{\Delta}_b \tag{11.77}
$$
$$
\mathbf{\Delta}_b = (\mathbf{b}_{i*}^{(k+1)} - \mathbf{b}_{i*}^{(k)})\mathbf{e}_{i*}^T
$$

where $\mathbf{b}_{i*}^{(k+1)}$ and $\mathbf{b}_{i*}^{(k)}$ are the $i*$th columns in $\mathbf{B}_{k+1}$ and $\mathbf{B}_k$, respectively. Efficient algorithms for updating the LU and QR factorizations of a matrix with a rank-one modification, which require only $O(n^2)$ flops, are available in the literature (see [3, Chap. 4], [5, Chap. 12], [7, Chap. 3], and Sect. A.12 for the details).

As a final remark on the matter, LP problems encountered in practice often involve a large number of parameters and the associated large-size system matrix $\mathbf{A}_{a_k}$ or $\mathbf{B}$ is often very *sparse*.[1] Sparse linear systems can be solved using specially designed algorithms that take full advantage of either particular patterns of sparsity that the system matrix exhibits or the general sparse nature of the matrix. Using these algorithms, reduction in the number of flops as well as the required storage space can be significant. (See Sect. 2.7 of [8] for an introduction to several useful methods and further references on the subject.)

---

[1] A matrix is said to be sparse if only a relatively small number of its elements are nonzero.

11.3.4.2 Performance in Terms of Number of Iterations

The number of iterations required for a given LP problem to converge depends on the data that specify the problem and on the initial point, and is difficult to predict accurately [3]. As far as the simplex method is concerned, there is a worse case analysis on the computational complexity of the method on the one hand, and observations on the algorithm's practical performance on the other hand.

Considering the alternative-form LP problem in Eqs. (11.2a) and (11.2b), in the worst case, the simplex method entails examining *every* vertex to find the minimizer. Consequently, the number of iterations would grow exponentially with the problem size. In 1972, Klee and Minty [9] described the following well-known LP problem

$$\text{maximize} \sum_{j=1}^{n} 10^{n-j} x_j \tag{11.78a}$$

$$\text{subject to:} \quad x_i + 2 \sum_{j=1}^{i-1} 10^{i-j} x_j \le 100^{i-1} \quad \text{for } i = 1, 2, \ldots, n \tag{11.78b}$$

$$x_j \ge 0 \quad \text{for } j = 1, 2, \ldots, n \tag{11.78c}$$

For each n, the LP problem involves $2n$ inequality constraints. By introducing n slack variables $s_1, s_2, \ldots, s_n$ and adding them to the constraints in Eq. ((11.78b)) to convert the constraints into equalities, it was shown that if we start with the initial point $s_i = 100^{i-1}$ and $x_i = 0$ for $i = 1, 2, \ldots, n$, then the simplex method has to perform $2^n - 1$ iterations to obtain the solution. However, the chances of encountering the worst case scenario in a real-life LP problem are extremely small. In fact, the simplex method is usually very efficient, and consistently requires a number of iterations that is a small multiple of the problem dimension [10], typically, 2 or 3 times.

Problems

11.1 (*a*) Develop a MATLAB function to generate the data matrices **A**, **b**, and **c** for the LP problem formulated in Prob. 10.1. Inputs of the function should include the order of polynomial $A(\omega)$, n, passband edge ω_p, stopband edge ω_a, number of grid points in the passband, N, and number of grid points in the stopband, M.

(*b*) Applying the MATLAB function obtained in part (*a*) with $n = 30$, $\omega_p = 0.45\pi$, $\omega_a = 0.55\pi$, and $M = N = 30$, obtain matrices **A**, **b**, and **c** for Prob. 10.1.

11.2 (*a*) Develop a MATLAB function that would find a vertex of the feasible
region defined by

$$\mathbf{Ax} \le \mathbf{b} \tag{P11.1}$$

The function may look like x=find_v(A,b,x0) and should accept
a general pair (**A, b**) that defines a nonempty feasible region through
(P11.1), and a feasible initial point $\mathbf{x}_0$.

(*b*) Test the MATLAB function obtained by applying it to the LP problem in
Example 11.2 using several different initial points.

(*c*) Develop a MATLAB function that would find a vertex of the feasible
region defined by $\mathbf{Ax} = \mathbf{b}$ and $\mathbf{x} \ge \mathbf{0}$.

11.3 (*a*) Develop a MATLAB function that would implement Algorithm 11.1. The
function may look like x=lp_and(A,b,c,x0) where $\mathbf{x}_0$ is a feasible
initial point.

(*b*) Apply the MATLAB function obtained to the LP problems in Examples
11.2 and 11.6.

11.4 (*a*) Develop a MATLAB function that would implement Algorithm 11.1
without requiring a feasible initial point. The code can be developed by
implementing the technique described in the first part of Sect. 11.2.3.4
using the code obtained from Prob. 11.3(*a*).

(*b*) Apply the MATLAB function obtained to the LP problems in Examples
11.2 and 11.6.

11.5 In connection with the LP problem in Eqs. (11.2a) and (11.2b), use Farkas'
Lemma (see Prob. 10.18) to show that if $\mathbf{x}$ is a feasible point but not a minimizer, then at $\mathbf{x}$ there always exists a feasible descent direction.

11.6 (*a*) Using a graphical approach, describe the feasible region $\mathcal{R}$ defined by

$$-x_1 \le 0$$
$$-x_2 \le 0$$
$$-x_1 - x_2 + 1 \le 0$$
$$-x_1 + 2x_2 - 4 \le 0$$
$$-x_1 + x_2 - 1 \le 0$$
$$5x_1 - 2x_2 - 15 \le 0$$
$$5x_1 - 6x_2 - 5 \le 0$$
$$x_1 + 4x_2 - 14 \le 0$$

(*b*) Identify the degenerate vertices of $\mathcal{R}$.

11.7 (*a*) By modifying the MATLAB function obtained in Prob. 11.3(*a*), implement Algorithm 11.2. The function may look like `x=lp_ad(A,b,c, x0)` where $\mathbf{x}_0$ is a feasible initial point.

(*b*) Apply the MATLAB function obtained to the LP problem

$$\text{minimize } f(\mathbf{x}) = x_1$$
$$\text{subject to: } \mathbf{x} \in \mathcal{R}$$

where $\mathcal{R}$ is the polygon described in Prob. 11.6(a).

11.8 Consider the LP problem

$$\text{minimize } f(\mathbf{x}) = -2x_1 - 3x_2 + x_3 + 12x_4$$

$$\text{subject to: } \quad -2x_1 - 9x_2 + x_3 + 9x_4 \leq 0$$
$$x_1/3 + x_2 - x_3/3 - 2x_4 \leq 0$$
$$-x_i \leq 0 \quad \text{for } i = 1, 2, 3, 4$$

(See [3, p. 351].)

(*a*) Show that this LP problem does not have finite minimizers.
Hint: Any points of the form $[r\ 0\ r\ 0]^T$ with $r \geq 0$ are feasible.
(*b*) Apply Algorithm 11.1 to the LP problem using $\mathbf{x}_0 = \mathbf{0}$ as a starting point and observe the results.
(*c*) Repeat part (*b*) for Algorithm 11.2.

11.9 Applying an appropriate LP algorithm, solve the problem

$$\text{minimize } f(\mathbf{x}) = -4x_1 - 8x_3$$

$$\text{subject to: } \quad 16x_1 - x_2 + 5x_3 \leq 1$$
$$2x_1 + 4x_3 \leq 1$$
$$10x_1 + x_2 \leq 1$$
$$x_i \leq 1 \quad \text{for } i = 1, 2, 3$$

11.10 Applying Algorithm 11.1, solve the LP problem

$$\text{minimize } f(\mathbf{x}) = x_1 - 4x_2$$

$$\text{subject to: } \quad x_1 - x_2 - 2 \leq 0$$
$$x_1 + x_2 - 6 \leq 0$$
$$-x_i \leq 0 \quad \text{for } i = 1, 2$$

Draw the path of the simplex steps using $\mathbf{x}_0 = [2\ 0]^T$ as a starting point.

11.11 Applying Algorithm 11.2, solve the LP problem

$$\text{minimize} \ f(\mathbf{x}) = 2x_1 - 6x_2 - x_3$$

$$\text{subject to:} \ \begin{aligned} 3x_1 - x_2 + 2x_3 - 7 &\leq 0 \\ -2x_1 + 4x_2 - 12 &\leq 0 \\ -4x_1 + 3x_2 + 3x_3 - 14 &\leq 0 \\ -x_i &\leq 0 \quad \text{for } i = 1, 2, 3 \end{aligned}$$

11.12 Applying Algorithm 11.2, solve the LP problem described in Prob. 10.1 with $n = 30$, $\omega_p = 0.45\pi$, $\omega_a = 0.55\pi$, and $M = N = 30$. Note that the matrices $\mathbf{A}$, $\mathbf{b}$, and $\mathbf{c}$ of the problem can be generated using the MATLAB function developed in Prob. 11.1.

11.13 (a) Develop a MATLAB function that would implement Algorithm 11.3.
       (b) Apply the MATLAB function obtained in part (a) to the LP problem in Example 11.8.

11.14 (a) Convert the LP problem in Prob. 11.10 to a standard-form LP problem by introducing slack variables.
       (b) Apply Algorithm 11.3 to the LP problem obtained in part (a) and compare the results with those obtained in Prob. 11.10.

11.15 (a) Convert the LP problem in Prob. 11.11 to a standard-form LP problem by introducing slack variables.
       (b) Apply Algorithm 11.3 to the LP problem obtained in part (a) and compare the results with those of Prob. 11.11.

11.16 Applying Algorithm 11.3, solve the LP problem

$$\text{minimize} \ f(\mathbf{x}) = x_1 + 1.5x_2 + x_3 + x_4$$

$$\text{subject to:} \ \begin{aligned} x_1 + 2x_2 + x_3 + 2x_4 &= 3 \\ x_1 + x_2 + 2x_3 + 4x_4 &= 5 \\ x_i &\geq 0 \quad \text{for } i = 1, 2, 3, 4 \end{aligned}$$

11.17 Applying Algorithm 11.3, solve the LP problem

$$\text{minimize} \ f(\mathbf{x}) = x_1 + 0.5x_2 + 2x_3$$

$$\text{subject to:} \ \begin{aligned} x_1 + x_2 + 2x_3 &= 3 \\ 2x_1 + x_2 + 3x_3 &= 5 \\ x_i &\geq 0 \quad \text{for } i = 1, 2, 3 \end{aligned}$$

11.18 Based on the remarks given at the end of Sect. 11.3.2, develop a step-by-step description of an algorithm that extends Algorithm 11.3 to the degenerate case.

11.19 Develop a MATLAB function to implement the algorithm developed in Prob. 11.18.

11.20 (a) Convert the LP problem in Prob. 11.8 to a standard-form LP problem. Note that only *two* slack variables need to be introduced.
(b) Apply Algorithm 11.3 to the problem formulated in part (a) using an initial point $\mathbf{x}_0 = \mathbf{0}$, and observe the results.
(c) Applying the algorithm developed in Prob. 11.18, solve the problem formulated in part (a) using an initial point $\mathbf{x}_0 = \mathbf{0}$.

11.21 Consider the nonlinear minimization problem

$$\text{minimize } f(\mathbf{x}) = -2x_1 - 2.5x_2$$

$$\text{subject to:} \quad x_1^2 + x_2^2 - 1 \leq 0$$
$$-x_1 \leq 0, \ -x_2 \leq 0$$

(a) Find an approximate solution of this problem by solving the LP problem with the same linear objective function subject to $\mathbf{x} \in \mathcal{P}$ where $\mathcal{P}$ is a polygon in the first quadrant of the (x_1, x_2) plane that contains the feasible region described above.
(b) Improve the approximate solution obtained in part (a) by using a polygon with an increased number of edges.

References

1. G. B. Dantzig, "Programming in a linear structure," in *Comptroller*. Washington, DC: USAF, 1948.
2. G. B. Dantzig, *Linear Programming and Extensions*. Princeton, NJ: Princeton University Press, 1963.
3. P. E. Gill, W. Murray, and M. H. Wright, *Numerical Linear Algebra and Optimization, Vol. I*. Reading: Addison-Wesley, 1991.
4. R. Saigal, *Linear Programming — A Modern Integrated Analysis*. Norwell, MA: Kluwer Academic, 1995.
5. G. H. Golub and C. F. Van Loan, *Matrix Computations*, 4th ed. Baltimore, MD: Johns Hopkins University Press, 2013.
6. R. G. Bland, "New finite pivoting rules for the simplex method," *Math. Operations Research*, vol. 2, pp. 103–108, 1977.

7. J. E. Dennis, Jr. and R. B. Schnabel, *Numerical Methods for Unconstrained Optimization and Nonlinear Equations*. Philadelphia, PA: SIAM, 1996.

8. W. H. Press, S. A. Teukolsky, W. T. Vetterling, and B. P. Flannery, *Numerical Recipes in C*, 2nd ed. Cambridge, UK: Cambridge University Press, 1992.

9. V. Klee and G. Minty, "How good is the simplex method?" in *Inequalities*, O. Shisha, Ed. New York: Academic Press, 1972, pp. 159–175.

10. M. H. Wright, "Interior methods for constrained optimization," *Acta Numerica, Cambridge University Press*, vol. 1, pp. 341–407, 1992.

Linear Programming Part II: Interior-Point Methods

12

12.1 Introduction

A paper by Karmarkar in 1984 [1] and substantial progress made since that time have led to the field of modern *interior-point methods* for linear programming (LP). Unlike the family of simplex methods considered in Chap. 11, which approach the solution through a sequence of iterates that move from vertex to vertex along the edges on the boundary of the feasible polyhedron, the iterates generated by interior-point algorithms approach the solution from the *interior* of a polyhedron. Although the claims about the efficiency of the algorithm in [1] have not been substantiated in general, extensive computational testing has shown that a number of interior-point algorithms are much more efficient than simplex methods for large-scale LP problems [2].

In this chapter, we study several representative interior-point methods. Our focus will be on algorithmic development rather than theoretical analysis of the methods. *Duality* is a concept of central importance in modern interior-point methods. In Sect. 12.2, we discuss several basic concepts of a duality theory for linear programming. These include primal-dual solutions and central path. Two important primal interior-point methods, namely, the primal affine scaling method and the primal Newton barrier method will be studied in Sects. 12.3 and 12.4, respectively. In Sect. 12.5, we present two primal-dual path-following methods. One of these methods, namely, Mehrotra's predictor-corrector algorithm [3], has been the basis of most interior-point software for LP developed since 1990.

© Springer Science+Business Media, LLC, part of Springer Nature 2021
A. Antoniou and W.-S. Lu, *Practical Optimization*, Texts in Computer Science,
https://doi.org/10.1007/978-1-0716-0843-2_12

12.2   Primal-Dual Solutions and Central Path

12.2.1   Primal-Dual Solutions

The concept of duality was first introduced in Sect. 10.9 for the general convex programming problem in Eq. (10.107) and the main results of the Lagrange dual, namely, the results of Sect. 10.9 as applied to LP problems, were briefly discussed in Example 10.16. In this section, we present several additional results concerning duality which are of importance for the development of modern interior-point methods.

Consider the standard-form LP problem

$$\text{minimize } f(\mathbf{x}) = \mathbf{c}^T\mathbf{x} \tag{12.1a}$$
$$\text{subject to: } \mathbf{A}\mathbf{x} = \mathbf{b} \tag{12.1b}$$
$$\mathbf{x} \geq \mathbf{0} \tag{12.1c}$$

where matrix $\mathbf{A} \in R^{p \times n}$ is of full row rank as the primal problem (see Sect. 10.9). From Example 10.16, we obtain the *dual* problem

$$\text{maximize } h(\boldsymbol{\lambda}) = -\mathbf{b}^T\boldsymbol{\lambda} \tag{12.2a}$$
$$\text{subject to: } -\mathbf{A}^T\boldsymbol{\lambda} + \boldsymbol{\mu} = \mathbf{c} \tag{12.2b}$$
$$\boldsymbol{\mu} \geq \mathbf{0} \tag{12.2c}$$

Two basic questions concerning the LP problems in Eqs. (12.1a)–(12.1c) and Eqs. (12.2a)–(12.2c) are as follows:

(*a*) Under what conditions will the solutions of these problems exist?
(*b*) How are the feasible points and solutions of the primal and dual related?

An LP problem is said to be *feasible* if its feasible region is not empty. The problem in Eqs. (12.1a)–(12.1c) is said to be *strictly feasible* if there exists an $\mathbf{x}$ that satisfies Eq. (12.1b) with $\mathbf{x} > \mathbf{0}$. Likewise, the LP problem in Eqs. (12.2a)–(12.2c) is said to be strictly feasible if there exist $\boldsymbol{\lambda}$ and $\boldsymbol{\mu}$ that satisfy Eq. (12.2b) with $\boldsymbol{\mu} > \mathbf{0}$. It is known that $\mathbf{x}^*$ is a minimizer of the problem in Eqs. (12.1a)–(12.1c) if and only if there exist $\boldsymbol{\lambda}^*$ and $\boldsymbol{\mu}^* \geq \mathbf{0}$ such that

$$-\mathbf{A}^T\boldsymbol{\lambda}^* + \boldsymbol{\mu}^* = \mathbf{c} \tag{12.3a}$$
$$\mathbf{A}\mathbf{x}^* = \mathbf{b} \tag{12.3b}$$
$$x_i^*\mu_i^* = 0 \quad \text{for } 1 \leq i \leq n \tag{12.3c}$$
$$\mathbf{x}^* \geq \mathbf{0}, \quad \boldsymbol{\mu}^* \geq \mathbf{0} \tag{12.3d}$$

For the primal problem, $\boldsymbol{\lambda}^*$ and $\boldsymbol{\mu}^*$ in Eqs. (12.3a)–(12.3d) are the Lagrange multipliers. It can be readily verified that a set of vectors $\{\boldsymbol{\lambda}^*, \boldsymbol{\mu}^*\}$ satisfying Eqs. (12.3a)–(12.3d) is a maximizer for the dual problem in Eqs. (12.2a)–(12.2c) and $\mathbf{x}^*$ in Eqs. (12.3a)–(12.3d) may be interpreted as the Lagrange multipliers for the dual problem. A set $\{\mathbf{x}^*, \boldsymbol{\lambda}^*, \boldsymbol{\mu}^*\}$ satisfying Eqs. (12.3a)–(12.3d) is called a *primal-dual solution*. It follows that $\{\mathbf{x}^*, \boldsymbol{\lambda}^*, \boldsymbol{\mu}^*\}$ is a primal-dual solution if and only if $\mathbf{x}^*$ solves the primal and $\{\boldsymbol{\lambda}^*, \boldsymbol{\mu}^*\}$ solves the dual [3]. The next two theorems address the existence and boundedness of primal-dual solutions.

Theorem 12.1 (Existence of a primal-dual solution) *A primal-dual solution exists if the primal and dual problems are both feasible.*

Proof If point $\mathbf{x}$ is feasible for the LP problem in Eqs. (12.1a)–(12.1c) and $\{\boldsymbol{\lambda}, \boldsymbol{\mu}\}$ is feasible for the LP problem in Eqs. (12.2a)–(12.2c), then set

$$
-\boldsymbol{\lambda}^T \mathbf{b} \leq -\boldsymbol{\lambda}^T \mathbf{b} + \boldsymbol{\mu}^T \mathbf{x} = -\boldsymbol{\lambda}^T \mathbf{A} \mathbf{x} + \boldsymbol{\mu}^T \mathbf{x}
$$
$$
= (-\mathbf{A}^T \boldsymbol{\lambda} + \boldsymbol{\mu})^T \mathbf{x} = \mathbf{c}^T \mathbf{x} \tag{12.4}
$$

Since $f(\mathbf{x}) = \mathbf{c}^T \mathbf{x}$ has a finite lower bound in the feasible region, there exists a set $\{\mathbf{x}^*, \boldsymbol{\lambda}^*, \boldsymbol{\mu}^*\}$ that satisfies Eqs. (12.3a)–(12.3d). Evidently, this $\mathbf{x}^*$ solves the problem in Eqs. (12.1a)–(12.1c). From Eq. (12.4), $h(\boldsymbol{\lambda})$ has a finite upper bound and $\{\boldsymbol{\lambda}^*, \boldsymbol{\mu}^*\}$ solves the problem in Eqs. (12.2a)–(12.2c). Consequently, the set $\{\mathbf{x}^*, \boldsymbol{\lambda}^*, \boldsymbol{\mu}^*\}$ is a primal-dual solution.   ∎

Theorem 12.2 (Strict feasibility of primal-dual solutions) *If the primal and dual problems are both feasible, then*

(a) solutions of the primal problem are bounded if the dual is strictly feasible;
(b) solutions of the dual problem are bounded if the primal is strictly feasible;
(c) primal-dual solutions are bounded if the primal and dual are both strictly feasible.

Proof The statement in (c) is an immediate consequence of (a) and (b). To prove (a), we first note that by virtue of Theorem 12.1 a solution of the primal exists. Below we follow [3] to show the boundedness. Let $\{\boldsymbol{\lambda}, \boldsymbol{\mu}\}$ be strictly feasible for the dual, $\mathbf{x}$ be feasible for the primal, and $\mathbf{x}^*$ be a solution of the primal. It follows that

$$
\boldsymbol{\mu}^T \mathbf{x}^* = (\mathbf{c} + \mathbf{A}^T \boldsymbol{\lambda})^T \mathbf{x}^*
$$
$$
= \mathbf{c}^T \mathbf{x}^* + \boldsymbol{\lambda}^T \mathbf{A} \mathbf{x}^* = \mathbf{c}^T \mathbf{x}^* + \boldsymbol{\lambda}^T \mathbf{b}
$$
$$
\leq \mathbf{c}^T \mathbf{x} + \boldsymbol{\lambda}^T \mathbf{b} = \boldsymbol{\mu}^T \mathbf{x}
$$

Since $\mathbf{x}^* \geq \mathbf{0}$ and $\boldsymbol{\mu} > \mathbf{0}$, we conclude that

$$
\mu_i^* x_i^* \leq \boldsymbol{\mu}^T \mathbf{x}^* \leq \boldsymbol{\mu}^T \mathbf{x}
$$

Hence

$$
x_i^* \leq \frac{1}{\mu_i^*} \boldsymbol{\mu}^T \mathbf{x} \leq \max_{1 \leq i \leq n} \left(\frac{1}{\mu_i^*} \right) \cdot \boldsymbol{\mu}^T \mathbf{x}
$$

and $\mathbf{x}^*$ is bounded.

Part (b) can be proved in a similar manner.   ∎

From Eqs. (12.3a)–(12.3d), we observe that

$$
\mathbf{c}^T \mathbf{x}^* = [(\boldsymbol{\mu}^*)^T - (\boldsymbol{\lambda}^*)^T \mathbf{A}] \mathbf{x}^* = -(\boldsymbol{\lambda}^*)^T \mathbf{A} \mathbf{x}^* = -(\boldsymbol{\lambda}^*)^T \mathbf{b} \tag{12.5}
$$

i.e.,

$$
f(\mathbf{x}^*) = h(\boldsymbol{\lambda}^*)
$$

If we define the *duality gap* as

$$\delta(\mathbf{x}, \ \boldsymbol{\lambda}) = \mathbf{c}^T \mathbf{x} + \mathbf{b}^T \boldsymbol{\lambda} \tag{12.6}$$

then Eqs. (12.4)–(12.5) imply that $\delta(\mathbf{x}, \ \boldsymbol{\lambda})$ is always nonnegative with $\delta(\mathbf{x}^*, \ \boldsymbol{\lambda}^*) = 0$. Moreover, for any feasible $\mathbf{x}$ and $\boldsymbol{\lambda}$, we have

$$\mathbf{c}^T \mathbf{x} \geq \mathbf{c}^T \mathbf{x}^* \geq -\mathbf{b}^T \boldsymbol{\lambda}^* \geq -\mathbf{b}^T \boldsymbol{\lambda}$$

Hence

$$0 \leq \mathbf{c}^T \mathbf{x} - \mathbf{c}^T \mathbf{x}^* \leq \delta(\mathbf{x}, \ \boldsymbol{\lambda}) \tag{12.7}$$

Equation 12.7 indicates that the duality gap can serve as a bound on the closeness of $f(\mathbf{x})$ to $f(\mathbf{x}^*)$ [2].

12.2.2   Central Path

Another important concept related to primal-dual solutions is central path. By virtue of Eqs. (12.3a)–(12.3d) , set $\{\mathbf{x}, \ \boldsymbol{\lambda}, \ \boldsymbol{\mu}\}$ with $\mathbf{x} \in R^n$, $\boldsymbol{\lambda} \in R^p$, and $\boldsymbol{\mu} \in R^n$ is a primal-dual solution if it satisfies the conditions

$$\mathbf{A}\mathbf{x} = \mathbf{b} \qquad \text{with } \mathbf{x} \geq \mathbf{0} \tag{12.8a}$$

$$-\mathbf{A}^T \boldsymbol{\lambda} + \boldsymbol{\mu} = \mathbf{c} \qquad \text{with } \boldsymbol{\mu} \geq \mathbf{0} \tag{12.8b}$$

$$\mathbf{X}\boldsymbol{\mu} = \mathbf{0} \tag{12.8c}$$

where $\mathbf{X} = \text{diag}\{x_1, \ x_2, \ \ldots, \ x_n\}$. The central path for a standard-form LP problem is defined as a set of vectors $\{\mathbf{x}(\tau), \ \boldsymbol{\lambda}(\tau), \ \boldsymbol{\mu}(\tau)\}$ that satisfy the conditions

$$\mathbf{A}\mathbf{x} = \mathbf{b} \qquad \text{with } \mathbf{x} > \mathbf{0} \tag{12.9a}$$

$$-\mathbf{A}^T \boldsymbol{\lambda} + \boldsymbol{\mu} = \mathbf{c} \qquad \text{with } \boldsymbol{\mu} > \mathbf{0} \tag{12.9b}$$

$$\mathbf{X}\boldsymbol{\mu} = \tau \mathbf{e} \tag{12.9c}$$

where τ is a strictly positive scalar parameter, and $\mathbf{e} = [1 \ 1 \ \cdots \ 1]^T$. For each fixed $\tau > 0$, the vectors in the set $\{\mathbf{x}(\tau), \ \boldsymbol{\lambda}(\tau), \ \boldsymbol{\mu}(\tau)\}$ satisfying Eqs. (12.9a)–(12.9c) can be viewed as sets of points in R^n, R^p, and R^n, respectively, and when τ varies, the corresponding points form a set of trajectories called the *central path*. On comparing Eqs. (12.9a)–(12.9c) with Eqs. (12.2a)–(12.2c), it is obvious that the central path is closely related to the primal-dual solutions. From Eqs. (12.9a) and (12.9b), every point on the central path is strictly feasible. Hence the central path lies in the interior of the feasible regions of the problems in Eqs. 12.1a12.1c and 12.2a12.2c and it approaches a primal-dual solution as $\tau \to 0$.

A more explicit relation of the central path with the primal-dual solution can be observed using the duality gap defined in Eq. (12.6). Given $\tau > 0$, let $\{\mathbf{x}(\tau), \ \boldsymbol{\lambda}(\tau), \ \boldsymbol{\mu}(\tau)\}$ be on the central path. From Eqs. (12.9a)–(12.9c), the duality gap $\delta[\mathbf{x}(\tau), \ \boldsymbol{\lambda}(\tau)]$ is given by

$$
\begin{aligned}
\delta[\mathbf{x}(\tau), \ \boldsymbol{\lambda}(\tau)] &= \mathbf{c}^T \mathbf{x}(\tau) + \mathbf{b}^T \boldsymbol{\lambda}(\tau) \\
&= [-\boldsymbol{\lambda}^T(\tau)\mathbf{A} + \boldsymbol{\mu}^T(\tau)]\mathbf{x}(\tau) + \mathbf{b}^T \boldsymbol{\lambda}(\tau) \\
&= \boldsymbol{\mu}^T(\tau)\mathbf{x}(\tau) = n\tau
\end{aligned}
\tag{12.10}
$$

Hence the duality gap along the central path converges linearly to zero as τ approaches zero. Consequently, as $\tau \to 0$ the objective function of the primal problem, $\mathbf{c}^T \mathbf{x}(\tau)$, and the objective function of the dual problem, $\mathbf{b}^T \boldsymbol{\lambda}(\tau)$, approach the same optimal value.

Example 12.1 Sketch the central path of the LP problem

$$\text{minimize} \ \ f(\mathbf{x}) = -2x_1 + x_2 - 3x_3$$
$$\text{subject to:} \ \ x_1 + x_2 + x_3 = 1$$
$$x_1 \geq 0, \ x_2 \geq 0, \ x_3 \geq 0$$

Solution With $\mathbf{c} = [-2 \ 1 \ -3]^T$, $\mathbf{A} = [1 \ 1 \ 1]$, and $\mathbf{b} = 1$, Eqs. (12.9a)–(12.9c) become

$$x_1 + x_2 + x_3 = 1 \tag{12.11a}$$
$$-\lambda + \mu_1 = -2 \tag{12.11b}$$
$$-\lambda + \mu_2 = 1 \tag{12.11c}$$
$$-\lambda + \mu_3 = -3 \tag{12.11d}$$
$$x_1 \mu_1 = \tau \tag{12.11e}$$
$$x_2 \mu_2 = \tau \tag{12.11f}$$
$$x_3 \mu_3 = \tau \tag{12.11g}$$

where $x_i > 0$ and $\mu_i > 0$ for $i = 1, \ 2, \ 3$. From Eqs. (12.11b)–(12.11d), we have

$$\mu_1 = -2 + \lambda \tag{12.12a}$$
$$\mu_2 = 1 + \lambda \tag{12.12b}$$
$$\mu_3 = -3 + \lambda \tag{12.12c}$$

Hence $\mu_i > 0$ for $1 \leq i \leq 3$ if

$$\lambda > 3 \tag{12.13}$$

If we assume that λ satisfies Eq. (12.13), then Eqs. (12.11e)–(12.11g) and Eq. (12.11a) yield

$$\frac{1}{\lambda - 2} + \frac{1}{\lambda + 1} + \frac{1}{\lambda - 3} = \frac{1}{\tau}$$

i.e.,

$$\frac{1}{\tau}\lambda^3 - \left(\frac{4}{\tau} + 3\right)\lambda^2 + \left(\frac{1}{\tau} + 8\right)\lambda + \left(\frac{6}{\tau} - 1\right) = 0 \tag{12.14}$$

The central path can now be constructed by finding a root of Eq. (12.14), $\hat{\lambda}$, that satisfies Eq. (12.13), by computing μ_i for $1 \leq i \leq 3$ using Eqs. (12.12a)–(12.12c) with $\lambda = \hat{\lambda}$, and then evaluating x_i for $1 \leq i \leq 3$ using Eqs. (12.11a)–(12.11g) with $\lambda = \hat{\lambda}$. Fig. 12.1 shows the $\mathbf{x}(\tau)$ component of the central path for $\tau_0 = 5$ and $\tau_f = 10^{-4}$. Note that the entire trajectory lies inside the triangle which is the feasible region of the problem, and approaches vertex $[0 \ 0 \ 1]^T$ which is the unique minimizer of the LP problem. ∎

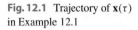

Fig. 12.1 Trajectory of $\mathbf{x}(\tau)$
in Example 12.1

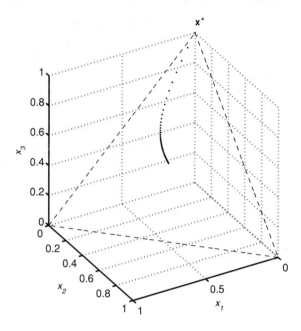

12.3   Primal Affine Scaling Method

The primal affine scaling (PAS) method was proposed in [4,5] as a modification of Karmarkar's interior-point algorithm [1] for LP problems. Though conceptually simple, the method has been found effective, particularly for large-scale problems.

Consider the standard-form LP problem in Eqs. (12.1a)–(12.1c) and let $\mathbf{x}_k$ be a strictly feasible point. The two major steps of the PAS method involve moving $\mathbf{x}_k$ along a *projected steepest-descent direction* and *scaling* the resulting point to center it in the feasible region in a transformed space.

For a linear objective function $f(\mathbf{x}) = \mathbf{c}^T\mathbf{x}$, the steepest-descent direction is $-\mathbf{c}$. At a feasible point $\mathbf{x}_k$, moving along $-\mathbf{c}$ does not guarantee the feasibility of the next iterate since $\mathbf{Ac}$ is most likely nonzero. The PAS method moves $\mathbf{x}_k$ along the direction that is the orthogonal projection of $-\mathbf{c}$ onto the null space of $\mathbf{A}$. This direction $\mathbf{d}_k$ is given by

$$\mathbf{d}_k = -\mathbf{Pc} \tag{12.15}$$

where $\mathbf{P}$ is the projection matrix given by

$$\mathbf{P} = \mathbf{I} - \mathbf{A}^T(\mathbf{AA}^T)^{-1}\mathbf{A} \tag{12.16}$$

It can be readily verified that $\mathbf{AP} = \mathbf{0}$. Hence if the next point is denoted as

$$\mathbf{x}_{k+1} = \mathbf{x}_k + \alpha_k\mathbf{d}_k$$

then $\mathbf{x}_{k+1}$ satisfies Eq. (12.1b), i.e.,

$$\mathbf{Ax}_{k+1} = \mathbf{Ax}_k + \alpha_k\mathbf{Ad}_k$$
$$= \mathbf{b} - \alpha_k\mathbf{APc} = \mathbf{b}$$

If matrix $\mathbf{A}$ is expressed in terms of its singular-value decomposition (SVD) as

$$\mathbf{A} = \mathbf{U}[\Sigma \ \mathbf{0}]\mathbf{V}^T$$

where $\mathbf{U} \in R^{p \times p}$ and $\mathbf{V} \in R^{n \times n}$ are orthogonal and Σ is positive definite and diagonal, then the projection matrix becomes

$$\mathbf{P} = \mathbf{V} \begin{bmatrix} \mathbf{0} & \mathbf{0} \\ \mathbf{0} & \mathbf{I}_{n-p} \end{bmatrix} \mathbf{V}^T$$

This gives

$$\mathbf{c}^T \mathbf{P} \mathbf{c} = \sum_{j=p+1}^{n} [(\mathbf{V}^T \mathbf{c})_j]^2 \qquad (12.17)$$

which is always nonnegative and strictly positive as long as one of the last $n - p$ components in $\mathbf{V}^T \mathbf{c}$ is nonzero. It follows that for any $\alpha_k > 0$

$$f(\mathbf{x}_{k+1}) = \mathbf{c}^T \mathbf{x}_{k+1} = \mathbf{c}^T \mathbf{x}_k - \alpha_k \mathbf{c}^T \mathbf{P} \mathbf{c}$$
$$\leq \mathbf{c}^T \mathbf{x}_k = f(\mathbf{x}_k)$$

and $f(\mathbf{x}_{k+1})$ will be strictly less than $f(\mathbf{x}_k)$ if at least one of the last $n - p$ components in $\mathbf{V}^T \mathbf{c}$ is nonzero.

The search direction, $\mathbf{d}_k$, determined by using Eqs. (12.15) and (12.16) is *independent* of the current point $\mathbf{x}_k$ and the progress that can be made along such a constant direction may become insignificant particularly when $\mathbf{x}_k$ is close to the boundary of the feasible region. A crucial step in the PAS method that overcomes this difficulty is to transform the original LP problem at the kth iteration from that in Eqs. (12.1a)–(12.1c) into an equivalent LP problem in which point $\mathbf{x}_k$ is at a more 'central' position so as to achieve significant reduction in $f(\mathbf{x})$ along the projected steepest-descent direction.

For the standard-form LP problem in Eqs. (12.1a)–(12.1c), the nonnegativity bounds in Eq. (12.1c) suggest that the point $\mathbf{e} = [1 \ 1 \ \cdots \ 1]^T$, which is situated at an equal distance from each x_i axis for $1 \leq i \leq n$, can be considered as a central point. The *affine scaling* transformation defined by

$$\bar{\mathbf{x}} = \mathbf{X}^{-1}\mathbf{x} \qquad (12.18a)$$

with

$$\mathbf{X} = \text{diag}\{(\mathbf{x}_k)_1, \ (\mathbf{x}_k)_2, \ \ldots, \ (\mathbf{x}_k)_n\} \qquad (12.18b)$$

maps point $\mathbf{x}_k$ to $\mathbf{e}$, and the equivalent LP problem given by this transformation is

$$\text{minimize } \bar{f}(\bar{\mathbf{x}}) = \bar{\mathbf{c}}^T \bar{\mathbf{x}} \qquad (12.19a)$$
$$\text{subject to: } \bar{\mathbf{A}}\bar{\mathbf{x}} = \mathbf{b} \qquad (12.19b)$$
$$\bar{\mathbf{x}} \geq \mathbf{0} \qquad (12.19c)$$

where $\bar{\mathbf{c}} = \mathbf{X}\mathbf{c}$ and $\bar{\mathbf{A}} = \mathbf{A}\mathbf{X}$. If the next point is generated along the projected steepest-descent direction from $\mathbf{x}_k$, then

$$\bar{\mathbf{x}}_{k+1} = \bar{\mathbf{x}}_k + \alpha_k \bar{\mathbf{d}}_k = \mathbf{e} + \alpha_k \bar{\mathbf{d}}_k \qquad (12.20)$$

where

$$\bar{\mathbf{d}}_k = -\bar{\mathbf{P}}\bar{\mathbf{c}} = -[\mathbf{I} - \bar{\mathbf{A}}^T(\bar{\mathbf{A}}\bar{\mathbf{A}}^T)^{-1}\bar{\mathbf{A}}]\bar{\mathbf{c}}$$
$$= -[\mathbf{I} - \mathbf{X}\mathbf{A}^T(\mathbf{A}\mathbf{X}^2\mathbf{A}^T)^{-1}\mathbf{A}\mathbf{X}]\mathbf{X}\mathbf{c} \qquad (12.21)$$

Equation (12.20) can be written in terms of the original variables as

$$\mathbf{x}_{k+1} = \mathbf{x}_k + \alpha_k\mathbf{d}_k \qquad (12.22a)$$

with

$$\mathbf{d}_k = \mathbf{X}\bar{\mathbf{d}}_k = -[\mathbf{X}^2 - \mathbf{X}^2\mathbf{A}^T(\mathbf{A}\mathbf{X}^2\mathbf{A}^T)^{-1}\mathbf{A}\mathbf{X}^2]\mathbf{c} \qquad (12.22b)$$

which is called the *primal affine scaling direction* [6]. In order to compare the two search directions given by Eqs. (12.15) and (12.22b), we write vector $\mathbf{d}_k$ in Eq. (12.22b) as

$$\mathbf{d}_k = -\mathbf{X}\bar{\mathbf{P}}\mathbf{X}\mathbf{c} \qquad (12.23)$$

with

$$\bar{\mathbf{P}} = \mathbf{I} - \mathbf{X}\mathbf{A}^T(\mathbf{A}\mathbf{X}^2\mathbf{A}^T)^{-1}\mathbf{A}\mathbf{X} \qquad (12.24)$$

Note that matrix $\mathbf{P}$ in Eq. (12.16) is the projection matrix for $\mathbf{A}$ while matrix $\bar{\mathbf{P}}$ in Eq. (12.24) is the projection matrix for $\mathbf{A}\mathbf{X}$, which depends on both $\mathbf{A}$ and the present point $\mathbf{x}_k$. Consequently, $\mathbf{A}\mathbf{X}\bar{\mathbf{P}} = \mathbf{0}$, which in conjunction with Eqs. (12.22a) and (12.23) implies that if $\mathbf{x}_k$ is strictly feasible, then $\mathbf{x}_{k+1}$ satisfies Eq. (12.1b)

$$\mathbf{A}\mathbf{x}_{k+1} = \mathbf{A}\mathbf{x}_k + \alpha_k\mathbf{A}\mathbf{d}_k = \mathbf{b} - \alpha_k\mathbf{A}\mathbf{X}\bar{\mathbf{P}}\mathbf{X}\mathbf{c} = \mathbf{b}$$

It can also be shown that for any $\alpha_k > 0$ in Eq. (12.22a),

$$f(\mathbf{x}_{k+1}) \le f(\mathbf{x}_k)$$

(see Prob. 12.2) and if at least one of the last $n - p$ components of $\mathbf{V}_k^T\mathbf{X}\mathbf{c}$ is nonzero, the above inequality becomes strict. Here matrix $\mathbf{V}_k$ is the $n \times n$ orthogonal matrix obtained from the SVD of $\mathbf{A}\mathbf{X}$, i.e.,

$$\mathbf{A}\mathbf{X} = \mathbf{U}_k[\Sigma_k \; \mathbf{0}]\mathbf{V}_k^T$$

Having calculated the search direction $\mathbf{d}_k$ using Eq. (12.22b), the step size α_k in Eq. (12.22a) can be chosen such that $\mathbf{x}_{k+1} > \mathbf{0}$. In practice, α_k is chosen as [6]

$$\alpha_k = \gamma\alpha_{max} \qquad (12.25a)$$

where $0 < \gamma < 1$ is a constant, usually close to unity, and

$$\alpha_{max} = \min_{i \text{ with } (\mathbf{d}_k)_i < 0} \left[-\frac{(\mathbf{x}_k)_i}{(\mathbf{d}_k)_i} \right] \qquad (12.25b)$$

The PAS algorithm can now be summarized as follows.

Table 12.1 Sequence of points $\{\mathbf{x}_k$ for $k = 0, 1, \ldots, 4\}$ in Example 12.2

$\mathbf{x}_0$	$\mathbf{x}_1$	$\mathbf{x}_2$	$\mathbf{x}_3$	$\mathbf{x}_4$
0.200000	0.099438	0.000010	0.000010	0.000008
0.700000	0.454077	0.383410	0.333357	0.333339
1.000000	0.290822	0.233301	0.333406	0.333336
1.000000	0.000100	0.000100	0.000100	0.000000
1.000000	0.624922	0.300348	0.000030	0.000029

Algorithm 12.1 Primal affine scaling algorithm for the standard-form LP problem

Step 1
Input $\mathbf{A}$, $\mathbf{c}$, and a strictly feasible initial point $\mathbf{x}_0$.
Set $k = 0$ and initialize the tolerance ε.
Evaluate $f(\mathbf{x}_k) = \mathbf{c}^T \mathbf{x}_k$.
Step 2
Form $\mathbf{X}$ at $\mathbf{x}_k$ and compute $\mathbf{d}_k$ using Eq. (12.22b).
Step 3
Calculate the step size α_k using Eqs. (12.25a) and (12.25b).
Step 4
Set $\mathbf{x}_{k+1} = \mathbf{x}_k + \alpha_k \mathbf{d}_k$ and evaluate $f(\mathbf{x}_{k+1}) = \mathbf{c}^T \mathbf{x}_{k+1}$.
Step 5
If

$$\frac{|f(\mathbf{x}_k) - f(\mathbf{x}_{k+1})|}{\max(1, \ |f(\mathbf{x}_k)|)} < \varepsilon$$

output $\mathbf{x}^* = \mathbf{x}_{k+1}$ and stop; otherwise, set $k = k + 1$ and repeat from Step 2.

Example 12.2 Solve the standard-form LP problem in Example 11.9 using the PAS algorithm.

Solution A strictly feasible initial point is $\mathbf{x}_0 = [0.2 \ 0.7 \ 1 \ 1 \ 1]^T$. With $\gamma = 0.9999$ and $\varepsilon = 10^{-4}$, Algorithm 12.1 converged to the solution

$$\mathbf{x}^* = \begin{bmatrix} 0.000008 \\ 0.333339 \\ 0.333336 \\ 0.000000 \\ 0.000029 \end{bmatrix}$$

after 4 iterations. The sequence of the iterates obtained is given in Table 12.1.   ■

12.4 Primal Newton Barrier Method

12.4.1 Basic Idea

In the primal Newton barrier (PNB) method [2,7], the inequality constraints in Eq. (12.1c) are incorporated in the objective function by adding a *logarithmic barrier function*. The subproblem obtained has the form

$$\text{minimize } f_\tau(\mathbf{x}) = \mathbf{c}^T \mathbf{x} - \tau \sum_{i=1}^{n} \ln x_i \tag{12.26a}$$

$$\text{subject to: } \mathbf{Ax} = \mathbf{b} \tag{12.26b}$$

where τ is a strictly positive scalar. The term $-\tau \sum_{i=1}^{n} \ln x_i$ in Eq. (12.26a) is called a *'barrier' function* for the reason that if we start with an initial $\mathbf{x}_0$ which is strictly inside the feasible region, then the term is well defined and acts like a barrier that prevents any component x_i from becoming zero. The scalar τ is known as the *barrier parameter*. The effect of the barrier function on the original LP problem depends largely on the magnitude of τ. If we start with an interior point, $\mathbf{x}_0$, then under certain conditions to be examined below for a given $\tau > 0$, a unique solution of the subproblem in Eqs. (12.26a)–(12.26b) exists. Thus, if we solve the subproblem in Eqs. (12.26a)–(12.26b) for a series of values of τ, a series of solutions are obtained that converge to the solution of the original LP problem as $\tau \to 0$. In effect, the PNB method solves the LP problem through the solution of a sequence of optimization problems [8] as in the minimax optimization methods of Chap. 8.

In a typical sequential optimization method, there are three issues that need to be addressed. These are:

(a) For each fixed $\tau > 0$ does a minimizer of the subproblem in Eqs. (12.26a)–(12.26b) exist?

(b) If $\mathbf{x}_\tau^*$ is a minimizer of the problem in Eqs. (12.26a)–(12.26b) and $\mathbf{x}^*$ is a minimizer of the problem in Eqs. (12.1a)–(12.1c), how close is $\mathbf{x}_\tau^*$ to $\mathbf{x}^*$ as $\tau \to 0$?

(c) For each fixed $\tau > 0$, how do we compute or estimate $\mathbf{x}_\tau^*$?

12.4.2 Minimizers of Subproblem

Throughout the rest of the section, we assume that the primal in Eqs. (12.1a)–(12.1c) and the dual in Eqs. (12.2a)–(12.2c) are both strictly feasible. Let $\tau > 0$ be fixed and $\mathbf{x}_0$ be a strictly feasible point for the problem in Eqs. (12.1a)–(12.1c). At $\mathbf{x}_0$ the objective function of Eq. (12.26a), $f_\tau(\mathbf{x}_0)$, is well defined. By virtue of Theorem 12.2, the above assumption implies that solutions of the primal exist and are bounded. Under these circumstances, it can be shown that for a given $\varepsilon > 0$ the set

$$S_0 = \{\mathbf{x} : \mathbf{x} \text{ is strictly feasible for problem (12.1)}; f_\tau(\mathbf{x}) \le f_\tau(\mathbf{x}_0) + \varepsilon\}$$

is compact for all $\tau > 0$ (see Theorem 4 in [2]). This implies that $f_\tau(\mathbf{x})$ has a local minimizer $\mathbf{x}_\tau^*$ at an interior point of $\mathcal{S}_0$. We can compute the gradient and Hessian of $f_\tau(\mathbf{x})$ as

$$\nabla f_\tau(\mathbf{x}) = \mathbf{c} - \tau \mathbf{X}^{-1}\mathbf{e} \tag{12.27a}$$

$$\nabla^2 f_\tau(\mathbf{x}) = \tau \mathbf{X}^{-2} \tag{12.27b}$$

with $\mathbf{X} = \mathrm{diag}\{x_1,\, x_2,\, \ldots,\, x_n\}$ and $\mathbf{e} = [1\ 1\ \cdots\ 1]^T$. Since $f_\tau(\mathbf{x})$ is convex, $\mathbf{x}_\tau^*$ in $\mathcal{S}_0$ is a global minimizer of the problem in Eqs. (12.26a) and (12.26b).

12.4.3   A Convergence Issue

Let $\{\tau_k\}$ be a sequence of barrier parameters that are monotonically decreasing to zero and $\mathbf{x}_k^*$ be the minimizer of the problem in Eqs. (12.26a) and (12.26b) with $\tau = \tau_k$. It follows that

$$\mathbf{c}^T\mathbf{x}_k^* - \tau_k \sum_{i=1}^{n} \ln(\mathbf{x}_k^*)_i \le \mathbf{c}^T\mathbf{x}_{k+1}^* - \tau_k \sum_{i=1}^{n} \ln(\mathbf{x}_{k+1}^*)_i$$

and

$$\mathbf{c}^T\mathbf{x}_{k+1}^* - \tau_{k+1} \sum_{i=1}^{n} \ln(\mathbf{x}_{k+1}^*)_i \le \mathbf{c}^T\mathbf{x}_k^* - \tau_{k+1} \sum_{i=1}^{n} \ln(\mathbf{x}_k^*)_i$$

These equations yield (see Prob. 12.11(a))

$$f(\mathbf{x}_{k+1}^*) = \mathbf{c}^T\mathbf{x}_{k+1}^* \le \mathbf{c}^T\mathbf{x}_k^* = f(\mathbf{x}_k^*) \tag{12.28}$$

i.e., the objective function of the original LP problem in Eqs. (12.1a)–(12.1c) is a monotonically decreasing function of sequence $\{\mathbf{x}_k^*$ for $k = 0,\ 1,\ \ldots\}$. An immediate consequence of Eq. (12.28) is that all the minimizers, $\mathbf{x}_k^*$, are contained in the compact set

$$\mathcal{S} = \{\mathbf{x} : \ \mathbf{x} \text{ is feasible for the problem in Eqs. (12.1a)–(12.1c) and } f(\mathbf{x}) \le f(\mathbf{x}_0)\}$$

Therefore, sequence $\{\mathbf{x}_k^*\}$ contains at least one convergent subsequence, which for the sake of simplicity is denoted again as $\{\mathbf{x}_k^*\}$, namely,

$$\lim_{k \to \infty} \mathbf{x}_k^* = \mathbf{x}^* \tag{12.29}$$

It can be shown that the limit vector $\mathbf{x}^*$ in Eq. (12.29) is a minimizer of the primal problem in Eqs. (12.1a)–(12.1c) [2,8]. Moreover, the closeness of $\mathbf{x}_k^*$ to $\mathbf{x}^*$ can be related to the magnitude of the barrier parameter τ_k as follows. The problem in Eqs. (12.1a)–(12.1c) is said to be *nondegenerate* if there are exactly p strictly positive components in $\mathbf{x}^*$ and is said to be *degenerate* otherwise. In [9,10], it was shown that

$$||\mathbf{x}_k^* - \mathbf{x}^*|| = O(\tau_k) \quad \text{if the problem in Eqs. (12.1a)–(12.1c) is nondegenerate}$$

and

$$||\mathbf{x}_k^* - \mathbf{x}^*|| = O(\tau_k^{1/2}) \quad \text{if the problem in Eqs. (12.1a)–(12.1c) is degenerate}$$

The sequence of minimizers for the subproblem in Eqs. (12.26a) and (12.26b) can also be related to the central path of the problems in Eqs. (12.1a)–(12.1c) and Eqs. (12.2a)–(12.2c). To see this, we write the Karush-Kuhn-Tucker (KKT) condition in Eq. (10.74) for the subproblem in Eqs. (12.26a) and (12.26b) at $\mathbf{x}_k^*$ as

$$-\mathbf{A}^T \boldsymbol{\lambda}_k + \tau_k \mathbf{X}^{-1} \mathbf{e} = \mathbf{c} \qquad (12.30)$$

where $\mathbf{X} = \mathrm{diag}\{(\mathbf{x}_k^*)_1,\ (\mathbf{x}_k^*)_2,\ \dots,\ (\mathbf{x}_k^*)_n\}$. If we let

$$\boldsymbol{\mu}_k = \tau_k \mathbf{X}^{-1} \mathbf{e} \qquad (12.31)$$

then with $\mathbf{x}_k^*$ being a strictly feasible point, Eqs. (12.30) and (12.31) lead to

$$\mathbf{A}\mathbf{x}_k^* = \mathbf{b} \qquad \text{with } \mathbf{x}_k^* > \mathbf{0} \qquad (12.32a)$$
$$-\mathbf{A}^T \boldsymbol{\lambda}_k + \boldsymbol{\mu}_k = \mathbf{c} \qquad \text{with } \boldsymbol{\mu}_k > \mathbf{0} \qquad (12.32b)$$
$$\mathbf{X}\boldsymbol{\mu}_k = \tau_k \qquad (12.32c)$$

On comparing Eqs. (12.32a)–(12.32c) with Eqs. (12.9a)–(12.9c), we conclude that the sequences of points $\{\mathbf{x}_k^*,\ \boldsymbol{\lambda}_k,\ \boldsymbol{\mu}_k\}$ are on the central path for the problems in Eqs. (12.1a)–(12.1c) and Eqs. (12.2a)–(12.2c). Further, since $\mathbf{x}^*$ is a minimizer of these problems, there exist $\boldsymbol{\lambda}^*$ and $\boldsymbol{\mu}^* \ge \mathbf{0}$ such that

$$\mathbf{A}\mathbf{x}^* = \mathbf{b} \qquad \text{with } \mathbf{x}^* \ge \mathbf{0} \qquad (12.33a)$$
$$-\mathbf{A}^T \boldsymbol{\lambda}^* + \boldsymbol{\mu}^* = \mathbf{c} \qquad \text{with } \boldsymbol{\mu}^* \ge \mathbf{0} \qquad (12.33b)$$
$$\mathbf{X}^* \boldsymbol{\mu}^* = \mathbf{0} \qquad (12.33c)$$

where $\mathbf{X}^* = \mathrm{diag}\{(\mathbf{x}^*)_1,\ (\mathbf{x}^*)_2,\ \dots,\ (\mathbf{x}^*)_n\}$. By virtue of Eq. (12.29) and $\tau_k \to 0$, Eqs. (12.32c) and (12.33c) imply that $\boldsymbol{\mu}_k \to \boldsymbol{\mu}^*$. From Eqs. (12.32b) and (12.33b), we have

$$\lim_{k \to \infty} \mathbf{A}^T (\boldsymbol{\lambda}_k - \boldsymbol{\lambda}^*) = \mathbf{0} \qquad (12.34)$$

Since $\mathbf{A}^T$ has full column rank, Eq. (12.34) implies that $\boldsymbol{\lambda}_k \to \boldsymbol{\lambda}^*$. Therefore, by letting $k \to \infty$ in Eqs. (12.32a)–(12.32c), we obtain Eqs. (12.33a)–(12.33c). In other words, as $k \to \infty$ the sequences of points $\{\mathbf{x}_k^*,\ \boldsymbol{\lambda}_k^*,\ \boldsymbol{\mu}_k^*\}$ converge to a primal-dual solution $\{\mathbf{x}^*,\ \boldsymbol{\lambda}^*,\ \boldsymbol{\mu}^*\}$ of the problems in Eqs. (12.1a)–(12.2c).

12.4.4   Computing a Minimizer of the Problem in Eqs. (12.26a) and (12.26b)

For a fixed $\tau > 0$, the PNB method starts with a strictly feasible point $\mathbf{x}_0$ and proceeds iteratively to find points $\mathbf{x}_k$ and

$$\mathbf{x}_{k+1} = \mathbf{x}_k + \alpha_k \mathbf{d}_k$$

such that the search direction satisfies the equality

$$\mathbf{A}\mathbf{d}_k = \mathbf{0} \qquad (12.35)$$

The constraints in Eq. (12.35) ensure that if $\mathbf{x}_k$ satisfies Eq. (12.26b), then so does $\mathbf{x}_{k+1}$, i.e.,

$$\mathbf{A}\mathbf{x}_{k+1} = \mathbf{A}\mathbf{x}_k + \alpha_k \mathbf{A}\mathbf{d}_k = \mathbf{b}$$

To find a descent direction, a second-order approximation of the problem in Eqs. (12.26a) and (12.26b) is employed using the gradient and Hessian of $f_\tau(\mathbf{x})$ in Eqs. (12.27a) and (12.27b), namely,

$$\text{minimize } \tfrac{1}{2}\tau \mathbf{d}^T \mathbf{X}^{-2}\mathbf{d} + \mathbf{d}^T (\mathbf{c} - \tau \mathbf{X}^{-1}\mathbf{e}) \tag{12.36a}$$

$$\text{subject to: } \mathbf{A}\mathbf{d} = \mathbf{0} \tag{12.36b}$$

For a strictly feasible $\mathbf{x}_k$, $\mathbf{X}^{-2}$ is positive definite. The problem in Eqs. Eqs. (12.36a) and (12.36b) is a convex programming problem whose solution $\mathbf{d}_k$ satisfies the KKT conditions

$$\tau \mathbf{X}^{-2}\mathbf{d}_k + \mathbf{c} - \tau \mathbf{X}^{-1}\mathbf{e} = -\mathbf{A}^T \boldsymbol{\lambda} \tag{12.37a}$$

$$\mathbf{A}\mathbf{d}_k = \mathbf{0} \tag{12.37b}$$

From Eqs. (12.37a) and (12.37b), we obtain

$$\mathbf{d}_k = \mathbf{x}_k - \frac{1}{\tau}\mathbf{X}^2(\mathbf{A}^T \boldsymbol{\lambda} + \mathbf{c}) \tag{12.38a}$$

and

$$\begin{aligned}\mathbf{A}\mathbf{X}^2\mathbf{A}^T \boldsymbol{\lambda} &= -\tau \mathbf{A}\mathbf{d}_k - \mathbf{A}\mathbf{X}^2\mathbf{c} + \tau \mathbf{A}\mathbf{x}_k \\ &= \mathbf{A}(\tau \mathbf{x}_k - \mathbf{X}^2\mathbf{c})\end{aligned} \tag{12.38b}$$

We see that the search direction $\mathbf{d}_k$ in the PNB method is determined by using Eq. (12.38a) with a $\boldsymbol{\lambda}$ obtained by solving the $p \times p$ symmetric positive-definite system in Eq. (12.38b).

Having determined $\mathbf{d}_k$, a line search along $\mathbf{d}_k$ can be carried out to determine a scalar $\alpha_k > 0$ such that $\mathbf{x}_k + \alpha \mathbf{d}_k$ remains strictly feasible and $\mathbf{f}_\tau(\mathbf{x}_k + \alpha \mathbf{d}_k)$ is minimized with respect to the range $0 \leq \alpha \leq \bar{\alpha}_k$ where $\bar{\alpha}_k$ is the largest possible scalar for $\mathbf{x}_k + \alpha \mathbf{d}_k$ to be strictly feasible. If we let

$$\mathbf{x}_k = \begin{bmatrix} x_1 \\ x_2 \\ \vdots \\ x_n \end{bmatrix} \quad \text{and} \quad \mathbf{d}_k = \begin{bmatrix} d_1 \\ d_2 \\ \vdots \\ d_n \end{bmatrix}$$

the strict feasibility of $\mathbf{x}_k + \alpha \mathbf{d}_k$ can be assured, i.e., $x_i + \alpha d_i > 0$ for $1 \leq i \leq n$, if $\alpha < x_i/(-d_i)$ for all $1 \leq i \leq n$. Hence point $\mathbf{x}_k + \alpha \mathbf{d}_k$ will remain strictly feasible if α satisfies the condition

$$\alpha < \min_{i \text{ with } d_i < 0} \left[\frac{x_i}{(-d_i)} \right]$$

In practice, the upper bound of α, namely,

$$\bar{\alpha}_k = 0.99 \times \min_{i \text{ with } d_i < 0} \left[\frac{x_i}{(-d_i)} \right] \tag{12.39}$$

gives satisfactory results. At $\mathbf{x}_k$, the line search for function $f_\tau(\mathbf{x}_k + \alpha\mathbf{d}_k)$ is carried out on the closed interval $[0, \bar{\alpha}_k]$. Since

$$\frac{d^2 f_\tau(\mathbf{x}_k + \alpha\mathbf{d}_k)}{d\alpha^2} = \tau \sum_{i=1}^{n} \frac{d_i}{(x_i + \alpha d_i)^2} > 0$$

$f_\tau(\mathbf{x}_k + \alpha\mathbf{d}_k)$ is strictly convex on $[0, \bar{\alpha}_k]$ and has a unique minimum. One of the search methods discussed in Chap. 4 can be used to find the minimizer, α_k, and the new point is obtained as $\mathbf{x}_{k+1} = \mathbf{x}_k + \alpha_k\mathbf{d}_k$.

The PNB algorithm can be summarized as follows.

Algorithm 12.2 Primal Newton barrier algorithm for the standard-form LP problem

Step 1
Input $\mathbf{A}$, $\mathbf{c}$, and a strictly feasible initial point $\mathbf{x}_0$.
Set $l = 0$, initialize the barrier parameter such that $\tau_0 > 0$, and input the outer-loop tolerance $\varepsilon_{\text{outer}}$.

Step 2
Set $k = 0$ and $\mathbf{x}_0^{(l)} = \mathbf{x}_l$, and input the inner-loop tolerance $\varepsilon_{\text{inner}}$.

   Step 3.1
   Use Eqs. (12.38a)–(12.38b) with $\tau = \tau_l$ to calculate $\mathbf{d}_k^{(l)}$ at $\mathbf{x}_k^{(l)}$.

   Step 3.2
   Use Eq. (12.39) to calculate $\bar{\alpha}_k$ where $\mathbf{x}_k = \mathbf{x}_k^{(l)}$ and $\mathbf{d}_k = \mathbf{d}_k^{(l)}$.

   Step 3.3
   Use a line search (e.g., a line search based on the golden-section method) to determine $\alpha_k^{(l)}$.

   Step 3.4
   Set $\mathbf{x}_{k+1}^{(l)} = \mathbf{x}_k^{(l)} + \alpha_k^{(l)}\mathbf{d}_k^{(l)}$.

   Step 3.5
   If $||\alpha_k^{(l)}\mathbf{d}_k^{(l)}|| < \varepsilon_{\text{inner}}$, set $\mathbf{x}_{l+1} = \mathbf{x}_{k+1}^{(l)}$ and go to Step 4; otherwise, set $k = k + 1$ and repeat from Step 3.1.

Step 4
If $||\mathbf{x}_l - \mathbf{x}_{l+1}|| < \varepsilon_{\text{outer}}$, output $\mathbf{x}^* = \mathbf{x}_{l+1}$, and stop; otherwise, choose $\tau_{l+1} < \tau_l$, set $l = l + 1$, and repeat from Step 2.

Example 12.3 Apply the PNB algorithm to the LP problem in Example 12.1.

Solution We start with $\mathbf{c} = [-2 \ 1 \ -3]^T$, $\mathbf{A} = [1 \ 1 \ 1]$, $\mathbf{b} = 1$, and $\mathbf{x}_0 = \left[\frac{1}{3} \ \frac{1}{3} \ \frac{1}{3}\right]^T$ which is strictly feasible, and employ the golden-section method (see Sect. 4.4) to perform the line search in Step 3.3. Parameter τ_i is chosen as $\tau_{l+1} = \sigma\tau_l$ with $\sigma = 0.1$.

Fig. 12.2 Iteration path in
Example 12.3

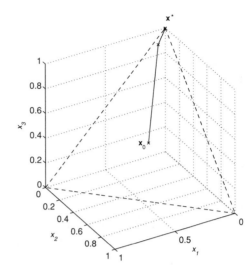

With $\tau_0 = 0.1$ and $\varepsilon_{\text{outer}} = 10^{-4}$, Algorithm 12.2 took six iterations to converge
to the solution

$$\mathbf{x}^* = \begin{bmatrix} 0.000007 \\ 0.000001 \\ 0.999992 \end{bmatrix}$$

The number of flops required was 5.194K. The path of the sequence $\{\mathbf{x}_l$ for $l =$
$0, 1, \ldots, 6\}$ is shown in Fig. 12.2.   ∎

12.5   Primal-Dual Interior-Point Methods

The methods studied in Sects. 12.3 and 12.4 are primal interior-point methods
in which the dual is not explicitly involved. Primal-dual methods, on the other
hand, solve the primal and dual LP problems simultaneously, and have emerged
as the most efficient interior-point methods for LP problems. In this section, we
examine two important primal-dual interior-point methods, namely, the primal-dual
path-following method and the nonfeasible-initialization primal-dual path-following
method.

12.5.1   Primal-Dual Path-Following Method

The path-following method to be discussed here is based on the work reported in
[11–13] . Consider the standard-form LP problem in Eqs. (12.1a)–(12.1c) and its dual
in Eqs. (12.2a)–(12.2c) and let $\mathbf{w}_k = \{\mathbf{x}_k, \lambda_k, \mu_k\}$ where $\mathbf{x}_k$ is strictly feasible for
the primal and $\{\lambda_k, \mu_k\}$ is strictly feasible for the dual. We need to find an increment

vector $\boldsymbol{\delta}_w = \{\boldsymbol{\delta}_x, \boldsymbol{\delta}_\lambda, \boldsymbol{\delta}_\mu\}$ such that the next iterate $\mathbf{w}_{k+1} = \{\mathbf{x}_{k+1}, \boldsymbol{\lambda}_{k+1}, \boldsymbol{\mu}_{k+1}\} = \{\mathbf{x}_k + \boldsymbol{\delta}_x, \boldsymbol{\lambda}_k + \boldsymbol{\delta}_\lambda, \boldsymbol{\mu}_k + \boldsymbol{\delta}_\mu\}$ remains strictly feasible and approaches the central path defined by Eqs. (12.9a)–(12.9c) with $\tau = \tau_{k+1} > 0$. In the path-following method, a suitable $\boldsymbol{\delta}_w$ is obtained as a first-order approximate solution of Eqs. (12.9a)–(12.9c). If $\mathbf{w}_{k+1}$ satisfies Eqs. (12.9a)–(12.9c) with $\tau = \tau_{k+1}$, then

$$\mathbf{A}\boldsymbol{\delta}_x = \mathbf{0} \tag{12.40a}$$

$$-\mathbf{A}^T\boldsymbol{\delta}_\lambda + \boldsymbol{\delta}_\mu = \mathbf{0} \tag{12.40b}$$

$$\Delta\mathbf{X}\boldsymbol{\mu}_k + \mathbf{X}\boldsymbol{\delta}_\mu + \Delta\mathbf{X}\boldsymbol{\delta}_\mu = \tau_{k+1}\mathbf{e} - \mathbf{X}\boldsymbol{\mu}_k \tag{12.40c}$$

where

$$\Delta\mathbf{X} = \text{diag}\{(\boldsymbol{\delta}_x)_1, \ (\boldsymbol{\delta}_x)_2, \ \ldots, \ (\boldsymbol{\delta}_x)_n\} \tag{12.41}$$

If the only second-order term in Eq. (12.40c), namely $\Delta\mathbf{X}\boldsymbol{\delta}_\mu$, is neglected, then Eqs. (12.40a)–(12.40c) is approximated by the system of linear equations

$$\mathbf{A}\boldsymbol{\delta}_x = \mathbf{0} \tag{12.42a}$$

$$-\mathbf{A}^T\boldsymbol{\delta}_\lambda + \boldsymbol{\delta}_\mu = \mathbf{0} \tag{12.42b}$$

$$\mathbf{M}\boldsymbol{\delta}_x + \mathbf{X}\boldsymbol{\delta}_\mu = \tau_{k+1}\mathbf{e} - \mathbf{X}\boldsymbol{\mu}_k \tag{12.42c}$$

where term $\Delta\mathbf{X}\boldsymbol{\mu}_k$ in Eq. (12.40c) has been replaced by $\mathbf{M}\boldsymbol{\delta}_x$ with

$$\mathbf{M} = \text{diag}\{(\boldsymbol{\mu}_k)_1, \ (\boldsymbol{\mu}_k)_2, \ \ldots, \ (\boldsymbol{\mu}_k)_n\} \tag{12.43}$$

Solving Eqs. (12.42a)–(12.42c) for $\boldsymbol{\delta}_w$, we obtain

$$\boldsymbol{\delta}_\lambda = -\mathbf{Y}\mathbf{A}\mathbf{y} \tag{12.44a}$$

$$\boldsymbol{\delta}_\mu = \mathbf{A}^T\boldsymbol{\delta}_\lambda \tag{12.44b}$$

$$\boldsymbol{\delta}_x = -\mathbf{y} - \mathbf{D}\boldsymbol{\delta}_\mu \tag{12.44c}$$

where

$$\mathbf{D} = \mathbf{M}^{-1}\mathbf{X} \tag{12.44d}$$

$$\mathbf{Y} = (\mathbf{A}\mathbf{D}\mathbf{A}^T)^{-1} \tag{12.44e}$$

and

$$\mathbf{y} = \mathbf{x}_k - \tau_{k+1}\mathbf{M}^{-1}\mathbf{e} \tag{12.44f}$$

Since $\mathbf{x}_k > \mathbf{0}$, $\boldsymbol{\mu}_k > \mathbf{0}$, and $\mathbf{A}$ has full row rank, matrix $\mathbf{Y}$ in Eq. (12.44e) is the inverse of a $p \times p$ positive definite matrix, and calculating $\mathbf{Y}$ is the major computation effort in the evaluation of $\boldsymbol{\delta}_w$ using Eqs. (12.44a)–(12.44f).

From Eqs. (12.42a) and (12.42b), the new iterate $\mathbf{w}_{k+1}$ satisfies Eqs. (12.9a) and (12.9b) but not necessarily Eq. (12.9c) because this is merely a linear approximation of Eq. (12.40c). If we define vector $\mathbf{f}(\mathbf{w}_k) = [f_1(\mathbf{w}_k) \ f_2(\mathbf{w}_k) \ \cdots \ f_n(\mathbf{w}_k)]^T$ with

$$f_i(\mathbf{w}_k) = (\boldsymbol{\mu}_k)_i \cdot (\mathbf{x}_k)_i \quad \text{for } 1 \le i \le n$$

then Eqs. (12.9c) and (12.10) suggest that the L_2 norm $\|\mathbf{f}(\mathbf{w}_k) - \tau_k\mathbf{e}\|$ can be viewed as a measure of the closeness of $\mathbf{w}_k$ to the central path. In [13], it was shown that if an initial point $\mathbf{w}_0 = \{\mathbf{x}_0, \boldsymbol{\lambda}_0, \boldsymbol{\mu}_0\}$ is chosen such that (a) $\mathbf{x}_0$ is strictly feasible for the primal and $\{\boldsymbol{\lambda}_0, \boldsymbol{\mu}_0\}$ is strictly feasible for the dual and (b)

$$||\mathbf{f}(\mathbf{w}_0) - \tau_0 \mathbf{e}|| \leq \theta \tau_0 \tag{12.45a}$$

where $\tau_0 = (\boldsymbol{\mu}_0^T \mathbf{x}_0)/n$ and θ satisfies the conditions

$$0 \leq \theta \leq \frac{1}{2} \tag{12.45b}$$

$$\frac{\theta^2 + \delta^2}{2(1-\theta)} \leq \left(1 - \frac{\delta}{\sqrt{n}}\right)\theta \tag{12.45c}$$

for some $\delta \in (0, \sqrt{n})$, then the iterate

$$\mathbf{w}_{k+1} = \mathbf{w}_k + \boldsymbol{\delta}_w \tag{12.46}$$

where $\boldsymbol{\delta}_w = \{\boldsymbol{\delta}_x, \boldsymbol{\delta}_\lambda, \boldsymbol{\delta}_\mu\}$ is given by Eqs. (12.44a)–(12.44f) with

$$\tau_{k+1} = \left(1 - \frac{\delta}{\sqrt{n}}\right)\tau_k \tag{12.47}$$

will remain strictly feasible and satisfy the conditions

$$||\mathbf{f}(\mathbf{w}_{k+1}) - \tau_{k+1}\mathbf{e}|| \leq \theta \tau_{k+1} \tag{12.48}$$

and

$$\boldsymbol{\mu}_{k+1}^T \mathbf{x}_{k+1} = n\tau_{k+1} \tag{12.49}$$

Since $0 < \delta/\sqrt{n} < 1$, it follows from Eq. (12.47) that $\tau_k = (1 - \delta/n)^k \tau_0 \to 0$ as $k \to \infty$. From Eqs. (12.49) and (12.10), the duality gap tends to zero, i.e., $\delta(\mathbf{x}_k, \boldsymbol{\lambda}_k) \to 0$, as $k \to \infty$. In other words, $\mathbf{w}_k$ converges to a primal-dual solution as $k \to \infty$. The above method can be implemented in terms of the following algorithm [13].

Algorithm 12.3   Primal-dual path-following algorithm for the standard-form LP problem

Step 1
Input $\mathbf{A}$ and a strictly feasible $\mathbf{w}_0 = \{\mathbf{x}_0, \boldsymbol{\lambda}_0, \boldsymbol{\mu}_0\}$ that satisfies Eqs. (12.45a)–(12.45c). Set $k = 0$ and initialize the tolerance ε for the duality gap.
Step 2
If $\boldsymbol{\mu}_k^T \mathbf{x}_k \leq \varepsilon$, output solution $\mathbf{w}^* = \mathbf{w}_k$ and stop; otherwise, continue with Step 3.
Step 3
Set τ_{k+1} using Eq. (12.47) and compute $\boldsymbol{\delta}_w = \{\boldsymbol{\delta}_x, \boldsymbol{\delta}_\lambda, \boldsymbol{\delta}_\mu\}$ using Eqs. (12.44a)–(12.44f).
Step 4
Set $\mathbf{w}_{k+1}$ using Eq. (12.46). Set $k = k + 1$ and repeat from Step 2.

A couple of remarks concerning Step 1 of the algorithm are in order. First, values of θ and δ that satisfy Eqs. (12.45b) and (12.45c) exist. For example, it can be readily verified that $\theta = 0.4$ and $\delta = 0.4$ meet Eqs. (12.45b) and (12.45c) for any $n \geq 2$. Second, in order to find an initial $\mathbf{w}_0$ that satisfies Eq. (12.45a), we can introduce an augmented pair of primal-dual LP problems such that (a) a strictly feasible initial point can be easily identified for the augmented problem and (b) a solution of the augmented problem will yield a solution of the original problem [13]. A more general remedy for dealing with this initialization problem is to develop a 'nonfeasible-initialization algorithm' so that a point $\mathbf{w}_0$ that satisfies $\mathbf{x}_0 > \mathbf{0}$ and $\boldsymbol{\mu}_0 > \mathbf{0}$ but not necessarily Eqs. (12.9a)–(12.9c) can be used as the initial point. Such a primal-dual path-following algorithm will be studied in Sect. 12.5.2.

It is important to stress that even for problems of moderate size, the choice $\delta = 0.4$ yields a factor $(1 - \delta/\sqrt{n})$ which is close to unity and, therefore, parameter τ_{k+1} determined using Eq. (12.47) converges to zero slowly and a large number of iterations are required to reach a primal-dual solution. In the literature, interior-point algorithms of this type are referred to as *short-step* path-following algorithms [3]. In practice, Algorithm 12.3 is modified to allow larger changes in parameter τ so as to accelerate the convergence [6, 14]. It was proposed in [14] that τ_{k+1} be chosen as

$$\tau_{k+1} = \frac{\boldsymbol{\mu}_k^T \mathbf{x}_k}{n + \rho} \tag{12.50}$$

with $\rho > \sqrt{n}$. In order to assume the strict feasibility of the next iterate, the modified path-following algorithm assigns

$$\mathbf{w}_{k+1} = \mathbf{w}_k + \alpha_k \boldsymbol{\delta}_w \tag{12.51}$$

where $\boldsymbol{\delta}_w = \{\boldsymbol{\delta}_x, \boldsymbol{\delta}_\lambda, \boldsymbol{\delta}_\mu\}$ is calculated using Eqs. (12.44a)–(12.44f), and

$$\alpha_k = (1 - 10^{-6}) \alpha_{\max} \tag{12.52a}$$

with $\alpha_{\max}$ being determined as

$$\alpha_{\max} = \min(\alpha_p, \ \alpha_d) \tag{12.52b}$$

where

$$\alpha_p = \min_{i \text{ with } (\delta_x)_i < 0} \left[-\frac{(\mathbf{x}_k)_i}{(\delta_x)_i} \right] \tag{12.52c}$$

$$\alpha_d = \min_{i \text{ with } (\delta_\mu)_i < 0} \left[-\frac{(\boldsymbol{\mu}_k)_i}{(\delta_\mu)_i} \right] \tag{12.52d}$$

The modified algorithm assumes the following form.

Algorithm 12.4 Modified version of Algorithm 12.3
Step 1
Input $\mathbf{A}$ and a strictly feasible $\mathbf{w}_0 = \{\mathbf{x}_0, \ \lambda_0, \ \boldsymbol{\mu}_0\}$.
Set $k = 0$ and $\rho > \sqrt{n}$, and initialize the tolerance ε for the duality gap.

Step 2
If $\boldsymbol{\mu}_k^T \mathbf{x}_k \leq \varepsilon$, output solution $\mathbf{w}^* = \mathbf{w}_k$ and stop; otherwise, continue with Step 3.

Step 3
Set τ_{k+1} using Eq. (12.50) and compute $\delta_w = \{\delta_x, \ \delta_\lambda, \ \delta_\mu\}$ using Eqs. (12.44a)–(12.44f).
Step 4
Compute step size α_k using Eqs. (12.52a)–(12.52d) and set $\mathbf{w}_{k+1}$ using Eq. (12.51).
Set $k = k + 1$ and repeat from Step 2.

Example 12.4 Apply Algorithm 12.4 to the LP problems in

(*a*) Example 12.3
(*b*) Example 12.2

Solution (*a*) In order to apply the algorithm to the LP problem in Example 12.3, we have used the method described in Example 12.1 to find an initial $\mathbf{w}_0$ on the central path with $\tau_0 = 5$. The vector $\mathbf{w}_0$ obtained is $\{\mathbf{x}_0, \ \lambda_0, \ \mu_0\}$ with

$$\mathbf{x}_0 = \begin{bmatrix} 0.344506 \\ 0.285494 \\ 0.370000 \end{bmatrix}, \quad \lambda_0 = 16.513519, \quad \mu_0 = \begin{bmatrix} 14.513519 \\ 17.513519 \\ 13.513519 \end{bmatrix}$$

With $\rho = 7\sqrt{n}$ and $\varepsilon = 10^{-6}$, Algorithm 12.4 converges after eight iterations to the solution

$$\mathbf{x}^* = \begin{bmatrix} 0.000000 \\ 0.000000 \\ 1.000000 \end{bmatrix}$$

The number of flops required was 858.

(*b*) In order to apply the algorithm to the LP problem in Example 12.2, we have to find a strictly feasible initial point $\mathbf{w}_0$ first. By using the method described in Sect. 10.4.1, a vector $\mathbf{x}$ that satisfies Eq. (12.9a) can be obtained as

$$\mathbf{x} = \mathbf{V}_r \boldsymbol{\phi} + \mathbf{A}^+ \mathbf{b}$$

where $\mathbf{V}_r$ is composed of the last $n - p$ columns of matrix $\mathbf{V}$ from the SVD of matrix $\mathbf{A}$ and $\boldsymbol{\phi} \in R^{(n-p)\times 1}$ is a free parameter vector. From Eq. (10.27), we have

$$\mathbf{x} = \begin{bmatrix} 0.5980 & 0.0000 & 0.0000 \\ 0.0608 & 0.1366 & 0.1794 \\ 0.6385 & 0.5504 & -0.2302 \\ -0.4358 & 0.8236 & 0.1285 \\ 0.2027 & -0.0039 & 0.9478 \end{bmatrix} \begin{bmatrix} \phi_1 \\ \phi_2 \\ \phi_3 \end{bmatrix} + \begin{bmatrix} -0.1394 \\ 0.2909 \\ 0.0545 \\ -0.0848 \\ -0.0303 \end{bmatrix} \quad (12.53)$$

The requirement $x_1 > 0$ is met if

$$\phi_1 > 0.2331 \quad (12.54a)$$

If we assume that $\phi_2 = \phi_3 > 0$, then $x_2 > 0$ and $x_3 > 0$. To satisfy the inequalities $x_4 > 0$ and $x_5 > 0$, we require

$$-0.4358\phi_1 + 0.9521\phi_2 > 0.0848 \tag{12.54b}$$

and

$$0.2027\phi_1 + 0.9439\phi_2 > 0.0303 \tag{12.54c}$$

Obviously, $\phi_1 = 0.5$ and $\phi_2 = 0.5$ satisfy Eqs. (12.54a)–(12.54c) and lead to a strictly feasible initial point

$$\mathbf{x}_0 = \begin{bmatrix} 0.1596 \\ 0.4793 \\ 0.5339 \\ 0.1733 \\ 0.5430 \end{bmatrix}$$

Next we can write Eq. (12.9b) as

$$\boldsymbol{\mu} = \mathbf{c} + \mathbf{A}^T\boldsymbol{\lambda} = \begin{bmatrix} 2 - 2\lambda_1 + \lambda_2 \\ 9 + 2\lambda_1 + 4\lambda_2 \\ 3 + \lambda_1 - \lambda_2 \\ -\lambda_1 \\ -\lambda_2 \end{bmatrix}$$

from which it is easy to verify that $\boldsymbol{\lambda}_0 = [-1 \; -1]^T$ leads to $\boldsymbol{\mu}_0 = [3\;3\;3\;1\;1]^T > \mathbf{0}$ and $\{\boldsymbol{\lambda}_0, \; \boldsymbol{\mu}_0\}$ satisfies Eq. (12.9b).

The application of Algorithm 12.4 using the above $\mathbf{w}_0$, $\rho = 12\sqrt{n}$, and $\varepsilon = 10^{-5}$ led to the solution

$$\mathbf{x}^* = \begin{bmatrix} 0.000000 \\ 0.333333 \\ 0.333333 \\ 0.000000 \\ 0.000000 \end{bmatrix}$$

in seven iterations. The number of flops required was 2.48K.   ∎

12.5.2   A Nonfeasible-Initialization Primal-Dual Path-Following Method

Both Algorithms 12.3 and 12.4 require an initial $\mathbf{w}_0 = \{\mathbf{x}_0, \; \boldsymbol{\lambda}_0, \; \boldsymbol{\mu}_0\}$ with $\mathbf{x}_0$ being strictly feasible for the primal and $\{\boldsymbol{\lambda}_0, \; \boldsymbol{\mu}_0\}$ being strictly feasible for the dual. As can be observed from Example 12.4, finding such an initial point is not straightforward, even for problems of small size, and it would certainly be highly desirable to start with an initial point $\mathbf{w}_0$ that is not necessarily feasible. In the literature, interior-point algorithms that accept nonfeasible initial points are often referred to as *nonfeasible-initialization* or *nonfeasible-start* algorithms. As described in [6], if $\mathbf{w}_k$ is nonfeasible in the sense that it does not satisfy Eqs. (12.1b) and (12.2b), then a reasonable way to generate the next point is to find a set of vector increments $\boldsymbol{\delta}_w = \{\boldsymbol{\delta}_x, \; \boldsymbol{\delta}_\lambda, \; \boldsymbol{\delta}_\mu\}$ such that $\mathbf{w}_k + \boldsymbol{\delta}_w$ satisfies Eqs. (12.1b) and (12.2b). Based on this approach, the basic idea presented in Sect. 12.5.1 can be used to construct a *nonfeasible-initialization primal-dual path-following* algorithm [15].

Let $\mathbf{w}_k = \{\mathbf{x}_k, \boldsymbol{\lambda}_k, \boldsymbol{\mu}_k\}$ be such that only the conditions $\mathbf{x}_k > \mathbf{0}$ and $\boldsymbol{\mu}_k > \mathbf{0}$ are assumed. We need to obtain the next iterate

$$\mathbf{w}_{k+1} = \mathbf{w}_k + \alpha_k \boldsymbol{\delta}_w$$

such that $\mathbf{x}_{k+1} > \mathbf{0}$ and $\boldsymbol{\mu}_{k+1} > \mathbf{0}$, and that $\boldsymbol{\delta}_w = \{\boldsymbol{\delta}_x, \boldsymbol{\delta}_\lambda, \boldsymbol{\delta}_\mu\}$ satisfies the conditions

$$\mathbf{A}(\mathbf{x}_k + \boldsymbol{\delta}_x) = \mathbf{b} \tag{12.55a}$$

$$-\mathbf{A}^T(\boldsymbol{\lambda}_k + \boldsymbol{\delta}_\lambda) + (\boldsymbol{\mu}_k + \boldsymbol{\delta}_\mu) = \mathbf{c} \tag{12.55b}$$

$$\mathbf{M}\boldsymbol{\delta}_x + \mathbf{X}\boldsymbol{\delta}_\mu = \tau_{k+1}\mathbf{e} - \mathbf{X}\boldsymbol{\mu}_k \tag{12.55c}$$

Note that Eq. (12.55c) is the same as Eq. (12.42c) which is a linear approximation of Eq. (12.40c) but Eqs. (12.55a) and (12.55b) differ from Eqs. (12.42a) and (12.42b) since in the present case the feasibility of $\mathbf{w}_k$ is not assumed. At the kth iteration, $\mathbf{w}_k$ is known; hence Eqs. (12.55a)–(12.55c) is a system of linear equations for $\{\boldsymbol{\delta}_x, \boldsymbol{\delta}_\lambda, \boldsymbol{\delta}_\mu\}$, which can be written as

$$\mathbf{A}\boldsymbol{\delta}_x = \mathbf{r}_p \tag{12.56a}$$

$$-\mathbf{A}^T\boldsymbol{\delta}_\lambda + \boldsymbol{\delta}_\mu = \mathbf{r}_d \tag{12.56b}$$

$$\mathbf{M}\boldsymbol{\delta}_x + \mathbf{X}\boldsymbol{\delta}_\mu = \tau_{k+1}\mathbf{e} - \mathbf{X}\boldsymbol{\mu}_k \tag{12.56c}$$

where $\mathbf{r}_p = \mathbf{b} - \mathbf{A}\mathbf{x}_k$ and $\mathbf{r}_d = \mathbf{c} + \mathbf{A}^T\boldsymbol{\lambda}_k - \boldsymbol{\mu}_k$ are the residuals for the primal and dual constraints, respectively. Solving Eqs. (12.56a)–(12.56c) for $\boldsymbol{\delta}_w$, we obtain

$$\boldsymbol{\delta}_\lambda = -\mathbf{Y}(\mathbf{A}\mathbf{y} + \mathbf{A}\mathbf{D}\mathbf{r}_d + \mathbf{r}_p) \tag{12.57a}$$

$$\boldsymbol{\delta}_\mu = \mathbf{A}^T\boldsymbol{\delta}_\lambda + \mathbf{r}_d \tag{12.57b}$$

$$\boldsymbol{\delta}_x = -\mathbf{y} - \mathbf{D}\boldsymbol{\delta}_\mu \tag{12.57c}$$

where $\mathbf{D}$, $\mathbf{Y}$, and $\mathbf{y}$ are defined by Eqs. (12.44d)–(12.44f), respectively. It should be stressed that if the new iterate $\mathbf{w}_{k+1}$ is set as in Eq. (12.51) with α_k determined using Eqs. (12.52a)–(12.52d), then $\mathbf{x}_{k+1}$ and $\boldsymbol{\mu}_{k+1}$ remain strictly positive but $\mathbf{w}_{k+1}$ is not necessarily strictly feasible unless α_k happens to be unity. As the iterations proceed, the new iterates generated get closer and closer to the central path and approach to a primal-dual solution. The nonfeasible-initialization interior-point algorithm is summarized as follows.

Algorithm 12.5   Nonfeasible-initialization primal-dual path-following algorithm for the standard-form LP problem
Step 1
Input $\mathbf{A}$, $\mathbf{b}$, $\mathbf{c}$, and $\mathbf{w}_0 = \{\mathbf{x}_0, \boldsymbol{\lambda}_0, \boldsymbol{\mu}_0\}$ with $\mathbf{x}_0 > \mathbf{0}$ and $\boldsymbol{\mu}_0 > \mathbf{0}$.
Set $k = 0$ and $\rho > \sqrt{n}$, and initialize the tolerance ε for the duality gap.
Step 2
If $\boldsymbol{\mu}_k^T\mathbf{x}_k \leq \varepsilon$, output solution $\mathbf{w}^* = \mathbf{w}_k$ and stop; otherwise, continue with Step 3.

Step 3
Set τ_{k+1} using Eq. (12.50) and compute $\delta_w = (\delta_x, \delta_\lambda, \delta_\mu)$ using Eqs. (12.57a)–(12.57c).

Step 4
Compute step size α_k using Eqs. (12.52a)–(12.52d) and set $\mathbf{w}_{k+1}$ using Eq. (12.51).
Set $k = k + 1$ and repeat from Step 2.

Example 12.5 Apply Algorithm 12.5 to the LP problems in

(a) Example 12.3
(b) Example 12.2

with nonfeasible initial points.

Solution (a) In order to apply the algorithm to the LP problem in Example 12.3, we can use $\mathbf{w}_0 = \{\mathbf{x}_0, \lambda_0, \mu_0\}$ with

$$\mathbf{x}_0 = \begin{bmatrix} 0.4 \\ 0.3 \\ 0.4 \end{bmatrix}, \quad \lambda_0 = -0.5, \quad \text{and} \quad \mu_0 = \begin{bmatrix} 1.0 \\ 0.5 \\ 1.0 \end{bmatrix}$$

So $\mathbf{x}_0 > \mathbf{0}$ and $\mu_0 > \mathbf{0}$ but $\mathbf{w}_0$ is not feasible. With $\varepsilon = 10^{-6}$ and $\rho = 7\sqrt{n}$, Algorithm 12.5 took eight iterations to converge to the solution

$$\mathbf{x}^* = \begin{bmatrix} 0.000000 \\ 0.000000 \\ 1.000000 \end{bmatrix}$$

The number of flops required was 1.21K. Fig. 12.3 shows point $\mathbf{x}_0$ and the first three iterates, i.e., $\mathbf{x}_k$ for $k = 0, 1, 2, 3$, as compared to the central path, which is shown as a dotted curve.

(b) For the LP problem in Example 12.2, we can use $\mathbf{w}_0 = \{\mathbf{x}_0, \lambda_0, \mu_0\}$ with

$$\mathbf{x}_0 = \begin{bmatrix} 1.0 \\ 0.1 \\ 0.1 \\ 2.0 \\ 5.0 \end{bmatrix}, \quad \lambda_0 = \begin{bmatrix} 1 \\ -1 \end{bmatrix}, \quad \text{and} \quad \mu_0 = \begin{bmatrix} 1.0 \\ 0.1 \\ 0.2 \\ 1.0 \\ 10.0 \end{bmatrix} \quad (12.58)$$

With $\varepsilon = 10^{-8}$ and $\rho = 12\sqrt{n}$, the algorithm took 13 iterations to converge to the solution

$$\mathbf{x}^* = \begin{bmatrix} 0.000000 \\ 0.333333 \\ 0.333333 \\ 0.000000 \\ 0.000000 \end{bmatrix}$$

The number of flops required was 6.96K.   ∎

Fig. 12.3 Iteration path in
Example 12.5(a) as
compared to the central path

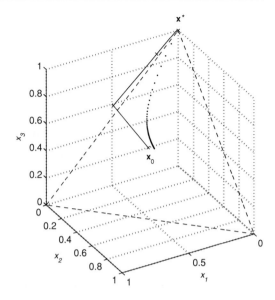

12.5.3   Predictor-Corrector Method

The predictor-corrector method (PCM) proposed by Mehrotra [16] can be viewed
as an important improved primal-dual path-following algorithm relative to the algorithms studied in Sects. 12.5.1 and 12.5.2. As a matter of fact, most interior-point software available since 1990 is based on Mehrotra's PCM algorithm [3]. Briefly speak-
ing, improvement is achieved by including the effect of the second-order term $\Delta \mathbf{X} \boldsymbol{\delta}_{\mu}$
in Eq. (12.40c) using a prediction-correction strategy rather than simply neglecting
it. In addition, in this method the parameter τ in Eq. (12.9c) is assigned a value
according to the relation

$$\tau = \sigma \hat{\tau}$$

where $\hat{\tau} = (\boldsymbol{\mu}^T \mathbf{x})/n$ and $0 < \sigma < 1$. The scalar σ, which is referred to as the
centering parameter, is determined *adaptively* in each iteration based on whether
good progress has been made in the prediction phase.

   At the kth iteration, there are three steps in the PCM algorithm that produce the
next iterate $\mathbf{w}_{k+1} = \{\mathbf{x}_{k+1}, \boldsymbol{\lambda}_{k+1}, \boldsymbol{\mu}_{k+1}\}$ with $\mathbf{x}_{k+1} > \mathbf{0}$ and $\boldsymbol{\mu}_{k+1} > \mathbf{0}$ as described
below (see [3, Chap. 10]).

1. *Generate an affine scaling 'predictor' direction $\boldsymbol{\delta}_w^{\text{aff}}$ using a linear approximation*
   of the KKT conditions in Eqs. 12.8a–12.8c.
   Let $\mathbf{w}_k = \{\mathbf{x}_k, \boldsymbol{\lambda}_k, \boldsymbol{\mu}_k\}$ with $\mathbf{x}_k > \mathbf{0}$ and $\boldsymbol{\mu}_k > \mathbf{0}$, and consider an increment
   $\boldsymbol{\delta}_w^{\text{aff}} = \{\boldsymbol{\delta}_x^{\text{aff}}, \boldsymbol{\delta}_\lambda^{\text{aff}}, \boldsymbol{\delta}_\mu^{\text{aff}}\}$ such that $\mathbf{w}_k + \boldsymbol{\delta}_w^{\text{aff}}$ linearly approximates the KKT
   conditions in Eqs. (12.8a)–(12.8c). Under these circumstances, $\boldsymbol{\delta}_w^{\text{aff}}$ satisfies the

equations

$$\mathbf{A}\boldsymbol{\delta}_x^{\text{aff}} = \mathbf{r}_p \tag{12.59a}$$

$$-\mathbf{A}^T\boldsymbol{\delta}_\lambda^{\text{aff}} + \boldsymbol{\delta}_\mu^{\text{aff}} = \mathbf{r}_d \tag{12.59b}$$

$$\mathbf{M}\boldsymbol{\delta}_x^{\text{aff}} + \mathbf{X}\boldsymbol{\delta}_\mu^{\text{aff}} = -\mathbf{X}\mathbf{M}\mathbf{e} \tag{12.59c}$$

where

$$\mathbf{r}_p = \mathbf{b} - \mathbf{A}\mathbf{x}_k \tag{12.59d}$$

$$\mathbf{r}_d = \mathbf{c} + \mathbf{A}^T\boldsymbol{\lambda}_k - \boldsymbol{\mu}_k \tag{12.59e}$$

Solving Eqs. (12.59a)–(12.59e) for $\boldsymbol{\delta}_w^{\text{aff}}$, we obtain

$$\boldsymbol{\delta}_\lambda^{\text{aff}} = -\mathbf{Y}(\mathbf{b} + \mathbf{A}\mathbf{D}\mathbf{r}_d) \tag{12.60a}$$

$$\boldsymbol{\delta}_\mu^{\text{aff}} = \mathbf{r}_d + \mathbf{A}^T\boldsymbol{\delta}_\lambda^{\text{aff}} \tag{12.60b}$$

$$\boldsymbol{\delta}_x^{\text{aff}} = -\mathbf{x}_k - \mathbf{D}\boldsymbol{\delta}_\mu^{\text{aff}} \tag{12.60c}$$

where

$$\mathbf{D} = \mathbf{M}^{-1}\mathbf{X} \tag{12.60d}$$

$$\mathbf{Y} = (\mathbf{A}\mathbf{D}\mathbf{A}^T)^{-1} \tag{12.60e}$$

Along the directions $\boldsymbol{\delta}_x^{\text{aff}}$ and $\boldsymbol{\delta}_\mu^{\text{aff}}$, two scalars α_p^{aff} and α_d^{aff} are determined as

$$\alpha_p^{\text{aff}} = \max_{0\le\alpha\le 1,\ \mathbf{x}_k+\alpha\boldsymbol{\delta}_x^{\text{aff}}\ge\mathbf{0}} (\alpha) \tag{12.61a}$$

and

$$\alpha_d^{\text{aff}} = \max_{0\le\alpha\le 1,\ \boldsymbol{\mu}_k+\alpha\boldsymbol{\delta}_\mu^{\text{aff}}\ge\mathbf{0}} (\alpha) \tag{12.61b}$$

A hypothetical value of τ_{k+1}, denoted as τ_{aff}, is then determined as

$$\tau_{\text{aff}} = \frac{1}{n}[(\boldsymbol{\mu}_k + \alpha_d^{\text{aff}}\boldsymbol{\delta}_\mu^{\text{aff}})^T (\mathbf{x}_k + \alpha_p^{\text{aff}}\boldsymbol{\delta}_x^{\text{aff}})] \tag{12.62}$$

2. *Determine the centering parameter σ_k.*
   A heuristic choice of σ_k, namely,

$$\sigma_k = \left(\frac{\tau_{\text{aff}}}{\hat{\tau}_k}\right)^3 \tag{12.63}$$

with

$$\hat{\tau}_k = \frac{1}{n}(\boldsymbol{\mu}_k^T\mathbf{x}_k) \tag{12.64}$$

was suggested in [16] and was found effective in extensive computational testing. Intuitively, if $\tau_{\text{aff}} \ll \hat{\tau}_k$, then the predictor direction $\boldsymbol{\delta}_w^{\text{aff}}$ given by Eqs. (12.60a)–(12.60e) is good and we should use a small centering parameter σ_k to substantially reduce the magnitude of parameter $\tau_{k+1} = \sigma_k\hat{\tau}_k$. If τ_{aff} is close to $\hat{\tau}_k$, then we should choose σ_k close to unity so as to move the next iterate $\mathbf{w}_{k+1}$ closer to the central path.

3. *Generate a 'corrector' direction to compensate for the nonlinearity in the affine scaling direction.*

The corrector direction $\boldsymbol{\delta}_w^c = \{\boldsymbol{\delta}_x^c, \boldsymbol{\delta}_\lambda^c, \boldsymbol{\delta}_\mu^c\}$ is determined using Eqs. (12.40a)–(12.40c) with the term $\mathbf{X}\boldsymbol{\mu}_k$ in Eq. (12.40c) neglected and the second-order term $\Delta\mathbf{X}\boldsymbol{\delta}_\mu$ in Eq. (12.40c) replaced by $\Delta\mathbf{X}^{\mathrm{aff}}\boldsymbol{\delta}_\mu^{\mathrm{aff}}$ where $\Delta\mathbf{X}^{\mathrm{aff}} = \mathrm{diag}\{(\delta_x^{\mathrm{aff}})_1,$ $(\delta_x^{\mathrm{aff}})_2, \ldots, (\delta_x^{\mathrm{aff}})_n\}$. The reason that term $\mathbf{X}\boldsymbol{\mu}_k$ is neglected is because it has been included in Eq. (12.59c) where $\mathbf{XMe} = \mathbf{X}\boldsymbol{\mu}_k$. In the primal-dual path-following algorithms studied in Sects. 12.5.1 and 12.5.2, the second-order term $\Delta\mathbf{X}\boldsymbol{\delta}_\mu$ was dropped to obtain the *linear* systems in Eqs. (12.42a)–(12.42c) and Eqs. (12.56a)–(12.56c). The PCM method approximates this second-order term with the increment vectors $\boldsymbol{\delta}_x$ and $\boldsymbol{\delta}_\mu$ obtained from the predictor direction. Having made the above modifications, the equations to be used to compute $\boldsymbol{\delta}_w^c$ become

$$\mathbf{A}\boldsymbol{\delta}_x^c = \mathbf{0} \tag{12.65a}$$

$$-\mathbf{A}^T\boldsymbol{\delta}_\lambda^c + \boldsymbol{\delta}_\mu^c = \mathbf{0} \tag{12.65b}$$

$$\mathbf{M}\boldsymbol{\delta}_x^c + \mathbf{X}\boldsymbol{\delta}_\mu^c = \tau_{k+1}\mathbf{e} - \Delta\mathbf{X}^{\mathrm{aff}}\boldsymbol{\delta}_\mu^{\mathrm{aff}} \tag{12.65c}$$

where

$$\tau_{k+1} = \sigma_k\hat{\tau}_k \tag{12.65d}$$

with σ_k and $\hat{\tau}_k$ given by Eqs. (12.63) and (12.64), respectively. Solving Eqs. (12.65a)–(12.65d) for $\boldsymbol{\delta}_w^c$, we obtain

$$\boldsymbol{\delta}_\lambda^c = -\mathbf{Y}\mathbf{A}\mathbf{y} \tag{12.66a}$$

$$\boldsymbol{\delta}_\mu^c = \mathbf{A}^T\boldsymbol{\delta}_\lambda^c \tag{12.66b}$$

$$\boldsymbol{\delta}_x^c = -\mathbf{y} - \mathbf{D}\boldsymbol{\delta}_\mu^c \tag{12.66c}$$

where $\mathbf{D}$ and $\mathbf{Y}$ are given by Eqs. (12.60d) and (12.60e), respectively, and

$$\mathbf{y} = \mathbf{M}^{-1}(\Delta\mathbf{X}^{\mathrm{aff}}\boldsymbol{\delta}_\mu^{\mathrm{aff}} - \tau_{k+1}\mathbf{e}) \tag{12.66d}$$

The predictor and corrector directions are now combined to obtain the search direction $\{\boldsymbol{\delta}_x, \boldsymbol{\delta}_\lambda, \boldsymbol{\delta}_\mu\}$ where

$$\boldsymbol{\delta}_x = \boldsymbol{\delta}_x^{\mathrm{aff}} + \boldsymbol{\delta}_x^c \tag{12.67a}$$

$$\boldsymbol{\delta}_\lambda = \boldsymbol{\delta}_\lambda^{\mathrm{aff}} + \boldsymbol{\delta}_\lambda^c \tag{12.67b}$$

$$\boldsymbol{\delta}_\mu = \boldsymbol{\delta}_\mu^{\mathrm{aff}} + \boldsymbol{\delta}_\mu^c \tag{12.67c}$$

and the new iterate is given by

$$\mathbf{w}_{k+1} = \mathbf{w}_k + \{\alpha_{k,p}\boldsymbol{\delta}_x, \ \alpha_{k,d}\boldsymbol{\delta}_\lambda, \ \alpha_{k,d}\boldsymbol{\delta}_\mu\} \tag{12.68}$$

where the step sizes for $\boldsymbol{\delta}_x$ and $(\boldsymbol{\delta}_\lambda, \boldsymbol{\delta}_\mu)$ are determined separately as

$$\alpha_{k,p} = \min(0.99\,\alpha_{\max}^{(p)}, \ 1) \tag{12.69a}$$

$$\alpha_{\max}^{(p)} = \max_{\alpha \geq 0, \ \mathbf{x}_k + \alpha\boldsymbol{\delta}_x \geq \mathbf{0}} (\alpha) \tag{12.69b}$$

$$\alpha_{k,d} = \min(0.99\,\alpha_{\max}^{(d)}, \ 1) \tag{12.69c}$$

$$\alpha_{\max}^{(d)} = \max_{\alpha \geq 0, \ \boldsymbol{\mu}_k + \alpha\boldsymbol{\delta}_\mu \geq \mathbf{0}} (\alpha) \tag{12.69d}$$

A note on the computational complexity of the method is in order. From Eqs. (12.67a)–(12.67c), we see that the search direction is obtained by computing δ_w^{aff} and δ_w^c; hence the two linear systems in Eqs. (12.59a)–(12.59e) and Eqs. (12.65a)–(12.65d) have to be solved. However, the system matrices for Eqs. (12.59a)–(12.59e) and Eqs. (12.65a)–(12.65d) are *identical* and, consequently, the computational effort required by the PCM algorithm is increased only slightly relative to that required by the primal-dual path-following algorithms discussed in the preceding sections. This can also be observed from the fact that matrices $\mathbf{Y}$ and $\mathbf{D}$ used to solve Eqs. (12.59a)–(12.59e) can also be used to solve Eqs. (12.65a)–(12.65d). A step-by-step summary of the PCM algorithm is given below.

Algorithm 12.6   Mehrotra's predictor-corrector algorithm for the standard-form LP problem

Step 1
Input $\mathbf{A}$, $\mathbf{b}$, $\mathbf{c}$, and $\mathbf{w}_0 = \{\mathbf{x}_0,\ \lambda_0,\ \mu_0\}$ with $\mathbf{x}_0 > 0$ and $\mu_0 > 0$.
Set $k = 0$ and $\hat{\tau}_0 = (\mu_0^T \mathbf{x}_0)/n$, and initialize the tolerance ε for the duality gap.

Step 2
If $\mu_k^T \mathbf{x}_k \leq \varepsilon$, output solution $\mathbf{w}^* = \mathbf{w}_k$ and stop; otherwise, go to Step 3.

Step 3
Compute predictor direction $\{\delta_x^{\text{aff}},\ \delta_\lambda^{\text{aff}},\ \delta_\mu^{\text{aff}}\}$ using Eqs. (12.60a)–(12.60e).

Step 4
Compute τ_{aff} using Eqs. (12.61a)–(12.62) and determine τ_{k+1} as

$$\tau_{k+1} = \sigma_k \hat{\tau}_k$$

where σ_k and $\hat{\tau}_k$ are evaluated using Eqs. (12.63) and (12.64).

Step 5
Compute corrector direction $\{\delta_x^c,\ \delta_\lambda^c,\ \delta_\mu^c\}$ using Eqs. (12.66a)–(12.66d).

Step 6
Obtain search direction $\{\delta_x,\ \delta_\lambda,\ \delta_\mu\}$ using Eqs. (12.67a)–(12.67c) and evaluate step sizes $\alpha_{k,p}$ and $\alpha_{k,d}$ using Eqs. (12.69a)–(12.69d).

Step 7
Set $\mathbf{w}_{k+1}$ using Eq. (12.68).
Set $k = k + 1$ and repeat from Step 2.

Example 12.6 Apply Algorithm 12.6 to the LP problems in

(a) Example 12.3
(b) Example 12.2

with nonfeasible initial points.

Fig. 12.4 Iteration path in
Example 12.6(*a*) as
compared to the central path

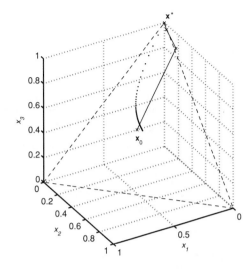

Solution (*a*) We can use the same $\mathbf{w}_0$ and ε as in Example 12.5(*a*) to start Algorithm 12.6. It took six iterations for the algorithm to converge to the solution

$$\mathbf{x}^* = \begin{bmatrix} 0.000000 \\ 0.000000 \\ 1.000000 \end{bmatrix}$$

The number of flops required was 1.268K, which entails a slight increase as compared to that in Example 12.5(*a*) but the solution $\mathbf{x}^*$ is more accurate. Figure 12.4 shows point $\mathbf{x}_0$ and the first three iterates, i.e., $\mathbf{x}_k$ for $k = 0, 1, 2, 3$ as compared to the central path, which is plotted as the dotted curve.

(*b*) The same $\mathbf{w}_0$ and ε as in Example 12.5(*b*) were used here. The algorithm took 11 iterations to converge to the solution

$$\mathbf{x}^* = \begin{bmatrix} 0.000000 \\ 0.333333 \\ 0.333333 \\ 0.000000 \\ 0.000000 \end{bmatrix}$$

The number of flops required was 7.564K. This is slightly larger than the number of flops in Example 12.5(b) but some improvement in the accuracy of the solution has been achieved.   ■

Problems

12.1 This problem concerns the central path of the LP problem described in Example 12.1.

(*a*) For a sample number of values τ ranging from 500 to 10^{-3}, use MATLAB command `roots` to evaluate the roots λ of Eq. (12.14) with $\lambda < -3$.

(b) Generate a trajectory (x_1, x_2, x_3) similar to that in Fig. 12.1.

(c) Change the range of τ from $[10^{-3}, 500]$ to $[10^{-2}, 200]$ and then to $[2.5 \times 10^{-2}, 20]$ and observe the trajectories (x_1, x_2, x_3) obtained.

12.2 Consider the LP problem in Eqs. (12.1a)–(12.1c) and let $\mathbf{x}_{k+1}$ be determined using Eqs. (12.22a) and (12.22b). Show that if $\alpha_k > 0$, then $f(\mathbf{x}_{k+1}) \leq f(\mathbf{x}_k)$ and this inequality holds strictly if at least one of the last $n - p$ components of $\mathbf{V}_k^T \mathbf{Xc}$ is nonzero, where $\mathbf{V}_k$ is the $n \times n$ orthogonal matrix obtained from the SVD of $\mathbf{AX} : \mathbf{AX} = \mathbf{U}_k[\boldsymbol{\Sigma}_k \ \mathbf{0}]\mathbf{V}_k^T$.

12.3 (a) Apply the PAS algorithm to solve the LP problem in Prob. 11.16. Compare the results with those obtained in Prob. 11.16.

(b) Apply the PAS algorithm to solve the LP problem in Prob. 11.17. Compare the results with those obtained in Prob. 11.17.

12.4 (a) Derive the KKT conditions for the minimizer of the problem in Eqs. (12.26a)–(12.26b).

(b) Relate the KKT conditions obtained in part (a) to the central path of the original LP problem in Eqs. (12.9a)–(12.9c).

12.5 (a) Apply the PNB algorithm to solve the LP problem in Prob. 11.16. Compare the results with those obtained in Prob. 12.3(a).

(b) Apply the PNB algorithm to solve the LP problem in Prob. 11.17. Compare the results with those obtained in Prob. 12.3(b).

12.6 Develop a PNB algorithm that is directly applicable to the LP problem in Eqs. (11.2a) and (11.2b).

Hint: Denote $\mathbf{A}$ and $\mathbf{b}$ in Eq. (11.2b) as

$$\mathbf{A} = \begin{bmatrix} \mathbf{a}_1^T \\ \mathbf{a}_2^T \\ \vdots \\ \mathbf{a}_p^T \end{bmatrix} \quad \text{and} \quad \mathbf{b} = \begin{bmatrix} b_1 \\ b_2 \\ \vdots \\ b_p \end{bmatrix}$$

and consider the logarithmic barrier function

$$f_\tau(\mathbf{x}) = \mathbf{c}^T \mathbf{x} - \tau \sum_{i=1}^{p} \ln(b_i - \mathbf{a}_i^T \mathbf{x})$$

where $\tau > 0$ is a barrier parameter.

12.7 Using the initial point $[x_1 \ x_2 \ x_3 \ s_1 \ s_2 \ s_3] = [0.5 \ 1 \ 10 \ 0.5 \ 98 \ 9870]^T$, solve the LP problem

$$\text{minimize} \quad 100x_1 + 10x_2 + x_3$$

$$\text{subject to: } s_1 + x_1 = 1$$
$$s_2 + 2x_1 + x_2 = 100$$
$$s_3 + 200x_1 + 20x_2 + x_3 = 10000$$
$$x_i \geq 0, \ s_i \geq 0 \text{ for } i = 1, 2, 3$$

by using

(a) the PAS algorithm.
(b) the PNB algorithm.

12.8 The primal Newton barrier method discussed in Sect. 12.4 is related to the primal LP problem in Eqs. (12.1a)–(12.1c). It is possible to develop a dual Newton barrier (DNB) method in terms of the following steps:

(a) Define a dual subproblem similar to that in Eqs. (12.26a) and (12.26b) for the LP problem in Eqs. (12.2a)–(12.2c).
(b) Derive the first-order optimality conditions for the subproblem obtained in part (a).
(c) Show that the points satisfying these first-order conditions are on the primal-dual central path.
(d) Develop a DNB algorithm for solving the dual problem in Eqs. (12.2a)–(12.2c).

12.9 Consider the standard-form LP problem in Eqs. (12.1a)–(12.2c). A strictly feasible point $\mathbf{x}^* > \mathbf{0}$ is said to be the *analytic center* of the feasible region if $\mathbf{x}^*$ is the farthest away from all the boundaries of the feasible region in the sense that $\mathbf{x}^*$ solves the problem

$$\text{minimize} \quad -\sum_{i=1}^{n} \ln x_i \qquad \qquad (\text{P12.1a})$$

$$\text{subject to: } \mathbf{Ax} = \mathbf{b} \qquad \qquad (\text{P12.1b})$$

(a) Derive the KKT conditions for the minimizer of the problem in Eqs. (P12.1a) and (P12.1b).
(b) Are the KKT conditions necessary and sufficient conditions?
(c) Use the KKT conditions obtained to find the analytic center for the LP problem in Example 12.1.

12.10 Generalize the concept of analytic center discussed in Prob. 12.9 to the feasible region given by $\mathbf{Ax} \leq \mathbf{b}$, where $\mathbf{A} \in R^{p \times n}$ with $p > n$, and rank$(\mathbf{A}) = n$.

12.11 (a) Prove the inequality in Eq. (12.28).
(b) Derive the formulas in Eqs. (12.38a) and (12.38b).

12.12 Develop a primal path-following interior-point algorithm for the primal LP problem in Eqs. (12.1a)–(12.1c) in several steps as described below.

(a) Formulate a subproblem by adding a logarithmic barrier function to the objective function, i.e.,

$$\text{minimize } f_\tau(\mathbf{x}) = \mathbf{c}^T\mathbf{x} - \tau \sum_{i=1}^{n} \ln x_i$$

$$\text{subject to:} \quad \mathbf{Ax} = \mathbf{b}$$

where $\tau > 0$ is a barrier parameter.

(b) Show that the KKT conditions for the minimizer of the above subproblem can be expressed as

$$\mathbf{Ax} = \mathbf{b}$$
$$\mathbf{c} + \mathbf{A}^T\boldsymbol{\lambda} - \tau\mathbf{X}^{-1}\mathbf{e} = \mathbf{0}$$

where $\mathbf{X} = \text{diag}(\mathbf{x})$ and $\mathbf{e} = [1\ 1\ \cdots\ 1]^T$.

(c) At the kth iteration, let $\mathbf{x}_{k+1} = \mathbf{x}_k + \mathbf{d}$ such that $\mathbf{x}_{k+1}$ would better approximate the above KKT conditions. Show that up to first-order approximation, we would require that $\mathbf{d}$ satisfy the equations

$$\mathbf{Ad} = \mathbf{0} \qquad\qquad\qquad \text{(P12.2a)}$$
$$\tau\mathbf{X}^{-2}\mathbf{d} + \mathbf{c} - \tau\mathbf{X}^{-1}\mathbf{e} + \mathbf{A}^T\boldsymbol{\lambda}_k = \mathbf{0} \qquad \text{(P12.2b)}$$

where $\mathbf{X} = \text{diag}\{\mathbf{x}_k\}$.

(d) Show that the search direction $\mathbf{d}$ in Eqs. (P12.2a) and (P12.2b) can be obtained as

$$\mathbf{d} = \mathbf{x}_k - \frac{1}{\tau}\mathbf{X}^2\boldsymbol{\mu}_k \qquad\qquad \text{(P12.3a)}$$

where

$$\boldsymbol{\mu}_k = \mathbf{c} + \mathbf{A}^T\boldsymbol{\lambda}_k \qquad\qquad\qquad \text{(P12.3b)}$$
$$\boldsymbol{\lambda}_k = (\mathbf{AX}^2\mathbf{A}^T)^{-1}\mathbf{AX}^2(\tau\mathbf{X}^{-1}\mathbf{e} - \mathbf{c}) \qquad \text{(P12.3c)}$$

(e) Based on the results obtained in parts (a)-(d), describe a primal path-following interior-point algorithm.

12.13 (a) Apply the algorithm developed in Prob. 12.12 to the LP problem in Prob. 11.16.

(b) Compare the results obtained in part (a) with those of Prob. 12.3(a) and Prob. 12.5(a).

12.14 (a) Apply the algorithm developed in Prob. 12.12 to the LP problem in Prob. 11.17.

(b) Compare the results obtained in part (a) with those of Prob. 12.3(b) and Prob. 12.5(b).

12.15 Show that the search direction determined by Eqs. (P12.3a)–(P12.3c) can be expressed as

$$\mathbf{d} = -\frac{1}{\tau}\mathbf{X\bar{P}Xc} + \mathbf{X\bar{P}e} \qquad (P12.4)$$

where $\mathbf{\bar{P}} = \mathbf{I} - \mathbf{XA}^T(\mathbf{AX}^2\mathbf{A}^T)^{-1}\mathbf{AX}$ is the projection matrix given by Eq. (12.24).

12.16 In the literature, the two terms on the right-hand side of Eq. (P12.4) are called the *primal affine scaling direction* and *centering direction*, respectively. Justify the use of this terminology.
Hint: Use the results of Sect. 12.3 and Prob. 12.9.

12.17 (a) Derive the formulas in Eqs. (12.44a)–(12.44f) using Eqs. (12.42a)–(12.42c).
(b) Derive the formulas in Eqs. (12.57a)–(12.57c) using Eqs. (12.56a)–(12.56c).

12.18 (a) Apply Algorithm 12.4 to the LP problem in Prob. 11.16. Compare the results obtained with those of Probs. 12.3(a), 12.5(a) and 12.13(a).
(b) Apply Algorithm 12.4 to the LP problem 11.17. Compare the results obtained with those of Probs. 12.3(b), 12.5(b), and 12.14(a).

12.19 (a) Apply Algorithm 12.5 to the LP problem in Prob. 11.16 with a nonfeasible initial point $\{\mathbf{x}_0, \lambda_0, \mu_0\}$ with $\mathbf{x}_0 > \mathbf{0}$ and $\mu_0 > \mathbf{0}$. Compare the results obtained with those of Prob. 12.18(a).
(b) Apply Algorithm 12.5 to the LP problem in Prob. 11.17 with a nonfeasible initial point $\{\mathbf{x}_0, \lambda_0, \mu_0\}$ with $\mathbf{x}_0 > \mathbf{0}$ and $\mu_0 > \mathbf{0}$. Compare the results obtained with those of Prob. 12.18(b).

12.20 (a) Derive the formulas in Eqs. (12.60a)–(12.60e) using Eqs. (12.59a)–(12.59e).
(b) Derive the formulas in Eqs. (12.66a)–(12.66d) using Eqs. (12.65a)–(12.65d).

12.21 (a) Apply Algorithm 12.6 to the LP problem in Prob. 11.16 with the same nonfeasible initial point used in Prob. 12.19(a). Compare the results obtained with those of Prob. 12.19(a).
(b) Apply Algorithm 12.6 to the LP problem in Prob. 11.17 with the same nonfeasible initial point used in Prob. 12.19(b). Compare the results obtained with those of Prob. 12.19(b).

12.22 Consider the nonstandard-form LP problem

$$\text{minimize } \mathbf{c}^T\mathbf{x}$$

$$\text{subject to: } \mathbf{Ax} \le \mathbf{b}$$

where $\mathbf{c} \in R^{n \times 1}$, $\mathbf{A} \in R^{p \times n}$, and $\mathbf{b} \in R^{p \times 1}$ with $p > n$. Show that its solution $\mathbf{x}^*$ can be obtained by solving the standard-form LP problem

$$\text{minimize } \mathbf{b}^T \mathbf{x}$$
$$\text{subject to: } -\mathbf{A}^T \mathbf{x} = \mathbf{c}$$
$$\mathbf{x} \geq \mathbf{0}$$

using a primal-dual algorithm and then taking the optimal Lagrange multiplier $\boldsymbol{\lambda}^*$ as $\mathbf{x}^*$.

References

1. N. K. Karmarkar, "A new polynomial time algorithm for linear programming," *Combinatorica*, vol. 4, pp. 373–395, 1984.
2. M. H. Wright, "Interior methods for constrained optimization," *Acta Numerica, Cambridge University Press*, vol. 1, pp. 341–407, 1992.
3. S. J. Wright, *Primal-Dual Interior-Point Methods*. Philadelphia, PA: SIAM, 1997.
4. E. R. Barnes, "A variation on Karmarkar's algorithm for solving linear programming problems," *Math. Programming*, vol. 36, pp. 174–182, 1986.
5. R. J. Vanderbei, M. S. Meketon, and B. A. Freedman, "A modification of Karmarkar's linear programming algorithm," *Algorithmica*, vol. 1, pp. 395–407, 1986.
6. S. G. Nash and A. Sofer, *Linear and Nonlinear Programming*. New York: McGraw-Hill, 1996.
7. P. E. Gill, W. Murray, M. A. Saunders, J. A. Tomlin, and M. H. Wright, "On projected Newton barrier methods for linear programming and an equivalence to Karmarkar's projective method," *Math. Programming*, vol. 36, pp. 183–209, 1986.
8. A. V. Fiacco and G. P. McCormick, *Nonlinear Programming: Sequential Unconstrained Minimization Techniques*. New York: Wiley, 1968, (republished by SIAM in 1990).
9. K. Jittorntrum, "Sequential algorithms in nonlinear programming," Ph.D. dissertation, Australian National University, 1978.
10. K. Jittorntrum and M. R. Osborne, "Trajectory analysis and extrapolation in barrier function methods," *J. Australian Math. Soc.*, vol. 20, Series B, pp. 352–369, 1978.
11. M. Kojima, S. Mizuno, and A. Yoshise, "A primal-dual interior point algorithm for linear programming," in *Progress in Mathematical Programming: Interior Point and Related Methods*, N. Megiddo, Ed. New York: Springer-Verlag, 1989, pp. 29–47.
12. N. Megiddo, "Pathways to the optimal set in linear programming," in *Progress in Mathematical Programming: Interior Point and Related Methods*, N. Megiddo, Ed. New York: Springer-Verlag, 1989, pp. 131–158.
13. R. D. C. Monteiro and I. Adler, "Interior path following primal-dual algorithms, Part I: Linear programming," *Math. Programming*, vol. 44, pp. 27–41, 1989.
14. Y. Ye, *Interior Point Algorithms: Theory and Analysis*. New York: Wiley, 1997.
15. I. J. Lustig, R. E. Marsten, and D. F. Shanno, "Computational experience with a primal-dual interior point method for linear programming," *Linear Algebra and Its Applications*, vol. 152, pp. 191–222, 1991.
16. S. Mehrotra, "On the implementation of a primal-dual interior point method," *SIAM J. Optim.*, vol. 2, pp. 575–601, 1992.

Quadratic, Semidefinite, and Second-Order Cone Programming

13

13.1 Introduction

Quadratic programming (QP) is a family of methods, techniques, and algorithms that can be used to minimize quadratic objective functions subject to linear constraints. QP shares many combinatorial features with linear programming (LP) and it is often used as the basis of constrained nonlinear programming. An important branch of QP is convex QP where the objective function is a convex quadratic function.

Semidefinite programming (SDP) is a branch of convex programming (CP) that has been a subject of intensive research since the early 1990s [1–8]. Continued interest in SDP has been motivated mainly by two factors. First, many important classes of optimization problems such as linear-programming (LP) and convex quadratic-programming (QP) problems can be formulated as SDP problems and many CP problems of practical usefulness that are neither LP nor QP problems can also be formulated as SDP problems. Second, several interior-point methods that have proven efficient for the solution of LP and convex QP problems have been extended to the solution of SDP problems in recent years.

Another important branch of convex programming known as *second-order cone programming* (SOCP) has emerged in recent years . Although quite specialized, this branch of optimization can deal effectively with many analysis and design problems in various disciplines. Furthermore, as for the case of SDP, efficient interior-point methods are available for the solution of SOCP problems.

In this chapter, we first explore the solution of convex QP problems with equality constraints and describe a QR-decomposition-based solution method. Next, we discuss in detail two active-set methods for strictly convex QP problems. These methods can be viewed as direct extensions of the simplex method discussed in Chap. 11. In Sect. 13.4, we extend the concepts of central path and duality gap to QP and study two primal-dual path-following methods. The chapter continues with the formulation of the primal and dual SDP problems. It then demonstrates that several useful CP problems can be formulated in an SDP setting. After an introduction of several basic properties of the primal-dual solutions of an SDP problem, a detailed account

© Springer Science+Business Media, LLC, part of Springer Nature 2021
A. Antoniou and W.-S. Lu, *Practical Optimization*, Texts in Computer Science,
https://doi.org/10.1007/978-1-0716-0843-2_13

on the primal-dual interior-point methods studied in [4–8] is provided. The last two sections of the chapter are devoted to the primal and dual SOCP formulations and their relations to corresponding LP, QP, and SDP formulations; an interior-point algorithm as well as several illustrative examples are also included.

13.2   Convex QP Problems with Equality Constraints

The problem we consider in this section is

$$\text{minimize } f(\mathbf{x}) = \tfrac{1}{2}\mathbf{x}^T \mathbf{H}\mathbf{x} + \mathbf{x}^T \mathbf{p} \tag{13.1a}$$
$$\text{subject to: } \mathbf{A}\mathbf{x} = \mathbf{b} \tag{13.1b}$$

where $\mathbf{A} \in R^{p \times n}$. We assume in the rest of this section that the Hessian $\mathbf{H}$ is symmetric and positive semidefinite, $\mathbf{A}$ is of full row rank, and $p < n$. From Sect. 10.4.1, the solutions of the system $\mathbf{A}\mathbf{x} = \mathbf{b}$ in Eq. (13.1b) assume the form

$$\mathbf{x} = \mathbf{V}_r \boldsymbol{\phi} + \mathbf{A}^+\mathbf{b} \tag{13.2}$$

where $\boldsymbol{\phi}$ is an arbitrary r-dimensional vector with $r = n - p$, $\mathbf{A}^+$ denotes the Moore-Penrose pseudo-inverse of $\mathbf{A}$, $\mathbf{V}_r$ is composed of the last r columns of $\mathbf{V}$, and $\mathbf{A}^+$ as well as $\mathbf{V}$ are obtained by performing the singular-value decomposition (SVD) of $\mathbf{A}$, namely, $\mathbf{U}\boldsymbol{\Sigma}\mathbf{V}^T$ (see Appendix A.9). By using Eq. (13.2), the constraints in Eq. (13.1b) can be eliminated to yield the unconstrained minimization problem

$$\text{minimize } \hat{f}(\boldsymbol{\phi}) = \tfrac{1}{2}\boldsymbol{\phi}^T \hat{\mathbf{H}}\boldsymbol{\phi} + \boldsymbol{\phi}^T \hat{\mathbf{p}} \tag{13.3a}$$

where

$$\hat{\mathbf{H}} = \mathbf{V}_r^T \mathbf{H}\mathbf{V}_r \tag{13.3b}$$

and

$$\hat{\mathbf{p}} = \mathbf{V}_r^T (\mathbf{H}\mathbf{A}^+\mathbf{b} + \mathbf{p}) \tag{13.3c}$$

The global minimizer of the problem in Eqs. (13.1a) and (13.1b) is given by

$$\mathbf{x}^* = \mathbf{V}_r \boldsymbol{\phi}^* + \mathbf{A}^+\mathbf{b} \tag{13.4a}$$

where $\boldsymbol{\phi}^*$ is a solution of the linear system of equations

$$\hat{\mathbf{H}}\boldsymbol{\phi} = -\hat{\mathbf{p}} \tag{13.4b}$$

An alternative and often more economical approach to obtain Eq. (13.2) is to use the QR decomposition of $\mathbf{A}^T$, i.e.,

$$\mathbf{A}^T = \mathbf{Q}\begin{bmatrix} \mathbf{R} \\ \mathbf{0} \end{bmatrix} \tag{13.5}$$

where $\mathbf{Q}$ is an $n \times n$ orthogonal matrix and $\mathbf{R}$ is a $p \times p$ upper triangular matrix (see Sec. A.12 and [9]). Using Eq. (13.5), the constraints in Eq. (13.1b) can be expressed as

$$\mathbf{R}^T \hat{\mathbf{x}}_1 = \mathbf{b}$$

where $\hat{\mathbf{x}}_1$ is the vector composed of the first p components of $\hat{\mathbf{x}}$ with

$$\hat{\mathbf{x}} = \mathbf{Q}^T \mathbf{x}$$

If we let

$$\hat{\mathbf{x}} = \begin{bmatrix} \hat{\mathbf{x}}_1 \\ \phi \end{bmatrix} \quad \text{and} \quad \mathbf{Q} = [\mathbf{Q}_1 \ \mathbf{Q}_2]$$

with $\phi \in R^{(n-p) \times 1}$, $\mathbf{Q}_1 \in R^{n \times p}$, and $\mathbf{Q}_2 \in R^{n \times (n-p)}$, then

$$\mathbf{x} = \mathbf{Q}\hat{\mathbf{x}} = \mathbf{Q}_2 \phi + \mathbf{Q}_1 \hat{\mathbf{x}}_1 = \mathbf{Q}_2 \phi + \mathbf{Q}_1 \mathbf{R}^{-T} \mathbf{b}$$

i.e.,

$$\mathbf{x} = \mathbf{Q}_2 \phi + \mathbf{Q}_1 \mathbf{R}^{-T} \mathbf{b} \tag{13.6a}$$

where Eq. (13.6a) is equivalent to Eq. (13.2). The parameterized solution in Eq. (13.6a) can be used to convert the problem in Eqs. (13.1a) and (13.1b) to the reduced-size unconstrained problem in Eqs. (13.3a)–(13.3c) where $\hat{\mathbf{H}}$ and $\hat{\mathbf{p}}$ are given by

$$\hat{\mathbf{H}} = \mathbf{Q}_2^T \mathbf{H} \mathbf{Q}_2 \tag{13.6b}$$

and

$$\hat{\mathbf{p}} = \mathbf{Q}_2^T (\mathbf{H} \mathbf{Q}_1 \mathbf{R}^{-T} \mathbf{b} + \mathbf{p})$$

respectively. The global minimizer of the problem in Eqs. (13.1a) and (13.1b) can be determined as

$$\mathbf{x}^* = \mathbf{Q}_2 \phi^* + \mathbf{Q}_1 \mathbf{R}^{-T} \mathbf{b} \tag{13.7}$$

where ϕ^* is a solution of Eq. (13.4b) with $\hat{\mathbf{H}}$ given by Eq. (13.6b).

In the above two approaches, if $\hat{\mathbf{H}}$ is positive definite then the system in Eq. (13.4b) can be solved efficiently through an LDL^T decomposition (see Chap. 5) or the Cholesky decomposition (see Sec. A.13).

Example 13.1 Solve the QP problem

$$\text{minimize } f(\mathbf{x}) = \tfrac{1}{2}(x_1^2 + x_2^2) + 2x_1 + x_2 - x_3 \tag{13.8a}$$
$$\text{subject to: } \mathbf{Ax} = \mathbf{b} \tag{13.8b}$$

where

$$\mathbf{A} = [0 \ 1 \ 1], \quad \mathbf{b} = 1$$

Solution Since matrix $\mathbf{H}$ is positive semidefinite, the SVD of $\mathbf{A}$ leads to

$$\mathbf{V}_r = \begin{bmatrix} 1 & 0 \\ 0 & \frac{1}{\sqrt{2}} \\ 0 & -\frac{1}{\sqrt{2}} \end{bmatrix} \quad \text{and} \quad \mathbf{A}^+ = \begin{bmatrix} 0 \\ \frac{1}{2} \\ \frac{1}{2} \end{bmatrix}$$

Since

$$\hat{\mathbf{H}} = \mathbf{V}_r^T \mathbf{H} \mathbf{V}_r = \begin{bmatrix} 1 & 0 \\ 0 & \frac{1}{\sqrt{2}} \end{bmatrix}$$

is positive definite, the use of Eq. (13.4a) yields the unique global minimizer

$$\mathbf{x}^* = \mathbf{V}_r \boldsymbol{\phi}^* + \mathbf{A}^+ \mathbf{b}$$

$$= \begin{bmatrix} 1 & 0 \\ 0 & \frac{1}{\sqrt{2}} \\ 0 & -\frac{1}{\sqrt{2}} \end{bmatrix} \begin{bmatrix} -2.0000 \\ -3.5355 \end{bmatrix} + \begin{bmatrix} 0 \\ \frac{1}{2} \\ \frac{1}{2} \end{bmatrix} = \begin{bmatrix} -2 \\ -2 \\ 3 \end{bmatrix}$$

Alternatively, the problem can be solved by using the QR decomposition of $\mathbf{A}^T$. From Eq. (13.5), we have

$$\mathbf{Q} = \begin{bmatrix} 0 & \frac{\sqrt{2}}{2} & \frac{\sqrt{2}}{2} \\ \frac{\sqrt{2}}{2} & -0.5 & 0.5 \\ \frac{\sqrt{2}}{2} & 0.5 & -0.5 \end{bmatrix}, \quad \mathbf{R} = \sqrt{2}$$

which leads to

$$\mathbf{Q}_1 = \begin{bmatrix} 0 \\ \frac{\sqrt{2}}{2} \\ \frac{\sqrt{2}}{2} \end{bmatrix}, \quad \mathbf{Q}_2 = \begin{bmatrix} \frac{\sqrt{2}}{2} & \frac{\sqrt{2}}{2} \\ -0.5 & 0.5 \\ 0.5 & -0.5 \end{bmatrix}$$

and

$$\hat{\mathbf{H}} = \begin{bmatrix} 0.75 & 0.25 \\ 0.25 & 0.75 \end{bmatrix}, \quad \hat{\mathbf{p}} = \begin{bmatrix} 0.1642 \\ 2.6642 \end{bmatrix}$$

Hence

$$\boldsymbol{\phi}^* = \begin{bmatrix} 1.0858 \\ -3.9142 \end{bmatrix}$$

The same solution, i.e., $\mathbf{x}^* = [-2 \ -2 \ 3]^T$, can be obtained by using Eq. (13.7).

Note that if the constraint matrix $\mathbf{A}$ is changed to

$$\mathbf{A} = [1 \ 0 \ 0] \tag{13.9}$$

then

$$\mathbf{V}_r = \begin{bmatrix} 0 & 0 \\ 1 & 0 \\ 0 & 1 \end{bmatrix}, \quad \hat{\mathbf{H}} = \begin{bmatrix} 1 & 0 \\ 0 & 0 \end{bmatrix}$$

and

$$\hat{\mathbf{p}} = \begin{bmatrix} 1 \\ -1 \end{bmatrix}$$

Obviously, $\hat{\mathbf{p}}$ cannot be expressed as a linear combination of the columns of $\hat{\mathbf{H}}$ in this case and hence the problem in Eqs. (13.8a) and (13.8b) with $\mathbf{A}$ given by Eq. (13.9) does not have a finite solution.

If the objective function is modified to

$$f(\mathbf{x}) = \tfrac{1}{2}(x_1^2 + x_2^2) + 2x_1 + x_2$$

then with $\mathbf{A}$ given by Eq. (13.9), we have

$$\hat{\mathbf{p}} = \begin{bmatrix} 1 \\ 0 \end{bmatrix}$$

In this case, $\hat{\mathbf{p}}$ is a linear combination of the columns of $\hat{\mathbf{H}}$ and hence there are infinitely many solutions. As a matter of fact, it can be readily verified that any vector $\mathbf{x}^* = [1 \ -1 \ x_3]^T$ with an arbitrary value for x_3 is a global minimizer of the problem. ∎

The problem in Eqs. (13.1a) and (13.1b) can also be solved by using the first-order necessary conditions described in Theorem 10.1, which are given by

$$\mathbf{H}\mathbf{x}^* + \mathbf{p} + \mathbf{A}^T\boldsymbol{\lambda}^* = \mathbf{0}$$
$$\mathbf{A}\mathbf{x}^* - \mathbf{b} = \mathbf{0}$$

i.e.,

$$\begin{bmatrix} \mathbf{H} & \mathbf{A}^T \\ \mathbf{A} & \mathbf{0} \end{bmatrix}\begin{bmatrix} \mathbf{x}^* \\ \boldsymbol{\lambda}^* \end{bmatrix} = \begin{bmatrix} -\mathbf{p} \\ \mathbf{b} \end{bmatrix} \tag{13.10}$$

If $\mathbf{H}$ is positive definite and $\mathbf{A}$ is of full row rank, then the system matrix in Eq. (13.10) is nonsingular (see Eq. (10.69)) and the solution $\mathbf{x}^*$ from Eq. (13.10) is the unique global minimizer of the problem. Hence solution $\mathbf{x}^*$ and Lagrange multipliers $\boldsymbol{\lambda}^*$ can be expressed as

$$\boldsymbol{\lambda}^* = -(\mathbf{A}\mathbf{H}^{-1}\mathbf{A}^T)^{-1}(\mathbf{A}\mathbf{H}^{-1}\mathbf{p} + \mathbf{b}) \tag{13.11a}$$
$$\mathbf{x}^* = -\mathbf{H}^{-1}(\mathbf{A}^T\boldsymbol{\lambda}^* + \mathbf{p}) \tag{13.11b}$$

It should be mentioned that the solution of the symmetric system in Eq. (13.10) can be obtained by using numerical methods that are often more reliable and efficient than the formulas in Eqs. (13.11a) and (13.11b) (see Chap. 10 of [9]).

13.3 Active-Set Methods for Strictly Convex QP Problems

The general form of a QP problem is to minimize a quadratic function subject to a set of linear equality and a set of linear inequality constraints. Using Eq. (13.2) or Eq. (13.6a), the equality constraints can be eliminated and without loss of generality the problem can be reduced to a QP problem subject to only linear inequality constraints, i.e.,

$$\text{minimize } f(\mathbf{x}) = \tfrac{1}{2}\mathbf{x}^T\mathbf{H}\mathbf{x} + \mathbf{x}^T\mathbf{p} \tag{13.12a}$$
$$\text{subject to: } \mathbf{A}\mathbf{x} \leq \mathbf{b} \tag{13.12b}$$

where $\mathbf{A} \in R^{p \times n}$. The Karush-Kuhn-Tucker (KKT) conditions of the problem at a minimizer $\mathbf{x}$ are given by

$$\mathbf{H}\mathbf{x} + \mathbf{p} + \mathbf{A}^T\boldsymbol{\mu} = \mathbf{0} \tag{13.13a}$$
$$(\mathbf{a}_i^T\mathbf{x} - b_i)\mu_i = 0 \quad \text{for } i = 1, 2, \ldots, p \tag{13.13b}$$
$$\mu_i \geq 0 \quad \text{for } i = 1, 2, \ldots, p \tag{13.13c}$$
$$\mathbf{A}\mathbf{x} \leq \mathbf{b} \tag{13.13d}$$

For the sake of simplicity, we assume in the rest of this section that $\mathbf{H}$ is positive definite and all vertices of the feasible region are nondegenerate (see Sect. 11.3.1.1). First, we consider the case where there is a solution $\{\mathbf{x}^*,\ \boldsymbol{\mu}^*\}$ for Eqs. (13.13a)–(13.13d) with $\mathbf{x}^*$ in the interior of the feasible region $\mathcal{R}$. In such a case, $\mathbf{A}\mathbf{x}^* < \mathbf{b}$ and Eq. (13.13b) implies that $\boldsymbol{\mu}^* = \mathbf{0}$, and Eq. (13.13a) gives

$$\mathbf{x}^* = -\mathbf{H}^{-1}\mathbf{p} \tag{13.14}$$

which would be the unique global minimizer of $f(\mathbf{x})$ if there were no constraints. Therefore, we conclude that solutions of the problem in Eqs. (13.12a) and (13.12b) are on the boundary of the feasible region $\mathcal{R}$ unless the unconstrained minimizer in Eq. (13.14) is an interior point of $\mathcal{R}$. In any given iteration, the search direction in an active-set method is determined by treating the constraints that are active at the iterate as a set of equality constraints while neglecting the rest of the constraints. In what follows, we describe first a primal active-set method [10, 11] and then a dual active-set method [12] for the problem in Eqs. (13.12a) and (13.12b).

13.3.1 Primal Active-Set Method

Let $\mathbf{x}_k$ be the feasible iterate obtained in the kth iteration and assume that $\mathcal{J}_k$ is the index set of the active constraints, often referred to as the *active-set*, at $\mathbf{x}_k$. The next iterate is given by

$$\mathbf{x}_{k+1} = \mathbf{x}_k + \alpha_k \mathbf{d}_k \tag{13.15}$$

The constraints that are active at $\mathbf{x}_k$ will remain active if

$$\mathbf{a}_j^T \mathbf{x}_{k+1} - b_j = 0 \quad \text{for } j \in \mathcal{J}_k$$

which leads to

$$\mathbf{a}_j^T \mathbf{d}_k = 0 \quad \text{for } j \in \mathcal{J}_k$$

Under these circumstances, the objective function at $\mathbf{x}_k + \mathbf{d}$ becomes

$$f_k(\mathbf{d}) = \tfrac{1}{2}\mathbf{d}^T \mathbf{H}\mathbf{d} + \mathbf{d}^T \mathbf{g}_k + c_k$$

where

$$\mathbf{g}_k = \mathbf{p} + \mathbf{H}\mathbf{x}_k \tag{13.16}$$

and c_k is a constant. A major step in the active-set method is to solve the QP subproblem

$$\text{minimize } \hat{f}(\mathbf{d}) = \tfrac{1}{2}\mathbf{d}^T \mathbf{H}\mathbf{d} + \mathbf{d}^T \mathbf{g}_k \tag{13.17a}$$

$$\text{subject to: } \mathbf{a}_j^T \mathbf{d} = 0 \quad \text{for } j \in \mathcal{J}_k \tag{13.17b}$$

and this can be accomplished by using one of the methods described in the preceding section.

If the solution of the problem in Eqs. (13.17a) and (13.17b) is denoted as $\mathbf{d}_k$, then there are two possibilities: either $\mathbf{d}_k = \mathbf{0}$ or $\mathbf{d}_k \neq \mathbf{0}$.

If $\mathbf{d}_k = \mathbf{0}$, then the first-order necessary conditions imply that there exists a μ_j for $j \in \mathcal{J}_k$ such that

$$\mathbf{Hx}_k + \mathbf{p} + \sum_{j \in \mathcal{J}_k} \mu_j \mathbf{a}_j = \mathbf{0} \tag{13.18}$$

i.e.,

$$\mathbf{Hx}_k + \mathbf{p} + \mathbf{A}_{a_k}^T \hat{\boldsymbol{\mu}} = \mathbf{0} \tag{13.19}$$

where $\mathbf{A}_{a_k}$ is the matrix composed of those rows of $\mathbf{A}$ that are associated with the constraints that are active at $\mathbf{x}_k$ and $\hat{\boldsymbol{\mu}}$ is the vector composed of the μ_i's in Eq. (13.18). If we augment vector $\hat{\boldsymbol{\mu}}$ to n-dimensional vector $\boldsymbol{\mu}$ by padding zeros at the places corresponding to those rows of $\mathbf{A}$ that are inactive at $\mathbf{x}_k$, then Eq. (13.19) can be written as

$$\mathbf{Hx}_k + \mathbf{p} + \mathbf{A}^T \boldsymbol{\mu} = \mathbf{0}$$

which is the same as Eq. (13.13a). Since $\mathbf{x}_k$ is a feasible point, it satisfies Eq. (13.13d). Moreover, because of the way vector $\boldsymbol{\mu}$ is constructed, the complementarity condition in Eq. (13.13b) is also satisfied. So the first-order necessary conditions in Eqs. (13.13a)–(13.13d), which are also sufficient conditions since the present problem is a convex QP problem, will be satisfied at $\mathbf{x}_k$ if $\hat{\boldsymbol{\mu}} \geq \mathbf{0}$. In such a case, $\mathbf{x}_k$ can be deemed to be the unique global solution and the iteration can be terminated. On the other hand, if one of the components of $\hat{\boldsymbol{\mu}}$, say, μ_i, is negative, then if point $\mathbf{x}$ moves along a feasible direction at $\mathbf{x}_k$, say, $\tilde{\mathbf{d}}$, where the ith constraint becomes inactive while all the other constraints that were active at $\mathbf{x}_k$ remain active, then the objective function will *decrease*. In fact, at $\mathbf{x}_k$ we have $\mathbf{a}_j^T \tilde{\mathbf{d}} = 0$ for $j \in \mathcal{J}_k$, $j \neq i$, and $\mathbf{a}_i^T \tilde{\mathbf{d}} < 0$ and from Eq. (13.19), we obtain

$$\nabla^T f(\mathbf{x}_k) \tilde{\mathbf{d}} = (\mathbf{Hx}_k + \mathbf{p})^T \tilde{\mathbf{d}} = -\hat{\boldsymbol{\mu}}^T \mathbf{A}_{a_k}^T \tilde{\mathbf{d}} = -\sum_{j \in \mathcal{J}_k} \mu_j \mathbf{a}_j^T \tilde{\mathbf{d}}$$

$$= -\mu_i (\mathbf{a}_i^T \tilde{\mathbf{d}}) < 0$$

Consequently, active-set $\mathcal{J}_k$ can be updated by removing index i from $\mathcal{J}_k$. For the sake of simplicity, the updated index set is again denoted as $\mathcal{J}_k$. If there is more than one negative Lagrange multiplier, then the index associated with the most negative Lagrange multiplier is removed.

It should be stressed that in an implementation of the method described, verifying whether or not $\mathbf{d}_k$ is zero can be carried out without solving the problem in Eqs. (13.17a) and (13.17b). At point $\mathbf{x}_k$, we can write Eq. (13.19) as

$$\mathbf{A}_{a_k}^T \hat{\boldsymbol{\mu}} = \mathbf{g}_k \tag{13.20}$$

where $\mathbf{g}_k$ is given by Eq. (13.16). It is well known that a solution $\hat{\boldsymbol{\mu}}$ exists if and only if

$$\text{rank } [\mathbf{A}_{a_k}^T \ \ \mathbf{g}_k] = \text{rank } (\mathbf{A}_{a_k}^T) \tag{13.21}$$

SVD- and QR-decomposition-based methods are available for checking the condition in Eq. (13.21) [9,13]. If the condition in Eq. (13.21) is satisfied, the components of $\hat{\boldsymbol{\mu}}$ are examined to determine whether $\mathbf{x}_k$ is the solution or whether $\mathcal{J}_k$ needs to be updated. If the condition in Eq. (13.21) is not satisfied, then we need to solve the subproblem in Eqs. (13.17a) and (13.17b) to see whether or not $\mathbf{d}_k = \mathbf{0}$.

If $\mathbf{d}_k \neq \mathbf{0}$, then parameter α_k in Eq. (13.15) needs to be determined to assure the feasibility of $\mathbf{x}_{k+1}$. Using Eq. (13.17b), the optimal α_k can be determined as

$$\alpha_k = \min \left\{ 1, \quad \min_{\substack{i \in \mathcal{J}_k \\ \mathbf{a}_i^T \mathbf{d}_k > 0}} \frac{-(\mathbf{a}_i^T \mathbf{x}_k - b_i)}{\mathbf{a}_i^T \mathbf{d}_k} \right\} \tag{13.22}$$

If $\alpha_k < 1$, then a new constraint becomes active at $\mathbf{x}_{k+1}$. The active-set $\mathcal{J}_{k+1}$ at $\mathbf{x}_{k+1}$ is obtained by adding the index of the new active constraint, j_k, to $\mathcal{J}_k$.

The active-set method can be implemented in terms of the following algorithm.

Algorithm 13.1 Primal active-set algorithm for QP problems with inequality constraints

Step 1
Input a feasible point, $\mathbf{x}_0$, identify the active-set $\mathcal{J}_0$, form matrix $\mathbf{A}_{a_0}$, and set $k = 0$.

Step 2
Compute $\mathbf{g}_k$ using Eq. (13.16).
Check the rank condition in Eq. (13.21); if Eq. (13.21) does not hold, go to Step 4.

Step 3
Solve Eq. (13.20) for $\hat{\boldsymbol{\mu}}$. If $\hat{\boldsymbol{\mu}} \geq \mathbf{0}$, output $\mathbf{x}_k$ as the solution and stop; otherwise, remove the index that is associated with the most negative Lagrange multiplier from $\mathcal{J}_k$.

Step 4
Solve the problem in Eqs. (13.17a) and (13.17b) for $\mathbf{d}_k$.

Step 5
Find α_k using Eq. (13.22) and set $\mathbf{x}_{k+1} = \mathbf{x}_k + \alpha_k \mathbf{d}_k$.

Step 6
If $\alpha_k < 1$, construct $\mathcal{J}_{k+1}$ by adding the index that yields the minimum in Eq. (13.22) to $\mathcal{J}_k$; otherwise, let $\mathcal{J}_{k+1} = \mathcal{J}_k$.

Step 7
Set $k = k + 1$ and repeat from Step 2.

Algorithm 13.1 requires a feasible initial point $\mathbf{x}_0$ that satisfies the constraints $\mathbf{A}\mathbf{x}_0 \leq \mathbf{b}$. Such a point can be identified by using, for example, the method described in Sect. 11.2.3.4. The method involves solving an LP problem of size $n + 1$ for which a feasible initial point can be easily identified.

Example 13.2 Find the shortest distance between triangles $\mathcal{R}$ and $\mathcal{S}$ shown in Fig. 13.1 and the points $\mathbf{r}^* \in \mathcal{R}$ and $\mathbf{s}^* \in \mathcal{S}$ that yield the minimum distance.

Fig. 13.1 Triangles $\mathcal{R}$ and $\mathcal{S}$ in Example 13.2

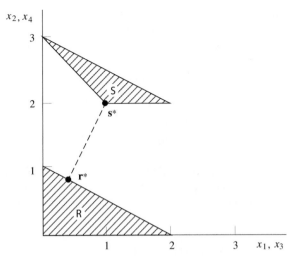

Solution Let $\mathbf{r} = [x_1\ x_2]^T \in \mathcal{R}$ and $\mathbf{s} = [x_3\ x_4]^T \in \mathcal{S}$. The square of the distance between $\mathbf{r}$ and $\mathbf{s}$ is given by

$$(x_1 - x_3)^2 + (x_2 - x_4)^2 = \mathbf{x}^T \mathbf{H} \mathbf{x}$$

where

$$\mathbf{H} = \begin{bmatrix} 1 & 0 & -1 & 0 \\ 0 & 1 & 0 & -1 \\ -1 & 0 & 1 & 0 \\ 0 & -1 & 0 & 1 \end{bmatrix} \tag{13.23}$$

and $\mathbf{x} = [x_1\ x_2\ x_3\ x_4]^T$ is constrained to satisfy the inequalities

$$x_1 \geq 0$$
$$x_2 \geq 0$$
$$x_1 + 2x_2 \leq 2$$
$$x_4 \geq 2$$
$$x_3 + x_4 \geq 3$$
$$x_3 + 2x_4 \leq 6$$

The problem can be formulated as the QP problem

$$\text{minimize } f(\mathbf{x}) = \tfrac{1}{2}\mathbf{x}^T \mathbf{H} \mathbf{x}$$
$$\text{subject to: } \mathbf{A}\mathbf{x} \leq \mathbf{b}$$

where

$$\mathbf{A} = \begin{bmatrix} -1 & 0 & 0 & 0 \\ 0 & -1 & 0 & 0 \\ 1 & 2 & 0 & 0 \\ 0 & 0 & 0 & -1 \\ 0 & 0 & -1 & -1 \\ 0 & 0 & 1 & 2 \end{bmatrix}, \quad \mathbf{b} = \begin{bmatrix} 0 \\ 0 \\ 2 \\ -2 \\ -3 \\ 6 \end{bmatrix}$$

Since matrix $\mathbf{H}$ in Eq. (13.23) is positive semidefinite, Algorithm 13.1 is not immediately applicable since it requires a positive definite $\mathbf{H}$. One approach to fix this
problem which has been found to be effective for convex QP problems of moderate
size with positive semidefinite Hessian is to introduce a small perturbation to the
Hessian to make it *positive definite*, i.e., we let

$$\tilde{\mathbf{H}} = \mathbf{H} + \delta \mathbf{I} \tag{13.24}$$

where $\mathbf{I}$ is the identity matrix and δ is a small positive scalar. Modifying matrix $\mathbf{H}$ in
Eq. (13.23) to $\tilde{\mathbf{H}}$ as in Eq. (13.24) with $\delta = 10^{-9}$ and then applying Algorithm 13.1
to the modified QP problem with an initial point $\mathbf{x}_0 = [2\ 0\ 0\ 3]^T$, the minimizer
was obtained in 5 iterations as $\mathbf{x}^* = [0.4\ 0.8\ 1.0\ 2.0]^T$. Hence $\mathbf{r}^* = [0.4\ 0.8]^T$
and $\mathbf{s}^* = [1\ 2]^T$ and the shortest distance between these points is 1.341641
(see Fig. 13.1). ∎

13.3.2   Dual Active-Set Method

A dual active-set method for the QP problem in Eqs. (13.12a) and (13.12b) with $\mathbf{H}$
positive definite was proposed by Goldfarb and Idnani [12]. The method is essentially
the active-set method described in the preceding section but applied to the dual of the
problem in Eqs. (13.12a) and (13.12b). According to Definition 10.6, the Lagrange
dual of the minimization problem in Eqs. (13.12a) and (13.12b) is the maximization
problem

$$\text{maximize} \quad -\tfrac{1}{2}\boldsymbol{\mu}^T \mathbf{A}\mathbf{H}^{-1}\mathbf{A}^T \boldsymbol{\mu} - \boldsymbol{\mu}^T(\mathbf{A}\mathbf{H}^{-1}\mathbf{p} + \mathbf{b})$$

$$\text{subject to:} \quad \boldsymbol{\mu} \geq \mathbf{0}$$

which is equivalent to

$$\text{minimize } h(\boldsymbol{\mu}) = \tfrac{1}{2}\boldsymbol{\mu}^T \mathbf{A}\mathbf{H}^{-1}\mathbf{A}^T \boldsymbol{\mu} + \boldsymbol{\mu}^T(\mathbf{A}\mathbf{H}^{-1}\mathbf{p} + \mathbf{b}) \tag{13.25a}$$

$$\text{subject to:} \quad \boldsymbol{\mu} \geq \mathbf{0} \tag{13.25b}$$

Once the minimizer of the problem in Eqs. (13.25a) and (13.25b), $\boldsymbol{\mu}^*$, is determined,
the minimizer of the primal problem is obtained from one of the KKT conditions,
i.e.,

$$\mathbf{H}\mathbf{x} + \mathbf{p} + \mathbf{A}^T\boldsymbol{\mu} = \mathbf{0}$$

which gives

$$\mathbf{x}^* = -\mathbf{H}^{-1}(\mathbf{A}^T\boldsymbol{\mu}^* + \mathbf{p})$$

The advantages of the dual problem in Eqs. (13.25a) and (13.25b) include the
following:

(a) A feasible initial point can be easily identified as any vector with nonnegative
    components, e.g., $\boldsymbol{\mu}_0 = \mathbf{0}$.
(b) The constraint matrix in Eq. (13.25b) is the $p \times p$ identity matrix. Consequently,
    the dual problem always satisfies the nondegeneracy assumption.

(c) The dual problem only involves bound-type inequality constraints, which considerably simplifies the computations required in Algorithm 13.1. For example, checking the rank condition in Eq. (13.21) for the dual problem entails examining whether the components of $\mathbf{g}_k$ that correspond to those indices not in the active-set are all zero.

As in the primal active-set method discussed in Sect. 13.3.1, a major step in the dual active-set method is to solve the QP subproblem which is the dual of the QP problem in Eqs. (13.17a) and (13.17b). This can be reduced to the unconstrained optimization problem

$$\text{minimize } \tfrac{1}{2}\tilde{\mathbf{d}}^T \tilde{\mathbf{H}} \tilde{\mathbf{d}} + \tilde{\mathbf{d}}^T \tilde{\mathbf{g}}_k$$

where $\tilde{\mathbf{d}}$ is the column vector obtained by deleting the components of $\mathbf{d}$ whose indices are in $\mathcal{J}_k$, $\tilde{\mathbf{H}}$ is the principal submatrix of $\mathbf{H}$ obtained by deleting the columns and rows associated with index set $\mathcal{J}_k$, and $\tilde{\mathbf{g}}_k$ is obtained by deleting the components of $\mathbf{g}_k$ whose indices are in $\mathcal{J}_k$.

13.4 Interior-Point Methods for Convex QP Problems

In this section, we discuss several interior-point methods for convex QP problems that can be viewed as natural extensions of the interior-point methods for LP problems discussed in Chap. 12.

13.4.1 Dual QP Problem, Duality Gap, and Central Path

As described in Sect. 11.2.1, linear constraints of the form $\mathbf{A}\mathbf{x} \leq \mathbf{b}$ can be converted to $\mathbf{A}\mathbf{x} = \mathbf{b}$ and $\mathbf{x} \geq \mathbf{0}$ by introducing slack variables and decomposing vector variable $\mathbf{x}$ as $\mathbf{x} = \mathbf{x}^+ - \mathbf{x}^-$ with $\mathbf{x}^+ \geq \mathbf{0}$ and $\mathbf{x}^- \geq \mathbf{0}$. Hence the QP problem in Eqs. (13.12a) and (13.12b) can be reformulated as

$$\text{minimize } f(\mathbf{x}) = \tfrac{1}{2}\mathbf{x}^T \mathbf{H}\mathbf{x} + \mathbf{x}^T \mathbf{p} \tag{13.26a}$$

$$\text{subject to: } \mathbf{A}\mathbf{x} = \mathbf{b} \tag{13.26b}$$

$$\mathbf{x} \geq \mathbf{0} \tag{13.26c}$$

where $\mathbf{H} \in R^{n \times n}$ is positive semidefinite and $\mathbf{A} \in R^{p \times n}$ is of full row rank. By applying Theorem 10.9 to Eqs. (13.26a)–(13.26c), the dual problem can be obtained as

$$\text{maximize } h(\mathbf{x}, \, \boldsymbol{\lambda}, \, \boldsymbol{\mu}) = -\tfrac{1}{2}\mathbf{x}^T \mathbf{H}\mathbf{x} - \boldsymbol{\lambda}^T \mathbf{b} \tag{13.27a}$$

$$\text{subject to: } -\mathbf{A}^T \boldsymbol{\lambda} + \boldsymbol{\mu} - \mathbf{H}\mathbf{x} = \mathbf{p} \tag{13.27b}$$

$$\boldsymbol{\mu} \geq \mathbf{0} \tag{13.27c}$$

The necessary and sufficient conditions for vector $\mathbf{x}$ to be the global minimizer of the problem in Eqs. (13.26a)–(13.26c) are the KKT conditions which are given by

$$\mathbf{Ax} - \mathbf{b} = \mathbf{0} \quad \text{for } \mathbf{x} \geq \mathbf{0} \tag{13.28a}$$

$$-\mathbf{A}^T\boldsymbol{\lambda} + \boldsymbol{\mu} - \mathbf{Hx} - \mathbf{p} = \mathbf{0} \quad \text{for } \boldsymbol{\mu} \geq \mathbf{0} \tag{13.28b}$$

$$\mathbf{X}\boldsymbol{\mu} = \mathbf{0} \tag{13.28c}$$

where $\mathbf{X} = \text{diag}\{x_1, x_2, \ldots, x_n\}$.

Let set $\{\mathbf{x}, \boldsymbol{\lambda}, \boldsymbol{\mu}\}$ be feasible for the problems in Eqs. (13.26a)–(13.26c) and Eqs. (13.27a)–(13.27c). The duality gap, which was defined in Eq. (12.6) for LP problems, can be obtained for $\{\mathbf{x}, \boldsymbol{\lambda}, \boldsymbol{\mu}\}$ as

$$\delta(\mathbf{x}, \boldsymbol{\lambda}, \boldsymbol{\mu}) = f(\mathbf{x}) - h(\mathbf{x}, \boldsymbol{\lambda}, \boldsymbol{\mu}) = \mathbf{x}^T\mathbf{Hx} + \mathbf{x}^T\mathbf{p} + \boldsymbol{\lambda}^T\mathbf{b}$$
$$= \mathbf{x}^T(-\mathbf{A}^T\boldsymbol{\lambda} + \boldsymbol{\mu}) + \boldsymbol{\lambda}^T\mathbf{b} = \mathbf{x}^T\boldsymbol{\mu}$$

which is always nonnegative and is equal to zero at solution $\{\mathbf{x}^*, \boldsymbol{\lambda}^*, \boldsymbol{\mu}^*\}$ because of the complementarity condition in Eq. (13.28c).

Based on Eqs. (13.28a)–(13.28c), the concept of central path, which was initially introduced for LP problems in Sect. 12.2.2, can be readily extended to the problems in Eqs. (13.26a)–(13.26c) and Eqs. (13.27a)–(13.27c) as the parameterized set $\mathbf{w}(\tau) = \{\mathbf{x}(\tau), \boldsymbol{\lambda}(\tau), \boldsymbol{\mu}(\tau)\}$ that satisfies the conditions

$$\mathbf{Ax} - \mathbf{b} = \mathbf{0} \quad \text{for } \mathbf{x} > \mathbf{0} \tag{13.29a}$$

$$-\mathbf{A}^T\boldsymbol{\lambda} + \boldsymbol{\mu} - \mathbf{Hx} - \mathbf{p} = \mathbf{0} \quad \text{for } \boldsymbol{\mu} > \mathbf{0} \tag{13.29b}$$

$$\mathbf{X}\boldsymbol{\mu} = \tau\mathbf{e} \tag{13.29c}$$

where τ is a strictly positive scalar parameter and $\mathbf{e} = [1 \; 1 \; \cdots \; 1]^T \in R^n$. It follows that every point on the central path is strictly feasible and the entire central path lies in the interior of the feasible regions described by Eqs. (13.26b), (13.26c), (13.27b), and (13.27c). On comparing Eqs. (13.29a)–(13.29c) with Eqs. (13.28a)–(13.28c), we see that as $\tau \to 0$ the central path approaches the set $\mathbf{w}^* = \{\mathbf{x}^*, \boldsymbol{\lambda}^*, \boldsymbol{\mu}^*\}$ which solves the problems in Eqs. (13.26a)–(13.26c) and Eqs. (13.27a)–(13.27c) simultaneously. This can also be seen by computing the duality gap on the central path, i.e.,

$$\delta[\mathbf{x}(\tau), \boldsymbol{\lambda}(\tau), \boldsymbol{\mu}(\tau)] = \mathbf{x}^T(\tau)\boldsymbol{\mu}(\tau) = n\tau$$

Hence the duality gap approaches zero linearly as $\tau \to 0$.

As in the LP case, the equations in Eqs. (13.29a)–(13.29c) that define the central path for the problem in Eqs. (13.26a)–(13.26c) and its dual can be interpreted as the KKT conditions for the modified minimization problem

$$\text{minimize } \hat{f}(\mathbf{x}) = \tfrac{1}{2}\mathbf{x}^T\mathbf{Hx} + \mathbf{x}^T\mathbf{p} - \tau\sum_{i=1}^{n}\ln x_i \tag{13.30a}$$

$$\text{subject to: } \mathbf{Ax} = \mathbf{b} \tag{13.30b}$$

where $\tau > 0$ is the barrier parameter (see Sect. 12.4). In order to ensure that $\hat{f}(\mathbf{x})$ in Eq. (13.30a) is well defined, we require that

$$\mathbf{x} > \mathbf{0} \tag{13.30c}$$

The KKT conditions for the problem in Eqs. (13.30a) and (13.30b) are given by

$$\mathbf{Ax} - \mathbf{b} = \mathbf{0} \quad \text{for } \mathbf{x} > \mathbf{0} \tag{13.31a}$$

$$-\mathbf{A}^T\boldsymbol{\lambda} + \tau\mathbf{X}^{-1}\mathbf{e} - \mathbf{Hx} - \mathbf{p} = \mathbf{0} \tag{13.31b}$$

If we let $\boldsymbol{\mu} = \tau\mathbf{X}^{-1}\mathbf{e}$, then the condition $\mathbf{x} > \mathbf{0}$ implies that $\boldsymbol{\mu} > \mathbf{0}$ and Eq. (13.31b) can be written as

$$-\mathbf{A}^T\boldsymbol{\lambda} + \boldsymbol{\mu} - \mathbf{Hx} - \mathbf{p} = \mathbf{0} \quad \text{for } \boldsymbol{\mu} > \mathbf{0} \tag{13.32a}$$

and

$$\mathbf{X}\boldsymbol{\mu} = \tau\mathbf{e} \tag{13.32b}$$

Consequently, Eqs. (13.31a), (13.32a), and (13.32b) are identical with Eqs. (13.29a), (13.29b), and (13.29c), respectively.

In what follows, we describe a primal-dual path-following method similar to the one proposed by Monteiro and Adler [14] which is an extension of their work on LP [15] described in Sect. 12.5.

13.4.2   A Primal-Dual Path-Following Method for Convex QP Problems

Consider the convex QP problem in Eqs. (13.26a)–(13.26c) and let $\mathbf{w}_k = \{\mathbf{x}_k, \boldsymbol{\lambda}_k, \boldsymbol{\mu}_k\}$ be such that $\mathbf{x}_k$ is strictly feasible for the primal problem in Eqs. (13.26a)–(13.26c) and $\mathbf{w}_k = \{\mathbf{x}_k, \boldsymbol{\lambda}_k, \boldsymbol{\mu}_k\}$ is strictly feasible for the dual problem in Eqs. (13.27a)–(13.27c). We require an increment set $\boldsymbol{\delta}_w = \{\boldsymbol{\delta}_x, \boldsymbol{\delta}_\lambda, \boldsymbol{\delta}_\mu\}$ such that the next iterate $\mathbf{w}_{k+1} = \{\mathbf{x}_{k+1}, \boldsymbol{\lambda}_{k+1}, \boldsymbol{\mu}_{k+1}\} = \mathbf{w}_k + \boldsymbol{\delta}_w$ remains strictly feasible and approaches the central path defined by Eqs. (13.29a)–(13.29c). If $\mathbf{w}_k$ were to satisfy Eqs. (13.29a)–(13.29c) with $\tau = \tau_{k+1}$, we would have

$$-\mathbf{H}\boldsymbol{\delta}_x - \mathbf{A}^T\boldsymbol{\delta}_\lambda + \boldsymbol{\delta}_\mu = \mathbf{0} \tag{13.33a}$$

$$\mathbf{A}\boldsymbol{\delta}_x = \mathbf{0} \tag{13.33b}$$

$$\Delta\mathbf{X}\boldsymbol{\mu}_k + \mathbf{X}\boldsymbol{\delta}_\mu + \Delta\mathbf{X}\boldsymbol{\delta}_\mu = \tau_{k+1}\mathbf{e} - \mathbf{X}\boldsymbol{\mu}_k \tag{13.33c}$$

where $\Delta\mathbf{X} = \text{diag}\{(\boldsymbol{\delta}_x)_1, (\boldsymbol{\delta}_x)_2, \ldots, (\boldsymbol{\delta}_x)_n\}$. If the second-order term in Eq. (13.33c), namely, $\Delta\mathbf{X}\boldsymbol{\delta}_\mu$, is neglected, then Eqs. (13.33a)–(13.33c) become the system of *linear* equations

$$-\mathbf{H}\boldsymbol{\delta}_x - \mathbf{A}^T\boldsymbol{\delta}_\lambda + \boldsymbol{\delta}_\mu = \mathbf{0} \tag{13.34a}$$

$$\mathbf{A}\boldsymbol{\delta}_x = \mathbf{0} \tag{13.34b}$$

$$\mathbf{M}\boldsymbol{\delta}_x + \mathbf{X}\boldsymbol{\delta}_\mu = \tau_{k+1}\mathbf{e} - \mathbf{X}\boldsymbol{\mu}_k \tag{13.34c}$$

where $\mathbf{M} = \text{diag}\{(\boldsymbol{\mu}_k)_1, (\boldsymbol{\mu}_k)_2, \ldots, (\boldsymbol{\mu}_k)_n\}$. These equations can be expressed in matrix form as

$$\begin{bmatrix} -\mathbf{H} & -\mathbf{A}^T & \mathbf{I} \\ \mathbf{A} & \mathbf{0} & \mathbf{0} \\ \mathbf{M} & \mathbf{0} & \mathbf{X} \end{bmatrix} \boldsymbol{\delta}_w = \begin{bmatrix} \mathbf{0} \\ \mathbf{0} \\ \tau_{k+1}\mathbf{e} - \mathbf{X}\boldsymbol{\mu}_k \end{bmatrix} \tag{13.35}$$

A good choice of parameter τ_{k+1} in Eqs. (13.34c) and (13.35) is

$$\tau_{k+1} = \frac{\mathbf{x}_k^T \boldsymbol{\mu}_k}{n + \rho} \quad \text{with } \rho \geq \sqrt{n} \tag{13.36}$$

It can be shown that for a given tolerance ε for the duality gap, this choice of τ_{k+1} will reduce the primal-dual potential function, which is defined as

$$\psi_{n+\rho}(\mathbf{x}, \boldsymbol{\mu}) = (n + \rho) \ln(\mathbf{x}^T \boldsymbol{\mu}) - \sum_{i=1}^{n} \ln(x_i \mu_i)$$

to a small but constant amount. This would lead to an iteration complexity of $O(\rho \ln(1/\varepsilon))$ (see [16, Chap. 4]).

The solution of Eqs. (13.34a)–(13.34c) can be obtained as

$$\boldsymbol{\delta}_\lambda = -\mathbf{Y}\mathbf{y} \tag{13.37a}$$
$$\boldsymbol{\delta}_x = -\boldsymbol{\Gamma}\mathbf{X}\mathbf{A}^T\boldsymbol{\delta}_\lambda - \mathbf{y} \tag{13.37b}$$
$$\boldsymbol{\delta}_\mu = \mathbf{H}\boldsymbol{\delta}_x + \mathbf{A}^T\boldsymbol{\delta}_\lambda \tag{13.37c}$$

where

$$\boldsymbol{\Gamma} = (\mathbf{M} + \mathbf{X}\mathbf{H})^{-1} \tag{13.37d}$$
$$\mathbf{Y} = (\mathbf{A}\boldsymbol{\Gamma}\mathbf{X}\mathbf{A}^T)^{-1}\mathbf{A} \tag{13.37e}$$

and

$$\mathbf{y} = \boldsymbol{\Gamma}(\mathbf{X}\boldsymbol{\mu}_k - \tau_{k+1}\mathbf{e}) \tag{13.37f}$$

Since $\mathbf{x}_k > \mathbf{0}$ and $\boldsymbol{\mu}_k > \mathbf{0}$, matrices $\mathbf{X}$ and $\mathbf{M}$ are positive definite. Therefore, $\mathbf{X}^{-1}\mathbf{M} + \mathbf{H}$ is also positive definite and the inverse of matrix

$$\mathbf{M} + \mathbf{X}\mathbf{H} = \mathbf{X}(\mathbf{X}^{-1}\mathbf{M} + \mathbf{H})$$

exists. Moreover, since $\mathbf{A}$ is of full row rank, the matrix

$$\mathbf{A}\boldsymbol{\Gamma}\mathbf{X}\mathbf{A}^T = \mathbf{A}(\mathbf{X}^{-1}\mathbf{M} + \mathbf{H})^{-1}\mathbf{A}^T$$

is also positive definite and hence nonsingular. Therefore, matrices $\boldsymbol{\Gamma}$ and $\mathbf{Y}$ in Eqs. (13.37a)–(13.37f) are well defined.

Once $\boldsymbol{\delta}_w$ is calculated, an appropriate value of α_k needs to be determined such that

$$\mathbf{w}_{k+1} = \mathbf{w}_k + \alpha_k \boldsymbol{\delta}_w \tag{13.38}$$

remains strictly feasible. A value of α_k can be chosen in the same way as in the primal-dual interior-point algorithm discussed in Sect. 12.5.1, i.e., we let

$$\alpha_k = (1 - 10^{-6})\alpha_{\max} \tag{13.39a}$$

where

$$\alpha_{\max} = \min(\alpha_p, \alpha_d) \tag{13.39b}$$

with

$$\alpha_p = \min_{i \text{ with } (\delta_x)_i < 0} \left[-\frac{(\mathbf{x}_k)_i}{(\delta_x)_i} \right] \tag{13.39c}$$

$$\alpha_d = \min_{i \text{ with } (\delta_\mu)_i < 0} \left[-\frac{(\mu_k)_i}{(\delta_\mu)_i} \right] \tag{13.39d}$$

The method described can be implemented in terms of the following algorithm.

Algorithm 13.2   Primal-dual path-following algorithm for convex QP problems

Step 1
Input a strictly feasible vector $\mathbf{w}_0 = \{\mathbf{x}_0, \, \lambda_0, \, \mu_0\}$.
Set $k = 1$ and $\rho \geq \sqrt{n}$, and initialize the tolerance ε for the duality gap.
Step 2
If $\mathbf{x}_k^T \mu_k \leq \varepsilon$, output solution $\mathbf{w}^* = \mathbf{w}_k$, and stop; otherwise, continue with Step 3.
Step 3
Set τ_{k+1} using Eq. (13.36) and compute $\delta_w = \{\delta_x, \, \delta_\lambda, \, \delta_\mu\}$ using Eqs. (13.37a)–(13.37c).
Step 4
Compute α_k using Eqs. (13.39a)–(13.39d) and update $\mathbf{w}_{k+1}$ using Eq. (13.38).
Set $k = k + 1$ and repeat from Step 2.

A value for tolerance ε in the range 10^{-9}–10^{-6} works well in practice.

13.4.3   Nonfeasible-initialization Primal-Dual Path-Following Method for Convex QP Problems

Algorithm 13.2 requires a strictly feasible $\mathbf{w}_0$, which might be difficult to obtain particularly for large-scale problems. However, the idea of using a set of vector increments $\delta_w = \{\delta_x, \, \delta_\lambda, \, \delta_\mu\}$ such that $\mathbf{w}_k + \delta_w$ becomes feasible, as described in Sect. 12.5.2, can be used to develop a nonfeasible-initialization primal-dual path-following algorithm for convex QP problems.

Let $\mathbf{w}_k = \{\mathbf{x}_k, \, \lambda_k, \, \mu_k\}$ be such that $\mathbf{x}_k > \mathbf{0}$ and $\mu_k > \mathbf{0}$ but may not satisfy Eqs. (13.29a) and (13.29b). We need to find the next iterate

$$\mathbf{w}_{k+1} = \mathbf{w}_k + \alpha_k \delta_w$$

such that $\mathbf{x}_{k+1} > \mathbf{0}$ and $\mu_{k+1} > \mathbf{0}$ and that $\delta_w = \{\delta_x, \, \delta_\lambda, \, \delta_\mu\}$ satisfies the equations

$$-\mathbf{H}(\mathbf{x}_k + \delta_x) - \mathbf{p} - \mathbf{A}^T(\lambda_k + \delta_\lambda) + (\mu_k + \delta_\mu) = \mathbf{0}$$
$$\mathbf{A}(\mathbf{x}_k + \delta_x) = \mathbf{b}$$
$$\mathbf{M}\delta_x + \mathbf{X}\delta_\mu = \tau_{k+1}\mathbf{e} - \mathbf{X}\mu_k$$

i.e.,

$$-\mathbf{H}\boldsymbol{\delta}_x - \mathbf{A}^T\boldsymbol{\delta}_\lambda + \boldsymbol{\delta}_\mu = \mathbf{r}_d$$
$$\mathbf{A}\boldsymbol{\delta}_x = \mathbf{r}_p$$
$$\mathbf{M}\boldsymbol{\delta}_x + \mathbf{X}\boldsymbol{\delta}_\mu = \tau_{k+1}\mathbf{e} - \mathbf{X}\boldsymbol{\mu}_k$$

where

$$\mathbf{r}_d = \mathbf{H}\mathbf{x}_k + \mathbf{p} + \mathbf{A}^T\boldsymbol{\lambda}_k - \boldsymbol{\mu}_k \tag{13.40a}$$
$$\mathbf{r}_p = \mathbf{b} - \mathbf{A}\mathbf{x}_k \tag{13.40b}$$

The above system of linear equations can be expressed as

$$\begin{bmatrix} -\mathbf{H} & -\mathbf{A}^T & \mathbf{I} \\ \mathbf{A} & \mathbf{0} & \mathbf{0} \\ \mathbf{M} & \mathbf{0} & \mathbf{X} \end{bmatrix} \boldsymbol{\delta}_w = \begin{bmatrix} \mathbf{r}_d \\ \mathbf{r}_p \\ \tau_{k+1}\mathbf{e} - \mathbf{X}\boldsymbol{\mu}_k \end{bmatrix} \tag{13.41}$$

On comparing Eq. (13.41) with Eq. (13.35), we see that $\boldsymbol{\delta}_w$ becomes the search direction determined by using Eq. (13.35) when the residual vectors $\mathbf{r}_p$ and $\mathbf{r}_d$ are reduced to zero. Note that in general the elimination of $\mathbf{r}_p$ and $\mathbf{r}_d$ cannot be accomplished in a single iteration because the next iterate also depends on α_k which may not be unity. The solution of Eq. (13.41) can be obtained as

$$\boldsymbol{\delta}_\lambda = -\mathbf{Y}_0(\mathbf{A}\mathbf{y}_d + \mathbf{r}_p) \tag{13.42a}$$
$$\boldsymbol{\delta}_x = -\boldsymbol{\Gamma}\mathbf{X}\mathbf{A}^T\boldsymbol{\delta}_\lambda - \mathbf{y}_d \tag{13.42b}$$
$$\boldsymbol{\delta}_\mu = \mathbf{H}\boldsymbol{\delta}_x + \mathbf{A}^T\boldsymbol{\delta}_\lambda + \mathbf{r}_d \tag{13.42c}$$

where

$$\boldsymbol{\Gamma} = (\mathbf{M} + \mathbf{X}\mathbf{H})^{-1} \tag{13.42d}$$
$$\mathbf{Y}_0 = (\mathbf{A}\boldsymbol{\Gamma}\mathbf{X}\mathbf{A}^T)^{-1} \tag{13.42e}$$
$$\mathbf{y}_d = \boldsymbol{\Gamma}[\mathbf{X}(\boldsymbol{\mu}_k + \mathbf{r}_d) - \tau_{k+1}\mathbf{e}] \tag{13.42f}$$
$$\tau_{k+1} = \frac{\mathbf{x}_k^T\boldsymbol{\mu}_k}{n + \rho} \quad \text{with } \rho \geq \sqrt{n} \tag{13.42g}$$

Obviously, if residual vectors $\mathbf{r}_p$ and $\mathbf{r}_d$ are reduced to zero, then the vector $\boldsymbol{\delta}_w = \{\boldsymbol{\delta}_x, \boldsymbol{\delta}_\lambda, \boldsymbol{\delta}_\mu\}$ determined by using Eqs. (13.42a)–(13.42g) is identical with that obtained using Eqs. (13.37a)–(13.37f). Once $\boldsymbol{\delta}_w$ is determined, α_k can be calculated using Eqs. (13.39a)–(13.39d). The above principles lead to the following algorithm.

Algorithm 13.3   Nonfeasible-initialization primal-dual path-following algorithm for convex QP problems
Step 1
Input a vector $\mathbf{w}_0 = \{\mathbf{x}_0, \boldsymbol{\lambda}_0, \boldsymbol{\mu}_0\}$ with $\mathbf{x}_0 > \mathbf{0}$ and $\boldsymbol{\mu}_0 > \mathbf{0}$.
Set $k = 0$ and $\rho \geq \sqrt{n}$, and initialize the tolerance ε for the duality gap.
Step 2
If $\mathbf{x}_k^T\boldsymbol{\mu}_k \leq \varepsilon$, output solution $\mathbf{w}^* = \mathbf{w}_k$ and stop; otherwise, continue with Step 3.
Step 3
Compute τ_{k+1} using Eq. (13.42g) and determine $\boldsymbol{\delta}_w = \{\boldsymbol{\delta}_x, \boldsymbol{\delta}_\lambda, \boldsymbol{\delta}_\mu\}$ using Eqs. (13.42a)–(13.42g).

Step 4
Compute α_k using Eqs. (13.39a)–(13.39d) and update $\mathbf{w}_{k+1}$ using Eq. (13.38).
Set $k = k + 1$ and repeat from Step 2.

Example 13.3 Solve the convex QP problem

$$\text{minimize } f(\mathbf{x}) = \tfrac{1}{2}\mathbf{x}^T \begin{bmatrix} 4 & 0 & 0 \\ 0 & 1 & -1 \\ 0 & -1 & 1 \end{bmatrix} \mathbf{x} + \mathbf{x}^T \begin{bmatrix} -8 \\ -6 \\ -6 \end{bmatrix}$$

$$\text{subject to: } x_1 + x_2 + x_3 = 3$$

$$\mathbf{x} \geq 0$$

Solution The problem can be solved by using either Algorithm 13.2 or Algorithm 13.3. Using a strictly feasible point $\mathbf{x}_0 = [1 \ 1 \ 1]^T$ and assigning $\lambda_0 = 7$ and $\boldsymbol{\mu}_0 = [3 \ 1 \ 1]^T$, it took Algorithm 13.2 11 iterations to converge to the solution

$$\mathbf{x}^* = \begin{bmatrix} 0.500000 \\ 1.250000 \\ 1.250000 \end{bmatrix}$$

On the other hand, using a nonfeasible initial point $\mathbf{x}_0 = [1 \ 2 \ 2]^T$ and assigning $\lambda_0 = 1$, $\boldsymbol{\mu}_0 = [0.2 \ 0.2 \ 0.2]^T$, $\rho = n + 2\sqrt{n}$, and $\varepsilon = 10^{-5}$, Algorithm 13.3 took 13 iterations to converge to the solution

$$\mathbf{x}^* = \begin{bmatrix} 0.500001 \\ 1.249995 \\ 1.249995 \end{bmatrix}$$

■

Example 13.4 Solve the shortest-distance problem described in Example 13.2 by using Algorithm 13.3.

Solution By letting $\mathbf{x} = \mathbf{x}^+ - \mathbf{x}^-$ where $\mathbf{x}^+ \geq \mathbf{0}$ and $\mathbf{x}^- \geq \mathbf{0}$, and then introducing slack vector $\boldsymbol{\eta} \geq \mathbf{0}$, the problem in Eqs. (13.12a) and (13.12b) can be converted into a QP problem of the type given in Eqs. (13.26a)–(13.26c), i.e.,

$$\text{minimize } \tfrac{1}{2}\hat{\mathbf{x}}^T \hat{\mathbf{H}}\hat{\mathbf{x}} + \hat{\mathbf{x}}^T \hat{\mathbf{p}}$$
$$\text{subject to: } \hat{\mathbf{A}}\hat{\mathbf{x}} = \mathbf{b}$$
$$\hat{\mathbf{x}} \geq \mathbf{0}$$

where

$$\hat{\mathbf{H}} = \begin{bmatrix} \mathbf{H} & -\mathbf{H} & \mathbf{0} \\ -\mathbf{H} & \mathbf{H} & \mathbf{0} \\ \mathbf{0} & \mathbf{0} & \mathbf{0} \end{bmatrix}, \quad \hat{\mathbf{p}} = \begin{bmatrix} \mathbf{p} \\ -\mathbf{p} \\ \mathbf{0} \end{bmatrix}, \quad \hat{\mathbf{x}} = \begin{bmatrix} \mathbf{x}^+ \\ \mathbf{x}^- \\ \boldsymbol{\eta} \end{bmatrix}$$

$$\hat{\mathbf{A}} = [\mathbf{A} \ -\mathbf{A} \ \mathbf{I}_p]$$

and $n = 14$, $p = 6$. We note that $\hat{\mathbf{H}}$ is positive semidefinite if $\mathbf{H}$ is positive semidefinite. Since a strictly feasible initial $\mathbf{w}_0$ is difficult to obtain in this example, Algorithm 13.3 was used with $\mathbf{x}_0 = \text{ones}\{14, 1\}$, $\boldsymbol{\lambda}_0 = \text{ones}\{6, 1\}$, $\boldsymbol{\mu}_0 = \text{ones}\{14, 1\}$, where ones$\{m, 1\}$ represents a column vector of dimension m whose elements are all equal to one. Assigning $\varepsilon = 10^{-5}$ and $\rho = n + 20\sqrt{n}$, the algorithm took 11 iterations to converge to $\hat{\mathbf{x}}^*$ whose first 8 elements were then used to obtain

$$\mathbf{x}^* = \begin{bmatrix} 0.400002 \\ 0.799999 \\ 1.000001 \\ 2.000003 \end{bmatrix}$$

The shortest distance can be obtained as 1.341644.

Note that we do not need to introduce a small perturbation to matrix $\mathbf{H}$ to make it positive definite in this example as was the case in Example 13.2.   ∎

13.4.4   Linear Complementarity Problems

In this section, we explore the class of monotone linear complementarity (LC) problems and a variant of this class known as the class of mixed LC problems, and recast convex QP problems as mixed LC problems (see [17, Chap. 8]).

LC problems involve finding a vector pair $\{\mathbf{x}, \boldsymbol{\mu}\}$ in R^n that satisfies the relations

$$\mathbf{Kx} + \mathbf{q} = \boldsymbol{\mu} \tag{13.43a}$$

$$\mathbf{x} \geq \mathbf{0} \quad \text{for } \boldsymbol{\mu} \geq \mathbf{0} \tag{13.43b}$$

$$\mathbf{x}^T \boldsymbol{\mu} = 0 \tag{13.43c}$$

where $\mathbf{K} \in R^{n \times n}$ and $\mathbf{q} \in R^n$ are given, and $\mathbf{K}$ is positive semidefinite. Such a problem can be formulated as the minimization problem

$$\text{minimize } f(\hat{\mathbf{x}}) = \hat{\mathbf{x}}_1^T \hat{\mathbf{x}}_2 \tag{13.44a}$$

$$\text{subject to: } \mathbf{A}\hat{\mathbf{x}} = \mathbf{b} \tag{13.44b}$$

$$\hat{\mathbf{x}} \geq \mathbf{0} \tag{13.44c}$$

where

$$\hat{\mathbf{x}} = \begin{bmatrix} \hat{\mathbf{x}}_1 \\ \hat{\mathbf{x}}_2 \end{bmatrix} = \begin{bmatrix} \mathbf{x} \\ \boldsymbol{\mu} \end{bmatrix}, \quad \mathbf{A} = [\mathbf{K} \ -\mathbf{I}_n], \quad \text{and} \quad \mathbf{b} = -\mathbf{q}$$

The objective function $f(\hat{\mathbf{x}})$ in Eq. (13.44a) can be expressed as

$$f(\hat{\mathbf{x}}) = \tfrac{1}{2}\hat{\mathbf{x}}^T \begin{bmatrix} \mathbf{0} & \mathbf{I}_n \\ \mathbf{I}_n & \mathbf{0} \end{bmatrix} \hat{\mathbf{x}}$$

and hence the problem in Eqs. (13.44a)–(13.44c) is a QP problem with an *indefinite* Hessian.

A variant of the LC class of problems that is well connected to the class of QP problems is the so-called class of *mixed* LC problems, which entail finding a vector pair $\{\mathbf{x}, \ \boldsymbol{\mu}\}$ in R^n and vector $\boldsymbol{\lambda} \in R^p$ such that

$$\begin{bmatrix} \mathbf{K}_{11} & \mathbf{K}_{12} \\ \mathbf{K}_{21} & \mathbf{K}_{22} \end{bmatrix} \begin{bmatrix} \mathbf{x} \\ \boldsymbol{\lambda} \end{bmatrix} + \begin{bmatrix} \mathbf{q}_1 \\ \mathbf{q}_2 \end{bmatrix} = \begin{bmatrix} \boldsymbol{\mu} \\ \mathbf{0} \end{bmatrix} \tag{13.45a}$$

$$\mathbf{x} \geq \mathbf{0}, \ \boldsymbol{\mu} \geq \mathbf{0} \tag{13.45b}$$

$$\mathbf{x}^T \boldsymbol{\mu} = 0 \tag{13.45c}$$

where matrix $\mathbf{K} \in R^{(n+p) \times (n+p)}$ given by

$$\begin{bmatrix} \mathbf{K}_{11} & \mathbf{K}_{12} \\ \mathbf{K}_{21} & \mathbf{K}_{22} \end{bmatrix}$$

is not necessarily symmetric but is positive semidefinite such that

$$\mathbf{y}^T \mathbf{K} \mathbf{y} \geq 0 \qquad \text{for any } \mathbf{y} \in R^{n+p} \tag{13.46}$$

The class of LC problems described by Eqs. (13.43a)–(13.43c) can be viewed as a special case of the class of mixed LC problems described by Eqs. (13.45a)–(13.45c) where dimension p is 0. As will become clear shortly, the class of mixed LC problems is closely related to the class of standard-form LP problems in Eqs. (11.1a)–(11.1c) as well as the class of convex QP problems in Eqs. (13.26a)–(13.26c). In order to see the relation of Eqs. (13.45a)–(13.45c) to Eqs. (11.1a)–(11.1c), note that the conditions in Eqs. (13.45b) and (13.45c) imply that

$$x_i \mu_i = 0 \qquad \text{for } i = 1, \ 2, \ \ldots, \ n$$

which is the complementarity condition in Eq. (11.5d). Hence the KKT conditions in Eqs. (11.5a)–(11.5d) can be restated as

$$\begin{bmatrix} \mathbf{0} & \mathbf{A}^T \\ -\mathbf{A} & \mathbf{0} \end{bmatrix} \begin{bmatrix} \mathbf{x} \\ \boldsymbol{\lambda} \end{bmatrix} + \begin{bmatrix} \mathbf{c} \\ \mathbf{b} \end{bmatrix} = \begin{bmatrix} \boldsymbol{\mu} \\ \mathbf{0} \end{bmatrix}$$

$$\mathbf{x} \geq \mathbf{0}, \ \boldsymbol{\mu} \geq \mathbf{0}$$

$$\mathbf{x}^T \boldsymbol{\mu} = 0$$

Since matrix

$$\mathbf{K} = \begin{bmatrix} \mathbf{0} & \mathbf{A}^T \\ -\mathbf{A} & \mathbf{0} \end{bmatrix}$$

is positive semidefinite according to Eq. (13.46) (see Prob. 13.11(*a*)), it follows that standard-form LP problems can be formulated as mixed LC problems.

For the class of convex QP problems in Eqs. (13.26a)–(13.26c), the KKT conditions given in Eqs. (13.28a)–(13.28c) can be written as

$$\begin{bmatrix} \mathbf{H} & \mathbf{A}^T \\ -\mathbf{A} & \mathbf{0} \end{bmatrix} \begin{bmatrix} \mathbf{x} \\ \boldsymbol{\lambda} \end{bmatrix} + \begin{bmatrix} \mathbf{p} \\ \mathbf{b} \end{bmatrix} = \begin{bmatrix} \boldsymbol{\mu} \\ \mathbf{0} \end{bmatrix} \tag{13.47a}$$

$$\mathbf{x} \geq \mathbf{0}, \quad \boldsymbol{\mu} \geq \mathbf{0} \tag{13.47b}$$

$$\mathbf{x}^T \boldsymbol{\mu} = 0 \tag{13.47c}$$

where

$$K = \begin{bmatrix} H & A^T \\ -A & 0 \end{bmatrix}$$

is positive semidefinite if H is positive semidefinite (see Prob. 13.11(b)). From the above analysis, we see that the class of mixed LC problems is a fairly general class that includes standard-form LP, convex QP, and LC problems.

Let

$$w_k = \{x_k, \, \lambda_k, \, \mu_k\} \text{ be the } k\text{th iterate with } x_k > 0,$$

$$\mu_k > 0, \text{ and}$$

$$w_{k+1} = w_k + \alpha_k \delta_w$$

where the search direction $\delta_w = \{\delta_x, \, \delta_\lambda, \, \delta_\mu\}$ is chosen to satisfy the relations

$$\begin{bmatrix} K_{11} & K_{12} \\ K_{21} & K_{22} \end{bmatrix} \begin{bmatrix} x_k + \delta_x \\ \lambda_k + \delta_\lambda \end{bmatrix} + \begin{bmatrix} q_1 \\ q_2 \end{bmatrix} = \begin{bmatrix} \mu_k + \delta_\mu \\ 0 \end{bmatrix}$$

$$(x_k + \delta_x)^T (\mu_k + \delta_\mu) \approx x_k^T \mu_k + \delta_x^T \mu_k + x_k^T \delta_\mu = \tau_{k+1} e$$

These equations can be expressed as

$$\begin{bmatrix} -K_{11} & -K_{12} & I \\ K_{21} & K_{22} & 0 \\ M & 0 & X \end{bmatrix} \begin{bmatrix} \delta_x \\ \delta_\lambda \\ \delta_\mu \end{bmatrix} = \begin{bmatrix} r_1 \\ r_2 \\ \tau_{k+1} e - X\mu_k \end{bmatrix} \qquad (13.48a)$$

where $M = \text{diag}\{(\mu_k)_1, \, (\mu_k)_2, \, \ldots, \, (\mu_k)_n\}$, $X = \text{diag}\{(x_k)_1, \, (x_k)_2, \, \ldots, \, (x_k)_n\}$, and

$$r_1 = K_{11}x_k + K_{12}\lambda_k - \mu_k + q_1 \qquad (13.48b)$$

$$r_2 = -K_{21}x_k - K_{22}\lambda_k - q_2 \qquad (13.48c)$$

$$\tau_{k+1} = \frac{x_k^T \mu_k}{n + \rho}, \quad \text{with } \rho \geq \sqrt{n} \qquad (13.48d)$$

Note that Eq. (13.48a) combines three equations and it can be readily verified that with $K_{11} = K_{22} = 0$, $K_{21} = -K_{12}^T = -A$, $q_1 = c$, and $q_2 = b$, Eq. (13.48a) becomes identical to Eqs. (12.56a)–(12.56c) which determines the search direction for the nonfeasible-initialization primal-dual path-following algorithm in Sect. 12.5.2. Likewise, with $K_{11} = H$, $K_{21} = -K_{12}^T = -A$, $K_{22} = 0$, $q_1 = p$, and $q_2 = b$, Eqs. (13.48a)–(13.48d) become identical to Eqs. (13.41), (13.40a), (13.40b), and (13.42g) which determine the search direction for the nonfeasible-initialization primal-dual path-following algorithm for the convex QP in Sect. 13.4.3. Once δ_w is determined by solving Eqs. (13.48a)–(13.48d), α_k can be calculated using Eqs. (13.39a)–(13.39d). The above method can be implemented in terms of the following algorithm.

Algorithm 13.4   Nonfeasible-initialization interior-point algorithm for mixed LC problems

Step 1
Input an initial point $\mathbf{w}_0 = \{\mathbf{x}_0,\ \boldsymbol{\lambda}_0,\ \boldsymbol{\mu}_0\}$ with $\mathbf{x}_0 > \mathbf{0}$ and $\boldsymbol{\mu}_0 > \mathbf{0}$.
Set $k = 0$ and $\rho > \sqrt{n}$ and initialize the tolerance ε for $\mathbf{x}_k^T \boldsymbol{\mu}_k$.

Step 2
If $\mathbf{x}_k^T \boldsymbol{\mu}_k \le \varepsilon$, output solution $\mathbf{w}^* = \mathbf{w}_k$ and stop; otherwise, continue with Step 3.

Step 3
Compute τ_{k+1} using Eq. (13.33c) and determine $\boldsymbol{\delta}_w = (\boldsymbol{\delta}_x,\ \boldsymbol{\delta}_\lambda,\ \boldsymbol{\delta}_\mu)$ by solving the system of equations in Eq. (13.48a).

Step 4
Compute α_k using Eqs. (13.39a)–(13.39d) and set $\mathbf{w}_{k+1} = \mathbf{w}_k + \alpha_k \boldsymbol{\delta}_w$.
Set $k = k + 1$ and repeat from Step 2.

13.5   Primal and Dual SDP Problems

13.5.1   Notation and Definitions

Let S^n be the space of real symmetric $n \times n$ matrices. The standard inner product on S^n is defined as

$$\mathbf{A} \cdot \mathbf{B} = \text{trace } (\mathbf{AB}) = \sum_{i=1}^{n} \sum_{j=1}^{n} a_{ij} b_{ij}$$

where $\mathbf{A} = \{a_{ij}\}$ and $\mathbf{B} = \{b_{ij}\}$ are two members of S^n.

The primal SDP problem is defined as

$$\text{minimize } \mathbf{C} \cdot \mathbf{X} \tag{13.49a}$$

$$\text{subject to: } \mathbf{A}_i \cdot \mathbf{X} = b_i \quad \text{for } i = 1,\ 2,\ \ldots,\ p \tag{13.49b}$$

$$\mathbf{X} \succeq \mathbf{0} \tag{13.49c}$$

where $\mathbf{C}$, $\mathbf{X}$, and $\mathbf{A}_i$ for $1 \le i \le p$ are members of S^n and the notation in Eq. (13.49c) denotes that $\mathbf{X}$ is positive semidefinite (see Sect. 10.2). It can be readily verified that the problem formulated in Eqs. (13.49a)–(13.49c) is a CP problem (see Prob. 13.13). An important feature of the problem is that the variable involved is a *matrix* rather than a vector. Despite this distinction, SDP is closely related to several important classes of optimization problems. For example, if matrices $\mathbf{C}$ and $\mathbf{A}_i$ for $1 \le i \le p$ are all diagonal matrices, i.e.,

$$\mathbf{C} = \text{diag } \{\mathbf{c}\}, \quad \mathbf{A}_i = \text{diag } \{\mathbf{a}_i\}$$

with $\mathbf{c} \in R^{n \times 1}$ and $\mathbf{a}_i \in R^{n \times 1}$ for $1 \le i \le p$, then the problem in Eqs. (13.49a)–(13.49c) is reduced to the standard-form LP problem

$$\text{minimize } \mathbf{c}^T \mathbf{x} \tag{13.50a}$$

$$\text{subject to:} \quad \mathbf{Ax} = \mathbf{b} \tag{13.50b}$$

$$\mathbf{x} \geq \mathbf{0} \tag{13.50c}$$

where $\mathbf{A} \in R^{p \times n}$ is a matrix with $\mathbf{a}_i^T$ as its ith row, $\mathbf{b} = [b_1\, b_2\, \cdots\, b_p]^T$, and vector $\mathbf{x} \in R^{n \times 1}$ is the diagonal of $\mathbf{X}$. The similarities between Eqs. (13.49a) and (13.50a) and between Eqs. (13.49b) and (13.50b) are quite evident. To see the similarity between Eq. (13.49c) and Eq. (13.50c), we need the concept of *convex cone*.

Definition 13.1 A convex cone $\mathcal{K}$ is a convex set such that $\mathbf{x} \in \mathcal{K}$ implies that $\alpha \mathbf{x} \in \mathcal{K}$ for any scalar $\alpha \geq 0$.

■

It can be readily verified that both sets $\{\mathbf{X} : \mathbf{X} \in R^{n \times n}, \mathbf{X} \succeq \mathbf{0}\}$ and $\{\mathbf{x} : \mathbf{x} \in R^{n \times 1}, \mathbf{x} \geq \mathbf{0}\}$ are convex cones (see Prob. 13.14). We now recall that the dual of the LP problem in Eqs. (13.50a)–(13.50c) is given by

$$\text{maximize} \quad \mathbf{b}^T \mathbf{y} \tag{13.51a}$$

$$\text{subject to:} \quad \mathbf{A}^T \mathbf{y} + \mathbf{s} = \mathbf{c} \tag{13.51b}$$

$$\mathbf{s} \geq \mathbf{0} \tag{13.51c}$$

(see Chap. 12) and, therefore, the *dual* SDP problem with respect to the primal SDP problem in Eqs. (13.49a)–(13.49c) can be obtained as

$$\text{maximize} \quad -\mathbf{b}^T \mathbf{y} \tag{13.52a}$$

$$\text{subject to:} \quad -\sum_{i=1}^{p} y_i \mathbf{A}_i + \mathbf{S} = \mathbf{C} \tag{13.52b}$$

$$\mathbf{S} \succeq \mathbf{0} \tag{13.52c}$$

where $\mathbf{S}$ is a slack matrix variable that can be regarded as the matrix counterpart of the slack vector $\mathbf{s}$ in Eqs. (13.51a)–(13.51c). To demonstrate that the maximization problem in Eqs. (13.52a)–(13.52c) is a dual of the problem in Eqs. (13.49a)–(13.49c), we assume that there exist $\mathbf{X} \in S^n$, $\mathbf{y} \in R^p$, and $\mathbf{S} \in S^n$ with $\mathbf{X} \succeq \mathbf{0}$ and $\mathbf{S} \succeq \mathbf{0}$ such that $\mathbf{X}$ is feasible for the primal and $\{\mathbf{y}, \mathbf{S}\}$ is feasible for the dual, and evaluate

$$\mathbf{C} \cdot \mathbf{X} + \mathbf{b}^T \mathbf{y} = \left(-\sum_{i=1}^{p} y_i \mathbf{A}_i + \mathbf{S}\right) \cdot \mathbf{X} + \mathbf{b}^T \mathbf{y}$$

$$= \mathbf{S} \cdot \mathbf{X} \geq 0 \tag{13.53}$$

where the first and second equalities follow from Eq. (13.52b) and the inequality is a consequence of the fact that both $\mathbf{S}$ and $\mathbf{X}$ are positive semidefinite (see Prob. 13.15). Later in Sect. 13.6, it will be shown that if $\mathbf{X}^*$ is a solution of the primal problem and $\mathbf{y}^*$ is a solution of the dual problem, then

$$\mathbf{S}^* \cdot \mathbf{X}^* = 0 \tag{13.54}$$

where $\mathbf{S}^*$ is determined from Eq. (13.52b), i.e.,

$$\mathbf{S}^* = \mathbf{C} + \sum_{i=1}^{p} y_i^* \mathbf{A}_i$$

From Eqs. (13.53) and (13.54), it follows that

$$\mathbf{C} \cdot \mathbf{X}^* + \mathbf{b}^T \mathbf{y}^* = 0 \tag{13.55}$$

Equations 13.53 13.55 suggest that a duality gap similar to that in Eq. (12.6) can be defined for the problems in Eqs. (13.49a)–(13.49c) and Eqs. (13.52a)–(13.52c) as

$$\delta(\mathbf{X}, \mathbf{y}) = \mathbf{C} \cdot \mathbf{X} + \mathbf{b}^T \mathbf{y} \tag{13.56}$$

for $\mathbf{X} \in \mathcal{F}_p$ and $\{\mathbf{y}, \mathbf{S}\} \in \mathcal{F}_d$ where $\mathcal{F}_p$ and $\mathcal{F}_d$ are the feasible sets for the primal and dual problems defined by

$$\mathcal{F}_p = \{\mathbf{X} : \mathbf{X} \succeq \mathbf{0}, \, \mathbf{A}_i \cdot \mathbf{X} = b_i \quad \text{for } 1 \le i \le p\}$$

$$\mathcal{F}_d = \left\{ \{\mathbf{y}, \mathbf{S}\} : -\sum_{i=1}^{p} y_i \mathbf{A}_i + \mathbf{S} = \mathbf{C}, \, \mathbf{S} \succeq \mathbf{0} \right\}$$

respectively. From Eqs. (13.53) and (13.55), it follows that for any $\mathbf{X} \in \mathcal{F}_p$ and $\{\mathbf{y}, \mathbf{S}\} \in \mathcal{F}_d$ the duality gap $\delta(\mathbf{X}, \mathbf{y})$ is nonnegative and it is reduced to zero at the solutions $\mathbf{X}^*$ and $\mathbf{S}^*$ of the primal and dual problems, respectively.

If we combine the constraints in Eqs. (13.52b) and (13.52c) into one inequality constraint, the dual SDP problem becomes

$$\text{maximize} \ -\mathbf{b}^T \mathbf{y}$$
$$\text{subject to:} \ -\mathbf{C} - \sum_{i=1}^{p} y_i \mathbf{A}_i \preceq \mathbf{0}$$

This is obviously equivalent to the following minimization problem

$$\text{minimize} \ \mathbf{c}^T \mathbf{x} \tag{13.57a}$$
$$\text{subject to:} \ \mathbf{F}(\mathbf{x}) \preceq \mathbf{0} \tag{13.57b}$$

where $\mathbf{c} \in R^{p \times 1}$, $\mathbf{x} \in R^{p \times 1}$, and

$$\mathbf{F}(\mathbf{x}) = \mathbf{F}_0 + \sum_{i=1}^{p} x_i \mathbf{F}_i$$

with $\mathbf{F}_i \in S^n$ for $0 \le i \le p$. Notice that the semidefinite constraint on matrix $\mathbf{F}(\mathbf{x})$ in Eq. (13.57b) is dependent on vector $\mathbf{x}$ in an *affine* manner. In the literature, the type of problems described by Eqs. (13.57a) and (13.57b) are often referred to as *convex optimization* problems with *linear matrix inequality* (LMI) constraints, and have found many applications in science and engineering [3,18]. Since the minimization problem in Eqs. (13.57a) and (13.57b) is equivalent to a dual SDP problem, the problem itself is often referred to as an SDP problem.

13.5.2 Examples

(i) *LP Problems* As we have seen in Sect. 13.5.1, standard-form LP problems can be viewed as a special class of SDP problems where the matrices $\mathbf{C}$ and $\mathbf{A}_i$ for $1 \le i \le p$ in Eqs. (13.49a)–(13.49c) are all diagonal.

The alternative-form LP problem

$$\text{minimize } \mathbf{c}^T \mathbf{x}$$

$$\text{subject to: } \mathbf{Ax} \le \mathbf{b}, \ \mathbf{A} \in R^{p \times n}$$

which was studied extensively in Chap. 11 (see Eqs. (11.2a) and (11.2b)) can be viewed as a linear minimization problem with LMI constraints. This can be demonstrated by expressing matrices $\mathbf{F}_i$ for $0 \le i \le n$ in Eq. (13.57b) as

$$\mathbf{F}_0 = -\text{ diag } \{\mathbf{b}\}, \quad \mathbf{F}_i = \text{ diag } \{\mathbf{a}_i\} \quad \text{for } i = 1, \ 2, \ \dots, \ n \qquad (13.58)$$

where $\mathbf{a}_i$ denotes the ith column of $\mathbf{A}$.

(ii) *Convex QP Problems* The general convex QP problem

$$\text{minimize } \mathbf{x}^T \mathbf{Hx} + \mathbf{p}^T \mathbf{x} \quad \text{with } \mathbf{H} \succeq \mathbf{0} \qquad (13.59a)$$

$$\text{subject to: } \mathbf{Ax} \le \mathbf{b} \qquad (13.59b)$$

which was studied in Sect. 13.4 can be formulated as

$$\text{minimize } \delta \qquad (13.60a)$$

$$\text{subject to: } \mathbf{x}^T \mathbf{Hx} + \mathbf{p}^T \mathbf{x} \le \delta \qquad (13.60b)$$

$$\mathbf{Ax} \le \mathbf{b} \qquad (13.60c)$$

where δ is an auxiliary scalar variable.

Since $\mathbf{H}$ is positive semidefinite, we can find a matrix $\hat{\mathbf{H}}$ such that $\mathbf{H} = \hat{\mathbf{H}}^T \hat{\mathbf{H}}$ (see proof of Theorem 7.2); hence the constraint in Eq. (13.60b) can be expressed as

$$\delta - \mathbf{p}^T \mathbf{x} - (\hat{\mathbf{H}} \mathbf{x})^T (\hat{\mathbf{H}} \mathbf{x}) \ge 0 \qquad (13.61)$$

It can be shown (see Prob. 13.16(a)) that the inequality in Eq. (13.61) holds if and only if

$$\mathbf{G}(\delta, \ \mathbf{x}) = \begin{bmatrix} -\mathbf{I}_n & -\hat{\mathbf{H}}\mathbf{x} \\ -(\hat{\mathbf{H}}\mathbf{x})^T & -\delta + \mathbf{p}^T \mathbf{x} \end{bmatrix} \preceq \mathbf{0} \qquad (13.62)$$

where $\mathbf{I}_n$ is the $n \times n$ identity matrix. Note that matrix $\mathbf{G}(\delta, \ \mathbf{x})$ is *affine* with respect to variables $\mathbf{x}$ and δ. In addition, the linear constraints in Eq. (13.60c) can be expressed as

$$\mathbf{F}(\mathbf{x}) = \mathbf{F}_0 + \sum_{i=1}^{n} x_i \mathbf{F}_i \preceq \mathbf{0} \qquad (13.63)$$

where the $\mathbf{F}_i$ for $0 \le i \le n$ are given by Eq. (13.58). Therefore, by defining an augmented vector

$$\hat{\mathbf{x}} = \begin{bmatrix} \delta \\ \mathbf{x} \end{bmatrix} \qquad (13.64)$$

the problem in Eqs. (13.59a) and (13.59b) can be reformulated as the SDP problem

$$\text{minimize } \hat{\mathbf{c}}^T \hat{\mathbf{x}}$$

$$\text{subject to: } \mathbf{E}(\hat{\mathbf{x}}) \preceq \mathbf{0}$$

where $\hat{\mathbf{c}} \in R^{n+1}$ with

$$\hat{\mathbf{c}} = [1\ 0\ \cdots\ 0]^T \qquad (13.65)$$

and

$$\mathbf{E}(\hat{\mathbf{x}}) = \text{diag}\ \{\mathbf{G}(\delta,\ \mathbf{x}),\ \mathbf{F}(\mathbf{x})\}$$

(iii) *Convex QP Problems with Quadratic Constraints* The class of CP problems represented by

$$\text{minimize}\ \mathbf{x}^T \mathbf{H}\mathbf{x} + \mathbf{p}^T \mathbf{x} \qquad (13.66a)$$

$$\text{subject to:}\ \ \mathbf{x}^T \mathbf{Q}_i \mathbf{x} + \mathbf{q}_i^T \mathbf{x} + r_i \le 0 \quad \text{for}\ i = 1,\ 2,\ \ldots,\ p \qquad (13.66b)$$

where $\mathbf{H} \succeq \mathbf{0}$ and $\mathbf{Q}_i \succeq \mathbf{0}$ for $1 \le i \le p$ covers the class of conventional convex QP problems represented by Eqs. (13.59a) and (13.59b) as a subclass if $\mathbf{Q}_i = \mathbf{0}$ for all i. Again, by introducing an auxiliary scalar variable δ, the problem in Eqs. (13.66a) and (13.66b) can be converted to

$$\text{minimize}\ \delta \qquad (13.67a)$$

$$\text{subject to:}\ \ \mathbf{x}^T \mathbf{H}\mathbf{x} + \mathbf{p}^T \mathbf{x} \le \delta \qquad (13.67b)$$

$$\mathbf{x}^T \mathbf{Q}_i \mathbf{x} + \mathbf{q}_i^T \mathbf{x} + r_i \le 0 \quad \text{for}\ 1 \le i \le p \qquad (13.67c)$$

As in the convex QP case, the constraint in Eq. (13.67b) is equivalent to the constraint in Eq. (13.62) and the constraints in Eq. (13.67c) are equivalent to

$$\mathbf{F}_i(\mathbf{x}) = \begin{bmatrix} -\mathbf{I}_n & -\hat{\mathbf{Q}}_i \mathbf{x} \\ -(\hat{\mathbf{Q}}_i \mathbf{x})^T & \mathbf{q}_i^T \mathbf{x} + r_i \end{bmatrix} \preceq \mathbf{0} \quad \text{for}\ 1 \le i \le p \qquad (13.68)$$

where $\hat{\mathbf{Q}}_i$ is related to $\mathbf{Q}_i$ by the equation $\mathbf{Q}_i = \hat{\mathbf{Q}}_i^T \hat{\mathbf{Q}}_i$. Consequently, the quadratically constrained convex QP problem in Eqs. (13.66a) and (13.66b) can be formulated as the SDP problem

$$\text{minimize}\ \hat{\mathbf{c}}^T \hat{\mathbf{x}} \qquad (13.69a)$$

$$\text{subject to:}\ \ \mathbf{E}(\hat{\mathbf{x}}) \preceq \mathbf{0} \qquad (13.69b)$$

where $\hat{\mathbf{x}}$ and $\hat{\mathbf{c}}$ are given by Eqs. (13.64) and (13.65), respectively, and

$$\mathbf{E}(\hat{\mathbf{x}}) = \text{diag}\ \{\mathbf{G}(\delta,\ \mathbf{x}),\ \mathbf{F}_1(\mathbf{x}),\ \mathbf{F}_2(\mathbf{x}),\ \ldots,\ \mathbf{F}_p(\mathbf{x})\}$$

where $\mathbf{G}(\delta,\ \mathbf{x})$ and $\mathbf{F}_i(\mathbf{x})$ are given by Eqs. (13.62) and (13.68), respectively.

There are many other types of CP problems that can be recast as SDP problems. One of them is the problem of minimizing the maximum eigenvalue of an affine matrix that can arise in structure optimization, control theory, and other areas [1, 19]. This problem can be formulated as an SDP problem of the form in Eqs. (13.52a)–(13.52c) (see Prob. 13.17). The reader is referred to [3, 18] for more examples.

13.6 Basic Properties of SDP Problems

13.6.1 Basic Assumptions

The feasible sets $\mathcal{F}_p$ and $\mathcal{F}_a$ for the primal and dual problems were defined in Sect. 13.5.1. A matrix $\mathbf{X}$ is said to be *strictly feasible* for the primal problem in Eqs. (13.49a)–(13.49c) if it satisfies Eq. (13.49b) and $\mathbf{X} \succ \mathbf{0}$. Such a matrix $\mathbf{X}$ can be viewed as an *interior point* of $\mathcal{F}_p$. If we let

$$\mathcal{F}_p^o = \{\mathbf{X} : \mathbf{X} \succ \mathbf{0}, \ \mathbf{A}_i \cdot \mathbf{X} = b_i \ \text{ for } \ 1 \le i \le p\}$$

then $\mathcal{F}_p^o$ is the set of all interior points of $\mathcal{F}_p$ and $\mathbf{X}$ is strictly feasible for the primal problem if $\mathbf{X} \in \mathcal{F}_p^o$. Similarly, we can define the set of all interior points of $\mathcal{F}_d$ as

$$\mathcal{F}_d^o = \left\{ \{\mathbf{y}, \ \mathbf{S}\} : -\sum_{i=1}^{p} y_i \mathbf{A}_i + \mathbf{S} = \mathbf{C}, \ \mathbf{S} \succ \mathbf{0} \right\}$$

and a pair $\{\mathbf{y}, \ \mathbf{S}\}$ is said to be *strictly feasible* for the dual problem in Eqs. (13.52a)–(13.52c) if $\{\mathbf{y}, \ \mathbf{S}\} \in \mathcal{F}_d^o$.

Unless otherwise stated, the following assumptions will be made in the rest of the chapter:

1. There exists a strictly feasible point $\mathbf{X}$ for the primal problem in Eqs. (13.49a)–(13.49c) and a strictly feasible pair $\{\mathbf{y}, \ \mathbf{S}\}$ for the dual problem in Eqs. (13.52a)–(13.52c). In other words, both $\mathcal{F}_p^o$ and $\mathcal{F}_d^o$ are nonempty.
2. Matrices $\mathbf{A}_i$ for $i = 1, \ 2, \ \ldots, \ p$ in Eq. (13.49b) are linearly independent, i.e., they span a p-dimensional linear space in $\mathcal{S}^n$.

The first assumption assures that the optimization problem at hand can be tackled by using an *interior-point* approach. The second assumption, on the other hand, can be viewed as a matrix counterpart of the assumption made for the LP problem in Eqs. (13.50a)–(13.50c) that the row vectors in matrix $\mathbf{A}$ in Eq. (13.50b) are linearly independent.

13.6.2 Karush-Kuhn-Tucker Conditions

The Karush-Kuhn-Tucker (KKT) conditions for the SDP problem in Eqs. (13.49a)–(13.49c) can be stated as follows: Matrix $\mathbf{X}^*$ is a minimizer of the problem in Eqs. (13.49a)–(13.49c) if and only if there exist a matrix $\mathbf{S}^* \in \mathcal{S}^n$ and a vector $\mathbf{y}^* \in R^p$ such that

$$-\sum_{i=1}^{p} y_i^* \mathbf{A}_i + \mathbf{S}^* = \mathbf{C} \tag{13.70a}$$

$$\mathbf{A}_i \cdot \mathbf{X}^* = b_i \quad \text{ for } 1 \le i \le p \tag{13.70b}$$

$$\mathbf{S}^* \mathbf{X}^* = \mathbf{0} \tag{13.70c}$$

$$\mathbf{X}^* \succeq \mathbf{0}, \ \mathbf{S}^* \succeq \mathbf{0} \tag{13.70d}$$

As noted in Sect. 13.5.1, if $\mathbf{A}_i = \text{diag}\{\mathbf{a}_i\}$ and $\mathbf{C} = \text{diag}\{\mathbf{c}\}$ with $\mathbf{a}_i \in R^n$ and $\mathbf{c} \in R^n$ for $1 \le i \le p$, the problem in Eqs. (13.49a)–(13.49c) becomes a standard-form LP problem. In such a case, matrix $\mathbf{X}^*$ in Eqs. (13.70a)–(13.70d) is also diagonal and the conditions in Eqs. (13.70a)–(13.70d) become identical with those in Eqs. (12.3a)–(12.3d), which are the KKT conditions for the LP problem in Eqs. (13.50a)–(13.50c).

While Eqs. (13.70a) and (13.70b) are linear, the complementarity constraint in Eq. (13.70c) is a nonlinear matrix equation. It can be shown that under the assumptions made in Sect. 13.6.2, the solution of Eqs. (13.70a)–(13.70d) exists (see Theorem 3.1 of [3]). Furthermore, if we denote a solution of Eqs. (13.70a)–(13.70d) as $\{\mathbf{X}^*, \ \mathbf{y}^*, \ \mathbf{S}^*\}$, then it can be readily verified that $\{\mathbf{y}^*, \ \mathbf{S}^*\}$ is a max-imizer of the dual problem in Eqs. (13.52a)–(13.52c). For these reasons, a set $\{\mathbf{X}^*, \ \mathbf{y}^*, \ \mathbf{S}^*\}$ satisfying Eqs. (13.70a)–(13.70d) is called a *primal-dual solution*. It follows that $\{\mathbf{X}^*, \ \mathbf{y}^*, \ \mathbf{S}^*\}$ is a primal-dual solution if and only if $\mathbf{X}^*$ solves the primal problem in Eqs. (13.49a)–(13.49c) and $\{\mathbf{y}^*, \ \mathbf{S}^*\}$ solves the dual problem in Eqs. (13.52a)–(13.52c).

13.6.3   Central Path

As we have seen in Chap. 12 and this chapter, the concept of central path plays an important role in the development of interior-point algorithms for LP and QP prob-lems. For the SDP problems in Eqs. (13.49a)–(13.49c) and Eqs. (13.52a)–(13.52c), the central path consists of a set $\{\mathbf{X}(\tau), \ \mathbf{y}(\tau), \ \mathbf{S}(\tau)\}$ such that for each $\tau > 0$ the equations

$$-\sum_{i=1}^{p} y_i(\tau)\mathbf{A}_i + \mathbf{S}(\tau) = \mathbf{C} \tag{13.71a}$$

$$\mathbf{A}_i \cdot \mathbf{X}(\tau) = b_i \quad \text{for } 1 \le i \le p \tag{13.71b}$$

$$\mathbf{X}(\tau)\mathbf{S}(\tau) = \tau\mathbf{I} \tag{13.71c}$$

$$\mathbf{S}(\tau) \succ \mathbf{0}, \ \mathbf{X}(\tau) \succ \mathbf{0} \tag{13.71d}$$

are satisfied.

Using Eq. (13.56) and Eqs. (13.71a)–(13.71d), the duality gap on the central path can be evaluated as

$$\delta[\mathbf{X}(\tau), \ \mathbf{y}(\tau)] = \mathbf{C} \cdot \mathbf{X}(\tau) + \mathbf{b}^T \mathbf{y}(\tau)$$
$$= \left[-\sum_{i=1}^{p} y_i(\tau)\mathbf{A}_i + \mathbf{S}(\tau)\right] \cdot \mathbf{X}(\tau) + \mathbf{b}^T \mathbf{y}(\tau)$$
$$= \mathbf{S}(\tau) \cdot \mathbf{X}(\tau) = \text{trace } [\mathbf{S}(\tau)\mathbf{X}(\tau)]$$
$$= \text{trace } (\tau\mathbf{I}) = n\tau$$

which implies that

$$\lim_{\tau \to 0} \delta[\mathbf{X}(\tau), \ \mathbf{y}(\tau)] = 0$$

Therefore, the limiting set $\{\mathbf{X}^*, \mathbf{y}^*, \mathbf{S}^*\}$ obtained when $\mathbf{X}(\tau) \rightarrow \mathbf{X}^*$, $\mathbf{y}(\tau) \rightarrow \mathbf{y}^*$, and $\mathbf{S}(\tau) \rightarrow \mathbf{S}^*$ as $\tau \rightarrow 0$ is a primal-dual solution. This claim can also be confirmed by examining Eqs. (13.71a)–(13.71d) which, as τ approaches zero, become the KKT conditions in Eqs. (13.70a)–(13.70d). In other words, as $\tau \rightarrow 0$, the central path approaches a primal-dual solution. In the sections that follow, several algorithms are developed for generating iterates that converge to a primal-dual solution by following the central path of the problem. Since $\mathbf{X}(\tau)$ and $\mathbf{S}(\tau)$ are positive semidefinite and satisfy Eqs. (13.70a) and (13.70b), respectively, $\mathbf{X}(\tau) \in \mathcal{F}_p$ and $\{\mathbf{y}(\tau), \mathbf{S}(\tau)\} \in \mathcal{F}_d$. Furthermore, the relaxed complementarity condition in Eq. (13.70c) implies that for each $\tau > 0$ both $\mathbf{X}(\tau)$ and $\mathbf{S}(\tau)$ are nonsingular; hence $\mathbf{X}(\tau) \succ \mathbf{0}$ and $\mathbf{S}(\tau) \succ \mathbf{0}$ which imply that $\mathbf{X}(\tau) \in \mathcal{F}_p^\circ$ and $\{\mathbf{y}(\tau), \mathbf{S}(\tau)\} \in \mathcal{F}_d^\circ$. In other words, for each $\tau > 0$, $\mathbf{X}(\tau)$ and $\{\mathbf{y}(\tau), \mathbf{S}(\tau)\}$ are in the *interior* of the feasible regions for the problems in Eqs. (13.49a)–(13.49c) and Eqs. (13.52a)–(13.52c), respectively. Therefore, a path-following algorithm that generates iterates that follow the central path is intrinsically an interior-point algorithm.

13.6.4 Centering Condition

On comparing Eqs. (13.70a)–(13.70d) and Eqs. (13.71a)–(13.71d), we see that the only difference between the two systems of equations is that the complementarity condition in Eq. (13.70c) is relaxed in Eq. (13.71c). This equation is often referred to as the *centering condition* since the central path is parameterized by introducing variable τ in Eq. (13.71c). Obviously, if $\mathbf{X}(\tau) = \text{diag} \{x_1(\tau), x_2(\tau), \ldots, x_n(\tau)\}$ and $\mathbf{S}(\tau) = \text{diag} \{s_1(\tau), s_2(\tau), \ldots, s_n(\tau)\}$ as in LP problems, the centering condition is reduced to n scalar equations, i.e.,

$$x_i(\tau)s_i(\tau) = \tau \quad \text{for } 1 \leq i \leq n \tag{13.72}$$

In general, the centering condition in Eq. (13.71c) involves n^2 nonlinear equations and, consequently, it is much 'more complicated than the condition in Eq. (13.72). The similarity between the general centering condition and the condition in Eq. (13.72) [7] is illustrated by the analysis in the next paragraph.

Since $\mathbf{X}(\tau)$ and $\mathbf{S}(\tau)$ are positive definite, their eigenvalues are strictly positive. Let $\delta_1(\tau) \geq \delta_2(\tau) \geq \cdots \geq \delta_n(\tau) > 0$ and $0 < \gamma_1(\tau) \leq \gamma_2(\tau) \leq \cdots \leq \gamma_n(\tau)$ be the eigenvalues of $\mathbf{X}(\tau)$ and $\mathbf{S}(\tau)$, respectively. There exists an orthogonal matrix $\mathbf{Q}(\tau)$ such that

$$\mathbf{X}(\tau) = \mathbf{Q}(\tau) \, \text{diag} \{\delta_1(\tau), \delta_2(\tau), \ldots, \delta_n(\tau)\}\mathbf{Q}^T(\tau)$$

From Eq. (13.71c), it follows that

$$\mathbf{S}(\tau) = \tau \mathbf{X}^{-1}(\tau)$$
$$= \mathbf{Q}(\tau) \, \text{diag} \left\{\frac{\tau}{\delta_1(\tau)}, \frac{\tau}{\delta_2(\tau)}, \ldots, \frac{\tau}{\delta_n(\tau)}\right\} \mathbf{Q}^T(\tau)$$
$$= \mathbf{Q}(\tau) \, \text{diag} \{\gamma_1(\tau), \gamma_2(\tau), \ldots, \gamma_n(\tau)\}\mathbf{Q}^T(\tau)$$

which leads to

$$\delta_i(\tau)\gamma_i(\tau) = \tau \quad \text{for } 1 \le i \le n \tag{13.73}$$

As $\tau \to 0$, we have $\delta_i(\tau) \to \delta_i^*$ and $\gamma_i(\tau) \to \gamma_i^*$ where $\delta_1^* \ge \delta_2^* \ge \cdots \ge \delta_n^* > 0$ and $0 \le \gamma_1^* \le \gamma_2^* \le \cdots \le \gamma_n^*$ are the eigenvalues of $\mathbf{X}^*$ and $\mathbf{S}^*$, respectively, and Eq. (13.73) becomes

$$\delta_i^* \gamma_i^* = 0 \quad \text{for } 1 \le i \le n \tag{13.74}$$

We note that the relations between the eigenvalues of $\mathbf{X}(\tau)$ and $\mathbf{S}(\tau)$ as specified by Eq. (13.73) resemble the scalar centering conditions in Eq. (13.72). In addition, there is an interesting similarity between Eq. (13.74) and the complementarity conditions in LP problems (see Eq. (12.3c)).

13.7 Primal-Dual Path-Following Method

13.7.1 Reformulation of Centering Condition

A primal-dual path-following algorithm for SDP usually generates iterates by obtaining approximate solutions of Eqs. (13.71a)–(13.71d) for a sequence of decreasing $\tau_k > 0$ for $k = 0, 1, \ldots$. If we let

$$\mathbf{G}(\mathbf{X}, \ \mathbf{y}, \ \mathbf{S}) = \begin{bmatrix} -\sum_{i=1}^{p} y_i \mathbf{A}_i + \mathbf{S} - \mathbf{C} \\ \mathbf{A}_1 \cdot \mathbf{X} - b_1 \\ \vdots \\ \mathbf{A}_p \cdot \mathbf{X} - b_p \\ \mathbf{X}\mathbf{S} - \tau \mathbf{I} \end{bmatrix} \tag{13.75}$$

then Eqs. (13.71a)–(13.71c) can be expressed as

$$\mathbf{G}(\mathbf{X}, \ \mathbf{y}, \ \mathbf{S}) = \mathbf{0} \tag{13.76}$$

We note that the domain of function $\mathbf{G}$ is in $\mathcal{S}^n \times R^p \times \mathcal{S}^n$ while the range of $\mathbf{G}$ is in $\mathcal{S}^n \times R^p \times R^{n \times n}$ simply because matrix $\mathbf{X}\mathbf{S} - \tau \mathbf{I}$ is not symmetric in general although both $\mathbf{X}$ and $\mathbf{S}$ are symmetric. This domain inconsistency would cause difficulties if, for example, the Newton method were to be applied to Eq. (13.76) to obtain an approximate solution. Several approaches that deal with this nonsymmetrical problem are available, see, for example, [4–7]. In [7], Eq. (13.71c) is rewritten in symmetric form as

$$\mathbf{X}\mathbf{S} + \mathbf{S}\mathbf{X} = 2\tau \mathbf{I} \tag{13.77}$$

Accordingly, function $\mathbf{G}$ in Eq. (13.75) is modified as

$$\mathbf{G}(\mathbf{X},\ \mathbf{y},\ \mathbf{S}) = \begin{bmatrix} -\sum_{i=1}^{p} y_i \mathbf{A}_i + \mathbf{S} - \mathbf{C} \\ \mathbf{A}_1 \cdot \mathbf{X} - b_1 \\ \vdots \\ \mathbf{A}_p \cdot \mathbf{X} - b_p \\ \mathbf{XS} + \mathbf{SX} - 2\tau\mathbf{I} \end{bmatrix} \tag{13.78}$$

and its range is now in $\mathcal{S}^n \times R^p \times \mathcal{S}^n$. It can be shown that if $\mathbf{X} \succeq \mathbf{0}$ or $\mathbf{S} \succeq 0$, then Eqs. (13.71c) and (13.77) are equivalent (see Prob. 13.18).

In the Newton method, we start with a given set $\{\mathbf{X},\ \mathbf{y},\ \mathbf{S}\}$ and find increments $\Delta\mathbf{X}$, $\Delta\mathbf{y}$, and $\Delta\mathbf{S}$ with $\Delta\mathbf{X}$ and $\Delta\mathbf{S}$ symmetric such that set $\{\Delta\mathbf{X},\ \Delta\mathbf{y},\ \Delta\mathbf{S}\}$ satisfies the linearized equations

$$-\sum_{i=1}^{p} \Delta y_i \mathbf{A}_i + \Delta\mathbf{S} = \mathbf{C} - \mathbf{S} + \sum_{i=1}^{p} y_i \mathbf{A}_i \tag{13.79a}$$

$$\mathbf{A}_i \cdot \Delta\mathbf{X} = b_i - \mathbf{A}_i \cdot \mathbf{X} \quad \text{for } 1 \le i \le p \tag{13.79b}$$

$$[5pt]\mathbf{X}\Delta\mathbf{S} + \Delta\mathbf{S}\mathbf{X} + \Delta\mathbf{X}\mathbf{S} + \mathbf{S}\Delta\mathbf{X} = 2\tau\mathbf{I} - \mathbf{XS} - \mathbf{SX} \tag{13.79c}$$

Eqs. (13.79a)–(13.79c) contain matrix equations with matrix variables $\Delta\mathbf{X}$ and $\Delta\mathbf{S}$. A mathematical operation known as *symmetric Kronecker product* [7] (see also Sec. A.14) turns out to be effective in dealing with this type of linear equations.

13.7.2 Symmetric Kronecker Product

Given matrices $\mathbf{K}$, $\mathbf{M}$, and $\mathbf{N}$ in $R^{n \times n}$, the general asymmetric Kronecker product $\mathbf{M} \otimes \mathbf{N}$ with $\mathbf{M} = \{m_{ij}\}$ is defined as

$$\mathbf{M} \otimes \mathbf{N} = \begin{bmatrix} m_{11}\mathbf{N} & \cdots & m_{1n}\mathbf{N} \\ \vdots & & \vdots \\ m_{n1}\mathbf{N} & \cdots & m_{nn}\mathbf{N} \end{bmatrix}$$

(see Sec. A.14). To deal with matrix variables, it is sometimes desirable to represent a matrix $\mathbf{K}$ as a vector, denoted as nvec($\mathbf{K}$), which stacks the columns of $\mathbf{K}$. It can be readily verified that

$$(\mathbf{M} \otimes \mathbf{N})\text{nvec}(\mathbf{K}) = \text{nvec}(\mathbf{NKM}^T) \tag{13.80}$$

The usefulness of Eq. (13.80) is that if a matrix equation involves terms like $\mathbf{NKM}^T$, where $\mathbf{K}$ is a matrix variable, then Eq. (13.80) can be used to convert $\mathbf{NKM}^T$ into a vector variable multiplied by a known matrix.

If a matrix equation contains a symmetric term given by $(\mathbf{NKM}^T + \mathbf{MKN}^T)/2$ where $\mathbf{K} \in \mathcal{S}^n$ is a matrix variable, then the term can be readily handled using the *symmetric Kronecker product* of $\mathbf{M}$ and $\mathbf{N}$, denoted as $\mathbf{M} \odot \mathbf{N}$, which is defined by the identity

$$(\mathbf{M} \odot \mathbf{N})\text{svec}(\mathbf{K}) = \text{svec}[\tfrac{1}{2}(\mathbf{NKM}^T + \mathbf{MKN}^T)] \tag{13.81}$$

where $\text{svec}(\mathbf{K})$ converts symmetric matrix $\mathbf{K} = \{k_{ij}\}$ into a vector of dimension $n(n+1)/2$ as

$$\text{svec}(\mathbf{K}) = [k_{11} \ \sqrt{2}k_{12} \ \cdots \ \sqrt{2}k_{1n} \ k_{22} \ \sqrt{2}k_{23} \ \cdots \ \sqrt{2}k_{2n} \ \cdots \ k_{nn}]^T$$

Note that the standard inner product of $\mathbf{A}$ and $\mathbf{B}$ in $\mathcal{S}^n$ can be expressed as the standard inner product of vectors $\text{svec}(\mathbf{A})$ and $\text{svec}(\mathbf{B})$, i.e.,

$$\mathbf{A} \cdot \mathbf{B} = \text{svec}(\mathbf{A})^T \text{svec}(\mathbf{B}) \tag{13.82}$$

If we use a matrix $\mathbf{K} = \{k_{ij}\}$ with only one nonzero element k_{ij} for $1 \le i \le j \le n$, then Eq. (13.81) can be used to obtain each column of $\mathbf{M} \odot \mathbf{N}$. Based on this observation, a simple algorithm can be developed to obtain the $n(n+1)/2$-dimensional matrix $\mathbf{M} \odot \mathbf{N}$ (see Prob. 13.20). The following lemma describes an explicit relation between the eigenvalues and eigenvectors of $\mathbf{M} \odot \mathbf{N}$ and the eigenvalues and eigenvectors of $\mathbf{M}$ and $\mathbf{N}$ (see Prob. 13.21).

Lemma 13.1 *If $\mathbf{M}$ and $\mathbf{N}$ are symmetric matrices satisfying the relation $\mathbf{MN} = \mathbf{NM}$, then the $n(n+1)/2$ eigenvalues of $\mathbf{M} \odot \mathbf{N}$ are given by*

$$\tfrac{1}{2}(\alpha_i \beta_j + \beta_i \alpha_j) \quad \text{for } 1 \le i \le j \le n$$

and the corresponding orthonormal eigenvectors are given by

$$\text{svec}(\mathbf{v}_i \mathbf{v}_i^T) \quad \text{if } i = j$$
$$\tfrac{1}{2}\text{svec}(\mathbf{v}_i \mathbf{v}_j^T + \mathbf{v}_j \mathbf{v}_i^T) \quad \text{if } i < j$$

where α_i for $1 \le i \le n$ and β_j for $1 \le j \le n$ are the eigenvalues of $\mathbf{M}$ and $\mathbf{N}$, respectively, and $\mathbf{v}_i$ for $1 \le i \le n$ is a common basis of orthonormal eigenvectors of $\mathbf{M}$ and $\mathbf{N}$.

13.7.3 Reformulation of Eqs. (13.79a)–(13.79c)

As the unknowns in Eqs. (13.79a)–(13.79c) are matrices instead of vectors, these equations are not in the standard form and, therefore, they are more difficult to solve. In what follows, Eqs. (13.79a)–(13.79c) are reformulated in terms of a system of linear equations with unknowns that are vectors and in this way they can be easily solved using readily available software.

Eq. (13.79c) can be expressed in terms of the symmetric Kronecker product as

$$(\mathbf{X} \odot \mathbf{I})\text{svec}(\mathbf{\Delta S}) + (\mathbf{S} \odot \mathbf{I})\text{svec}(\mathbf{\Delta X}) = \text{svec}[\tau\mathbf{I} - \tfrac{1}{2}(\mathbf{XS} + \mathbf{SX})] \tag{13.83}$$

For the sake of simplicity, we let

$$\text{svec}(\mathbf{\Delta X}) = \mathbf{\Delta x} \tag{13.84a}$$
$$\text{svec}(\mathbf{\Delta S}) = \mathbf{\Delta s} \tag{13.84b}$$
$$\mathbf{S} \odot \mathbf{I} = \mathbf{E} \tag{13.84c}$$
$$\mathbf{X} \odot \mathbf{I} = \mathbf{F} \tag{13.84d}$$
$$\text{svec}[\tau\mathbf{I} - \tfrac{1}{2}(\mathbf{XS} + \mathbf{SX})] = \mathbf{r}_c \tag{13.84e}$$

With this notation, Eq. (13.83) becomes

$$\mathbf{E}\boldsymbol{\Delta}\mathbf{x} + \mathbf{F}\boldsymbol{\Delta}\mathbf{s} = \mathbf{r}_c$$

In order to simplify Eqs. (13.79a) and (13.79b), we let

$$\mathbf{A} = \begin{bmatrix} [\text{svec}(\mathbf{A}_1)]^T \\ [\text{svec}(\mathbf{A}_2)]^T \\ \vdots \\ [\text{svec}(\mathbf{A}_p)]^T \end{bmatrix} \qquad (13.85a)$$

$$\mathbf{x} = \text{svec}(\mathbf{X}) \qquad (13.85b)$$

$$\mathbf{y} = [y_1 \ y_2 \ \cdots \ y_p]^T \qquad (13.85c)$$

$$\boldsymbol{\Delta}\mathbf{y} = [\Delta y_1 \ \Delta y_2 \ \cdots \ \Delta y_p]^T \qquad (13.85d)$$

$$\mathbf{r}_p = \mathbf{b} - \mathbf{A}\mathbf{x} \qquad (13.85e)$$

$$\mathbf{r}_d = \text{svec}[\mathbf{C} - \mathbf{S} + \text{mat}(\mathbf{A}^T\mathbf{y})] \qquad (13.85f)$$

where $\text{mat}(\cdot)$ is the inverse of $\text{svec}(\cdot)$. With the use of Eqs. (13.85a)–(13.85f), Eqs. (13.79a) and (13.79b) can be written as

$$-\mathbf{A}^T\boldsymbol{\Delta}\mathbf{y} + \boldsymbol{\Delta}\mathbf{s} = \mathbf{r}_d$$

$$\mathbf{A}\boldsymbol{\Delta}\mathbf{x} = \mathbf{r}_p$$

and, therefore, Eqs. (13.79a)–(13.79c) can be reformulated as

$$\mathbf{J}\begin{bmatrix} \boldsymbol{\Delta}\mathbf{x} \\ \boldsymbol{\Delta}\mathbf{y} \\ \boldsymbol{\Delta}\mathbf{s} \end{bmatrix} = \begin{bmatrix} \mathbf{r}_d \\ \mathbf{r}_p \\ \mathbf{r}_c \end{bmatrix} \qquad (13.86)$$

where

$$\mathbf{J} = \begin{bmatrix} \mathbf{0} & -\mathbf{A}^T & \mathbf{I} \\ \mathbf{A} & \mathbf{0} & \mathbf{0} \\ \mathbf{E} & \mathbf{0} & \mathbf{F} \end{bmatrix}$$

It can be readily verified that the solution of Eq. (13.86) is given by

$$\boldsymbol{\Delta}\mathbf{x} = -\mathbf{E}^{-1}[\mathbf{F}(\mathbf{r}_d + \mathbf{A}^T\boldsymbol{\Delta}\mathbf{y}) - \mathbf{r}_c] \qquad (13.87a)$$

$$\boldsymbol{\Delta}\mathbf{s} = \mathbf{r}_d + \mathbf{A}^T\boldsymbol{\Delta}\mathbf{y} \qquad (13.87b)$$

$$\mathbf{M}\boldsymbol{\Delta}\mathbf{y} = \mathbf{r}_p + \mathbf{A}\mathbf{E}^{-1}(\mathbf{F}\mathbf{r}_d - \mathbf{r}_c) \qquad (13.87c)$$

where matrix $\mathbf{M}$, which is known as the *Schur complement* matrix, is given by

$$\mathbf{M} = \mathbf{A}\mathbf{E}^{-1}\mathbf{F}\mathbf{A}^T$$

From Eqs. (13.87a)–(13.87c), we see that solving the system of linear equations in Eq. (13.86) involves evaluating $\mathbf{E}^{-1}$ and computing $\boldsymbol{\Delta}\mathbf{y}$ from the linear system in Eq. (13.37c). In effect, the computational complexity is mainly determined by the computations required to solve the system in Eq. (13.37c) [7]. Matrix $\mathbf{J}$ in Eq. (13.86) is actually the Jacobian matrix of function $\mathbf{G}$ defined by Eq. (13.78). From Eqs. (13.87a)–(13.87c), it can be shown that $\mathbf{J}$ is nonsingular (i.e., Eq. (13.86) has a unique solution) if and only if $\mathbf{M}$ is nonsingular. It can also be shown that if $\mathbf{XS} + \mathbf{SX} \succ \mathbf{0}$, then $\mathbf{M}$ is nonsingular [20]. Therefore, $\mathbf{XS} + \mathbf{SX} \succ \mathbf{0}$ is a sufficient condition for Eqs. (13.87a)–(13.87c) to have a unique solution set $\{\boldsymbol{\Delta}\mathbf{x}, \ \boldsymbol{\Delta}\mathbf{y}, \ \boldsymbol{\Delta}\mathbf{s}\}$.

13.7.4 Primal-Dual Path-Following Algorithm

The above analysis leads to the following algorithm.

Algorithm 13.5 Primal-dual path-following algorithm for SDP problems
Step 1
Input $\mathbf{A}_i$ for $1 \leq i \leq p$, $\mathbf{b} \in R^p$, $\mathbf{C} \in R^{n \times n}$, and a strictly feasible set $\{\mathbf{X}_p, \mathbf{y}_0, \mathbf{S}_0\}$ that satisfies Eqs. (13.49b) and (13.52b) with $\mathbf{X}_0 \succ \mathbf{0}$ and $\mathbf{S}_0 \succ \mathbf{0}$.
Choose a scalar σ in the range $0 \leq \sigma < 1$.
Set $k = 0$ and initialize the tolerance ε for the duality gap δ_k.
Step 2
Compute
$$\delta_k = \frac{\mathbf{X}_k \cdot \mathbf{S}_k}{n}$$

Step 3
If $\delta_k \leq \varepsilon$, output solution $\{\mathbf{X}_k, \mathbf{y}_k, \mathbf{S}_k\}$ and stop; otherwise, set
$$\tau_k = \sigma \frac{\mathbf{X}_k \cdot \mathbf{S}_k}{n} \tag{13.88}$$
and continue with Step 4.
Step 4
Solve Eq. (13.86) using Eqs. (13.87a)–(13.87c) where $\mathbf{X} = \mathbf{X}_k, \mathbf{y} = \mathbf{y}_k, \mathbf{S} = \mathbf{S}_k$, and $\tau = \tau_k$.
Convert the solution $\{\Delta\mathbf{x}, \Delta\mathbf{y}, \Delta\mathbf{s}\}$ into $\{\Delta\mathbf{X}, \Delta\mathbf{y}, \Delta\mathbf{S}\}$ with $\Delta\mathbf{X} = \mathrm{mat}(\Delta\mathbf{x})$ and $\Delta\mathbf{S} = \mathrm{mat}(\Delta\mathbf{s})$.
Step 5
Choose a parameter γ in the range $0 < \gamma < 1$ and determine parameters α and β as
$$\alpha = \min(1, \gamma\hat{\alpha}) \tag{13.89a}$$
$$\beta = \min(1, \gamma\hat{\beta}) \tag{13.89b}$$
where
$$\hat{\alpha} = \max_{\mathbf{X}_k + \bar{\alpha}\Delta\mathbf{X} \succeq \mathbf{0}} (\bar{\alpha}) \quad \text{and} \quad \hat{\beta} = \max_{\mathbf{S}_k + \bar{\beta}\Delta\mathbf{S} \succeq \mathbf{0}} (\bar{\beta})$$
Step 6 Set
$$\mathbf{X}_{k+1} = \mathbf{X}_k + \alpha\Delta\mathbf{X}$$
$$\mathbf{y}_{k+1} = \mathbf{y}_k + \beta\Delta\mathbf{y}$$
$$\mathbf{S}_{k+1} = \mathbf{S}_k + \beta\Delta\mathbf{S}$$
Set $k = k + 1$ and repeat from Step 2.

A couple of remarks on Step 5 of the algorithm are in order. First, it follows from Eqs. (13.89a) and (13.89b) that if the increments $\Delta\mathbf{X}, \Delta\mathbf{y}$, and $\Delta\mathbf{S}$ obtained in Step 4 are such that $\mathbf{X}_k + \Delta\mathbf{X} \in \mathcal{F}_p^o$ and $\{\mathbf{y}_k + \Delta\mathbf{y}, \mathbf{S}_k + \Delta\mathbf{S}\} \in \mathcal{F}_d^o$, then we should use $\alpha = 1$ and $\beta = 1$. Otherwise, we should use $\alpha = \gamma\hat{\alpha}$ and $\beta = \gamma\hat{\beta}$ where $0 < \gamma < 1$

to ensure that $\mathbf{X}_{k+1} \in \mathcal{F}_p^o$ and $\{\mathbf{y}_{k+1}, \ \mathbf{S}_{k+1}\} \in \mathcal{F}_d^o$. Typically, a value of γ in the range $0.9 \leq \gamma \leq 0.99$ works well in practice. Second, the numerical values of $\hat{\alpha}$ and $\hat{\beta}$ can be determined by using the eigendecomposition of symmetric matrices as follows. Since $\mathbf{X}_k \succ \mathbf{0}$, the Cholesky decomposition (see Sec. A.13) of $\mathbf{X}_k$ gives

$$\mathbf{X}_k = \hat{\mathbf{X}}_k^T \hat{\mathbf{X}}_k$$

Now if we perform an eigendecomposition of the symmetric matrix

$$(\hat{\mathbf{X}}_k^T)^{-1} \boldsymbol{\Delta X} \hat{\mathbf{X}}_k^{-1}$$

as

$$(\hat{\mathbf{X}}_k^T)^{-1} \boldsymbol{\Delta X} \hat{\mathbf{X}}_k^{-1} = \mathbf{U}^T \boldsymbol{\Lambda} \mathbf{U}$$

where $\mathbf{U}$ is orthogonal and $\boldsymbol{\Lambda} = \mathrm{diag}\{\lambda_1, \ \lambda_2, \ \ldots, \ \lambda_n\}$, we get

$$\begin{aligned}
\mathbf{X}_k + \bar{\alpha} \boldsymbol{\Delta X} &= \hat{\mathbf{X}}_k^T [\mathbf{I} + \bar{\alpha}(\hat{\mathbf{X}}_k^T)^{-1} \boldsymbol{\Delta X} \hat{\mathbf{X}}_k^{-1}] \hat{\mathbf{X}}_k \\
&= \hat{\mathbf{X}}_k^T (\mathbf{I} + \bar{\alpha} \mathbf{U}^T \boldsymbol{\Lambda} \mathbf{U}) \hat{\mathbf{X}}_k \\
&= (\mathbf{U}\hat{\mathbf{X}}_k)^T (\mathbf{I} + \bar{\alpha} \boldsymbol{\Lambda})(\mathbf{U}\hat{\mathbf{X}}_k)
\end{aligned}$$

Hence $\mathbf{X}_k \bar{\alpha} \boldsymbol{\Delta X} \succeq \mathbf{0}$ if and only if $\mathbf{I} + \bar{\alpha} \boldsymbol{\Lambda} = \mathrm{diag}\{1 + \bar{\alpha}\lambda_1, \ 1 + \bar{\alpha}\lambda_2, \ \ldots, \ 1 + \bar{\alpha}\lambda_n\} \succeq \mathbf{0}$. If $\min\{\lambda_i\} \geq 0$, then $\mathbf{I} + \bar{\alpha} \boldsymbol{\Lambda} \succeq \mathbf{0}$ holds true for any $\bar{\alpha} \geq 0$; otherwise, the largest $\bar{\alpha}$ to assure the positive definiteness of $\mathbf{I} + \bar{\alpha} \boldsymbol{\Lambda}$ is given by

$$\hat{\alpha} = \frac{1}{\max_i(-\lambda_i)} \tag{13.90}$$

Therefore, the numerical value of α in Eq. (13.89a) can be obtained as

$$\alpha = \begin{cases} 1 & \text{if all } \lambda_i \geq 0 \\ \min(1, \ \gamma\hat{\alpha}) & \text{otherwise} \end{cases} \tag{13.91}$$

where $\hat{\alpha}$ is determined using Eq. (13.90). Similarly, the numerical value of β in Eq. (13.89b) can be obtained as

$$\beta = \begin{cases} 1 & \text{if all } \mu_i \geq 0 \\ \min(1, \ \gamma\hat{\beta}) & \text{otherwise} \end{cases} \tag{13.92}$$

where the μ_i's are the eigenvalues of $(\hat{\mathbf{S}}_k^T)^{-1} \boldsymbol{\Delta S} \hat{\mathbf{S}}_k^{-1}$ with $\mathbf{S}_k = \hat{\mathbf{S}}_k^T \hat{\mathbf{S}}_k$ and

$$\hat{\beta} = \frac{1}{\max_i(-\mu_i)}$$

The numerical value of the centering parameter σ should be in the range of $[0, 1)$. For small-scale applications, the choice

$$\sigma = \frac{n}{15\sqrt{n} + n}$$

is usually satisfactory.

Example 13.5 Find scalars α_1, α_2, and α_3 such that the maximum eigenvalue of $\mathbf{F} = \mathbf{A}_0 + \alpha_1\mathbf{A}_1 + \alpha_2\mathbf{A}_2 + \alpha_3\mathbf{A}_3$ is minimized where

$$\mathbf{A}_0 = \begin{bmatrix} 2 & -0.5 & -0.6 \\ -0.5 & 2 & 0.4 \\ -0.6 & 0.4 & 3 \end{bmatrix}, \quad \mathbf{A}_1 = \begin{bmatrix} 0 & 1 & 0 \\ 1 & 0 & 0 \\ 0 & 0 & 0 \end{bmatrix}$$

$$\mathbf{A}_2 = \begin{bmatrix} 0 & 0 & 1 \\ 0 & 0 & 0 \\ 1 & 0 & 0 \end{bmatrix}, \quad \mathbf{A}_3 = \begin{bmatrix} 0 & 0 & 0 \\ 0 & 0 & 1 \\ 0 & 1 & 0 \end{bmatrix}$$

Solution As matrix $\mathbf{F}$ is symmetric, there exists an orthogonal matrix $\mathbf{U}$ such that $\mathbf{U}^T\mathbf{F}\mathbf{U} = \text{diag}\{\lambda_1, \lambda_2, \lambda_3\}$ with $\lambda_1 \geq \lambda_2 \geq \lambda_3$. Hence we can write

$$\mathbf{U}^T(\alpha_4\mathbf{I} - \mathbf{F})\mathbf{U} = \alpha_4\mathbf{I} - \mathbf{U}^T\mathbf{F}\mathbf{U} = \text{diag}\{\alpha_4 - \lambda_1, \ \alpha_4 - \lambda_2, \ \alpha_4 - \lambda_3\}$$

and conclude that matrix $\alpha_4\mathbf{I} - \mathbf{F}$ will remain positive semidefinite as long as α_4 is not less than the maximum eigenvalue of $\mathbf{F}$. Consequently, the problem at hand can be solved by minimizing α_4 subject to the condition $\alpha_4\mathbf{I} - \mathbf{F} \succeq \mathbf{0}$. Obviously, this problem can be formulated as

$$\text{maximize} \ -\mathbf{b}^T\mathbf{y} \tag{13.93a}$$

$$\text{subject to:} \ -\sum_{i=1}^{4} y_i\mathbf{A}_i + \mathbf{S} = \mathbf{C} \tag{13.93b}$$

$$\mathbf{S} \succeq \mathbf{0} \tag{13.93c}$$

where $\mathbf{b} = [0 \ 0 \ 0 \ 1]^T$, $\mathbf{C} = -\mathbf{A}_0$, $\mathbf{A}_4 = \mathbf{I}$, and $\mathbf{y} = [y_1 \ y_2 \ y_3 \ y_4]^T$ where $y_1 = -\alpha_1$, $y_2 = -\alpha_2$, $y_3 = -\alpha_3$, and $y_4 = \alpha_4$ is the maximum eigenvalue of matrix $\mathbf{F}$ (see Prob. 13.17). We observe that the optimization problem in Eqs. (13.93a)–(13.93c) is of the type described by Eqs. (13.52a)–(13.52c) with $n = 3$ and $p = 4$.

It is easy to verify that the set $\{\mathbf{X}_0, \mathbf{y}_0, \mathbf{S}_0\}$ with

$$\mathbf{X}_0 = \tfrac{1}{3}\mathbf{I}, \ \mathbf{y}_0 = [-0.2 \ -0.2 \ -0.2 \ 4.0]^T \ \text{ and } \ \mathbf{S}_0 = \begin{bmatrix} 2 & 0.3 & 0.4 \\ 0.3 & 2 & -0.6 \\ 0.4 & -0.6 & 1 \end{bmatrix}$$

is strictly feasible for the associated primal-dual problems. Matrix $\mathbf{A}$ in Eq. (13.85a) is in this case a 4×6 matrix given by

$$\mathbf{A} = \begin{bmatrix} 0 & \sqrt{2} & 0 & 0 & 0 & 0 \\ 0 & 0 & \sqrt{2} & 0 & 0 & 0 \\ 0 & 0 & 0 & 0 & \sqrt{2} & 0 \\ 1 & 0 & 0 & 1 & 0 & 1 \end{bmatrix}$$

At the initial point, the maximum eigenvalue of $\mathbf{F}$ is 3.447265. With $\sigma = n/(15\sqrt{n} + n) = 0.1035$, $\gamma = 0.9$, and $\varepsilon = 10^{-3}$, it took Algorithm 13.5 four iterations to converge to the solution set $\{\mathbf{X}^*, \mathbf{y}^*, \mathbf{S}^*\}$ where

$$\mathbf{y}^* = \begin{bmatrix} -0.392921 \\ -0.599995 \\ 0.399992 \\ 3.000469 \end{bmatrix}$$

i.e., $\alpha_1 = 0.392921$, $\alpha_2 = 0.599995$, $\alpha_3 = -0.399992$, and $\alpha_4 = 3.000469$. The minimized maximum eigenvalue of $\mathbf{F}$ is, therefore, obtained as 3.000469.   ∎

13.8 Predictor-Corrector Method

Algorithm 13.5 can be improved by incorporating the predictor-corrector rule proposed by Mehrotra [21] for LP problems (see Sect. 12.5.3).

As in the LP case, the predictor-corrector method involves two steps. Consider the kth iteration of Algorithm 13.5. In the first step, a *predictor direction* $\{\Delta\mathbf{X}^{(p)}, \Delta\mathbf{y}^{(p)}, \Delta\mathbf{S}^{(p)}\}$ is first identified by using a linear approximation of the KKT conditions. This set $\{\Delta\mathbf{X}^{(p)}, \Delta\mathbf{y}^{(p)}, \Delta\mathbf{S}^{(p)}\}$ can be obtained by letting $\tau = 0$ in Eq. (13.84e) to obtain

$$\mathbf{r}_c = \text{svec} \left[-\tfrac{1}{2}(\mathbf{X}_k\mathbf{S}_k + \mathbf{S}_k\mathbf{X}_k) \right] \tag{13.94}$$

and then using Eqs. (13.87a)–(13.87c). Next, the numerical values of α_p and β_p can be determined in a way similar to that described in Eqs. (13.91) and (13.92) as

$$\alpha_p = \min(1, \gamma\hat{\alpha}) \tag{13.95a}$$

$$\beta_p = \min(1, \gamma\hat{\beta}) \tag{13.95b}$$

where

$$\hat{\alpha} = \max_{\mathbf{X}_k + \bar{\alpha}\Delta\mathbf{X}^{(p)} \succeq 0} (\bar{\alpha})$$

$$\hat{\beta} = \max_{\mathbf{S}_k + \bar{\beta}\Delta\mathbf{S}^{(p)} \succeq 0} (\bar{\beta})$$

The centering parameter σ_k is then computed as

$$\sigma_k = \left[\frac{(\mathbf{X}_k + \alpha_p\Delta\mathbf{X}^{(p)}) \cdot (\mathbf{S}_k + \beta_p\Delta\mathbf{S}^{(p)})}{\mathbf{X}_k \cdot \mathbf{S}_k} \right]^3 \tag{13.96}$$

and is used to determine the value of τ_k in Eq. (13.88), i.e.,

$$\tau_k = \sigma_k \frac{\mathbf{X}_k \cdot \mathbf{S}_k}{n} \tag{13.97}$$

In the second step, the parameter τ_k in Eq. (13.97) is utilized to obtain vector

$$\mathbf{r}_c = \tau_k\mathbf{I} - \tfrac{1}{2}(\mathbf{X}_k\mathbf{S}_k + \mathbf{S}_k\mathbf{X}_k + \Delta\mathbf{X}^{(p)}\Delta\mathbf{S}^{(p)} + \Delta\mathbf{S}^{(p)}\Delta\mathbf{X}^{(p)}) \tag{13.98}$$

which is then used in Eqs. (13.87a)–(13.87c) to obtain the *corrector direction* $\{\Delta \mathbf{X}^{(c)}, \Delta \mathbf{y}^{(c)}, \Delta \mathbf{S}^{(c)}\}$. The set $\{\mathbf{X}_k, \mathbf{y}_k, \mathbf{S}_k\}$ is then updated as

$$\mathbf{X}_{k+1} = \mathbf{X}_k + \alpha_c \Delta \mathbf{X}^{(c)} \tag{13.99a}$$

$$\mathbf{y}_{k+1} = \mathbf{y}_k + \beta_c \Delta \mathbf{y}^{(c)} \tag{13.99b}$$

$$\mathbf{S}_{k+1} = \mathbf{S}_k + \beta_c \Delta \mathbf{S}^{(c)} \tag{13.99c}$$

where α_c and β_c are given by

$$\alpha_c = \min(1, \ \gamma \hat{\alpha}) \tag{13.100a}$$

$$\beta_c = \min(1, \ \gamma \hat{\beta}) \tag{13.100b}$$

where

$$\hat{\alpha} = \max_{\mathbf{X}_k + \bar{\alpha} \Delta \mathbf{X}^{(c)} \succeq 0} (\bar{\alpha})$$

$$\hat{\beta} = \max_{\mathbf{S}_k + \bar{\beta} \Delta \mathbf{S}^{(c)} \succeq 0} (\bar{\beta})$$

The predictor-corrector method can be incorporated in Algorithm 13.5 as follows.

Algorithm 13.6 Predictor-corrector algorithm for SDP problems
Step 1
Input $\mathbf{A}_i$ for $1 \leq i \leq p$, $\mathbf{b} \in R^p$, and $\mathbf{C} \in R^{n \times n}$, and a strictly feasible set $\{\mathbf{X}_0, \mathbf{y}_0, \mathbf{S}_0\}$ that satisfies Eqs. (13.49b) and (13.52b) with $\mathbf{X}_0 \succ 0$ and $\mathbf{S}_0 \succ 0$. Set $k = 0$ and initialize the tolerance ε for the duality gap δ_k.
Step 2
Compute

$$\delta_k = \frac{\mathbf{X}_k \cdot \mathbf{S}_k}{n}$$

Step 3
If $\delta_k \leq \varepsilon$, output solution $\{\mathbf{X}_k, \mathbf{y}_k, \mathbf{S}_k\}$ and stop; otherwise, continue with Step 4.
Step 4
Compute $\{\Delta \mathbf{X}^{(p)}, \Delta \mathbf{y}^{(p)}, \Delta \mathbf{S}^{(p)}\}$ using Eqs. (13.87a)–(13.87c) with $\mathbf{X} = \mathbf{X}_k$, $\mathbf{y} = \mathbf{y}_k$, $\mathbf{S} = \mathbf{S}_k$, and $\mathbf{r}_c$ given by Eq. (13.74).
Choose a parameter γ in the range $0 < \gamma \leq 1$ and compute α_p and β_p using Eqs. (13.95a) and (13.95b) and evaluate σ_k using Eq. (13.96).
Compute τ_k using Eq. (13.97).
Step 5
Compute $\{\Delta \mathbf{X}^{(c)}, \Delta \mathbf{y}^{(c)}, \Delta \mathbf{S}^{(c)}\}$ using Eqs. (13.87a)–(13.87c) with $\mathbf{X} = \mathbf{X}_k$, $\mathbf{y} = \mathbf{y}_k$, $\mathbf{S} = \mathbf{S}_k$, and $\mathbf{r}_c$ given by Eq. (13.98).
Step 6
Compute α_c and β_c using Eqs. (13.100a) and (13.100b).
Step 7
Obtain set $\{\mathbf{X}_{k+1}, \mathbf{y}_{k+1}, \mathbf{S}_{k+1}\}$ using Eqs. (13.99a)–(13.99c).
Set $k = k + 1$ and repeat from Step 2.

Fig. 13.2 Distance between
two ellipses (Example 13.6)

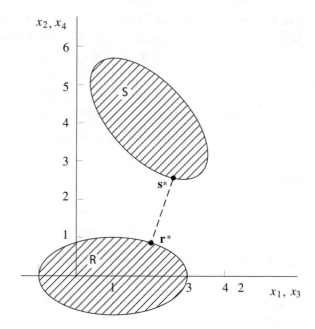

Example 13.6 The two ellipses in Fig. 13.2 are described by

$$c_1(\mathbf{x}) = [x_1 \ x_2] \begin{bmatrix} \frac{1}{4} & 0 \\ 0 & 1 \end{bmatrix} \begin{bmatrix} x_1 \\ x_2 \end{bmatrix} - [x_1 \ x_2] \begin{bmatrix} \frac{1}{2} \\ 0 \end{bmatrix} - \frac{3}{4} \le 0$$

$$c_2(\mathbf{x}) = \frac{1}{8}[x_3 \ x_4] \begin{bmatrix} 5 & 3 \\ 3 & 5 \end{bmatrix} \begin{bmatrix} x_3 \\ x_4 \end{bmatrix} - [x_3 \ x_4] \begin{bmatrix} \frac{11}{2} \\ \frac{13}{2} \end{bmatrix} + \frac{35}{2} \le 0$$

where $\mathbf{x} = [x_1 \ x_2 \ x_3 \ x_4]^T$. Find the shortest distance between them using
Algorithm 13.6.

Solution From Sect. 13.5.2, we can first formulate the problem as a CP problem of
the form given by Eqs. (13.67a)–(13.67c), i.e.,

$$\text{minimize } \delta$$

$$\text{subject to: } \mathbf{x}^T \mathbf{H} \mathbf{x} \le \delta$$

$$\mathbf{x}^T \mathbf{Q}_1 \mathbf{x} + \mathbf{q}_1^T \mathbf{x} + r_1 \le 0$$

$$\mathbf{x}^T \mathbf{Q}_2 \mathbf{x} + \mathbf{q}_2^T \mathbf{x} + r_2 \le 0$$

with

$$
\mathbf{x} = \begin{bmatrix} x_1 \\ x_2 \\ x_3 \\ x_4 \end{bmatrix}, \quad
\mathbf{H} = \begin{bmatrix} 1 & 0 & -1 & 0 \\ 0 & 1 & 0 & -1 \\ -1 & 0 & 1 & 0 \\ 0 & -1 & 0 & 1 \end{bmatrix}
$$

$$
\mathbf{Q}_1 = \begin{bmatrix} \frac{1}{4} & 0 \\ 0 & 1 \end{bmatrix}, \quad
\mathbf{Q}_2 = \begin{bmatrix} 5 & 3 \\ 3 & 5 \end{bmatrix}, \quad
\mathbf{q}_1 = \begin{bmatrix} -\frac{1}{2} \\ 0 \end{bmatrix}, \quad
\mathbf{q}_2 = \begin{bmatrix} -44 \\ -52 \end{bmatrix}
$$

$$
r_1 = -\tfrac{3}{4}, \quad \text{and} \quad r_2 = 140
$$

The above CP problem can be converted into the SDP problem in Eqs. (13.69a) and (13.69b) with

$$
\hat{\mathbf{c}} = \begin{bmatrix} 0 \\ 0 \\ 0 \\ 0 \\ 1 \end{bmatrix}, \quad
\hat{\mathbf{x}} = \begin{bmatrix} x_1 \\ x_2 \\ x_3 \\ x_4 \\ \delta \end{bmatrix}
$$

and

$$
\mathbf{E}(\hat{\mathbf{x}}) = \mathrm{diag}\{\mathbf{G}(\delta, \mathbf{x}), \; \mathbf{F}_1(\mathbf{x}), \; \mathbf{F}_2(\mathbf{x})\}
$$

with

$$
\mathbf{G}(\delta, \mathbf{x}) = \begin{bmatrix} -\mathbf{I}_4 & -\hat{\mathbf{H}}\mathbf{x} \\ -(\hat{\mathbf{H}}\mathbf{x})^T & -\delta \end{bmatrix}, \quad
\hat{\mathbf{H}} = \begin{bmatrix} 1 & 0 & -1 & 0 \\ 0 & -1 & 0 & 1 \\ 0 & 0 & 0 & 0 \\ 0 & 0 & 0 & 0 \end{bmatrix}
$$

$$
\mathbf{F}_1(\mathbf{x}) = \begin{bmatrix} -\mathbf{I}_2 & -\hat{\mathbf{Q}}_1\mathbf{x}_a \\ -(\hat{\mathbf{Q}}_1\mathbf{x}_a)^T & \mathbf{q}_1^T\mathbf{x}_a + r_1 \end{bmatrix}, \quad
\hat{\mathbf{Q}}_1 = \begin{bmatrix} \frac{1}{2} & 0 \\ 0 & 1 \end{bmatrix}, \quad
\mathbf{x}_a = \begin{bmatrix} x_1 \\ x_2 \end{bmatrix}
$$

$$
\mathbf{F}_2(\mathbf{x}) = \begin{bmatrix} -\mathbf{I}_2 & -\hat{\mathbf{Q}}_2\mathbf{x}_b \\ -(\hat{\mathbf{Q}}_2\mathbf{x}_b)^T & \mathbf{q}_2^T\mathbf{x}_b + r_2 \end{bmatrix}, \quad
\hat{\mathbf{Q}}_2 = \begin{bmatrix} 2 & 2 \\ -1 & 1 \end{bmatrix}, \quad
\mathbf{x}_b = \begin{bmatrix} x_3 \\ x_4 \end{bmatrix}
$$

The SDP problem in Eqs. (13.69a) and (13.69b) is equivalent to the standard SDP problem in Eqs. (13.52a)–(13.52c) with $p = 5$, $n = 11$, and

$$
\mathbf{y} = \begin{bmatrix} x_1 \\ x_2 \\ x_3 \\ x_4 \\ \delta \end{bmatrix}, \quad
\mathbf{b} = \begin{bmatrix} 0 \\ 0 \\ 0 \\ 0 \\ 1 \end{bmatrix}
$$

Matrices $\mathbf{A}_i$ for $1 \le i \le 5$ and $\mathbf{C}$ are given by

$$\mathbf{A}_1 = \text{diag}\left\{ \begin{bmatrix} \mathbf{0}_4 & \mathbf{h}_1 \\ \mathbf{h}_1^T & 0 \end{bmatrix}, \begin{bmatrix} \mathbf{0}_2 & \mathbf{q}_{11} \\ \mathbf{q}_{11}^T & -\mathbf{q}_1(1) \end{bmatrix}, \mathbf{0}_3 \right\}$$

$$\mathbf{A}_2 = \text{diag}\left\{ \begin{bmatrix} \mathbf{0}_4 & \mathbf{h}_2 \\ \mathbf{h}_2^T & 0 \end{bmatrix}, \begin{bmatrix} \mathbf{0}_2 & \mathbf{q}_{12} \\ \mathbf{q}_{12}^T & -\mathbf{q}_1(2) \end{bmatrix}, \mathbf{0}_3 \right\}$$

$$\mathbf{A}_3 = \text{diag}\left\{ \begin{bmatrix} \mathbf{0}_4 & \mathbf{h}_3 \\ \mathbf{h}_3^T & 0 \end{bmatrix}, \mathbf{0}_3, \begin{bmatrix} \mathbf{0}_2 & \mathbf{q}_{21} \\ \mathbf{q}_{21}^T & -\mathbf{q}_2(1) \end{bmatrix} \right\}$$

$$\mathbf{A}_4 = \text{diag}\left\{ \begin{bmatrix} \mathbf{0}_4 & \mathbf{h}_4 \\ \mathbf{h}_4^T & 0 \end{bmatrix}, \mathbf{0}_3, \begin{bmatrix} \mathbf{0}_2 & \mathbf{q}_{22} \\ \mathbf{q}_{22}^T & -\mathbf{q}_2(2) \end{bmatrix} \right\}$$

$$\mathbf{A}_5 = \text{diag}\left\{ \begin{bmatrix} \mathbf{0}_4 & \mathbf{0} \\ \mathbf{0} & 1 \end{bmatrix}, \mathbf{0}_3, \mathbf{0}_3 \right\}$$

$$\mathbf{C} = -\text{diag}\left\{ \begin{bmatrix} \mathbf{I}_4 & \mathbf{0} \\ \mathbf{0} & 0 \end{bmatrix}, \begin{bmatrix} \mathbf{I}_2 & \mathbf{0} \\ \mathbf{0} & -r_1 \end{bmatrix}, \begin{bmatrix} \mathbf{I}_2 & \mathbf{0} \\ \mathbf{0} & -r_2 \end{bmatrix} \right\}$$

where $\mathbf{h}_i$, $\mathbf{q}_{1j}$, and $\mathbf{q}_{2j}$ for $1 \le i \le 4$ and $1 \le j \le 2$ are the ith and jth columns of $\hat{\mathbf{H}}$, $\hat{\mathbf{Q}}_1$, and $\hat{\mathbf{Q}}_2$, respectively, $\mathbf{I}_k$ is the $k \times k$ identity matrix, and $\mathbf{0}_k$ is the $k \times k$ zero matrix. A strictly feasible initial set $\{\mathbf{X}_0, \mathbf{y}_0, \mathbf{S}_0\}$ can be identified as

$$\mathbf{X}_0 = \text{diag}\{\mathbf{I}_5, \mathbf{X}_{02}, \mathbf{X}_{03}\}$$

$$\mathbf{y}_0 = [1\ 0\ 2\ 4\ 20]^T$$

$$\mathbf{S}_0 = \mathbf{C} + \sum_{i=1}^{5} \mathbf{y}_0(i)\mathbf{A}_i$$

where

$$\mathbf{X}_{02} = \begin{bmatrix} 1 & 0 & -0.5 \\ 0 & 1 & 0 \\ -0.5 & 0 & 1 \end{bmatrix} \quad \text{and} \quad \mathbf{X}_{03} = \begin{bmatrix} 180 & 0 & -12 \\ 0 & 60 & -2 \\ -12 & -2 & 1 \end{bmatrix}$$

With $\gamma = 0.9$ and $\varepsilon = 10^{-3}$, it took Algorithm 13.6 six iterations to converge to the solution $\{\mathbf{X}^*, \mathbf{y}^*, \mathbf{S}^*\}$ where

$$\mathbf{y}^* = \begin{bmatrix} 2.044717 \\ 0.852719 \\ 2.544895 \\ 2.485678 \\ 2.916910 \end{bmatrix}$$

This corresponds to the solution points $\mathbf{r}^* \in \mathcal{R}$ and $\mathbf{s}^* \in \mathcal{S}$ (see Fig. 13.2) with

$$\mathbf{r}^* = \begin{bmatrix} 2.044717 \\ 0.852719 \end{bmatrix} \quad \text{and} \quad \mathbf{s}^* = \begin{bmatrix} 2.544895 \\ 2.485678 \end{bmatrix}$$

which yield the shortest distance between $\mathcal{R}$ and $\mathcal{S}$ as $||\mathbf{r}^* - \mathbf{s}^*|| = 1.707845$.   ∎

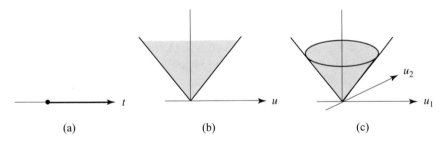

Fig. 13.3 Second-order cones of dimension **a** $n = 1$, **b** $n = 2$, and **c** $n = 3$

13.9 Second-Order Cone Programming

13.9.1 Notation and Definitions

The concept of convex cone has been defined in Sect. 13.5 as a convex set where any element multiplied by any nonnegative scalar still belongs to the set defining the cone (see Def. 13.1). Here we are interested in a special class of convex cones known as *second-order cones*.

Definition 13.2 A second-order cone (also known as *quadratic* or *Lorentz* cone) of dimension n is defined as

$$\mathcal{K} = \left\{ \begin{bmatrix} t \\ \mathbf{u} \end{bmatrix} : t \in R, \ \mathbf{u} \in R^{n-1} \text{ for } \|\mathbf{u}\|_2 \leq t \right\}$$

For $n = 1$, a second-order cone degenerates into a ray on the t axis starting at $t = 0$, as shown in Fig. 13.3a. Second-order cones for $n = 2$ and 3 are illustrated in Fig. 13.3b and c, respectively.

Note that the second-order cone $\mathcal{K}$ is a convex set in R^n because for any two points in $\mathcal{K}$, $[t_1 \ \mathbf{u}_1^T]^T$ and $[t_2 \ \mathbf{u}_2^T]^T$, and $\lambda \in [0, \ 1]$, we have

$$\lambda \begin{bmatrix} t_1 \\ \mathbf{u}_1 \end{bmatrix} + (1 - \lambda) \begin{bmatrix} t_2 \\ \mathbf{u}_2 \end{bmatrix} = \begin{bmatrix} \lambda t_1 + (1 - \lambda) t_2 \\ \lambda \mathbf{u}_1 + (1 - \lambda) \mathbf{u}_2 \end{bmatrix}$$

where

$$\|\lambda \mathbf{u}_1 + (1 - \lambda) \mathbf{u}_2\|_2 \leq \lambda \|\mathbf{u}_1\|_2 + (1 - \lambda) \|\mathbf{u}_2\|_2 \leq \lambda t_1 + (1 - \lambda) t_2$$

The primal second-order cone-programming (SOCP) problem is a constrained optimization problem that can be formulated as

$$\text{minimize} \ \sum_{i=1}^{q} \hat{\mathbf{c}}_i^T \mathbf{x}_i \tag{13.101a}$$

$$\text{subject to:} \ \sum_{i=1}^{q} \hat{\mathbf{A}}_i \mathbf{x}_i = \mathbf{b} \tag{13.101b}$$

$$\mathbf{x}_i \in \mathcal{K}_i \quad \text{for } i = 1, \ 2, \ \ldots, \ q \tag{13.101c}$$

where $\hat{c}_i \in R^{n_i \times 1}$, $\mathbf{x}_i \in R^{n_i \times 1}$, $\hat{\mathbf{A}}_i \in R^{m \times n_i}$, $\mathbf{b} \in R^{m \times 1}$, and $\mathcal{K}_i$ is the second-order cone of dimension n_i. It is interesting to note that there exists an analogy between the SOCP problem in Eqs. (13.101a)–(13.101c) and the LP problem in Eqs. (12.1a)–(12.1c): both problems involve a linear objective function and a linear equality constraint. While the variable vector $\mathbf{x}$ in an LP problem is constrained to the region $\{\mathbf{x} \geq \mathbf{0}, \mathbf{x} \in R^n\}$, which is a convex cone (see Def. 13.1), each variable vector $\mathbf{x}_i$ in an SOCP problem is constrained to the second-order cone $\mathcal{K}_i$.

The dual of the SOCP problem in Eqs. (13.101a)–(13.101c) will be referred to hereafter as the *dual SOCP problem* and it can be shown to be of the form

$$\text{maximize}\ \mathbf{b}^T\mathbf{y} \tag{13.102a}$$

$$\text{subject to:}\ \hat{\mathbf{A}}_i^T\mathbf{y} + \mathbf{s}_i = \hat{\mathbf{c}}_i \tag{13.102b}$$

$$\mathbf{s}_i \in \mathcal{K}_i \quad \text{for } i = 1,\ 2,\ \dots,\ q \tag{13.102c}$$

where $\mathbf{y} \in R^{m \times 1}$ and $\mathbf{s}_i \in R^{n_i \times 1}$ (see Prob. 13.24).

Note that a similar analogy exists between the dual SOCP problem in Eqs. (13.102a)–(13.102c) and the dual LP problem in Eqs. (12.2a)–(12.2c). If we let

$$\mathbf{x} = -\mathbf{y}, \quad \hat{\mathbf{A}}_i^T = \begin{bmatrix} \mathbf{b}_i^T \\ \mathbf{A}_i^T \end{bmatrix} \quad \text{and} \quad \hat{\mathbf{c}}_i = \begin{bmatrix} d_i \\ \mathbf{c}_i \end{bmatrix} \tag{13.103}$$

where $\mathbf{b}_i \in R^{m \times 1}$ and d_i is a scalar, then the SOCP problem in Eqs. (13.102a)–(13.102c) can be expressed as

$$\text{minimize}\ \mathbf{b}^T\mathbf{x} \tag{13.104a}$$

$$\text{subject to:}\ \|\mathbf{A}_i^T\mathbf{x} + \mathbf{c}_i\|_2 \leq \mathbf{b}_i^T\mathbf{x} + d_i \quad \text{for } i = 1,\ 2,\ \dots,\ q \tag{13.104b}$$

(see Prob. 13.25). As we will see next, this SOCP formulation turns out to have a direct connection to many convex-programming problems in engineering and science.

13.9.2 Relations Among LP, QP, SDP, and SOCP Problems

The class of SOCP problems is large enough to include both LP and convex QP problems as will be shown next. If $\mathbf{A}_i^T = \mathbf{0}$ and $\mathbf{c}_i = \mathbf{0}$ for $i = 1,\ 2,\ \dots,\ q$, then the SOCP problem in Eqs. (13.104a) and (13.104b) becomes

$$\text{minimize}\ \mathbf{b}^T\mathbf{x}$$

$$\text{subject to:}\quad -\mathbf{b}_i^T\mathbf{x} - d_i \leq 0 \quad \text{for } i = 1,\ 2,\ \dots,\ q$$

which is obviously an LP problem.

Now consider the convex QP problem

$$\text{minimize}\ f(\mathbf{x}) = \mathbf{x}^T\mathbf{H}\mathbf{x} + 2\mathbf{x}^T\mathbf{p} \tag{13.105a}$$

$$\text{subject to:}\ \mathbf{A}\mathbf{x} \leq \mathbf{b} \tag{13.105b}$$

where $\mathbf{H}$ is positive definite. If we write matrix $\mathbf{H}$ as $\mathbf{H} = \mathbf{H}^{T/2}\mathbf{H}^{1/2}$ and let $\tilde{\mathbf{p}} = \mathbf{H}^{-T/2}\mathbf{p}$, then the objective function in Eq. (13.105a) can be expressed as

$$f(\mathbf{x}) = \|\mathbf{H}^{1/2}\mathbf{x} + \tilde{\mathbf{p}}\|_2^2 - \mathbf{p}^T\mathbf{H}^{-1}\mathbf{p}$$

Since the term $\mathbf{p}^T\mathbf{H}^{-1}\mathbf{p}$ is a constant, minimizing $f(\mathbf{x})$ is equivalent to minimizing $\|\mathbf{H}^{1/2}\mathbf{x} + \tilde{\mathbf{p}}\|_2$ and thus the problem at hand can be converted to the problem

$$\text{minimize } \delta$$

$$\text{subject to: } \|\mathbf{H}^{1/2}\mathbf{x} + \tilde{\mathbf{p}}\|_2 \le \delta$$

$$\mathbf{A}\mathbf{x} \le \mathbf{b}$$

where δ is an upper bound for $\|\mathbf{H}^{1/2}\mathbf{x} + \tilde{\mathbf{p}}\|_2$ that can be treated as an auxiliary variable of the problem. By letting

$$\tilde{\mathbf{x}} = \begin{bmatrix} \delta \\ \mathbf{x} \end{bmatrix}, \quad \tilde{\mathbf{b}} = \begin{bmatrix} 1 \\ \mathbf{0} \end{bmatrix}, \quad \tilde{\mathbf{H}} = [\mathbf{0} \; \mathbf{H}^{1/2}], \quad \tilde{\mathbf{A}} = [\mathbf{0} \; \mathbf{A}]$$

the problem becomes

$$\text{minimize } \tilde{\mathbf{b}}^T\tilde{\mathbf{x}}$$

$$\text{subject to: } \|\tilde{\mathbf{H}}\tilde{\mathbf{x}} + \tilde{\mathbf{p}}\|_2 \le \tilde{\mathbf{b}}^T\tilde{\mathbf{x}}$$

$$\tilde{\mathbf{A}}\tilde{\mathbf{x}} \le \mathbf{b}$$

which is an SOCP problem. On the other hand, it can be shown that every SOCP problem can be formulated as an SDP problem. To see this, note that the constraint $\|\mathbf{u}\|_2 \le t$ implies that

$$\begin{bmatrix} -t\mathbf{I} & -\mathbf{u} \\ -\mathbf{u}^T & -t \end{bmatrix} \preceq \mathbf{0}$$

(see Prob. 13.26). Thus a second-order cone can be embedded into a cone of positive semidefinite matrices, and the SOCP problem in Eqs. (13.104a) and (13.104b) can be formulated as

$$\text{minimize } \mathbf{b}^T\mathbf{x}$$

$$\text{subject to: } \begin{bmatrix} -(\mathbf{b}_i^T\mathbf{x} + d_i)\mathbf{I} & -\mathbf{A}_i^T\mathbf{x} - \mathbf{c}_i \\ -(\mathbf{A}_i^T\mathbf{x} + \mathbf{c}_i)^T & -(\mathbf{b}_i^T\mathbf{x} + d_i) \end{bmatrix} \preceq \mathbf{0} \quad \text{for } i = 1, 2, \ldots, q$$

which is an SDP problem.

The above analysis has demonstrated that the branch of nonlinear programming known as CP can be subdivided into a series of nested branches of optimization, namely, SDP, SOCP, convex QP, and LP as illustrated in Fig. 13.4.

Fig. 13.4 Relations among
LP, convex QP, SOCP, SDP,
and CP problems

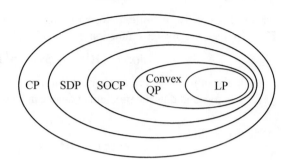

13.9.3 Examples

In this section, we present several examples to demonstrate that a variety of interesting optimization problems can be formulated as SOCP problems [22].

(i) *QP problems with quadratic constraints* A general QP problem with quadratic constraints can be expressed as

$$\text{minimize} \ \ \mathbf{x}^T \mathbf{H}_0 \mathbf{x} + 2\mathbf{p}_0^T \mathbf{x} \tag{13.106a}$$
$$\text{subject to:} \ \ \mathbf{x}^T \mathbf{H}_i \mathbf{x} + 2\mathbf{p}_i^T \mathbf{x} + r_i \le 0 \ \ \text{for} \ i = 1, \ 2, \ \dots, \ q$$
$$\tag{13.106b}$$

where $\mathbf{H}_i$ for $i = 0, \ 1, \ \dots, \ q$ are assumed to be positive-definite matrices. Using the matrix decomposition $\mathbf{H}_i = \mathbf{H}_i^{T/2} \mathbf{H}_i^{1/2}$, the problem in Eqs. (13.106a) and (13.106b) can be expressed as

$$\text{minimize} \ \ \|\mathbf{H}_0^{1/2}\mathbf{x} + \tilde{\mathbf{p}}_0\|_2^2 - \mathbf{p}_0^T \mathbf{H}_0^{-1}\mathbf{p}_0$$
$$\text{subject to:} \ \ \|\mathbf{H}_i^{1/2}\mathbf{x} + \tilde{\mathbf{p}}_i\|_2^2 - \mathbf{p}_i^T \mathbf{H}_i^{-1}\mathbf{p}_i + r_i \le 0 \ \ \text{for} \ i = 1, \ 2, \ \dots, \ q$$

where $\tilde{\mathbf{p}}_i = \mathbf{H}_i^{-T/2}\mathbf{p}_i$ for $i = 0, \ 1. \ \dots, \ q$. The term $-\mathbf{p}_0^T \mathbf{H}_0^{-1}\mathbf{p}_0$ in the objective function is a constant and it can be neglected. Hence the above problem is equivalent to the SOCP problem

$$\text{minimize} \ \ \delta$$
$$\text{subject to:} \ \ \|\mathbf{H}^{1/2}\mathbf{x} + \tilde{\mathbf{p}}_0\|_2 \le \delta$$
$$\|\mathbf{H}^{1/2}\mathbf{x} + \tilde{\mathbf{p}}_i\|_2 \le (\mathbf{p}_i^T \mathbf{H}_i^{-1}\mathbf{p}_i - r_i)^{1/2} \ \text{for} \ i = 1, \ 2, \ \dots, \ q$$

(ii) *Minimization of a sum of L_2 norms* Unconstrained minimization problems of the form

$$\text{minimize} \ \sum_{i=1}^{N} \|\mathbf{A}_i \mathbf{x} + \mathbf{c}_i\|_2$$

occur in many applications. By introducing an upper bound for each L_2-norm term in the objective function, the problem can be converted to

$$\text{minimize} \ \sum_{i=1}^{N} \delta_i \tag{13.107a}$$

$$\text{subject to:} \quad \|\mathbf{A}_i \mathbf{x} + \mathbf{c}_i\|_2 \leq \delta_i \quad \text{for } i = 1, 2, \ldots, N$$

$$(13.107b)$$

If we define an augmented variable vector

$$\tilde{\mathbf{x}} = \begin{bmatrix} \delta_1 \\ \vdots \\ \delta_N \\ \mathbf{x} \end{bmatrix}$$

and let

$$\tilde{\mathbf{b}} = \begin{bmatrix} 1 \\ \vdots \\ 1 \\ \mathbf{0} \end{bmatrix}, \quad \tilde{\mathbf{A}}_i = [\mathbf{0} \ \mathbf{A}_i], \quad \tilde{\mathbf{b}}_i = \begin{bmatrix} 0 \\ \vdots \\ 0 \\ 1 \\ 0 \\ \vdots \\ 0 \end{bmatrix} \leftarrow i\text{th component}$$

then Eqs. (13.107a) and (13.107b) become

$$\text{minimize} \ \tilde{\mathbf{b}}^T \tilde{\mathbf{x}}$$

$$\text{subject to:} \quad \|\tilde{\mathbf{A}}_i \tilde{\mathbf{x}} + \mathbf{c}_i\|_2 \leq \tilde{\mathbf{b}}_i^T \tilde{\mathbf{x}} \quad \text{for } i = 1, 2, \ldots, N$$

which is an SOCP problem.

Another unconstrained problem related to the problem in Eqs. (13.107a) and (13.107b) is the minimax problem

$$\underset{\mathbf{x}}{\text{minimize}} \ \underset{1 \leq i \leq N}{\text{maximize}} \ \|\mathbf{A}_i \mathbf{x} + \mathbf{c}_i\|_2 \tag{13.108}$$

where

$$\underset{1 \leq i \leq N}{\text{maximize}} \ \|\mathbf{A}_i \mathbf{x} + \mathbf{c}_i\|_2$$

maximizes function $\|\mathbf{A}_i \mathbf{x} + \mathbf{c}_i\|_2$ with respect to i while variable $\mathbf{x}$ is kept fixed. The problem in Eq. (13.108) can be re-formulated as the SOCP problem

$$\text{minimize} \ \delta$$

$$\text{subject to:} \quad \|\mathbf{A}_i \mathbf{x} + \mathbf{c}_i\|_2 \leq \delta \quad \text{for } i = 1, 2, \ldots, N$$

(iii) *Complex L_1-norm approximation problem* An interesting special case of the sum-of-norms problem is the complex L_1-norm approximation problem whereby a complex-valued approximate solution for the linear equation $\mathbf{A}\mathbf{x} = \mathbf{b}$ is required where $\mathbf{A}$ and $\mathbf{b}$ are complex-valued such that $\mathbf{x}$ solves the unconstrained problem

$$\text{minimize} \ \|\mathbf{A}\mathbf{x} - \mathbf{c}\|_1$$

where $\mathbf{A} \in C^{m \times n}$, $\mathbf{c} \in C^{m \times 1}$, $\mathbf{x} \in C^{n \times 1}$, and the L_1 norm of $\mathbf{x}$ is defined as $\|\mathbf{x}\|_1 = \sum_{k=1}^{n} |x_k|$. If we let $\mathbf{A} = [\mathbf{a}_1 \; \mathbf{a}_2 \; \cdots \; \mathbf{a}_m]^T$ and $\mathbf{c} = [c_1 \; c_2 \; \cdots \; c_m]^T$ where $\mathbf{a}_k = \mathbf{a}_{kr} + j\mathbf{a}_{ki}$, $c_k = c_{kr} + jc_{ki}$, $\mathbf{x} = \mathbf{x}_r + j\mathbf{x}_i$, and $j = \sqrt{-1}$, then we have

$$\|\mathbf{A}\mathbf{x} - \mathbf{c}\|_1 = \sum_{k=1}^{m} |\mathbf{a}_k^T \mathbf{x} - c_k|$$

$$= \sum_{k=1}^{m} [(\mathbf{a}_{kr}^T \mathbf{x}_r - \mathbf{a}_{ki}^T \mathbf{x}_i - c_{kr})^2 + (\mathbf{a}_{kr}^T \mathbf{x}_i + \mathbf{a}_{ki}^T \mathbf{x}_r - c_{ki})^2]^{1/2}$$

$$= \sum_{k=1}^{m} \left\| \underbrace{\begin{bmatrix} \mathbf{a}_{kr}^T & -\mathbf{a}_{ki}^T \\ \mathbf{a}_{ki}^T & \mathbf{a}_{kr}^T \end{bmatrix}}_{\mathbf{A}_k} \underbrace{\begin{bmatrix} \mathbf{x}_r \\ \mathbf{x}_i \end{bmatrix}}_{\hat{\mathbf{x}}} - \underbrace{\begin{bmatrix} c_{kr} \\ c_{ki} \end{bmatrix}}_{\mathbf{c}_k} \right\|_2 = \sum_{k=1}^{m} \|\mathbf{A}_k \hat{\mathbf{x}} - \mathbf{c}_k\|_2$$

Hence the problem under consideration can be converted to

$$\text{minimize} \sum_{k=1}^{m} \delta_k \tag{13.109a}$$

$$\text{subject to:} \quad \|\mathbf{A}_k \hat{\mathbf{x}} - \mathbf{c}_k\|_2 \le \delta_k \quad \text{for } k = 1, \; 2, \; \ldots, \; m \tag{13.109b}$$

By letting

$$\tilde{\mathbf{x}} = \begin{bmatrix} \delta_1 \\ \vdots \\ \delta_m \\ \hat{\mathbf{x}} \end{bmatrix}, \quad \tilde{\mathbf{b}}_0 = \begin{bmatrix} 1 \\ \vdots \\ 1 \\ \mathbf{0} \end{bmatrix}, \quad \tilde{\mathbf{A}}_k = [\mathbf{0} \; \mathbf{A}_k], \quad \text{and}$$

$$\tilde{\mathbf{b}}_k = \begin{bmatrix} 0 \\ \vdots \\ 0 \\ 1 \\ 0 \\ \vdots \\ 0 \end{bmatrix} \leftarrow \text{the } k\text{th component}$$

the problem in Eqs. (13.109a) and (13.109b) becomes

$$\text{minimize} \; \tilde{\mathbf{b}}_0^T \tilde{\mathbf{x}}$$

$$\text{subject to:} \quad \|\tilde{\mathbf{A}}_k \tilde{\mathbf{x}} - \mathbf{c}_k\|_2 \le \tilde{\mathbf{b}}_k^T \tilde{\mathbf{x}} \quad \text{for } k = 1, \; 2, \; \ldots, \; m$$

which is obviously an SOCP problem.

(iv) *Linear fractional problem* The linear fractional problem can be described as

$$\text{minimize} \sum_{i=1}^{p} \frac{1}{\mathbf{a}_i^T \mathbf{x} + c_i} \tag{13.110a}$$

$$\text{subject to:} \quad \mathbf{a}_i^T \mathbf{x} + c_i > 0 \quad \text{for } i = 1, 2, \ldots, p \qquad (13.110b)$$

$$\mathbf{b}_j^T \mathbf{x} + d_j \geq 0 \quad \text{for } j = 1, 2, \ldots, q \qquad (13.110c)$$

It can be readily verified that subject to the constraints in Eq. (13.110b), each term in the objective function is convex and hence the objective function itself is also convex. It, therefore, follows that the problem in Eqs. (13.110a)–(13.110c) is a CP problem. By introducing the auxiliary constraints

$$\frac{1}{\mathbf{a}_i^T \mathbf{x} + c_i} \leq \delta_i \qquad \text{for } i = 1, 2, \ldots, p$$

i.e.,

$$\delta_i (\mathbf{a}_i^T \mathbf{x} + c_i) \geq 1$$

and

$$\delta_i \geq 0$$

the problem in Eqs. (13.110a)–(13.110c) can be expressed as

$$\text{minimize} \sum_{i=1}^{p} \delta_i \qquad (13.111a)$$

$$\text{subject to:} \quad \delta_i (\mathbf{a}_i^T \mathbf{x} + c_i) \geq 1 \quad \text{for } i = 1, 2, \ldots, p \qquad (13.111b)$$

$$\delta_i \geq 0 \qquad (13.111c)$$

$$\mathbf{b}_j^T \mathbf{x} + d_j \geq 0 \quad \text{for } j = 1, 2, \ldots, q \qquad (13.111d)$$

Furthermore, we note that $w^2 \leq uv$, $u \geq 0$, $v \geq 0$ if and only if

$$\left\| \begin{bmatrix} 2w \\ u - v \end{bmatrix} \right\|_2 \leq u + v$$

(see Prob. 13.27) and hence the constraints in Eqs. (13.111b) and (13.111c) can be written as

$$\left\| \begin{bmatrix} 2 \\ \mathbf{a}_i^T \mathbf{x} + c_i - \delta_i \end{bmatrix} \right\|_2 \leq \mathbf{a}_i^T \mathbf{x} + c_i + \delta_i \qquad \text{for } i = 1, 2, \ldots, p$$

Hence the problem in Eqs. (13.111a)–(13.111d) can be formulated as

$$\text{minimize} \sum_{i=1}^{p} \delta_i$$

$$\text{subject to:} \quad \left\| \begin{bmatrix} 2 \\ \mathbf{a}_i^T \mathbf{x} + c_i - \delta_i \end{bmatrix} \right\|_2 \leq \mathbf{a}_i^T \mathbf{x} + c_i + \delta_i \quad \text{for } i = 1, 2, \ldots, p$$

$$\mathbf{b}_j^T \mathbf{x} + d_j \geq 0 \quad \text{for } j = 1, 2, \ldots, q$$

which is an SOCP problem.

13.10   A Primal-Dual Method for SOCP Problems

13.10.1   Assumptions and KKT Conditions

If we let

$$\mathbf{c} = \begin{bmatrix} \hat{\mathbf{c}}_1 \\ \hat{\mathbf{c}}_2 \\ \vdots \\ \hat{\mathbf{c}}_q \end{bmatrix}, \quad \mathbf{x} = \begin{bmatrix} \mathbf{x}_1 \\ \mathbf{x}_2 \\ \vdots \\ \mathbf{x}_q \end{bmatrix}, \quad \mathbf{s} = \begin{bmatrix} \mathbf{s}_1 \\ \mathbf{s}_2 \\ \vdots \\ \mathbf{s}_q \end{bmatrix}$$

$$\mathbf{A} = [\hat{\mathbf{A}}_1 \ \hat{\mathbf{A}}_2 \ \cdots \ \hat{\mathbf{A}}_q], \quad \text{and} \quad \mathcal{K} = \mathcal{K}_1 \times \mathcal{K}_2 \times \cdots \times \mathcal{K}_q$$

where $\mathcal{K}_1$, $\mathcal{K}_2$, ..., $\mathcal{K}_q$ are the second-order cones in Eqs. (13.101a)–(13.101c) and Eqs. (13.102a)–(13.102c), and $\mathcal{K} = \mathcal{K}_1 \times \mathcal{K}_2 \times \cdots \times \mathcal{K}_q$ represents a second-order cone whose elements are of the form $\mathbf{x} = [\mathbf{x}_1^T \ \mathbf{x}_2^T \ \cdots \ \mathbf{x}_q^T]^T$ with $\mathbf{x}_i \in \mathcal{K}_i$ for $i = 1, 2, \ldots, q$, then the primal and dual SOCP problems in Eqs. (13.101a)–(13.101c) and Eqs. (13.102a)–(13.102c) can be expressed as

$$\text{minimize} \ \mathbf{c}^T \mathbf{x} \tag{13.112a}$$

$$\text{subject to:} \ \ \mathbf{A}\mathbf{x} = \mathbf{b}, \ \mathbf{x} \in \mathcal{K} \tag{13.112b}$$

and

$$\text{maximize} \ - \mathbf{b}^T \mathbf{y} \tag{13.113a}$$

$$\text{subject to:} \ \ - \mathbf{A}^T \mathbf{y} + \mathbf{s} = \mathbf{c}, \ \mathbf{s} \in \mathcal{K} \tag{13.113b}$$

respectively. The feasible sets for the primal and dual SOCP problems are defined as

$$\mathcal{F}_p = \{\mathbf{x} : \ \mathbf{A}\mathbf{x} = \mathbf{b}, \ \mathbf{x} \in \mathcal{K}\}$$

and

$$\mathcal{F}_d = \{(\mathbf{s}, \ \mathbf{y}) : \ - \mathbf{A}^T \mathbf{y} + \mathbf{s} = \mathbf{c}, \ \mathbf{s} \in \mathcal{K}\}$$

respectively. The duality gap between $\mathbf{x} \in \mathcal{F}_p$ and $(\mathbf{s}, \ \mathbf{y}) \in \mathcal{F}_d$ assumes the form

$$\delta(\mathbf{x}, \mathbf{s}, \mathbf{y}) = \mathbf{c}^T \mathbf{x} + \mathbf{b}^T \mathbf{y} = -(\mathbf{A}^T \mathbf{y} + \mathbf{s})^T \mathbf{x} + \mathbf{b}^T \mathbf{y} = \mathbf{s}^T \mathbf{x}$$

A vector $\mathbf{x}_i = [t_i \ \mathbf{u}_i^T]^T$ in space $R^{n_i \times 1}$ is said to be an interior point of the second-order cone $\mathcal{K}_i$ if $\|\mathbf{u}_i\|_2 < t_i$. If we denote the set of all interior points of $\mathcal{K}_i$ as $\mathcal{K}_i^o$ and let

$$\mathcal{K}^o = \mathcal{K}_1^o \times \mathcal{K}_2^o \times \cdots \times \mathcal{K}_q^o$$

then a strictly feasible vector for the problem in Eqs. (13.112a) and (13.112b) is a vector $\mathbf{x} \in \mathcal{K}_i^o$ satisfying the constraint in Eq. (13.112b). Based on these ideas, the strictly feasible sets of the primal and dual SOCP problems are given by

$$\mathcal{F}_p^o = \{\mathbf{x} : \ \mathbf{A}\mathbf{x} = \mathbf{b}, \ \mathbf{x} \in \mathcal{K}^o\}$$

and

$$\mathcal{F}_d^o = \{(\mathbf{s}, \ \mathbf{y}) : \ - \mathbf{A}^T \mathbf{y} + \mathbf{s} = \mathbf{c}, \ \mathbf{s} \in \mathcal{K}^o\}$$

respectively.

In the rest of the chapter, we make the following two assumptions:

1. There exists a strictly feasible point $\mathbf{x}$ for the primal problem in Eqs. (13.112a) and (13.112b) and a strictly feasible pair $(\mathbf{s}, \mathbf{y})$ for the dual problem in Eqs. (13.113a) and (13.113b), i.e., both $\mathcal{F}_p^o$ and $\mathcal{F}_d^o$ are nonempty.
2. The rows of matrix $\mathbf{A}$ are linearly independent.

Under these assumptions, solutions for the primal and dual SOCP problems exist and finding these solutions is equivalent to finding a vector set $(\mathbf{x}, \mathbf{s}, \mathbf{y}) \in \mathcal{K} \times \mathcal{K} \times R^m$ that satisfies the KKT conditions [23]

$$\mathbf{A}\mathbf{x} = \mathbf{b} \tag{13.114a}$$

$$-\mathbf{A}^T\mathbf{y} + \mathbf{s} = \mathbf{c} \tag{13.114b}$$

$$\mathbf{x}^T\mathbf{s} = 0 \tag{13.114c}$$

where the condition in Eq. (13.114c) is referred to as the *complementarity* condition. From Eq. (13.114c), we note that the duality gap $\delta(\mathbf{x}, \mathbf{s}, \mathbf{y})$ at the primal and dual solution points becomes zero.

13.10.2   A Primal-Dual Interior-Point Algorithm

In this section, we introduce a primal-dual interior-point algorithm for SOCP, which is a slightly modified version of an algorithm proposed in [23]. In the kth iteration of the algorithm, the vector set $(\mathbf{x}_k, \mathbf{s}_k, \mathbf{y}_k)$ is updated to

$$(\mathbf{x}_{k+1}, \mathbf{s}_{k+1}, \mathbf{y}_{k+1}) = (\mathbf{x}_k, \mathbf{s}_k, \mathbf{y}_k) + \alpha_k(\Delta\mathbf{x}, \Delta\mathbf{s}, \Delta\mathbf{y}) \tag{13.115}$$

where $(\Delta\mathbf{x}, \Delta\mathbf{s}, \Delta\mathbf{y})$ is obtained by solving the linear system of equations

$$\mathbf{A}\Delta\mathbf{x} = \mathbf{b} - \mathbf{A}\mathbf{x} \tag{13.116a}$$

$$-\mathbf{A}^T\Delta\mathbf{y} + \Delta\mathbf{s} = \mathbf{c} - \mathbf{s} + \mathbf{A}^T\mathbf{y} \tag{13.116b}$$

$$\mathbf{S}\Delta\mathbf{x} + \mathbf{X}\Delta\mathbf{s} = \sigma\mu\mathbf{e} - \mathbf{X}\mathbf{s} \tag{13.116c}$$

where $\mathbf{e} = [1\ 1\ \cdots\ 1]^T$,

$$\mathbf{X} = \text{diag}\{\mathbf{X}_1, \ldots, \mathbf{X}_q\} \quad \text{with } \mathbf{X}_i = \begin{bmatrix} t_i & \mathbf{u}_i^T \\ \mathbf{u}_i & t_i\mathbf{I}_i \end{bmatrix} \tag{13.116d}$$

$$\mathbf{S} = \text{diag}\{\mathbf{S}_1, \ldots, \mathbf{S}_q\} \tag{13.116e}$$

$$\mu = \mathbf{x}^T\mathbf{s}/q \tag{13.116f}$$

σ is a small positive scalar, and $(\mathbf{x}, \mathbf{s}, \mathbf{y})$ assumes the value of $(\mathbf{x}_k, \mathbf{s}_k, \mathbf{y}_k)$. In Eq. (13.116d), t_i and $\mathbf{u}_i$ are the first component and the remaining part of vector $\mathbf{x}_i$, respectively, and $\mathbf{I}_i$ is the identity matrix of dimension $n_i - 1$. Matrices $\mathbf{S}_i$ for $i = 1, 2, \ldots, q$ in Eq. (13.116e) are defined in a similar manner. On comparing Eqs. (13.116a)–(13.116f) with Eqs. (13.11a) and (13.11b), it is evident that the vector set $(\mathbf{x}_k, \mathbf{s}_k, \mathbf{y}_k)$ is updated so that the new vector set $(\mathbf{x}_{k+1}, \mathbf{s}_{k+1}, \mathbf{y}_{k+1})$ better approximates the KKT conditions in Eqs. (13.114a)–(13.114c).

In Eq. (13.115), α_k is a positive scalar that is determined using the line search

$$\alpha_k = 0.75 \min(\alpha_{k1}, \; \alpha_{k2}, \; \alpha_{k3}) \tag{13.117a}$$

$$\alpha_{k1} = \max_{0<\alpha\le1} \; (\mathbf{x}_k + \alpha\Delta\mathbf{x} \in \mathcal{F}_p^o) \tag{13.117b}$$

$$\alpha_{k2} = \max_{0<\alpha\le1} \; (\mathbf{s}_k + \alpha\Delta\mathbf{s} \in \mathcal{F}_d^o) \tag{13.117c}$$

$$\alpha_{k3} = \max_{0<\alpha\le1} \; [\mathbf{c} + \mathbf{A}^T(\mathbf{y}_k + \alpha\Delta\mathbf{y}) \in \mathcal{F}_d^o] \tag{13.117d}$$

It follows from Eq. (13.115) and Eqs. (13.101a)–(13.101c) that the updated vector set $(\mathbf{x}_{k+1}, \; \mathbf{s}_{k+1}, \; \mathbf{y}_{k+1})$ will remain strictly feasible.

An algorithm based on the above approach is as follows.

Algorithm 13.7 Primal-dual interior-point algorithm for SOCP problems

Step 1
Input data set $(\mathbf{A}, \; \mathbf{b}, \; \mathbf{c})$, parameters q and n_i for $i = 1, \; 2, \; \ldots, \; q$, and tolerance ε.
Input an initial vector set $(\mathbf{x}_0, \mathbf{s}_0, \mathbf{y}_0)$ with $\mathbf{x}_0 \in \mathcal{F}_p^o$ and $(\mathbf{s}_0, \mathbf{y}_0) \in \mathcal{F}_d^o$.
Set $\mu_0 = \mathbf{x}_0^T \mathbf{s}_0/q$, $\sigma = 10^{-5}$, and $k = 0$.
Step 2
Compute the solution $(\Delta\mathbf{x}, \; \Delta\mathbf{s}, \; \Delta\mathbf{y})$ of Eqs. (13.116a)–(13.116c) where $(\mathbf{x}, \; \mathbf{s}, \; \mathbf{y}) = (\mathbf{x}_k, \mathbf{s}_k, \mathbf{y}_k)$ and $\mu = \mu_k$.
Step 3
Compute α_k using Eqs. (13.117a)–(13.117d).
Step 4
Set $(\mathbf{x}_{k+1}, \mathbf{s}_{k+1}, \mathbf{y}_{k+1}) = (\mathbf{x}_k + \alpha_k\Delta\mathbf{x}, \; \mathbf{s}_k + \alpha_k\Delta\mathbf{s}, \; \mathbf{y}_k + \alpha_k\Delta\mathbf{y})$
Step 5
Compute $\mu_{k+1} = \mathbf{x}_{k+1}^T\mathbf{s}_{k+1}/q$. If $\mu_{k+1} \le \varepsilon$, output solution $(\mathbf{x}^*, \mathbf{s}^*, \mathbf{y}^*) = (\mathbf{x}_{k+1}, \mathbf{s}_{k+1}, \mathbf{y}_{k+1})$ and stop; otherwise, set $k = k + 1$ and repeat from Step 2.

Example 13.7 Apply Algorithm 13.7 to solve the shortest-distance problem discussed in Example 13.6.

Solution The problem can be formulated as

$$\text{minimize } \delta$$

$$\text{subject to: } [(x_1 - x_3)^2 + (x_2 - x_4)^2]^{1/2} \le \delta$$

$$[x_1 \; x_2]\begin{bmatrix}1/4 & 0\\0 & 1\end{bmatrix}\begin{bmatrix}x_1\\x_2\end{bmatrix} - [x_1 \; x_2]\begin{bmatrix}1/2\\0\end{bmatrix} \le \frac{3}{4}$$

$$[x_3 \; x_4]\begin{bmatrix}5/8 & 3/8\\3/8 & 5/8\end{bmatrix}\begin{bmatrix}x_3\\x_4\end{bmatrix} - [x_3 \; x_4]\begin{bmatrix}11/2\\13/2\end{bmatrix} \le -\frac{35}{2}$$

If we let $\mathbf{x} = [\delta \; x_1 \; x_2 \; x_3 \; x_4]^T$, the above problem can be expressed as

$$\text{minimize } \mathbf{b}^T\mathbf{x} \tag{13.118a}$$

$$\text{subject to:} \quad \|A_i^T x + c_i\|_2 \leq b_i^T x + d_i \quad \text{for } i = 1, 2, 3$$

$$(13.118b)$$

where

$$b = [1\ 0\ 0\ 0\ 0]^T$$

$$A_1^T = \begin{bmatrix} 0 & -1 & 0 & 1 & 0 \\ 0 & 0 & 1 & 0 & -1 \end{bmatrix}, \quad A_2^T = \begin{bmatrix} 0 & 0.5 & 0 & 0 & 0 \\ 0 & 0 & 1 & 0 & 0 \end{bmatrix}$$

$$A_3^T = \begin{bmatrix} 0 & 0 & 0 & -0.7071 & -0.7071 \\ 0 & 0 & 0 & -0.3536 & 0.3536 \end{bmatrix}$$

$$b_1 = b, \quad b_2 = 0, \quad b_3 = 0$$

$$c_1 = 0, \quad c_2 = [-0.5\ 0]^T, \quad c_3 = [4.2426\ -0.7071]^T$$

$$d_1 = 0, \quad d_2 = 1, \quad \text{and} \quad d_3 = 1$$

The SOCP formulation in Eqs. (13.118a) and (13.118b) is the same as that in Eqs. (13.104a) and (13.104b). Hence by using Eq. (13.103), the problem at hand can be converted into the primal and dual formulations in Eqs. (13.101a)–(13.101c) and Eqs. (13.102a)–(13.102c), respectively.

In order to generate a strictly feasible vector set (x_0, s_0, y_0), we note in Fig. 13.2 that point $[x_1\ x_2\ x_3\ x_4] = [1.5\ 0.5\ 2.5\ 4]$ is feasible. This suggests an initial $y_0 = [\beta\ 1.5\ 0.5\ 2.5\ 4]^T$ where β is a scalar to ensure that $s = c + A_0^T y_0 \in \mathcal{F}_d^o$. Since

$$s_0 = c - A^T y_0 = [-\beta\ 1\ 3.5\ 0\ 0\ 1\ 0.25\ 0.5\ 1\ -0.3535\ -0.1767]^T$$

$n_1 = 5$, $n_2 = 3$, and $n_3 = 3$, choosing $\beta = -3.7$ guarantees that $s_0 \in \mathcal{F}_d^o$. This gives

$$y_0 = [-3.7\ 1.5\ 0.5\ 2.5\ 4.0]^T$$

and

$$s_0 = [3.7\ 1\ 3.5\ 0\ 0\ 1\ 0.25\ 0.5\ 1\ -0.3535\ -0.1767]^T$$

Moreover, it can be readily verified that

$$x_0 = [1\ 0\ 0\ 0\ 0\ 0.1\ 0\ 0\ 0.1\ 0\ 0]^T \in \mathcal{F}_p^o$$

With $\varepsilon = 10^{-4}$, it took Algorithm 13.7 15 iterations to converge to vector set (x^*, s^*, y^*) where

$$y^* = \begin{bmatrix} 1.707791 \\ 2.044705 \\ 0.852730 \\ 2.544838 \\ 2.485646 \end{bmatrix}$$

which corresponds to the solution points $r^* \in \mathcal{R}$ and $s^* \in \mathcal{S}$ given by

$$r^* = \begin{bmatrix} 2.044705 \\ 0.852730 \end{bmatrix} \quad \text{and} \quad s^* = \begin{bmatrix} 2.544838 \\ 2.485646 \end{bmatrix}$$

These points give the shortest distance between $\mathcal{R}$ and $\mathcal{S}$ as $\|r^* - s^*\|_2 = 1.707790$. ∎

Problems

13.1 Show that the constrained optimization problem

$$\text{minimize} \ f(\mathbf{x}) = \int_0^\pi W(\omega)|H(\omega) - H_d(\omega)|^2 d\omega$$

$$\text{subject to:} \ |H(\omega_k) - H_d(\omega_k)| \leq \delta_k \quad \text{for } k = 1, 2, \ldots, K$$

where $H(\omega) = \sum_{i=0}^{N} a_i \cos i\omega$ and $\mathbf{x} = [a_0 \ a_1 \ \cdots \ a_N]^T$ is a convex QP problem.
$H_d(\omega)$ and $W(\omega)$ are given real-valued functions, $W(\omega) \geq 0$ is a weighting function, $\{\omega : \omega_k, \ k = 1, 2, \ldots, K\}$ is a set of grid points on $[0, \pi]$, and $\delta_k > 0$ for $1 \leq k \leq K$ are constants.

13.2 (*a*) Solve the QP problem

$$\text{minimize} \ f(\mathbf{x}) = 2x_1^2 + x_2^2 + x_1 x_2 - x_1 - x_2$$

$$\text{subject to:} \ x_1 + x_2 = 1$$

by using the SVD method.
(*b*) Repeat part (*a*) by using the QR-decomposition method.
(*c*) Repeat part (*a*) by using the Lagrange-multiplier method.

13.3 (*a*) Solve the QP problems

$$\text{minimize} \ f(\mathbf{x}) = 1.5x_1^2 - x_1 x_2 + x_2^2 - x_2 x_3 + 0.5x_3^2 + x_1 + x_2 + x_3$$

$$\text{subject to:} \ x_1 + 2x_2 + x_3 = 4$$

by using the SVD method.
(*b*) Repeat part (*a*) by using the QR-decomposition method.
(*c*) Repeat part (*a*) by using the Lagrange-multiplier method.

13.4 (*a*) By applying Algorithm 13.1, solve the following QP problem

$$\text{minimize} \ f(\mathbf{x}) = 3x_1^2 + 3x_2^2 - 10x_1 - 24x_2$$

$$\text{subject to:} \ 2x_1 + x_2 \leq 4$$

$$\mathbf{x} \geq \mathbf{0}$$

where $\mathbf{x}_0 = [0 \ 0]^T$.

(*b*) Repeat part (*a*) for the problem

$$\text{minimize} \ f(\mathbf{x}) = x_1^2 - x_1 x_2 + x_2^2 - 3x_1$$

$$\text{subject to:} \ x_1 + x_2 \leq 2$$

$$\mathbf{x} \geq \mathbf{0}$$

where $\mathbf{x}_0 = [0 \ 0]^T$.

(c) Repeat part (a) for the problem

$$\text{minimize } f(\mathbf{x}) = x_1^2 + 0.5x_2^2 - x_1x_2 - 3x_1 - x_2$$
$$\text{subject to: } x_1 + x_2 \leq 2$$
$$2x_1 - x_2 \leq 2$$
$$\mathbf{x} \geq \mathbf{0}$$

where $\mathbf{x}_0 = [0 \; 0]^T$.

(d) Repeat part (a) for the problem

$$\text{minimize } f(\mathbf{x}) = x_1^2 + x_2^2 + 0.5x_3^2 + x_1x_2 + x_1x_3 - 4x_1 - 3x_2 - 2x_3$$
$$\text{subject to: } x_1 + x_2 + x_3 \leq 3$$
$$\mathbf{x} \geq \mathbf{0}$$

where $\mathbf{x}_0 = [0 \; 0 \; 0]^T$.

13.5 Verify that the solution of Eq. (13.35) is given by Eqs. (13.37a)–(13.37f).

13.6 (a) Convert the QP problems in Prob. 13.4 into the form in Eqs. (13.26a)–(13.26c).

(b) Solve the QP problems obtained in part (a) by applying Algorithm 13.2.

13.7 Verify that the solution of Eq. (13.41) is given by Eqs. (13.42a)–(13.42g).

13.8 (a) Solve the QP problems obtained in Prob. 13.6(a) by applying Algorithm 13.3.

(b) Compare the results obtained in part (a) with those obtained in Prob. 13.6(b).

13.9 Show that if $\mathbf{H}$ is positive definite, $\mathbf{A}$ is of full row rank, $\boldsymbol{\mu}_k > \mathbf{0}$, and $\mathbf{x}_k > \mathbf{0}$, then the system in Eq. (13.35) has a unique solution for $\boldsymbol{\delta}_w$.

13.10 (a) By applying Algorithm 13.2, solve the QP problem

$$\text{minimize } \tfrac{1}{2}\mathbf{x}^T (\mathbf{hh}^T)\mathbf{x} + \mathbf{x}^T \mathbf{p}$$
$$\text{subject to: } \mathbf{Ax} = \mathbf{b}$$
$$\mathbf{x} \geq \mathbf{0}$$

where

$$\mathbf{h} = \begin{bmatrix} 1 \\ -4 \\ 2 \\ 1 \end{bmatrix}, \quad \mathbf{A} = [1 \; 1 \; 1 \; 1], \quad b = 4, \quad \text{and } \mathbf{p} = \begin{bmatrix} -1 \\ 0 \\ 7 \\ 4 \end{bmatrix}$$

with $\mathbf{x}_0 = [1 \; 1 \; 1 \; 1]^T$, $\lambda_0 = 2$, and $\boldsymbol{\mu}_0 = [1 \; 2 \; 9 \; 6]^T$.

(b) By applying Algorithm 13.3, solve the QP problem in part (a) with $\mathbf{x}_0 = [3 \; 3 \; 3 \; 3]^T$, $\lambda_0 = -1$, and $\boldsymbol{\mu}_0 = [1 \; 1 \; 1 \; 1]^T$. Compare the solution obtained with that obtained in part (a).

13.11 (a) Show that

$$K = \begin{bmatrix} 0 & A^T \\ -A & 0 \end{bmatrix}$$

is positive semidefinite in the sense of Eq. (13.46).
   (b) Show that if H is positive semidefinite, then

$$K = \begin{bmatrix} H & A^T \\ -A & 0 \end{bmatrix}$$

is positive semidefinite in the sense of Eq. (13.46).

13.12 (a) Convert the QP problem in Prob. 13.10(a) using the initial values for x_0, λ_0, and μ_0 given in Prob. 13.10(b) to a mixed LC problem.
   (b) Solve the LC problem obtained in part (a) by applying Algorithm 13.4.
   (c) Compare the solutions obtained with those obtained in Prob. 13.10(b).

13.13 Prove that the minimization problem in Eqs. (13.49a)–(13.49c) is a CP problem.

13.14 Show that the sets $\{X : X \in R^{n \times n}, X \succeq 0\}$ and $\{x : x \in R^{n \times 1}, x \geq 0\}$ are convex cones.

13.15 Given that $X \succeq 0$ and $S \succeq 0$, prove that

$$S \cdot X \geq 0$$

where $S \cdot X$ denotes the standard inner product in space $\mathcal{S}^n$ defined in Sect. 13.5.1.

13.16 (a) Prove that the inequality in Eq. (13.61) holds if and only if matrix $G(\delta, x)$ in Eq. (13.62) is positive semidefinite.
   (b) Specify matrices F_0 and F_i ($1 \leq i \leq n$) in Eq. (13.63) so that $F(x) \preceq 0$ reformulates the constraints in Eq. (13.60c).

13.17 The problem of minimizing the maximum eigenvalue of an affine matrix can be stated as follows: given matrices $A_0, A_1, \ldots, A_p$ in $\mathcal{S}^n$, find scalars $y_1, y_2, \ldots, y_p$ such that the maximum eigenvalue of

$$A_0 + \sum_{i=1}^{p} y_i A_i$$

is minimized. Formulate this optimization problem as an SDP problem.

13.18 (a) Prove that if $X \succ 0$ or $S \succ 0$, then

$$XS = \tau I$$

is equivalent to

$$XS + SX = 2\tau I$$

(b) Give a numerical example to demonstrate that if the condition $\mathbf{X} \succ \mathbf{0}$ or $\mathbf{S} \succ \mathbf{0}$ is removed, then the equalities in part (a) are no longer equivalent.

13.19 (a) Verify the identity in Eq. (13.80).

(b) Verify the identity in Eq. (13.82).

13.20 By using Eq. (13.81) with a special symmetric matrix $\mathbf{K}$, each row of matrix $\mathbf{M} \odot \mathbf{N}$ can be determined for given matrices $\mathbf{M}$ and $\mathbf{N}$.

(a) Develop an algorithm that computes the symmetric Kronecker product $\mathbf{M} \odot \mathbf{N}$ for given $\mathbf{M}, \mathbf{N} \in R^{n \times n}$.

(b) Write a MATLAB program that implements the algorithm developed in part (a).

13.21 Prove Lemma 13.1.

13.22 Using Lemma 13.1, show that if matrices $\mathbf{M}$ and $\mathbf{N}$ commute then the solution of the Lyapunov equation

$$\mathbf{MXN}^T + \mathbf{NXM}^T = \mathbf{B}$$

where $\mathbf{B} \in \mathcal{S}^n$, $\mathbf{M}$, and $\mathbf{N}$ are given, can be expressed as

$$\mathbf{X} = \mathbf{VCV}^T$$

where $\mathbf{V} = [\mathbf{v}_1 \ \mathbf{v}_2 \ \cdots \ \mathbf{v}_n]$, $\mathbf{v}_i$ for $1 \leq i \leq n$ are the eigenvectors in Lemma 13.1, and $\mathbf{C}$ is obtained by calculating $\mathbf{V}^T \mathbf{B} \mathbf{V}$ and dividing its elements by $(\alpha_i \beta_j + \beta_i \alpha_j)$ componentwise.

13.23 Verify that $\{\Delta \mathbf{x}, \ \Delta \mathbf{y}, \ \Delta \mathbf{s}\}$ obtained from Eqs. (13.87a)–(13.87c) solves Eq. (13.86).

13.24 Show that the dual of the primal SOCP problem in Eqs. (13.101a)–(13.101c) assumes the form in Eqs. (13.102a)–(13.102c).

13.25 Show that the SOCP problem in Eqs. (13.102a)–(13.102c) is equivalent to the SOCP problem in Eqs. (13.104a) and (13.104b).

13.26 Given a column vector $\mathbf{u}$ and a nonnegative scalar t such that $\|\mathbf{u}\|_2 \leq t$, the matrix

$$\begin{bmatrix} t\mathbf{I} & \mathbf{u} \\ \mathbf{u}^T & t \end{bmatrix}$$

can be constructed. Show that the matrix is positive semidefinite.

13.27 Show that $w^2 \leq uv$, $u \geq 0$, and $v \geq 0$ if and only if

$$\left\| \begin{bmatrix} 2w \\ u - v \end{bmatrix} \right\|_2 \leq u + v$$

13.28 Consider the least-squares minimization problem with quadratic inequality (LSQI) constraints, which arises in cases where the solution to the ordinary least-squares problem needs to be regularized [9], namely,

$$\text{minimize } ||\mathbf{Ax} - \mathbf{b}||_2$$

$$\text{subject to: } ||\mathbf{Bx}||_2 \leq \delta$$

where $\mathbf{A} \in R^{m \times n}$, $\mathbf{b} \in R^{m \times 1}$, $\mathbf{B} \in R^{n \times n}$ with $\mathbf{B}$ nonsingular and $\delta \geq 0$.

(a) Convert the LSQI problem to an SOCP problem.
(b) By applying Algorithm 13.7, solve the LSQI problem if

$$\mathbf{A} = \begin{bmatrix} 2 & 0 \\ 0 & 1 \\ 0 & 0 \end{bmatrix}, \quad \mathbf{b} = \begin{bmatrix} 4 \\ 2 \\ 3 \end{bmatrix}, \quad \mathbf{B} = \begin{bmatrix} 1 & 0 \\ 0 & 1 \end{bmatrix}, \quad and \quad \delta = 0.1$$

(c) Apply the algorithm used in part (b) to the case where δ is increased to 1. Compare the solution obtained with that obtained in part (b).

References

1. F. Alizadeh, "Combinational optimization with interior point methods and semidefinite matrices," Ph.D. dissertation, University of Minnesota, Oct. 1991.
2. Y. Nesterov and A. Nemirovskii, *Interior-Point Polynomial Algorithms in Convex Programming.* Philadelphia, PA: SIAM, 1994.
3. L. Vandenberghe and S. Boyd, "Semidefinite programming," *SIAM Review*, vol. 38, pp. 49–95, Mar. 1996.
4. C. Helmberg, F. Rendl, R. Vanderbei, and H. Wolkowicz, "An interior-point method for semidefinite programming," *SIAM J. Optim.*, vol. 6, pp. 342–361, 1996.
5. M. Kojima, S. Shindoh, and S. Hara, "Interior-point methods for the monotone linear complementarity problem in symmetric matrices," *SIAM J. Optim.*, vol. 7, pp. 86–125, Nov. 1997.
6. Y. Nesterov and M. Todd, "Primal-dual interior-point method for self-scaled cones," *SIAM J. Optim.*, vol. 8, pp. 324–364, 1998.
7. F. Alizadeh, J. A. Haeberly, and M. L. Overton, "Primal-dual interior-point methods for semidefinite programming: Convergence rates, stability and numerical results," *SIAM J. Optim.*, vol. 8, pp. 746–768, 1998.
8. R. Monteiro, "Polynomial convergence of primal-dual algorithm for semidefinite programming based on Monteiro and Zhang family of directions," *SIAM J. Optim.*, vol. 8, pp. 797–812, 1998.
9. G. H. Golub and C. F. Van Loan, *Matrix Computations*, 4th ed. Baltimore, MD: Johns Hopkins University Press, 2013.
10. R. Fletcher, *Practical Methods of Optimization*, 2nd ed. New York: Wiley, 1987.
11. P. E. Gill, W. Murray, and M. H. Wright, *Practical Optimization.* New York: Academic Press, 1981.
12. D. Goldfarb and A. Idnani, "A numerically stable dual method for solving strictly convex quadratic programs," *Math. Programming*, vol. 27, pp. 1–33, 1983.
13. C. L. Lawson and R. J. Hanson, *Solving Least Squares Problems.* Englewood Cliffs, NJ: Prentice-Hall, 1974.

14. R. D. C. Monteiro and I. Adler, "Interior path following primal-dual algorithms, Part II: Convex quadratic programming," *Math. Programming*, vol. 44, pp. 45–66, 1989.

15. R. D. C. Monteiro and I. Adler, "Interior path following primal-dual algorithms, Part I: Linear programming," *Math. Programming*, vol. 44, pp. 27–41, 1989.

16. Y. Ye, *Interior Point Algorithms: Theory and Analysis*. New York: Wiley, 1997.

17. S. J. Wright, *Primal-Dual Interior-Point Methods*. Philadelphia, PA: SIAM, 1997.

18. S. Boyd, L. El Ghaoui, E. Feron, and V. Balakrishnan, *Linear Matrix Inequalities in System and Control Theory*. Philadelphia, PA: SIAM, 1994.

19. A. S. Lewis and M. L. Overton, "Eigenvalue optimization," *Acta Numerica*, vol. 5, pp. 149–190, 1996.

20. M. Shida, S. Shindoh, and M. Kojima, "Existence and uniqueness of search directions in interior-point algorithms for the SDP and the monotone SDLCP," *SIAM J. Optim.*, vol. 8, pp. 387–398, 1998.

21. S. Mehrotra, "On the implementation of a primal-dual interior point method," *SIAM J. Optim.*, vol. 2, pp. 575–601, 1992.

22. M. S. Lobo, L. Vandenberghe, S. Boyd, and H. Lebret, "Applications of second-order cone programming," *Linear Algebra and Its Applications*, vol. 284, pp. 193–228, Nov. 1998.

23. R. D. C. Monteiro and T. Tsuchiya, "Polynomial convergence of primal-dual algorithms for the second-order cone program based on the MZ-family of directions," *Math. Programming*, vol. 88, pp. 61–83, 2000.

Algorithms for General Convex Problems

14

14.1 Introduction

In Chap. 13, we have addressed several special classes of convex problems that have
found widespread applications in engineering and explored a variety of algorithms
for the solution of these problems. There are, however, certain convex problems that
do not belong to any one of these classes. This chapter is devoted to the study of methods for the general convex problem whereby the objective function or constraints
do not satisfy all the required properties imposed by the problem types described
in Chap. 13. The algorithms considered include Newton algorithms, proximal-point
algorithms for composite convex functions, and alternating direction algorithms.
Several concepts and properties of convex functions that are needed in the development of these algorithms are introduced in Sect. 14.2. These include the concept of subgradient, convex functions with Lipschitz continuous gradients, strongly
convex functions, conjugate functions, and proximal operators. Newton algorithms
for unconstrained and constrained convex problems are addressed in Sect. 14.3.
In Sect. 14.4, several algorithms for minimizing composite convex functions are
studied. The $l_1 - l_2$ minimization problem, which finds applications in digital signal
processing and machine learning, is covered as a special case. Alternating direction
methods are becoming increasingly important nowadays because of their capability
in dealing with large-scale convex problems. In Sect. 14.5, we present two representative alternating direction methods known as the *alternating direction method
of multipliers* and *the alternating minimization method*.

14.2 Concepts and Properties of Convex Functions

The notion of a convex function and the elementary properties of convex functions
have been examined in Chap. 2. Here we introduce several additional concepts and
properties of convex functions that are of use in the development of efficient algorithms for convex problems.

© Springer Science+Business Media, LLC, part of Springer Nature 2021
A. Antoniou and W.-S. Lu, *Practical Optimization*, Texts in Computer Science,
https://doi.org/10.1007/978-1-0716-0843-2_14

14.2.1 Subgradient

Practical optimization problems involving nondifferentiable objective functions and/or constraints are pervasive. The nonsmoothness of the functions in an optimization problem may be encountered in several ways. In some optimization problems, the objective function and/or constraints are inherently nondifferentiable, for example, in the problem

$$\text{minimize } \|\mathbf{x}\|_1$$

$$\text{subject to: } \|\mathbf{A} - \mathbf{b}\|_2 \leq \varepsilon$$

where the l_1 norm of variable $\mathbf{x}$ is minimized subject to an l_2 norm constraint. Obviously, the objective function $\|\mathbf{x}\|_1 = \sum_{i=1}^{n} |x_i|$ is continuous and convex but *not everywhere differentiable*. There are also scenarios where the individual functions involved are differentiable but the objective function itself is not, for example, if the objective function is of the form

$$f(\mathbf{x}) = \max_{1 \leq j \leq p} \{\phi_j(\mathbf{x})\}$$

where the $\phi_j(\mathbf{x})$ for $j = 1, 2, \ldots, p$ are smooth convex functions and the objective function is also convex but is not everywhere differentiable. A specific objective function of this type that involves three linear $\phi_j(\mathbf{x})$ is illustrated in Fig. 14.1.

The gradient of a function plays a key role for the optimization of functions that are everywhere differentiable. The concept of subgradient is a natural generalization of the concept of gradient that allows us to deal with optimization problems involving

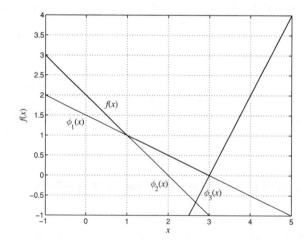

Fig. 14.1 The pointwise maximum of three affine functions yields a piecewise affine function which is convex but not differentiable

convex functions that are not everywhere differentiable [1]. Recall that a function $f(\mathbf{x})$ that is differentiable at point $\mathbf{x} = \tilde{\mathbf{x}}$ is convex if

$$f(\mathbf{x}) \geq f(\tilde{\mathbf{x}}) + \nabla^T f(\tilde{\mathbf{x}})(\mathbf{x} - \tilde{\mathbf{x}}) \quad \text{for } \mathbf{x}, \tilde{\mathbf{x}} \in \text{dom}(f) \tag{14.1}$$

where dom(f) is the domain of function $f(\mathbf{x})$ that defines the set of points $\mathbf{x}$ where $f(\mathbf{x})$ assumes finite values. In geometric terms, Eq. (14.1) states that at any point $\mathbf{x}$ in the domain of a convex function $f(\mathbf{x})$, the tangent line, plane, or hyperplane in general defined by $y = f(\mathbf{x})$ always lies below the curve, surface, or hypersurface defined by the function $y = f(\mathbf{x})$.

Definition 14.1 If function $f(\mathbf{x})$ is convex but not necessarily differentiable at a point $\tilde{\mathbf{x}}$, then vector $\mathbf{g} \in R^n$ is said to be a subgradient of $f(\mathbf{x})$ at $\tilde{\mathbf{x}}$ if

$$f(\mathbf{x}) \geq f(\tilde{\mathbf{x}}) + \mathbf{g}^T(\mathbf{x} - \tilde{\mathbf{x}}) \quad \text{for } \tilde{\mathbf{x}} \in \text{dom}(f) \tag{14.2}$$

The subgradient, $\mathbf{g}$, of a function $f(\mathbf{x})$ is not unique if the function is not differentiable at point $\tilde{\mathbf{x}}$. The set of all subgradients of $f(\mathbf{x})$ at a point $\tilde{\mathbf{x}}$ is called a *subdifferential* of $f(\mathbf{x})$ and is denoted as $\partial f(\tilde{\mathbf{x}})$. ∎

The right-hand side of Eq. (14.2) may be viewed as a linear lower bound of $f(\mathbf{x})$ and the subgradients at a point $\tilde{\mathbf{x}}$ where the convex function $f(\mathbf{x})$ is not differentiable correspond to different tangent hyperplanes at $\tilde{\mathbf{x}}$.

Definition 14.1 is illustrated in Fig. 14.2 for the case of a function $f(x)$ which is not differentiable at point $\tilde{x}$ where the two subgradients of $f(x)$ are given by $g_1 = \tan\theta_1$ and $g_2 = \tan\theta_2$. From the figure, it is obvious that any tangent line at $\tilde{x}$ with a slope between g_2 and g_1 satisfies Eq. (14.2) and, therefore, any value $g \in [g_2, g_1]$ is a subgradient of $f(x)$ at $\tilde{x}$.

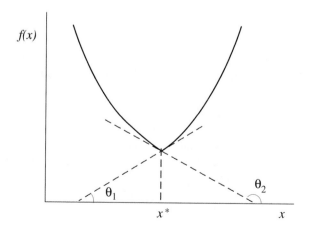

Fig. 14.2 Two subgradients of $f(x)$ at x^* where $f(x)$ is not differentiable

From Eq. (14.2), it follows that $f(\mathbf{x}) \geq f(\tilde{\mathbf{x}})$ as long as $\mathbf{g}^T(\mathbf{x} - \tilde{\mathbf{x}}) \geq 0$. Note that for a given point $\tilde{\mathbf{x}}$, $\mathbf{g}^T(\mathbf{x} - \tilde{\mathbf{x}})$ defines a hyperplane which passes through point $\tilde{\mathbf{x}}$ with $\mathbf{g}$ as its normal. This hyperplane divides space R^n into two parts with the hyperplane as boundary. In the part of the space where $\mathbf{x}$ satisfies the condition $\mathbf{g}^T(\mathbf{x} - \tilde{\mathbf{x}}) > 0$, no minimizers exist because Eq. (14.2) implies that $f(\mathbf{x}) > f(\tilde{\mathbf{x}})$. Consequently, a minimizer of $f(\mathbf{x})$ can only be found in the part of the space characterized by $\{\mathbf{x} : \mathbf{g}^T(\mathbf{x} - \tilde{\mathbf{x}}) \leq 0\}$. Evidently, the subgradient enables the construction of a cutting plane in the parameter space that reduces the search region significantly.

There are several important special cases in which the computation of a subgradient of a convex $f(\mathbf{x})$ can be readily carried out, as follows:

(a) If $f(\mathbf{x})$ is convex and differentiable at $\mathbf{x}$, then the subdifferential $\partial f(\mathbf{x})$ contains only one subgradient, which is the same as the gradient $\nabla f(\mathbf{x})$;

(b) If $\alpha > 0$, a subgradient of $\alpha f(\mathbf{x})$ is given by $\alpha \mathbf{g}$ where $\mathbf{g}$ is a subgradient of $f(\mathbf{x})$;

(c) If $f(\mathbf{x}) = f_1(\mathbf{x}) + f_2(\mathbf{x}) + \cdots + f_r(\mathbf{x})$ where $f_i(\mathbf{x})$ for $1 \leq i \leq r$ are convex, then $\mathbf{g} = \mathbf{g}_1 + \mathbf{g}_2 + \cdots + \mathbf{g}_r$ is a subgradient of $f(\mathbf{x})$ where $\mathbf{g}_i$ is a subgradient of $f_i(\mathbf{x})$;

(d) If $f(\mathbf{x}) = \max[f_1(\mathbf{x}), f_2(\mathbf{x}), \ldots, f_r(\mathbf{x})]$ where $f_i(\mathbf{x})$ for $i = 1, 2, \ldots, r$ are convex, then at a given point $\mathbf{x}$ there exists at least one index, say i^*, such that $f(\mathbf{x}) = f_{i^*}(\mathbf{x})$. In this situation, a subgradient of $f_{i^*}(\mathbf{x})$ is a subgradient of $f(\mathbf{x})$.

(e) If $h(\mathbf{x}) = f(\mathbf{A}\mathbf{x} + \mathbf{b})$ where $f(\mathbf{x})$ is convex, then

$$\partial h(\mathbf{x}) = \mathbf{A}^T \partial f(\mathbf{A}\mathbf{x} + \mathbf{b})$$

Example 14.1 Show that the subdifferential of function $f(x) = |x|$ is given by

$$\partial|x| = \begin{cases} 1 & \text{for } x > 0 \\ \text{any } g \in [-1, 1] & \text{for } x = 0 \\ -1 & \text{for } x < 0 \end{cases} \tag{14.3}$$

Solution Function $f(x) = |x|$ is convex because for $0 \leq \alpha \leq 1$ we have

$$f(\alpha x_1 + (1-\alpha)x_2) = |\alpha x_1 + (1-\alpha)x_2| \leq \alpha|x_1| + (1-\alpha)|x_2| = \alpha f(x_1) + (1-\alpha)f(x_2)$$

Consider an arbitrary point $\tilde{x}$. If $\tilde{x} > 0$, we have $f(\tilde{x}) = \tilde{x}$ which is obviously differentiable. Hence the differential of $f(\tilde{x})$ is equal to the derivative of $f(\tilde{x})$, i.e., $f'(\tilde{x}) = \tilde{x}' = 1$ which verifies the first constraint in Eq. (14.3). If $\tilde{x} < 0$, we have $f(\tilde{x}) = -\tilde{x}$ which is differentiable. Hence the differential of $f(\tilde{x})$ is equal to $f'(\tilde{x}) = (-\tilde{x})' = -1$ which verifies the third constraint in Eq. (14.3). Finally, if $\tilde{x} = 0$, Eq. (14.2) is reduced to $|x| \geq gx$ which holds for any value of g between -1 and 1. This verifies the second constraint in Eq. (14.3).   ∎

The next two theorems concern optimization problems where the functions involved are convex but not necessarily differentiable. These theorems may be regarded as extensions of the well-known first-order optimality condition and KKT conditions to their differentiable counterparts that were studied earlier in Chaps. 2 and 10, respectively.

Theorem 14.1 *First-order   optimality   condition   for   nondifferentiable unconstrained convex problems* *Point* $\mathbf{x}^*$ *is a global solution of the minimization problem*

$$\begin{aligned} minimize \quad & f(\mathbf{x}) \\ \mathbf{x} \in & \, domf \end{aligned}$$

where $f(\mathbf{x})$ *is convex, if and only if* $\mathbf{0} \in \partial f(\mathbf{x}^*)$.

Proof Suppose that $\mathbf{0} \in \partial f(\mathbf{x}^*)$. By letting $\tilde{\mathbf{x}} = \mathbf{x}^*$ and $\mathbf{g} = \mathbf{0}$ in Eq. (14.2), we obtain $f(\mathbf{x}) \geq f(\mathbf{x}^*)$ for all $\mathbf{x}$ in dom(f) and hence $\mathbf{x}^*$ is a global minimizer of $f(\mathbf{x})$. Conversely, if $\mathbf{x}^*$ is a global minimizer, then $f(\mathbf{x}) \geq f(\mathbf{x}^*)$ for all $\tilde{\mathbf{x}} \in$ dom(f) and thus $f(\mathbf{x}) \geq f(\mathbf{x}^*) + \mathbf{0}^T(\mathbf{x} - \mathbf{x}^*)$ for all $\mathbf{x}$ in dom(f), which implies that $\mathbf{0} \in \partial f(\mathbf{x}^*)$. ∎

Theorem 14.2 *KKT conditions for nondifferentiable constrained convex problems* *Consider constrained convex problem*

$$\begin{aligned} minimize \quad & f(\mathbf{x}) \\ subject \ to: \quad & \mathbf{a}_i^T \mathbf{x} = b_i \quad for \ 1 \leq i \leq p \\ & c_j(\mathbf{x}) \leq 0 \quad for \ 1 \leq j \leq q \end{aligned} \qquad (14.4)$$

where $f(\mathbf{x})$ *and* $c_j(\mathbf{x})$ *are convex but not necessarily differentiable. A regular point* $\mathbf{x}^*$ *is a solution of the problem in Eq. (14.4) if and only if*

(a) $\mathbf{a}_i^T \mathbf{x}^* = b_i$ *for* $1 \leq i \leq p$.
(b) $c_j(\mathbf{x}^*) \leq 0$ *for* $1 \leq j \leq q$.
(c) *there exist* λ_i^*'s *and* μ_j^*'s *such that*

$$\mathbf{0} \in \partial f(\mathbf{x}^*) + \sum_{i=1}^{p} \lambda_i^* \mathbf{a}_i + \sum_{j=1}^{q} \mu_j^* \partial c_j(\mathbf{x}^*)$$

(d) $\mu_j^* c_j(\mathbf{x}^*) = 0$ *for* $j = 1, \ 2, \ \ldots, \ q$.
(e) $\mu_j^* \geq 0$ *for* $j = 1, \ 2, \ \ldots, \ q$.

Proof Below we prove the sufficiency of these conditions for point $\mathbf{x}^*$ to be a global solution of the problem in Eq. (14.4) and leave the necessity part as an exercise for the reader.

Suppose that point $\mathbf{x}^*$ satisfies conditions (a)–(e). Conditions (a) and (b) imply that $\mathbf{x}^*$ is a feasible point of the problem in Eq. (14.4). If $\mathbf{x}$ is an arbitrary feasible point, then both $\mathbf{x}$ and $\mathbf{x}^*$ are feasible and hence we can write

$$\sum_{i=1}^{p} \lambda_i^* \mathbf{a}_i^T (\mathbf{x} - \mathbf{x}^*) = \sum_{i=1}^{p} \lambda_i^* (\mathbf{a}_i^T \mathbf{x} - b_i) - \sum_{i=1}^{p} \lambda_i^* (\mathbf{a}_i^T \mathbf{x}^* - b_i) = 0 \qquad (14.5)$$

As $c_j(\mathbf{x}) \leq 0$ and $\mu_j^* \geq 0$ for $j = 1, 2, \ldots, q$, we have

$$\sum_{j=1}^{q} \mu_j^* c_j(\mathbf{x}) \leq 0$$

which in conjunction with condition (d) implies that

$$\sum_{j=1}^{q} \mu_j^* (c_j(\mathbf{x}) - c_j(\mathbf{x}^*)) \leq 0 \qquad (14.6)$$

The convexity of functions $c_j(\mathbf{x})$ gives

$$c_j(\mathbf{x}) - c_j(\mathbf{x}^*) \geq \partial c_j(\mathbf{x}^*)^T (\mathbf{x} - \mathbf{x}^*)$$

and from Eq. (14.6), we have

$$\sum_{j=1}^{q} \mu_j^* \partial c_j(\mathbf{x}^*)^T (\mathbf{x} - \mathbf{x}^*) \leq 0 \qquad (14.7)$$

From Eqs. (14.5) and (14.7) and the convexity of $f(\mathbf{x})$, we deduce

$$f(\mathbf{x}) - f(\mathbf{x}^*)$$

$$\geq \partial f(\mathbf{x}^*)^T (\mathbf{x} - \mathbf{x}^*) + \sum_{i=1}^{p} \lambda_i^* \mathbf{a}_i^T (\mathbf{x} - \mathbf{x}^*) + \sum_{j=1}^{q} \mu_j^* \partial c_j(\mathbf{x}^*)^T (\mathbf{x} - \mathbf{x}^*)$$

$$= \left[\partial f(\mathbf{x}^*) + \sum_{i=1}^{p} \lambda_i^* \mathbf{a}_i + \sum_{j=1}^{q} \mu_j^* \partial c_j(\mathbf{x}^*) \right]^T (\mathbf{x} - \mathbf{x}^*)$$

Now from condition (c), the expression in the square brackets has a zero value and, therefore, $f(\mathbf{x}^*) \leq f(\mathbf{x})$. ∎

14.2.2 Convex Functions with Lipschitz-Continuous Gradients

A continuously differentiable function $f(\mathbf{x})$ is said to have a *Lipschitz continuous gradient* [2] if

$$\|\nabla f(\mathbf{x}) - \nabla f(\mathbf{y})\|_2 \leq L \|\mathbf{x} - \mathbf{y}\|_2 \qquad (14.8)$$

for any $\mathbf{x}$ and $\mathbf{y} \in \mathrm{dom}(f)$ where $L > 0$ is called a *Lipschitz* constant.

Example 14.2 Show that the gradient of $f(\mathbf{x}) = \frac{1}{2}\|\mathbf{A}\mathbf{x} - \mathbf{b}\|_2^2$ is Lipschitz continuous.

Solution We can write function $f(\mathbf{x})$ as

$$f(\mathbf{x}) = \frac{1}{2}\mathbf{x}^T\mathbf{A}^T\mathbf{A}\mathbf{x} - \mathbf{x}^T\mathbf{A}^T\mathbf{b} + \frac{1}{2}\mathbf{b}^T\mathbf{b}$$

Hence the gradient of $f(\mathbf{x})$ is given by

$$\nabla f(\mathbf{x}) = \mathbf{A}^T\mathbf{A}\mathbf{x} - \mathbf{A}^T\mathbf{b}$$

and

$$\|\nabla f(\mathbf{x}) - \nabla f(\mathbf{y})\|_2 = \|\mathbf{A}^T\mathbf{A}(\mathbf{x} - \mathbf{y})\|_2 \leq \lambda_{\max}(\mathbf{A}^T\mathbf{A})\|\mathbf{x} - \mathbf{y}\|_2$$

where $\lambda_{\max}(\mathbf{A}^T\mathbf{A})$ denotes the largest eigenvalue of $\mathbf{A}^T\mathbf{A}$. Therefore, $\nabla f(\mathbf{x})$ is Lipschitz continuous with $L = \lambda_{\max}(\mathbf{A}^T\mathbf{A})$. ∎

Now recall the property of a convex function $f(\mathbf{x})$ expressed in Eq. (14.2). Essentially it says that a convex function is bounded from below by its supporting planes meaning that it is tangent to the surface defined by the function. The theorem below shows that a convex function with a Lipschitz continuous gradient is, in addition, bounded from above by a family of simple convex quadratic functions.

Theorem 14.3 *Upper bound of a convex function with Lipschitz continuous gradients* If $f(\mathbf{x})$ *is convex with a Lipschitz continuous gradient, then* $f(\mathbf{x})$ *is bounded from above by a family of convex quadratic functions with Hessian* $L \cdot \mathbf{I}$, *namely,*

$$f(\mathbf{x}) \leq f(\mathbf{y}) + \nabla f(\mathbf{y})^T(\mathbf{x} - \mathbf{y}) + \frac{L}{2}\|\mathbf{x} - \mathbf{y}\|_2^2 \tag{14.9}$$

Proof By expressing $f(\mathbf{x})$ as

$$f(\mathbf{x}) = f(\mathbf{y}) + \int_0^1 \nabla f(\mathbf{y} + \tau(\mathbf{x} - \mathbf{y}))^T(\mathbf{x} - \mathbf{y})d\tau$$

$$= f(\mathbf{y}) + \nabla f(\mathbf{y})^T(\mathbf{x} - \mathbf{y}) + \int_0^1 [\nabla f(\mathbf{y} + \tau(\mathbf{x} - \mathbf{y})) - \nabla f(\mathbf{y})]^T(\mathbf{x} - \mathbf{y})d\tau$$

we have

$$|f(\mathbf{x}) - f(\mathbf{y}) - \nabla^T f(\mathbf{y})(\mathbf{x} - \mathbf{y})| =$$

$$= \left| \int_0^1 \{\nabla f[\mathbf{y} + \tau(\mathbf{x} - \mathbf{y})] - \nabla f(\mathbf{y})\}^T(\mathbf{x} - \mathbf{y})d\tau \right|$$

$$\leq \int_0^1 |\{\nabla f[\mathbf{y} + \tau(\mathbf{x} - \mathbf{y})] - \nabla f(\mathbf{y})\}^T(\mathbf{x} - \mathbf{y})|d\tau$$

$$\leq \int_0^1 \|\nabla f[\mathbf{y} + \tau(\mathbf{x} - \mathbf{y})] - \nabla f(\mathbf{y})\|_2 \cdot \|\mathbf{x} - \mathbf{y}\|_2 d\tau$$

$$\leq \int_0^1 \tau L\|\mathbf{x} - \mathbf{y}\|_2^2 d\tau = \frac{L}{2}\|\mathbf{x} - \mathbf{y}\|_2^2$$

which implies Eq. (14.9) [2]. ∎

Note that the quadratic upper bound in Eq. (14.9) has a simple constant Hessian $L \cdot \mathbf{I}$ and, therefore, it is straightforward to compute its minimizer as a function of x. The upper bound of $f(\mathbf{x})$ can be written as

$$h_U(\mathbf{x}, \mathbf{y}) = f(\mathbf{y}) + \nabla f(\mathbf{y})^T (\mathbf{x} - \mathbf{y}) + \frac{L}{2} \|\mathbf{x} - \mathbf{y}\|_2^2 = \frac{L}{2} \left\| \mathbf{x} - \left(\mathbf{y} - \frac{1}{L} f(\mathbf{y})\right) \right\|_2^2 + \text{const}$$

and hence its global minimizer is given by

$$\hat{\mathbf{x}} = \mathbf{y} - \frac{1}{L} \nabla f(\mathbf{y}) \tag{14.10}$$

As a simple application of the above analysis, consider a convex function $f(\mathbf{x})$ with Lipschitz continuous gradient and assume that the kth iterate $\mathbf{x}_k$ is available. Instead of minimizing $f(\mathbf{x})$ itself, $\mathbf{x}_k$ can be updated by minimizing the upper bound of $f(\mathbf{x})$ in Eq. (14.9) with y taken to be $\mathbf{x}_k$. In doing so, Eq. (14.10) can be used to generate the next iterate $\mathbf{x}_{k+1}$ as

$$\mathbf{x}_{k+1} = \mathbf{x}_k - \frac{1}{L} \nabla f(\mathbf{x}_k) \tag{14.11}$$

It is interesting to note that Eq. (14.11) is a steepest-descent formula with a *constant* line search step $\alpha_k = 1/L$. Now if we combine Eqs. (14.2) and (14.9), we get

$$f(\mathbf{y}) + \nabla f(\mathbf{y})^T (\mathbf{x} - \mathbf{y}) \le f(\mathbf{x}) \le f(\mathbf{y}) + \nabla f(\mathbf{y})^T (\mathbf{x} - \mathbf{y}) + \frac{L}{2} \|\mathbf{x} - \mathbf{y}\|_2^2 \tag{14.12}$$

Example 14.3

(a) By using a graphical construction, demonstrate the validity of the property in Eq. (14.12) for the scalar variable function $f(x) = \frac{1}{2}(x-1)^2 - \frac{1}{2} \log(x^2+1) + 3$ where y is set to 3.

(b) With an initial guess $x_0 = 3$, apply the formula in Eq. (14.11) to generate three iterates and evaluate how close they are to the minimizer of $f(x)$.

Solution (a) The first- and second-order derivatives of $f(x)$ are given by

$$f'(x) = x - 1 - \frac{x}{x^2 + 1} \quad \text{and} \quad f''(x) = \frac{x^2(x^2 + 3)}{(x^2 + 1)^2}$$

respectively. Since $f''(x) \ge 0$, $f(x)$ is convex. Moreover, it can be verified that

$$|f'(x) - f'(y)| \le 1.25|x - y|$$

and hence $f'(x)$ is Lipschitz continuous with $L = 1.25$. At $y = 3$, the lower and upper bounds in Eq. (14.12) are given by

$$h_L(x, 3) = 3.8487 + 1.7(x - 3)$$

and

$$h_U(x, 3) = 3.8487 + 1.7(x - 3) + 0.625(x - 3)^2$$

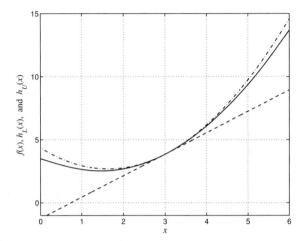

Fig. 14.3 Function $f(x)$ (solid curve), its linear lower bound (dashed line), and quadratic upper bound (dot-dashed curve) over interval $[0, 6]$

respectively. The above bounds as well as the function $f(x)$ itself over the range $[0, 6]$ are illustrated in Fig. 14.3.

(*b*) By setting $f'(x) = 0$, we obtain a third-order equation $x^3 - x^2 - 1 = 0$ which has only one real root $x^* = 1.4656$. Since $f(x)$ is strictly convex, x^* is the unique global minimizer of $f(x)$. By applying Eq. (14.11) with

$$\nabla f(x_k) = x_k - 1 - \frac{x_k}{x_k^2 + 1}$$

$L = 1.25$, and $x_0 = 3$, the first three iterates were found to be $x_1 = 1.64$, $x_2 = 1.4838$, and $x_3 = 1.4675$ and, in effect, x_3 is a good approximation for minimizer x^*.   ∎

14.2.3 Strongly Convex Functions

Definition 14.2 A continuously differentiable function $f(\mathbf{x})$ is said to be *strongly convex* if there exists an $m > 0$ such that

$$f(\mathbf{x}_1) \geq f(\mathbf{x}) + \nabla f(\mathbf{x})^T (\mathbf{x}_1 - \mathbf{x}) + \frac{m}{2} \|\mathbf{x}_1 - \mathbf{x}\|_2^2 \tag{14.13}$$

holds for any $\mathbf{x}_1$ and $\mathbf{x}$.   ∎

As seen in the right-hand side of Eq. (14.13), a strongly convex function has a quadratic lower bound which is obviously tighter than the linear lower bound given in Eq. (14.2) because the second-order term in Eq. (14.13) is always nonnegative. One would, therefore, expect that the class of strongly convex functions possesses additional desirable properties relative to the class of conventional convex functions. Below we summarize some of these properties [3].

Properties of strongly convex functions

(a) A twice continuously differentiable function $f(\mathbf{x})$ is strongly convex if and only if

$$\nabla^2 f(\mathbf{x}) - m\mathbf{I} \succeq \mathbf{0} \tag{14.14}$$

for some $m > 0$.

Proof Suppose that Eq. (14.14) holds. The Taylor expansion of $f(\mathbf{x})$ at point $\mathbf{x}_1$ gives

$$f(\mathbf{x}_1) = f(\mathbf{x}) + \nabla f(\mathbf{x})^T (\mathbf{x}_1 - \mathbf{x}) + \tfrac{1}{2}(\mathbf{x}_1 - \mathbf{x})^T \nabla^2 f(\mathbf{y})(\mathbf{x}_1 - \mathbf{x}) \tag{14.15}$$

for some $\mathbf{y}$ on the line between $\mathbf{x}$ and $\mathbf{x}_1$, which in conjunction with Eq. (14.14) leads to Eq. (14.13). Therefore, $f(\mathbf{x})$ is strongly convex. Conversely, suppose $f(\mathbf{x})$ is strongly convex, then Eqs. (14.13) and (14.15) imply that

$$(\mathbf{x}_1 - \mathbf{x})^T \nabla^2 f(\mathbf{y})(\mathbf{x}_1 - \mathbf{x}) \ge m\|\mathbf{x}_1 - \mathbf{x}\|_2^2$$

By choosing $\mathbf{x}_1 = \mathbf{x} + t\boldsymbol{\delta}$ with t a scalar and $\boldsymbol{\delta}$ an arbitrary vector in R^n, the above expression becomes $\boldsymbol{\delta}^T [\nabla^2 f(\mathbf{y}) - m\mathbf{I}]\boldsymbol{\delta} \ge 0$ which leads to

$$\boldsymbol{\delta}^T [\nabla^2 f(\mathbf{x}) - m\mathbf{I}]\boldsymbol{\delta} \ge 0$$

as $t \to 0$. Hence $\nabla^2 f(\mathbf{x}) \succeq m\mathbf{I}$. This completes the proof.   ∎

(b) If $f(\mathbf{x})$ is strongly convex and f^* is the minimum value of $f(\mathbf{x})$, then

$$f(\mathbf{x}) - f^* \le \frac{1}{2m}\|\nabla f(\mathbf{x})\|_2^2 \tag{14.16}$$

Proof Note that the convex quadratic function of $\mathbf{x}_1$ on the right-hand side of Eq. (14.13) achieves its minimum at $\mathbf{x}_1^* = \mathbf{x} - \frac{1}{m}\nabla f(\mathbf{x})$. This implies that

$$f(\mathbf{x}_1) \ge f(\mathbf{x}) + \nabla f(\mathbf{x})^T (\mathbf{x}_1 - \mathbf{x}) + \frac{m}{2}\|\mathbf{x}_1 - \mathbf{x}\|_2^2$$

$$\ge f(\mathbf{x}) + \nabla f(\mathbf{x})^T (\mathbf{x}_1^* - \mathbf{x}) + \frac{m}{2}\|\mathbf{x}_1^* - \mathbf{x}\|_2^2$$

$$= f(\mathbf{x}) - \frac{1}{2m}\|\nabla f(\mathbf{x})\|_2^2 \tag{14.17}$$

Equation 14.17 also holds when $f(\mathbf{x}_1)$ is replaced by its infimum, i.e., $f^* \ge f(\mathbf{x}) - \frac{1}{2m}\|\nabla f(\mathbf{x})\|_2^2$; hence Eq. (14.16) follows.   ∎

The upper bound in Eq. (14.16) quantifies the difference between $f(\mathbf{x})$ and its minimum in terms of the gradient of $f(\mathbf{x})$. From Eq. (14.16), we can readily deduce that

$$f(\mathbf{x}) - f^* \le \varepsilon \quad \text{if } \|\nabla f(\mathbf{x})\|_2 \le \sqrt{2m\varepsilon}$$

The next property provides an upper bound for the difference between $\mathbf{x}$ and the minimizer $\mathbf{x}^*$ of $f(\mathbf{x})$ in terms of the gradient of $f(\mathbf{x})$.

(c) If $f(\mathbf{x})$ is strongly convex and $\mathbf{x}^*$ is its minimizer, then

$$\|\mathbf{x} - \mathbf{x}^*\|_2 \le \frac{2}{m}\|\nabla f(\mathbf{x})\|_2 \tag{14.18}$$

Proof By Eq. (14.13) with $\mathbf{x}_1 = \mathbf{x}^*$, we have

$$\nabla f(\mathbf{x})^T (\mathbf{x}^* - \mathbf{x}) + \frac{m}{2}\|\mathbf{x}^* - \mathbf{x}\|_2^2 \le f(\mathbf{x}^*) - f(\mathbf{x}) \le 0$$

Hence

$$\frac{m}{2}\|\mathbf{x}^* - \mathbf{x}\|_2^2 \le -\nabla f(\mathbf{x})^T (\mathbf{x}^* - \mathbf{x}) \le \|\nabla f(\mathbf{x})\|_2 \cdot \|\mathbf{x}^* - \mathbf{x}\|_2$$

which yields Eq. (14.18).   ∎

(d) If $f(\mathbf{x})$ is strongly convex, then

$$(\nabla f(\mathbf{x}_1) - \nabla f(\mathbf{x}))^T (\mathbf{x}_1 - \mathbf{x}) \ge m\|\mathbf{x}_1 - \mathbf{x}\|_2^2 \tag{14.19}$$

Proof Since the inequality in Eq. (14.13), i.e.,

$$f(\mathbf{x}_1) \ge f(\mathbf{x}) + \nabla f(\mathbf{x})^T (\mathbf{x}_1 - \mathbf{x}) + \frac{m}{2}\|\mathbf{x}_1 - \mathbf{x}\|_2^2$$

holds true for any pair of points $\mathbf{x}_1$ and $\mathbf{x}$, it follows that $\mathbf{x}_1$ and $\mathbf{x}$ can be interchanged to give the inequality

$$f(\mathbf{x}) \ge f(\mathbf{x}_1) + \nabla f(\mathbf{x}_1)^T (\mathbf{x} - \mathbf{x}_1) + \frac{m}{2}\|\mathbf{x} - \mathbf{x}_1\|_2^2$$

The above two inequalities imply that

$$f(\mathbf{x}_1) + f(\mathbf{x}) \ge f(\mathbf{x}) + f(\mathbf{x}_1) + \nabla f(\mathbf{x})^T (\mathbf{x}_1 - \mathbf{x}) + \nabla f(\mathbf{x}_1)^T (\mathbf{x} - \mathbf{x}_1) + m\|\mathbf{x} - \mathbf{x}_1\|_2^2$$

or

$$0 \ge \nabla f(\mathbf{x})^T (\mathbf{x}_1 - \mathbf{x}) + \nabla f(\mathbf{x}_1)^T (\mathbf{x} - \mathbf{x}_1) + m\|\mathbf{x} - \mathbf{x}_1\|_2^2$$

and, therefore,

$$(\nabla f(\mathbf{x}_1) - \nabla f(\mathbf{x}))^T (\mathbf{x}_1 - \mathbf{x}) \ge m\|\mathbf{x}_1 - \mathbf{x}\|_2^2$$

                                                                                  ∎

If $f(\mathbf{x})$ is strongly convex and possesses a Lipschitz continuous gradient (see Sec. 14.2.2), then Eqs. (14.8) and (14.19) imply that

$$m\|\mathbf{x} - \mathbf{y}\|_2^2 \le (\mathbf{x} - \mathbf{y})^T (\nabla f(\mathbf{x}) - \nabla f(\mathbf{y})) \le L\|\mathbf{x} - \mathbf{y}\|_2^2$$

The value $L/m \ge 1$ is called the *condition number* of function $f(\mathbf{x})$.

14.2.4 Conjugate Functions

An important notion in convex analysis is that of conjugate functions [1]. We start by introducing several concepts relevant to conjugate functions.

Definition 14.3 The *epigraph* of a real-valued function $f(\mathbf{x})$ is defined as the set

$$\text{epi}(f) = \{(\mathbf{x}, u) : \mathbf{x} \in \text{dom}(f), \ u \geq f(\mathbf{x})\}$$

∎

It can be shown that function $f(\mathbf{x})$ is convex if and only if set epi(f) is convex [2]. Figure 14.4 illustrates the connection between the convexity of epi(f) and the convexity of $f(\mathbf{x})$.

Definition 14.4 Function $f(\mathbf{x})$ is said to be *lower-semicontinuous* at point $\mathbf{x}_0$ if for every $\varepsilon > 0$ there exists a neighborhood $\mathcal{U}$ of $\mathbf{x}_0$ such that $f(\mathbf{x}) \geq f(\mathbf{x}_0) - \varepsilon$ for all $\mathbf{x} \in \mathcal{U}$ [4]. ∎

Figure 14.5 illustrates the concept of lower-semicontinuity where at point $\mathbf{x}_0$ the function in Fig. 14.5a is lower-semicontinuous while the function in Fig. 14.5b is not.

It can be shown that a function is lower-semicontinuous if its epigraph is a closed set [1]. This is illustrated in Fig. 14.6a, b which show a function that is lower-semicontinuous in Fig. 14.6a and a function that is not lower-semicontinuous in Fig. 14.6b. The epigraph epi(f) of the function in Fig. 14.6a is a closed set whereas that of the function in Fig. 14.6b is *not* a closed set.

Definition 14.5 Function $f(\mathbf{x})$ is said to be proper if dom(f) is nonempty.   ∎

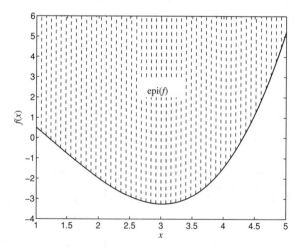

Fig. 14.4 Epigraph of a convex function

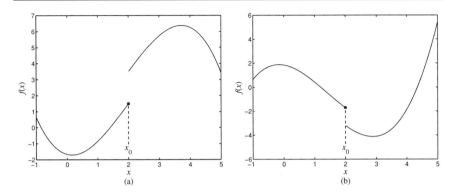

Fig. 14.5 a A function that is lower-semicontinuous at x_0; **b** a function that is not lower-semicontinuous at x_0

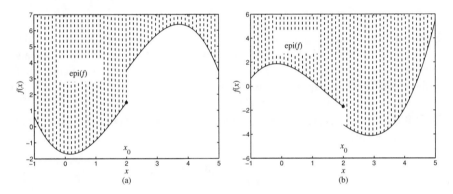

Fig. 14.6 a Epigraph of a lower-semicontinuous function is a closed set; **b** epigraph of function that is not lower-semicontinuous is not a closed set

We recall that by definition the domain of $f(\mathbf{x})$ is defined by $\mathrm{dom}(f) = \{\mathbf{x} : f(\mathbf{x}) < +\infty\}$. The domain of a proper function $f(\mathbf{x})$ is usually extended to the entire space R^n such that at any point $\mathbf{x}$ outside $\mathrm{dom}(f)$ it is assumed that $f(\mathbf{x}) = +\infty$.

Definition 14.6 The class of all lower-semicontinuous convex functions is commonly referred to as *class* Γ_0. ∎

Definition 14.7 The *indicator function* associated with a nonempty closed convex set in R^n C is defined by

$$I_C(\mathbf{x}) = \begin{cases} 0 & \text{if } \mathbf{x} \in C \\ +\infty & \text{if } \mathbf{x} \notin C \end{cases} \tag{14.20}$$

∎

Example 14.4 Verify that $I_C \in \Gamma_0$.

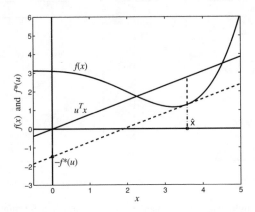

Fig. 14.7 Conjugate function $f^*(\mathbf{u})$ as the largest difference between $\mathbf{u}^T\mathbf{x}$ and $f(\mathbf{x})$

Solution Since the domain $\text{dom}(I_C) = C$ is nonempty, $I_C(\mathbf{x})$ is proper. By definition the epigraph of $I_C(\mathbf{x})$ is the closed cylinder whose base is the closed convex set C and hence $\text{epi}(I_C)$ is closed and convex. This implies that $I_C(\mathbf{x})$ is lower-semicontinuous and convex. Therefore, $I_C(\mathbf{x})$ belongs to class Γ_0. ∎

Definition 14.8 If $f(\mathbf{x})$ is a proper function, the conjugate of $f(\mathbf{x})$, denoted as $f^*(\mathbf{u})$, is defined by the equation

$$f^*(\mathbf{u}) = \sup_{\mathbf{x}\in dom(f)} (\mathbf{u}^T\mathbf{x} - f(\mathbf{x})) \tag{14.21}$$

∎

Given a $\mathbf{u} \in R^n$, Eq. (14.21) defines the value $f^*(\mathbf{u})$ as the largest difference between the linear function $\mathbf{u}^T\mathbf{x}$ and function $f(\mathbf{x})$. As can be observed from Fig. 14.7, for a smooth $f(\mathbf{x})$ the largest difference occurs at a point $\hat{\mathbf{x}}$ where the gradient $\nabla f(\mathbf{x})$ is equal to $\mathbf{u}$. From the figure, it can also be observed that the intercept of the tangent of $f(\mathbf{x})$ at $\hat{\mathbf{x}}$ is equal to $-f^*(\mathbf{u})$.

Properties of conjugate functions

(a) Conjugate function $f^*(\mathbf{u})$ is convex regardless of whether $f(\mathbf{x})$ is convex or not.
Proof By definition, $f^*(\mathbf{u})$ as a function of $\mathbf{u}$ is the supremum of a family of affine functions which are convex. Since the supremum of a family of convex functions is convex, $f^*(\mathbf{u})$ is convex. ∎
(b) For any $\mathbf{x}$ and $\mathbf{u}$,

$$f(\mathbf{x}) + f^*(\mathbf{u}) \geq \mathbf{x}^T\mathbf{u} \tag{14.22}$$

Equation (14.22) is known as the *Fenchel inequality*.

Proof From Eq. (14.21), we can write

$$f^*(\mathbf{u}) = \sup_{\mathbf{x} \in dom(f)} (\mathbf{u}^T \mathbf{x} - f(\mathbf{x})) \geq \mathbf{u}^T \mathbf{x} - f(\mathbf{x})$$

which immediately leads to Eq. (14.22).   ■

(c) If $f(\mathbf{x})$ is a proper and convex function and $\mathbf{u} \in \partial f(\mathbf{x})$, then

$$f(\mathbf{x}) + f^*(\mathbf{u}) = \mathbf{x}^T \mathbf{u} \tag{14.23}$$

Proof The convexities of $f(\mathbf{x})$ and $\mathbf{u} \in \partial f(\mathbf{x})$ imply that

$$f(\mathbf{x}_1) \geq f(\mathbf{x}) + \mathbf{u}^T(\mathbf{x}_1 - \mathbf{x}) \quad \text{for any } \mathbf{x}_1 \in \text{dom}(f)$$

i.e.,

$$\mathbf{u}^T \mathbf{x} - f(\mathbf{x}) \geq \mathbf{u}^T \mathbf{x}_1 - f(\mathbf{x}_1) \quad \text{for any } \mathbf{x}_1 \in \text{dom}(f)$$

Hence

$$\mathbf{u}^T \mathbf{x} - f(\mathbf{x}) \geq \sup_{\mathbf{x}_1 \in dom(f)} (\mathbf{u}^T \mathbf{x}_1 - f(\mathbf{x}_1))$$

i.e.,

$$f(\mathbf{x}) + f^*(\mathbf{u}) \leq \mathbf{x}^T \mathbf{u}$$

which in conjunction with Eq. (14.22) implies that Eq. (14.23) holds.   ■

(d) If $f(\mathbf{x}) \in \Gamma_0$, then the conjugate of the conjugate function $f^*(\mathbf{u})$ is equal to $f(\mathbf{x})$. In other words,

$$f^{**}(\mathbf{x}) = f(\mathbf{x}) \quad \text{for} \quad f(\mathbf{x}) \in \Gamma_0 \tag{14.24}$$

Proof By definition

$$f^{**}(\mathbf{x}) = \sup_{\mathbf{u}} [\mathbf{x}^T \mathbf{u} - f^*(\mathbf{u})] = \mathbf{x}^T \hat{\mathbf{u}} - f^*(\hat{\mathbf{u}})$$

where $\hat{\mathbf{u}}$ achieves the above supremum, and

$$f^*(\hat{\mathbf{u}}) = \sup_{\hat{x}} (\hat{\mathbf{u}}^T \hat{\mathbf{x}} - f(\hat{\mathbf{x}})) \geq \hat{\mathbf{u}}^T \mathbf{x} - f(\mathbf{x})$$

By combining the above two expressions, we obtain

$$f^{**}(\mathbf{x}) \leq f(\mathbf{x}) \tag{14.25}$$

By applying the Fenchel inequality to $f^*(\mathbf{u})$, we have

$$f^*(\mathbf{u}) + f^{**}(\mathbf{x}) \geq \mathbf{u}^T \mathbf{x}$$

If $\mathbf{u} \in \partial f(\mathbf{x})$, then Eq. (14.23) and the above inequality imply that

$$f^*(\mathbf{u}) + f^{**}(\mathbf{x}) \geq f(\mathbf{x}) + f^*(\mathbf{u})$$

i.e., $f^{**}(\mathbf{x}) \geq f(\mathbf{x})$ which in conjunction with Eq. (14.25) proves Eq. (14.24).   ■

(e) If $f(\mathbf{x}) \in \Gamma_0$, then $\mathbf{u} \in \partial f(\mathbf{x})$ if and only if $\mathbf{x} \in \partial f^*(\mathbf{u})$.

Proof If $\mathbf{u} \in \partial f(\mathbf{x})$ for $\mathbf{x} \in \text{dom}(f)$, then

$$f(\mathbf{x}_1) \geq f(\mathbf{x}) + \mathbf{u}^T(\mathbf{x}_1 - \mathbf{x}) \text{ for any } \mathbf{x}_1 \in \text{dom}(f)$$

Hence

$$\mathbf{u}^T\mathbf{x} - f(\mathbf{x}) \geq \mathbf{u}^T\mathbf{x}_1 - f(\mathbf{x}_1) \text{ for any } \mathbf{x}_1 \in \text{dom}(f)$$

By taking the supremum with respect to $\mathbf{x}_1$ in the above inequality, we obtain $\mathbf{u}^T\mathbf{x} - f(\mathbf{x}) \geq f^*(\mathbf{u})$, i.e., $-f(\mathbf{x}) \geq f^*(\mathbf{u}) - \mathbf{u}^T\mathbf{x}$. By adding term $\mathbf{u}_1^T\mathbf{x}$ with an arbitrary $\mathbf{u}_1$ to both sides of the above inequality, we obtain

$$\mathbf{u}_1^T\mathbf{x} - f(\mathbf{x}) \geq f^*(\mathbf{u}) + \mathbf{x}^T(\mathbf{u}_1 - \mathbf{u})$$

Since

$$f^*(\mathbf{u}_1) = \sup_{\hat{\mathbf{x}}}(\mathbf{u}_1^T\hat{\mathbf{x}} - f(\hat{\mathbf{x}})) \geq \mathbf{u}_1^T\mathbf{x} - f(\mathbf{x})$$

combining the last two inequalities yields

$$f^*(\mathbf{u}_1) \geq f^*(\mathbf{u}) + \mathbf{x}^T(\mathbf{u}_1 - \mathbf{u}) \qquad \text{for any } \mathbf{u}_1$$

Therefore, $\mathbf{x} \in \partial f^*(\mathbf{u})$. Conversely, if $\mathbf{x} \in \partial f^*(\mathbf{u})$, by using a similar argument in conjunction with the property $f^{**}(\mathbf{x}) = f(\mathbf{x})$, we can show that $\mathbf{u} \in \partial f(\mathbf{x})$. The details are left as an exercise for the reader. ∎

14.2.5   Proximal Operators

The proximal operator of a closed, proper, convex function $f(\mathbf{x})$ is defined as

$$\text{prox}_{\tau f}(\mathbf{v}) = \arg\min_{\mathbf{x}} \left(f(\mathbf{x}) + \frac{1}{2\tau}\|\mathbf{x} - \mathbf{v}\|_2^2 \right) \tag{14.26}$$

where $\tau > 0$ is a scaling parameter. With a fixed τ, $\text{prox}_{\tau f}(\mathbf{v})$ maps a vector $\mathbf{v} \in R^n$ to the unique minimizer of the objective function on the right-hand side of Eq. (14.26). We call $\text{prox}_{\tau f}(\mathbf{v})$ an *operator* because obtaining $\text{prox}_{\tau f}(\mathbf{v})$ involves a procedure of solving a minimization problem for strongly convex function $f(\mathbf{x}) + \frac{1}{2\tau}\|\mathbf{x} - \mathbf{v}\|_2^2$ rather than straightforward algebraic manipulations of vector $\mathbf{v}$. A proximal operator essentially generates a series of points that approach the minimizer of $f(\mathbf{x})$ and parameter τ controls how close $\text{prox}_{\tau f}(\mathbf{v})$ is to the minimizer of $f(\mathbf{x})$. In effect, under mild assumptions $\text{prox}_{\tau f}(\mathbf{v})$ can be shown to act as a descent gradient step $\mathbf{v} - \tau\nabla f(\mathbf{v})$ when τ is small [5].

Properties of proximal operators   Several properties are associated with proximal operators, as detailed below [5].

(a) Point $\mathbf{x}^*$ is a minimizer of $f(\mathbf{x})$ if and only if $\mathbf{x}^*$ is a fixed point of $\mathrm{prox}_{\tau f}(\mathbf{v})$, namely,

$$\mathbf{x}^* = \mathrm{prox}_{\tau f}(\mathbf{x}^*) \tag{14.27}$$

Proof A minimizer of $f(\mathbf{x})$, $\mathbf{x}^*$, minimizes both $f(\mathbf{x})$ and $\frac{1}{2\tau}\|\mathbf{x} - \mathbf{x}^*\|_2^T$ and hence it minimizes $f(\mathbf{x}) + \frac{1}{2\tau}\|\mathbf{x} - \mathbf{x}^*\|_2^T$, which implies that Eq. (14.27) holds. Conversely, if $\mathbf{x}^*$ is a fixed point of $\mathrm{prox}_{\tau f}(\mathbf{v})$, i.e., Eq. (14.27) holds, then $\mathbf{x}^*$ minimizes $f(\mathbf{x}) + \frac{1}{2\tau}\|\mathbf{x} - \mathbf{x}^*\|_2^T$. Hence

$$\nabla(f(\mathbf{x}) + \frac{1}{2\tau}\|\mathbf{x} - \mathbf{x}^*\|_2^2)\big|_{\mathbf{x}=\mathbf{x}^*} = 0$$

which implies that $\nabla f(\mathbf{x}^*) = 0$. This means that $\mathbf{x}^*$ is a minimizer of $f(\mathbf{x})$. ∎
It can be shown that if $f(\mathbf{x})$ possesses a minimizer, then the sequence of iterates produced by

$$\mathbf{x}_{k+1} = \mathrm{prox}_{\tau f}(\mathbf{x}_k) \quad \text{for } k = 0, 1, \ldots \tag{14.28}$$

converges to a fixed point $\mathbf{x}^*$ of $\mathrm{prox}_{\tau f}(\mathbf{v})$ [5], which is a minimizer of $f(\mathbf{x})$ by virtue of the above property. In the literature, algorithms based on Eq. (14.28) are called *proximal-point* algorithms.
(b) For a fully separable function $f(\mathbf{x}) = \sum_{i=1}^n f_i(x_i)$,

$$[\mathrm{prox}_{\tau f}(\mathbf{v})]_i = \mathrm{prox}_{\tau f}(v_i) \tag{14.29}$$

Proof Since

$$\mathrm{prox}_{\tau f}(\mathbf{v}) = \arg\min_{\mathbf{x}} \left[\sum_{i=1}^n f_i(x_i) + \frac{1}{2\tau}\sum_{i=1}^n (x_i - v_i)^2 \right]$$

$$= \arg\min_{\mathbf{x}} \sum_{i=1}^n \left[f_i(x_i) + \frac{1}{2\tau}(x_i - v_i)^2 \right]$$

it follows that the minimizer of $f(\mathbf{x})$, $\mathbf{x}^*$, can be obtained by minimizing $f_i(x_i) + \frac{1}{2\tau}(x_i - v_i)^2$ with respect to each component x_i, i.e., by obtaining

$$\arg\min_{x_i} \left(f_i(x_i) + \frac{1}{2\tau}(x_i - v_i)^2 \right) = \mathrm{prox}_{\tau f_i}(v_i)$$

for $i = 1, 2, \ldots, n$. This demonstrates the validity of Eq. (14.29). ∎
(c) The proximal operator can be interpreted as

$$\mathrm{prox}_{\tau f} = (\mathbf{I} + \tau \partial f)^{-1} \tag{14.30}$$

where $(\mathbf{I} + \tau \partial f)^{-1}$ is called the *resolvent* of subgradient ∂f.
Proof We demonstrate the validity of Eq. (14.30) by showing that $\mathrm{prox}_{\tau f}(\mathbf{v}) = (\mathbf{I} + \tau \partial f)^{-1}\mathbf{v}$ for an arbitrary $\mathbf{v}$. We need to show that if $\mathbf{u} \in (\mathbf{I} + \tau \partial f)^{-1}\mathbf{v}$, then $\mathbf{u} =$

$\text{prox}_{\tau f}(\mathbf{v})$. Since $\mathbf{u} \in (\mathbf{I} + \tau \partial f)^{-1}\mathbf{v}$ implies that $\mathbf{v} \in (\mathbf{I} + \tau \partial f)\mathbf{u} = \mathbf{u} + \tau \partial f(\mathbf{u})$, we have

$$\mathbf{0} \in \mathbf{u} - \mathbf{v} + \partial f(\mathbf{u}) = \tau \partial \left[f(\mathbf{u}) + \frac{1}{2\tau}\|\mathbf{u} - \mathbf{v}\|_2^2 \right]$$

and hence $\mathbf{u}$ minimizes $f(\mathbf{x}) + \frac{1}{2\tau}\|\mathbf{x} - \mathbf{v}\|_2^2$, i.e.,

$$\mathbf{u} = \arg \min_{\mathbf{x}} (f(\mathbf{x}) + \frac{1}{2\tau}\|\mathbf{x} - \mathbf{v}\|_2^2) = \text{prox}_{\tau f}(\mathbf{v})$$

which completes the proof.   ∎

14.3   Extension of Newton Method to Convex Constrained and Unconstrained Problems

In Sects. 5.3 and 5.4, the Newton method has been applied for the solution of unconstrained optimization problems in which the Hessian matrix is nonsingular and the approximation in Eq. (5.17) is valid. The optimization problems that can be solved by using the algorithms presented in Chap. 5 may or may not be convex. In this section, we use the Newton method to develop algorithms for the solution of unconstrained *convex* problems. In Sects. 14.3.2–14.3.3, we develop algorithms for the solution of constrained convex problems with equality constraints and also for constrained general convex problems [3].

14.3.1   Minimization of Smooth Convex Functions Without Constraints

If $f(\mathbf{x})$ is a smooth strictly convex function, the Newton method can be used to solve the unconstrained problem

$$\text{minimize}   f(\mathbf{x})$$

by iteratively minimizing a quadratic approximation of $f(\mathbf{x})$ by solving the unconstrained problem

$$\text{minimize}   f(\mathbf{x}_k) + \mathbf{g}_k^T(\mathbf{x} - \mathbf{x}_k) + \tfrac{1}{2}(\mathbf{x} - \mathbf{x}_k)^T \mathbf{H}_k(\mathbf{x} - \mathbf{x}_k) \qquad (14.31)$$

for $k = 0, 1, \ldots$ where $\mathbf{H}_k = \nabla^2 f(\mathbf{x}_k)$ and $\mathbf{g}_k = \nabla f(\mathbf{x}_k)$. As $\mathbf{H}_k$ is positive definite, the quadratic function in Eq. (14.31) is convex and hence the solution of the above problem can be obtained by setting the gradient of the quadratic function to zero, which yields the linear equation

$$\mathbf{H}_k(\mathbf{x} - \mathbf{x}_k) = -\mathbf{g}_k$$

If we let $\mathbf{d}_k = \mathbf{x} - \mathbf{x}_k$, then the above equation becomes

$$\mathbf{H}_k \mathbf{d}_k = -\mathbf{g}_k \tag{14.32}$$

and the next iterate is given by $\mathbf{x}_{k+1} = \mathbf{x}_k + \alpha_k \mathbf{d}_k$ where $\mathbf{d}_k$ satisfies Eq. (14.32) and scalar $\alpha_k > 0$ can be obtained by using a line search.

Algorithm 5.3 uses the stopping criterion

$$||\mathbf{x}_{k+1} - \mathbf{x}_k||_2 < \varepsilon$$

where ε is a convergence tolerance in conjunction with one of several available line searches as described in Chap. 4. Below we develop a Newton algorithm using a stopping criterion based on the Newton decrement in conjunction with the back-tracking line search described in Sect. 5.3.4. These techniques offer some important computational advantages. A desirable property of the Newton decrement is that it is independent of linear changes in the problem variables. The backtracking line search requires a reduced amount of computation relative to that required by the line searches described in Chap. 4, including the inexact line search, because it requires the gradient of the objective function to be evaluated only once per iteration at the starting point $\mathbf{x}_k$. Another property which makes the backtracking line search espe-cially suitable for a Newton algorithm is that as the iterates approach the solution point, α_k approaches unity, the quadratic approximation used as the basis of the Newton method becomes increasingly accurate, and as a consequence the Newton directions $\mathbf{d}_k = -\mathbf{H}_k^{-1}\mathbf{g}_k$ become optimal updates without the need for a line search. Note that the backtracking search always starts with the initial value $\alpha_k = 1$ and if the problem is quadratic, the solution is obtained in one iteration.

The minimizer of the quadratic function in Eq. (14.31) is given by point $\tilde{\mathbf{x}} = \mathbf{x}_k - \mathbf{H}_k^{-1}\mathbf{g}_k$ at which the quadratic function assumes the minimum value

$$\begin{aligned}
\tilde{f} &= f(\mathbf{x}_k) + \mathbf{g}_k^T(\tilde{\mathbf{x}} - \mathbf{x}_k) + \tfrac{1}{2}(\tilde{\mathbf{x}} - \mathbf{x}_k)^T \mathbf{H}_k(\tilde{\mathbf{x}} - \mathbf{x}_k) \\
&= f(\mathbf{x}_k) - \mathbf{g}_k^T \mathbf{H}_k^{-1}\mathbf{g}_k + \tfrac{1}{2}\mathbf{g}_k^T \mathbf{H}_k^{-1}\mathbf{g}_k \\
&= f(\mathbf{x}_k) - \tfrac{1}{2}\mathbf{g}_k^T \mathbf{H}_k^{-1}\mathbf{g}_k
\end{aligned}$$

Since $\tilde{f}$ is a good estimate of the minimum of $f(\mathbf{x}_k)$, $f(\mathbf{x}^*)$, the above equation gives

$$f(\mathbf{x}_k) - f(\mathbf{x}^*) \approx f(\mathbf{x}_k) - \tilde{f} = \tfrac{1}{2}\mathbf{g}_k^T \mathbf{H}_k^{-1}\mathbf{g}_k = \tfrac{1}{2}\lambda(\mathbf{x}_k)^2$$

where $\lambda(\mathbf{x}_k) = (\mathbf{g}_k^T \mathbf{H}_k^{-1}\mathbf{g}_k)^{1/2}$ is the Newton decrement at point $\mathbf{x}_k$ (see Sect. 5.3.3). From Eq. (14.32) it follows that

$$\lambda(\mathbf{x}_k) = (-\mathbf{g}_k^T \mathbf{d}_k)^{1/2} \tag{14.33}$$

A Newton algorithm based on the above principles can be constructed as follows:

Algorithm 14.1 Newton algorithm for unconstrained minimization
of convex functions

Step 1
Input $\mathbf{x}_0$, initialize tolerance ε, and set $k = 0$.
Step 2
Compute $\mathbf{g}_k$ and $\mathbf{H}_k$.
Step 3
Compute $\mathbf{d}_k$ by solving Eq. (14.32).
Step 4
Compute $\lambda(\mathbf{x}_k)$ using Eq. (14.33).
If $\frac{1}{2}\lambda^2(\mathbf{x}_k) < \varepsilon$, output $\mathbf{x}_k$ as the solution and stop; otherwise, find α_k, the value of α that minimizes $f(\mathbf{x}_k + \alpha\mathbf{d}_k)$, using the backtracking line search in Sect. 5.3.4.
Step 5
Set $\mathbf{x}_{k+1} = \mathbf{x}_k + \alpha_k\mathbf{d}_k$.
Set $k = k + 1$ and repeat from Step 2.

For problems in which the Hessian $\mathbf{H}_k$ is of large size, solving Eq. (14.32) by computing the inverse of $\mathbf{H}_k$ would require a large amount of computation. An alternative approach for obtaining $\mathbf{H}_k$ for such problems is to use the conjugate-gradient method described in Sect. 6.9.

Note that vectors $\mathbf{g}_k$ and $\mathbf{d}_k$ are both available in the kth iteration and hence the amount of computation required to obtain $\lambda(\mathbf{x}_k)$ in Step 4 of the algorithm is insignificant.

14.3.2 Minimization of Smooth Convex Functions Subject to Equality Constraints

In a convex programming (CP) problem, all equality constraints must be linear and hence the problem assumes the form

$$\begin{aligned} \text{minimize} \quad & f(\mathbf{x}) \\ \text{subject to:} \quad & \mathbf{Ax} = \mathbf{b} \end{aligned} \tag{14.34}$$

where $f(\mathbf{x})$ is a convex twice continuously differentiable function and $\mathbf{A}$ is of full row rank. It is well known that a point $\mathbf{x}$ is a minimizer of the above problem if and only if the KKT conditions

$$\mathbf{g}(\mathbf{x}) + \mathbf{A}^T\boldsymbol{\lambda} = \mathbf{0} \tag{14.35a}$$

$$\mathbf{Ax} = \mathbf{b} \tag{14.35b}$$

are satisfied for some Lagrange multiplier $\boldsymbol{\lambda}$.

Let us suppose that iterate $\{\mathbf{x}_k, \boldsymbol{\lambda}_k\}$ with primal $\mathbf{x}_k$ feasible, i.e., $\mathbf{A}\mathbf{x}_k = \mathbf{b}$, is known and that $\{\mathbf{x}_k, \boldsymbol{\lambda}_k\}$ is to be updated to $\{\mathbf{x}_{k+1}, \boldsymbol{\lambda}_{k+1}\}$ with $\mathbf{x}_{k+1} = \mathbf{x}_k + \mathbf{d}_k$ so as to better satisfy the KKT conditions in Eqs. (14.35a) and (14.35b). To satisfy Eq. (14.35b), namely, $\mathbf{A}\mathbf{x}_{k+1} = \mathbf{b}$, we must have

$$\mathbf{A}\mathbf{d}_k = \mathbf{0} \tag{14.36}$$

To deal with Eq. (14.35a), we replace gradient $\mathbf{g}(\mathbf{x})$ by its first-order Taylor approximation at $\mathbf{x}_k$, i.e.,

$$\mathbf{g}(\mathbf{x}) \approx \mathbf{g}_k + \mathbf{H}_k \mathbf{d}_k$$

in which case the KKT condition in Eq. (14.35a) can be approximated as

$$\mathbf{g}_k + \mathbf{H}_k \mathbf{d}_k + \mathbf{A}^T \boldsymbol{\lambda}_{k+1} = \mathbf{0} \tag{14.37}$$

Now on combining Eqs. (14.36) and (14.37), we get

$$\begin{bmatrix} \mathbf{H}_k & \mathbf{A}^T \\ \mathbf{A} & \mathbf{0} \end{bmatrix} \begin{bmatrix} \mathbf{d}_k \\ \boldsymbol{\lambda}_{k+1} \end{bmatrix} = \begin{bmatrix} -\mathbf{g}_k \\ \mathbf{0} \end{bmatrix} \tag{14.38}$$

Note that if there are no constraints, Eq. (14.38) is reduced to Eq. (14.32) as expected. For this reason, the vector $\mathbf{d}_k$ that satisfies Eq. (14.38) is regarded as a Newton direction. The solution of Eq. (14.38) is obtained as

$$\mathbf{d}_k = -\mathbf{H}_k^{-1}(\mathbf{A}^T \boldsymbol{\lambda}_{k+1} + \mathbf{g}_k) \tag{14.39a}$$

where

$$\boldsymbol{\lambda}_{k+1} = -(\mathbf{A}\mathbf{H}_k^{-1}\mathbf{A}^T)\mathbf{A}\mathbf{H}_k^{-1}\mathbf{g}_k \tag{14.39b}$$

This solution always satisfies Eq. (14.36) and thus the second KKT condition in Eq. (14.35b) is also satisfied. Consequently, the Newton iteration can in practice be terminated as long as the KKT condition in Eq. (14.35a) is satisfied to within a prescribed tolerance. Based on this idea, a stopping criterion can be formulated as

$$\|\mathbf{g}_{k+1} + \mathbf{A}^T \boldsymbol{\lambda}_{k+1}\|_2 < \varepsilon \tag{14.40}$$

where ε is a prescribed tolerance.

A Newton algorithm for the solution of the problem in Eq. (14.34) can now be constructed as follows:

Algorithm 14.2 Newton algorithm for convex problems with equality constraints

Step 1

Input a feasible initial point $\mathbf{x}_0 \in \text{dom}(f)$ and a termination tolerance $\varepsilon > 0$ and set $k = 0$.

Step 2

Compute $\mathbf{d}_k$ and $\boldsymbol{\lambda}_{k+1}$ using Eqs. (14.39a) and (14.39b).

Step 3

Find α_k, the value of α that minimizes $f(\mathbf{x}_k + \alpha\mathbf{d}_k)$, using the backtracking line search in Sect. 5.3.4.

Step 4

Set $\mathbf{x}_{k+1} = \mathbf{x}_k + \alpha_k\mathbf{d}_k$ and compute $\mathbf{g}_{k+1}$.

Step 5

If Eq. (14.40) is satisfied, stop and output $\mathbf{x}_{k+1}$ as solution; otherwise set $k = k + 1$ and repeat from Step 2.

14.3.3   Newton Algorithm for Problem in Eq. (14.34) with a Nonfeasible $\mathbf{x}_0$

Suppose the kth iterate $\{\mathbf{x}_k, \boldsymbol{\lambda}_k\}$ is known but $\mathbf{x}_k$ is not necessarily feasible, and we want to update $\{\mathbf{x}_k, \boldsymbol{\lambda}_k\}$ to $\{\mathbf{x}_{k+1}, \boldsymbol{\lambda}_{k+1}\}$ to ensure that $\{\mathbf{x}_{k+1}, \boldsymbol{\lambda}_{k+1}\}$ better satisfies the KKT conditions in Eqs. (14.35a) and (14.35b). To satisfy the condition in Eq. (14.35b), we must have

$$\mathbf{A}\mathbf{d}_k = -(\mathbf{A}\mathbf{x}_k - \mathbf{b})$$

and hence the Newton direction must satisfy the equation

$$\begin{bmatrix} \mathbf{H}_k & \mathbf{A}^T \\ \mathbf{A} & \mathbf{0} \end{bmatrix} \begin{bmatrix} \mathbf{d}_k \\ \boldsymbol{\lambda}_{k+1} \end{bmatrix} = \begin{bmatrix} -\mathbf{g}_k \\ \mathbf{b} - \mathbf{A}\mathbf{x}_k \end{bmatrix} \tag{14.41}$$

Note that $\mathbf{d}_k$ is not necessarily a descent direction because of the additional requirement that $\mathbf{A}\mathbf{d}_k = -(\mathbf{A}\mathbf{x}_k - \mathbf{b})$. As a result, the line search and stopping criterion used in Algorithm 14.2 are not appropriate for the present case. We proceed by letting $\boldsymbol{\lambda}_{k+1} = \boldsymbol{\lambda}_k + \boldsymbol{\delta}_k$ and writing Eq. (14.41) as

$$\begin{bmatrix} \mathbf{H}_k & \mathbf{A}^T \\ \mathbf{A} & \mathbf{0} \end{bmatrix} \begin{bmatrix} \mathbf{d}_k \\ \boldsymbol{\delta}_k \end{bmatrix} = -\begin{bmatrix} \mathbf{r}_d \\ \mathbf{r}_p \end{bmatrix}$$

where $\mathbf{r}_d$ and $\mathbf{r}_p$ are called *dual* and *primal residuals* and are given by

$$\mathbf{r}_d = \mathbf{g}_k + \mathbf{A}^T \boldsymbol{\lambda}_k$$

and

$$\mathbf{r}_p = \mathbf{A}\mathbf{x}_k - \mathbf{b}$$

respectively,

The solution of the above equation is given by

$$\mathbf{d}_k = -\mathbf{H}_k^{-1}(\mathbf{A}^T \boldsymbol{\delta}_k + \mathbf{r}_d) \tag{14.42a}$$

and

$$\boldsymbol{\delta}_k = -(\mathbf{A}\mathbf{H}_k^{-1}\mathbf{A}^T)^{-1}(\mathbf{A}\mathbf{H}_k^{-1}\mathbf{r}_d - \mathbf{r}_p) \tag{14.42b}$$

It is reasonable to terminate the algorithm if

$$\|\mathbf{r}(\mathbf{x}_k, \boldsymbol{\lambda}_k)\|_2 < \varepsilon \tag{14.43}$$

for a prescribed tolerance ε where

$$\mathbf{r}(\mathbf{x}_k, \boldsymbol{\lambda}_k) \triangleq \begin{bmatrix} \mathbf{r}_d \\ \mathbf{r}_p \end{bmatrix} = \begin{bmatrix} \mathbf{g}_k + \mathbf{A}^T \boldsymbol{\lambda}_k \\ \mathbf{A}\mathbf{x}_k - \mathbf{b} \end{bmatrix}$$

A value of $\alpha_k > 0$ that can be used in the backtracking line search can be found by minimizing $\|\mathbf{r}(\mathbf{x}_k + \alpha\mathbf{d}_k, \boldsymbol{\lambda}_k + \alpha\boldsymbol{\delta}_k)\|_2$ where $\{\mathbf{d}_k, \boldsymbol{\delta}_k\}$ are given by Eqs. (14.42a) and (14.42b) as detailed below.

Algorithm 14.3 Backtracking line search for minimizing
$$\| \mathbf{r}(\mathbf{x}_k + \alpha\mathbf{d}_k, \boldsymbol{\lambda}_k + \alpha\boldsymbol{\delta}_k)\|_2$$
Step 1
Select constants $\rho \in (0, 0.5)$ and $\gamma \in (0, 1)$, say $\rho = 0.1$ and $\gamma = 0.5$, and set $\alpha = 1$.
Step 2
If $\|\mathbf{r}(\mathbf{x}_k + \alpha\mathbf{d}_k, \boldsymbol{\lambda}_k + \alpha\boldsymbol{\delta}_k)\|_2 \le (1 - \rho\alpha)\|\mathbf{r}(\mathbf{x}_k, \boldsymbol{\lambda}_k)\|_2$, output $\alpha_k = \alpha$ and stop.
Step 3
Set $\alpha = \gamma\alpha$ and repeat from Step 2.

A Newton algorithm for convex problems with equality constraints and a nonfeasible initial point can be constructed as follows:

Algorithm 14.4 Newton algorithm for convex problems with equality constraints and a nonfeasible initial point
Step 1
Input initial point $\mathbf{x}_0 \in \text{dom}(f)$, initial $\boldsymbol{\lambda}_0$, and tolerance $\varepsilon > 0$ and set $k = 0$.
Step 2
If Eq. (14.43) holds, output $\mathbf{x}_k$ as solution and stop.
Step 3
Compute a Newton directon $\{\mathbf{d}_k, \boldsymbol{\delta}_k\}$ by using Eqs. (14.42a) and (14.42b).
Step 4
Apply Algorithm 14.3 to find α_k that minimizes $\|\mathbf{r}(\mathbf{x}_k + \alpha\mathbf{d}_k, \boldsymbol{\lambda}_k + \alpha\boldsymbol{\delta}_k)\|_2$.
Step 5
Set $\mathbf{x}_{k+1} = \mathbf{x}_k + \alpha_k\mathbf{d}_k$ and $k = k + 1$ and repeat from Step 2.

Example 14.5 Apply (a) Algorithm 14.2 and (b) Algorithm 14.4 to the analytic center problem

$$\text{minimize } f(\mathbf{x}) = -\sum_{i=1}^{n} \log x_i$$

$$\text{subject to: } \mathbf{e}^T \mathbf{x} = b$$

where $\mathbf{e}$ is a column vector of length n whose components are all ones and $b = 1$.

Solution (a) The gradient and Hessian of $f(\mathbf{x})$ are given by $\mathbf{g}(\mathbf{x}) = [-1/x_1 - 1/x_2 \ldots -1/x_n]^T$ and $\mathbf{H}(\mathbf{x}) = \text{diag}\{1/x_1^2, 1/x_2^2, \ldots, 1/x_n^2\}$, respectively. By applying Eqs. (14.39a) and (14.39b), we obtain

$$\lambda_{k+1} = \frac{\mathbf{e}^T \mathbf{x}_k}{\mathbf{x}_k^T \mathbf{x}_k}, \quad \mathbf{d}_k = \mathbf{x}_k - \lambda_{k+1} \mathbf{x}_k^{(2)}$$

where $\mathbf{x}_k^{(2)}$ is a vector whose ith component is equal to $x_k^2(i)$. Algorithm 14.2 was run for five instances with problem size $n = 10, 50, 100, 200$, and 500. For each n, five feasible points were tried. Each of these initial points was generated randomly. The convergence tolerance was set to $\varepsilon = 10^{-5}$ for all 25 cases tested and the algorithm was found to converge in fewer than 10 iterations in each case. Figure 14.8 shows a plot of $f(\mathbf{x}_k) - f^*$ versus the number of iterations for the case with $n = 200$.

(b) To apply Algorithm 14.4, we use Eqs. (14.42a) and (14.42b) to compute the components of $\mathbf{d}_k$ and δ_k, which are found to be

$$\mathbf{d}_k(i) = -(\delta_k - \mathbf{r}_d(i))\mathbf{x}_k^2(i)$$

$$\delta_k = -\frac{1}{\mathbf{x}_k^T \mathbf{x}_k} \left(\sum_{i=1}^{n} \mathbf{r}_d(i)\mathbf{x}_k^2(i) - r_p \right)$$

where

$$\mathbf{r}_d(i) = \lambda_k - \frac{1}{\mathbf{x}_k(i)}$$

$$r_p = \mathbf{e}^T \mathbf{x}_k - 1$$

As for the setup in part (a), Algorithm 14.4 was evaluated for five instances with $n = 10, 50, 100, 200$, and 500. For each n, five nonfeasible points were generated randomly with strictly positive entries to ensure that they were in the domain of the objective function. The convergence tolerance was set to $\varepsilon = 10^{-6}$ as in part (a) and the algorithm was found to converge in fewer than 25 iterations in all cases. Figure 14.9a shows a plot of $\|\mathbf{r}(\mathbf{x}_k, \lambda_k)\|_2$ versus the number of iterations while Fig. 14.9b shows a plot of $f(\mathbf{x}_k) - f^*$ versus the number of iterations for the case where $n = 200$. We remark that in the implementation of the algorithm a "projection back to dom(f)" step was executed in each iteration to project the newly produced iterate $\mathbf{x}_k$ back to the domain of the objective function dom(f) = $\{\mathbf{x} : \mathbf{x} > 0\}$. This was done by verifying the components of $\mathbf{x}_k$ and simply setting any component that was less than 10^{-6} to 10^{-6}. The projection step was found helpful to make the algorithm less sensitive to the choice of a nonfeasible initial point. ∎

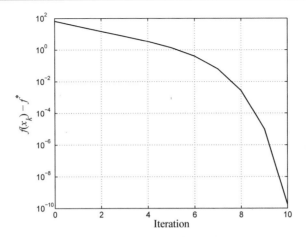

Fig. 14.8 $f(\mathbf{x}_k) - f^*$ versus number of iterations for the analytic center problem with $n = 200$

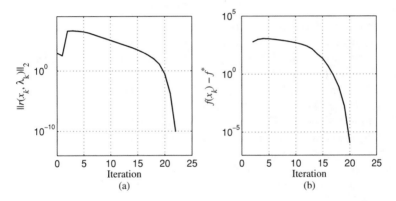

Fig. 14.9 **a** Residual $\|\mathbf{r}(\mathbf{x}_k, \boldsymbol{\lambda}_k)\|_2$ and **b** $f(\mathbf{x}_k) - f^*$ versus the number of iterations for the analytic center problem with $n = 200$. The initial point was in dom(f) but was not feasible

14.3.4   A Newton Barrier Method for General Convex Programming Problems

We now consider the general convex problem

$$\text{minimize } f(\mathbf{x}) \tag{14.44a}$$

$$\text{subject to: } \mathbf{Ax} = \mathbf{b} \tag{14.44b}$$

$$c_j(\mathbf{x}) \leq 0 \quad \text{for } j = 1, \ 2, \ \ldots, \ q \tag{14.44c}$$

where $f(\mathbf{x})$ and $c_j(\mathbf{x})$ are smooth convex functions. This problem is equivalent to

$$\text{minimize } f(\mathbf{x}) + \sum_{j=1}^{q} I_-(c_j(\mathbf{x}))$$

(14.45)

$$\text{subject to: } \mathbf{Ax} = \mathbf{b}$$

where $I_-(u)$ is the indicator function for nonpositive real u, which is consistent with the definition of a general indicator function in Eq. (14.20). An effective way to deal with the nonsmoothness of the objective function in Eq. (14.45) is to use a logarithmic function $h_\tau(u) = -(1/\tau)\log(-u)$ to approximate $I_-(u)$, which is smooth and convex in its domain $u < 0$. Parameter $\tau > 0$ controls the approximation error in the sense that $\lim_{\tau\to\infty} h_\tau(u) = I_-(u)$ for $u < 0$ (see Probs. 14.9 and 14.10). With these observations in mind, we proceed by examining the smooth convex problem

$$\text{minimize } F_\tau(\mathbf{x}) = f(\mathbf{x}) - \frac{1}{\tau}\sum_{j=1}^{q}\log(-c_j(\mathbf{x}))$$

(14.46)

$$\text{subject to: } \mathbf{Ax} = \mathbf{b}$$

which is to be solved sequentially with a fixed τ and the value of τ increasing as the sequential minimization proceeds. Under these circumstances, the above problem can be considered to be a *subproblem* whose solution is utilized as initial point in solving the next subproblem. The second term of the objective function in Eq. (14.46), namely, $b(\mathbf{x}) = -\sum_{j=1}^{q}\log(-c(\mathbf{x}))$, is known as the *logarithmic barrier*. Note that $b(\mathbf{x})$ is convex as long as all $c_j(\mathbf{x})$ are convex. Therefore, the problem in Eq. (14.46) is convex. The motivation behind this formulation is that the inclusion of the logarithmic barrier term as part of the objective function prevents $c_j(\mathbf{x})$ from becoming positive which, in turn, assures that the constraints in Eq. (14.44c) are satisfied. At the same time with an increasing τ, as the algorithm proceeds, we see a gradually weakening logarithmic barrier relative to function $f(\mathbf{x})$. As a result after a few iterations, minimizing $F_\tau(\mathbf{x})$ is practically the same as minimizing $f(\mathbf{x})$. Another formulation equivalent to that in Eq. (14.46) can be obtained by rescaling the objective function as

$$\text{minimize } F_\tau(\mathbf{x}) = \tau f(\mathbf{x}) + b(\mathbf{x})$$
$$\text{subject to: } \mathbf{Ax} = \mathbf{b}$$

(14.47)

With a fixed τ, the problem in Eq. (14.47) is a smooth convex problem subject to linear equalities, which can be solved by applying Algorithm 14.2. To apply this algorithm, the gradient and Hessian of $F_\tau(\mathbf{x})$ can be evaluated as $\nabla F_\tau(\mathbf{x}) = \tau\nabla f(\mathbf{x}) + \nabla b(\mathbf{x})$ and $\nabla^2 F_\tau(\mathbf{x}) = \tau\nabla^2 f(\mathbf{x}) + \nabla^2 b(\mathbf{x})$, respectively, where

$$\nabla b(\mathbf{x}) = -\sum_{j=1}^{q}\frac{\nabla c_j(\mathbf{x})}{c_j(\mathbf{x})}$$

(14.48)

$$\nabla^2 b(\mathbf{x}) = \sum_{j=1}^{q}\frac{\nabla c_j(\mathbf{x})\nabla^T c_j(\mathbf{x})}{c_j(\mathbf{x})^2} - \sum_{j=1}^{q}\frac{\nabla^2 c_j(\mathbf{x})}{c_j(\mathbf{x})}$$

We now address a convergence issue for the Newton barrier method where, for the sake of simplicity, the problem in Eq. (14.47) is assumed to have a unique solution $\mathbf{x}^*(\tau)$. The KKT conditions of the problem are given by

$$\mathbf{A}\mathbf{x}^* = \mathbf{b}, \quad c_j(\mathbf{x}^*(\tau)) < 0 \text{ for } 1 \le j \le q$$

$$\tau \nabla f(\mathbf{x}^*(\tau)) + \mathbf{A}^T \hat{\boldsymbol{\lambda}} + \nabla b(\mathbf{x}^*(\tau)) = \mathbf{0}$$

Using Eq. (14.48), the second equation can be expressed as

$$\nabla f(\mathbf{x}^*(\tau)) + \mathbf{A}^T \boldsymbol{\lambda}^*(\tau) + \sum_{j=1}^{q} \mu_j^*(\tau) \nabla c_j(\mathbf{x}^*(\tau)) = \mathbf{0} \qquad (14.49a)$$

where

$$\boldsymbol{\lambda}^*(\tau) = \hat{\boldsymbol{\lambda}}/\tau, \quad \mu_j^*(\tau) = -\frac{1}{\tau c_j(\mathbf{x}^*(\tau))} \qquad (14.49b)$$

Now recall that the Lagrangian for the problem in Eqs. (14.44a)–(14.44c) is defined by

$$L(\mathbf{x}, \boldsymbol{\lambda}, \boldsymbol{\mu}) = f(\mathbf{x}) + \boldsymbol{\lambda}^T (\mathbf{A}\mathbf{x} - \mathbf{b}) + \sum_{j=1}^{q} \mu_j c_j(\mathbf{x}) \qquad (14.50)$$

On comparing Eq. (14.50) with Eqs. (14.49a) and (14.49b), we see that $\mathbf{x}^*(\tau)$ minimizes the Lagrangian for $\boldsymbol{\lambda} = \boldsymbol{\lambda}^*(\tau)$, $\mu_j = \mu_j^*(\tau)$ with $1 \le j \le q$ which means that the dual function at $\boldsymbol{\lambda} = \boldsymbol{\lambda}^*(\tau)$ and $\mu_j = \mu_j^*(\tau)$ for $1 \le j \le q$ is given by

$$q(\boldsymbol{\lambda}^*(\tau), \boldsymbol{\mu}^*(\tau)) = f(\mathbf{x}^*(\tau)) + \boldsymbol{\lambda}^{*T}(\mathbf{A}\mathbf{x}^*(\tau) - \mathbf{b}) + \sum_{j=1}^{q} \mu_j^* c_j(\mathbf{x}^*(\tau))$$

$$= f(\mathbf{x}^*(\tau)) - q/\tau \qquad (14.51)$$

(see Eq. (10.110) for the definition of dual function). Since the dual function is bounded from above by the minimum value f^* of the objective function in the primal problem in Eqs. (14.44a)–(14.44c), Eq. (14.51) implies that

$$f(\mathbf{x}^*(\tau)) - f^* \le q/\tau \qquad (14.52)$$

which shows that $\mathbf{x}^*(\tau)$ converges to the optimal solution of the problem in Eqs. (14.44a)–(14.44c) as τ goes to infinity. Moreover, Eq. (14.52) also suggests the stopping criterion for the Newton barrier algorithm being q/τ is less than a prescribed tolerance. The Newton barrier algorithm can now be outlined as follows.

Algorithm 14.5 Newton barrier algorithm for general convex problem in Eqs. (14.44a)–(14.44c)

Step 1
Input strictly feasible initial point x_0, $\tau > 0$ (say, $\tau = 1$), $\gamma > 1$ (say, $\gamma = 10$), and tolerance $\varepsilon > 0$ and set $k = 0$.

Step 2
Solve the problem in Eq. (14.47) using Algorithm 14.2 to obtain solution $\mathbf{x}^*(\tau)$.

Step 3
If $q/\tau < \varepsilon$, stop and output $\mathbf{x}^*(\tau)$ as the solution; otherwise, set $\tau = \gamma\tau$, $k = k + 1$, and $x_0 = \mathbf{x}^*(\tau)$, and repeat from Step 2.

A value of γ in the range [10, 20] usually works well.

The Newton barrier algorithm requires a strictly feasible initial point x_0 and finding such a point is not always trivial. Below we describe a method that solves this *feasibility problem* by converting the problem into a convex problem for which a strictly feasible initial point is easy to find. The feasibility problem at hand is to find a point $\mathbf{x}$ such that $\mathbf{A}\mathbf{x} = \mathbf{b}$ and $c_j(\mathbf{x}) < 0$ for $j = 1, 2, \ldots, q$. First, we find a special solution of $\mathbf{A}\mathbf{x} = \mathbf{b}$ as $\mathbf{x}_s = \mathbf{A}^+\mathbf{b}$ where $\mathbf{A}^+$ is the Moore-Penrose pseudo-inverse of $\mathbf{A}$. Next, we find $c_0 = \max\{c_j(\mathbf{x}_s), j = 1, 2, \ldots, q\}$. If $c_0 < 0$, then obviously $\mathbf{x}_s$ is a strictly feasible point. If $c_0 \geq 0$, then the augmented convex problem

$$
\begin{aligned}
\text{minimize } & \delta \\
\text{subject to: } & \mathbf{A}\mathbf{x} = \mathbf{b} \\
& c_j(\mathbf{x}) \leq \delta \text{ for } 1 \leq j \leq q
\end{aligned}
\tag{14.53}
$$

where δ is treated as an additional variable is solved. In the literature, Eq. (14.53) is referred to as the *phase-one optimization problem*. Note that a strictly feasible initial point for the problem in Eq. (14.53) can be identified easily, for example, as $\{\delta_0, \mathbf{x}_0\} = \{\hat{c}_0, \mathbf{x}_s\}$ where $\hat{c}_0$ is a scalar $\hat{c}_0 > c_0$. Let the solution of the problem in Eq. (14.53) be denoted by $\{\delta^*, \mathbf{x}^*\}$. If $\delta^* < 0$, then $\mathbf{x}^*$ can be used as a strictly feasible initial point for the problem in Eqs. (14.44a)–(14.44c). If $\delta^* \geq 0$, then no strictly feasible point exists for the problem.

Example 14.6 Apply Algorithm 14.5 to the LP problem

$$
\begin{aligned}
\text{minimize } & f(\mathbf{x}) = \mathbf{c}^T\mathbf{x} \\
\text{subject to: } & \mathbf{A}\mathbf{x} \leq \mathbf{b}
\end{aligned}
\tag{14.54}
$$

where data $\mathbf{A} \in R^{100\times 50}$, $\mathbf{b} \in R^{100\times 1}$, $\mathbf{c} \in R^{50\times 1}$ are randomly generated such that both the primal and dual problems are strictly feasible and the optimal value of $f(\mathbf{x})$ is $f^* = 1$.

Solution Denote the ith row of $\mathbf{A}$ and ith entry of $\mathbf{b}$ as $\mathbf{a}_i^T$ and b_i, respectively. The solution of the problem in Eq. (14.54) is obtained by sequentially solving the subproblem

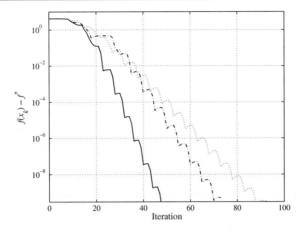

Fig. 14.10 $f(\mathbf{x}_k) - f^*$ versus cumulative number of Newton iterations for the problem in Example 14.6 where γ was set to 5 (dotted), 10 (dashed-dotted), and 20 (solid)

$$\text{minimize } F_\tau(\mathbf{x}) = \tau \mathbf{c}^T \mathbf{x} - \sum_{i=1}^{100} \log(b_i - \mathbf{a}_i^T \mathbf{x}) \qquad (14.55)$$

We start by finding a strictly feasible initial point $\mathbf{x}_0$ such that $\mathbf{A}\mathbf{x}_0 - \mathbf{b} < 0$. This is done by solving the LP problem

$$\text{minimize } \delta$$

$$\text{subject to: } \mathbf{A}\mathbf{x} - \mathbf{b} \le \delta \mathbf{e}$$

where $\mathbf{e}$ denotes a vector of dimension 100 whose components are all 1s. If we denote $\tilde{\mathbf{x}} = [\delta \; \mathbf{x}^T]^T$, it can readily be verified that point $\tilde{\mathbf{x}}_0 = [\delta_0 \; \mathbf{0}]^T$ with $\delta_0 = \max(-\mathbf{b}) + 10^{-3}$ is strictly feasible for the above LP problem. Now suppose that the LP problem is solved and the first component of the solution, δ^*, is found to be negative, then the remaining part of the solution can be taken as a strictly feasible initial point for the problem in Eq. (14.55). We applied Algorithm 14.5 to the problem in Eq. (14.55) with an initial $\tau = 1$ and three choices of γ, namely, $\gamma = 5, 10,$ and 50. Since the problem in Eq. (14.55) is unconstrained, Step 2 of the algorithm was implemented using Algorithm 14.1 in which the convergence tolerance was set to 10^{-2} as higher accuracy of the solution for these subproblems is unnecessary. The convergence tolerance ε was set to 10^{-11}. Figure 14.10 shows the algorithm's performance in terms of $f(\mathbf{x}_k) - f^*$ versus cumulative number of Newton steps used. ∎

14.4   Minimization of Composite Convex Functions

In this section, we study a class of unconstrained convex problems which assume
the form

$$\text{minimize } F(\mathbf{x}) = f(\mathbf{x}) + \Psi(\mathbf{x}) \tag{14.56}$$

where function $f(\mathbf{x})$ is differentiable and convex while $\Psi(\mathbf{x})$ is convex but possibly
nondifferentiable. The problem in Eq. (14.56) arises in several applications in statistics, engineering, and other fields and has also motivated recent developments in the
theoretical analysis and design of new algorithms for nonsmooth convex optimization
in general and the minimization of composite convex functions in particular.

14.4.1   Proximal-Point Algorithm

Below we develop a proximal-point algorithm for the solution of a class of problems
that assume the formulation in Eq. (14.56) where $f(\mathbf{x})$ is convex with a Lipschitz
continuous gradient. From Eq. (14.9),

$$f(\mathbf{x}) + \Psi(\mathbf{x}) \le f(\mathbf{v}) + \nabla f(\mathbf{v})^T (\mathbf{x} - \mathbf{v}) + \Psi(\mathbf{x}) + \frac{L}{2}\|\mathbf{x} - \mathbf{v}\|_2^2 \tag{14.57}$$

for any $\mathbf{x}$ and $\mathbf{v}$ where L is a Lipschitz constant for the gradient of $f(\mathbf{x})$ according
to Eq. (14.8). If we let

$$\hat{F}(\mathbf{x}, \mathbf{v}) = f(\mathbf{v}) + \nabla f(\mathbf{v})^T (\mathbf{x} - \mathbf{v}) + \Psi(\mathbf{x}) \tag{14.58}$$

then Eq. (14.57) becomes

$$f(\mathbf{x}) + \Psi(\mathbf{x}) \le \hat{F}(\mathbf{x}, \mathbf{v}) + \frac{1}{2\tau}\|\mathbf{x} - \mathbf{v}\|_2^2 \tag{14.59}$$

with $\tau = 1/L$ which leads to the proximal-point operator

$$\text{prox}_{\tau \hat{F}}(\mathbf{v}) = \arg \min_{\mathbf{x}} \left(\hat{F}(\mathbf{x}, \mathbf{v}) + \frac{1}{2\tau}\|\mathbf{x} - \mathbf{v}\|_2^2 \right) \tag{14.60}$$

Note that this proximal operator differs slightly from that defined in Eq. (14.26) as
function $\hat{F}$ depends on both $\mathbf{x}$ and $\mathbf{v}$. Nevertheless, like a conventional proximal
operator, for a given point $\mathbf{v}$, $\text{prox}_{\tau \hat{F}}(\mathbf{v})$ maps $\mathbf{v}$ to a point that is closer to the minimizer of $F(\mathbf{x})$ relative to point $\mathbf{v}$. This suggests a series of proximal-point iterations
$\mathbf{x}_{k+1} = \text{prox}_{\tau \hat{F}}(\mathbf{x}_k)$ for $k = 0, 1, \ldots,$ which in conjunction with Eq. (14.60) lead
to

$$\mathbf{x}_{k+1} = \arg \min_{\mathbf{x}} \left(\hat{F}(\mathbf{x}, \mathbf{x}_k) + \frac{1}{2\tau}\|\mathbf{x} - \mathbf{x}_k\|_2^2 \right) \tag{14.61}$$

Theorem 14.4 *Point* $\mathbf{x}^*$ *is a minimizer of* $F(\mathbf{x})$ *if and only if* $\mathbf{x}^*$ *is a fixed point of* $\mathrm{prox}_{\tau\hat{F}}(\mathbf{v})$, *that is*

$$\mathbf{x}^* = \mathrm{prox}_{\tau\hat{F}}(\mathbf{x}^*) \tag{14.62}$$

Proof Suppose $\mathbf{x}^*$ is a minimizer of $F(\mathbf{x})$. If $\mathbf{v} = \mathbf{x}^*$, the inequality in Eq. (14.59) can be expressed as

$$f(\mathbf{x}) + \Psi(\mathbf{x}) \leq \hat{F}(\mathbf{x}, \mathbf{x}^*) + \frac{1}{2\tau}\|\mathbf{x} - \mathbf{x}^*\|_2^2 \tag{14.63}$$

At point $\mathbf{x} = \mathbf{x}^*$, the right-hand side of Eq. (14.63) becomes $f(\mathbf{x}^*) + \Psi(\mathbf{x}^*)$ which is the minimum of the lower bound on the left-hand side of Eq. (14.63). Therefore, $\mathbf{x}^*$ minimizes $\hat{F}(\mathbf{x}, \mathbf{x}^*) + \frac{1}{2\tau}\|\mathbf{x} - \mathbf{x}^*\|_2^2$ and Eq. (14.62) holds. Conversely, if $\mathbf{x}^*$ is a fixed point of $\mathrm{prox}_{\tau\hat{F}}$, i.e., Eq. (14.62) holds, then $\mathbf{x}^*$ minimizes $\hat{F}(\mathbf{x}, \mathbf{x}^*) + \frac{1}{2\tau}\|\mathbf{x} - \mathbf{x}^*\|_2^2$. Hence, from Theorem 14.1, we have

$$\mathbf{0} \in \partial\left(\hat{F}(\mathbf{x}, \mathbf{x}^*) + \frac{1}{2\tau}\|\mathbf{x} - \mathbf{x}^*\|_2^2\right)\Big|_{\mathbf{x}=\mathbf{x}^*}$$

which in conjunction with Eq. (14.58) implies that $\mathbf{0} \in \nabla f(\mathbf{x}^*) + \partial\Psi(\mathbf{x}^*)$. This means that $\mathbf{x}^*$ is a minimizer of $F(\mathbf{x})$.   ■

From the definition of $\hat{F}(\mathbf{x}, \mathbf{x}^*)$, the proximal-point iteration given by Eq. (14.61) can be made more explicit as

$$\mathbf{x}_{k+1} = \arg\min_{\mathbf{x}}\left(f(\mathbf{x}_k) + \nabla f(\mathbf{x}_k)^T(\mathbf{x} - \mathbf{x}_k) + \frac{L}{2}\|\mathbf{x} - \mathbf{x}_k\|_2^2 + \Psi(\mathbf{x})\right)$$

which is the same as

$$\mathbf{x}_{k+1} = \arg\min_{\mathbf{x}}\left(\frac{L}{2}\|\mathbf{x} - (\mathbf{x}_k - \frac{1}{L}\nabla f(\mathbf{x}_k))\|_2^2 + \Psi(\mathbf{x})\right)$$

or

$$\mathbf{x}_{k+1} = \arg\min_{\mathbf{x}}\left(\frac{L}{2}\|\mathbf{x} - \mathbf{b}_k\|_2^2 + \Psi(\mathbf{x})\right) \tag{14.64a}$$

where

$$\mathbf{b}_k = \mathbf{x}_k - \frac{1}{L}\nabla f(\mathbf{x}_k) \tag{14.64b}$$

Algorithm 14.6 Proximal-point algorithm for the problem in Eq. (14.56)
Step 1
Input initial point $\mathbf{x}_0$, Lipschitz constant L for $\nabla f(\mathbf{x})$, and tolerance ε and set $k = 0$.
Step 2
Compute $\mathbf{x}_{k+1}$ by solving the convex problem in Eqs. (14.64a) and (14.64b).
Step 3
If $\|\mathbf{x}_{k+1} - \mathbf{x}_k\|_2/\sqrt{n} < \varepsilon$, output solution $\mathbf{x}^* = \mathbf{x}_{k+1}$ and stop; otherwise, set $k = k + 1$ and repeat from Step 2.

A problem of particular interest that often arises in several statistics and signal processing applications which fits well into the formulation in Eq. (14.56) is the $l_1 - l_2$ minimization problem

$$\text{minimize } F(\mathbf{x}) = \tfrac{1}{2}\|\mathbf{Ax} - \mathbf{b}\|_2^2 + \mu\|\mathbf{x}\|_1 \tag{14.65}$$

where data $\mathbf{A} \in R^{m \times n}$, $\mathbf{b} \in R^{m \times 1}$, and *regularization* parameter $\mu > 0$ are given. The two component functions in this case are $f(\mathbf{x}) = \tfrac{1}{2}\|\mathbf{Ax} - \mathbf{b}\|_2^2$ which is convex with Lipschitz continuous gradient $L = \lambda_{\max}(\mathbf{A}^T\mathbf{A})$ (see Example 14.2) and $\Psi(\mathbf{x}) = \mu\|\mathbf{x}\|_1$ which is convex but nondifferentiable. Applying Algorithm 14.6, the iteration step in Eqs. (14.64a) and (14.64b) becomes

$$\mathbf{x}_{k+1} = \arg\min_{\mathbf{x}} \left(\frac{1}{2}\|\mathbf{x} - \mathbf{b}_k\|_2^2 + \frac{\mu}{L}\|\mathbf{x}\|_1 \right) \tag{14.66a}$$

where

$$\mathbf{b}_k = \frac{1}{L}\mathbf{A}^T(\mathbf{b} - \mathbf{Ax}_k) + \mathbf{x}_k \tag{14.66b}$$

It is well known that the minimizer of the problem in Eqs. (14.66a) and (14.66b) can be obtained by applying soft shrinkage by an amount of μ/L to $\mathbf{b}_k$ [6], i.e.,

$$\mathbf{x}_{k+1} = S_{\mu/L}\left\{ \frac{1}{L}\mathbf{A}^T(\mathbf{b} - \mathbf{Ax}_k) + \mathbf{x}_k \right\} \tag{14.67}$$

where operator S_δ is defined by

$$S_\delta(\mathbf{z}) \triangleq \text{sign}(\mathbf{z}) \cdot \max\{|\mathbf{z}| - \delta, \mathbf{0}\} \tag{14.68}$$

and the operations of "sign", "max" and "$|\mathbf{z}| - \delta$" are performed component by component.

Algorithm 14.7 Proximal-point algorithm for the problem in Eq. (14.65)
Step 1
Input data $\{\mathbf{A}, \mathbf{b}\}$, regularization parameter μ, initial point $\mathbf{x}_0$, Lipschitz constant L, tolerance ε and set $k = 0$.
Step 2
Compute $\mathbf{x}_{k+1}$ using Eq. (14.67).
Step 3
If $\|\mathbf{x}_{k+1} - \mathbf{x}_k\|_2/\sqrt{n} < \varepsilon$, output solution $\mathbf{x}^* = \mathbf{x}_{k+1}$ and stop; otherwise, set $k = k + 1$ and repeat from Step 2.

14.4.2   Fast Algorithm For Solving the Problem in Eq. (14.56)

In his seminal paper, Nesterov proposed an algorithm for the solution of problems of the type in Eq. (14.56) [7], which have a rate of convergence $O(k^{-2})$ in the sense that

$$F(\mathbf{x}_k) - F(\mathbf{x}_{optimal}) \leq \frac{2L\|\mathbf{x}_0 - \mathbf{x}_{optimal}\|_2^2}{(k+1)^2} \tag{14.69}$$

Nesterov reviewed his algorithm in 2004 in [2, Chap.2] and later on, in 2013, he proposed several new gradient methods for the minimization of composite functions such as that in Eq. (14.56) [8]. A modified version of Nesterov's algorithm that uses a subtly modified previous iterate instead of soft-thresholding and which has the same rate of convergence as Nesterov's algorithm was published by Beck and Teboullein in [6]. This is commonly referred to as the *fast iterative shrinkage-thresholding algorithm* (FISTA) for short, and the steps involved are as follows:

Algorithm 14.8 FISTA for the problem in Eq. (14.56)
Step 1
Input initial point $\mathbf{x}_0$, Lipschitz constant L for $\nabla f(\mathbf{x})$, tolerance ε and set $\mathbf{z}_1 = \mathbf{x}_0$, $t_1 = 1$, and $k = 1$.
Step 2
Compute $\mathbf{x}_k = \arg\min_{\mathbf{x}}(\frac{L}{2}\|\mathbf{x} - (\mathbf{z}_k - \frac{1}{L}\nabla f(\mathbf{z}_k))\|_2^2 + \Psi(\mathbf{x}))$.
Step 3
Compute $t_{k+1} = \frac{1+\sqrt{1+4t_k^2}}{2}$.
Step 4
Update $\mathbf{z}_{k+1} = \mathbf{x}_k + \left(\frac{t_k-1}{t_{k+1}}\right)(\mathbf{x}_k - \mathbf{x}_{k-1})$.
Step 5
If $\|\mathbf{x}_k - \mathbf{x}_{k-1}\|_2/\sqrt{n} < \varepsilon$, output $\mathbf{x}^* = \mathbf{x}_k$ as the solution and stop; otherwise, set $k = k + 1$ and repeat from Step 2.

By applying FISTA to the $l_1 - l_2$ minimization problem in Eq. (14.65), the following algorithm can be constructed:

Algorithm 14.9 FISTA for the problem in Eq. (14.65)
Step 1
Input data $\{\mathbf{A}, \mathbf{b}\}$, regularization parameter μ, initial point $\mathbf{x}_0$, Lipschitz constant L, and tolerance ε and set $\mathbf{z}_1 = \mathbf{x}_0$, $t_1 = 1$, and $k = 1$.
Step 2
Compute $\mathbf{x}_k = S_{\mu/L}\left\{\frac{1}{L}\mathbf{A}^T(\mathbf{b} - \mathbf{A}\mathbf{z}_k) + \mathbf{z}_k\right\}$.
Step 3
Compute $t_{k+1} = \frac{1+\sqrt{1+4t_k^2}}{2}$.
Step 4
Update $\mathbf{z}_{k+1} = \mathbf{x}_k + \left(\frac{t_k-1}{t_{k+1}}\right)(\mathbf{x}_k - \mathbf{x}_{k-1})$.
Step 5
If $\|\mathbf{x}_k - \mathbf{x}_{k-1}\|_2/\sqrt{n} < \varepsilon$, output $\mathbf{x}^* = \mathbf{x}_k$ as the solution and stop; otherwise, set $k = k + 1$ and repeat from Step 2.

Note that the complexities of Algorithms 14.8 and 14.9 are practically the same as those of Algorithms 14.6 and 14.7, respectively, because the amount of additional computation required by Steps 3 and 4 of FISTA is insignificant.

Example 14.7 *Deconvolution of distorted and noise-contaminated signals*   Consider a communication channel where the signal received suffers both channel distortion and noise contamination. We assume an FIR channel of length N characterized by the transfer function

$$H(z) = \sum_{i=0}^{N-1} h_i z^{-i}$$

The output from the channel is modeled by

$$y(k) = \sum_{i=0}^{N-1} h_i x(k-i) + w(k) \tag{14.70}$$

where $\{w(k), k = 0, 1, \ldots, N-1\}$ are additive Gaussian noise components with zero mean and variance σ^2. A *deconvolution* problem associated with the model in Eq. (14.70) is to estimate transmitted samples $\{x(k), k = 0, 1, \ldots, n-1\}$ using m distorted and noisy measurements $\{y(k), k = 0, 1, \ldots, m-1\}$. In this example, we consider a typical case where both n and m are considerably larger than the channel length N. In addition, we assume that there are fewer measurements than the signal length, i.e., $m \leq n$. Formulate and solve the deconvolution problem as an $l_1 - l_2$ minimization problem.

Solution From Eq. (14.70), we can write

$$\begin{bmatrix} y_0 \\ y_1 \\ \vdots \\ \vdots \\ \vdots \\ \vdots \\ y_{m-1} \end{bmatrix} = \begin{bmatrix} h_0 & 0 & \cdots & 0 & 0 & \cdots & 0 \\ h_1 & h_0 & \cdots & 0 & 0 & \cdots & 0 \\ \vdots & \vdots & \ddots & \vdots & \vdots & \ddots & \vdots \\ h_{N-1} & h_{N-2} & \cdots & h_0 & 0 & \cdots & 0 \\ 0 & h_{N-1} & \cdots & h_1 & h_0 & \cdots & 0 \\ \vdots & \vdots & \ddots & \vdots & \vdots & \ddots & \vdots \\ 0 & 0 & \cdots & h_{N-1} & h_{N-2} & \cdots & h_0 \end{bmatrix} \begin{bmatrix} x_0 \\ x_1 \\ \vdots \\ \vdots \\ \vdots \\ \vdots \\ x_{n-1} \end{bmatrix} + \begin{bmatrix} w_0 \\ w_1 \\ \vdots \\ \vdots \\ \vdots \\ \vdots \\ w_{m-1} \end{bmatrix}$$

i.e.,

$$\mathbf{y} = \mathbf{Hx} + \mathbf{w} \tag{14.71}$$

where $\mathbf{H}$ is a Toeplitz matrix   which can be constructed by using its first column and first row.

Typically a signal of interest, like vector $\mathbf{x}$ above, is compressible under a certain transform such as the Fourier or cosine or wavelet transform. In analytic terms, this

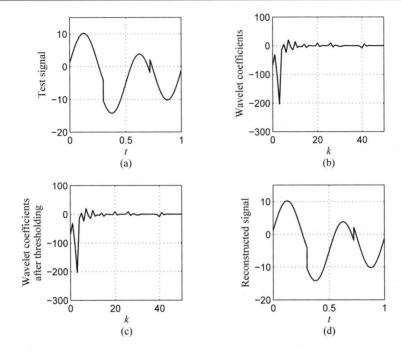

Fig. 14.11 a Heavisine test signal of length 1024; **b** first 50 wavelet coefficients of the heavisine signal; **c** first 50 wavelet coefficients after hard-thresholding that sets coefficients with magnitude less than 0.05 to zero. This leaves only 10.25% of the coefficients that are nonzero; **d** reconstructed signal using the sparse wavelet coefficients from part (**c**). The relative reconstruction error in l_2-norm was found to be 3.71×10^{-4}

means that there exists an orthogonal matrix $\mathbf{T}$ that converts a nonsparse $\mathbf{x}$ into an approximately sparse $\boldsymbol{\theta}$, namely, $\boldsymbol{\theta} = \mathbf{Tx}$. Here an approximately sparse signal is a discrete signal having a small number of significant nonzero components relative to the signal length. The signal used in this example is the standard test signal known as the *heavisine* signal of length 1024, which is illustrated in Fig. 14.11. With $\boldsymbol{\theta} = \mathbf{Tx}$, the model in Eq. (14.71) in terms of $\boldsymbol{\theta}$ becomes

$$\mathbf{y} = \mathbf{A}\boldsymbol{\theta} + \mathbf{w} \tag{14.72}$$

where $\mathbf{A} = \mathbf{HT}^T$ and the deconvolution problem can be tackled by solving the convex problem

$$\text{minimize } \|\boldsymbol{\theta}\|_1 \tag{14.73a}$$

$$\text{subject to: } \|\mathbf{A}\boldsymbol{\theta} - \mathbf{y}\|_2 \leq \varepsilon \tag{14.73b}$$

The bound ε in Eq. (14.73b) is related to channel noise component $\mathbf{w}$ and can be estimated using Eq. (14.72) as

$$\varepsilon = \|\mathbf{A}\boldsymbol{\theta} - \mathbf{y}\|_2 = \|\mathbf{w}\|_2 \approx \sqrt{m}\sigma \qquad (14.74)$$

We remark that the estimation of ε in Eq. (14.74) is useful only if the noise variance of $\mathbf{w}$ is available.

With an appropriate weight $\mu > 0$, the problem in Eqs. (14.73a) and (14.73b) is equivalent to the $l_1 - l_2$ minimization problem

$$\text{minimize } F(\boldsymbol{\theta}) = \tfrac{1}{2}\|\mathbf{A}\boldsymbol{\theta} - \mathbf{y}\|_2^2 + \mu\|\boldsymbol{\theta}\|_1 \qquad (14.75)$$

The original signal model associated with Eq. (14.75) is given by Eq. (14.71) where $\mathbf{x}$ is the heavisine signal of length 1024 (see Fig. 14.12a), $\mathbf{H}$ is a Toeplitz matrix of size $m = n = 36$ which was generated by using a narrow-band lowpass filter with a passband of $[0, 0.1\pi]$, and $\mathbf{w}$ is a noise vector with components that are Gaussian white with mean 0 and variance $\sigma^2 = 0.25$.

The problem in Eq. (14.75) was solved using Algorithm 14.7. The solution $\boldsymbol{\theta}^*$ was used to construct signal $\mathbf{x}^* = \mathbf{T}^T\boldsymbol{\theta}^*$. With $\varepsilon = 2.4 \times 10^{-4}$ and $\mu = 0.224$, it took the algorithm 176 iterations to converge to solution $\boldsymbol{\theta}^*$. The signal-to-noise ratio (SNR) of the deconvolved signal was found to be 26.6023 dB. For comparison, the SNR of the distorted and noise-corrupted signal $\mathbf{y}$ was 5.7825 dB. The original, distorted, and recovered signals are depicted in Fig. 14.12.

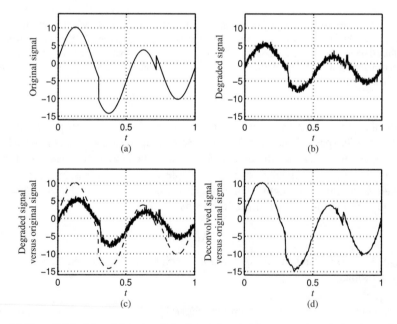

Fig. 14.12 a Original heavisine signal; **b** distorted and noise-contaminated signal; **c** original (dashed) versus degraded (solid) signal, SNR = 5.7825 dB; **d** original (dashed) versus reconstructed (solid) signal, SNR = 26.6023 dB

Algorithm 14.9 was also applied to the same deconvolution problem. With $\varepsilon = 2.68 \times 10^{-3}$ and $\mu = 0.224$, it took the algorithm 29 iterations to converge to solution $\boldsymbol{\theta}^*$. The SNR of the deconvolved signal was found to be 26.5980 dB which is practically the same as that obtained by Algorithm 14.7. As expected, Algorithm 14.9 was considerably faster: the average CPU time it required was approximately 19% of that required by Algorithm 14.7.   ■

14.5   Alternating Direction Methods

Alternating direction methods have become increasingly important because of their capability in dealing with large-scale convex problems. This section presents two representative alternating direction methods, the *alternating direction method of multipliers* and the *alternating minimization method*.

The alternating direction method of multipliers is so-called in view of the fact that it is critically dependent on the use of Lagrange multipliers although the alternating minimization method also uses Lagrange multipliers.

14.5.1   Alternating Direction Method of Multipliers

The basic alternating direction method of multipliers (ADMM) has been developed to solve convex minimization problems of the type [9]

$$\begin{aligned} \text{minimize } & f(\mathbf{x}) + h(\mathbf{y}) \\ \text{subject to: } & \mathbf{Ax} + \mathbf{By} = \mathbf{c} \end{aligned} \tag{14.76}$$

where $\mathbf{x} \in R^n$ and $\mathbf{y} \in R^m$ are variables, $\mathbf{A} \in R^{p\times n}$, $\mathbf{B} \in R^{p\times m}$, $\mathbf{c} \in R^{p\times 1}$, and $f(\mathbf{x})$ as well as $h(\mathbf{y})$ are convex functions. Note that in Eq. (14.76) the variables in both the objective function and the constraints are split into two parts, each involving only one set of different variables. By definition, the Lagrangian for the problem in Eq. (14.76) is given by

$$L(\mathbf{x}, \mathbf{y}, \boldsymbol{\lambda}) = f(\mathbf{x}) + h(\mathbf{y}) + \boldsymbol{\lambda}^T (\mathbf{Ax} + \mathbf{By} - \mathbf{c})$$

If both $f(\mathbf{x})$ and $h(\mathbf{y})$ are differentiable, the KKT conditions for the problem in Eq. (14.76) are given by

$$\mathbf{Ax} + \mathbf{By} = \mathbf{c} \tag{14.77a}$$

$$\nabla f(\mathbf{x}) + \mathbf{A}^T \boldsymbol{\lambda} = \mathbf{0} \tag{14.77b}$$

$$\nabla h(\mathbf{y}) + \mathbf{B}^T \boldsymbol{\lambda} = \mathbf{0} \tag{14.77c}$$

The Lagrange dual of Eq. (14.76) assumes the form

$$\text{maximize } q(\boldsymbol{\lambda}) \tag{14.78}$$

where

$$q(\lambda) = \inf_{\mathbf{x},\mathbf{y}}[f(\mathbf{x}) + h(\mathbf{y}) + \lambda^T(\mathbf{A}\mathbf{x} + \mathbf{B}\mathbf{y} - \mathbf{c})]$$

which in conjunction with Eq. (14.21) leads to

$$
\begin{aligned}
q(\lambda) &= \inf_{\mathbf{x}}[f(\mathbf{x}) + \lambda^T\mathbf{A}\mathbf{x}] + \inf_{\mathbf{y}}[h(\mathbf{y}) + \lambda^T\mathbf{B}\mathbf{y}] - \lambda^T\mathbf{c} \\
&= -\sup_{\mathbf{x}}[(-\mathbf{A}^T\lambda)^T\mathbf{x} - f(\mathbf{x})] - \sup_{\mathbf{y}}[(-\mathbf{B}^T\lambda)^T\mathbf{x} - h(\mathbf{y})] - \lambda^T\mathbf{c} \quad (14.79) \\
&= -f^*(-\mathbf{A}^T\lambda) - h^*(-\mathbf{B}^T\lambda) - \mathbf{c}^T\lambda
\end{aligned}
$$

Based on the properties that $\mathbf{u} = \nabla f(\mathbf{v})$ if and only if $\mathbf{v} = \nabla f^*(\mathbf{u})$ and that $\nabla f^*(-\mathbf{A}^T\lambda) = -\mathbf{A}\nabla f^*(\lambda)$ (see Sect. 14.2.4), Eq. (14.79) implies that

$$\nabla q(\lambda) = \mathbf{A}\mathbf{x} + \mathbf{B}\mathbf{y} - \mathbf{c} \qquad (14.80)$$

where $\{\mathbf{x}, \mathbf{y}\}$ minimizes $L(\mathbf{x}, \mathbf{y}, \lambda)$ for a given λ.

If in addition, we assume that $f(\mathbf{x})$ and $h(\mathbf{y})$ are strictly convex, a solution of the problem in Eq. (14.76) can be found by minimizing the Lagrangian $L(\mathbf{x}, \mathbf{y}, \lambda^*)$ with respect to primal variables $\mathbf{x}$ and $\mathbf{y}$ where λ^* maximizes the dual function $q(\lambda)$ in Eq. (14.79). This in conjunction with Eq. (14.80) suggests the *dual ascent iterations* for the problem in Eq. (14.76) given by

$$\mathbf{x}_{k+1} = \arg\min_{\mathbf{x}} L(\mathbf{x}, \mathbf{y}_k, \lambda_k) = \arg\min_{\mathbf{x}}[f(\mathbf{x}) + \lambda_k^T\mathbf{A}\mathbf{x}] \qquad (14.81a)$$

$$\mathbf{y}_{k+1} = \arg\min_{\mathbf{y}} L(\mathbf{x}_k, \mathbf{y}, \lambda_k) = \arg\min_{\mathbf{y}}[h(\mathbf{y}) + \lambda_k^T\mathbf{B}\mathbf{y}] \qquad (14.81b)$$

$$\lambda_{k+1} = \lambda_k + \alpha_k(\mathbf{A}\mathbf{x}_{k+1} + \mathbf{B}\mathbf{y}_{k+1} - \mathbf{c}) \qquad (14.81c)$$

where scalar α_k, such that $\alpha_k > 0$, is chosen to maximize $q(\lambda)$ (see Eq. (14.78)) along the direction $\mathbf{A}\mathbf{x}_{k+1} + \mathbf{B}\mathbf{y}_{k+1} - \mathbf{c}$.

It is well known that convex problems of the form in Eq. (14.76) with less restrictive functions $f(\mathbf{x})$ and $h(\mathbf{y})$ and data matrices $\mathbf{A}$ and $\mathbf{B}$ can be handled by using an augmented dual based on the *augmented* Lagrangian [9]

$$L_\alpha(\mathbf{x}, \mathbf{y}, \lambda) = f(\mathbf{x}) + h(\mathbf{y}) + \lambda^T(\mathbf{A}\mathbf{x} + \mathbf{B}\mathbf{y} - \mathbf{c}) + \frac{\alpha}{2}\|\mathbf{A}\mathbf{x} + \mathbf{B}\mathbf{y} - \mathbf{c}\|_2^2 \quad (14.82)$$

which includes the conventional Lagrangian $L(\mathbf{x}, \mathbf{y}, \lambda)$ as a special case when parameter α is set to zero. The introduction of the augmented Lagrangian may be understood by considering the following [9]: if we modify the objective function in Eq. (14.76) by adding a penalty term $\frac{\alpha}{2}\|\mathbf{A}\mathbf{x} + \mathbf{B}\mathbf{y} - \mathbf{c}\|_2^2$ for violation of the equality constraint, namely,

$$
\begin{aligned}
&\text{minimize} \ \ f(\mathbf{x}) + h(\mathbf{y}) + \frac{\alpha}{2}\|\mathbf{A}\mathbf{x} + \mathbf{B}\mathbf{y} - \mathbf{c}\|_2^2 \\
&\text{subject to:} \ \mathbf{A}\mathbf{x} + \mathbf{B}\mathbf{y} = \mathbf{c}
\end{aligned}
\qquad (14.83)
$$

then the conventional Lagrangian of the problem in Eq. (14.83) is exactly equal to $L_\alpha(\mathbf{x}, \mathbf{y}, \boldsymbol{\lambda})$ in Eq. (14.82). The augmented dual problem associated with $L_\alpha(\mathbf{x}, \mathbf{y}, \boldsymbol{\lambda})$ is given by

$$\text{minimize} \quad q_\alpha(\boldsymbol{\lambda})$$

where

$$q_\alpha(\boldsymbol{\lambda}) = \inf_{\mathbf{x},\mathbf{y}} \left[f(\mathbf{x}) + h(\mathbf{y}) + \boldsymbol{\lambda}^T(\mathbf{A}\mathbf{x} + \mathbf{B}\mathbf{y} - \mathbf{c}) + \frac{\alpha}{2} \|\mathbf{A}\mathbf{x} + \mathbf{B}\mathbf{y} - \mathbf{c}\|_2^2 \right]$$

The augmented Lagrangian is no longer separable in terms of variables $\mathbf{x}$ and $\mathbf{y}$ as in Eqs. (14.81a)–(14.81c) because of the presence of the penalty term. In the ADMM this issue is addressed by using *alternating* updates of the primal variables $\mathbf{x}$ and $\mathbf{y}$, namely,

$$\mathbf{x}_{k+1} = \arg\min_{\mathbf{x}} \left[f(\mathbf{x}) + \boldsymbol{\lambda}_k^T \mathbf{A}\mathbf{x} + \frac{\alpha}{2} \|\mathbf{A}\mathbf{x} + \mathbf{B}\mathbf{y}_k - \mathbf{c}\|_2^2 \right] \tag{14.84a}$$

$$\mathbf{y}_{k+1} = \arg\min_{\mathbf{y}} \left[h(\mathbf{y}) + \boldsymbol{\lambda}_k^T \mathbf{B}\mathbf{y} + \frac{\alpha}{2} \|\mathbf{A}\mathbf{x}_{k+1} + \mathbf{B}\mathbf{y} - \mathbf{c}\|_2^2 \right] \tag{14.84b}$$

$$\boldsymbol{\lambda}_{k+1} = \boldsymbol{\lambda}_k + \alpha(\mathbf{A}\mathbf{x}_{k+1} + \mathbf{B}\mathbf{y}_{k+1} - \mathbf{c}) \tag{14.84c}$$

Note that parameter α associated with the quadratic penalty term is used in Eq. (14.84c) to update Lagrange multiplier $\boldsymbol{\lambda}_k$ thereby eliminating the need for a line search step to compute α_k as required in Eq. (14.81c). To justify Eqs. (14.84a)–(14.84c), note that $\mathbf{y}_{k+1}$ minimizes $h(\mathbf{y}) + \boldsymbol{\lambda}_k^T \mathbf{B}\mathbf{y} + \frac{\alpha}{2} \|\mathbf{A}\mathbf{x}_{k+1} + \mathbf{B}\mathbf{y} - \mathbf{c}\|_2^2$ and hence

$$\begin{aligned}
\mathbf{0} &= \nabla h(\mathbf{y}_{k+1}) + \mathbf{B}^T \boldsymbol{\lambda}_k + \alpha \mathbf{B}^T(\mathbf{A}\mathbf{x}_{k+1} + \mathbf{B}\mathbf{y}_{k+1} - \mathbf{c}) \\
&= \nabla h(\mathbf{y}_{k+1}) + \mathbf{B}^T[\boldsymbol{\lambda}_k + \alpha(\mathbf{A}\mathbf{x}_{k+1} + \mathbf{B}\mathbf{y}_{k+1} - \mathbf{c})]
\end{aligned}$$

which in conjunction with Eq. (14.84c) leads to

$$\nabla h(\mathbf{y}_{k+1}) + \mathbf{B}^T \boldsymbol{\lambda}_{k+1} = 0$$

Therefore, the KKT condition in Eq. (14.77c) is forced by using the ADMM iterations in Eqs. (14.94a)–(14.94c). In addition, since $\mathbf{x}_{k+1}$ minimizes $f(\mathbf{x}) + \boldsymbol{\lambda}_k^T \mathbf{A}\mathbf{x} + \frac{\alpha}{2} \|\mathbf{A}\mathbf{x} + \mathbf{B}\mathbf{y}_k - \mathbf{c}\|_2^2$, we have

$$\begin{aligned}
\mathbf{0} &= \nabla f(\mathbf{x}_{k+1}) + \mathbf{A}^T \boldsymbol{\lambda}_k + \alpha \mathbf{A}^T(\mathbf{A}\mathbf{x}_{k+1} + \mathbf{B}\mathbf{y}_k - \mathbf{c}) \\
&= \nabla f(\mathbf{x}_{k+1}) + \mathbf{A}^T[\boldsymbol{\lambda}_k + \alpha(\mathbf{A}\mathbf{x}_{k+1} + \mathbf{B}\mathbf{y}_k - \mathbf{c})] \\
&= \nabla f(\mathbf{x}_{k+1}) + \mathbf{A}^T \boldsymbol{\lambda}_{k+1} - \alpha \mathbf{A}^T \mathbf{B}(\mathbf{y}_{k+1} - \mathbf{y}_k)
\end{aligned}$$

i.e.,

$$\nabla f(\mathbf{x}_{k+1}) + \mathbf{A}^T \boldsymbol{\lambda}_{k+1} = \alpha \mathbf{A}^T \mathbf{B}(\mathbf{y}_{k+1} - \mathbf{y}_k) \tag{14.85}$$

On comparing Eq. (14.85) with Eq. (14.77b), a *dual residual* in the kth iteration can be obtained as

$$\mathbf{d}_k = \alpha \mathbf{A}^T \mathbf{B}(\mathbf{y}_{k+1} - \mathbf{y}_k) \tag{14.86}$$

and from Eq. (14.77a), the *primal residual* in the kth iteration is given by

$$\mathbf{r}_k = \mathbf{A}\mathbf{x}_{k+1} + \mathbf{B}\mathbf{y}_{k+1} - \mathbf{c} \tag{14.87}$$

The set $\{\mathbf{r}_k, \mathbf{d}_k\}$ measures the closeness of set $\{\mathbf{x}_k, \mathbf{y}_k, \boldsymbol{\lambda}_k\}$ to the solution set of the problem in Eq. (14.76) and thus a reasonable criterion for terminating the ADMM iterations is

$$\|\mathbf{r}_k\|_2 \leq \varepsilon_p \quad \text{and} \quad \|\mathbf{d}_k\|_2 \leq \varepsilon_d \tag{14.88}$$

where ε_p and ε_d are prescribed tolerances for the primal and dual residuals, respectively.

The convergence properties of the ADMM iterations in Eqs. (14.84a) and (14.84c) have been investigated under various assumptions in [9, 10] and the references cited therein. If both $f(\mathbf{x})$ and $h(\mathbf{y})$ are strongly convex with parameters m_f and m_h, respectively, and parameter α is chosen to satisfy the inequality

$$\alpha^3 \leq \frac{m_f m_h^2}{\rho(\mathbf{A}^T \mathbf{A})\rho(\mathbf{B}^T \mathbf{B})^2}$$

where $\rho(\mathbf{M})$ denotes the largest eigenvalue of symmetric matrix $\mathbf{M}$, then both primal and dual residuals vanish at the rate $O(1/k)$ [10], namely,

$$\|\mathbf{r}_k\|_2 \leq O(1/k) \quad \text{and} \quad \|\mathbf{d}_k\|_2 \leq O(1/k)$$

These ideas for solving the problem in Eq. (14.76) lead to the following ADMM algorithm.

Algorithm 14.10 ADMM for the problem in Eq. (14.76)
Step 1
Input parameter $\alpha > 0$, $\mathbf{y}_0$, $\boldsymbol{\lambda}_0$, and tolerances $\varepsilon_p > 0$, $\varepsilon_d > 0$ and set $k = 0$.
Step 2
Compute $\{\mathbf{x}_{k+1}, \mathbf{y}_{k+1}, \boldsymbol{\lambda}_{k+1}\}$ using Eqs. (14.84a)–(14.84c).
Step 3
Compute $\mathbf{d}_k$ and $\mathbf{r}_k$ using Eqs. (14.86) and (14.87), respectively.
Step 4
If the conditions in Eq. (14.88) are satisfied, output $(\mathbf{x}_{k+1}, \mathbf{y}_{k+1})$ as the solution and stop; otherwise, set $k = k + 1$ and repeat from Step 2.

Several variants of the ADMM algorithm are available, one of them is the *scaled* form of ADMM [9] which will now be described.

By letting

$$\mathbf{r} = \mathbf{Ax} + \mathbf{By} - \mathbf{c} \quad \text{and} \quad \boldsymbol{v} = \boldsymbol{\lambda}/\alpha,$$

we can write the augmented Lagrangian as

$$L_\alpha(\mathbf{x}, \mathbf{y}, \boldsymbol{\lambda}) = f(\mathbf{x}) + h(\mathbf{y}) + \boldsymbol{\lambda}^T \mathbf{r} + \frac{\alpha}{2}\|\mathbf{r}\|_2^2$$

$$= f(\mathbf{x}) + h(\mathbf{y}) + \frac{\alpha}{2}\|\mathbf{r} + \boldsymbol{v}\|_2^2 - \frac{\alpha}{2}\|\boldsymbol{v}\|_2^2$$

$$= f(\mathbf{x}) + h(\mathbf{y}) + \frac{\alpha}{2}\|\mathbf{Ax} + \mathbf{By} - \mathbf{c} + \boldsymbol{v}\|_2^2 - \frac{\alpha}{2}\|\boldsymbol{v}\|_2^2$$

Consequently, the scaled ADMM algorithm can be constructed as follows.

Algorithm 14.11 Scaled ADMM for the problem in Eq. (14.76)
Step 1
Input parameter $\alpha > 0$, $\mathbf{y}_0$, $\boldsymbol{v}_0$, and tolerances $\varepsilon_p > 0$, $\varepsilon_d > 0$ and set $k = 0$.
Step 2
Compute

$$\mathbf{x}_{k+1} = \arg\min_{\mathbf{x}} \left[f(\mathbf{x}) + \frac{\alpha}{2}\|\mathbf{Ax} + \mathbf{By}_k - \mathbf{c} + \boldsymbol{v}_k\|_2^2 \right]$$

$$\mathbf{y}_{k+1} = \arg\min_{\mathbf{y}} \left[h(\mathbf{y}) + \frac{\alpha}{2}\|\mathbf{Ax}_{k+1} + \mathbf{By} - \mathbf{c} + \boldsymbol{v}_k\|_2^2 \right]$$

$$\boldsymbol{v}_{k+1} = \boldsymbol{v}_k + \mathbf{Ax}_{k+1} + \mathbf{By}_{k+1} - \mathbf{c}$$

Step 3
Compute $\mathbf{d}_k$ and $\mathbf{r}_k$ using Eqs. (14.86) and (14.87), respectively.
Step 4
If the conditions in Eq. (14.88) are satisfied, output $(\mathbf{x}_{k+1}, \mathbf{y}_{k+1})$ as the solution
and stop; otherwise, set $k = k + 1$ and repeat from Step 2.

A variant of the scaled ADMM algorithm is the *over-relaxed* ADMM algorithm
where the term $\mathbf{Ax}_{k+1}$ in the y- and v-updates can be replaced by $\tau\mathbf{Ax}_{k+1} - (1 - \tau)(\mathbf{By}_k - \mathbf{c})$ with $\tau \in (0, 2]$. Thus the iterations in Step 2 of Algorithm 14.11 can be
replaced by

$$\mathbf{x}_{k+1} = \arg\min_{\mathbf{x}} \left[f(\mathbf{x}) + \frac{\alpha}{2}\|\mathbf{Ax} + \mathbf{By}_k - \mathbf{c} + \boldsymbol{v}_k\|_2^2 \right]$$

$$\mathbf{y}_{k+1} = \arg\min_{\mathbf{y}} \left[h(\mathbf{y}) + \frac{\alpha}{2}\|\tau\mathbf{Ax}_{k+1} - (1 - \tau)\mathbf{By}_k + \mathbf{By} - \tau\mathbf{c} + \boldsymbol{v}_k\|_2^2 \right]$$

$$\boldsymbol{v}_{k+1} = \boldsymbol{v}_k + \tau\mathbf{Ax}_{k+1} - (1 - \tau)\mathbf{By}_k + \mathbf{By}_{k+1} - \tau\mathbf{c}$$

An ADMM version of Nesterov's accelerated gradient descent algorithm [2] has
also been proposed in [10], as detailed below, which converges if functions $f(\mathbf{x})$ and
$h(\mathbf{y})$ are both strongly convex.

Algorithm 14.12 Accelerated ADMM for the problem in Eq. (14.76)
Step 1
Input parameter $\alpha > 0$, $\hat{\mathbf{y}}_0$, $\hat{\boldsymbol{\lambda}}_0$, and tolerances $\varepsilon_p > 0$, $\varepsilon_d > 0$.
Set $\mathbf{y}_{-1} = \hat{\mathbf{y}}_0$, $\boldsymbol{\lambda}_{-1} = \hat{\boldsymbol{\lambda}}_0$, $t_0 = 1$, and $k = 0$.
Step 2
Compute

$$\mathbf{x}_k = \arg\min_{\mathbf{x}} \left[f(\mathbf{x}) + \hat{\boldsymbol{\lambda}}_k^T \mathbf{A}\mathbf{x} + \frac{\alpha}{2} \|\mathbf{A}\mathbf{x} + \mathbf{B}\hat{\mathbf{y}}_k - \mathbf{c}\|_2^2 \right]$$

$$\mathbf{y}_k = \arg\min_{\mathbf{y}} \left[h(\mathbf{y}) + \hat{\boldsymbol{\lambda}}_k^T \mathbf{B}\mathbf{y} + \frac{\alpha}{2} \|\mathbf{A}\mathbf{x}_k + \mathbf{B}\mathbf{y} - \mathbf{c}\|_2^2 \right]$$

$$\boldsymbol{\lambda}_k = \hat{\boldsymbol{\lambda}}_k + \alpha(\mathbf{A}\mathbf{x}_k + \mathbf{B}\mathbf{y}_k - \mathbf{c})$$

$$t_{k+1} = \frac{1}{2}\left(1 + \sqrt{1 + 4t_k^2}\right)$$

$$\hat{\mathbf{y}}_{k+1} = \mathbf{y}_k + \frac{t_k - 1}{t_{k+1}}(\mathbf{y}_k - \mathbf{y}_{k-1})$$

$$\hat{\boldsymbol{\lambda}}_{k+1} = \boldsymbol{\lambda}_k + \frac{t_k - 1}{t_{k+1}}(\boldsymbol{\lambda}_k - \boldsymbol{\lambda}_{k-1})$$

Step 3
If $\|\alpha \mathbf{A}^T \mathbf{B}(\mathbf{y}_k - \mathbf{y}_{k-1})\|_2 < \varepsilon_d$ and $\|\mathbf{A}\mathbf{x}_k + \mathbf{B}\mathbf{y}_k - \mathbf{c}\|_2 < \varepsilon_p$, output $(\mathbf{x}_k, \mathbf{y}_k)$ as
the solution and stop; otherwise, set $k = k + 1$ and repeat from Step 2.

Example 14.8

(a) Derive the formulas for the iterations in Step 2 of Algorithm 14.11 for the solution
   of the $L_1 - L_2$ problem

$$\text{minimize} \ \tfrac{1}{2}\|\mathbf{A}\mathbf{x} - \mathbf{b}\|_2^2 + \mu\|\mathbf{x}\|_1 \tag{14.89}$$

(b) Using the algorithm obtained in part (a), solve the deconvolution problem in
   Example 14.7.

Solution (*a*) The problem in Eq. (14.89) can be formulated as [9]

$$\text{minimize} \ \tfrac{1}{2}\|\mathbf{A}\mathbf{x} - \mathbf{b}\|_2^2 + \mu\|\mathbf{y}\|_1$$

$$\text{subject to: } \mathbf{x} - \mathbf{y} = \mathbf{0}$$

which is consistent with the formulation in Eq. (14.76) with $f(\mathbf{x}) = \tfrac{1}{2}\|\mathbf{A}\mathbf{x} - \mathbf{b}\|_2^2$
and $h(\mathbf{y}) = \mu\|\mathbf{y}\|_1$. The required iterations for Step 2 of the algorithm can, therefore,
be obtained as

$$\mathbf{x}_{k+1} = \arg\min_{\mathbf{x}} \left[\frac{1}{2}\|\mathbf{A}\mathbf{x} - \mathbf{b}\|_2^2 + \frac{\alpha}{2}\|\mathbf{x} - (\mathbf{y}_k - \boldsymbol{\lambda}_k/\alpha)\|_2^2 \right]$$

$$\mathbf{y}_{k+1} = \arg\min_{\mathbf{y}} \left[\mu\|\mathbf{y}\|_1 + \frac{\alpha}{2}\|\mathbf{y} - (\mathbf{x}_{k+1} + \boldsymbol{\lambda}_k/\alpha)\|_2^2 \right]$$

$$\boldsymbol{\lambda}_{k+1} = \boldsymbol{\lambda}_k + \alpha(\mathbf{x}_{k+1} - \mathbf{y}_{k+1})$$

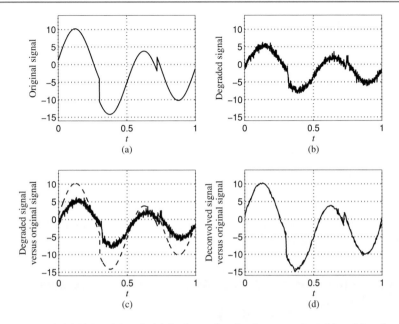

Fig. 14.13 a Original heavisine signal; **b** distorted and noise-contaminated heavisine signal; **c** original (dashed) versus degraded (solid) heavisine signal, SNR = 5.7825 dB; **d** original (dashed) versus reconstructed (solid) heavisine signal, SNR = 26.6029 dB (Example 14.8)

Evidently, updating $\mathbf{x}_k$ amounts to minimizing a convex quadratic function and updating $\mathbf{y}_k$ can be done by soft shrinkage of $\mathbf{x}_{k+1} + \lambda_k/\alpha$ by μ/α (see Sect. 14.4.1). Thus we have

$$\begin{aligned}
\mathbf{x}_{k+1} &= (\mathbf{A}^T\mathbf{A} + \alpha\mathbf{I})^{-1}(\mathbf{A}^T\mathbf{b} + \alpha\mathbf{y}_k - \lambda_k) \\
\mathbf{y}_{k+1} &= \mathcal{S}_{\mu/\alpha}(\mathbf{x}_{k+1} + \lambda_k/\alpha) \\
\lambda_{k+1} &= \lambda_k + \alpha(\mathbf{x}_{k+1} - \mathbf{y}_{k+1})
\end{aligned} \qquad (14.90)$$

where operator $\mathcal{S}_{\mu/\alpha}$ is given in Eq. (14.68). Note that matrix $(\mathbf{A}^T\mathbf{A} + \alpha\mathbf{I})^{-1}$ as well as vector $\mathbf{A}^T\mathbf{b}$ in Eq. (14.90) are independent of iterations and hence they need to be computed only once.

(*b*) By applying Eq. (14.90) to the data set described in Example 14.7 with $\mathbf{y}_0 = \mathbf{0}$, $\lambda_0 = \mathbf{0}$, $\mu = 0.223$, and $\alpha = 0.1$, it took Algorithm 14.11 36 iterations to yield a satisfactory estimate of the signal. The SNR of the estimated signal was found to be 26.6029 dB. The original, distorted, and recovered signals are plotted in Fig. 14.13. The profiles of the primal and dual residuals in terms of $\|\mathbf{r}_k\|_2$ and $\|\mathbf{d}_k\|_2$ are shown in Fig. 14.14. It is observed that both $\|\mathbf{r}_k\|_2$ and $\|\mathbf{d}_k\|_2$ assume values less than 5×10^{-3} after 36 iterations. On comparing with the proximal-point algorithms in Sect. 14.4, the average CPU time required using Algorithm 14.11 was found to be practically the same as that required by Algorithm 14.9.   ■

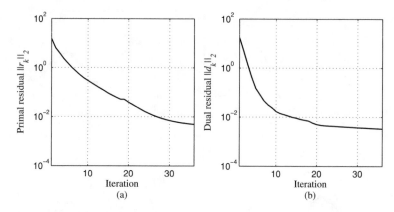

Fig. 14.14 **a** Primal residual $\|\mathbf{r}_k\|_2$ versus the number of iterations; **b** dual residual $\|\mathbf{d}_k\|_2$ versus the number of iterations (Example 14.8)

14.5.2 Application of ADMM to General Constrained Convex Problem

Consider the general constrained convex problem

$$\begin{aligned} \text{minimize } & f(\mathbf{x}) \\ \text{subject to: } & \mathbf{x} \in \mathcal{C} \end{aligned} \tag{14.91}$$

where $f(\mathbf{x})$ is a convex function and $\mathcal{C}$ is a convex set representing the feasible region of the problem. The problem in Eq. (14.91) can be formulated as

$$\text{minimize } f(\mathbf{x}) + I_c(\mathbf{x}) \tag{14.92}$$

where

$$I_C(\mathbf{x}) = \begin{cases} 0 & \text{if } \mathbf{x} \in \mathcal{C} \\ +\infty & \text{otherwise} \end{cases}$$

is the so-called *indicator function* associated with set $\mathcal{C}$. The problem in Eq. (14.92) can be written as

$$\begin{aligned} \text{minimize } & f(\mathbf{x}) + I_C(\mathbf{y}) \\ \text{subject to: } & \mathbf{x} - \mathbf{y} = \mathbf{0} \end{aligned} \tag{14.93}$$

which fits into the ADMM formulation in Eq. (14.76) [9]. The iterations in Step 2 of the scaled ADMM algorithm, i.e., Algorithm 14.11, for this problem are given by

$$\begin{aligned} \mathbf{x}_{k+1} &= \arg\min_{\mathbf{x}} \left[f(\mathbf{x}) + \frac{\alpha}{2} \|\mathbf{x} - \mathbf{y}_k + \boldsymbol{v}_k\|_2^2 \right] \\ \mathbf{y}_{k+1} &= \arg\min_{\mathbf{y}} \left[I_C(\mathbf{y}) + \frac{\alpha}{2} \|\mathbf{y} - (\mathbf{x}_{k+1} + \boldsymbol{v}_k)\|_2^2 \right] \\ \boldsymbol{v}_{k+1} &= \boldsymbol{v}_k + \mathbf{x}_{k+1} - \mathbf{y}_{k+1} \end{aligned}$$

where the y-minimization is obtained by minimizing $\|y - (x_{k+1} + v_k)\|_2$ subject to $y \in C$. This means that y_{k+1} can be obtained by projecting $x_{k+1} + v_k$ onto set C. Therefore, the iterations in Step 2 of Algorithm 14.11 become

$$x_{k+1} = \arg\min_{x} \left[f(x) + \frac{\alpha}{2} \|x - y_k + v_k\|_2^2 \right] \tag{14.94a}$$

$$y_{k+1} = P_C(x_{k+1} + v_k) \tag{14.94b}$$

$$v_{k+1} = v_k + x_{k+1} - y_{k+1} \tag{14.94c}$$

where $P_C(z)$ denotes the projection of point z onto convex set C. The projection can be accomplished by solving the convex problem

$$\text{minimize } \|y - z\|_2$$
$$\text{subject to: } y \in C$$

Example 14.9 A sparse solution of an undetermined system of linear equations $Ax = b$ can be obtained by solving the constrained convex problem

$$\text{minimize } \|x\|_1$$
$$\text{subject to: } Ax = b \tag{14.95}$$

(a) Formulate the problem in the form of the general constrained convex problem of Eq. (14.91).
(b) Solve the problem in Eq. (14.95) for the case where A is a randomly generated matrix of size 20×50 and vector b is given by $b = Ax_s$ where sparse column vector x_s is a specific solution of the problem of length 50 obtained by placing six randomly generated nonzero numbers in a zero vector of length 50 for six randomly selected coordinates.

Solution *(a)* The problem fits into the formulation in Eq. (14.91) with $f(x) = \|x\|_1$ and $C = \{x : Ax = b\}$. The x-minimization step in Eq. (14.94a) becomes

$$x_{k+1} = \arg\min_{x} \left[\|x\|_1 + \frac{\alpha}{2} \|x - y_k + v_k\|_2^2 \right]$$

and hence

$$x_{k+1} = S_{1/\alpha}(y_k - v_k) \tag{14.96}$$

where operator S is defined by Eq. (14.68). The y-minimization step in Eq. (14.94b) can be carried out by solving the simple convex QP problem

$$\text{minimize } \|y - (x_{k+1} + v_k)\|_2$$
$$\text{subject to: } Ay = b$$

whose solution is given by

$$y_{k+1} = A^+ b + P_A(x_{k+1} + v_k) \tag{14.97}$$

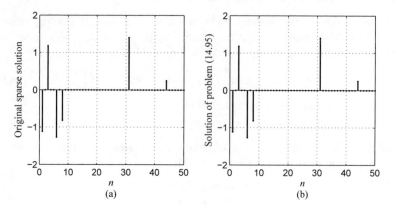

Fig. 14.15 a Original sparse solution $\mathbf{x}_s$ and **b** solution obtained by solving the problem in Eq. (14.95)

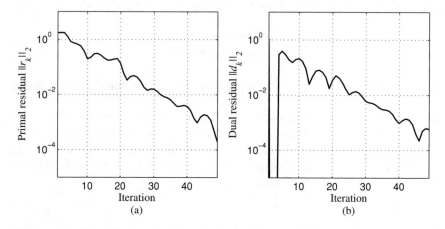

Fig. 14.16 a Primal and **b** dual residuals (Example 14.9)

where $\mathbf{A}^+ = \mathbf{A}^T (\mathbf{A}\mathbf{A}^T)^{-1}$ and $P_A = \mathbf{I} - \mathbf{A}^T (\mathbf{A}\mathbf{A}^T)^{-1}\mathbf{A}$. In effect, the iterations in Eqs. (14.94a)–(14.94c) can be implemented using Eqs. (14.96), (14.97), and (14.94c).

(*b*) Using $\alpha = 0.5$ and the above iterations in Step 2 of Algorithm 14.11, it took the algorithm 49 iterations to converge to solution $\mathbf{x}^*$, which is plotted in Fig. 14.15b. As can be seen, the ADMM-based solution recovered the original sparse signal $\mathbf{x}_s$ accurately. The l_2 reconstruction error of the solution was $\|\mathbf{x}^* - \mathbf{x}_s\|_2 = 4.1282 \times 10^{-4}$.

The primal and dual residuals are given by $\mathbf{r}_k = \mathbf{x}_k - \mathbf{y}_k$ and $\mathbf{d}_k = -\alpha(\mathbf{y}_k - \mathbf{y}_{k-1})$, respectively. The profiles of the residuals are plotted versus the number of iterations in Fig. 14.16.

Note that in problems where sparse solutions of underdetermined systems of linear equations are required, like the one in Example 14.9, vector $\mathbf{b}$ is known and vector $\mathbf{x}$

needs to be obtained using optimization. We started with a known vector $\mathbf{x}_s$ just for the purpose of evaluating the quality of the solution that can be obtained by using the above version of Algorithm 14.11. ∎

14.5.3   Alternating Minimization Algorithm (AMA)

Alternating minimization algorithms (AMAs) are developed for the solution of the problem in Eq. (14.76), namely,

$$\text{minimize } f(\mathbf{x}) + h(\mathbf{y})$$
$$\text{subject to: } \mathbf{A}\mathbf{x} + \mathbf{B}\mathbf{y} = \mathbf{c}$$

where one of the parts of the objective components, say, $f(\mathbf{x})$, is strongly convex [11]. Under this assumption, updating $\mathbf{x}_k$ in an AMA iteration as follows

$$\mathbf{x}_{k+1} = \arg\min_{\mathbf{x}}[f(\mathbf{x}) + \boldsymbol{\lambda}_k^T \mathbf{A}\mathbf{x}] \tag{14.98a}$$

$$\mathbf{y}_{k+1} = \arg\min_{\mathbf{y}} \left[h(\mathbf{y}) + \boldsymbol{\lambda}_k^T \mathbf{B}\mathbf{y} + \frac{\alpha}{2} \|\mathbf{A}\mathbf{x}_{k+1} + \mathbf{B}\mathbf{y} - \mathbf{c}\|_2^2 \right] \tag{14.98b}$$

$$\boldsymbol{\lambda}_{k+1} = \boldsymbol{\lambda}_k + \alpha(\mathbf{A}\mathbf{x}_{k+1} + \mathbf{B}\mathbf{y}_{k+1} - \mathbf{c}) \tag{14.98c}$$

would result in a simplified iteration relative to the corresponding ADMM iteration. On the basis of an argument similar to that used for the ADMM iterations in Eqs. (14.94a)–(14.94c), Eqs. (14.98b) and (14.98c) can be shown to imply that

$$\nabla h(\mathbf{y}_{k+1}) + \mathbf{B}^T \boldsymbol{\lambda}_{k+1} = \mathbf{0}$$

Therefore, the KKT condition in Eq. (14.77c) is satisfied by the above AMA iterations. Since $\mathbf{x}_{k+1}$ minimizes $f(\mathbf{x}) + \boldsymbol{\lambda}_k^T \mathbf{A}\mathbf{x}$, we have

$$\mathbf{0} = \nabla f(\mathbf{x}_{k+1}) + \mathbf{A}^T \boldsymbol{\lambda}_k = \nabla f(\mathbf{x}_{k+1}) + \mathbf{A}^T \boldsymbol{\lambda}_{k+1} - \mathbf{A}^T (\boldsymbol{\lambda}_{k+1} - \boldsymbol{\lambda}_k)$$

i.e.,

$$\nabla f(\mathbf{x}_{k+1}) + \mathbf{A}^T \boldsymbol{\lambda}_{k+1} = \mathbf{A}^T (\boldsymbol{\lambda}_{k+1} - \boldsymbol{\lambda}_k)$$

and on comparing this equation with Eq. (14.77b), the dual residual

$$\mathbf{d}_k = \mathbf{A}^T (\boldsymbol{\lambda}_{k+1} - \boldsymbol{\lambda}_k) \tag{14.99}$$

is obtained. The *primal residual* for AMA is the same as for ADMM, namely,

$$\mathbf{r}_k = \mathbf{A}\mathbf{x}_{k+1} + \mathbf{B}\mathbf{y}_{k+1} - \mathbf{c} \tag{14.100}$$

The AMA iterations can be terminated when the conditions

$$\|\mathbf{r}_k\|_2 < \varepsilon_p \quad \text{and} \quad \|\mathbf{d}_k\|_2 < \varepsilon_d \tag{14.101}$$

are satisfied where ε_p and ε_d are prescribed tolerances for the primal and dual residuals, respectively. Based on these ideas, an algorithm for the solution of the problem in Eq. (14.76) can be constructed as follows.

Algorithm 14.13 AMA for the problem in Eq. (14.76)
Step 1
Input parameter $\alpha > 0$, $\mathbf{y}_0$, λ_0, and tolerances $\varepsilon_p > 0$ and $\varepsilon_d > 0$.
Set $k = 0$.
Step 2
Compute $\{\mathbf{x}_{k+1}, \mathbf{y}_{k+1}, \lambda_{k+1}\}$ using Eqs. (14.98a)–(14.98c).
Step 3
Compute $\mathbf{d}_k$ and $\mathbf{r}_k$ using Eqs. (14.99) and (14.100), respectively.
Step 4
If the conditions in Eq. (14.101) are satisfied, output $(\mathbf{x}_{k+1}, \mathbf{y}_{k+1})$ as the solution
and stop; otherwise, set $k = k + 1$ and repeat from Step 2.

It can be shown that if $f(\mathbf{x})$ is strongly convex and $\alpha < m/\rho(\mathbf{A}^T\mathbf{A})$ where m is
the parameter associated with the strong convexity of $f(\mathbf{x})$, as seen in Eq. (14.13),
and $\rho(\mathbf{A}^T\mathbf{A})$ denotes the largest eigenvalue of $\mathbf{A}^T\mathbf{A}$, then the fast AMA algorithm
outlined below converges to the solution of Eq. (14.76) (see Ref. [10]).

Algorithm 14.14 Accelerated AMA for the problem Eq. (14.76)
Step 1
Input $\alpha < m/\rho(\mathbf{A}^T\mathbf{A})$, $\hat{\lambda}_0$ and tolerances $\varepsilon_p > 0$ and $\varepsilon_d > 0$.
Set $\lambda_{-1} = \hat{\lambda}_0$, $t_0 = 1$, and $k = 0$.
Step 2
Compute

$$\mathbf{x}_k = \arg\min_{\mathbf{x}}[f(\mathbf{x}) + \hat{\lambda}_k^T\mathbf{A}\mathbf{x}]$$

$$\mathbf{y}_k = \arg\min_{\mathbf{y}}\left[h(\mathbf{y}) + \hat{\lambda}_k^T\mathbf{B}\mathbf{y} + \frac{\alpha}{2}\|\mathbf{A}\mathbf{x}_k + \mathbf{B}\mathbf{y} - \mathbf{c}\|_2^2\right]$$

$$\lambda_k = \hat{\lambda}_k + \alpha(\mathbf{A}\mathbf{x}_k + \mathbf{B}\mathbf{y} - \mathbf{c})$$

$$t_{k+1} = \frac{1}{2}\left(1 + \sqrt{1 + 4t_k^2}\right)$$

$$\hat{\lambda}_{k+1} = \lambda_k + \frac{t_k - 1}{t_{k+1}}(\lambda_k - \lambda_{k-1})$$

Step 3
If $\|\mathbf{A}^T(\lambda_k - \lambda_{k-1})\|_2 < \varepsilon_d$ and $\|\mathbf{A}\mathbf{x}_k + \mathbf{B}\mathbf{y}_k - \mathbf{c}\|_2 < \varepsilon_p$, output $(\mathbf{x}_k, \mathbf{y}_k)$ as the
solution and stop; otherwise, set $k = k + 1$ and repeat from Step 2.

Problems

14.1 *(a)* Find the subdifferential of $f(\mathbf{x}) = \|\mathbf{x}\|_1$ (Hint: Use the result in
        Example 14.1).
    (b) Find the subdifferential of $f(\mathbf{x}) = \|\mathbf{x}\|_2$.
    (c) Show that the subdifferential of $f(\mathbf{x}) = \sum_{i=1}^{m}|\mathbf{a}_i^T\mathbf{x} - b_i|$ is given by

$$\partial f(\mathbf{x}) = \sum_{i \in I_+(\mathbf{x})} \mathbf{a}_i - \sum_{i \in I_-(\mathbf{x})} \mathbf{a}_i + \sum_{i \in I_0(\mathbf{x})} \text{any } \mathbf{g}_i \in [-\mathbf{a}_i, \mathbf{a}_i] \qquad \text{(P14.1)}$$

where the index sets are defined by

$$I_+(\mathbf{x}) = \{i : \mathbf{a}_i^T \mathbf{x} - b_i > 0\}$$
$$I_-(\mathbf{x}) = \{i : \mathbf{a}_i^T \mathbf{x} - b_i < 0\}$$
$$I_0(\mathbf{x}) = \{i : \mathbf{a}_i^T \mathbf{x} - b_i = 0\}$$

In Eq. (P14.1), $\mathbf{g}_i \in [-\mathbf{a}_i, \mathbf{a}_i]$ stands for $\mathbf{g}_i = t\mathbf{a}_i$ for some scalar t between -1 and 1 for all i.

(d) Assuming that $f(\mathbf{x})$ is convex, show that if $h(\mathbf{x}) = f(\mathbf{Ax} + \mathbf{b})$, then $\partial h(\mathbf{x}) = \mathbf{A}^T \partial f(\mathbf{Ax} + \mathbf{b})$.

14.2 Show that subgradient of a convex function is a monotone operator, namely,

$$(\mathbf{g}_x - \mathbf{g}_y)^T (\mathbf{x} - \mathbf{y}) \geq 0$$

where $\mathbf{g}_x$ and $\mathbf{g}_y$ are any subgradients of $f(\mathbf{x})$ and $f(\mathbf{y})$ at $\mathbf{x}$ and $\mathbf{y}$, respectively.

14.3 Show that

$$\begin{bmatrix} \partial f(\mathbf{x}) \\ -1 \end{bmatrix}^T \left(\begin{bmatrix} \mathbf{y} \\ t \end{bmatrix} - \begin{bmatrix} \mathbf{x} \\ f(\mathbf{x}) \end{bmatrix} \right) \leq 0 \quad \forall (\mathbf{y}, t) \in \text{epi}(f)$$

where function $f(\mathbf{x})$ is convex but not necessarily differentiable.

14.4 The *projection* of a point $\mathbf{z} \in R^n$ onto a closed convex set $\mathcal{C}$, to be denoted as $P_{\mathcal{C}}(\mathbf{z})$, is defined as a point in $\mathcal{C}$ that is located at the smallest Euclidean distance from point $\mathbf{z}$ among all points in $\mathcal{C}$, namely,

$$P_{\mathcal{C}}(\mathbf{z}) = \arg \min_{\mathbf{x} \in \mathcal{C}} \|\mathbf{x} - \mathbf{z}\|_2$$

(a) Given $\mathcal{C} = \{x : a \leq x \leq b\}$ and $z \in R$, derive a closed-form expression for projection $P_{\mathcal{C}}(z)$.

(b) Given $\mathcal{C} = \left\{ \mathbf{x} = \begin{bmatrix} x_1 \\ x_2 \end{bmatrix} : \frac{(x_1-4)^2}{9} + \frac{(x_2-5)^2}{4} \leq 1 \right\}$ and $\mathbf{z} = \begin{bmatrix} 2 \\ 1 \end{bmatrix}$, evaluate $P_{\mathcal{C}}(\mathbf{z})$.

14.5 Consider the convex set $\mathcal{C} = \{\mathbf{x} : \mathbf{Ax} = \mathbf{b}\}$ where $\mathbf{A} \in R^{p \times n}$, $\mathbf{b} \in R^{p \times l}$, $p < n$, and $\mathbf{A}$ is of full row rank.

(a) Given a point $\mathbf{z} \notin \mathcal{C}$, formulate the problem of projecting $\mathbf{z}$ onto $\mathcal{C}$ as a convex problem.

(b) Deduce a closed-form formula for $P_{\mathcal{C}}(\mathbf{z})$ by solving the problem in part (a).

(c) Given $\mathcal{C} = \{\mathbf{x} : \mathbf{Ax} = \mathbf{b}\}$ where $\mathbf{A} = \begin{bmatrix} 1 & 2 & -1 \\ 0 & -1 & 3 \end{bmatrix}$, $\mathbf{b} = \begin{bmatrix} 1 \\ 2 \end{bmatrix}$, and a point $\mathbf{z} = \begin{bmatrix} 1 & -1 & 2 \end{bmatrix}^T$, apply the result of part (b) to evaluate $P_{\mathcal{C}}(\mathbf{z})$.

14.6 *(a)* Given $\mathcal{C} = \{x : \mathbf{Ax} = \mathbf{b}\}$, formulate the problem of finding $P_{\mathcal{C}}(\mathbf{z})$ as a convex minimization problem.

   (b) Given $\mathcal{C} = \{x : \mathbf{Ax} = \mathbf{b}\}$ where

$$\mathbf{A} = \begin{bmatrix} -1 & 0 & 0 & 0 \\ 0 & -1 & 0 & 0 \\ 1 & 2 & 0 & 0 \\ 0 & 0 & 0 & -1 \\ 0 & 0 & -1 & -1 \\ 0 & 0 & 1 & 2 \end{bmatrix}, \quad \mathbf{b} = \begin{bmatrix} 0 \\ 0 \\ 2 \\ -2 \\ -3 \\ 6 \end{bmatrix}$$

and a point $\mathbf{z} = \begin{bmatrix} 2 & 1 & 4 & 3 \end{bmatrix}^T$, evaluate $P_{\mathcal{C}}(\mathbf{z})$ by solving the convex problem formulated in part (a).

14.7 Identify which of the following quadratic functions $f_i(\mathbf{x}) = \frac{1}{2}\mathbf{x}^T \mathbf{H}_i \mathbf{x} + \mathbf{x}^T \mathbf{p}_i$ are (a) convex with Lipschitz continuous gradients and (b) strongly convex:

- $\mathbf{H}_1 = \begin{bmatrix} 2 & -2 & 0 \\ -2 & 2 & 0 \\ 0 & 0 & 1 \end{bmatrix}$, $\mathbf{p}_1 = \begin{bmatrix} 1 \\ 1 \\ 1 \end{bmatrix}$

- $\mathbf{H}_2 = \begin{bmatrix} 2 & -2.01 & 0 \\ -2.01 & 2 & 0 \\ 0 & 0 & 1 \end{bmatrix}$, $\mathbf{p}_2 = \begin{bmatrix} 1 \\ 1 \\ 1 \end{bmatrix}$

- $\mathbf{H}_3 = \begin{bmatrix} 2 & -2 & 0 \\ -2 & 2.01 & 0 \\ 0 & 0 & 1 \end{bmatrix}$, $\mathbf{p}_3 = \begin{bmatrix} 1 \\ 1 \\ 1 \end{bmatrix}$

14.8 This problem examines several additional properties of conjugate functions which are closely related to the notion of Lagrange dual function. Recall that the *conjugate function* $f^*(\mathbf{u})$ of $f(\mathbf{x})$ is defined as

$$f^*(\mathbf{u}) = \sup_{\mathbf{x} \in \text{dom}(f)} \left(\mathbf{u}^T \mathbf{x} - f(\mathbf{x}) \right)$$

(a) The concept of conjugate function is related to the inverse of the gradient in the following way: one can regard the gradient $\nabla f(\mathbf{x})$ as a mapping from a subset of R^n to a subset of R^n. Now consider the 'inverse' question: given a vector $\mathbf{y}$ from R^n, find a vector $\mathbf{x}$ such that $\nabla f(\mathbf{x}) = \mathbf{u}$. In effect, this question is about finding the inverse mapping of $\nabla f(\mathbf{x})$. Show that the solution of this inverse gradient mapping problem (if it exists) is given by

$$\mathbf{x} = \arg\max_{\hat{\mathbf{x}}} (\mathbf{u}^T \hat{\mathbf{x}} - f(\hat{\mathbf{x}}))$$

(b) Show that the conjugate function is differentiable and

$$\nabla f^*(\mathbf{u}) = \arg\max_{\hat{\mathbf{x}}} (\mathbf{u}^T \hat{\mathbf{x}} - f(\hat{\mathbf{x}})) \tag{P14.2}$$

Based on the result of part (a) and Eq. (P14.2), show that

$$\nabla f^*(\mathbf{u}) = \nabla^{-1} f(\mathbf{u})$$

where $\nabla^{-1} f$ denotes the inverse of the graduate mapping of $f(\mathbf{x})$.

(c) Show that the Lagrange dual and conjugate functions of the problems

$$\text{minimize } f(\mathbf{x})$$
$$\text{subject to: } \mathbf{x} = \mathbf{0}$$

are related by

$$q(\boldsymbol{\lambda}) = -f^*(-\boldsymbol{\lambda})$$

(d) Show that the Lagrange dual function $q(\boldsymbol{\lambda}, \boldsymbol{\mu})$ and conjugate functions of the convex problem

$$\text{minimize } f(\mathbf{x})$$
$$\text{subject to: } \mathbf{Ax} \leq \mathbf{b}, \ \mathbf{Cx} = \mathbf{d}$$

are related by

$$q(\boldsymbol{\lambda}, \boldsymbol{\mu}) = -\mathbf{b}^T \boldsymbol{\mu} - \mathbf{d}^T \boldsymbol{\lambda} - f^*(-\mathbf{A}^T \boldsymbol{\mu} - \mathbf{C}^T \boldsymbol{\lambda})$$

14.9 Consider the general convex minimization problem

$$\text{minimize } f(\mathbf{x}) \tag{P14.3a}$$
$$\text{subject to: } \quad \mathbf{Ax} = \mathbf{b} \tag{P14.3b}$$
$$c_j(\mathbf{x}) \leq 0 \quad \text{for } j = 1, 2, \ldots, q \tag{P14.3c}$$

where functions $f(\mathbf{x})$ and $c_j(\mathbf{x})$ are all convex.

(a) Show that the problem in Eqs. (P14.3a)–(P14.3c) is equivalent to the problem

$$\text{minimize } F(\mathbf{x}) = f(\mathbf{x}) + \sum_{j=1}^{q} I_-(c_j(\mathbf{x})) \tag{P14.4a}$$

$$\text{subject to: } \mathbf{Ax} = \mathbf{b} \tag{P14.4b}$$

where

$$I_-(u) = \begin{cases} 0 & \text{if } u \leq 0 \\ \infty & \text{if } u > 0 \end{cases} \tag{P14.5}$$

denotes the *indicator function* for nonpositive real u.

(b) Show that the objective function $F(\mathbf{x})$ in Eq. (P14.4a) is convex.

14.10 (a) Evaluate and compare function $h(u) = -(1/\tau) \log(-u)$ over $u < 0$ with the indicator function $I_-(u)$ defined in Eq. (P14.5) where $\tau > 0$ is a parameter that controls the accuracy of the approximation.

(b) Show that for a fixed $\tau > 0$, $h(u)$ is convex with respect to u.

(c) Plot the indicator function $I_-(u)$ over $u < 0$.

(d) In the same graph, plot function $h(u)$ over the interval $[-2, 0)$ for $\tau = 0.5$, 1, and 2. What can you conclude from the plots in (c) and (d)?

(e) Formulate a convex optimization problem whose objective function is a differentiable function that approximates the previous objective function in Eqs. (P14.4a) and (P14.4b).

14.11 Using Algorithm 14.5, solve the convex problem

$$\text{minimize } f(\mathbf{x}) = \frac{\|\mathbf{Ax} + \mathbf{b}\|_2^2}{\mathbf{c}^T \mathbf{x} + d}$$

$$\text{subject to: } \mathbf{c}^T \mathbf{x} + d \geq \delta$$

where

$$\mathbf{A} = \begin{bmatrix} 2 & -1 \\ 0 & 1 \end{bmatrix}, \quad \mathbf{b} = \begin{bmatrix} 1 \\ -1 \end{bmatrix}, \quad \mathbf{c} = \begin{bmatrix} 1 \\ 1 \end{bmatrix}, \quad d = -1,$$

and δ is a small positive constant which is set to $\delta = 0.01$.

14.12 This is a *signal separation* problem which is to be solved using an *alternating $l_1 - l_2$ minimization* strategy. Suppose the signal measurement that can be accessed, $\mathbf{u}$, is an additive mixture of two different types of signals, $\mathbf{x}_1$ and $\mathbf{x}_2$, and some measurement noise, $\mathbf{w}$, such that

$$\mathbf{u} = \mathbf{x}_1 + \mathbf{x}_2 + \mathbf{w} \tag{P14.6}$$

and there is a requirement to separate the two signals from each other. An effective technique to solve this problem is to use an alternating $l_1 - l_2$ minimization technique as will now be described. First, we choose a coordinate transformation $\mathbf{T}_i$ for each $\mathbf{x}_i$ such that $\mathbf{x}_i$ becomes sparse under $\mathbf{T}_i$. That is, we find orthogonal $\mathbf{T}_1$ and $\mathbf{T}_2$ such that $\boldsymbol{\theta}_1 = \mathbf{T}_1 \mathbf{x}_1$ and $\boldsymbol{\theta}_2 = \mathbf{T}_2 \mathbf{x}_2$ with $\boldsymbol{\theta}_1$ and $\boldsymbol{\theta}_2$ sparse. Under these circumstances, the signal model in Eq. (P14.6) becomes

$$\mathbf{u} = \mathbf{T}_1 \boldsymbol{\theta}_1 + \mathbf{T}_2 \boldsymbol{\theta}_2 + \mathbf{w} \tag{P14.7}$$

A natural way to recover the two signals in Eq. (P14.7) is to solve the constrained problem

$$\text{minimize} \quad \|\boldsymbol{\theta}_1\|_1 + \|\boldsymbol{\theta}_2\|_1$$

$$\text{subject to: } \|\mathbf{T}_1^T \boldsymbol{\theta}_1 + \mathbf{T}_2^T \boldsymbol{\theta}_2 - \mathbf{u}\|_2 \leq \varepsilon$$

which can be converted into the unconstrained problem

$$\text{minimize } \tfrac{1}{2}\|\mathbf{T}_1^T \boldsymbol{\theta}_1 + \mathbf{T}_2^T \boldsymbol{\theta}_2 - \mathbf{u}\|_2^2 + \mu\|\boldsymbol{\theta}_1\|_1 + \mu\|\boldsymbol{\theta}_2\|_1 \tag{P14.8}$$

where $\mu > 0$ is an appropriate regularization parameter. The second step is to solve the $l_1 - l_2$ minimization problem in Eq. (P14.8) by using alternating shrinkage.

(a) Show that for a fixed $\boldsymbol{\theta}_2$, the global minimizer of the problem in Eq. (P14.8) is given by

$$\boldsymbol{\theta}_1 = S_\mu[\mathbf{T}_1(\mathbf{u} - \mathbf{T}_2^T\boldsymbol{\theta}_2)] \qquad \text{(P14.9)}$$

(b) Show that for a fixed $\boldsymbol{\theta}_1$, the global minimizer of the problem in Eq. (P14.8) is given by

$$\boldsymbol{\theta}_2 = S_\mu[\mathbf{T}_2(\mathbf{u} - \mathbf{T}_1^T\boldsymbol{\theta}_1)] \qquad \text{(P14.10)}$$

Based on Eqs. (P14.9) and (P14.10), develop an iterative formula that updates the kth iterate $\{\boldsymbol{\theta}_1^{(k)}, \boldsymbol{\theta}_2^{(k)}\}$ to $\{\boldsymbol{\theta}_1^{(k+1)}, \boldsymbol{\theta}_2^{(k+1)}\}$.

(c) Based on the formula developed in part (b), solve the signal separation problem where the data in model (P14.7) are generated in MATLAB as follows:

(i) Signals $\mathbf{x}_1$ and $\mathbf{x}_2$ are generated as t = 0:1/511:1;
```
x1 = -0.9*sin(2*pi*1.2*t)+0.6*cos(2*pi*1.8*t)+...
     -0.5*sin(2*pi*3.1*t)-0.4;
x1 = x1(:);
x2 = zeros(512,1);
rand('state',14);
r = randperm(512);
randn('state',28);
x2(r(1:13)) = randn(13,1);
```
(ii) Matrix $\mathbf{T}_1$ is a DCT matrix and matrix $\mathbf{T}_2$ is an identity matrix which are generated as T1 = dctmtx(512);
```
T2 = eye(512);
```
(iii) Measurement noise $\mathbf{w}$ is generated as
```
randn('state',10)
w = 0.01*randn(512,1);
```
(iv) Using the above data, you can generate the measurement vector $\mathbf{u}$ using Eq. (P14.6).

Now start with $\{\mathbf{x}_1^{(0)}, \mathbf{x}_2^{(0)}\} = \{\mathbf{0}, \mathbf{0}\}$, perform a total of 31 iterations based on part (c) to obtain estimates $\tilde{\mathbf{x}}_1 \triangleq \mathbf{x}_1^{(31)}$ and $\tilde{\mathbf{x}}_2 \triangleq \mathbf{x}_2^{(31)}$. Use $\mu = 0.01$ in your iterations. The separation performance can be evaluated as follows:

- Construct a plot that displays the data curve $\mathbf{u}$ versus t;
- Construct a plot that displays both $\mathbf{x}_1$ and its estimate $\tilde{\mathbf{x}}_1$;
- Construct a plot that displays both $\mathbf{x}_2$ and its estimate $\tilde{\mathbf{x}}_2$;
- Compute the average 2-norm estimation errors $\|\tilde{\mathbf{x}}_1 - \mathbf{x}_1\|_2/\sqrt{n}$ and $\|\tilde{\mathbf{x}}_2 - \mathbf{x}_2\|_2/\sqrt{n}$ with $n = 512$;
- Comment on the results you obtained.

14.13 This problem involves finding a point in the intersection of two given closed nonempty convex sets S_1 and S_2. An ADMM formulation of such a problem is given by [9]

$$\text{minimize } I_{S_1}(\mathbf{x}) + I_{S_2}(\mathbf{y})$$
$$\text{subject to: } \mathbf{x} - \mathbf{y} = \mathbf{0}$$

(P14.11)

where I_{S_1} and I_{S_2} are the indicator functions of S_1 and S_2, respectively.

(a) Write down the scaled form of ADMM iterations for the problem in Eq. (P14.11). (Hint: the $\mathbf{x}$- and $\mathbf{y}$-minimization steps can be expressed explicitly in term of projections onto sets S_1 and S_2, respectively.)

(b) Apply the ADMM iterations obtained from part (a) to obtain a point in the intersection of sets

$$S_1 = \{\mathbf{x} : x_1 \leq 0; \ -x_2 \leq 0; \ 0.5x_1 + x_2 - 1 \leq 0\}$$

and

$$S_2 = \{\mathbf{x} : 2x_1 - x_2 - 1.6 \leq 0; \ -2x_1 - x_2 + 2.4 \leq 0; \ x_2 - 3 \leq 0\}$$

14.14 Consider the problem of finding a point that lies in the intersection of K closed convex sets $\{S_i, \ i = 1, \ 2, \ \ldots, \ K\}$ with $K > 2$.

(a) Extend the ADMM-based method developed in Problem 14.13 to the problem of finding a point that lies in the intersection of K closed convex sets $\{S_i, \ i = 1, \ 2, \ \ldots, \ K\}$ with $K > 2$.

(b) Apply the method developed in part (a) to find a feasible point for linear inequality constraints $\mathbf{Ax} \leq \mathbf{b}$ where $\mathbf{A}$ and $\mathbf{b}$ are given in Problem 14.6(b).

(c) Apply the technique in part (b) to find a strictly feasible point for linear inequality constraints $\mathbf{Ax} \leq \mathbf{b}$ where $\mathbf{A}$ and $\mathbf{b}$ are given in Problem 14.6(b).

14.15 Consider the convex problem

$$\text{minimize } f(\mathbf{x}) = \sum_{i=1}^{K} f_i(\mathbf{x})$$

(P14.12)

where each function $f_i(\mathbf{x})$ is convex and assumes the value $+\infty$ for any $\mathbf{x}$ that violates a constraint. The problem in Eq. (P14.12) arises in several applications, especially in the area of modeling, data analysis, and data processing where number K in the objective function is often large. The problem at hand can be reformulated as

$$\text{minimize } f(\mathbf{x}) = \sum_{i=1}^{K} f_i(\mathbf{x}_i)$$
$$\text{subject to: } \mathbf{x}_i - \mathbf{y} = \mathbf{0} \quad \text{for } i = 1, \ 2, \ \ldots, \ K$$

(P14.13)

which is referred to as the *global consensus problem* in that all local variables $\mathbf{x}_i$ are constrained to be equal [9]. This reformulation is of interest as it enables the solution of the problem in Eq. (P14.12) in a way that each term in $f(\mathbf{x})$ is dealt with by a local processing unit when ADMM is applied to the problem in Eq. (P14.13). Write the ADMM iterations for the problem in Eq. (P14.13).

References

1. R. T. Rockafellar, *Convex Analysis*. Princeton, NJ: Princeton University Press, 1970.
2. Y. E. Nesterov, *Introductory Lectures on Convex Optimization — A Basic Course*. Norwell, MA: Kluwer Academic Publishers, 2004.
3. S. Boyd and L. Vandenberghe, *Convex Optimization*. Cambridge, UK: Cambridge University Press, 2004.
4. N. Komodakis and J.-C. Pesquet, "Playing with duality," *IEEE Signal Processing Magazine*, vol. 21, no. 6, pp. 31–54, 2015.
5. N. Parikh and S. Boyd, "Proximal algorithms," *Foundations and Trends in Optimization*, vol. 1, no. 3, pp. 123–231, 2013.
6. A. Beck and M. Teboulle, "A fast shrinkage-thresholding algorithm for linear inverse problems," *SIAM J. Imaging Sciences*, vol. 2, no. 1, pp. 183–202, 2009.
7. Y. Nesterov, "A method for solving a convex programming problem with rate of convergence $O(1/k^2)$," *Soviet Math. Doklady*, vol. 269, no. 3, pp. 543–547, 1983, in Russian.
8. Y. Nesterov, "Gradient methods for minimizing composite functions," *Math. Programming*, vol. 140, no. 1, pp. 125–161, 2013.
9. S. Boyd, N. Parikh, E. Chu, B. Peleato, and J. Eckstein, "Distributed optimization and statistical learning via the alternating direction method of multipliers," *Foundations and Trends in Machine Learning*, vol. 3, no. 1, pp. 1–122, 2010.
10. T. Goldstein, B. O'Donoghue, S. Setzer, and R. Baraniuk, "Fast alternating direction optimization methods," *SIAM J. Imaging Sciences*, vol. 7, no. 3, pp. 1588–1623, 2014.
11. P. Tseng, "Applications of splitting algorithm to decomposition in convex programming and variational inequalities," *SIAM J. Control and Optimization*, vol. 29, no. 1, pp. 119–138, 1991.

Algorithms for General Nonconvex Problems

15

15.1 Introduction

In many optimization problems of practical importance, the objective function or constraints or both are nonconvex. For these problems, the algorithms developed for convex problems in previous chapters are not directly applicable. In this chapter we refer to these problems *as general optimization problems*. A methodology that is particularly effective in dealing with a general optimization problem is to tackle it sequentially whereby a convex model is formulated, often called a *convex subproblem*, that well approximates the original problem in a local region of the parameter space. The subproblem is then solved using an appropriate convex algorithm and the solution obtained can then be used as the initialization of the next convex subproblem. In this way, a sequence of convex models is created where each model is valid in the local area of interest of the parameter space. Through this sequence of convex models, a sequence of iterates is generated that will hopefully converge to a solution of the original problem. Since the original problem is not convex, an algorithm based on such sequential convex approximation is heuristic in nature and is not guaranteed to work. In practice, however, heuristic methodologies of this type often work well for many general optimization problems, especially when a good initial point can be identified.

In Sects. 15.2 and 15.3, we explore two sequential methods for general optimization problems, known as *sequential convex programming* (SCP) and *sequential quadratic programming* (SQP), respectively. In Sect. 15.4, we study a specific SCP method that is based on a 'convexification' technique known as the *convex-concave procedure*. Finally, in Sect. 15.5, the alternating direction method of multipliers (ADMM) studied in Chap. 14 is extended to general optimization problems.

© Springer Science+Business Media, LLC, part of Springer Nature 2021
A. Antoniou and W.-S. Lu, *Practical Optimization*, Texts in Computer Science,
https://doi.org/10.1007/978-1-0716-0843-2_15

15.2 Sequential Convex Programming

SCP is one of the optimization techniques that implement the methodology outlined in the introduction. In the following sections, we assume that the objective function as well as the functions associated with the constraints are twice continuously differentiable.

15.2.1 Principle of SCP

Consider the constrained problem

$$\begin{aligned}
&\text{minimize } f(\mathbf{x})\\
&\text{subject to: } a_i(\mathbf{x}) = 0 \quad \text{for } i = 1,\, 2,\, \ldots,\, p\\
&\qquad\quad c_j(\mathbf{x}) \le 0 \quad \text{for } j = 1,\, 2,\, \ldots,\, q
\end{aligned} \tag{15.1}$$

where $f(\mathbf{x})$ and $c_j(\mathbf{x})$ are twice continuously differentiable but may not be convex, and the constraints $a_i(\mathbf{x})$ are continuously differentiable but may not be affine. Let $\mathbf{x}_k$ be the kth iterate generated by the SCP algorithm, which is to be updated to the next iterate $\mathbf{x}_{k+1}$. We begin by defining a region surrounding point $\mathbf{x}_k$ in which a convex subproblem is to be solved to produce point $\mathbf{x}_{k+1}$. In the literature, such a local region is called a *trust region* and is usually characterized either as a bounding box defined by

$$\mathcal{R}_\rho^{(k)} = \{\mathbf{x} : |x_i - x_i^{(k)}| \le \rho_i \quad \text{for } i = 1,\, 2,\, \ldots,\, n\} \tag{15.2}$$

where $x_i^{(k)}$ denotes the ith component of $\mathbf{x}_k$, or as a 'ball' and is usually characterized in the n-dimensional Euclidean space defined by

$$\mathcal{R}_\rho^{(k)} = \{\mathbf{x} : \|\mathbf{x} - \mathbf{x}_k\|_2 \le \rho\} \tag{15.3}$$

In the trust region, functions $f(\mathbf{x})$ and $c_j(\mathbf{x})$ are approximated by convex functions $\tilde{f}(\mathbf{x})$ and $\tilde{c}_j(\mathbf{x})$, respectively, and constraint functions $a_i(\mathbf{x})$ are approximated by affine functions and are usually denoted as $\tilde{a}_i(\mathbf{x})$. It is reasonable to expect that within $\mathcal{R}_\rho^{(k)}$ the *convex* subproblem

$$\begin{aligned}
&\text{minimize } \tilde{f}(\mathbf{x})\\
&\text{subject to: } \tilde{a}_i(\mathbf{x}) \le 0 \quad \text{for } i = 1,\, 2,\, \ldots,\, p\\
&\qquad\quad \tilde{c}_j(\mathbf{x}) \le 0 \quad \text{for } j = 1,\, 2,\, \ldots,\, q\\
&\qquad\quad \mathbf{x} \in \mathcal{R}_\rho^{(k)}
\end{aligned} \tag{15.4}$$

approximates the original problem in Eq. (15.1). Hence it is reasonable to take the solution of the convex subproblem in Eq. (15.4) as the next iterate $\mathbf{x}_{k+1}$. Having generated $\mathbf{x}_{k+1}$, a new trust region $\mathcal{R}_\rho^{(k+1)}$ centered at $\mathbf{x}_{k+1}$ can be defined by Eq. (15.2) or (15.3), and convex approximations for functions $f(\mathbf{x})$, $c_j(\mathbf{x})$, and $a_i(\mathbf{x})$ in region $\mathcal{R}_\rho^{(k+1)}$ can be obtained. In this way, a convex subproblem similar to that in Eq. (15.4) can be formulated and its solution is taken to be the new iterate $\mathbf{x}_{k+2}$. This procedure is continued until a certain convergence criterion is met.

15.2.2   Convex Approximations for $f(\mathbf{x})$ and $c_j(\mathbf{x})$ and Affine Approximation of $a_i(\mathbf{x})$

Convex approximation based on Taylor expansion
To facilitate the description of the process of generating a convex approximation for a differentiable function, we need a concept called the *nonnegative part* of a symmetric matrix. Let $\mathbf{M}$ be an $n \times n$ symmetric matrix. The eigen-decomposition of $\mathbf{M}$ is $\mathbf{M} = \mathbf{U}\Lambda\mathbf{U}^T$ where $\mathbf{U}$ is an orthogonal matrix whose columns are the eigenvectors of $\mathbf{M}$ and Λ is a diagonal matrix whose diagonal elements are the eigenvalues of matrix $\mathbf{M}$ arranged in descending order, namely, $\Lambda = \text{diag}\{\lambda_1, \lambda_2, \ldots, \lambda_n\}$ with $\lambda_1 \geq \lambda_2 \geq \cdots \geq \lambda_n$. If all eigenvalues of $\mathbf{M}$ are nonnegative, then the nonnegative part of $\mathbf{M}$ can be denoted as $(\mathbf{M})_+ = \mathbf{M}$. If $\mathbf{M}$ has r negative eigenvalues, i.e., $0 > \lambda_{n-r+1} \geq \cdots \geq \lambda_n$, then $(\mathbf{M})_+ = \mathbf{U}\tilde{\Lambda}\mathbf{U}^T$ where $\tilde{\Lambda} = \text{diag}\{\lambda_1, \ldots, \lambda_{n-r}, 0, \ldots, 0\}$.

Using the above notion, convex approximations for $f(\mathbf{x})$ and $c_j(\mathbf{x})$, denoted as $\tilde{f}(\mathbf{x})$ and $\tilde{c}_j(\mathbf{x})$, respectively, can be obtained by using the Taylor expansions of their respective functions at point $\mathbf{x}_k$ and, if necessary, modifying the Hessian by using its nonnegative part. This approach yields

$$\tilde{f}(\mathbf{x}) = f(\mathbf{x}_k) + \nabla^T f(\mathbf{x}_k)(\mathbf{x} - \mathbf{x}_k) + \tfrac{1}{2}(\mathbf{x} - \mathbf{x}_k)^T \mathbf{H}_k(\mathbf{x} - \mathbf{x}_k) \quad (15.5\text{a})$$

$$\tilde{c}(\mathbf{x}) = c_j(\mathbf{x}_k) + \nabla^T c_j(\mathbf{x}_k)(\mathbf{x} - \mathbf{x}_k) + \tfrac{1}{2}(\mathbf{x} - \mathbf{x}_k)^T \mathbf{P}_{j,k}(\mathbf{x} - \mathbf{x}_k) \quad (15.5\text{b})$$

where $\mathbf{H}_k = (\nabla^2 f(\mathbf{x}_k))_+$ and $\mathbf{P}_{j,k} = (\nabla^2 c_j(\mathbf{x}_k))_+$ are the nonnegative parts of Hessians $\nabla^2 f(\mathbf{x}_k)$ and $\nabla^2 c_j(\mathbf{x}_k)$, respectively.

An affine approximation for function $a_i(\mathbf{x})$, denoted as $\tilde{a}_i(\mathbf{x})$, can be obtained by using the first-order Taylor expansion of $a_i(\mathbf{x})$ at $\mathbf{x}_k$, namely,

$$\tilde{a}_i(\mathbf{x}) = a_i(\mathbf{x}_k) + \nabla^T a_i(\mathbf{x}_k)(\mathbf{x} - \mathbf{x}_k) \quad (15.5\text{c})$$

It is intuitively clear that the size of the feasible region $\mathcal{R}_\rho^{(k)}$, which is determined as $\{\rho_i, i = 1, 2, \ldots, n\}$ in the case of Eq. (15.2) or as a single scalar ρ in the case of Eq. (15.3), affects the accuracy of the approximation obtained by using Eqs. (15.5a)–(15.5c) and, in consequence, the performance of the SCP algorithm: if $\mathcal{R}_\rho^{(k)}$ is larger, the convex models in Eqs. (15.5a)–(15.5c) are less accurate and, as a result, the solution of the convex subproblem will be less accurate; if $\mathcal{R}_\rho^{(k)}$ is smaller, the convex models will be more accurate, but more iterations will be required for the SCP algorithm to converge.

Particle method   Some other convex approximations for $f(\mathbf{x})$ and $c_j(\mathbf{x})$ and affine approximations for $a_i(\mathbf{x})$ are possible. One of them is the so-called *particle method*. In the particle method, a function $f(\mathbf{x})$ is approximated by a convex (or affine) function by using the following steps:

1. Choose a set of points $\mathbf{z}_1, \mathbf{z}_2, \ldots, \mathbf{z}_K$ from region $\mathcal{R}_\rho^{(k)}$.
2. Evaluate $y_i = f(\mathbf{z}_i)$ for $i = 1, 2, \ldots, K$.
3. Fit a convex or affine function to the data $\{(\mathbf{z}_i, y_i), i = 1, 2, \ldots, K\}$ by using convex optimization.

The last step can be implemented by solving the SDP problem

$$
\underset{\mathbf{H},\, \mathbf{p},\, r}{\text{minimize}} \ \sum_{i=1}^{K} \left[(\mathbf{z}_i - \mathbf{x}_k)^T \mathbf{H}(\mathbf{z}_i - \mathbf{x}_k) + \mathbf{p}^T (\mathbf{z}_i - \mathbf{x}_k) + r - y_i \right]^2
$$
$$
\text{subject to: } \mathbf{H} \succeq \mathbf{0}
$$

where the variables are $\mathbf{H}$, $\mathbf{p}$, and r. The convex approximation for $f(\mathbf{x})$ in this case is taken to be

$$
\tilde{f}(\mathbf{x}) = (\mathbf{x} - \mathbf{x}_k)^T \mathbf{H}(\mathbf{x} - \mathbf{x}_k) + \mathbf{p}^T (\mathbf{x} - \mathbf{x}_k) + r \tag{15.6a}
$$

Evidently, this approximation depends on the current iterate $\mathbf{x}_k$ as well as the trust region $\mathcal{R}_\rho^{(k)}$. Note that unlike the Taylor expansion-based method, the particle method can handle nonsmooth functions as well as smooth functions whose derivatives are difficult to compute. For function $a_i(\mathbf{x})$, an affine approximation of the form

$$
\tilde{a}_i(\mathbf{x}) = \mathbf{p}^T (\mathbf{x} - \mathbf{x}_k) + r \tag{15.6b}
$$

can be obtained by solving the least-squares problem

$$
\underset{\mathbf{p},\, r}{\text{minimize}} \ \sum_{i=1}^{K} \left[\mathbf{p}^T (\mathbf{z}_i - \mathbf{x}_k) + r - y_i \right]^2
$$

Other options

There are other techniques that can be used to approximate nonconvex functions by convex ones. If a nonconvex function $f(\mathbf{x})$ assumes the form $f(\mathbf{x}) = f_c(\mathbf{x}) + f_{nc}(\mathbf{x})$ where $f_c(\mathbf{x})$ is convex while $f_{nc}(\mathbf{x})$ is not convex, then a natural convex approximation for $f(\mathbf{x})$ is

$$
\tilde{f}_{nc}(\mathbf{x}) = f_c(\mathbf{x}) + \tilde{f}_{nc}(\mathbf{x}) \tag{15.7a}
$$

which is a convex quadratic approximation for $f_{nc}(\mathbf{x})$ that can be obtained by using the Taylor expansion of $f_{nc}(\mathbf{x})$. In doing so, we obtain

$$
\tilde{f}_{nc}(\mathbf{x}) = f_{nc}(\mathbf{x}_k) + \nabla^T f_{nc}(\mathbf{x})(\mathbf{x} - \mathbf{x}_k) + \tfrac{1}{2}(\mathbf{x} - \mathbf{x}_k)^T \mathbf{H}_k(\mathbf{x} - \mathbf{x}_k) \tag{15.7b}
$$

where $\mathbf{H}_k = (\nabla^2 f_{nc}(\mathbf{x}_k))_+$. In the case where the nonnegative part of $f_{nc}(\mathbf{x})$ is zero, $\tilde{f}_{nc}(\mathbf{x})$ is reduced to a linear approximation given by

$$
\tilde{f}_{nc}(\mathbf{x}) = f_{nc}(\mathbf{x}_k) + \nabla^T f_{nc}(\mathbf{x}_k)(\mathbf{x} - \mathbf{x}_k) \tag{15.8}
$$

Approximating a nonconvex function by using a Taylor expansion based on the linear function in Eq. (15.8) is the starting point of a method known as the *convex-concave procedure* which will be studied in Sect. 15.4. Note that the convex approximation $\tilde{f}(\mathbf{x})$ in Eq. (15.7a) is not necessarily quadratic because of the presence of $f_c(\mathbf{x})$.

An SCP algorithm based on the above principles is as follows.

Algorithm 15.1 Sequential convex programming for the problem
in Eq. (15.1)

Step 1
Input feasible initial point for Eq. (15.1), $\{\rho_i, i = 1, 2, \ldots, n\}$, and a tolerance $\varepsilon > 0$ and set $k = 0$.

Step 2
Using Eqs. (15.5a)–(15.5c), construct functions $\tilde{f}(\mathbf{x})$, $\tilde{a}_i(\mathbf{x})$, for $1 \le i \le p$, and $\tilde{c}_j(\mathbf{x})$ for $1 \le j \le q$.

Step 3
Using $\mathbf{x}_k$ as initial point, solve the convex problem in Eq. (15.4) where $\mathcal{R}_\rho^{(k)}$ is defined by Eq. (15.2). Denote the solution obtained as $\mathbf{x}_{k+1}$.

Step 4
If $\|\mathbf{x}_{k+1} - \mathbf{x}_k\|_2 < \varepsilon$, stop and output $\mathbf{x}_{k+1}$ as the solution; otherwise, set $k = k + 1$ and repeat from Step 2.

Variants of the above algorithm can be obtained by using different convex approximations for the functions involved and different ways of constructing the trust region.

Example 15.1 Applying Algorithm 15.1, solve the constrained optimization problem

$$\text{minimize } f(\mathbf{x}) = \ln(1 + x_1^2) + x_2$$
$$\text{subject to: } a(\mathbf{x}) = (1 + x_1^2)^2 + x_2^2 - 4 = 0$$
$$c(\mathbf{x}) = x_1 + x_2^2 + 0.3 \le 0$$

Solution From Eq. (15.5a), we have

$$\tilde{f}(\mathbf{x}) = f(\mathbf{x}_k) + \nabla^T f(\mathbf{x}_k)(\mathbf{x} - \mathbf{x}_k) + \tfrac{1}{2}(\mathbf{x} - \mathbf{x}_k)^T \mathbf{H}_k(\mathbf{x} - \mathbf{x}_k)$$

where

$$\nabla f(\mathbf{x}_k) = \begin{bmatrix} \frac{2x_1^{(k)}}{1+(x_1^{(k)})^2} \\ 1 \end{bmatrix}, \quad \mathbf{H}_k = \begin{bmatrix} \frac{2(1-(x_1^{(k)})^2)_+}{(1+(x_1^{(k)})^2)^2} & 0 \\ 0 & 0 \end{bmatrix}$$

and $\tilde{c}(\mathbf{x}) = c(\mathbf{x})$ because $c(\mathbf{x})$ is quadratic and convex. An affine approximation of $a(\mathbf{x})$ is given by Eq. (15.5b), namely,

$$\tilde{a}(\mathbf{x}) = z(\mathbf{x}_k) + \nabla^T a(\mathbf{x}_k)(\mathbf{x} - \mathbf{x}_k)$$

where $\nabla a(\mathbf{x}_k) = [4x_1^{(k)}(1+(x_1^{(k)})^2) \ 2x_2^{(k)}]^T$. With $\mathbf{x}_0 = [-1.5 \ 1]^T$, $\rho_i = 0.4$ for $i = 1, 2$, and $\varepsilon = 10^{-6}$, it took Algorithm 15.1 eight iterations to converge to the solution $\mathbf{x}$. The value of the objective function at the solution point was $f^* = -1.755174 \times 10^{-1}$. The value of $\mathbf{a}(\mathbf{x}^*)$ was less than 10^{-15} and $c(\mathbf{x}^*) = -2.060264 \times 10^{-9}$ which means that, for all practical purposes, the specified constraints are satisfied at the solution. Figure 15.1 shows the trajectory of the first eight iterates and a profile of the objective function versus the number of iterations. ∎

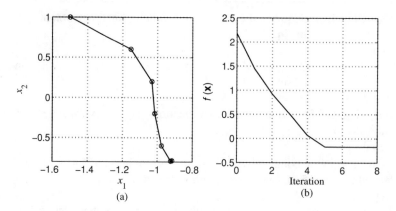

Fig. 15.1 **a** Trajectory of the solution produced by Algorithm 15.1 **b** objective function versus the number of iterations (Example 15.1)

15.2.3 Exact Penalty Formulation

A weak point of Algorithm 15.1 is that the convex formulation in Eq. (15.4) is not guaranteed to be feasible. This difficulty can be eliminated by using the exact penalty formulation described below [1]. The method consists of two steps. First, the problem in Eq. (15.1) is reformulated as the unconstrained problem

$$\text{minimize } F_\tau(\mathbf{x}) = f(\mathbf{x}) + \tau\left(\sum_{i=1}^{p} |a_i(\mathbf{x})| + \sum_{j=1}^{q} c_j(\mathbf{x})_+\right) \qquad (15.9)$$

where $c_j(\mathbf{x})_+ = \max\{c_j(\mathbf{x}), 0\}$ and $\tau > 0$ is a weighting constant for the second term which represents the degree to which the constraints are violated. The size of τ determines the magnitude of the penalty. It can readily be shown that the solution of the problem in Eq. (15.9) is also the solution of the original problem in Eq. (15.1) if τ is sufficiently large. Second, the problem in Eq. (15.9) is solved in a sequential manner where each subproblem is given by

$$\text{minimize } \tilde{F}_\tau(\mathbf{x}) = \tilde{f}(\mathbf{x}) + \tau\left(\sum_{i=1}^{p} |\tilde{a}_i(\mathbf{x})| + \sum_{j=1}^{q} \tilde{c}_j(\mathbf{x})_+\right)$$
$$\text{subject to: } \mathbf{x} \in \mathcal{R}_\rho^{(k)} \qquad (15.10)$$

and τ is assigned some fixed value. The solution of the problem is then used as the initialization of the next subproblem in Eq. (15.10) with an increased value of τ. Note that the formulation in Eq. (15.10) is always convex and entails no feasibility issues because the choice of the initial point ensures that it is always inside the current trust region $\mathcal{R}_\rho^{(k+1)}$.

As mentioned earlier, it is of importance that the trust region used in each subproblem is of the appropriate size so that the iterates generated by solving these subproblems converge to a local solution of the initial problem at a satisfactory convergence rate. A technique for updating the size of the trust region in each subproblem can be implemented in terms of the following algorithm.

Algorithm 15.2   Technique for updating the size of the trust region
                    in Eq. (15.2)

Step 1
Input iterate $\mathbf{x}_k$ and current ρ_k. Set constants $\alpha = 0.1$, $\beta_s = 1.1$, and $\beta_f = 0.5$ and assume that the solution of the problem in Eq. (15.10) is $\tilde{\mathbf{x}}$.

Step 2
Evaluate the predicted decrease

$$\tilde{\delta} = F_\tau(\mathbf{x}_k) - \tilde{F}_\tau(\tilde{\mathbf{x}})$$

and also the actual decrease in the objective function

$$\delta = F_\tau(\mathbf{x}_k) - F_\tau(\tilde{\mathbf{x}})$$

Step 3
If $\delta \geq \alpha\tilde{\delta}$, indicating that the actual decrease is greater than a fraction α of the predicted decrease, then set $\mathbf{x}_{k+1} = \tilde{\mathbf{x}}$, and $\rho_{k+1} = \beta_s\rho_k$ to increase the size of the trust region.

Step 4
If $\delta < \alpha\tilde{\delta}$, indicating that the actual decrease is less than a fraction α of the predicted decrease, then set $\mathbf{x}_{k+1} = \mathbf{x}_k$ and $\rho_{k+1} = \beta_f\rho_k$ to decrease the size of the trust region.

An algorithm based on the exact penalty formulation in Eq. (15.10) is summarized below.

Algorithm 15.3   Sequential convex programming for the problem
                    in Eq. (15.1) based on an exact penalty function

Step 1
Input feasible initial point $\mathbf{x}_0$, $\rho_0 = [\rho_1 \ \rho_2 \ \cdots \ \rho_n]^T$, and the number of iterations K. Let $k = 0$ and set $\alpha = 0.1$, $\beta_s = 1.1$, and $\beta_f = 0.5$.

Step 2
Using Eqs. (15.5a)–(15.5c), construct functions $\tilde{f}(\mathbf{x})$, $\tilde{a}_i(\mathbf{x})$ for $1 \leq i \leq p$, and $\tilde{c}_j(\mathbf{x})$ for $1 \leq j \leq q$.

Step 3
Using $\mathbf{x}_k$ as the initial point, solve the convex subproblem in Eq. (15.10) where $\mathcal{R}_\rho^{(k)}$ is defined by Eq. (15.2). Denote the solution obtained as $\tilde{\mathbf{x}}$.

Step 4
Apply Algorithm 15.2 to obtain $\mathbf{x}_{k+1}$ and ρ_{k+1}.

Step 5
If $k + 1 = K$, stop and output $\mathbf{x}_{k+1}$ as the solution; otherwise, set $k = k + 1$ and repeat from Step 2.

Example 15.2 Applying Algorithm 15.3, solve the constrained optimization problem in Example 15.1, i.e.,

$$\text{minimize } f(\mathbf{x}) = \ln(1 + x_1^2) + x_2$$

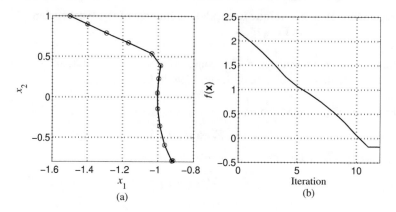

Fig. 15.2 a Trajectory of the solution produced by Algorithm 15.3, **b** objective function versus the number of iterations (Example 15.2)

$$\text{subject to: } a(\mathbf{x}) = (1 + x_1^2)^2 + x_2^2 - 4 = 0$$
$$c(\mathbf{x}) = x_1 + x_2^2 + 0.3 \le 0$$

Solution With $\boldsymbol{\rho}_0 = [0.1 \ \ 0.1]^T$, $\tau = 20$, and the same initial point as in Example 15.1, $\mathbf{x}_0 = [-1.5 \ \ 1]^T$, it took Algorithm 15.3 twelve iterations to converge to the solution $\mathbf{x}^* = [-0.9162810 \ \ -0.7850357]^T$. The objective function at the solution point was found to be $f^* = -1.755033 \times 10^{-1}$. The values of $a(\mathbf{x}^*)$ and $c(\mathbf{x}^*)$ were found to be $a(\mathbf{x}^*) = 3.020135 \times 10^{-4}$ and $c(\mathbf{x}^*) = -9.936028 \times 10^{-9}$, respectively, which means that, for all practical purposes, the specified constraints are satisfied at the solution. Figure 15.2 shows the trajectory of the solution and the value of the objective function versus the number of iterations.   ∎

15.2.4 Alternating Convex Optimization

In this section, a method known as *alternating convex optimization* [2] is described, which is often used in a number of applications.

Suppose that the number of variables in the problem of Eq. (15.1) is n. Index set $\{1, 2, \ldots, n\}$ can be divided into k nonoverlapping subsets I_i such that

$$\bigcup_{i=1}^{k} I_i = \{1, 2, \ldots, n\} \tag{15.11}$$

The components of variable $\mathbf{x} = [x_1 \ \ x_2 \ \ \cdots \ \ x_n]^T$ can then be divided into k groups with each group forming a vector which can be denoted as $\mathbf{x}^{(i)}$ for $i = 1, 2, \ldots, k$.

In other words, the variable vector $\mathbf{x}$ can be arranged into a vector of the form

$$\mathbf{x} = \begin{bmatrix} \mathbf{x}^{(1)} \\ \mathbf{x}^{(2)} \\ \vdots \\ \mathbf{x}^{(k)} \end{bmatrix}$$

We assume that there exists an index assignment in Eq. (15.11) such that if any specific variable, say $\mathbf{x}^{(i)}$, is considered as the only variable with the rest of the variables assumed to be fixed, the problem in Eq. (15.1) is *convex with respect to* $\mathbf{x}^{(i)}$. Such an index assignment, say $\{I_i, \text{ for } i = 1, 2, \ldots, k\}$, is said to be *regular* for the problem of Eq. (15.1). A natural SCP strategy under these circumstances is to solve the problem in Eq. (15.1) with respect to one specific variable, say $\mathbf{x}^{(i)}$, with the rest of the variables remaining fixed to ensure that the problem becomes convex. The procedure is then repeated with i cycling through enough times until variations in all variables $\{\mathbf{x}^{(i)} \text{ for } i = 1, 2, \ldots, k\}$ or in the objective function or in both are less than a given tolerance. Evidently, the values of the objective function at the iterates produced by this method are always nonincreasing. On the other hand, due to the nonconvex nature of the problem of Eq. (15.1), the method is not guaranteed to converge to the global minimizer and the quality of the local solution tends to depend to a large extent on the initialization of the algorithm. The above principles lead to the following SCP algorithm.

Algorithm 15.4   Alternating convex optimization for the problem in Eq. (15.1)

Step 1
Input a regular index assignment $\{I_i, \text{ for } i = 1, 2, \ldots, k\}$, a tolerance $\varepsilon > 0$, an initial point $\mathbf{x}_0$ arranged in accordance with the index assignment as

$$\mathbf{x}_0 = \begin{bmatrix} \mathbf{x}_0^{(1)} \\ \mathbf{x}_0^{(2)} \\ \vdots \\ \mathbf{x}_0^{(k)} \end{bmatrix}.$$

Set $l = 0$.
Step 2
Solve the problem in Eq. (15.1) with respect to variable $\mathbf{x}^{(i)}$ for $i = 1, 2, \ldots, k$, with $\mathbf{x}_l$ as initial point by using an appropriate convex-problem solver. Denote the solution obtained as $\mathbf{x}_{l+1}$.
Step 3
If $\left| \frac{f(\mathbf{x}_{l+1}) - f(\mathbf{x}_l)}{f(\mathbf{x}_l)} \right| < \varepsilon$, stop and output $\mathbf{x}_{l+1}$ as the solution; otherwise, set $l = l + 1$ and repeat from Step 2.

Applications of the alternating convex optimization technique typically involve a matrix factorization of the form $\mathbf{X} \approx \mathbf{W}\mathbf{H}$ where matrices $\mathbf{X}$, $\mathbf{W}$, and $\mathbf{H}$ are so-called *nonnegative matrices* in that their components are all nonnegative. In view of the nonnegativity property of the matrices involved, such a factorization is referred to as a *nonnegative matrix factorization* (NMF) [3,4].

In facial image representation, each column of the given data matrix $\mathbf{X}$ is a vector representation of a facial image that is generated by stacking the columns of the image. Each component of matrix $\mathbf{X}$ is the light intensity of a corresponding image pixel. In a typical application, matrix $\mathbf{X}$ involves a large number of images and hence its size $n \times m$ is large. An NMF of matrix $\mathbf{X}$ is required to be of a reduced rank such that $\mathbf{X} \approx \mathbf{WH}$ where $\mathbf{W} \in R^{n \times r}$ and $\mathbf{H} \in R^{r \times m}$ are nonnegative and $r \ll \min\{n, m\}$. Note that an approximation of this type is not unique because if $\mathbf{X} \approx \mathbf{WH}$, then $\mathbf{X} \approx \tilde{\mathbf{W}}\tilde{\mathbf{H}}$ also holds for $\tilde{\mathbf{W}} = \mathbf{WD}$ and $\tilde{\mathbf{H}} = \mathbf{D}^{-1}\mathbf{H}$ with any positive-definite diagonal matrix $\mathbf{D}$. It has been shown [3] that for a facial data matrix $\mathbf{X}$, there exist nonnegative matrices $\mathbf{W}$ and $\mathbf{H}$ with a small value of r such that $\mathbf{WH}$ provides a good approximation for $\mathbf{X}$. The columns of $\mathbf{W}$, which are often regarded as basis vectors, are visually interpretable as parts of human faces such as eyes, noses, lips, etc., whereas the components of a column of matrix $\mathbf{H}$ are nonnegative weights that linearly combine the basis vectors to reconstruct the corresponding facial image for input data matrix $\mathbf{X}$.

For $r \ll \min\{n, m\}$, the total number of components of $\mathbf{W}$ and $\mathbf{H}$, which is equal to $r \cdot (n + m)$, is much less than that of $\mathbf{X}$, which is equal to $n \cdot m$. Therefore, the NMF of $\mathbf{X}$ is a more economical representation of the original data matrix $\mathbf{X}$. More importantly, matrices $\mathbf{W}$ and $\mathbf{H}$ can reveal explicit structural information about $\mathbf{X}$ that is of immediate use in applications such as face recognition [4]. The example below describes an application of the alternating convex optimization method for the construction of an NMF for a large-size facial data matrix $\mathbf{X}$.

Example 15.3 Find nonnegative matrices $\mathbf{W} = \{w_{i,j}\}$ and $\mathbf{H} = \{h_{i,j}\}$ of sizes $n \times r$ and $r \times m$, respectively, that solve the problem

$$\text{minimize } \tfrac{1}{2}\|\mathbf{WH} - \mathbf{X}\|_F^2 + \mu_1 \sum_{i=1}^{n} \sum_{j=1}^{r} |w_{i,j}| + \mu_2 \sum_{i=1}^{r} \sum_{j=1}^{m} |h_{i,j}| \qquad (15.12a)$$

$$\text{subject to: } \mathbf{W} \geq \mathbf{0}, \ \mathbf{H} \geq \mathbf{0} \qquad (15.12b)$$

where $\mathbf{X}$ is a given data matrix of size $n \times m$ and $r \ll \min\{n, m\}$. Parameters μ_1 and μ_2 are *regularization parameters* that are selected to promote sparsity in matrices $\mathbf{W}$ and $\mathbf{H}$, respectively. The constraints in Eq. (15.12b) are imposed in a component-wise sense such that every component of $\mathbf{W}$ and $\mathbf{H}$ is constrained to be nonnegative. Illustrate the alternating convex optimization technique by applying it to a matrix $\mathbf{X}$ of size 361×2429 where each column of $\mathbf{X}$ is generated by stacking a facial image of size 19×19 pixels from the database of the Center for Biological and Computational Learning at MIT [5]. Assume that parameter r is set to 49.

Solution With $n = 361$, $m = 2429$, and $r = 49$, there is a total of $n \cdot r + r \cdot m = 136,710$ variables in the optimization problem under consideration. In order to optimize such a large-scale problem with the convex optimization technique, matrix $\mathbf{H}$ is first optimized by solving the problem in Eqs. (15.12a) and (15.12b) with matrix $\mathbf{W}$ fixed; then matrix $\mathbf{W}$ is optimized by solving Eqs. (15.12a) and (15.12b) with matrix $\mathbf{H}$ fixed; this alternating optimization procedure is then repeated as many times as necessary until matrices $\mathbf{H}$ and $\mathbf{W}$ satisfy a certain convergence criterion. The subprograms involved are convex but their sizes, which involve 119,021 and 17,689 variables, respectively, are too large to be solved efficiently. To improve the efficiency of the optimization process, we let

$$
\mathbf{W} = \begin{bmatrix} \mathbf{w}_1^T \\ \mathbf{w}_2^T \\ \vdots \\ \mathbf{w}_n^T \end{bmatrix}, \quad \mathbf{H} = \begin{bmatrix} \mathbf{h}_1 & \mathbf{h}_2 & \cdots & \mathbf{h}_m \end{bmatrix}, \quad \mathbf{X} = \begin{bmatrix} \mathbf{x}_1 & \mathbf{x}_2 & \cdots & \mathbf{x}_m \end{bmatrix} = \begin{bmatrix} \hat{\mathbf{x}}_1^T \\ \hat{\mathbf{x}}_2^T \\ \vdots \\ \hat{\mathbf{x}}_n^T \end{bmatrix}
$$

and note that in the subproblem for which $\mathbf{H}$ is held fixed, the third term of the objective function in Eq. (15.12a) becomes a constant and, consequently, it can be taken to be zero and hence the objective function can be expressed as

$$
\tfrac{1}{2}\|\mathbf{WH} - \mathbf{X}\|_F^2 + \mu_1 \sum_{i=1}^{n} \sum_{j=1}^{r} |w_{i,j}| = \sum_{i=1}^{n} \left[\tfrac{1}{2}\|\mathbf{H}^T \mathbf{w}_i - \hat{\mathbf{x}}_i\|_2^2 + \mu_1 \|\mathbf{w}_i\|_1 \right]
$$

where each of the variable vectors $\{\mathbf{w}_i, \ i = 1, \ 2, \ \ldots, \ n\}$ appears in only one of the n-term sums on the right-hand side of the above expression. As a result, the minimization of the objective function can be performed by optimizing with respect to each vector $\mathbf{w}_i$. This can be done by solving a standard $L_1 - L_2$ minimization problem subject to the nonnegativity of $\mathbf{w}_i$, namely,

$$
\text{minimize} \ \ \tfrac{1}{2}\|\mathbf{H}^T \mathbf{w}_i - \hat{\mathbf{x}}_i\|_2^2 + \mu_1 \|\mathbf{w}_i\|_1
$$

$$
\text{subject to: } \mathbf{w}_i \geq \mathbf{0}
$$

which is a convex problem involving only r variables.

Similarly, in the subproblem where $\mathbf{W}$ is held fixed, the second term of the objective function in Eq. (15.12a) becomes a constant and it can be taken to be zero. Hence the objective function can be expressed as

$$
\tfrac{1}{2}\|\mathbf{WH} - \mathbf{X}\|_F^2 + \mu_2 \sum_{i=1}^{r} \sum_{j=1}^{m} |h_{i,j}| = \sum_{j=1}^{m} \left[\tfrac{1}{2}\|\mathbf{Wh}_j - \mathbf{x}_j\|_2^2 + \mu_2 \|\mathbf{h}_j\|_1 \right]
$$

where each of the variable vectors $\{\mathbf{h}_j, \ j = 1, \ 2, \ \ldots, \ m\}$ appears in only one of the m-term sums on the right-hand side of the above expression. Consequently, the minimization of the objective function can be performed by optimizing with respect to each vector $\mathbf{h}_j$. This can be done by solving a standard $L_1 - L_2$ minimization problem subject to the nonnegativity of $\mathbf{h}_j$, namely,

$$\text{minimize} \ \tfrac{1}{2}\|\mathbf{Wh}_j - \mathbf{x}_j\|_2^2 + \mu_2\|\mathbf{h}_j\|_1$$

$$\text{subject to: } \mathbf{h}_j \geq \mathbf{0}$$

which is a convex problem involving only r variables. In this example, the regularization parameters were set to $\mu_1 = 0.015$ and $\mu_2 = 0.1$ and a random nonnegative matrix of size 49×2429 was used as the initial $\mathbf{H}$ to start the alternating convex optimization. The $L_1 - L_2$ optimization involved in the subproblems was carried out by performing 20 iterations of the alternating direction method of multipliers (ADMM) with $\alpha = 0.02$ (see Sect. 14.5, especially Example 14.8, for technical details associated with ADMM). Once the minimizer of the $L_1 - L_2$ objective function is obtained, the nonnegativity constraints are satisfied by projecting each component of the minimizer onto the domain of nonnegative real numbers: if the component is nonnegative, the component is preserved; otherwise, it is set to zero. Using NNF, it took 387 iterations to yield a solution $\{\mathbf{W}^*, \ \mathbf{H}^*\}$ which leads to the approximation

$$\mathbf{X} \approx \mathbf{W}^*\mathbf{H}^*$$

for the data set $\mathbf{X}$. Because of the inclusion of the two regularization terms in the objective function in Eq. (15.12a) both matrices $\mathbf{W}^*$ and $\mathbf{H}^*$ are sparse, i.e., each contains a large number of negligible components. Figure 15.3 shows 49 images, each of which can be obtained by converting a column of matrix $\mathbf{W}^*$ into a square matrix of size 19×19. We note that these matrices are sparse and they are also interpretable as various 'face parts'. Consequently, they can be utilized as basis images for reconstructing the facial images from input data matrix $\mathbf{X}$. Figure 15.4 displays 49 components of a typical column of $\mathbf{H}^*$ where the sparsity of $\mathbf{H}^*$ is clearly evident. The reconstruction of the jth input facial image $\mathbf{x}_j$ can readily be obtained by using the r components of the jth column of matrix $\mathbf{W}^*$ as weights to linearly combine the r basis images from $\mathbf{W}^*$. The sparsity and nonnegativity of $\mathbf{H}^*$ imply that a facial image can be reconstructed by selecting a small number of basic face parts and combining them with a set of appropriate weights. A set of 49 input facial images and their approximations by NMF are illustrated in Fig. 15.5a, b, respectively.   ∎

15.3 Sequential Quadratic Programming

In this section, we explore a class of algorithms, known as *sequential quadratic programming* (SQP) algorithms [6,7], for the general constrained optimization problem in Eq. (15.1) i.e.,

$$\text{minimize} \ f(\mathbf{x})$$

$$\text{subject to: } a_i(\mathbf{x}) = 0 \quad \text{for } i = 1, \ 2, \ \ldots, \ p$$

$$c_j(\mathbf{x}) \leq 0 \quad \text{for } i = 1, \ 2, \ \ldots, \ q$$

where functions $f(\mathbf{x})$, $a_i(\mathbf{x})$ for $i = 1, \ 2, \ \ldots, \ p$, and $c_j(\mathbf{x})$ for $i = 1, \ 2, \ \ldots, \ q$ are twice continuously differentiable.

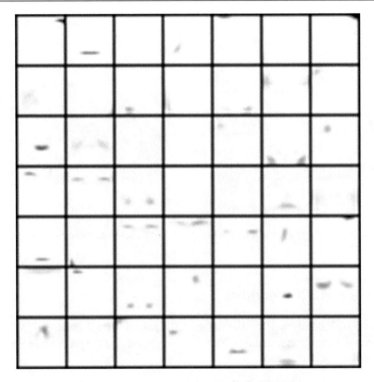

Fig. 15.3 The 49 columns of $\mathbf{W}^*$, here each column is converted into an image of 19×19 pixels (Example 15.3)

Fig. 15.4 The 49 components of a typical column of matrix $\mathbf{H}^*$ (Example 15.3)

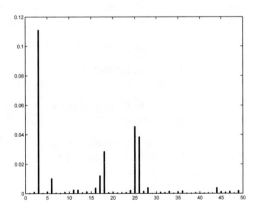

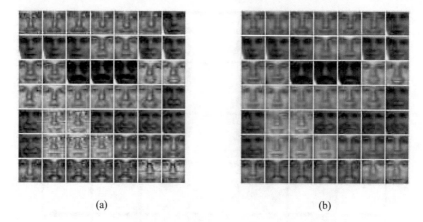

(a)	(b)

Fig. 15.5 a 49 input facial images, each of size 19×19, and **b** their approximations by NMF (Example 15.3)

15.3.1 Basic SQP Algorithm

The basic SQP algorithm is similar to the SCP algorithms studied in Sect. 15.2 except that it has several distinct features. The KKT conditions for the problem in Eq. (15.1) are given by

$$\nabla_x L(\mathbf{x}, \boldsymbol{\lambda}, \boldsymbol{\mu}) = \mathbf{0}$$
$$a_i(\mathbf{x}) = 0 \quad \text{for } i = 1, 2, \ldots, p$$
$$c_j(\mathbf{x}) \leq 0 \quad \text{for } i = 1, 2, \ldots, q \qquad (15.13)$$
$$\boldsymbol{\mu} \geq \mathbf{0}$$
$$\mu_j c_j(\mathbf{x}) = 0 \quad \text{for } j = 1, 2, \ldots, q$$

where the Lagrangian assumes the form

$$L(\mathbf{x}, \boldsymbol{\lambda}, \boldsymbol{\mu}) = f(\mathbf{x}) = \sum_{i=1}^{p} \lambda_i a_i(\mathbf{x}) + \sum_{j=1}^{q} \mu_j c_j(\mathbf{x})$$

In the kth iteration, we update a known iterate $\{\mathbf{x}_k, \boldsymbol{\lambda}_k, \boldsymbol{\mu}_k\}$ to a new iterate such that the KKT conditions in Eq. (15.13) are satisfied at $\{\mathbf{x}_{k+1}, \boldsymbol{\lambda}_{k+1}, \boldsymbol{\mu}_{k+1}\}$ approximately. To this end, we let $\mathbf{x}_{k+1} = \mathbf{x}_k + \boldsymbol{\delta}_k$ and examine an affine approximation for Eq. (15.13) at $\{\mathbf{x}_{k+1}, \boldsymbol{\lambda}_{k+1}, \boldsymbol{\mu}_{k+1}\}$ given by

$$\mathbf{Z}_k \boldsymbol{\delta}_k + \mathbf{g}_k + \mathbf{A}_{ek}^T \boldsymbol{\lambda}_{k+1} + \mathbf{A}_{ik}^T \boldsymbol{\mu}_{k+1} = \mathbf{0}$$
$$\mathbf{A}_{ek} \boldsymbol{\delta}_x = -\mathbf{a}_k$$
$$\mathbf{A}_{ik} \boldsymbol{\delta}_x \leq -\mathbf{c}_k \qquad (15.14)$$
$$\boldsymbol{\mu}_{k+1} \geq \mathbf{0}$$
$$(\mu_{k+1})(\mathbf{A}_{ik}\boldsymbol{\delta}_k + \mathbf{c}_k)_j = 0 \quad \text{for } j = 1, 2, \ldots, q$$

where

$$\mathbf{Z} = \nabla_x^2 f(\mathbf{x}_k) + \sum_{i=1}^{p} (\lambda_k)_i \nabla_x^2 a_i(\mathbf{x}_k) + \sum_{j=1}^{q} (\boldsymbol{\mu})_j \nabla_x^2 c_j(\mathbf{x}_k) \qquad (15.15\text{a})$$

$$\mathbf{g}_k = \nabla_x f(\mathbf{x}_k) \qquad (15.15\text{b})$$

$$\mathbf{A}_{ek} = \begin{bmatrix} \nabla_x^T a_1(\mathbf{x}_k) \\ \nabla_x^T a_2(\mathbf{x}_k) \\ \vdots \\ \nabla_x^T a_p(\mathbf{x}_k) \end{bmatrix}, \ \mathbf{A}_{ik} = \begin{bmatrix} \nabla_x^T c_1(\mathbf{x}_k) \\ \nabla_x^T c_2(\mathbf{x}_k) \\ \vdots \\ \nabla_x^T c_q(\mathbf{x}_k) \end{bmatrix}, \ \mathbf{a}_k = \begin{bmatrix} a_1(\mathbf{x}_k) \\ a_2(\mathbf{x}_k) \\ \vdots \\ a_p(\mathbf{x}_k) \end{bmatrix}, \ \mathbf{c}_k = \begin{bmatrix} c_1(\mathbf{x}_k) \\ c_2(\mathbf{x}_k) \\ \vdots \\ c_q(\mathbf{x}_k) \end{bmatrix} \qquad (15.15\text{c})$$

Equation (15.14) may be interpreted as the exact set of KKT conditions of the QP problem

$$\text{minimize } \tfrac{1}{2}\boldsymbol{\delta}^T \mathbf{Z}_k \boldsymbol{\delta} + \boldsymbol{\delta}^T \mathbf{g}_k \qquad (15.16\text{a})$$

$$\text{subject to: } \mathbf{A}_{ek}\boldsymbol{\delta} = -\mathbf{a}_k \qquad (15.16\text{b})$$

$$\mathbf{A}_{ik}\boldsymbol{\delta} \leq -\mathbf{c}_k \qquad (15.16\text{c})$$

Let $\boldsymbol{\delta}_x$ be the solution of the above QP problem and $\{\lambda_{qp}, \boldsymbol{\mu}_{qp}\}$ be the associated set of Lagrange multipliers. If we let

$$\boldsymbol{\delta}_\lambda = \lambda_{qp} - \lambda_k \quad \text{and} \quad \boldsymbol{\delta}_\mu = \boldsymbol{\mu}_{qp} - \boldsymbol{\mu}_k \qquad (15.17)$$

the new iterate $\{\mathbf{x}_{k+1}, \lambda_{k+1}, \boldsymbol{\mu}_{k+1}\}$ can be obtained as

$$\mathbf{x}_{k+1} = \mathbf{x}_k + \alpha_k \boldsymbol{\delta}_x$$
$$\lambda_{k+1} = \lambda_k + \alpha_k \boldsymbol{\delta}_\lambda$$
$$\boldsymbol{\mu}_{k+1} = \boldsymbol{\mu}_k + \alpha_k \boldsymbol{\delta}_\mu \qquad (15.18)$$

where α_k is determined by using a line search that involves minimizing a *merit function*. Popular merit functions in the context of SQP include the L_1 merit function [8,9] and the augmented Lagrangian merit function [10–12]. The L_1 merit function is defined by

$$\Psi_k(\alpha) = f(\mathbf{x}_k + \alpha\boldsymbol{\delta}_x) + \tau \sum_{i=1}^{p} |a_i(\mathbf{x}_k + \alpha\boldsymbol{\delta}_x)| + \tau \sum_{j=1}^{q} c_j(\mathbf{x}_k + \alpha\boldsymbol{\delta}_x)_+ \qquad (15.19)$$

where $c_j(\mathbf{x}_k + \alpha\boldsymbol{\delta}_x)_+ = \max\{c_j(\mathbf{x}_k + \alpha\boldsymbol{\delta}_x), 0\}$ and $\tau > 0$ is a sufficiently large scalar. Evidently, minimizing a merit function with respect to α reduces the objective function along the search direction $\boldsymbol{\delta}_x$ and at the same time reduces the degree of violation of both the equality and inequality constraints. Therefore, a good choice of α_k for Eq. (15.18) is obtained by minimizing $\Psi_k(\alpha)$ over a reasonable interval, say $0 \leq \alpha \leq 1$, namely,

$$\alpha_k = \arg \min_{0 \leq \alpha \leq 1} \Psi_k(\alpha)$$

Because of the presence of the second and third terms in Eq. (15.19), the merit function is nondifferentiable in general and may be nonconvex. A straightforward way to obtain a good estimate of α_k is to evaluate $\Psi_k(\alpha)$ over a set of uniformly spaced grid points $\mathcal{A} = \{\alpha_l, l = 1, 2, \ldots, L\}$ over the range 0 to 1 and then identify a grid point at which $\Psi_k(\alpha)$ is minimized, i.e.,

$$\alpha_k \approx \arg\min_{\alpha \in \mathcal{A}} \Psi_k(\alpha) \tag{15.20}$$

The above approach is quite effective in practice especially when the number of constraints is not large. The basic SQP technique can be implemented in terms of the following algorithm.

Algorithm 15.5 Basic SQP algorithm for the problem in Eq. (15.1)
Step 1
Input initial point $\{\mathbf{x}_0, \lambda_0, \mu_0\}$, $\tau > 0$, and tolerance ε and set $k = 0$.
Step 2
Solve the QP problem in Eqs. (15.16a)–(15.16c) to obtain δ_x and $\{\lambda_{qp}, \mu_{qp}\}$ and then use Eq. (15.17) to obtain δ_λ and δ_μ.
Step 3
Choose α_k using Eq. (15.20).
Step 4
Set $\{\mathbf{x}_{k+1}, \lambda_{k+1}, \mu_{k+1}\}$ using Eq. (15.18).
Step 5
If

$$\left\| \begin{bmatrix} \mathbf{x}_{k+1} - \mathbf{x}_k \\ \lambda_{k+1} - \lambda_k \\ \mu_{k+1} - \mu_k \end{bmatrix} \right\|_2 < \varepsilon,$$

stop and output $\mathbf{x}_{k+1}$ as the solution; otherwise, set $k = k + 1$ and repeat from Step 2.

15.3.2 Positive Definite Approximation of Hessian

As can be seen in Eq. (15.15a), computing matrix $\mathbf{Z}_k$ is quite involved and, more importantly, $\mathbf{Z}_k$ is not guaranteed to be positive definite or positive semidefinite, especially when the current iterate $\mathbf{x}_k$ is not sufficiently close to a solution point $\mathbf{x}^*$. This can, in turn, cause difficulties in solving the QP subproblem in Eqs. (15.16a)–(15.16c). One way to handle this problem is to use an approximate Hessian, denoted again as $\mathbf{Z}_k$, which is updated using a modified BFGS formula [13] as

$$\mathbf{Z}_{k+1} = \mathbf{Z}_k + \frac{\eta_k \eta_k^T}{\delta_x^T \eta_k} - \frac{\mathbf{Z}_k \delta_x \delta_x^T \mathbf{Z}_k}{\delta_x^T \mathbf{Z}_k \delta_x} \tag{15.21}$$

where

$$\eta_k = \theta \gamma_k + (1 - \theta)\mathbf{Z}_k \delta_x$$
$$\gamma_k = (\mathbf{g}_{k+1} - \mathbf{g}_k) + (\mathbf{A}_{e,k+1} - \mathbf{A}_{ek})^T \lambda_{k+1} + (\mathbf{A}_{i,k+1} - \mathbf{A}_{i,k})^T \mu_{k+1}$$
$$\theta = \begin{cases} 1 & \text{if } \delta_k^T \geq 0.2\delta_x^T \mathbf{Z}_k \delta_x \\ \frac{0.8\delta_x^T \mathbf{Z}_k \delta_x}{\delta_k^T \mathbf{Z}_k \delta_x - \delta_x^T \gamma_k} & \text{otherwise} \end{cases}$$

The initial $\mathbf{Z}_0$ is usually set to $\mathbf{I}_n$. A desirable property of the above BFGS approximation is that $\mathbf{Z}_{k+1}$ is positive definite if $\mathbf{Z}_k$ is positive definite. As a result, the QP problem in Eqs. (15.16a)–(15.16c) where $\mathbf{Z}_k$ is generated by Eq. (15.21) is a convex QP problem.

15.3.3   Robustness and Solvability of QP Subproblem of Eqs. (15.16a)–(15.16c)

In the subproblem of Eqs. (15.16a)–(15.16c), it is important to stress that the development of Algorithm 15.5 is based on the affine approximation of the KKT conditions in Eq. (15.13), and the validity of such an approximation is based on the assumption that the variable δ_x is sufficiently small in magnitude. In many practical problems, difficulties associated with the use of an initial point that is far from the solution can be avoided by including a bound constraint such as $\|\delta_x\|_2 \leq \rho$.

Another technical issue that may occur in implementing an SQP algorithm is that the QP subproblem of Eqs. (15.16a)–(15.16c) may be unsolvable even if matrix $\mathbf{Z}_k$ in Eq. (15.16a) is positive definite. This will indeed be the case if the feasible region defined by Eqs. (15.16b) and (15.16c) turns out to be empty. This problem can be fixed by introducing nonnegative slack variables $\mathbf{u}, \mathbf{v} \in R^p$, $\mathbf{w} \in R^q$, and $r \in R$ in the QP subproblem as

$$\text{minimize } \tfrac{1}{2}\delta^T \mathbf{Z}_k \delta + \delta^T \mathbf{g}_k + \eta \left(r + \sum_{i=1}^{p}(u_i + v_i) + \sum_{j=1}^{p} w_j \right) \tag{15.22a}$$

$$\text{subject to: } \mathbf{A}_{ek}\delta + \mathbf{a}_k = \mathbf{u} - \mathbf{v} \tag{15.22b}$$

$$\mathbf{A}_{ik}\delta + \mathbf{c}_k \leq \mathbf{w} \tag{15.22c}$$

$$\|\delta\|_2 \leq \rho + r \tag{15.22d}$$

$$\mathbf{u} \geq \mathbf{0}, \mathbf{v} \geq \mathbf{0}, \mathbf{w} \geq \mathbf{0}, r \geq 0 \tag{15.22e}$$

where $\eta > 0$ is a sufficiently large scalar [14]. With this modification, the feasible region is always nonempty. In fact, if we arbitrarily assign $\delta = \delta_0$ and let $\mathbf{u}_0 = \max\{\mathbf{A}_{ek}\delta_p + \mathbf{a}_k, \mathbf{0}\}$, $\mathbf{v}_0 = \max\{-(\mathbf{A}_{ek}\delta_0 + \mathbf{a}_k), \mathbf{0}\}$, $\mathbf{w}_0 = \max\{\mathbf{A}_{ik}\delta_0 + \mathbf{c}_k, \mathbf{0}\}$, and $r_0 = \max\{\|\delta_0\|_2 - \rho, 0\}$, it can be readily verified that point $\{\delta_0, \mathbf{u}_0, \mathbf{v}_0, \mathbf{w}_0, r_0\}$ is in the feasible region. The last term in the objective function is a penalty term that, with a sufficiently large $\eta > 0$, helps reduce the values of the slack variables. Let $\{\delta_x, \mathbf{u}^*, \mathbf{v}^*, \mathbf{w}^*, r^*\}$ be a solution of the above problem. If the variables in set $\{\mathbf{u}^*, \mathbf{v}^*, \mathbf{w}^*, r^*\}$ assume zero values, then vector δ_x is taken to be a solution of the original QP problem subject to an additional bound constraint $\|\delta\|_2 \leq \rho$. On the other hand, if the variables in set $\{\mathbf{u}^*, \mathbf{v}^*, \mathbf{w}^*, r^*\}$ assume nonzero values, then the constraints must be relaxed for the problem to be solvable. In such a case, the QP subproblem in Eqs. (15.16a)–(15.16c) is relaxed to

$$\text{minimize } \tfrac{1}{2}\delta^T \mathbf{Z}_k \delta + \delta^T \mathbf{g}_k \tag{15.23a}$$

$$\text{subject to: } \mathbf{A}_{ek}\boldsymbol{\delta} = -\tilde{\mathbf{a}}_k \tag{15.23b}$$

$$\mathbf{A}_{ik}\boldsymbol{\delta} = -\tilde{\mathbf{c}}_k \tag{15.23c}$$

$$\|\boldsymbol{\delta}\|_2 \le \tilde{\rho} \tag{15.23d}$$

where

$$\tilde{\mathbf{a}}_k = \mathbf{a}_k - \mathbf{u}^* + \mathbf{v}^*, \ \tilde{\mathbf{c}}_k = \mathbf{c}_k - \mathbf{w}^*, \ \tilde{\rho} = \rho + r^* \tag{15.24}$$

The optimal Lagrange multipliers associated with the constraints in Eqs. (15.23b) and (15.23c) are denoted as $\boldsymbol{\lambda}_{qp}$ and $\boldsymbol{\mu}_{qp}$, respectively, and are used in Eq. (15.17) to generate $\boldsymbol{\delta}_\lambda$ and $\boldsymbol{\delta}_\mu$.

15.3.4   Practical SQP Algorithm for the Problem of Eq. (15.1)

An SQP algorithm that incorporates the positive definite approximation technique in Sect. 15.3.2 and the solution technique for the QP subproblem in Sect. 15.3.3 is outlined below.

Algorithm 15.6 Practical SQP algorithm for the problem in Eq. (15.1)
Step 1
Input initial point $\{\mathbf{x}_0, \boldsymbol{\lambda}_0, \boldsymbol{\mu}_0\}$, $\eta > 0$, $\tau > 0$, $\rho > 0$, $\mathbf{Z}_0 = \mathbf{I}_n$, and tolerance ε and set $k = 0$.
Step 2
Solve the QP problem in Eqs. (15.22a)–(15.22e) to obtain $\{\boldsymbol{\delta}_x, \mathbf{u}^*, \mathbf{v}^*, \mathbf{w}^*, r^*\}$. If $\mathbf{u}^* \approx \mathbf{0}$, $\mathbf{v}^* \approx \mathbf{0}$, and $\mathbf{w}^* \approx \mathbf{0}$, then take the optimal Lagrange multipliers associated with the constraints in Eqs. (15.22b) and (15.22c) as $\boldsymbol{\lambda}_{qp}$ and $\boldsymbol{\mu}_{qp}$; otherwise, solve the QP problem in Eqs. (15.23a)–(15.23d) to obtain $\{\boldsymbol{\delta}_x, \boldsymbol{\lambda}_{qp}, \boldsymbol{\mu}_{qp}\}$.
Step 3
Choose α_k using Eq. (15.20).
Step 4
Using Eq. (15.17), obtain $\boldsymbol{\delta}_\lambda$ and $\boldsymbol{\delta}_\mu$ and set $\{\mathbf{x}_{k+1}, \boldsymbol{\lambda}_{k+1}, \boldsymbol{\mu}_{k+1}\}$ using Eq. (15.18).
Step 5
If

$$\left\| \begin{bmatrix} \mathbf{x}_{k+1} - \mathbf{x}_k \\ \boldsymbol{\lambda}_{k+1} - \boldsymbol{\lambda}_k \\ \boldsymbol{\mu}_{k+1} - \boldsymbol{\mu}_k \end{bmatrix} \right\|_2 < \varepsilon,$$

stop and output $\mathbf{x}_{k+1}$ as the solution; otherwise, update $\mathbf{Z}_k$ to $\mathbf{Z}_{k+1}$ using Eq. (15.21), set $k = k + 1$, and repeat from Step 2.

Example 15.4 Applying Algorithm 15.6, solve the constrained problem in Example 15.1, i.e.,

$$\begin{aligned} \text{minimize } & f(\mathbf{x}) = \ln(1 + x_1^2) + x_2 \\ \text{subject to: } & a(\mathbf{x}) = (1 + x_1^2)^2 + x_2^2 - 4 = 0 \\ & c(\mathbf{x}) = x_1 + x_2^2 + 0.3 \le 0 \end{aligned}$$

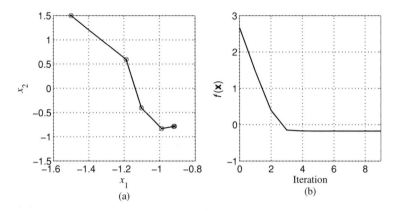

Fig. 15.6 a Trajectory of the solution produced by Algorithm 15.6, **b** objective function versus the number of iterations (Example 15.4)

Solution From Eqs. (15.15a)–(15.15c),

$$\mathbf{g}_k = \begin{bmatrix} 2x_1^{(k)} \\ 1+x_1^{(k)2} \\ 1 \end{bmatrix}, \quad \mathbf{A}_{ek} = \nabla^T a(\mathbf{x}_k) = \begin{bmatrix} 4x_1^{(k)}(1+x_1^{(k)2}) \\ 2x_2^{(k)} \end{bmatrix}^T, \quad \text{and } \mathbf{A}_{ik} = \nabla^T c(\mathbf{x}_k) = \begin{bmatrix} 1 \\ 2x_2^{(k)} \end{bmatrix}^T$$

In Example 15.1, the initial point was taken to be $\mathbf{x}_0 = [-1.5 \ \ 1]^T$ which violates the equality constraint $a(\mathbf{x}) = 0$ while satisfying the inequality constraint $c(\mathbf{x}) \leq 0$. In the present example, the initial point was taken to be $\mathbf{x}_0 = [-1.5 \ \ 1.5]^T$ which violates both the equality and inequality constraints. With $\lambda_0 = 1$, $\mu_0 = 1$, $\eta = 20$, $\tau = 1$, $\rho = 1$, and $\varepsilon = 10^{-6}$, it took Algorithm 15.6 nine iterations to converge to the solution $\mathbf{x}^* = [-0.9162420 \ \ -0.7850108]^T$. The objective function at the solution point was found to be $f^* = -1.755174 \times 10^{-1}$. The values of constraints $a(\mathbf{x}^*)$ and $c(\mathbf{x}^*)$ at the solution were $-9.059420 \times 10^{-14}$ and $-3.363848 \times 10^{-10}$, respectively, which means that, for all practical purposes, the specified constraints are satisfied at the solution. Figure 15.6 shows the trajectory of the solution and the value of the objective function versus the number of iterations.   ∎

15.4 Convex-Concave Procedure

In this section, we develop an algorithm for the solution of the general optimization problem in Eq. (15.1) by using a convexification technique known as the convex-concave (or concave-convex) procedure [15].

15.4.1 Basic Convex-Concave Procedure

The convex-concave procedure (CCP) is a heuristic method of convexifying non-convex problems based on the fact that a twice continuously differentiable function can be expressed as a difference of two convex functions [16]. Furthermore, if the function in question has a bounded Hessian, the difference of two convex functions can be explicitly constructed [15]. We are thus motivated to consider a nonconvex problem of the form

$$\text{minimize } f(\mathbf{x}) = u_0(\mathbf{x}) - v_0(\mathbf{x}) \tag{15.25a}$$

$$\text{subject to: } c_j(\mathbf{x}) = u_j(\mathbf{x}) - v_j(\mathbf{x}) \le 0 \quad \text{for } j = 1, 2, \dots, q \tag{15.25b}$$

where functions $u_j(\mathbf{x})$ and $v_j(\mathbf{x})$ for $j = 0, 1, \dots, q$ are convex.

Consider the kth iteration for the problem in Eqs. (15.25a) and (15.25b) where iterate $\mathbf{x}_k$ is known. The CCP consists of two steps. The first step is to convexify the objective function and constraints in Eqs. (15.25a) and (15.25b) by replacing each $v_j(\mathbf{x})$ with the affine approximation

$$\hat{v}_j(\mathbf{x}, \mathbf{x}_k) = v_j(\mathbf{x}_k) + \nabla^T v_j(\mathbf{x}_k)(\mathbf{x} - \mathbf{x}_k) \quad \text{for } j = 0, 1, \dots, q \tag{15.26}$$

As a result, the modified objective function $\mathbf{u}_0(\mathbf{x}) - \hat{v}_0(\mathbf{x} - \mathbf{x}_k)$ as well as constraint functions $\mathbf{u}_j(\mathbf{x}) - \hat{v}_j(\mathbf{x} - \mathbf{x}_k)$ are all convex. The second step is to solve the convex subproblem

$$\begin{aligned} \text{minimize } &\hat{f}(\mathbf{x}, \mathbf{x}_k) = u_0(\mathbf{x}) - \hat{v}_0(\mathbf{x}, \mathbf{x}_k) \\ \text{subject to: } &\hat{c}_j(\mathbf{x}, \mathbf{x}_k) = u_j(\mathbf{x}) - \hat{v}_j(\mathbf{x}, \mathbf{x}) \le 0 \text{ for } j = 1, 2, \dots, q \end{aligned} \tag{15.27}$$

whose solution is denoted as $\mathbf{x}_{k+1}$. These two steps are repeated for the new iterate $\mathbf{x}_{k+1}$ and in this way a sequence of iterates $\{\mathbf{x}_k, k = 0, 1, \dots, q\}$ is generated. There are two issues arising from the above procedure.

The first issue is whether each and every iterate in the sequence generated is feasible for the initial problem, i.e., it satisfies the constraints in Eq. (15.25b). We deal with this issue by mathematical induction as follows. Suppose we start the CCP with a point $\mathbf{x}_0$ that is feasible in the initial problem, that is, $\mathbf{x}_0$ satisfies the inequalities

$$u_j(\mathbf{x}_0) - v_j(\mathbf{x}_0) \le 0 \quad \text{for } j = 1, 2, \dots, q \tag{15.28}$$

From Eq. (15.26), we have $\hat{v}_j(\mathbf{x}_0, \mathbf{x}_0) = v_j(\mathbf{x}_0)$ which in conjunction with Eq. (15.28) gives

$$\hat{c}_j(\mathbf{x}_0, \mathbf{x}_0) = u_j(\mathbf{x}_0) - \hat{v}_j(\mathbf{x}_0, \mathbf{x}_0) = u_j(\mathbf{x}_0) - v_j(\mathbf{x}_0) \le 0$$

for $j = 1, 2, \dots, q$. This means that point $\mathbf{x}_0$ is also feasible for the convex problem in Eq. (15.27) and, as a consequence, the CCP can be performed. Now if iterate $\mathbf{x}_k$ is feasible for the initial problem and point $\mathbf{x}_{k+1}$ is obtained by solving the convex problem in Eq. (15.27), then we need to show that $\mathbf{x}_{k+1}$ is also feasible for the initial problem. We note that as $\mathbf{x}_{k+1}$ is a solution of the problem in Eq. (15.27), it is a feasible point of the problem and hence

$$u_j(\mathbf{x}_{k+1}) - \hat{v}_j(\mathbf{x}_{k+1}, \mathbf{x}_k) \le 0 \quad \text{for } j = 1, 2, \dots, q \tag{15.29}$$

In addition, since function $v_j(\mathbf{x})$ is convex, we can write

$$v_j(\mathbf{x}) \geq v_j(\mathbf{x}_k) + \nabla^T v_j(\mathbf{x}_k)(\mathbf{x} - \mathbf{x}_k) = \hat{v}_j(\mathbf{x}, \mathbf{x}_k)$$

which implies that $v_j(\mathbf{x}_{k+1}) \geq \hat{v}_j(\mathbf{x}_{k+1}, \mathbf{x}_k)$. This in conjunction with Eq. (15.29) leads to the inequality

$$u_j(\mathbf{x}_{k+1}) - v_j(\mathbf{x}_{k+1}) \leq u_j(\mathbf{x}_{k+1}) - \hat{v}_j(\mathbf{x}_{k+1}, \mathbf{x}_k) \leq 0$$

for $j = 1, 2, \ldots, q$. Therefore, $\mathbf{x}_{k+1}$ satisfies the constraints in Eq. (15.25b) and hence it is feasible for the initial problem.

The second issue is whether the sequence generated, $\{\mathbf{x}_k, k = 0, 1, \ldots\}$, converges. It can be shown that the objective function of the initial problem, namely, $u_0(\mathbf{x}) - v_0(\mathbf{x})$, decreases monotonically for $\{\mathbf{x}_k, k = 0, 1, \ldots\}$ and thus the CCP is a descent method [17]. To show this, let $f_k = u_0(\mathbf{x}_k) - v_0(\mathbf{x}_k)$ and note that

$$f_k = u_0(\mathbf{x}_k) - v_0(\mathbf{x}_k) = u_0(\mathbf{x}_k) - \hat{v}_0(\mathbf{x}_k, \mathbf{x}_k) \geq u_0(\mathbf{x}_{k+1}) - \hat{v}_0(\mathbf{x}_{k+1}, \mathbf{x}_k) \quad (15.30)$$

Since $v_0(\mathbf{x})$ is convex, we have $v_0(\mathbf{x}_{k+1}) \geq \hat{v}_0(\mathbf{x}_{k+1}, \mathbf{x}_k)$ and hence

$$u_0(\mathbf{x}_{k+1}) - \hat{v}_0(\mathbf{x}_{k+1}, \mathbf{x}_k) \geq u_0(\mathbf{x}_{k+1}) - v_0(\mathbf{x}_{k+1}) = f_{k+1}$$

which in conjunction with Eq. (15.30) gives $f_{k+1} \leq f_k$. In addition, it can be shown that the sequence $\{\mathbf{x}_k, k = 0, 1, \ldots\}$ converges to a critical point[1] of the initial problem in Eqs. (15.25a) and (15.25b) [18]. Based on the above considerations, a CCP algorithm for the general problem in Eqs. (15.25a) and (15.25b) can be constructed as follows.

Algorithm 15.7 CCP for the problem in Eqs. (15.25a) and (15.25b)
Step 1
Input a feasible initial point $\mathbf{x}_0$ and a tolerance $\varepsilon > 0$ and set $k = 0$.
Step 2
Solve the problem in Eq. (15.27) and denote the solution as $\mathbf{x}_{k+1}$.
Step 3
If $\left| \frac{f(\mathbf{x}_{k+1}) - f(\mathbf{x}_k)}{f(\mathbf{x}_k)} \right| < \varepsilon$, stop and output $\mathbf{x}_{k+1}$ as the solution; otherwise, set $k = k + 1$ and repeat from Step 2.

15.4.2 Penalty Convex-Concave Procedure

In Algorithm 15.7, an initialization point $\mathbf{x}_0$ is required that satisfies the constraints in Eq. (15.25b). For many practical problems, finding a suitable initial point $\mathbf{x}_0$ is part of the formulation of Eqs. (15.25a) and (15.25b) and for problems with a large number of constraints, the task can be challenging. In what follows, we describe a modified CCP, known as *penalty CCP* (PCCP) [17] that accepts an initial point $\mathbf{x}_0$ which does not need to be feasible for the problem in Eqs. (15.25a) and (15.25b). This can be done by introducing nonnegative slack variables $\{s_j, j = 1, 2, \ldots, q\}$

---

[1] In the sense that it satisfies the KKT conditions (see Chap. 10).

into the formulation, which modifies the CCP formulation in Eqs. (15.25a) and (15.25b) to

$$\text{minimize } u_0(\mathbf{x}) - \hat{v}_0(\mathbf{x}, \mathbf{x}_k) + \tau_k \sum_{j=1}^{q} s_j \tag{15.31a}$$

$$\text{subject to: } u_j(\mathbf{x}) - \hat{v}_j(\mathbf{x}, \mathbf{x}_k) \le s_j \text{ for } j = 1, 2, \ldots, q \tag{15.31b}$$

$$s_j \ge 0 \text{ for } j = 1, 2, \ldots, q \tag{15.31c}$$

To explain the role the slack variables play, consider an initial point $\mathbf{x}_0$ which violates only one of the constraints in Eq. (15.25b), say the $\tilde{j}$-th constraint, i.e.,

$$z_{\tilde{j}} \triangleq u_{\tilde{j}}(\mathbf{x}_0) - v_{\tilde{j}}(\mathbf{x}_0) = u_{\tilde{j}}(\mathbf{x}_0) - \hat{v}_{\tilde{j}}(\mathbf{x}_0, \mathbf{x}_0) > 0$$

For the modified formulation in Eqs. (15.31a)–(15.31c), point $\mathbf{x}_0$ is acceptable because Eqs. (15.31a) and (15.31c) involve both $\mathbf{x}$ and $\mathbf{s} = [s_1 \ s_2 \ \cdots \ s_q]^T$ as the design variables and an initial point $\{\mathbf{x}_0, \mathbf{s}_0\}$ is obviously feasible as long as $\mathbf{s}_0 \ge \mathbf{0}$ and its $\tilde{j}$-th component is no less than $z_{\tilde{j}}$. Once the algorithm is started with an easy-to-find feasible initial point, the values of $\{s_j, j = 1, 2, \ldots, q\}$ have to be reduced, gradually approaching zero, so that the iterates obtained by solving the convex problem in Eqs. (15.31a)–(15.31c) approach the solution of the problem in Eq. (15.27). The gradual reduction of slack variable s_j is achieved by including a penalty term, $\tau_k \sum_{j=1}^{q} s_j$, in the objective function in Eq. (15.31a) where $\tau_k > 0$ is a scalar weight that controls the trade-off between the original objective function and the penalty term. The algorithm can be started with a reasonable value τ_0, say $\tau_0 = 1$, and the value of τ increased monotonically as the algorithm proceeds until it reaches an upper limit $\tau_{\max}$. In this way, the penalty term in Eq. (15.31a) forces a quick reduction of $\{s_j, j = 1, 2, \ldots, q\}$. As soon as all s_j's are reduced to zero, they remain zero because of the presence of the penalty term in the objective function. From this point on, the problem in Eqs. (15.31a)–(15.31c) is practically identical to that in Eq. (15.27). The upper limit $\tau_{\max}$ is imposed to avoid numerical difficulties that may occur if τ_k becomes too large [17]. These considerations lead to the following PCCP-based algorithm for the problem in Eqs. (15.25a) and (15.25b).

Algorithm 15.8 Penalty convex-concave procedure for the problem in Eqs. (15.25a) and (15.25b)

Step 1
Input an initial point $\mathbf{x}_0$, $\tau_0 = 1$, $\tau_{\max} > 0$, $\mu > 1$, and a tolerance $\varepsilon > 0$ and set $k = 0$.

Step 2
Generate a feasible initial point $\{\mathbf{x}_0, \mathbf{s}_0\}$ where $\mathbf{s}_0 = [s_1^{(0)} \ s_2^{(0)} \ \cdots \ s_q^{(0)}]^T$ with $s_j^{(0)} = \max\{0.1, u_j(\mathbf{x}_0) - v_j(\mathbf{x}_0) + 0.1\}$ for $j = 1, 2, \ldots, q$.

Step 3
Solve the problem in Eqs. (15.31a)–(15.31c) and denote the solution as $\mathbf{x}_{k+1}$.

Step 4
If $\left| \frac{f(\mathbf{x}_{k+1}) - f(\mathbf{x}_k)}{f(\mathbf{x}_k)} \right| < \varepsilon$, stop and output $\mathbf{x}_{k+1}$ as the solution; otherwise, set $\tau_{k+1} = \min\{\mu \tau_k, \tau_{\max}\}$, $k = k + 1$, and repeat from Step 2.

Example 15.5 Solve the constrained problem in Example 15.1, i.e.,

$$\text{minimize } f(\mathbf{x}) = \ln(1 + x_1^2) + x_2$$
$$\text{subject to: } a(\mathbf{x}) = (1 + x_1^2)^2 + x_2^2 - 4 = 0$$
$$c(\mathbf{x}) = x_1 + x_2^2 + 0.3 \le 0$$

by using Algorithm 15.8.

Solution First, we reformulate the problem at hand into the form of Eqs. (15.25a)–(15.25b) as

$$\text{minimize } f(\mathbf{x}) = \ln(1 + x_1^2) + x_2 \tag{15.32a}$$

$$c_1(\mathbf{x}) = 4 - (1 + x_1^2)^2 - x_2^2 \le 0 \tag{15.32b}$$
$$c_2(\mathbf{x}) = (1 + x_1^2)^2 + x_2^2 - 4 \le 0 \tag{15.32c}$$
$$c_3(\mathbf{x}) = x_1 + x_2^2 + 0.3 \le 0 \tag{15.32d}$$

The objective and constraint functions $f(\mathbf{x})$ and $c_1(\mathbf{x})$ are not convex. The objective function can be expressed as $f(\mathbf{x}) = u_0(\mathbf{x}) - v_0(\mathbf{x})$ where $u_0(\mathbf{x}) = \ln(1 + x_1^2) + \frac{1}{2}x_1^2 + x_2$ and $v_0(\mathbf{x}) = \frac{1}{2}x_1^2 + x_2$ are convex because the Hessians

$$\nabla^2 u_0(\mathbf{x}) = \text{diag}\left\{\frac{x_1^4 + 3}{(x_1^2 + 1)^2}, 0\right\} \quad \text{and} \quad \nabla^2 v_0(\mathbf{x}) = \text{diag}\{1, 0\}$$

are positive semidefinite. The constraint function $c_1(\mathbf{x})$ is in the form $c_1(\mathbf{x}) = u_1(\mathbf{x}) - v_1(\mathbf{x}) \le 0$ with $u_1(\mathbf{x}) = 4$ and $v_1(\mathbf{x}) = (1 + x_1^2)^2 + x_2^2$ both of which are obviously convex. In the kth iteration of the algorithm, affine approximations for $v_0(\mathbf{x})$ and $v_1(\mathbf{x})$ at $\mathbf{x}_k$ are given by

$$\hat{v}_0(\mathbf{x}, \mathbf{x}_k) = \frac{1}{2}(x_1^{(k)})^2 + [x_1^{(k)} \ 0](\mathbf{x} - \mathbf{x}_k)$$

and

$$\hat{v}_1(\mathbf{x}, \mathbf{x}_k) = v_1(\mathbf{x}_k) + [4x_1^{(k)}(1 + (x_1^{(k)})^2) \ 2x_2^{(k)}](\mathbf{x} - \mathbf{x}_k)$$

respectively. Because of the convexity of $c_2(\mathbf{x})$ and $c_3(\mathbf{x})$, the last two constraints do not require convex approximations and we simply let $u_2(\mathbf{x}) = c_2(\mathbf{x})$, $v_2(\mathbf{x}) = 0$, $u_3(\mathbf{x}) = c_3(\mathbf{x})$, and $v_3(\mathbf{x}) = 0$. Consequently, the convex problem to be solved in the kth iteration assumes the form

$$\text{minimize } \ln(1 + x_1^2) + \frac{1}{2}x_1^2 + x_2 - \frac{1}{2}(x_1^{(k)})^2 - [x_1^{(k)} \ 0](\mathbf{x} - \mathbf{x}_k) + \tau_k \sum_{j=1}^{3} s_j$$
$$\text{subject to: } 4 - v_1(\mathbf{x}_k) - [4x_1^{(k)}(1 + (x_1^{(k)})^2) \ 2x_2^{(k)}](\mathbf{x} - \mathbf{x}_k) \le s_1$$
$$(1 + x_1^2)^2 + x_2^2 - 4 \le s_2$$
$$x_1 + x_2^2 + 0.3 \le s_3$$
$$s_j \ge 0 \quad \text{for } j = 1, 2, 3.$$

$$\tag{15.33}$$

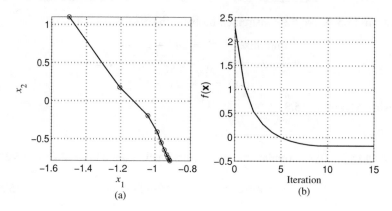

Fig. 15.7 a Trajectory of the solution produced by Algorithm 15.8, **b** objective function versus the number of iterations (Example 15.5)

To examine how the algorithm handles an initial point $\mathbf{x}_0$ that violates some of the constraints in the problem of Eqs. (15.32b)–(15.32d), we chose $\mathbf{x}_0 = [1.5\ 1.2]^T$ which satisfies the constraint in Eq. (15.32b) but violates the constraints in Eqs. (15.32c) and (15.32d). With the specified initial point $\mathbf{x}_0$, Step 2 of the algorithm gave an initial slack variable $\mathbf{s}_0 = [0.1\ 7.8725\ 0.11]^T$ which assured that the initial point $\{\mathbf{x}_0, \mathbf{s}_0\}$ was feasible for the convex problem in Eq. (15.33). With $\tau_0 = 1$, $\tau_{\max} = 10^3$, $\mu = 1.5$, and $\varepsilon = 10^{-8}$, it took Algorithm 15.8 fourteen iterations to converge to the solution $\mathbf{x}^* = [-0.9162420\ -0.7850108]^T$ and the value of the objective function at the solution was $f^* = -1.755174 \times 10^{-1}$. The values of $a(\mathbf{x}^*)$ and $c(\mathbf{x}^*)$ were $-1.775460 \times 10^{-10}$ and -6.922224×10^{-9}, respectively, which means that the specified constraints are, for all practical purposes, satisfied at the solution. Figure 15.7 shows the trajectory of the solution and the value of the objective function versus the number of iterations. ∎

15.5 ADMM Heuristic Technique for Nonconvex Problems

Some alternating direction methods of multipliers (ADMM) for convex problems have been studied in Sect. 14.5. In this section, ADMM is used as a heuristic technique for some nonconvex problems. We consider the class of constrained problems of the form

$$\begin{aligned} \text{minimize} \quad & f(\mathbf{x}) \\ \text{subject to:} \quad & \mathbf{x} \in C \end{aligned} \qquad (15.34)$$

where function $f(\mathbf{x})$ is convex but the feasible region C is nonconvex [19, Sect. 9.1]. In effect, Eq. (15.34) represents a class of nonconvex problems. On comparing the problem formulations in Eq. (15.34) and Eq. (14.91), we note that the two formulations are the same except for the convexity of the feasible region involved: set C is convex in Eq. (14.91) but not in Eq. (15.34). It is, therefore, reasonable to expect that an ADMM heuristic technique can be developed by extending the techniques used for the problem in Eq. (14.91) to the problem in Eq. (15.34).

To start with, the problem in Eq. (15.34) can be reformulated as

$$\text{minimize } f(\mathbf{x}) + I_C(\mathbf{x})$$

where $I_C(\mathbf{x})$ is the indicator function associated with set C (see Sect. 14.5.2). Next, a new variable $\mathbf{y}$ can be introduced in order to split the objective function into two parts. After these modifications, the problem under consideration would be converted to

$$\begin{aligned} \text{minimize } & f(\mathbf{x}) + I_C(\mathbf{y}) \\ \text{subject to: } & \mathbf{x} - \mathbf{y} = \mathbf{0} \end{aligned} \tag{15.35}$$

and as in the scaled ADMM method in Sect. 14.5.2, Eq. (15.35) suggests the scaled ADMM iterations

$$\mathbf{x}_{k+1} = \arg\min_{\mathbf{x}}[f(\mathbf{x}) + \frac{\alpha}{2}\|\mathbf{x} - \mathbf{y}_k + \mathbf{v}_k\|_2^2]$$

$$\mathbf{y}_{k+1} = \arg\min_{\mathbf{y}}[I_C(\mathbf{y}) + \frac{\alpha}{2}\|\mathbf{y} - (\mathbf{x}_{k+1} + \mathbf{v})\|_2^2]$$

$$\mathbf{v}_{k+1} = \mathbf{v}_k + \mathbf{x}_{k+1} - \mathbf{y}_{k+1}$$

where the $\mathbf{x}$-minimization is obviously a convex problem because $f(\mathbf{x})$ is convex while the $\mathbf{y}$-minimization can be performed by minimizing $\|\mathbf{y} - (\mathbf{x}_{k+1} + \mathbf{v}_k)\|_2$ subject to $\mathbf{y} \in C$. This means that $\mathbf{y}_{k+1}$ can be computed by projecting $\mathbf{x}_{k+1} + \mathbf{v}_k$ onto set C, and hence the ADMM iterations can be expressed as

$$\mathbf{x}_{k+1} = \arg\min_{\mathbf{x}}[f(\mathbf{x}) + \frac{\alpha}{2}\|\mathbf{x} - \mathbf{y}_k + \mathbf{v}_k\|_2^2] \tag{15.36a}$$

$$\mathbf{y}_{k+1} = P_C(\mathbf{x}_{k+1} + \mathbf{v}_k) \tag{15.36b}$$

$$\mathbf{v}_{k+1} = \mathbf{v}_k + \mathbf{x}_{k+1} - \mathbf{y}_{k+1} \tag{15.36c}$$

where $P_C(\mathbf{z})$ denotes the projection of point $\mathbf{z}$ onto nonconvex set C. The difference between Eq. (15.36b) and Eq. (14.94b) is the projection in Eq. (15.36b), which is difficult to calculate in general as it involves a nonconvex feasible region C. However, as demonstrated in [19, Sect. 9.1], there are several important cases where the projection in Eq. (15.36b) can be carried out precisely.

The preceding analysis leads to the ADMM-based algorithm for the nonconvex problem in Eq. (15.34) as detailed below.

Algorithm 15.9 Scaled ADMM for the Problem in Eq. (15.1)
Step 1
Input parameter $\alpha > 0$, $\mathbf{y}_0$, $\mathbf{v}_0$, and tolerances $\varepsilon_p > 0$ and $\varepsilon_d > 0$ and set $k = 0$.
Step 2
Compute

$$\mathbf{x}_{k+1} = \arg\min_{\mathbf{x}}[f(\mathbf{x}) + \frac{\alpha}{2}\|\mathbf{x} - \mathbf{y}_k + \mathbf{v}_k\|_2^2]$$

$$\mathbf{y}_{k+1} = P_C(\mathbf{x}_{k+1} + \mathbf{v}_k)$$

$$\mathbf{v}_{k+1} = \mathbf{v}_k + \mathbf{x}_{k+1} - \mathbf{y}_{k+1}$$

Step 3
Compute the dual and primal residuals as

$$\mathbf{d}_k = -\alpha(\mathbf{y}_k - \mathbf{y}_{k-1})$$

and

$$\mathbf{r}_k = \mathbf{x}_{k+1} - \mathbf{y}_{k+1}$$

Step 4
If

$$\|\mathbf{r}_k\|_2 < \varepsilon_p \quad \text{and} \quad \|\mathbf{d}_k\|_2 < \varepsilon_d$$

output $(\mathbf{x}_{k+1}, \mathbf{y}_{k+1})$ as the solution and stop; otherwise, set $k = k + 1$ and repeat from Step 2.

Example 15.6 Solve the constrained problem

$$\text{minimize } f(\mathbf{x}) = x_2^2 - 2x_1 + x_2$$
$$\text{subject to: } x_1^2 + x_2^2 - 9 = 0$$

by using Algorithm 15.9.

Solution We start by writing the objective function in Eq. (15.36a) as

$$x_2^2 - 2x_1 + x_2 + \frac{\alpha}{2}\|\mathbf{x} - \mathbf{y}_k + \mathbf{v}_k\|_2^2$$

$$= \mathbf{x}^T \begin{bmatrix} \alpha/2 & 0 \\ 0 & 1 + \alpha/2 \end{bmatrix} \mathbf{x} - \mathbf{x}^T \left(\alpha(\mathbf{y}_k - \mathbf{v}_k) - \begin{bmatrix} 2 \\ -1 \end{bmatrix}\right)$$

By minimizing the above objective function, we obtain

$$\mathbf{x}_{k+1} = \begin{bmatrix} 1/\alpha & 0 \\ 0 & 1/(2 + \alpha) \end{bmatrix} \left(\alpha(\mathbf{y}_k - \mathbf{v}_k) - \begin{bmatrix} 2 & -1 \end{bmatrix}\right)$$

The feasible region C is a circle with center at the origin and radius 3, namely,

$$C = \{\mathbf{x} : x_1^2 + x_2^2 = 9\} \tag{15.37}$$

The step in Eq. (15.36b) is carried out by projecting point $\mathbf{x}_{k+1} + \mathbf{v}_k$ onto circle C. If we let the two coordinates of $\mathbf{x}_{k+1} + \mathbf{v}_k$ be p_1 and p_2, respectively, and the two

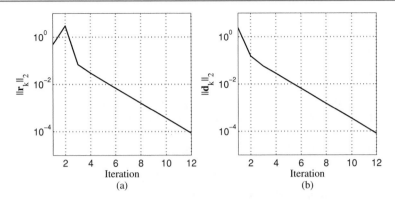

Fig. 15.8 **a** 2-norm of primal residual $\|\mathbf{r}_k\|_2$, **b** 2-norm of dual residual $\|\mathbf{d}_k\|_2$ (Example 15.6)

coordinates of the projection $P_C(\mathbf{x}_{k+1} + \mathbf{v}_k)$ be q_1 and q_2, then it can readily be verified that (i) if $p_1 = 0$ and $p_2 > 0$, then $q_1 = 0$ and $q_2 = 3$; (ii) if $p_1 = 0$ and $p_2 < 0$, then $q_1 = 0$ and $q_2 = -3$; (iii) if $p_1 > 0$, then $q_1 = t$ and $q_2 = t \cdot p_2/p_1$; and (iv) if $p_1 < 0$, then $q_1 = -t$ and $q_2 = -t \cdot p_2/p_1$, where $t = 3/\sqrt{1 + (p_2/p_1)^2}$.

With $\alpha = 0.8$, $\varepsilon_p = 10^{-4}$, and $\varepsilon_d = 10^{-4}$, it took Algorithm 15.9 twelve iterations to converge to a vector $\hat{\mathbf{x}}^*$, which was projected onto the feasible region C to yield the solution $\mathbf{x}^* = [2.976563 \ -0.3742591]^T$. The 2-norms of the primal residual $\mathbf{r}_k$ and dual residual $\mathbf{d}_k$ versus the number of iterations are plotted in Fig. 15.8. ∎

Problems

15.1 Applying Algorithm 15.3, solve the problem

$$\text{minimize } f(\mathbf{x}) = \ln(1 + x_1^2) + x_2$$
$$\text{subject to: } a(\mathbf{x}) = (1 + x_1^2)^2 + x_2^2 - 4 = 0$$

Start with $\mathbf{x}_0 = [1 \ 1]^T$ and discuss the solution obtained in terms of reduction of the value of the objective function relative to $f(\mathbf{x}_0)$ and satisfaction of the constraint.

15.2 Applying Algorithm 15.6, solve the problem in Problem 15.1 with the same initial point. Discuss the solution obtained in terms of reduction of the value of the objective function relative to $f(\mathbf{x}_0)$ and satisfaction of the constraint.

15.3 Applying Algorithm 15.3, solve the problem

$$\text{minimize } f(\mathbf{x}) = x_1^4 + 2x_2^4 + x_3^4 - x_1^2 x_2^2 - x_1^2 x_3^2$$
$$\text{subject to: } a_1(\mathbf{x}) = x_1^4 + x_2^4 + x_3^4 - 25 = 0$$
$$a_2(\mathbf{x}) = 8x_1^2 + 14x_2^2 + 7x_3^2 - 56 = 0$$

Start with $\mathbf{x}_0 = [3\ 2\ 3]^T$ and discuss the solution obtained in terms of reduction of the value of the objective function relative to $f(\mathbf{x}_0)$ and satisfaction of the constraints.

15.4 Applying Algorithm 15.6, solve the problem in Problem 15.3 with the same initial point and discuss the solution obtained in terms of reduction of the value of the objective function relative to $f(\mathbf{x}_0)$ and satisfaction of the constraints.

15.5 Applying Algorithm 15.3, solve the problem [14, Chap. 18]

$$\text{minimize } f(\mathbf{x}) = e^{x_1 x_2 x_3 x_4 x_5} - \tfrac{1}{2}(x_1^3 + x_2^3 + 1)^2$$
$$\text{subject to: } a_1(\mathbf{x}) = x_1^2 + x_2^2 + x_3^2 + x_4^2 + x_5^2 - 11 = 0$$
$$a_2(\mathbf{x}) = x_2 x_3 - 5 x_4 x_5 = 0$$
$$a_3(\mathbf{x}) = x_1^3 + x_2^3 + 1 = 0$$

Start with $\mathbf{x}_0 = [-1.71\ \ 1.59\ \ 1.82\ \ -0.763\ \ -0.763]^T$ and discuss the solution obtained in terms of reduction of the value of the objective function relative to $f(\mathbf{x}_0)$ and satisfaction of the constraints.

15.6 Applying Algorithm 15.6, solve the problem in Problem 15.5 with the same initial point and discuss the solution obtained in terms of reduction of the value of the objective function relative to $f(\mathbf{x}_0)$ and satisfaction of the constraints.

15.7 Applying Algorithm 15.7, solve the constrained problem

$$\text{minimize } f(\mathbf{x}) = -2 x_1 x_2$$
$$\text{subject to: } c_1(\mathbf{x}) = 0.2 x_1^2 + 0.4 x_2^2 - 1 \le 0$$

(a) Start with $\mathbf{x}_0 = [0\ 0]^T$ and discuss the solution obtained in terms of reduction of the value of the objective function relative to $f(\mathbf{x}_0)$ and satisfaction of the constraint.

(b) Start with $\mathbf{x}_0 = [-2\ \ -2]^T$ and discuss the solution obtained in terms of reduction of the value of the objective function relative to $f(\mathbf{x}_0)$ and satisfaction of the constraint.

15.8 Applying Algorithm 15.8, solve the constrained problem

$$\text{minimize } f(\mathbf{x}) = (x_1 - 1)^2 + (x_2 - 0.5)^2$$
$$\text{subject to: } c_1(\mathbf{x}) = 2 x_1^2 - x_2 \le 0$$
$$c_2(\mathbf{x}) = x_1 - 2 x_2^2 \le 0$$

Start with $\mathbf{x}_0 = [0.2\ \ -0.2]^T$ and discuss the solution obtained in terms of reduction of the value of the objective function relative to $f(\mathbf{x}_0)$ and satisfaction of the constraints.

15.9 Applying Algorithm 15.8, solve the constrained problem

$$\text{minimize} \ \ f(\mathbf{x}) = x_1^2 + 4x_1 x_2 + 3x_2^2$$
$$\text{subject to:} \ c_1(\mathbf{x}) = -x_1^2 - x_2 \le 0$$
$$c_2(\mathbf{x}) = x_1^2 + x_2^2 - 1 \le 0$$

Start with $\mathbf{x}_0 = [-0.5 \ \ 4]^T$ and discuss the solution obtained in terms of reduction of the value of the objective function relative to $f(\mathbf{x}_0)$ and satisfaction of the constraints.

15.10 Verify that the KKT conditions of the QP problem in Eqs. (15.16a)–(15.16c) are given by Eq. (15.14).

15.11 Applying Algorithm 15.6, solve the problem

$$\text{minimize} \ \ f(\mathbf{x}) = 0.01x_1^2 + x_2^2$$
$$\text{subject to:} \ c_1(\mathbf{x}) = x_1 x_2 - 25 \le 0$$
$$c_2(\mathbf{x}) = -x_1^2 - x_2^2 + 25 \le 0$$
$$c_3(\mathbf{x}) = -x_1 + 2 \le 0$$

Start with $\mathbf{x}_0 = [15 \ \ 5]^T$ and discuss the solution obtained in terms of reduction of the value of the objective function relative to $f(\mathbf{x}_0)$ and satisfaction of the constraints.

15.12 Applying Algorithm 15.8, solve the problem in Problem 15.11 with the same initial point as in Problem 15.11 and discuss the solution obtained in terms of reduction of the value of the objective function relative to $f(\mathbf{x}_0)$ and satisfaction of the constraints.

15.13 Applying Algorithm 15.6, solve the constrained problem

$$\text{minimize} \ \ f(\mathbf{x}) = 2x_1 + x_2^2$$
$$\text{subject to:} \ a_1(\mathbf{x}) = 4x_1^2 + x_2^2 - 9 = 0$$
$$\mathbf{A}\mathbf{x} \le \mathbf{b}$$

where

$$\mathbf{A} = \begin{bmatrix} 0 & -1 \\ 0 & 1 \\ -2 & 0 \\ 2 & 0 \end{bmatrix} \quad \text{and} \quad \mathbf{b} = \begin{bmatrix} -1 \\ 5 \\ -2 \\ 4 \end{bmatrix}$$

Start with $\mathbf{x}_0 = [6 \ -2]^T$ and discuss the solution obtained in terms of reduction of the value of the objective function relative to $f(\mathbf{x}_0)$ and satisfaction of the constraints.

15.14 Applying Algorithm 15.8, solve the problem in Problem 15.13 with the same initial point as in Problem 15.13 and discuss the solution obtained in terms of reduction of the value of the objective function relative to $f(\mathbf{x}_0)$ and satisfaction of the constraints.

15.15 Applying Algorithm 15.6, solve the constrained problem

$$\text{minimize} \ \ f(\mathbf{x}) = 2x_1^2 + 3x_2^2$$
$$\text{subject to: } a_1(\mathbf{x}) = x_1^2 + x_2^2 - 16 = 0$$
$$\mathbf{Ax} \le \mathbf{b}$$

where

$$\mathbf{A} = \begin{bmatrix} -1 & 0 \\ 1 & 0 \\ 0 & -1 \\ 0 & 1 \end{bmatrix} \quad \text{and} \quad \mathbf{b} = \begin{bmatrix} -2 \\ 5 \\ -1 \\ 5 \end{bmatrix}$$

Start with $\mathbf{x}_0 = [-1 \ \ -2]^T$ and discuss the solution obtained in terms of reduction of the value of the objective function relative to $f(\mathbf{x}_0)$ and satisfaction of the constraints.

15.16 Applying Algorithm 15.8, solve the problem in Problem 15.15 with the same initial point as in Problem 15.15 and discuss the solution obtained in terms of reduction of the value of the objective function relative to $f(\mathbf{x}_0)$ and satisfaction of the constraints.

15.17 Applying Algorithm 15.9, solve the least-squares problem

$$\text{minimize} \ \ f(\mathbf{x}) = \tfrac{1}{2}\|\mathbf{Ax} - \mathbf{b}\|_2^2$$
$$\text{subject to: } x_i \in \{1, -1\} \ \ \text{for } i = 1, \ 2, \ \ldots, \ n$$

where $\mathbf{A} \in R^{m \times n}$ and $\mathbf{b} \in R^{m \times 1}$ with $m = 50$ and $n = 12$ can be generated using the following MATLAB code:

```
randn('state',17)
A = randn(50,12);
randn('state',6)
b = 5*randn(50,1);
```

References

1. S. Boyd and J. Duchi, "EE364b: Convex Optimization II, Lectures, Stanford University." [Online]. Available: http://stanford.edu/class/ee364b/
2. G. B. Dantzig, *Linear Programming and Extensions*. Princeton, NJ: Princeton University Press, 1963.

3. D. D. Lee and H. S. Seung, "Learning the parts of objects by non-negative matrix factorization," *Nature*, vol. 401, pp. 788–791, Oct. 21 1999.

4. N. Gillis, "The why and how of nonnegative matrix factorization," in *Regularization, Optimization, Kernels, and Support Vector Machines*, M. S. J. A. K. Suykens and A. Argyriou, Eds. Boca Raton, FL: CRC Press, 2014, pp. 257–291.

5. "Pattern classification database of the center for biological and computational learning at MIT." [Online]. Available: https://github.com/HyTruongSon/Pattern-Classification/tree/master/MIT-CBCL-database

6. R. B. Wilson, "A simplicial algorithm for concave programming," Ph.D. dissertation, Graduate School of Business Administration, Harvard University, Cambridge, MA, 1963.

7. P. T. Boggs and J. W. Tolle, "Sequential quadratic programming," *Acta Numerica*, vol. 4, pp. 1–52, 1995.

8. S. P. Han, "A globally convergent method for nonlinear programming," *J. Optimization Theory and Applications*, vol. 22, pp. 297–309, Jul. 1977.

9. R. H. Byrd and J. Nocedal, "An analysis of reduced Hessian methods for constrained optimization," *Math. Programming*, vol. 49, pp. 285–323, 1991.

10. R. Fletcher, "A class of methods for nonlinear programming, III: Rates of convergence," in *Numerical Methods for Nonlinear Optimization*, F. A. Lootsma, Ed. New York: Academic Press, 1972, pp. 371–382.

11. M. J. D. Powell and Y. Yuan, "A recursive quadratic programming algorithm that uses differentiable exact penalty functions," *Math. Programming*, vol. 35, pp. 265–278, 1986.

12. P. T. Boggs and J. W. Tolle, "A strategy for global convergence in sequential quadratic programming algorithm," *SIAM J. Numerical Analysis*, vol. 21, pp. 600–623, 1989.

13. M. J. D. Powell, "A fast algorithm for nonlinearly constrained optimization calculations," in *Numerical Analysis*, 1978, vol. 630 of the series Lecture Notes in Mathematics, pp. 144–157.

14. J. N. Nocedal and S. J. Wright, *Numerical Optimization*, 2nd ed. New York: Springer, 2006.

15. A. L. Yuille and A. Rangarajan, "The concave-convex procedure," *Neural Computation*, vol. 15, no. 4, pp. 915–936, 2003.

16. P. Hartman, "On functions representable as a difference of convex functions," *Pacific J. of Math.*, vol. 9, no. 3, pp. 707–713, 1959.

17. T. Lipp and S. Boyd, "Variations and extensions of the convex-concave procedure," *Optimization and Engineering*, vol. 17, no. 2, pp. 263–287, 2016.

18. G. R. Lanckreit and B. K. Sriperumbudur, "On the convergence of the concave-convex procedure," in *Advances in Neural Information Processing Systems*, 2009, pp. 1759–1767.

19. S. Boyd, N. Parikh, E. Chu, B. Peleato, and J. Eckstein, "Distributed optimization and statistical learning via the alternating direction method of multipliers," *Foundations and Trends in Machine Learning*, vol. 3, no. 1, pp. 1–122, 2010.

Applications of Constrained Optimization

16

16.1 Introduction

Constrained optimization provides a general framework in which a variety of design criteria and specifications can be readily imposed on the required solution. Usually, a multivariable objective function that quantifies a performance measure of a design can be identified. This objective function may be linear, quadratic, or highly nonlinear, and usually it is differentiable so that its gradient and sometimes Hessian can be evaluated. In a real-life design problem, the design is carried out under certain physical limitations with limited resources. If these limitations can be quantified as equality or inequality constraints on the design variables, then a constrained optimization problem can be formulated whose solution leads to an optimal design that satisfies the limitations imposed. Depending on the degree of nonlinearity of the objective function and constraints, the problem at hand can be a linear programming (LP), quadratic programming (QP), convex programming, semidefinite programming (SDP), second-order cone programming (SOCP), or general nonlinear constrained optimization problem.

This chapter is devoted to several applications of some of the constrained optimization algorithms studied in Chaps. 11–15 in the areas of digital signal processing, control, robotics, and telecommunications. In Sect. 16.2, we show how constrained algorithms of various types, e.g., LP, QP, CP, SDP, and SOCP algorithms, can be utilized for the design of FIR and IIR digital filters. Filters that would satisfy multiple specifications such as maximum passband gain, minimum stopband gain, maximum transition-band gain, and maximum pole radius can be designed with these methods. The authors draw from their extensive research experience on the subject [1,2]. Section 16.3 introduces several models for uncertain dynamic systems and develops an effective control strategy known as model predictive control for this class of systems, which involves the use of SDP. In Sect. 16.4, LP and SDP are applied to solve a problem that entails optimizing the grasping force distribution for dextrous robotic hands. In Sect. 16.5, an SDP-based method for multiuser detection and a CP approach to minimize bit-error rate for wireless communication systems is described.

© Springer Science+Business Media, LLC, part of Springer Nature 2021
A. Antoniou and W.-S. Lu, *Practical Optimization*, Texts in Computer Science,
https://doi.org/10.1007/978-1-0716-0843-2_16

16.2   Design of Digital Filters

16.2.1   Design of Linear-Phase FIR Filters Using QP

In many applications of digital filters in communication systems, it is often desirable to design linear-phase finite-duration impulse response (FIR) digital filters with a specified maximum passband error, δ_p, and/or a specified maximum stopband gain, δ_a [3] (see Sect. B.9.1). FIR filters of this class can be designed relatively easily by using a QP approach as described below.

For the sake of simplicity, we consider the problem of designing a linear-phase lowpass FIR filter of even order N (odd length $N+1$) with normalized passband and stopband edges ω_p and ω_a, respectively (see Sect. B.9.1). The frequency response of such a filter can be expressed as

$$H(e^{j\omega}) = e^{-j\omega N/2} A(\omega)$$

as in Eq. (9.34) and the desired amplitude response, $A_d(\omega)$, can be assumed to be of the form

$$A_d(\omega) = \begin{cases} 1 & \text{for } 0 \le \omega \le \omega_p \\ 0 & \text{for } \omega_a \le \omega \le \pi \end{cases}$$

If we use the piecewise-constant weighting function defined by

$$W(\omega) = \begin{cases} 1 & \text{for } 0 \le \omega \le \omega_p \\ \gamma & \text{for } \omega_a \le \omega \le \pi \\ 0 & \text{elsewhere} \end{cases} \tag{16.1}$$

then the objective function $e_l(\mathbf{x})$ in Eq. (9.36) becomes

$$e_l(\mathbf{x}) = \int_0^{\omega_p} [A(\omega) - 1]^2 \, d\omega + \gamma \int_{\omega_a}^{\pi} A^2(\omega) \, d\omega \tag{16.2a}$$

$$= \mathbf{x}^T \mathbf{Q}_l \mathbf{x} - 2\mathbf{x}^T \mathbf{b}_l + \kappa \tag{16.2b}$$

where $\mathbf{Q}_l$ and $\mathbf{b}_l$ are given by Eqs. (9.38a)–(9.38b), respectively. If the weight γ in Eq. (16.2a) is much greater than 1, then minimizing $e_l(\mathbf{x})$ would yield an FIR filter with a minimized least-squares error in the stopband but the passband error would be left largely unaffected. This problem can be fixed by imposing the constraint

$$|A(\omega) - 1| \le \delta_p \quad \text{for } \omega \in [0, \ \omega_p] \tag{16.3}$$

where δ_p is the upper bound on the amplitude of the passband error. With $A(\omega) = \mathbf{c}_l^T(\omega)\mathbf{x}$ where $\mathbf{c}_l(\omega)$ is given by

$$\mathbf{c}_l(\omega) = [1 \ \cos\omega \ \dots \ \cos N\omega/2]^T$$

Equation (16.3) can be written as

$$\mathbf{c}_l^T(\omega)\mathbf{x} \le 1 + \delta_p \quad \text{for } \omega \in [0, \ \omega_p] \tag{16.4a}$$

and

$$- \mathbf{c}_I^T(\omega)\mathbf{x} \leq -1 + \delta_p \quad \text{for } \omega \in [0, \, \omega_p] \tag{16.4b}$$

Note that the frequency variable ω in Eqs. (16.4a)–(16.4b) can assume an infinite set of values in the range 0 to ω_p. A realistic way to implement these constraints is to impose the constraints on a finite set of sample frequencies $\mathcal{S}_p = \{\omega_i^{(p)} : i = 1, 2, \ldots, M_p\}$ in the passband. Under these circumstances, the above constraints can be expressed in matrix form as

$$\mathbf{A}_p \mathbf{x} \leq \mathbf{b}_p$$

where

$$\mathbf{A}_p = \begin{bmatrix} \mathbf{c}_I^T(\omega_1^{(p)}) \\ \vdots \\ \mathbf{c}_I^T(\omega_{M_p}^{(p)}) \\ -\mathbf{c}_I^T(\omega_1^{(p)}) \\ \vdots \\ -\mathbf{c}_I^T(\omega_{M_p}^{(p)}) \end{bmatrix} \quad \text{and} \quad \mathbf{b}_p = \begin{bmatrix} 1 + \delta_p \\ \vdots \\ 1 + \delta_p \\ -1 + \delta_p \\ \vdots \\ -1 + \delta_p \end{bmatrix} \tag{16.5}$$

Additional constraints can be imposed to ensure that the maximum stopband gain, δ_a, is also well controlled. To this end, we impose the constraint

$$|A(\omega)| \leq \delta_a \quad \text{for } \omega \in [\omega_a, \, \pi]$$

A discretized version of the above constraint is given by

$$\mathbf{c}_I^T(\omega)\mathbf{x} \leq \delta_a \quad \text{for } \omega \in \mathcal{S}_a \tag{16.6a}$$
$$-\mathbf{c}_I^T(\omega)\mathbf{x} \leq \delta_a \quad \text{for } \omega \in \mathcal{S}_a \tag{16.6b}$$

where $\mathcal{S}_a = \{\omega_i^{(a)} : i = 1, 2, \ldots, M_a\}$ is a set of sample frequencies in the stopband. The inequality constraints in Eqs. (16.6a)–(16.6b) can be expressed in matrix form as

$$\mathbf{A}_a \mathbf{x} \leq \mathbf{b}_a$$

where

$$\mathbf{A}_a = \begin{bmatrix} \mathbf{c}_I^T(\omega_1^{(a)}) \\ \vdots \\ \mathbf{c}_I^T(\omega_{M_a}^{(a)}) \\ -\mathbf{c}_I^T(\omega_1^{(a)}) \\ \vdots \\ -\mathbf{c}_I^T(\omega_{M_a}^{(a)}) \end{bmatrix} \quad \text{and} \quad \mathbf{b}_a = \delta_a \begin{bmatrix} 1 \\ \vdots \\ 1 \end{bmatrix} \tag{16.7}$$

The design problem can now be formulated as the optimization problem

$$\text{minimize } e(\mathbf{x}) = \mathbf{x}^T \mathbf{Q}_l \mathbf{x} - 2\mathbf{b}_l^T \mathbf{x} + \kappa \tag{16.8a}$$

$$\text{subject to: } \begin{bmatrix} \mathbf{A}_p \\ \mathbf{A}_a \end{bmatrix} \mathbf{x} \leq \begin{bmatrix} \mathbf{b}_p \\ \mathbf{b}_a \end{bmatrix} \tag{16.8b}$$

where $\mathbf{A}_p$ and $\mathbf{b}_p$ are defined in Eq. (16.5), and $\mathbf{A}_a$ and $\mathbf{b}_a$ are defined in Eq. (16.7). There are $(N+2)/2$ design variables in vector $\mathbf{x}$ and $2(M_p + M_a)$ linear inequality constraints in Eq. (16.8b). Since matrix $\mathbf{Q}_l$ is positive definite, the problem under consideration is a convex QP problem that can be solved using the algorithms studied in Chap. 13.

Example 16.1 Applying the above method, design a linear-phase lowpass FIR digital filter that would satisfy the following specifications: passband edge $= 0.45\pi$, stopband edge $= 0.5\pi$, maximum passband error $\delta_p = 0.025$, minimum stopband attenuation $= 40$ dB. Assume idealized passband and stopband gains of 1 and 0, respectively, and a normalized sampling frequency of 2π.

Solution The design was carried out by solving the QP problem in Eqs. (16.8a)–(16.8b) using Algorithm 13.1. We have used a weighting constant $\gamma = 3 \times 10^3$ in Eq. (16.2a) and $\delta_p = 0.025$ in Eq. (16.3). The maximum stopband gain, δ_a, in Eqs. (16.6a)–(16.6b) can be deduced from the minimum stopband attenuation, A_a, as

$$\delta_a = 10^{-0.05 A_a} = 10^{-2}$$

(see Sect. B.9.1). We assumed 80 uniformly distributed sample frequencies with respect to the passband $[0,\ 0.45\pi]$ and 10 sample frequencies in the lower one-tenth of the stopband $[0.5\pi,\ \pi]$, which is usually the most critical part of the stopband, i.e., $M_p = 80$ and $M_a = 10$ in sets $\mathcal{S}_p$ and $\mathcal{S}_a$, respectively.

Unfortunately, there are no analytical methods for predicting the filter order N that would yield a filter which would meet the required specifications but a trial-and-error approach can often be used. Such an approach has resulted in a filter order of 84.

The amplitude of the passband ripple and the minimum stopband attenuation achieved were 0.025 and 41.65 dB, respectively. The amplitude response of the filter is plotted in Fig. 16.1. It is interesting to note that an equiripple error has been achieved with respect to the passband, which is often a desirable feature. ∎

16.2.2 Minimax Design of FIR Digital Filters Using SDP

Linear-phase FIR filters are often designed very efficiently using the so-called *weighted-Chebyshev* method, which is essentially a minimax method based on the *Remez exchange algorithm* [1, Chap. 15]. These filters can also be designed using a minimax method based on SDP, as will be illustrated in this section. In fact,

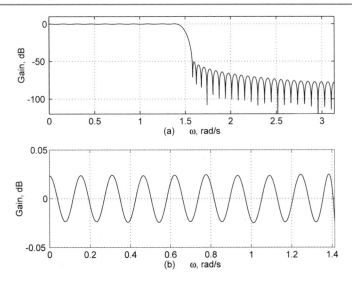

Fig. 16.1 Amplitude response of the filter for Example 16.1: **a** For baseband $0 \le \omega \le \pi$, **b** for passband $0 \le \omega \le \omega_p$

the SDP approach can be used to design FIR filters with arbitrary amplitude and phase responses including certain types of filters that cannot be designed with the weighted-Chebyshev method [4].

Consider an FIR filter of order N characterized by the general transfer function

$$H(z) = \sum_{n=0}^{N} h_n z^{-n}$$

The frequency response of such a filter can be expressed as

$$H(e^{j\omega}) = \sum_{n=0}^{N} h_n e^{-jn\omega} = \mathbf{h}^T[\mathbf{c}(\omega) - j\mathbf{s}(\omega)]$$

where

$$\mathbf{c}(\omega) = [1 \ \cos\omega \ \cdots \ \cos N\omega]^T$$
$$\mathbf{s}(\omega) = [0 \ \sin\omega \ \cdots \ \sin N\omega]^T$$

and $\mathbf{h} = [h_0 \ h_1 \ \cdots \ h_N]^T$. Let $H_d(\omega)$ be the desired frequency response and assume a normalized sampling frequency of 2π. In a minimax design, we need to find a coefficient vector $\mathbf{h}$ that solves the optimization problem

$$\begin{array}{ll} \underset{\mathbf{h}}{\text{minimize}} & \underset{\omega \in \Omega}{\max}[W(\omega)|H(e^{j\omega}) - H_d(\omega)|] \end{array} \qquad (16.9)$$

where Ω is a frequency region of interest over the positive half of the baseband $[0, \pi]$ and $W(\omega)$ is a given weighting function.

If δ denotes the upper bound of the squared weighted error in Eq. (16.9), i.e.,

$$W^2(\omega)|H(e^{j\omega}) - H_d(\omega)|^2 \le \delta \qquad \text{for } \omega \in \Omega \tag{16.10}$$

then a minimax problem equivalent to that in Eq. (16.9) can be reformulated as

$$\text{minimize } \delta \tag{16.11a}$$

$$\text{subject to: } W^2(\omega)|H(e^{j\omega}) - H_d(\omega)|^2 \le \delta \qquad \text{for } \omega \in \Omega \tag{16.11b}$$

Now let $H_r(\omega)$ and $H_i(\omega)$ be the real and imaginary parts of $H_d(\omega)$, respectively. We can write

$$W^2(\omega)|H(e^{j\omega}) - H_d(\omega)|^2 = W^2(\omega)\{[\mathbf{h}^T \mathbf{c}(\omega) - H_r(\omega)]^2 + [\mathbf{h}^T \mathbf{s}(\omega) + H_i(\omega)]^2\}$$
$$= \alpha_1^2(\omega) + \alpha_2^2(\omega) \tag{16.12}$$

where

$$\alpha_1(\omega) = \mathbf{h}^T \mathbf{c}_w(\omega) - H_{rw}(\omega)$$
$$\alpha_2(\omega) = \mathbf{h}^T \mathbf{s}_w(\omega) + H_{iw}(\omega)$$
$$\mathbf{c}_w(\omega) = W(\omega)\mathbf{c}(\omega)$$
$$\mathbf{s}_w(\omega) = W(\omega)\mathbf{s}(\omega)$$
$$H_{rw}(\omega) = W(\omega)H_r(\omega)$$
$$H_{iw}(\omega) = W(\omega)H_i(\omega)$$

Using Eq. (16.12), the constraint in Eq. (16.11b) becomes

$$\delta - \alpha_1^2(\omega) - \alpha_2^2(\omega) \ge 0 \qquad \text{for } \omega \in \Omega \tag{16.13}$$

It can be shown that the inequality in Eq. (16.13) holds if and only if

$$\mathbf{D}(\omega) = \begin{bmatrix} \delta & \alpha_1(\omega) & \alpha_2(\omega) \\ \alpha_1(\omega) & 1 & 0 \\ \alpha_2(\omega) & 0 & 1 \end{bmatrix} \succeq \mathbf{0} \qquad \text{for } \omega \in \Omega \tag{16.14}$$

(see Problem 16.3) i.e., $\mathbf{D}(\omega)$ is positive semidefinite for the frequencies of interest. If we write

$$\mathbf{x} = \begin{bmatrix} \delta \\ \mathbf{h} \end{bmatrix} = \begin{bmatrix} x_1 \\ x_2 \\ \vdots \\ x_{N+2} \end{bmatrix} \tag{16.15}$$

where $x_1 = \delta$ and $\mathbf{h} = [x_2 \ x_3 \ \cdots \ x_{N+2}]^T$, then matrix $\mathbf{D}(\omega)$ is *affine* with respect to $\mathbf{x}$. If $\mathcal{S} = \{\omega_i : i = 1, \ 2, \ \ldots, \ M\} \subset \Omega$ is a set of frequencies which is sufficiently dense on Ω, then a discretized version of Eq. (16.14) is given by

$$\mathbf{F(x)} \succeq \mathbf{0}$$

where

$$\mathbf{F(x)} = \text{diag}\{\mathbf{D}(\omega_1), \ \mathbf{D}(\omega_2), \ \ldots, \ \mathbf{D}(\omega_M)\}$$

and the minimization problem in Eqs. (16.11a)–(16.11b) can be converted into the optimization problem

$$\text{minimize } \mathbf{c}^T \mathbf{x} \tag{16.16a}$$

$$\text{subject to:} \ \ \mathbf{F(x)} \succeq \mathbf{0} \tag{16.16b}$$

where $\mathbf{c} = [1 \ 0 \ \cdots \ 0]^T$. Upon comparing Eqs. (16.16a)–(16.16b) with Eqs. (13.57a)–(13.57b), we conclude that this problem belongs to the class of SDP problems studied in Chap. 13.

Example 16.2 Assuming idealized passband and stopband gains of 1 and 0, respectively, and a normalized sampling frequency of 2π, apply the SDP-based minimax approach described in Sect. 16.2.2 to design a lowpass FIR filter of order 84 with a passband edge $\omega_p = 0.45\pi$ and a stopband edge $\omega_a = 0.5\pi$.

Solution The design was carried out by solving the SDP problem in Eqs. (16.16a)–(16.16b) using Algorithm 13.5. The desired specifications can be achieved by assuming an idealized frequency response of the form

$$
\begin{aligned}
H_d(\omega) &= \begin{cases} e^{-j42\omega} & \text{for } \omega \in [0, \ \omega_p] \\ 0 & \text{for } \omega \in [\omega_a, \ \omega_s/2] \end{cases} \\
&= \begin{cases} e^{-j42\omega} & \text{for } \omega \in [0, \ 0.45\pi] \\ 0 & \text{for } \omega \in [0.5\pi, \ \pi] \end{cases}
\end{aligned}
$$

For a filter order $N = 84$, the variable vector $\mathbf{x}$ has 86 elements as can be seen in Eq. (16.15). We assumed 300 sample frequencies that were uniformly distributed in $\Omega = [0, \ 0.45] \cup [0.5\pi, \ \pi]$, i.e., $M = 300$ in set $\mathcal{S}$. Consequently, matrix $\mathbf{F(x)}$ in Eq. (16.16b)is of dimension 900×900. Using a piecewise constant representation for the weighting function $W(\omega)$ defined in Eq. (16.1) with $\gamma = 1.5$, a filter was obtained that has an equiripple amplitude response as can be seen in the plots of Fig. 16.2a and b. The maximum passband error and minimum stopband attenuation were 0.0098 and 43.72 dB, respectively.

The existence of a unique equiripple linear-phase FIR-filter design for a given set of amplitude-response specifications is guaranteed by the so-called *alternation theorem* (see [1, p. 677]). This design has a constant group delay of $N/2$ s. Interestingly, the FIR filter designed here has a constant group delay of $N/2 = 42$, as can be seen in the delay characteristic of Fig. 16.2c, and this feature along with the equiripple amplitude response achieved suggests that the SDP minimax approach actually obtained the unique best equiripple linear-phase design. The SDP approach is much more demanding than the Remez exchange algorithm in terms of computation effort. However, it can be used to design FIR filter types that cannot be designed with the Remez exchange algorithm, for example, low-delay FIR filters with approximately constant passband group delay.   ■

16.2.3   Minimax Design of IIR Digital Filters Using SDP

16.2.3.1   Introduction
Infinite-duration impulse response (IIR) digital filters offer improved selectivity and computational efficiency and reduced system delay compared to what can be achieved using FIR filters of comparable approximation accuracy [1]. The major drawbacks of an IIR design are that linear phase response can be achieved only approximately and the design must deal with the stability problem, which does not exist in the FIR case.

A linear phase response is often required in digital filters for the purpose of avoiding phase distortion in the signals to be processed. Since signal components transmitted through stopbands are usually heavily attenuated, the linearity of the phase response is typically unimportant in stopbands. Consequently, IIR filters which have an approximately linear phase response in passbands and possibly a nonlinear phase response in stopbands are often entirely satisfactory particularly if they are also more economical in terms of computational complexity. Several methods are available for the design of IIR filters with approximately linear phase response in passbands [5–8].

The stability problem can be handled in several ways, see, for example, [1,5–8]. A popular approach is to impose *stability constraints* that establish a class of stable IIR filters from which the best solution for the design problem can be obtained. Obviously, this leads to a constrained optimization formulation for the design. However, technical difficulties can often occur if we attempt to implement a stability constraint that is explicitly related to the design variables. This is because the locations of the poles of the transfer function, which determine the stability of the filter, are related to the filter coefficients in a highly nonlinear and implicit way even for filters of moderate orders. Linear stability constraints that are affine with respect to the design variables were proposed in [6,8]. These constraints depend on the frequency variable ω, which can vary from 0 to π. Their linearity makes it possible to formulate the design of stable IIR filters as LP or convex QP problems. It should be mentioned, however, that constraints of this class that are *sufficient* conditions for stability are

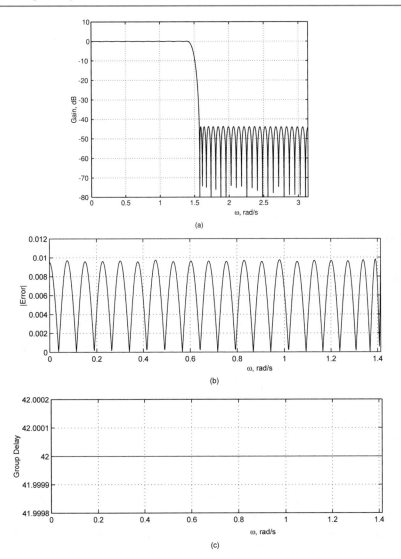

Fig. 16.2 Performance of the lowpass FIR filter designed in Example 16.2: **a** Amplitude response, **b** passband error, **c** group-delay characteristic

often too restrictive to permit a satisfactory design, especially the linear constraint proposed in [6].

Below we formulate the design of IIR filters as SDP problems. The stability of the filter to be designed is assured by using a *single* linear matrix inequality (LMI) constraint, which fits nicely into an SDP formulation and does not depend on continuous parameters other than the design variables.

The transfer function of the IIR filter is assumed to be of the form

$$H(z) = \frac{A(z)}{B(z)} \tag{16.17a}$$

where

$$A(z) = \sum_{i=0}^{N} a_i z^{-i} \tag{16.17b}$$

$$B(z) = 1 + \sum_{i=1}^{K} b_i z^{-i} \tag{16.17c}$$

and K is an integer between 1 and N. The particular form of the denominator polynomial $B(z)$ in Eq. (16.17c) has $N - K$ poles at the origin which, as was recently observed in [8], can be beneficial in the design of certain types of digital filters.

16.2.3.2 LMI Constraint for Stability

The stability of a filter represented by transfer function $H(z)$ such as that in Eq. (16.17a) is guaranteed if the zeros of polynomial $B(z)$ in Eq. (16.17c) are strictly inside the unit circle as was stated earlier. It can be shown that the zeros of $B(z)$ are the eigenvalues of the matrix

$$\mathbf{D} = \begin{bmatrix} -b_1 & -b_2 & \cdots & -b_{K-1} & -b_K \\ 1 & 0 & \cdots & 0 & 0 \\ 0 & 1 & \cdots & 0 & 0 \\ \vdots & \vdots & \ddots & \vdots & \vdots \\ 0 & 0 & \cdots & 1 & 0 \end{bmatrix} \tag{16.18}$$

(see Problem 16.4(a)). Consequently, the filter is stable if the moduli of the eigenvalues are all strictly less than one. The well-known Lyapunov theory [9] states that $\mathbf{D}$ represents a stable filter if and only if there exists a positive definite matrix $\mathbf{P}$ such that $\mathbf{P} - \mathbf{D}^T \mathbf{P} \mathbf{D}$ is positive definite, i.e.,

$$\mathcal{F} = \{\mathbf{P} : \ \mathbf{P} \succ \mathbf{0} \ \text{and} \ \mathbf{P} - \mathbf{D}^T \mathbf{P} \mathbf{D} \succ \mathbf{0}\} \tag{16.19}$$

is nonempty. Using simple linear algebraic manipulations, it can be verified that the set $\mathcal{F}$ in Eq. (16.19) can be characterized by

$$\mathcal{F} = \left\{ \mathbf{P} : \ \begin{bmatrix} \mathbf{P}^{-1} & \mathbf{D} \\ \mathbf{D}^T & \mathbf{P} \end{bmatrix} \succ \mathbf{0} \right\} \tag{16.20}$$

(see Problem 16.4b).

Note that the constraint matrix in Eq. (16.20) is linear with respect to matrix $\mathbf{D}$, which is not the case for the constraint matrix in Eq. (16.19).

16.2.3.3 SDP Formulation of the Design Problem

Given a desired frequency response $H_d(\omega)$, a minimax design of a stable IIR filter can be obtained by finding a transfer function $H(z)$ such as that in Eqs. (16.17a)–(16.17b) which solves the constrained optimization problem

$$\text{minimize} \max_{0 \leq \omega \leq \pi} [W(\omega)|H(e^{j\omega}) - H_d(\omega)|] \tag{16.21a}$$

$$\text{subject to: } B(z) \neq 0 \quad \text{for } |z| \geq 1 \tag{16.21b}$$

(see Sect. B.7). The frequency response of the filter can be expressed as

$$H(e^{j\omega}) = \frac{A(\omega)}{B(\omega)}$$

where

$$A(\omega) = \sum_{n=0}^{N} a_n e^{-jn\omega} = \mathbf{a}^T \mathbf{c}(\omega) - j\mathbf{a}^T \mathbf{s}(\omega)$$

$$B(\omega) = 1 + \sum_{n=1}^{K} b_n e^{-jn\omega} = 1 + \mathbf{b}^T \hat{\mathbf{c}}(\omega) - j\mathbf{b}^T \hat{\mathbf{s}}(\omega)$$

$$\mathbf{a} = [a_0 \ a_1 \ \cdots \ a_N]^T$$

$$\mathbf{b} = [b_1 \ b_2 \ \cdots \ b_K]^T$$

$$\mathbf{c}(\omega) = [1 \ \cos\omega \ \cdots \ \cos N\omega]^T$$

$$\mathbf{s}(\omega) = [0 \ \sin\omega \ \cdots \ \sin N\omega]^T$$

$$\hat{\mathbf{c}}(\omega) = [\cos\omega \ \cos 2\omega \ \cdots \ \cos K\omega]^T$$

$$\hat{\mathbf{s}}(\omega) = [\sin\omega \ \sin 2\omega \ \cdots \ \sin K\omega]^T$$

and $H_d(\omega)$ can be written as

$$H_d(\omega) = H_r(\omega) + jH_i(\omega)$$

where $H_r(\omega)$ and $H_i(\omega)$ denote the real and imaginary parts of $H_d(\omega)$, respectively.

Following the reformulation step in the FIR case (see Sect. 16.2.2), the problem in Eqs. (16.21a)–(16.21b) can be expressed as

$$\text{minimize } \delta \tag{16.22a}$$

$$\text{subject to: } W^2(\omega)|H(e^{j\omega}) - H_d(\omega)|^2 \leq \delta \quad \text{for } \omega \in \Omega \tag{16.22b}$$

$$B(z) \neq 0 \quad \text{for } |z| \geq 1 \tag{16.22c}$$

where $\Omega = [0, \pi]$, we can write

$$W^2(\omega)|H(e^{j\omega}) - H_d(\omega)|^2 = \frac{W^2(\omega)}{|B(\omega)|^2}|A(\omega) - B(\omega)H_d(\omega)|^2$$

which suggests the following iterative scheme: In the kth iteration, we seek to find polynomials $A_k(z)$ and $B_k(z)$ that solve the constrained optimization problem

$$\text{minimize } \delta \tag{16.23a}$$

$$\text{subject to: } \frac{W^2(\omega)}{|B_{k-1}(\omega)|^2}|A(\omega) - B(\omega)H_d(\omega)|^2 \leq \delta \quad \text{for } \omega \in \Omega$$

$$\tag{16.23b}$$

$$B(z) \neq 0 \quad \text{for } |z| \geq 1 \tag{16.23c}$$

where $B_{k-1}(\omega)$ is obtained in the $(k-1)$th iteration. An important difference between the problems in Eqs. (16.22b) and (16.23b) is that the constraint in Eq. (16.22b) is highly nonlinear because of the presence of $B(\omega)$ as the denominator of $H(e^{j\omega})$ while the constraint in Eq. (16.23b) is a *quadratic* function with respect to the components of $\mathbf{a}$ and $\mathbf{b}$ and $W^2(\omega)/|B_{k-1}(\omega)|^2$ is a weighting function. Using arguments similar to those in Sect. 16.2.2, it can be shown that the constraint in Eq. (16.23b) is equivalent to

$$\boldsymbol{\Gamma}(\omega) \succeq \mathbf{0} \quad \text{for } \omega \in \Omega \tag{16.24}$$

where

$$\boldsymbol{\Gamma}(\omega) = \begin{bmatrix} \delta & \alpha_1(\omega) & \alpha_2(\omega) \\ \alpha_1(\omega) & 1 & 0 \\ \alpha_2(\omega) & 0 & 1 \end{bmatrix}$$

with

$$\alpha_1(\omega) = \hat{\mathbf{x}}^T \mathbf{c}_k - H_{rw}(\omega)$$
$$\alpha_2(\omega) = \hat{\mathbf{x}}^T \mathbf{s}_k + H_{iw}(\omega)$$
$$\hat{\mathbf{x}} = \begin{bmatrix} \mathbf{a} \\ \mathbf{b} \end{bmatrix}, \quad \mathbf{c}_k = \begin{bmatrix} \mathbf{c}_w \\ \mathbf{u}_w \end{bmatrix}, \quad \mathbf{s}_k = \begin{bmatrix} \mathbf{s}_w \\ \mathbf{v}_w \end{bmatrix}$$
$$w_k = \frac{W(\omega)}{|B_{k-1}(\omega)|}$$
$$\mathbf{c}_w = w_k \mathbf{c}(\omega)$$
$$\mathbf{s}_w = w_k \mathbf{s}(\omega)$$
$$H_{rw}(\omega) = w_k H_r(\omega)$$
$$H_{iw}(\omega) = w_k H_i(\omega)$$
$$\mathbf{u}_w = w_k[-H_i(\omega)\hat{\mathbf{s}}(\omega) - H_r(\omega)\hat{\mathbf{c}}(\omega)]$$
$$\mathbf{v}_w = w_k[-H_i(\omega)\hat{\mathbf{c}}(\omega) + H_r(\omega)\hat{\mathbf{s}}(\omega)]$$

As for the stability constraint in Eq. (16.23c), we note from Sect. 16.2.3.2 that for a stable filter there exists a $\mathbf{P}_{k-1} \succ \mathbf{0}$ that solves the Lyapunov equation [9]

$$\mathbf{P}_{k-1} - \mathbf{D}_{k-1}^T \mathbf{P}_{k-1} \mathbf{D}_{k-1} = \mathbf{I} \tag{16.25}$$

where $\mathbf{I}$ is the $K \times K$ identity matrix and $\mathbf{D}_{k-1}$ is a $K \times K$ matrix of the form in Eq. (16.18) with $-\mathbf{b}_{k-1}^T$ as its first row. Eq. (16.20) suggests a stability constraint for the digital filter as

$$\begin{bmatrix} \mathbf{P}_{k-1}^{-1} & \mathbf{D} \\ \mathbf{D}^T & \mathbf{P}_{k-1} \end{bmatrix} \succ \mathbf{0}$$

or

$$\mathbf{Q}_k = \begin{bmatrix} \mathbf{P}_{k-1}^{-1} - \tau\mathbf{I} & \mathbf{D} \\ \mathbf{D}^T & \mathbf{P}_{k-1} - \tau\mathbf{I} \end{bmatrix} \succeq \mathbf{0} \tag{16.26}$$

where $\mathbf{D}$ is given by Eq. (16.18) and $\tau > 0$ is a scalar that can be used to control the stability margin of the IIR filter. We note that (a) $\mathbf{Q}_k$ in Eq. (16.26) is affine with respect to $\mathbf{D}$ (and hence with respect to $\hat{\mathbf{x}}$) and (b) because of Eq. (16.25), positive definite matrix $\mathbf{P}_{k-1}$ in Eq. (16.26) is *constrained*. Consequently, Eq. (16.26) is a *sufficient* (but not necessary) constraint for stability. However, if the iterative algorithm described above converges, then the matrix sequence $\{\mathbf{D}_k\}$ also converges. Since the existence of a $\mathbf{P}_{k-1} \succ \mathbf{0}$ in Eq. (16.25) is a necessary and sufficient condition for the stability of the filter, the LMI constraint in Eq. (16.26) becomes less and less restrictive as the iterations continue.

Combining a discretized version of Eq. (16.24) with the stability constraint in Eq. (16.26), the constrained optimization problem in Eqs. (16.23a)–(16.23c) can now be formulated as

$$\text{minimize } \mathbf{c}^T \mathbf{x} \tag{16.27a}$$

$$\text{subject to: } \begin{bmatrix} \boldsymbol{\Gamma}_k & \mathbf{0} \\ \mathbf{0} & \mathbf{Q}_k \end{bmatrix} \succeq \mathbf{0} \tag{16.27b}$$

where

$$\mathbf{x} = \begin{bmatrix} \delta \\ \hat{\mathbf{x}} \end{bmatrix} = \begin{bmatrix} \delta \\ \mathbf{a} \\ \mathbf{b} \end{bmatrix}, \quad \mathbf{c} = \begin{bmatrix} 1 \\ 0 \\ \vdots \\ 0 \end{bmatrix}$$

and

$$\boldsymbol{\Gamma}_k = \text{diag}\{\boldsymbol{\Gamma}(\omega_1), \ \boldsymbol{\Gamma}(\omega_2), \ \ldots, \ \boldsymbol{\Gamma}(\omega_M)\}$$

In the above equation, $\{\omega_i : 1 \le i \le M\}$ is a set of frequencies in the range of interest. Since both $\boldsymbol{\Gamma}_k$ and $\mathbf{Q}_k$ are affine with respect to variable vector $\mathbf{x}$, the problem in Eqs. (16.27a)–(16.27b) is an SDP problem.

16.2.3.4 Iterative SDP Algorithm

Given a desired frequency response $H_d(\omega)$, a weighting function $W(\omega)$, and the orders of $A(z)$ and $B(z)$, namely, N and K, respectively, we can start the design with an initial point $\hat{\mathbf{x}}_0 = [\mathbf{a}_0^T \ \mathbf{b}_0^T]^T$ with $\mathbf{b}_0 = \mathbf{0}$. Coefficient vector $\mathbf{a}_0$ is obtained by designing an FIR filter assuming a desired frequency response $H_d(\omega)$ using a routine design algorithm [1]. The SDP problem formulated in Eqs. (16.27a)–(16.27b) is solved for $k = 1$. If $||\hat{\mathbf{x}}_k - \hat{\mathbf{x}}_{k-1}||_2$ is less than a prescribed tolerance ε, then $\hat{\mathbf{x}}_k$ is deemed to be a solution for the design problem. Otherwise, the SDP problem in Eqs. (16.27a)–(16.27b) is solved for $k = 2$, etc. This algorithm is illustrated by the following example.

Example 16.3 Assuming idealized passband and stopband gains of 1 and 0, respectively, and a normalized sampling frequency of 2π, apply the above iterative minimax approach to design an IIR lowpass digital filter that would meet the following specifications: passband edge $\omega_p = 0.5\pi$, stopband edge $\omega_a = 0.6\pi$, maximum passband error $\delta_p \le 0.02$, minimum stopband attenuation $A_a \ge 34$ dB, group delay in passband $= 9$ s with a maximum deviation of less than 1 s.

Solution The required IIR filter was designed by solving the SDP problem in Eqs. (16.27a)–(16.27b) using Algorithm 13.5. The desired specifications were achieved by using an idealized frequency of the form

$$
H_d(\omega) = \begin{cases} e^{-j9\omega} & \text{for } \omega \in [0, \ \omega_p] \\ 0 & \text{for } \omega \in [\omega_a, \ \omega_s/2] \end{cases}
$$

$$
= \begin{cases} e^{-j9\omega} & \text{for } \omega \in [0, \ 0.5\pi] \\ 0 & \text{for } \omega \in [0.6\pi, \ \pi] \end{cases}
$$

along with $N = 12$, $K = 6$, $W(\omega) = 1$ on $[0, \ 0.5\pi] \cup [0.6\pi, \ \pi]$ and zero elsewhere, $\tau = 10^{-4}$, and $\varepsilon = 5 \times 10^{-3}$. The constraint $\Gamma(\omega) \succeq 0$ was discretized over a set of 240 equally-spaced sample frequencies on $[0, \ \omega_p] \cup [\omega_a, \ \pi]$. It took the algorithm 50 iterations to converge to a solution. The poles and zeros of the filter obtained are given in Table 16.1, and $a_0 = 0.00789947$. The largest pole magnitude is 0.944.

The performance of the filter obtained can be compared with that of an alternative design reported by Deczky as Example 1 in [5], which has the same passband and stopband edges and filter order. As can be seen in Table 16.2 and Fig. 16.3, the present design offers improved performance as well as a reduced group delay. In addition, the present filter has only six nonzero poles, which would lead to reduced computational complexity in the implementation of the filter.

■

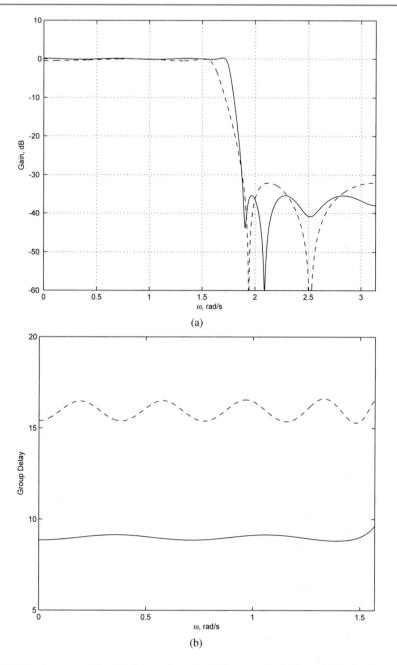

Fig. 16.3 Performance of the IIR filter designed (solid lines) and the filter in [5] (dashed lines) for Example 16.3: **a** Amplitude response, **b** passband group delay characteristic

Table 16.1 Zeros and poles of the transfer function for Example 16.3

Zeros	Poles
-2.12347973	
-1.22600378	$-0.15960464 \pm j0.93037014$
$1.49482238 \pm j0.55741991$	$-0.03719150 \pm j0.55679595$
$0.75350472 \pm j1.37716837$	$0.24717453 \pm j0.18656749$
$-0.89300316 \pm j0.65496710$	Plus 6 poles
$-0.32277491 \pm j0.93626367$	At the origin
$-0.49091195 \pm j0.86511412$	

Table 16.2 Performance comparisons for Example 16.3

Filter	Iterative SDP design	Deczky's design [5]
Maximum passband Error in magnitude	0.0171	0.0549
Minimum stopband Attenuation, dB	34.7763	31.5034
Maximum ripple Of group delay In passband, s	0.8320	1.3219

16.2.4 Minimax Design of FIR and IIR Digital Filters Using SOCP

An alternative approach to the design problems addressed in Sects. 16.2.2 and 16.2.3 that has evolved in recent years is through the use of SOCP.

To solve the optimization problem in Eq. (16.9), let δ be the upper bound of the weighted error in Eq. (16.9), i.e.,

$$W(\omega)|H(e^{j\omega}) - H_d(\omega)| \leq \delta \quad \text{for } \omega \in \Omega$$

A minimax problem that is equivalent to the problem in Eq. (16.9) is given by

$$\text{minimize } \delta \tag{16.28a}$$

$$\text{subject to: } W(\omega)|H(e^{j\omega}) - H_d(\omega)| \leq \delta \quad \text{for } \omega \in \Omega \tag{16.28b}$$

From Eq. (16.12), we have

$$W(\omega)|H(e^{j\omega}) - H_d(e^{j\omega})| = \sqrt{\alpha_1^2(\omega) + \alpha_2^2(\omega)} = \left\| \begin{bmatrix} \alpha_1(\omega) \\ \alpha_2(\omega) \end{bmatrix} \right\|_2$$

where the last term can be expressed as

$$\left\|\begin{bmatrix}\alpha_1(\omega)\\\alpha_2(\omega)\end{bmatrix}\right\|_2 = \left\|\begin{bmatrix}\mathbf{c}_w^T\\\mathbf{s}_w^T\end{bmatrix}\mathbf{h} + \begin{bmatrix}-H_{rw}(\omega)\\H_{iw}(\omega)\end{bmatrix}\right\|_2 = \left\|\begin{bmatrix}\mathbf{0}&\mathbf{c}_w^T\\\mathbf{0}&\mathbf{s}_w^T\end{bmatrix}\mathbf{x} + \begin{bmatrix}-H_{rw}(\omega)\\H_{iw}(\omega)\end{bmatrix}\right\|_2$$

Therefore, for $S = \{\omega_i : i = 1, 2, \ldots, M\} \in \Omega$, the minimization problem in Eqs. (16.28a)–(16.28b) can be expressed as the SOCP problem

$$\text{minimize}\ \mathbf{c}^T\mathbf{x} \tag{16.29a}$$

$$\text{subject to:}\|\mathbf{A}_i\mathbf{x} + \mathbf{c}_i\|_2 \le \mathbf{c}^T\mathbf{x}\ \ \text{for}\ i = 1, 2, \ldots, M \tag{16.29b}$$

where $\mathbf{c} = [1\ 0\ \cdots\ 0]^T$,

$$\mathbf{A}_i = \begin{bmatrix}\mathbf{0}&\mathbf{c}_w^T\\\mathbf{0}&\mathbf{s}_w^T\end{bmatrix}\ \ \text{and}\ \ \mathbf{c}_i = \begin{bmatrix}-H_{rw}(\omega)\\H_{iw}(\omega)\end{bmatrix}$$

The above approach is also applicable to the design of minimax IIR filters except that the design procedure is more complicated as we also have to deal with the stability issue. Two design steps that can be used to deal with the stability issue are as follows:

1. Replace the constraint in Eq. (16.23b) by

$$\frac{W(\omega)}{|B_{k-1}(\omega)|}|A(\omega) - B(\omega)H_d(\omega)| \le \delta\ \ \text{for}\ \omega \in \Omega$$

   so that the constraint assumes the same form as that in Eq. (16.28b).
2. Replace the stability constraint in Eq. (16.23c) by a set of linear inequalities using one of the techniques proposed in [6,7].

In this way, the problem in Eqs. (16.23a)–(16.23c) can be converted into an SOCP problem.

16.2.5   Minimax Design of IIR Digital Filters Satisfying Multiple Specifications

In this section, a minimax method for the design of IIR digital filters based on constrained optimization is described. Arbitrary amplitude response specifications as well as nearly linear phase response can be achieved and a variety of constraints can be imposed such as maximum passband ripple, maximum stopband gain, maximum transition-band gain, and prescribed stability margin [10]. The filter to be designed is assumed to consist of a cascade arrangement of second-order sections and has a transfer function of the form

$$H(z) = H_0 \prod_{m=1}^{J} \frac{a_{0m} + a_{1m}z + z^2}{b_{0m} + b_{1m}z + z^2}$$

The group delay of the filter is given by

$$\tau_h(\omega) = \sum_{i=1}^{J} \left[\frac{\alpha_d(\omega, i)}{\beta_d(\omega, i)} - \frac{\alpha_n(\omega, i)}{\beta_n(\omega, i)} \right]$$

where

$$\alpha_d(\omega, i) = 1 - b_{0i}^2 + b_{1i}(1 - b_{0i})\cos\omega$$
$$\beta_d(\omega, i) = b_{0i}^2 + b_{1i}^2 + 1 + 2b_{0i}\cos 2\omega + 2b_{1i}(b_{0i} + 1)\cos\omega$$
$$\alpha_n(\omega, i) = 1 - a_{0i}^2 + a_{1i}(1 - a_{0i})\cos\omega$$
$$\beta_n(\omega, i) = a_{0i}^2 + a_{1i}^2 + 1 + 2a_{0i}\cos 2\omega + 2a_{1i}(a_{0i} + 1)\cos\omega$$

The vector of the design variables is given by

$$\mathbf{x} = [a_{01}\ a_{11}\ b_{00}\ b_{01}\ \cdots\ b_{0J}\ b_{1J}\ H_0\ \tau]^T$$

and includes all the coefficients of the transfer function and the overall group delay of the filter τ. The group-delay deviation at frequency ω is given by

$$e_g(\mathbf{x}, \omega) = \tau_h(\omega) - \tau$$

Let us consider the design of a 2-band filter such as a lowpass or highpass filter and let $\mathbf{x}_k$ be the kth iterate of $\mathbf{x}$ and $\boldsymbol{\delta}$ be the update of $\mathbf{x}_k$. The L_p norm of the passband group-delay deviation is given by

$$E_p^{(gd)} = \left[\int_{\omega_{pl}}^{\omega_{pu}} |e_g(\mathbf{x}_k + \boldsymbol{\delta}, \omega)|^p d\omega \right]^{1/p} \approx \left[\sum_{i=1}^{N} |e_g(\mathbf{x}_k, \omega_i) + \nabla e_g(\mathbf{x}_k, \omega_i)^T \boldsymbol{\delta}|^p \Delta\omega \right]^{1/p}$$

where ω_{pl} and ω_{pu} are the lower and upper passband edges and $\{\omega_i,\ i = 1, 2, \ldots, N\}$ is a set of frequencies evenly distributed over the passband $\Psi_p = [\omega_{pl}, \omega_{pu}]$. The above expression can be written in matrix form as

$$E_p^{(gd)} \approx \|\mathbf{C}_k\boldsymbol{\delta} + \mathbf{d}_k\|_p$$

where matrix $\mathbf{C}_k$ and vector $\mathbf{d}_k$ are known.

Without loss of generality, we can assume that the desired amplitude response with respect to the passband is unity in which case the passband error function becomes

$$e^{(pb)}(\omega, \mathbf{x}_k) = |H(e^{j\omega})|^2 - 1 \qquad \omega \in \Psi_p$$

The L_p-norm of the passband error function in matrix form can be approximated as

$$E_p^{(pb)} \approx \|\mathbf{D}_k^{(pb)}\boldsymbol{\delta} + \mathbf{f}_k^{(pb)}\|_p$$

where matrix $\mathbf{D}_k^{(pb)}$ and vector $\mathbf{f}_k^{(pb)}$ are known.

Similarly, the L_p norm of the error of the amplitude response in the stopband and transition bands can be approximated as

$$E_p^{(sb)} \approx \|\mathbf{D}_k^{(sb)}\boldsymbol{\delta} + \mathbf{f}_k^{(sb)}\|_p \qquad \omega \in \Psi_s$$

and

$$E_p^{(tb)} \approx \|\mathbf{D}_k^{(tb)}\boldsymbol{\delta} + \mathbf{f}_k^{(tb)}\|_p \qquad \omega \in \Psi_t$$

respectively, where Ψ_s and Ψ_t are vectors whose components are the sample frequencies with respect to the stopband and transition band, respectively, and matrices $\mathbf{D}_k^{(sb)}, \mathbf{D}_k^{(tb)}$, and vectors $\mathbf{f}_k^{(sb)}$ and $\mathbf{f}_k^{(tb)}$ are known.

To assure the stability of the filter, the poles of the transfer function are constrained to be inside the unit circle. To ensure that the filter continues to be stable in situations where quantization errors are introduced in the transfer-function coefficients, we define a stability margin as $\epsilon_s = 1 - r_{max}^{(p)}$ where $r_{max}^{(p)}$ is the largest pole radius allowed. In order to achieve a stability margin ε_s, the coefficients of the denominator should satisfy the stability conditions (see Example 10.7)

$$b_{0m} \leq 1 - \gamma$$
$$b_{1m} - b_{0m} \leq 1 - \gamma$$
$$-b_{1m} - b_{0m} \leq 1 - \gamma$$

where $\gamma = 1 - (1 - \varepsilon_s)^2$.

The above stability conditions for the kth iteration can be expressed in matrix form as

$$\mathbf{B}\boldsymbol{\delta} \leq \mathbf{b}^{(k)}$$

where

$$\mathbf{B} = \begin{bmatrix} \mathcal{B} & \mathbf{0} & \cdots & \mathbf{0} & \mathbf{0} & \mathbf{0} \\ \vdots & \vdots & \ddots & \vdots & \vdots & \vdots \\ \mathbf{0} & \mathbf{0} & \cdots & \mathcal{B} & \mathbf{0} & \mathbf{0} \end{bmatrix}, \quad \mathcal{B} = \begin{bmatrix} 1 & 0 \\ -1 & 1 \\ -1 & -1 \end{bmatrix}$$

$$\mathbf{b}^{(k)} = \begin{bmatrix} \mathbf{b}_1^{(k)} \\ \vdots \\ \mathbf{b}_J^{(k)} \end{bmatrix}, \quad \mathbf{b}_m^{(k)} = \begin{bmatrix} 1 - \varepsilon_s - b_{0m}^{(k)} \\ 1 - \varepsilon_s - b_{1m}^{(k)} + b_{0m}^{(k)} \\ 1 - \varepsilon_s + b_{1m}^{(k)} + b_{0m}^{(k)} \end{bmatrix}$$

Based on the above analysis, the problem to be solved in the kth iteration can be formulated as

$$\text{minimize} \quad \|\mathbf{C}_k \boldsymbol{\delta} + \mathbf{d}_k\|_p \tag{16.30a}$$

$$\text{subject to:} \quad \|\mathbf{D}_k^{(pb)} \boldsymbol{\delta} + \mathbf{f}_k^{(pb)}\|_p \leq \Gamma_{pb} \tag{16.30b}$$

$$\|\mathbf{D}_k^{(sb)} \boldsymbol{\delta} + \mathbf{f}_k^{(sb)}\|_p \leq \Gamma_{sb} \tag{16.30c}$$

$$\|\mathbf{D}_k^{(tb)} \boldsymbol{\delta} + \mathbf{f}_k^{(tb)}\|_p \leq \Gamma_{tb} \tag{16.30d}$$

$$\|\boldsymbol{\delta}\|_2 \leq \Gamma_{\text{small}} \tag{16.30e}$$

$$\mathbf{B}\boldsymbol{\delta} \leq \mathbf{b}^{(k)} \tag{16.30f}$$

where Γ_{pb}, Γ_{sb}, and Γ_{tb} are the maximun prescribed levels for the passband error, stopband gain, and transition-band gain, respectively. The bound on the norm of $\boldsymbol{\delta}$ in Eq. (16.30e) assures the validity of the linear approximation and at the same time eliminates the need for a line-search step. The formulation in Eqs. (16.30a)–(16.30f) covers the most important cases of least-squares design when $p = 2$ and minimax design when $p = \infty$.

The above formulation can be readily extended to the case of 3-band filters like bandpass and bandstop filters. In the case of the design of bandpass filters with one passband, two stopbands, and two transition bands, the L_p-norm of the passband group delay and passband error can be approximated in the same way as for the case of 2-band filters whereas the amplitude response for the two stopbands and two transition bands can be approximated as

$$E_p^{(sb)} \approx \|\mathbf{D}_k^{(sb)} \boldsymbol{\delta} + \mathbf{f}_k^{(sb)}\|_p \quad \omega \in \Psi_s^{(1)} \text{ and } \Psi_s^{(2)}$$

and

$$E_p^{(tb)} \approx \|\mathbf{D}_k^{(tb)} \boldsymbol{\delta} + \mathbf{f}_k^{(tb)}\|_p \quad \omega \in \Psi_t^{(1)} \text{ and } \Psi_t^{(2)}$$

respectively, where $\Psi_s^{(1)}$ and $\Psi_s^{(2)}$ are vectors whose components are sample frequencies which are evenly distributed over the two stopbands, and $\Psi_t^{(1)}$ and $\Psi_t^{(2)}$ are vectors whose components are sample frequencies which are evenly distributed over the two transition bands. The design of bandstop filters can similarly be accomplished except that we have two passbands and one stopband for bandstop filters.

The problem in Eqs. (16.30a)–(16.30f) can be expressed as an SOCP problem (see Sect. 13.9). Additional information about the method and extensive experimental results and comparisons can be found in [1, Chap. 17].

16.3 Model Predictive Control of Dynamic Systems

One of the challenges encountered in modeling and control of real-life dynamic systems is the development of controllers whose performance remains robust with respect to various uncertainties that exist due to modeling errors, sensor noise, interference induced by the use of alternating-current power supply, and finite word length effects of the controller itself. Model predictive control (MPC) is a popular open-loop control methodology that has proven effective for the control of slow-varying dynamic systems such as process control in pulp and paper, chemical, and oil-refinement industries [11,12]. At each control instant, a model predictive controller performs online optimization to generate an optimal control input based on a model that describes the dynamics of the system to be controlled and the available input and output measurements. In [12], it is shown that robust MPC that takes into account model uncertainty and various constraints on the control input and plant output can be designed using SDP techniques. In this section, we follow the methodology used in [12] to illustrate several SDP-based techniques for the design of robust model predictive controllers.

16.3.1 Polytopic Model for Uncertain Dynamic Systems

A linear discrete-time time-varying dynamic system can be modeled in terms of a state-space formulation as [9]

$$\mathbf{x}(k + 1) = \mathbf{A}(k)\mathbf{x}(k) + \mathbf{B}(k)\mathbf{u}(k) \tag{16.31a}$$

$$\mathbf{y}(k) = \mathbf{C}\mathbf{x}(k) \tag{16.31b}$$

where $\mathbf{y}(k) \in R^{m \times 1}$ is the output vector, $\mathbf{u}(k) \in R^{p \times 1}$ is the input vector, and $\mathbf{x}(k) \in R^{n \times 1}$ is the state vector at time instant k. The system matrix $\mathbf{A}(k) \in R^{n \times n}$ and the input-to-state matrix $\mathbf{B}(k) \in R^{n \times m}$ are *time dependent*. The time dependence of matrices $\mathbf{A}(k)$ and $\mathbf{B}(k)$ can be utilized to describe systems whose dynamic characteristics vary with time. In order to incorporate modeling uncertainties into the model in Eqs. (16.31a)–(16.31b), the pair $[\mathbf{A}(k)\ \mathbf{B}(k)]$ is allowed to be a member of polytope $\mathcal{M}$ defined by

$$\mathcal{M} = \text{Co}\{[\mathbf{A}_1\ \mathbf{B}_1],\ [\mathbf{A}_2\ \mathbf{B}_2],\ \ldots,\ [\mathbf{A}_L\ \mathbf{B}_L]\}$$

where Co denotes the convex hull spanned by $[A_i \ B_i]$ for $1 \leq i \leq L$, which is defined as

$$M = \{[A \ B] : [A \ B] = \sum_{i=1}^{L} \lambda_i [A_i \ B_i], \ \lambda_i \geq 0, \ \sum_{i=1}^{L} \lambda_i = 1\} \qquad (16.32)$$

(see Sect. A.16).

The linear model in Eqs. (16.31a)–(16.31b) subject to the constraint $[A(k) \ B(k)] \in M$ can be used to describe a wide variety of real-life dynamic systems. As an example, consider the angular positioning system illustrated in Fig. 16.4 [13]. The control problem is to use the input voltage to the motor to rotate the antenna such that the antenna angle, θ, relative to some reference tracks the angle of the moving target, θ_r. The discrete-time equation of the motion of the antenna can be derived from its continuous-time counterpart by discretization using a sampling period of 0.1 s. A first-order approximation of the derivative can be obtained as

$$\mathbf{x}(k+1) = \begin{bmatrix} \theta(k) \\ \dot{\theta}(k+1) \end{bmatrix} = \begin{bmatrix} 1 & 0.1 \\ 0 & 1-0.1\alpha(k) \end{bmatrix} \mathbf{x}(k) + \begin{bmatrix} 0 \\ 0.1\eta \end{bmatrix} u(k)$$

$$= \mathbf{A}(k)\mathbf{x}(k) + \mathbf{B}u(k) \qquad (16.33a)$$

$$y(k) = [1 \ 0]\mathbf{x}(k) = \mathbf{C}\mathbf{x}(k) \qquad (16.33b)$$

where $\eta = 0.787$. The parameter $\alpha(k)$ in matrix $\mathbf{A}(k)$ is proportional to the coefficient

of viscous friction in the rotating parts of the antenna, and is assumed to be arbitrarily time-varying in the range $0.1 \leq \alpha(k) \leq 10$. It follows that Eqs. (16.33a)–(16.33b) represent a polytopic model with $\mathbf{A}(k) \in$ Co $\{\mathbf{A}_1, \mathbf{A}_2\}$ where

$$\mathbf{A}_1 = \begin{bmatrix} 1 & 0.10 \\ 0 & 0.99 \end{bmatrix}, \quad \mathbf{A}_2 = \begin{bmatrix} 1 & 0.1 \\ 0 & 0 \end{bmatrix} \qquad (16.33c)$$

Below, we deal with several aspects of MPCs.

16.3.2   Introduction to Robust MPC

At sampling instant k, a robust MPC uses plant measurements and a model, such as the polytopic model in Eqs. (16.33a)–(16.33b), to predict future outputs of the system. These measurements are utilized to compute m control inputs, $\mathbf{u}(k+i|k)$ for $i = 0, 1, \ldots, m-1$, by solving the minimax optimization problem

$$\underset{\mathbf{u}(k+i|k), \ 0 \leq i \leq m-1}{\text{minimize}} \quad \underset{[A(k+i) \ B(k+i)] \in M, i \geq 0}{\text{max}} \quad J(k) \qquad (16.34)$$

Fig. 16.4 Angular
positioning system

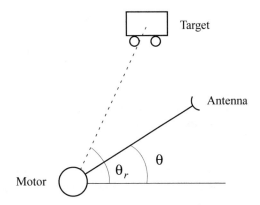

The notation $\mathbf{u}(k+i|k)$ denotes the control input at instant $k+i$ based on the measurements available at instant k; $J(k)$ is an objective function that measures system performance and is given by

$$J(k) = \sum_{i=0}^{\infty} [\mathbf{x}^T(k+i|k)\mathbf{Q}\mathbf{x}(k+i|k) + \mathbf{u}^T(k+i|k)\mathbf{R}\mathbf{u}(k+i|k)] \qquad (16.35)$$

where $\mathbf{Q} \succeq \mathbf{0}$ and $\mathbf{R} \succ \mathbf{0}$ are constant weighting matrices; and $\mathbf{x}(k+i|k)$ denotes the system state at instant $k+i$, which is predicted using the measurements at instant k. It follows that the control input obtained by solving the problem in Eq. (16.34) takes into account the system's uncertainties by minimizing the worst-case value of $J(k)$ for all possible plant models included in set $\mathcal{M}$. The control input so computed is, therefore, *robust* against model uncertainties. At any given sampling instant k, the solution of the optimization problem in Eq. (16.34) provides a total of m control actions $\mathbf{u}(k|k), \mathbf{u}(k+1|k), \ldots, \mathbf{u}(k+m-1|k)$, but in a model predictive controller only the first control action, $\mathbf{u}(k|k)$, is implemented. At the next sampling instant, new measurements are obtained and on the basis of the new measurements the problem in Eq. (16.34) is solved again to provide a new set of m control actions $\mathbf{u}(k|k), \mathbf{u}(k+1|k), \ldots, \mathbf{u}(k+m-1|k)$. The first control action, $\mathbf{u}(k|k)$, is then implemented.

Frequently, the solution of the above minimax problem is computationally too demanding to implement and in the MPC literature the problem in Eq. (16.34) has been addressed by deriving an upper bound of the objective function $J(k)$ and then minimizing this upper bound with a constant state-feedback control law

$$\mathbf{u}(k+i|k) = \mathbf{F}\mathbf{x}(k+i|k) \qquad \text{for } i \geq 0 \qquad (16.36)$$

Let us assume that there exists a quadratic function $V(\mathbf{x}) = \mathbf{x}^T\mathbf{P}\mathbf{x}$ with $\mathbf{P} \succ \mathbf{0}$ such that for all $\mathbf{x}(k+i|k)$ and $\mathbf{u}(k+i|k)$ satisfying Eq. (16.31a) and for $[\mathbf{A}(k+i)\ \mathbf{B}(k+i)] \in \mathcal{M}$, $i \geq 0$, $V(\mathbf{x})$ satisfies the inequality

$$V[\mathbf{x}(k+i+1|k)] - V[\mathbf{x}(k+i|k)] \leq -[\mathbf{x}^T(k+i|k)\mathbf{Q}\mathbf{x}(k+i|k)$$

$$+\mathbf{u}^T(k+i|k)\mathbf{R}\mathbf{u}(k+i|k)] \tag{16.37}$$

If the objective function is finite, then the series in Eq. (16.35) must converge and, consequently, we have $\mathbf{x}(\infty|k) = \mathbf{0}$ which implies that $V[\mathbf{x}(\infty|k)] = 0$. By summing the inequality in Eq. (16.37) from $i = 0$ to ∞, we obtain $J(k) \leq V[\mathbf{x}(k|k)]$. This means that $V[\mathbf{x}(k|k)]$ is an upper bound of the objective function, which is considerably easier to deal with than $J(k)$. In the next section, we study the condition under which a positive definite matrix $\mathbf{P}$ exists such that $V(\mathbf{x})$ satisfies the condition in Eq. (16.37); we then formulate a modified optimization problem that can be solved by using SDP algorithms.

16.3.3   Robust Unconstrained MPC by Using SDP

If γ is an upper bound of $V[\mathbf{x}(k|k)]$, namely,

$$V[\mathbf{x}(k|k)] = \mathbf{x}^T(k|k)\mathbf{P}\mathbf{x}(k|k) \leq \gamma \tag{16.38}$$

then minimizing $V[\mathbf{x}(k|k)]$ is equivalent to minimizing γ. If we let

$$\mathbf{S} = \gamma\mathbf{P}^{-1} \tag{16.39}$$

then $\mathbf{P} \succ \mathbf{0}$ implies that $\mathbf{S} \succ \mathbf{0}$ and Eq. (16.38) becomes

$$1 - \mathbf{x}^T(k|k)\mathbf{S}^{-1}\mathbf{x}(k|k) \geq 0$$

which is equivalent to

$$\begin{bmatrix} 1 & \mathbf{x}^T(k|k) \\ \mathbf{x}(k|k) & \mathbf{S} \end{bmatrix} \succeq \mathbf{0} \tag{16.40}$$

At sampling instant k, the state vector $\mathbf{x}(k|k)$ is assumed to be a known measurement which is used in a state feedback control $\mathbf{u}(k|k) = \mathbf{F}\mathbf{x}(k|k)$ (see Eq. (16.36)). Recall that for $V[\mathbf{x}(k|k)]$ to be an upper bound of $J(k)$, $V[\mathbf{x}(k|k)]$ is required to satisfy the condition in Eq. (16.37). By substituting Eqs. (16.36) and (16.31a) into Eq. (16.37), we obtain

$$\mathbf{x}^T(k+i|k)\mathbf{W}\mathbf{x}(k+i|k) \leq 0 \quad \text{for } i \geq 0 \tag{16.41}$$

where

$$\mathbf{W} = [\mathbf{A}(k + i) + \mathbf{B}(k + i)\mathbf{F}]^T \mathbf{P}[\mathbf{A}(k + i) + \mathbf{B}(k + i)\mathbf{F}]$$
$$-\mathbf{P} + \mathbf{F}^T \mathbf{RF} + \mathbf{Q}$$

Evidently, the inequality in Eq. (16.41) holds if $\mathbf{W} \preceq \mathbf{0}$. Now if we let $\mathbf{Y} = \mathbf{FS}$ where $\mathbf{S}$ is related to matrix $\mathbf{P}$ by Eq. (16.39), we note that the matrix inequality

$$\begin{bmatrix} \mathbf{D} & \mathbf{H} \\ \mathbf{H}^T & \mathbf{G} \end{bmatrix} \succ \mathbf{0} \tag{16.42}$$

is equivalent to

$$\mathbf{G} \succ \mathbf{0} \quad \text{and} \quad \mathbf{D} - \mathbf{HG}^{-1}\mathbf{H}^T \succ \mathbf{0} \tag{16.43}$$

or

$$\mathbf{D} \succ \mathbf{0} \quad \text{and} \quad \mathbf{G} - \mathbf{H}^T \mathbf{D}^{-1}\mathbf{H} \succ \mathbf{0} \tag{16.44}$$

Based on this equivalence, it can be shown that $\mathbf{W} \preceq \mathbf{0}$ can be expressed as

$$\begin{bmatrix} \mathbf{S} & \mathbf{SA}_{k+i}^T + \mathbf{Y}^T \mathbf{B}_{k+i}^T & \mathbf{SQ}^{1/2} & \mathbf{Y}^T \mathbf{R}^{1/2} \\ \mathbf{A}_{k+i}\mathbf{S} + \mathbf{B}_{k+i}\mathbf{Y} & \mathbf{S} & \mathbf{0} & \mathbf{0} \\ \mathbf{Q}^{1/2}\mathbf{S} & \mathbf{0} & \gamma \mathbf{I}_n & \mathbf{0} \\ \mathbf{R}^{1/2}\mathbf{Y} & \mathbf{0} & \mathbf{0} & \gamma \mathbf{I}_p \end{bmatrix} \succeq \mathbf{0} \tag{16.45}$$

where $\mathbf{A}_{k+i}$ and $\mathbf{B}_{k+i}$ stand for $\mathbf{A}(k + i)$ and $\mathbf{B}(k + i)$, respectively (see Probs. 16.7 and 16.8). Since the matrix inequality in Eq. (16.45) is affine with respect to $[\mathbf{A}(k + i) \ \mathbf{B}(k + i)]$, Eq. (16.45) is satisfied for all $[\mathbf{A}(k + i) \ \mathbf{B}(k + i)] \in \mathcal{M}$ defined by Eq. (16.32) if there exist $\mathbf{S} \succ \mathbf{0}$, $\mathbf{Y}$, and scalar γ such that

$$\begin{bmatrix} \mathbf{S} & \mathbf{SA}_j^T + \mathbf{Y}^T \mathbf{B}_j^T & \mathbf{SQ}^{1/2} & \mathbf{Y}^T \mathbf{R}^{1/2} \\ \mathbf{A}_j\mathbf{S} + \mathbf{B}_j\mathbf{Y} & \mathbf{S} & \mathbf{0} & \mathbf{0} \\ \mathbf{Q}^{1/2}\mathbf{S} & \mathbf{0} & \gamma \mathbf{I}_n & \mathbf{0} \\ \mathbf{R}^{1/2}\mathbf{Y} & \mathbf{0} & \mathbf{0} & \gamma \mathbf{I}_p \end{bmatrix} \succeq \mathbf{0} \tag{16.46}$$

for $j = 1, 2, \ldots, L$. Therefore, the unconstrained robust MPC can be formulated in terms of the constrained optimization problem

$$\underset{\gamma, \mathbf{S}, \mathbf{Y}}{\text{minimize}} \quad \gamma \tag{16.47a}$$

$$\text{subject to: constraints in (16.40) and (16.46)} \tag{16.47b}$$

There are a total of $L + 1$ matrix inequality constraints in this problem and they are affine with respect to variables γ, $\mathbf{S}$, and $\mathbf{Y}$. Therefore, this is an SDP problem and the algorithms studied in Chap. 13 can be used to solve it. Once the optimal matrices $\mathbf{S}^*$ and $\mathbf{Y}^*$ are obtained, the optimal feedback matrix can be computed as

$$\mathbf{F}^* = \mathbf{Y}^* \mathbf{S}^{*-1} \tag{16.48}$$

Example 16.4 Design a robust MPC for the angular positioning system discussed in Sect. 16.3.1. Assume that the initial angular position and velocity of the antenna are $\theta(0) = 0.12$ rad and $\dot{\theta}(0) = -0.1$ rad/s, respectively. The goal of the MPC is to steer the antenna to the desired position $\theta_r = 0$. The weighting matrix $\mathbf{R}$ in $J(k)$ in this case is a scalar and is set to $R = 2 \times 10^{-5}$.

Solution Note that $\theta(k)$ is related to $\mathbf{x}(k)$ through the equation $\theta(k) = [1 \ 0]\mathbf{x}(k)$; hence Eq. (16.31b) implies that $y(k) = \theta(k)$, and

$$y^2(k + i|k) = \mathbf{x}^T(k + i|k) \begin{bmatrix} 1 & 0 \\ 0 & 0 \end{bmatrix} \mathbf{x}(k + i|k)$$

The objective function can be written as

$$J(k) = \sum_{i=0}^{\infty} [y^2(k + i|k) + Ru^2(k + i|k)]$$

$$= \sum_{i=0}^{\infty} [\mathbf{x}^T(k + i|k)\mathbf{Q}\mathbf{x}(k + i|k) + Ru^2(k + i|k)]$$

where

$$\mathbf{Q} = \begin{bmatrix} 1 & 0 \\ 0 & 0 \end{bmatrix} \quad \text{and} \quad R = 2 \times 10^{-5}$$

Since the control system under consideration has only one scalar input, namely, the voltage applied to the motor, $u(k + i|k)$ is a scalar. Consequently, the feedback gain $\mathbf{F}$ is a row vector of dimension 2. Other known quantities in the constraints in Eqs. (16.40) and (16.46) are

$$\mathbf{x}(0|0) = \begin{bmatrix} 0.12 \\ -0.10 \end{bmatrix}, \quad \mathbf{Q}^{1/2} = \begin{bmatrix} 1 & 0 \\ 0 & 0 \end{bmatrix}, \quad R^{1/2} = 0.0045$$

$$\mathbf{B}_j = \begin{bmatrix} 0 \\ 0.0787 \end{bmatrix} \quad \text{for } j = 1, 2$$

and matrices $\mathbf{A}_1$ and $\mathbf{A}_2$ are given by Eq. (16.33c).

With the above data and a sampling period of 0.1 s, the solution of the SDP problem in Eqs. (16.47a)–(16.47b), $\{\mathbf{Y}^*, \mathbf{S}^*\}$, can be obtained using Algorithm 13.5, and by using Eq. (16.48) $\mathbf{F}^*$ can be deduced. The optimal MPC can then be computed using the state feedback control law

$$u(k + 1|k) = \mathbf{F}^*\mathbf{x}(k + 1|k) \quad \text{for } k = 0, 1, \ldots \quad (16.49)$$

The state $\mathbf{x}(k + 1|k)$ in Eq. (16.49) is calculated using the model in Eqs. (16.31a)–(16.31b) where $\mathbf{A}(k)$ is selected randomly from the set

$$\mathcal{M} = \text{Co}\{\mathbf{A}_1, \mathbf{A}_2\}$$

Figure 16.5a and b depicts the angular position $\theta(k)$ and velocity $\dot{\theta}(k)$ obtained over the first 2 s. It is observed that both $\theta(k)$ and $\dot{\theta}(k)$ are steered to the desired value of zero within 1 s. The corresponding MPC profile $u(k)$ for $k = 1, 2, \ldots, 20$ is shown in Fig. 16.5c.

∎

16.3.4   Robust Constrained MPC by Using SDP

Frequently, it is desirable to design an MPC subject to certain constraints on the system's input and/or output. For example, constraints on the control input may become necessary in order to represent limitations on control equipment (such as value saturation in a process control scenario). The need for constraints can be illustrated in terms of Example 16.4. In Fig. 16.5c, we observe that at instants 0.1, 0.2, and 0.3 s, the magnitude of the control voltage exceeds 2 V. In such a case, the controller designed would become nonfeasible if the maximum control magnitude were to be limited to 2 V. In the rest of the section we develop robust model-predictive controllers with L_2 norm and componentwise input constraints using SDP.

16.3.4.1 L_2-Norm Input Constraint

As in the unconstrained MPC studied in Sect. 16.3.4, the objective function considered here is also the upper bound γ in Eq. (16.38), and the state feedback control $\mathbf{u}(k + i|k) = \mathbf{F}\mathbf{x}(k + i|k)$ is assumed throughout this section. From Sect. 16.3.4, we know that the matrix inequality constraint in Eq. (16.46) implies the inequality in Eq. (16.45), which, in conjunction with Eqs. (16.31a)–(16.31b), leads to

$$\begin{aligned} \mathbf{x}^T(k + i + 1|k)\mathbf{P}\mathbf{x}(k + i + 1|k) &- \mathbf{x}^T(k + i|k)\mathbf{P}\mathbf{x}(k + i|k) \\ &\leq -\mathbf{x}^T(k + i|k)(\mathbf{F}^T\mathbf{R}\mathbf{F} + \mathbf{S})\mathbf{x}(k + i|k) < 0 \end{aligned}$$

Hence

$$\mathbf{x}^T(k + i + 1|k)\mathbf{P}\mathbf{x}(k + i + 1|k) < \mathbf{x}^T(k + i|k)\mathbf{P}\mathbf{x}(k + i|k)$$

Fig. 16.5 Performance of
MPC in Example 16.4 with
$R = 2 \times 10^{-5}$: **a** Angular
position $\theta(k)$, **b** angular
velocity $\dot{\theta}(k)$, **c** profile of the
MPC

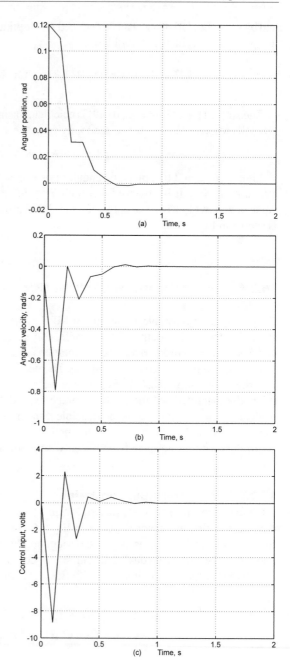

By repeating the above argument for $i = 0, 1, \ldots$, we conclude that

$$\mathbf{x}^T(k+i|k)\mathbf{P}\mathbf{x}(k+i|k) < \mathbf{x}^T(k|k)\mathbf{P}\mathbf{x}(k|k) \qquad \text{for } i \geq 1$$

Therefore,

$$\mathbf{x}^T(k|k)\mathbf{P}\mathbf{x}(k|k) \leq \gamma$$

implies that

$$\mathbf{x}^T(k+i|k)\mathbf{P}\mathbf{x}(k+i|k) \leq \gamma \qquad \text{for } i \geq 1$$

So if we define set $\mathcal{E}$ as

$$\mathcal{E} = \{\mathbf{z} : \mathbf{z}^T\mathbf{P}\mathbf{z} \leq \gamma\} = \{\mathbf{z} : \mathbf{z}^T\mathbf{S}^{-1}\mathbf{z} \leq 1\} \qquad (16.50)$$

then from the above analysis

$$\mathbf{x}(k|k) \in \mathcal{E} \text{ implies that } \mathbf{x}(k+i|k) \in \mathcal{E} \qquad \text{for } i \geq 1$$

In other words, set $\mathcal{E}$ is an *invariant ellipsoid* for the predicted states of the uncertain system.

Now let us consider the Euclidean norm constraint on the control input at sampling instant k, i.e.,

$$||\mathbf{u}(k+i|k)||_2 \leq u_{max} \qquad \text{for } i \geq 0 \qquad (16.51)$$

where u_{max} is a given upper bound. In a state feedback MPC, the control is given by

$$\mathbf{u}(k+i|k) = \mathbf{F}\mathbf{x}(k+i|k) = \mathbf{Y}\mathbf{S}^{-1}\mathbf{x}(k+i|k)$$

Since set $\mathcal{E}$ is invariant for the predicted state, we have

$$\max_{i \geq 0} ||\mathbf{u}(k+i|k)||_2^2 = \max_{i \geq 0} ||\mathbf{Y}\mathbf{S}^{-1}\mathbf{x}(k+i|k)||_2^2$$

$$\leq \max_{\mathbf{z} \in \mathcal{E}} ||\mathbf{Y}\mathbf{S}^{-1}\mathbf{z}||_2^2 \qquad (16.52)$$

It can be shown that

$$\max_{\mathbf{z} \in \mathcal{E}} ||\mathbf{Y}\mathbf{S}^{-1}\mathbf{z}||_2^2 = \lambda_{max}(\mathbf{Y}\mathbf{S}^{-1}\mathbf{Y}^T) \qquad (16.53)$$

where $\lambda_{max}(\mathbf{M})$ denotes the largest eigenvalue of matrix $\mathbf{M}$ (see Problem 16.9). Further, by using the equivalence between the matrix inequality in Eq. (16.42) and that in Eq. (16.43) or Eq. (16.44), it can be shown that the matrix inequality

$$\begin{bmatrix} u_{max}^2\mathbf{I} & \mathbf{Y} \\ \mathbf{Y}^T & \mathbf{S} \end{bmatrix} \succeq \mathbf{0} \tag{16.54}$$

implies that

$$\lambda_{max}(\mathbf{YS}^{-1}\mathbf{Y}^T) \leq u_{max}^2 \tag{16.55}$$

(see Problem 16.10). Therefore, the L_2-norm input constraint in Eq. (16.51) holds if the matrix inequality in Eq. (16.54) is satisfied. Thus a robust MPC that allows an L_2-norm input constraint can be formulated by adding the constraint in Eq. (16.54) to the SDP problem in Eqs. (16.47a)–(16.47b)) as

$$\underset{\gamma,\mathbf{S},\mathbf{Y}}{\text{minimize}} \quad \gamma \tag{16.56a}$$

$$\text{subject to: constraints in (16.40), (16.46) and (16.54)} \tag{16.56b}$$

Since the constraint matrix in Eq. (16.54) is affine with respect to variables $\mathbf{S}$ and $\mathbf{Y}$, the above is also an SDP problem.

16.3.4.2 Componentwise Input Constraints
Another type of commonly used input constraint is an upper bound for the magnitude of each component of the control input, i.e.,

$$|u_j(k+i|k)| \leq u_{j,max} \quad \text{for } i \geq 0, \ j = 1, 2, \ldots, p \tag{16.57}$$

It follows from Eq. (16.52) that

$$\max_{i\geq 0} |u_j(k+i|k)|^2 = \max_{i\geq 0} |[\mathbf{YS}^{-1}\mathbf{x}(k+i|k)]_j|^2$$

$$\leq \max_{\mathbf{z}\in\mathcal{E}} |(\mathbf{YS}^{-1}\mathbf{z})_j|^2 \quad \text{for } j = 1, 2, \ldots, p$$

$$\tag{16.58}$$

Note that the set $\mathcal{E}$ defined in Eq. (16.50) can be expressed as

$$\mathcal{E} = \{\mathbf{z} : \mathbf{z}^T\mathbf{S}^{-1}\mathbf{z} \leq 1\} = \{\mathbf{w} : ||\mathbf{w}||_2 \leq 1, \ \mathbf{w} = \mathbf{S}^{-1/2}\mathbf{z}\}$$

which, in conjunction with the use of the Cauchy-Schwarz inequality, modifies Eq. (16.57) into

$$\max_{i \geq 0} |u_j(k+i|k)|^2 \leq \max_{||\mathbf{w}||_2 \leq 1} |(\mathbf{YS}^{-1/2}\mathbf{w})_j|^2 \leq ||(\mathbf{YS}^{-1/2})_j||_2^2$$

$$= (\mathbf{YS}^{-1}\mathbf{Y}^T)_{j,j} \quad \text{for } j = 1, 2, \ldots, p \quad (16.59)$$

It can be readily verified that if there exists a symmetric matrix $\mathbf{X} \in R^{p \times p}$ such that

$$\begin{bmatrix} \mathbf{X} & \mathbf{Y} \\ \mathbf{Y}^T & \mathbf{S} \end{bmatrix} \succeq \mathbf{0} \tag{16.60a}$$

where the diagonal elements of $\mathbf{X}$ satisfy the inequalities

$$X_{jj} \leq u_{j,max}^2 \quad \text{for } j = 1, 2, \ldots, p \tag{16.60b}$$

then

$$(\mathbf{YS}^{-1}\mathbf{Y}^T)_{j,j} \leq u_{j,max}^2 \quad \text{for } j = 1, 2, \ldots, p$$

which, by virtue of Eq. (16.59), implies the inequalities in Eq. (16.57) (see Problem (16.11)). Therefore, the componentwise input constraints in Eq. (16.57) hold if there exists a symmetric matrix $\mathbf{X}$ that satisfies the inequalities in Eqs. (16.60a)–(16.60b). Hence a robust MPC with the input constraints in Eq. (16.57) can be formulated by

modifying the SDP problem in Eqs. (16.47a) and (16.47b) to

$$\underset{\gamma, \mathbf{S}, \mathbf{X}, \mathbf{Y}}{\text{minimize}} \quad \gamma$$

subject to the constraints in Eqs. (16.40), (16.46), (16.60a) and (16.60b)

Example 16.5 Design an MPC for the angular positioning system discussed in Sect. 16.3.1 with input constraint

$$|u(k|k+i)| \leq 2 \quad \text{for } i \geq 0$$

The initial state $\mathbf{x}(0)$ and other parameters are the same as in Example 16.4.

Solution Since $u(k|k+i)$ is a scalar, $||u(k+i|k)|| = |u(k+i|k)|$. Hence the L_2-norm input and the componentwise input constraints become identical. With $u_{max} = 2$ V, the MPC can be obtained by solving the SDP problem in Eqs. (16.56a)–(16.56b) using Algorithm 13.5 for each sampling instant k. The angular position $\theta(k)$ and velocity $\dot{\theta}(k)$ over the first 2 seconds are plotted in Fig. 16.6a and b, respectively. The control profile is depicted in Fig. 16.6c where we note that the magnitude of the control voltage has been kept within the 2V bound.

It is interesting to note that the magnitude of the MPC commands can also be reduced by using a larger value of weighting factor R in $J(k)$. For the angular positioning system in question with $R = 0.0035$, the unconstrained MPC developed in Example 16.4 generates the control profile shown in Fig. 16.7a, which obviously satisfies the constraint $|u(k|k + i)| \leq 2$. However, the corresponding $\theta(k)$ and $\dot{\theta}(k)$ plotted in Fig. 16.7b and c, respectively, indicate that such an MPC takes a longer time to steer the system from the same initial state to the desired zero state compared to what is achieved by the constrained MPC.   ∎

16.4   Optimal Force Distribution for Robotic Systems with Closed Kinematic Loops

Because of their use in a wide variety of applications ranging from robotic surgery to space exploration, robotic systems with closed kinematic loops such as multiple manipulators handling a single workload, dextrous hands with fingers holding the workload (see Fig. 16.8), and multilegged vehicles with kinematic chains holding the workload (see Fig. 16.9) have become an increasingly important subject of study in the past several years [14–19]. An issue of central importance for this class of robotic systems is the *force distribution* that determines the joint torques and forces to generate the desired motion of the workload [15].

In Sect. 16.4.1, the force distribution problem for multifinger dextrous hands is described and two models for the contact forces are studied. The optimal force distribution problem is then formulated and solved using LP and SDP in Sects. 16.4.2 and 16.4.3, respectively.

16.4.1   Force Distribution Problem in Multifinger Dextrous Hands

Consider a dextrous hand with m fingers grasping an object such as that depicted in Fig. 16.10 for $m = 3$. The contact force $\mathbf{c}_i$ of the ith finger is supplied by the finger's n_j joint torques τ_{ij} for $j = 1, 2, \ldots, n_j$, and $\mathbf{f}_{ext}$ is an external force exerted on the object. The force distribution problem involves finding vectors $\mathbf{c}_i$ for $i = 1, 2, \ldots, m$ for the problem at hand whose components $c_{i1}, c_{i2}, \ldots c_{ik}$ are the force components required at the contact points so as to assure a stable grasp. The value of integer k depends on the application. See below for typical examples.

The dynamics of the system can be represented by the equation

$$\mathbf{W}\mathbf{c} = -\mathbf{f}_{ext} \qquad (16.61)$$

Fig. 16.6 Performance of MPC in Example 16.5. **a** Angular position $\theta(k)$, **b** angular velocity $\dot{\theta}(k)$, **c** profile of the constrained MPC

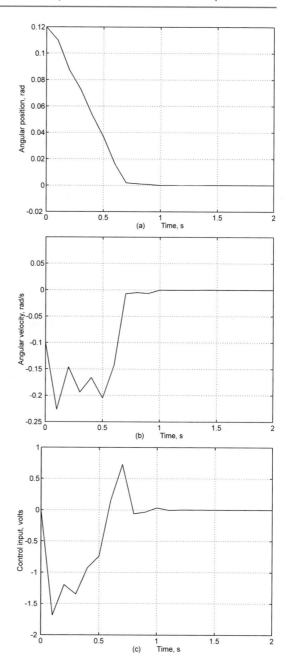

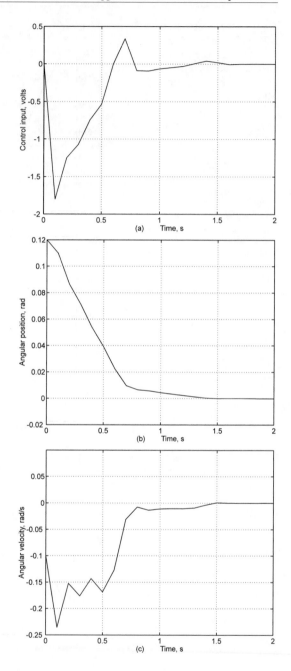

Fig. 16.7 Performance of
MPC in Example 16.4 with
$R = 0.0035$ **a** Profile of the
MPC, **b** angular position
$\theta(k)$, **c** angular velocity $\dot{\theta}(k)$

where **c** is a vector whose components $\mathbf{c}_i$ for $1 \le i \le m$ are the m contact forces, i.e., $\mathbf{c} = [\mathbf{c}_1^T \ \mathbf{c}_2^T \ \dots \ \mathbf{c}_m^T]^T$, and $\mathbf{W} \in R^{6 \times 3m}$ is a matrix whose columns comprise the directions of the m contact forces. The product vector $\mathbf{Wc}$ in Eq. (16.61) is a

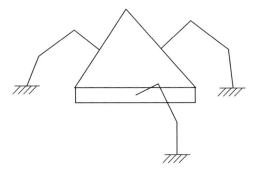

Fig. 16.8 Three coordinated manipulators (also known as a three-finger dextrous hand) grasping an object

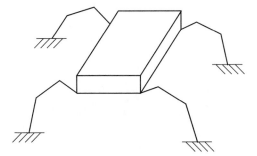

Fig. 16.9 Multilegged vehicle

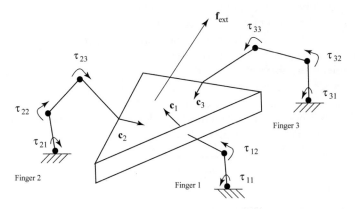

Fig. 16.10 A three-finger hand grasping an object

six-dimensional vector whose first three components represent the overall contact force and last three components represent the overall contact torque relative to a frame of reference with the center of mass of the object as its origin [15, 18].

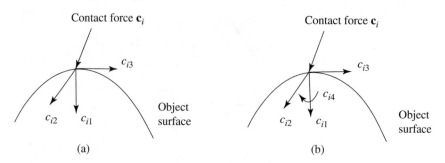

Fig. 16.11 a Point-contact model, **b** soft-finger contact model

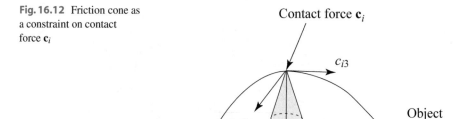

Fig. 16.12 Friction cone as a constraint on contact force $\mathbf{c}_i$

To maintain a stable grasp, the contact forces whose magnitudes are less than the friction force limit must remain positive towards the object surface. There are two commonly used models to describe a contact force, namely, the *point-contact* and *soft-finger contact model*. In the point-contact model, the contact force $\mathbf{c}_i$ has three components, a component c_{i1} that is orthogonal and two components c_{i2} and c_{i3} that are tangential to the object surface as shown in Fig. 16.11a. In the soft-finger contact model, $\mathbf{c}_i$ has an additional component c_{i4}, as shown in Fig. 16.11b, that describes the torsional moment around the normal on the object surface [18].

Friction force plays an important role in stable grasping. In a point-contact model, the friction constraint can be expressed as

$$\sqrt{c_{i2}^2 + c_{i3}^2} \le \mu_i c_{i1} \tag{16.62}$$

where c_{i1} is the normal force component, c_{i2} and c_{i3} are the tangential components of the contact force $\mathbf{c}_i$, and $\mu_i > 0$ denotes the friction coefficient at the contact point. It follows that for a given friction coefficient $\mu_i > 0$, the constraint in Eq. (16.62) describes a *friction cone* as illustrated in Fig. 16.12.

Obviously, the friction force modeled by Eq. (16.62) is nonlinear: for a fixed μ_i and c_{i1}, the magnitude of the tangential force is constrained to within a circle of radius $\mu_i c_{i1}$. A linear constraint for the friction force can be obtained by approximating

Fig. 16.13 Linear approximation for friction cone constraint

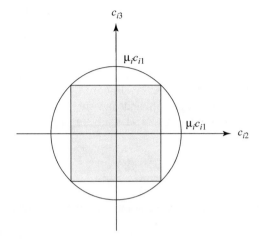

the circle with a square as shown in Fig. 16.13. The approximation involved can be described in terms of the linear constraints [15] illustrated in Fig. 16.13.

$$c_{i1} \geq 0 \tag{16.63a}$$

$$-\frac{\mu_i}{\sqrt{2}} c_{i1} \leq c_{i2} \leq \frac{\mu_i}{\sqrt{2}} c_{i1} \tag{16.63b}$$

$$-\frac{\mu_i}{\sqrt{2}} c_{i1} \leq c_{i3} \leq \frac{\mu_i}{\sqrt{2}} c_{i1} \tag{16.63c}$$

The friction limits in a soft-finger contact model depend on both the torsion and shear forces, and can be described by a linear or an elliptical approximation [18]. The linear model is given by

$$\frac{1}{\mu_i} f_t + \frac{1}{\hat{\mu}_{ti}} |c_{i4}| \leq c_{i1} \tag{16.64}$$

where $\hat{\mu}_{ti}$ is a constant between the torsion and shear limits, μ_i is the tangential friction coefficient, and $f_t = \sqrt{c_{i2}^2 + c_{i3}^2}$. The elliptical model, on the other hand, is described by

$$c_{i1} \geq 0 \tag{16.65a}$$

$$\frac{1}{\mu_i} (c_{i2}^2 + c_{i3}^2) + \frac{1}{\mu_{ti}} c_{i4}^2 \leq c_{i1}^2 \tag{16.65b}$$

where μ_{ti} is a constant.

16.4.2   Solution of Optimal Force Distribution Problem by Using LP

The problem of finding the optimal force distribution of an m-finger dextrous hand is to find the contact forces $\mathbf{c}_i$ for $1 \leq i \leq m$ that optimize a performance index subject to the force balance constraint in Eq. (16.61) and friction-force constraints in one of Eqs. (16.62)–(16.65b).

A typical performance measure in this case is the weighted sum of the m normal force components c_{i1} $(1 \leq i \leq m)$, i.e.,

$$p = \sum_{i=1}^{m} w_i c_{i1} \tag{16.66}$$

If we employ the point-contact model and let

$$\mathbf{c} = \begin{bmatrix} \mathbf{c}_1 \\ \vdots \\ \mathbf{c}_m \end{bmatrix}, \quad \mathbf{c}_i = \begin{bmatrix} c_{i1} \\ c_{i2} \\ c_{i3} \end{bmatrix}, \quad \mathbf{w} = \begin{bmatrix} \mathbf{w}_1 \\ \vdots \\ \mathbf{w}_m \end{bmatrix}, \quad \mathbf{w}_i = \begin{bmatrix} w_i \\ 0 \\ 0 \end{bmatrix}$$

then the objective function in Eq. (16.66) can be expressed as

$$p(\mathbf{c}) = \mathbf{w}^T \mathbf{c} \tag{16.67}$$

and the friction-force constraints in Eqs. (16.63a)–(16.63c) can be written as

$$\mathbf{Ac} \geq \mathbf{0} \tag{16.68}$$

where

$$\mathbf{A} = \begin{bmatrix} \mathbf{A}_1 & & \mathbf{0} \\ & \ddots & \\ \mathbf{0} & & \mathbf{A}_m \end{bmatrix} \quad \text{and} \quad \mathbf{A}_i = \begin{bmatrix} 1 & 0 & 0 \\ \mu_i/\sqrt{2} & -1 & 0 \\ \mu_i/\sqrt{2} & 1 & 0 \\ \mu_i/\sqrt{2} & 0 & -1 \\ \mu_i/\sqrt{2} & 0 & 1 \end{bmatrix}$$

Obviously, the problem of minimizing function $p(\mathbf{c})$ in Eq. (16.67) subject to the linear inequality constraints in Eq. (16.68) and linear equality constraints

$$\mathbf{Wc} = -\mathbf{f}_{ext} \tag{16.69}$$

is an LP problem and many algorithms studied in Chaps. 11 and 12 are applicable. In what follows, the above LP approach is illustrated using a four-finger robot hand grasping a rectangular object. The same robot hand was used in [18] to demonstrate a gradient-flow-based optimization method.

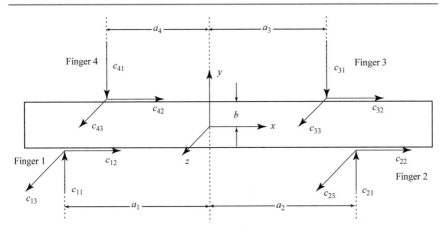

Fig. 16.14 Grasping a rectangular object with four fingers

Example 16.6 Find the optimal contact forces $\mathbf{c}_i$ for $i = 1, 2, \ldots, 4$, that minimize the objective function in Eq. (16.67) subject to the constraints in Eqs. (16.68)–(16.69) for a four-finger robot hand grasping the rectangular object illustrated in Fig. 16.14.

Solution The input data of the problem are given by

$$\mathbf{W}^T = \begin{bmatrix} 0 & 1 & 0 & 0 & 0 & -a_1 \\ 1 & 0 & 0 & 0 & 0 & b \\ 0 & 0 & 1 & -b & a_1 & 0 \\ 0 & 1 & 0 & 0 & 0 & a_2 \\ 1 & 0 & 0 & 0 & 0 & b \\ 0 & 0 & 1 & -b & -a_2 & 0 \\ 0 & -1 & 0 & 0 & 0 & -a_3 \\ 1 & 0 & 0 & 0 & 0 & -b \\ 0 & 0 & 1 & b & -a_3 & 0 \\ 0 & -1 & 0 & 0 & 0 & a_4 \\ 1 & 0 & 0 & 0 & 0 & -b \\ 0 & 0 & 1 & b & a_4 & 0 \end{bmatrix}$$

where $a_1 = 0.1$, $a_2 = 0.15$, $a_3 = 0.05$, $a_4 = 0.065$, and $b = 0.02$. The weights, μ_i, and $\mathbf{f}_{ext}$ are given by $w_i = 1$, $\mu_i = 0.4$ for $1 \le i \le 4$, and

$$\mathbf{f}_{ext} = [0 \ 0 \ -1 \ 0 \ 0 \ 0]^T$$

The rank of matrix $\mathbf{W}$ is 6; hence the solutions of Eq. (16.69) can be characterized by the equation

$$\mathbf{c} = -\mathbf{W}^+ \mathbf{f}_{ext} + \mathbf{V}_\eta \boldsymbol{\phi} \tag{16.70}$$

where $\mathbf{W}^+$ denotes the Moore-Penrose pseudoinverse of $\mathbf{W}$, $\mathbf{V}_\eta$ is the matrix formed using the last 6 columns of $\mathbf{V}$ obtained from the singular-value decomposition $\mathbf{W} = \mathbf{U\Sigma V}^T$, and $\boldsymbol{\phi} \in R^{6\times1}$ is the free parameter vector (see Sect. 10.4). Using Eq. (16.70), the above LP problem is reduced to

$$\text{minimize } \hat{\mathbf{w}}^T \boldsymbol{\phi} \tag{16.71a}$$

$$\text{subject to } \hat{\mathbf{A}}\boldsymbol{\phi} \geq \hat{\mathbf{b}} \tag{16.71b}$$

where

$$\hat{\mathbf{w}} = \mathbf{V}_\eta^T \mathbf{w}, \quad \hat{\mathbf{A}} = \mathbf{A}\mathbf{V}_\eta, \quad \hat{\mathbf{b}} = \mathbf{A}\mathbf{W}^+\mathbf{f}_{ext}$$

The reduced LP problem was solved by using Algorithm 11.1. In [15], this solution method is referred to as the *compact LP method*. If $\boldsymbol{\phi}^*$ is the minimizer of the LP problem in Eqs. (16.71a)–(16.71b), then the minimizer of the original LP problem is given by

$$\mathbf{c}^* = -\mathbf{W}^+\mathbf{f}_{ext} + \mathbf{V}_\eta\boldsymbol{\phi}^*$$

which leads to

$$\mathbf{c}^* = \begin{bmatrix} \mathbf{c}_1^* \\ \mathbf{c}_2^* \\ \mathbf{c}_3^* \\ \mathbf{c}_4^* \end{bmatrix}$$

with

$$\mathbf{c}_1^* = \begin{bmatrix} 1.062736 \\ 0.010609 \\ 0.300587 \end{bmatrix}, \quad \mathbf{c}_2^* = \begin{bmatrix} 0.705031 \\ 0.015338 \\ 0.199413 \end{bmatrix}$$

$$\mathbf{c}_3^* = \begin{bmatrix} 1.003685 \\ -0.038417 \\ 0.283885 \end{bmatrix}, \quad \mathbf{c}_4^* = \begin{bmatrix} 0.764082 \\ 0.012470 \\ 0.216115 \end{bmatrix}$$

The minimum value of $p(\mathbf{c})$ at $\mathbf{c}^*$ was found to be 3.535534.

∎

16.4.3 Solution of Optimal Force Distribution Problem by Using SDP

The LP-based solution discussed in Sect. 16.4.2 is an approximate solution because it was obtained for the case where the quadratic friction-force constraint in Eq. (16.62) is approximated using a linear model. An improved solution can be obtained by formulating the problem at hand as an SDP problem. To this end, we need to convert the friction-force constraints into linear matrix inequalities [18].

For the point-contact case, the friction-force constraint in Eq. (16.62) yields

$$\mu_i c_{i1} \geq 0$$

and

$$\mu_i^2 c_{i1}^2 - (c_{i2}^2 + c_{i3}^2) \geq 0$$

Hence Eq. (16.62) is equivalent to

$$\mathbf{P}_i = \begin{bmatrix} \mu_i c_{i1} & 0 & c_{i2} \\ 0 & \mu_i c_{i1} & c_{i3} \\ c_{i2} & c_{i3} & \mu_i c_{i1} \end{bmatrix} \succeq \mathbf{0} \tag{16.72}$$

(see Problem 16.12a). For an m-finger robot hand, the constraint on point-contact friction forces is given by

$$\mathbf{P}(\mathbf{c}) = \begin{bmatrix} \mathbf{P}_1 & & \mathbf{0} \\ & \ddots & \\ \mathbf{0} & & \mathbf{P}_m \end{bmatrix} \succeq \mathbf{0} \tag{16.73}$$

where $\mathbf{P}_i$ is defined by Eq. (16.72). Similarly, the constraint on the soft-finger friction forces of an m-finger robot hand can be described by Eq. (16.73) where matrix $\mathbf{P}_i$ is given by

$$\mathbf{P}_i = \left[\begin{array}{cccc|ccc} c_{i1} & 0 & 0 & 0 & & & \\ 0 & \alpha_i & 0 & c_{i2} & & \mathbf{0} & \\ 0 & 0 & \alpha_i & c_{i3} & & & \\ 0 & c_{i2} & c_{i3} & \alpha_i & & & \\ \hline & & & & \beta_i & 0 & c_{i2} \\ & \mathbf{0} & & & 0 & \beta_i & c_{i3} \\ & & & & c_{i2} & c_{i3} & \beta_i \end{array} \right] \tag{16.74}$$

with $\alpha_i = \mu_i(c_{i1} + c_{i4}/\hat{\mu}_{ti})$ and $\beta_i = \mu_i(c_{i1} - c_{i4}/\hat{\mu}_{ti})$ for the linear model in Eq. (16.64) or

$$\mathbf{P}_i = \begin{bmatrix} c_{i1} & 0 & 0 & \alpha_i c_{i2} \\ 0 & c_{i1} & 0 & \alpha_i c_{i3} \\ 0 & 0 & c_{i1} & \beta_i c_{i4} \\ \alpha_i c_{i2} & \alpha_i c_{i3} & \beta_i c_{i4} & c_{i1} \end{bmatrix} \tag{16.75}$$

with $\alpha_i = 1/\sqrt{\mu_i}$ and $\beta_i = 1/\sqrt{\mu_{ti}}$ for the elliptical model in Eqs. (16.65a)–(16.65b) (see Problem 16.12(b) and (c)).

Note that matrix $\mathbf{P}(\mathbf{c})$ for both point-contact and soft-finger models is *linear* with respect to parameters c_{i1}, c_{i2}, c_{i3}, and c_{i4}.

The optimal force distribution problem can now be formulated as

$$\text{minimize } p = \mathbf{w}^T \mathbf{c} \tag{16.76a}$$
$$\text{subject to: } \mathbf{Wc} = -\mathbf{f}_{ext} \tag{16.76b}$$
$$\mathbf{P}(\mathbf{c}) \succeq \mathbf{0} \tag{16.76c}$$

where $\mathbf{c} = [\mathbf{c}_1^T \ \mathbf{c}_2^T \ \cdots \ \mathbf{c}_m^T]^T$ with $\mathbf{c}_i = [c_{i1} \ c_{i2} \ c_{i3}]^T$ for the point-contact case or $\mathbf{c}_i = [c_{i1} \ c_{i2} \ c_{i3} \ c_{i4}]^T$ for the soft-finger case, and $\mathbf{P}(\mathbf{c})$ is given by Eq. (16.73) with $\mathbf{P}_i$ defined by Eq. (16.72) for the point-contact case or Eq. (16.75) for the soft-finger case. By using the variable elimination method discussed in Sect. 10.4, the solutions of Eq. (16.76b) can be expressed as

$$\mathbf{c} = \mathbf{V}_\eta \boldsymbol{\phi} + \mathbf{c}_0 \tag{16.77}$$

with $\mathbf{c}_0 = -\mathbf{W}^+ \mathbf{f}_{ext}$ where $\mathbf{W}^+$ is the Moore-Penrose pseudoinverse of $\mathbf{W}$. Thus the problem in Eqs. (16.76a)–(16.76c) reduces to

$$\text{minimize } \hat{p} = \hat{\mathbf{w}}^T \boldsymbol{\phi} \tag{16.78a}$$
$$\text{subject to: } \mathbf{P}(\mathbf{V}_\eta \boldsymbol{\phi} + \mathbf{c}_0) \succeq \mathbf{0} \tag{16.78b}$$

Since $\mathbf{P}(\mathbf{V}_\eta \boldsymbol{\phi} + \mathbf{c}_0)$ is affine with respect to vector $\boldsymbol{\phi}$, the optimization problem in Eqs. (16.78a)–(16.78b) is a standard SDP problem of the type studied in Chap. 13.

Example 16.7 Find the optimal contact forces $\mathbf{c}_i$ for $1 \leq i \leq 4$ that would solve the minimization problem in Eqs. (16.76a)–(16.76c) for the 4-finger robot hand grasping the rectangular object illustrated in Fig. 16.14 using the soft-finger model in Eqs. (16.65a)–(16.65b) with $\mu_i = 0.4$ and $\mu_{ti} = \sqrt{0.2}$ for $1 \leq i \leq 4$.

Solution The input data are given by

$$\mathbf{w} = [1\ 0\ 0\ 0\ 1\ 0\ 0\ 0\ 1\ 0\ 0\ 0\ 1\ 0\ 1\ 0\ 0\ 0\ 1\ 0\ 0\ 0]^T$$

$$\mathbf{f}_{ext} = \begin{bmatrix} 1 & 1 & -1 & 0 & 0.5 & 0.5 \end{bmatrix}^T$$

$$\mathbf{W}^T = \begin{bmatrix}
0 & 1 & 0 & 0 & 0 & -a_1 \\
1 & 0 & 0 & 0 & 0 & b \\
0 & 0 & 1 & -b & a_1 & 0 \\
0 & 0 & 0 & 0 & -1 & 0 \\
0 & 1 & 0 & 0 & 0 & a_2 \\
1 & 0 & 0 & 0 & 0 & b \\
0 & 0 & 1 & -b & -a_2 & 0 \\
0 & 0 & 0 & 0 & -1 & 0 \\
0 & -1 & 0 & 0 & 0 & -a_3 \\
1 & 0 & 0 & 0 & 0 & -b \\
0 & 0 & 1 & b & -a_3 & 0 \\
0 & 0 & 0 & 0 & 1 & 0 \\
0 & -1 & 0 & 0 & 0 & a_4 \\
1 & 0 & 0 & 0 & 0 & -b \\
0 & 0 & 1 & b & a_4 & 0 \\
0 & 0 & 0 & 0 & 1 & 0
\end{bmatrix}$$

where the numerical values of a_1, a_2, a_3, and b are the same as in Example 16.6.

By applying Algorithm 13.5 to the SDP problem in Eqs. (16.78a)–(16.78b), the minimizer $\boldsymbol{\phi}^*$ was found to be

$$\boldsymbol{\phi}^* = \begin{bmatrix}
-2.419912 \\
-0.217252 \\
3.275539 \\
0.705386 \\
-0.364026 \\
-0.324137 \\
-0.028661 \\
0.065540 \\
-0.839180 \\
0.217987
\end{bmatrix}$$

Eq. (16.77) then yields

$$\mathbf{c}^* = \mathbf{V}_\eta \boldsymbol{\phi}^* + \mathbf{c}_0 = \begin{bmatrix} \mathbf{c}_1^* \\ \mathbf{c}_2^* \\ \mathbf{c}_3^* \\ \mathbf{c}_4^* \end{bmatrix}$$

where

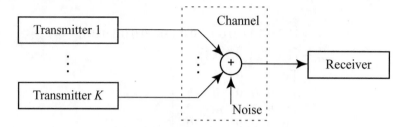

Fig. 16.15 A multiuser communication system

$$
\mathbf{c}_1^* = \begin{bmatrix} 2.706396 \\ -1.636606 \\ 0.499748 \\ -0.015208 \end{bmatrix}, \quad
\mathbf{c}_2^* = \begin{bmatrix} 0.003041 \\ -0.000633 \\ 0.000252 \\ -0.000172 \end{bmatrix}
$$

$$
\mathbf{c}_3^* = \begin{bmatrix} 3.699481 \\ 0.638543 \\ 0.500059 \\ -0.541217 \end{bmatrix}, \quad
\mathbf{c}_4^* = \begin{bmatrix} 0.009955 \\ -0.001303 \\ -0.000059 \\ 0.000907 \end{bmatrix}
$$

The minimum value of $p(\mathbf{c})$ at $\mathbf{c}^*$ is $p(\mathbf{c})^* = 6.418873$.

∎

16.5 Multiuser Detection in Wireless Communication Channels

Multiuser communication systems are systems that enable users to transmit information through a common channel [20] as illustrated in Fig. 16.15. A typical system of this type is a cellular communication system where a number of mobile users in a cell send their information to the receiver at the base station of the cell.

There are three basic multiple access methods for multiuser communication, namely, *frequency-division multiple access* (FDMA), *time-division multiple access* (TDMA), and *code-division multiple access* (CDMA). In FDMA, the available channel bandwidth is divided into a number of nonoverlapping subchannels and each subchannel is assigned to a user. In TDMA, the unit time duration known as the frame duration is divided into several nonoverlapping time intervals, and each time interval is assigned to a user. In CDMA, each user is assigned a distinct *code sequence*, which spreads the user's information signal across the assigned frequency band. These code sequences have small cross-correlation with each other so that signals from different users can be separated at the receiver using a bank of match filters, each performing cross-correlation of the received signal with a particular code sequence.

In order to accommodate asynchronous users in CDMA channels, practical code sequences are not orthogonal [21]. This nonorthogonality leads to nonzero cross-correlation between each pair of code sequences. Therefore, users interfere with each other and any interferer with sufficient power at the receiver can cause significant performance degradation. Multiuser detection is a demodulation technique that can perform quite effectively in the presence of multiple access interference. The purpose of this section is to demonstrate that several multiuser detection problems can be addressed using modern optimization methods. In Sect. 16.5.1, the CDMA channel model and the maximum-likelihood (ML) multiuser detector [22] are reviewed. A near-optimal multiuser detector for direct-sequence CDMA (DS-CDMA) channels using SDP relaxation [23] is described in Sect. 16.5.2. In digital communication systems, performance is usually measured in terms of the probability that a signal bit is in error at the receiver output, and this probability is referred to as the *bit-error rate* (BER). In Sect. 16.5.3, we describe a linear multiuser detection algorithm based on minimizing the BER subject to a set of reasonable con-straints [24].   ∎

16.5.1   Channel Model and ML Multiuser Detector

16.5.1.1   CDMA Channel Model

We consider a DS-CDMA system where K users transmit information bits through a common channel. The bit interval of each user is T_b seconds and each information bit belongs to the set $\{1, -1\}$. Each signal is assigned a signature waveform $s(t)$, often called the *spreading sequence*, given by

$$s(t) = \sum_{i=1}^{N} (-1)^{c_i} p_{T_c}[t - (i - 1)T_c] \quad \text{for } t \in [0, \ T_b]$$

where $p_{T_c}(t)$ is a rectangular pulse which takes the value of one for $0 \le t \le T_c$ and zero elsewhere, $\{c_1, c_2, \ldots, c_N\}$ is a binary sequence, and $N = T_b/T_c$ is the length of the signature waveform, which is often referred to as the *spreading gain*. Typically, the waveform of $p_{T_c}(t)$ is common to all the users, and it is the binary sequence $\{c_1, c_2, \ldots, c_N\}$ assigned to each user that distinguishes the different signature waveforms. One of the commonly used binary sequences is the Gold sequence, which has low crosscorrelations for all possible cyclic shifts [20]. The signature waveforms are normalized to have unit energy, i.e., $||s_k(t)||^2 = 1$ for $1 \le k \le K$. The received baseband signal is given by

$$y(t) = \sum_{i=0}^{\infty} \sum_{k=1}^{K} A_k^i b_k^i s_k(t - iT_b - \tau_k) + n(t) \tag{16.79}$$

where b_k^i is an information bit, τ_k is the transmission delay, A_k^i is the signal amplitude of the kth user, and $n(t)$ is additive white Gaussian noise (AWGN) with variance

σ^2. A DS-CDMA system is said to be *synchronous* if τ_k in Eq. (16.79) is zero for $1 \leq k \leq K$, and thus

$$y(t) = \sum_{k=1}^{K} A_k b_k s_k(t) + n(t) \tag{16.80}$$

where t can assume values in the bit interval $[0, \ T_b]$.

Demodulation is achieved by filtering the received signal $y(t)$ with a bank of matched filters. The filter bank consists of K filters, each matched to a signature waveform, and the filtered signals are sampled at the end of each bit interval. The outputs of the matched filters are given by

$$y_k = \int_0^{T_b} y(t) s_k(t) \, dt \quad \text{for } 1 \leq k \leq K \tag{16.81}$$

Using Eq. (16.80), Eq. (16.81) can be expressed as

$$y_k = A_k b_k + \sum_{j \neq k} A_j b_j \rho_{jk} + n_k \quad \text{for } 1 \leq k \leq K \tag{16.82}$$

where

$$\rho_{jk} = \int_0^{T_b} s_j(t) s_k(t) \, dt$$

and

$$n_k = \int_0^{T_b} n(t) s_k(t) \, dt$$

The discrete-time synchronous model in Eq. (16.82) can be described in matrix form as

$$\mathbf{y} = \mathbf{R} \mathbf{A} \mathbf{b} + \mathbf{n} \tag{16.83}$$

where $\mathbf{y} = [y_1 \ y_2 \ \cdots \ y_K]^T$, $\mathbf{A} = \text{diag}\{A_1, \ A_2, \ \ldots, \ A_K\}$, $\mathbf{b} = [b_1 \ b_2 \ \cdots \ b_K]^T$, $\mathbf{R}_{ij} = \rho_{ij}$, and $\mathbf{n} = [n_1 \ n_2 \ \cdots \ n_K]^T$. Since $n(t)$ in Eq. (16.80) is an AWGN with variance σ^2, the term $\mathbf{n}$ in Eq. (16.83) is a zero-mean Gaussian noise vector with covariance matrix $\sigma^2 \mathbf{R}$.

If we consider an ideal channel which is free of background noise and the signature waveforms are orthogonal to each other, then Eq. (16.82) assumes the form $y_k = A_k b_k$. In such a case the information bit b_k can be perfectly detected based on the output of the kth matched filter, y_k. In a realistic CDMA channel, however, the signature waveforms are nonorthogonal [21] and hence the second term at the right-hand side of Eq. (16.82), which quantifies the multiple access interference (MAI), is always nonzero. The MAI in conjunction with the noise represented by term n_k in Eq. (16.82) can in many cases be so large that it is difficult to estimate the transmitted information based on the outputs of the matched filters without further processing. A *multiuser detector* is essentially a digital signal processing algorithm or processor that takes **y** as its input to estimate the transmitted information vector **b** such that a low probability of error is achieved.

16.5.1.2   ML Multiuser Detector

The goal of the optimal multiuser detector is to generate an estimate of the information vector **b** in Eq. (16.83) that maximizes the log-likelihood function defined by

$$f(\mathbf{b}) = \exp\left(-\frac{1}{2\sigma^2}\int_0^{T_b}[y(t) - \sum_{k=1}^{K}A_k b_k s_k(t)]^2\,dt\right)$$

which is equivalent to maximizing the quadratic function

$$\Omega(\mathbf{b}) = \left[\sum_{k=1}^{K}A_k b_k s_k(t)\right]^2 = 2\mathbf{b}^T\mathbf{Ay} - \mathbf{b}^T\mathbf{ARAb}$$

By defining the unnormalized crosscorrelation matrix as $\mathbf{H} = \mathbf{ARA}$ and letting $\mathbf{p} = -2\mathbf{Ay}$, the ML detector is characterized by the solution of the combinatorial optimization problem [22]

$$\text{minimize } \mathbf{x}^T\mathbf{Hx} + \mathbf{x}^T\mathbf{p} \tag{16.84a}$$

$$\text{subject to: } \mathbf{x}_i \in \{1, -1\} \quad \text{for } i = 1, 2, \ldots, K \tag{16.84b}$$

Because of the binary constraints in Eq. (16.84b), the optimization problem in Eqs. (16.84a)–(16.84b) is an integer programming (IP) problem. Its solution can be obtained by exhaustive evaluation of the objective function over 2^K possible values of **x**. However, the amount of computation involved becomes prohibitive even for a moderate number of users, say, $K = 100$.

16.5.2   Near-Optimal Multiuser Detector Using SDP Relaxation

The near-optimal multiuser detector described in [23] is based on a relaxation of the so-called *MAX-CUT problem* as detailed below.

16.5.2.1 SDP Relaxation of MAX-CUT Problem

We begin by examining the MAX-CUT problem, which is a well-known IP problem in graph theory. It can be formulated as

$$\text{maximize } \tfrac{1}{2}\sum_{i<j}\sum w_{ij}(1 - x_i x_j) \qquad (16.85\text{a})$$

$$\text{subject to: } \quad x_i \in \{1, -1\} \quad \text{ for } 1 \le i \le n \qquad (16.85\text{b})$$

where w_{ij} denotes the weight from node i to node j in the graph. The constraints in Eq. (16.85b) can be expressed as $x_i^2 = 1$ for $1 \le i \le n$. If we define a symmetric matrix $\mathbf{W} = \{w_{ij}\}$ with $w_{ii} = 0$ for $1 \le i \le n$, then the objective function in Eq. (16.85a) can be expressed as

$$\tfrac{1}{2}\sum_{i<j}\sum w_{ij}(1 - x_i x_j) = \tfrac{1}{4}\sum\sum w_{ij} - \tfrac{1}{4}\mathbf{x}^T \mathbf{W}\mathbf{x}$$

$$= \tfrac{1}{4}\sum\sum w_{ij} - \tfrac{1}{4}\text{trace}(\mathbf{W}\mathbf{X})$$

where trace$(\cdot)$ denotes the trace of the matrix and $\mathbf{X} = \mathbf{x}\mathbf{x}^T$ with $\mathbf{x} = [x_1 \ x_2 \ \cdots \ x_n]^T$ (see Sect. A.7). Note that the set of matrices $\{\mathbf{X} : \mathbf{X} = \mathbf{x}\mathbf{x}^T \text{ with } x_i^2 = 1 \text{ for } 1 \le i \le n\}$ can be characterized by $\{\mathbf{X} : x_{ii} = 1 \text{ for } 1 \le i \le n, \mathbf{X} \succeq \mathbf{0}, \text{ and } \text{rank}(\mathbf{X}) = 1\}$ where x_{ii} denotes the ith diagonal element of $\mathbf{X}$. Hence the problem in Eqs. (16.85a)–(16.85b) can be expressed as

$$\text{minimize trace}(\mathbf{W}\mathbf{X}) \qquad (16.86\text{a})$$
$$\text{subject to: } \quad \mathbf{X} \succeq \mathbf{0} \qquad (16.86\text{b})$$
$$x_{ii} = 1 \quad \text{ for } 1 \le i \le n \qquad (16.86\text{c})$$
$$\text{rank}(\mathbf{X}) = 1 \qquad (16.86\text{d})$$

In [25], Geomans and Williamson proposed a relaxation of the above problem by removing the rank constraint in Eq. (16.86d), which leads to

$$\text{minimize trace}(\mathbf{W}\mathbf{X}) \qquad (16.87\text{a})$$
$$\text{subject to : } \quad \mathbf{X} \succeq \mathbf{0} \qquad (16.87\text{b})$$
$$x_{ii} = 1 \quad \text{ for } 1 \le i \le n \qquad (16.87\text{c})$$

Note that the objective function in Eq. (16.87a) is a linear function of $\mathbf{X}$ and the constraints in Eqs. (16.87b)–(16.87c) can be combined into an LMI as

$$\sum_{i>j}\sum x_{ij}\mathbf{F}_{ij} + \mathbf{I} \succeq \mathbf{0}$$

where, for each (i, j) with $i > j$, $\mathbf{F}_{ij}$ is a symmetric matrix whose (i, j)th and (j, i)th components are equal to one while the other components are equal to zero, for example, for $n = 4$, $i = 3$, and $j = 1$ we would have

$$\mathbf{F}_{31} = \begin{bmatrix} 0\,0\,1\,0 \\ 0\,0\,0\,0 \\ 1\,0\,0\,0 \\ 0\,0\,0\,0 \end{bmatrix}$$

The problem in Eqs. (16.87a)–(16.87c) fits into the formulation in Eqs. (13.49a)–(13.49c) and, therefore, is an SDP problem. For this reason, it is often referred to as an *SDP relaxation* of the IP problem in Eqs. (16.85a)–(16.85b) and, equivalently, of the problem in Eqs. (16.86a)–(16.86d).

If we denote the minimum values of the objective functions in the problems of Eqs. (16.86a)–(16.86d) and Eqs. (16.87a)–(16.87c) as μ^* and ν^*, respectively, then since the feasible region of the problem in Eqs. (16.86a)–(16.86d) is a subset of the feasible region of the problem in Eqs. (16.87a)–(16.87c), we have $\nu^* \le \mu^*$. Further, it has been shown that if the weights w_{ij} are all nonnegative, then $\nu^* \ge 0.87856\mu^*$ [26]. Therefore, we have

$$0.87856\mu^* \le \nu^* \le \mu^* \tag{16.88}$$

This indicates that the solution of the SDP problem in Eqs. (16.87a)–(16.87c) is in general a good approximation of the solution of the problem in Eqs. (16.86a)–(16.86d) . It is the good quality of the approximation in conjunction with the SDP's polynomial-time computational complexity that makes the Geomans-Williamson SDP relaxation an attractive optimization tool for combinatorial minimization problems. As a consequence, this approach has found applications in graph optimization, network management, and scheduling [27,28]. In what follows, we present an SDP-relaxation-based algorithm for multiuser detection.

16.5.2.2   An SDP-Relaxation-Based Multiuser Detector
Let

$$\hat{\mathbf{X}} = \begin{bmatrix} \mathbf{xx}^T & \mathbf{x} \\ \mathbf{x}^T & 1 \end{bmatrix} \quad \text{and} \quad \mathbf{C} = \begin{bmatrix} \mathbf{H} & \mathbf{p}/2 \\ \mathbf{p}^T/2 & 1 \end{bmatrix} \tag{16.89}$$

By using the property that $\text{trace}(\mathbf{AB}) = \text{trace}(\mathbf{BA})$, the objective function in Eq. (16.84a) can be expressed as

$$\mathbf{x}^T \mathbf{Hx} + \mathbf{x}^T \mathbf{p} = \text{trace}(\mathbf{C}\hat{\mathbf{X}})$$

(see Problem 16.14(*a*)). Using an argument similar to that in Sect. 16.5.2.1, the constraint in Eq. (16.84b) can be converted to

$$\hat{\mathbf{X}} \succeq \mathbf{0}, \quad \hat{x}_{ii} = 1 \quad \text{for } 1 \le i \le K+1 \tag{16.90a}$$
$$\text{rank}(\hat{\mathbf{X}}) = 1 \tag{16.90b}$$

where $\hat{x}_{ii}$ denotes the ith diagonal element of $\hat{\mathbf{X}}$ (see Problem 16.14(b)). By removing the rank constraint in Eq. (16.90b), we obtain an SDP relaxation of the optimization problem in Eqs. (16.84a)–(16.84b) as

$$\text{minimize trace}(\mathbf{C}\hat{\mathbf{X}}) \qquad (16.91a)$$

$$\text{subject to:} \quad \hat{\mathbf{X}} \succeq \mathbf{0} \qquad (16.91b)$$
$$\hat{x}_{ii} = 1 \quad \text{for } i = 1, 2, \ldots, K+1 \qquad (16.91c)$$

The variables in the original problem in Eqs. (16.84a)–(16.84b) assume only the values of 1 or -1 while the variable $\hat{\mathbf{X}}$ in the SDP minimization problem Eqs. (16.91a)–(16.91c) has real-valued components. In what follows, we describe two approaches that can be used to generate a binary solution for the problem in Eqs. (16.84a)–(16.84b) based on the solution $\hat{\mathbf{X}}$ of the SDP problem in Eqs. (16.91a)–(16.91c).

Let the solution of the problem in Eqs. (16.91a)–(16.91c) be denoted as $\hat{\mathbf{X}}^*$. It follows from Eq. (16.89) that $\hat{\mathbf{X}}^*$ is a $(K+1) \times (K+1)$ symmetric matrix of the form

$$\hat{\mathbf{X}}^* = \begin{bmatrix} \mathbf{X}^* & \mathbf{x}^* \\ \mathbf{x}^{*T} & 1 \end{bmatrix} \qquad (16.92)$$

with

$$\hat{x}_{ii}^* = 1 \quad \text{for } i = 1, 2, \ldots, K.$$

In view of Eq. (16.92), our first approach is simply to apply operator sgn($\cdot$) to $\mathbf{x}^*$ in Eq. (16.92), namely,

$$\hat{\mathbf{b}} = \text{sgn}(\mathbf{x}^*) \qquad (16.93)$$

where $\mathbf{x}^*$ denotes the vector formed by the first K components in the last column of $\hat{\mathbf{X}}^*$.

At the cost of more computation, a better binary solution can be obtained by using the eigendecomposition of matrix $\hat{\mathbf{X}}^*$, i.e., $\hat{\mathbf{X}}^* = \mathbf{U}\mathbf{S}\mathbf{U}^T$, where $\mathbf{U}$ is an orthogonal and $\mathbf{S}$ is a diagonal matrix with the eigenvalues of $\hat{\mathbf{X}}^*$ as its diagonal components in decreasing order (see Sect. A.9). It is well known that an optimal rank-one approximation of $\hat{\mathbf{X}}^*$ in the L_2 norm sense is given by $\lambda_1 \mathbf{u}_1 \mathbf{u}_1^T$, where λ_1 is the largest eigenvalue of $\hat{\mathbf{X}}^*$ and $\mathbf{u}_1$ is the eigenvector associated with λ_1 [29] . If we denote the vector formed by the first K components of $\mathbf{u}_1$ as $\tilde{\mathbf{u}}$, and the last component of $\mathbf{u}_1$ by u_{K+1}, i.e.,

$$\mathbf{u}_1 = \begin{bmatrix} \tilde{\mathbf{u}} \\ u_{K+1} \end{bmatrix}$$

then the optimal rank-one approximation of $\hat{\mathbf{X}}^*$ can be written as

$$\hat{\mathbf{X}}^* \approx \lambda_1 \mathbf{u}_1 \mathbf{u}_1^T = \lambda_1 \begin{bmatrix} \tilde{\mathbf{u}}\tilde{\mathbf{u}}^T & u_{K+1}\tilde{\mathbf{u}} \\ u_{K+1}\tilde{\mathbf{u}}^T & u_{K+1}^2 \end{bmatrix}$$

$$= \frac{\lambda_1}{u_{K+1}^2} \begin{bmatrix} \tilde{\mathbf{x}}_1 \tilde{\mathbf{x}}_1^T & \tilde{\mathbf{x}}_1 \\ \tilde{\mathbf{x}}_1^T & 1 \end{bmatrix} \qquad (16.94)$$

where $\tilde{\mathbf{x}}_1 = u_{K+1}\tilde{\mathbf{u}}$. Since $\lambda_1 > 0$, on comparing Eqs. (16.92) and (16.94) we note that the signs of the components of vector $\tilde{\mathbf{x}}_1$ are likely to be the same as the signs of the corresponding components in vector $\mathbf{x}^*$. Therefore, a binary solution of the problem in Eqs. (16.84a)–(16.84b) can be generated as

$$\hat{\mathbf{b}} = \begin{cases} \text{sgn}(\tilde{\mathbf{u}}) & \text{if } u_{K+1} > 0 \\ -\text{sgn}(\tilde{\mathbf{u}}) & \text{if } u_{K+1} < 0 \end{cases} \qquad (16.95)$$

16.5.2.3   Solution Suboptimality

Because of the relaxation involved, the detector described is *suboptimal* but, as mentioned in Sect. 16.5.2.1, the SDP relaxation of the MAX-CUT problem yields a good suboptimal solution. However, there are two important differences between the SDP problems in Eqs. (16.87a)–(16.87c) and Eqs. (16.91a)–(16.91c) : The diagonal components of $\mathbf{W}$ in Eq. (16.87a) are all zero whereas those of $\mathbf{C}$ in Eq. (16.91a) are all strictly positive; and although the off-diagonal components in $\mathbf{W}$ are assumed to be nonnegative, matrix $\mathbf{C}$ may contain negative off-diagonal components. Consequently, the bounds in Eq. (16.88) do not always hold for the SDP problem in Eqs. (16.91a)–(16.91c). However, as will be demonstrated in terms of some experimental results presented below, the near-optimal detector offers comparable performance to that of the optimal ML detector.

In the next section, we describe an alternative but more efficient SDP-relaxation-based detector.

16.5.2.4   Efficient-Relaxation-Based Detector Via Duality

Although efficient interior-point algorithms such as those in [28, 30] (see Sects. 13.7–13.8) can be applied to solve the SDP problem in Eqs. (16.91a)–(16.91c), numerical difficulties can arise because the number of variables can be quite large even for the case of a moderate number of users. For example, if $K = 20$, the dimension of vector $\mathbf{x}$ in Eq. (16.89) is 20 and the number of variables in $\hat{\mathbf{X}}$ becomes $K(K+1)/2 = 210$. In this section, we present a more efficient approach for the solution of the SDP problem under consideration. Essentially, we adopt an indirect approach by first solving the dual SDP problem, which involves a much smaller number of variables, and then convert the solution of the dual problem to that of the primal SDP problem.

We begin by rewriting the SDP problem in Eqs. (16.91a)–(16.91c) as

$$\text{minimize trace}(\mathbf{C}\hat{\mathbf{X}}) \tag{16.96a}$$

$$\text{subject to:} \quad \hat{\mathbf{X}} \succeq \mathbf{0} \tag{16.96b}$$

$$\text{trace}(\mathbf{A}_i \mathbf{X}) = 1 \quad \text{for } i = 1, 2, \ldots, K+1 \tag{16.96c}$$

where $\mathbf{A}_i$ is a diagonal matrix whose diagonal components are all zero except for the ith component, which is 1. It follows from Chap. 13 that the dual of the problem in Eqs. (16.96a)–(16.96c) is given by

$$\text{minimize} \; -\mathbf{b}^T \mathbf{y}_{K+1} \tag{16.97a}$$

$$\text{subject to:} \quad \mathbf{S} = \mathbf{C} - \sum_{i=1}^{K+1} y_i \mathbf{A}_i \tag{16.97b}$$

$$\mathbf{S} \succeq \mathbf{0} \tag{16.97c}$$

where $\mathbf{y} = [y_1 \; y_2 \; \cdots \; y_{K+1}]^T$ and $\mathbf{b} = [1 \; 1 \; \cdots \; 1]^T \in C^{(K+1)\times 1}$. Evidently, the dual problem in Eqs. (16.97a)–(16.97c) involves only $K+1$ variables and it is, therefore, much easier to solve than the primal problem. Any efficient interior-point algorithm can be used for the solution such as the projective algorithm proposed by Nemirovski and Gahinet [31].

In order to obtain the solution of the primal SDP problem in Eqs. (16.96a)–(16.96c), we need to carry out some analysis on the Karush-Kuhn-Tucker (KKT) conditions for the solutions of the problems in Eqs. (16.96a)–(16.96c) and Eqs. (16.97a)–(16.97c). The KKT conditions state that the set $\{\hat{\mathbf{X}}^*, \; \mathbf{y}^*\}$ solves the problems in Eqs. (16.96a)–(16.96c) and Eqs. (16.97a)–(16.97c) if and only if they satisfy the conditions

$$\sum_{i=1}^{K+1} y_i^* \mathbf{A}_i + \mathbf{S}^* = \mathbf{C} \tag{16.98a}$$

$$\text{trace}(\mathbf{A}_i \hat{\mathbf{X}}^*) = 1 \quad \text{for } i = 1, 2, \ldots, K+1 \tag{16.98b}$$

$$\mathbf{S}^* \hat{\mathbf{X}}^* = \mathbf{0} \tag{16.98c}$$

$$\hat{\mathbf{X}}^* \succeq \mathbf{0} \text{ and } \mathbf{S}^* \succeq \mathbf{0} \tag{16.98d}$$

From Eq. (16.98a),

we have

$$\mathbf{S}^* = \mathbf{C} - \sum_{i=1}^{K+1} y_i^* \mathbf{A}_i \tag{16.99}$$

Since the solution $\mathbf{y}^*$ is typically obtained by using an *iterative* algorithm, e.g., the projective algorithm of Nemirovski and Gahinet, $\mathbf{y}^*$ can only be a good approximate solution of the problem in Eqs. (16.98a)–(16.98d), which means that $\mathbf{y}^*$ is in the *interior* of the feasible region. Consequently, matrix $\mathbf{S}^*$ remains *positive definite*. Therefore, the set $\{\mathbf{y}^*,\ \mathbf{S}^*,\ \hat{\mathbf{X}}^*\}$ can be regarded as a point in the feasible region that is sufficiently close to the limiting point of the central path for the problems in Eqs. (16.96a)–(16.96c) and Eqs. (16.97a)–(16.97c). Recall that the central path is defined as a parameterized set $\{\mathbf{y}(\tau),\ \mathbf{S}(\tau),\ \hat{\mathbf{X}}(\tau)\ \text{for}\ \tau > 0\}$ that satisfies the modified KKT conditions

$$\sum_{i=1}^{K+1} y_i(\tau)\mathbf{A}_i + \mathbf{S}(\tau) = \mathbf{C} \tag{16.100a}$$

$$\text{tr}(\mathbf{A}_i\hat{\mathbf{X}}(\tau)) = 1 \quad \text{for}\ i = 1,\ 2,\ \ldots,\ K+1 \tag{16.100b}$$

$$\mathbf{S}(\tau)\hat{\mathbf{X}}(\tau) = \tau\mathbf{I} \tag{16.100c}$$

$$\hat{\mathbf{X}}(\tau) \succeq \mathbf{0} \ \text{and}\ \mathbf{S}(\tau) \succeq \mathbf{0} \tag{16.100d}$$

The relation between Eqs. (16.98a)–(16.98d) and Eqs. (16.100a)–(16.100d) becomes transparent since the entire central path defined by Eqs. (16.100a)–(16.100d) lies in the interior of the feasible region and as $\tau \to 0$, the path converges to the solution set $\{\mathbf{y}^*,\ \mathbf{S}^*,\ \hat{\mathbf{X}}^*\}$ that satisfies Eqs. (16.98a)–(16.98d).

From Eq. (16.100c), it follows that

$$\hat{\mathbf{X}}(\tau) = \tau\mathbf{S}^{-1}(\tau)$$

which suggests an approximate solution of Eqs. (16.96a)–(16.96c) as

$$\hat{\mathbf{X}} = \tau(\mathbf{S}^*)^{-1} \tag{16.101}$$

for some sufficiently small $\tau > 0$, where $\mathbf{S}^*$ is given by Eq. (16.99). In order for matrix $\hat{\mathbf{X}}$ in Eq. (16.101) to satisfy the equality constraints in Eq. (16.96c), $\hat{\mathbf{X}}$ needs to be slightly modified using a scaling matrix $\boldsymbol{\Pi}$ as

$$\hat{\mathbf{X}}^* = \boldsymbol{\Pi}(\mathbf{S}^*)^{-1}\boldsymbol{\Pi} \tag{16.102a}$$

where

$$\boldsymbol{\Pi} = \text{diag}\{\xi_1^{1/2}\ \xi_2^{1/2}\ \cdots\ \xi_{K+1}^{1/2}\} \tag{16.102b}$$

and ξ_i is the ith diagonal component of $(\mathbf{S}^*)^{-1}$. In Eq. (16.102a) we have pre- and post-multiplied $(\mathbf{S}^*)^{-1}$ by $\boldsymbol{\Pi}$ so that matrix $\hat{\mathbf{X}}^*$ remains *symmetric* and *positive definite*. It is worth noting that by imposing the equality constraints in Eq. (16.96c) on $\hat{\mathbf{X}}$, the parameter τ in Eq. (16.101) is absorbed in the scaling matrix $\boldsymbol{\Pi}$.

In summary, an approximate solution $\hat{\mathbf{X}}$ of the SDP problem in Eqs. (16.96a)–(16.96c) can be efficiently obtained by using the following algorithm.

Algorithm 16.1 SDP-relaxation algorithm based on dual problem
Step 1
Form matrix $\mathbf{C}$ using Eq. (16.89).
Step 2
Solve the dual SDP problem in Eqs. (16.97a)–(16.97c) and let its solution be
$\mathbf{y}^*$.
Step 3
Compute $\mathbf{S}^*$ using Eq. (16.99).
Step 4
Compute $\hat{\mathbf{X}}^*$ using Eqs. (16.102a)–(16.102b).
Step 5
Compute $\hat{\mathbf{b}}$ using Eq. (16.93) or Eq. (16.95).

We conclude this section with two remarks on the computational complexity of
the above algorithm and the accuracy of the solution obtained. To a large extent, the
mathematical complexity of the algorithm is determined by Steps 2 and 4 where a
$(K + 1)$-variable SDP problem is solved and a $(K + 1) \times (K + 1)$ positive definite matrix is inverted, respectively. Consequently, the dual approach reduces the
amount of computation required considerably compared to that required to solve the
$K(K + 1)/2$-variable SDP problem in Eqs. (16.96a)–(16.96c) directly. Concerning
the accuracy of the solution, we note that it is the binary solution that determines
the performance of the multiuser detector. Since the binary solution is the output of
the sign operation (see Eqs. (16.93) and (16.95)), the approximation introduced in
Eqs. (16.102a)–(16.102b) is expected to have an insignificant negative effect on the
solution.

Example 16.8 Apply the primal and dual SDP-relaxation-based multiuser detectors
to a six-user synchronous system and compare their performance with that of the ML
detector described in Sect. 16.5.1.2 in terms of bit-error rate (BER) and computational
complexity.

Solution For the sake of convenience, we refer to the detectors based on the primal
and dual problems of Sect. 16.5.2.2 and Sect. 16.5.2.4 as the SDP relaxation primal
(SDPR-P) and the SDP relaxation dual (SDPR-D) detectors, respectively. The SDP
problems in Eqs. (16.91a)–(16.91c) and Eqs. (16.97a)–(16.97c) for the SDPR-P and
SDPR-D detectors were solved by using Algorithms 14.1 and 14.4, respectively. The
user signatures used in the simulations were 15-chip Gold sequences. The received
signal powers of the six users were set to 5, 3, 1.8, 0.6, 0.3, and 0.2, respectively. The
last (weakest) user with power 0.2 was designated as the desired user. The average
BERs for the SDPR-P, SDPR-D, and ML detectors are plotted versus the signal-to-
noise ratio (SNR) in Fig. 16.16, and as can be seen the demodulation performance
of the SDPR-P and SDPR-D detectors is consistently very close to that of the ML
detector.

The computational complexity of the detectors was evaluated in terms of CPU time
and the results for the SDPR-P, SDPR-D, and ML detectors are plotted in Fig. 16.17a
and b versus the number of active users. As expected, the amount of computation

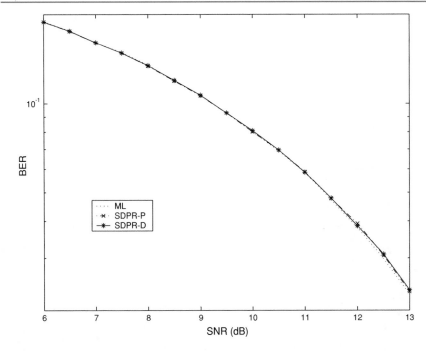

Fig. 16.16 BER of six-user synchronous DS-CDMA system in AWGN channel

required by the ML detector increases exponentially with the number of users as shown in Fig. 16.17a. The SDPR detectors reduce the amount of computation to less than 1 percent and between the SDPR-P and SDPR-D detectors, the latter one, namely, the one based on the dual problem, is significantly more efficient as can be seen in Fig. 16.17b.   ∎

16.5.3   A Constrained Minimum-BER Multiuser Detector

16.5.3.1   Problem Formulation
Although the SDP-based detectors described in Sect. 16.5.2 achieve near-optimal performance with reduced computational complexity compared to that of the ML detector, the amount of computation they require is still too large for real-time applications. A more practical solution is to develop *linear* multiuser detectors that estimate the users' information bits by processing the observation data with an FIR digital filter. Several linear detectors that offer satisfactory performance have been recently developed [21]. However, in general, these detectors do not provide the lowest BER and, therefore, it is of interest to develop a constrained minimum-BER detector that minimizes the BER directly.

We consider a DS-CDMA channel with K synchronous users whose continuous-time model is given by Eq. (16.80). Within the observation window, the critically

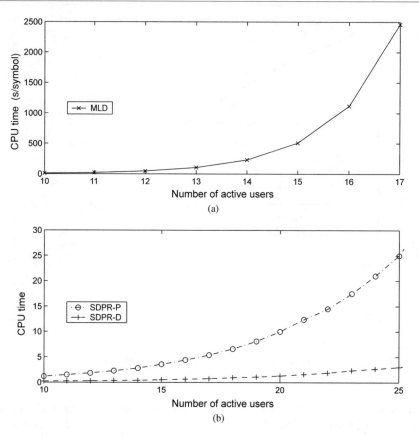

Fig. 16.17 Computational complexity of **a** ML detector and **b** SDPR-P and SDPR-D detectors

sampled version of the received signal $\mathbf{r} = [y(0)\ y(\Delta)\ \cdots\ y[(N-1)\Delta]]^T$, where Δ denotes the sampling period, can be expressed as

$$\mathbf{r} = \mathbf{Sb} + \mathbf{n} \qquad (16.103)$$

where

$$\begin{aligned}
\mathbf{S} &= [A_1\mathbf{s}_1\ A_2\mathbf{s}_2\ \cdots\ A_K\mathbf{s}_K] \\
\mathbf{s}_k &= [s_k(0)\ s_k(\Delta)\ \cdots\ s_k[(N-1)\Delta]]^T \\
\mathbf{b} &= [b_1\ b_2\ \cdots\ b_K]^T \\
\mathbf{n} &= [n(0)\ n(\Delta)\ \cdots\ n[(N-1)\Delta]]^T
\end{aligned}$$

In Eq. (16.103), $\mathbf{n}$ is an AWGN signal with zero mean and variance $\sigma^2\mathbf{I}$, and $\mathbf{s}_k \in R^{N \times 1}$ is the signature signal of the kth user.

The linear multiuser detector to be investigated in this section can be regarded as an FIR filter of length N that is characterized by the coefficient vector $\mathbf{c} \in R^{N \times 1}$. From the channel model in Eq. (16.103), it follows that the output of the detector is given by

$$\mathbf{c}^T \mathbf{r} = \mathbf{c}^T \mathbf{S} \mathbf{b} + \mathbf{c}^T \mathbf{n}$$

Let the kth user be the desired user. We want to detect the user's information bit with minimum error regardless of the information bits of the other $K - 1$ users. If $\hat{\mathbf{b}}_i$ for $1 \leq i \leq 2^{K-1}$ is the possible information vector with the kth entry $b_k = 1$, then for each $\hat{\mathbf{b}}_i$ the output of the detector is given by $\mathbf{c}^T \mathbf{r} = \mathbf{c}^T \hat{\mathbf{v}}_i + \mathbf{c}^T \mathbf{n}$ where $\hat{\mathbf{v}}_i = \mathbf{S}\hat{\mathbf{b}}_i$. The BER of the kth user can be shown to be [20]

$$P(\mathbf{c}) = \frac{1}{2^{K-1}} \sum_{i=1}^{2^{K-1}} Q\left(\frac{\mathbf{c}^T \hat{\mathbf{v}}_i}{||\mathbf{c}||\sigma}\right) \tag{16.104}$$

with

$$Q(x) = \frac{1}{\sqrt{2\pi}} \int_x^{\infty} e^{-v^2/2} \, dv \tag{16.105}$$

A detector whose coefficient vector $\mathbf{c}^*$ minimizes $P(\mathbf{c})$ in Eq. (16.104) can be referred to as a *constrained minimum-BER* (CMBER) *detector*. An optimal linear detector is an unconstrained optimization algorithm that minimizes the BER objective function $P(\mathbf{c})$ in Eq. (16.104). A difficulty associated with the above unconstrained problem is that function $P(\mathbf{c})$ is highly nonlinear and there may exist more than one local minimum. Consequently, convergence to $\mathbf{c}^*$ cannot be guaranteed for most optimization algorithms. In what follows, we present a *constrained optimization* formulation of the problem that can be used to implement a CMBER detector [24].

It can be shown that any local minimizer of the BER objective function in Eq. (16.104) subject to constraints

$$\mathbf{c}^T \hat{\mathbf{v}}_i \geq 0 \quad \text{for } 1 \leq i \leq 2^{K-1} \tag{16.106}$$

is a global minimizer. Furthermore, with the constraint $||\mathbf{c}||_2 = 1$, the global minimizer is unique (see Problem 16.15).

Before proceeding to the problem formulation, it should be mentioned that the constraints in Eq. (16.106) are reasonable in the sense that they will not exclude good local minimizers. This can be seen from Eqs. (16.104)–(16.105) which indicate that nonnegative inner products $\mathbf{c}^T \mathbf{v}_i$ for $1 \leq i \leq 2^{K-1}$ tend to reduce $P(\mathbf{c})$ compared with negative inner products. Now if we define the set

$$I = \{\mathbf{c} : \mathbf{c} \text{ satisfies Eq. (16.106) and} ||\mathbf{c}||_2 = 1\} \tag{16.107}$$

then it can be readily shown that as long as vectors $\{\mathbf{s}_i : 1 \leq i \leq K\}$ are linearly independent, set I contains an infinite number of elements (see Problem 16.16). Under these circumstances, we can formulate the multiuser detection problem at hand as the constrained minimization problem

$$\text{minimize } P(\mathbf{c}) \tag{16.108a}$$

$$\text{subject to: } \mathbf{c} \in I \tag{16.108b}$$

16.5.3.2   Conversion of the Problem in Eqs. (16.108a)–(16.108b) Into a CP Problem

We start with a simple conversion of the problem in Eqs. (16.108a)–(16.108b) into the following problem

$$\text{minimize } P(\mathbf{c}) = \frac{1}{2^{K-1}} \sum_{i=1}^{2^{K-1}} Q(\mathbf{c}^T \mathbf{v}_i) \tag{16.109a}$$

$$\text{subject to: } \mathbf{c}^T \mathbf{v}_i \geq 0 \quad \text{for } 1 \leq i \leq 2^{K-1} \tag{16.109b}$$

$$||\mathbf{c}||_2 = 1 \tag{16.109c}$$

where

$$\mathbf{v}_i = \frac{\hat{\mathbf{v}}_i}{\sigma} \quad \text{for } 1 \leq i \leq w^{K-1}$$

Note that the problem in Eqs. (16.109a)–(16.109c) is *not* a CP problem because the feasible region characterized by Eqs. (16.109b)–(16.109c) is not convex. However, it can be readily verified that the solution of Eqs. (16.109a)–(16.109c) coincides with the solution of the constrained optimization problem

$$\text{minimize } P(\mathbf{c}) \tag{16.110a}$$

$$\text{subject to: } \mathbf{c}^T \mathbf{v}_i \geq 0 \quad \text{for } 1 \leq i \leq 2^{K-1} \tag{16.110b}$$

$$||\mathbf{c}||_2 \leq 1 \tag{16.110c}$$

This is because for any $\mathbf{c}$ with $||\mathbf{c}||_2 < 1$, we always have $P(\hat{\mathbf{c}}) \leq P(\mathbf{c})$ where $\hat{\mathbf{c}} = \mathbf{c}/||\mathbf{c}||_2$. In other words, the minimizer $\mathbf{c}^*$ of the problem in Eqs. (16.110a)–(16.110c) always satisfies the constraint $||\mathbf{c}^*||_2 = 1$. A key distinction between the problems in Eqs. (16.109a)–(16.109c) and Eqs. (16.110a)–(16.110c) is that the latter one is a CP problem for which a number of efficient algorithms are available (see Chap. 14).

16.5.3.3 Newton-Barrier Method

The optimization algorithm described below fits into the class of barrier function methods studied in Chap. 12 but it has several additional features that are uniquely associated with the present problem. These include a closed-form formula for evaluating the Newton direction and an efficient line search.

By adopting a barrier function approach, we can drop the nonlinear constraint in Eq. (16.110c) and convert the problem in Eqs. (16.110a)–(16.110c) into the form

$$\text{minimize } F_\mu(\mathbf{c}) = P(\mathbf{c}) - \mu \ln(1 - \mathbf{c}^T \mathbf{c}) \tag{16.111a}$$

$$\text{subject to: } \mathbf{c}^T \mathbf{v}_i \geq 0 \quad \text{for } 1 \leq i \leq 2^{K-1} \tag{16.111b}$$

where $\mu > 0$ is the barrier parameter. With a strictly feasible initial point $\mathbf{c}_0$, which strictly satisfies the constraints in Eqs. (16.110b)–(16.110c), the logarithmic term in Eq. (16.111a) is well defined. The gradient and Hessian of $F_\mu(\mathbf{c})$ are given by

$$\nabla F_\mu(\mathbf{c}) = -\sum_{i=1}^{M} \frac{1}{M} e^{-\beta_i^2/2} \mathbf{v}_i + \frac{2\mu \mathbf{c}}{1 - \|\mathbf{c}\|_2^2}$$

$$\nabla^2 F_\mu(\mathbf{c}) = \sum_{i=1}^{M} \frac{1}{M} e^{-\beta_i^2/2} \beta_i \mathbf{v}_i \mathbf{v}_i^T + \frac{2\mu}{1 - \|\mathbf{c}\|_2^2} \mathbf{I} + \frac{4\mu}{(1 - \|\mathbf{c}\|)_2^2} \mathbf{c}\mathbf{c}^T$$

where $M = 2^{K-1}$ and $\beta_i = \mathbf{c}^T \mathbf{v}_i$ for $1 \leq i \leq M$. Note that the Hessian in the interior of the feasible region, i.e., $\mathbf{c}$ with $\beta_i = \mathbf{c}^T \mathbf{v}_i > 0$ and $\|\mathbf{c}\|_2 < 1$, is positive definite. This suggests that at the $(k+1)$th iteration, $\mathbf{c}_{k+1}$ can be obtained as

$$\mathbf{c}_{k+1} = \mathbf{c}_k + \alpha_k \mathbf{d}_k \tag{16.112}$$

where the search direction $\mathbf{d}_k$ is given by

$$\mathbf{d}_k = -[\nabla^2 F_\mu(\mathbf{c}_k)]^{-1} \nabla F_\mu(\mathbf{c}_k)$$

The positive scalar α_k in Eq. (16.112) can be determined by using a line search as follows. First, we note that the one-variable function $F_\mu(\mathbf{c}_k + \alpha \mathbf{d}_k)$ is strictly convex on the interval $[0, \bar{\alpha}]$ where $\bar{\alpha}$ is the largest positive scalar such that $\mathbf{c}_k + \alpha \mathbf{d}_k$ remains feasible for $0 \leq \alpha \leq \bar{\alpha}$. Once $\bar{\alpha}$ is determined, $F_\mu(\mathbf{c}_k + \alpha \mathbf{d}_k)$ is a unimodal function on $[0, \bar{\alpha}]$ and the search for the minimizer of the function can be carried out using one of the well-known methods such as quadratic or cubic interpolation or the Golden-section method (see Chap. 4). To find $\bar{\alpha}$, we note that a point $\mathbf{c}_k + \alpha \mathbf{d}_k$ satisfies the constraints in Eq. (16.110b) if

$$(\mathbf{c}_k + \alpha \mathbf{d}_k)^T \mathbf{v}_i \geq 0 \quad \text{for } 1 \leq i \leq M \tag{16.113}$$

Since $\mathbf{c}_k$ is feasible, we have $\mathbf{c}_k^T \mathbf{v}_i \geq 0$ for $1 \leq i \leq M$. Hence for any indices i such that $\mathbf{d}_k^T \mathbf{v}_i \geq 0$, any nonnegative α will satisfy Eq. (16.113). In other words, only those constraints in Eq. (16.110b) whose indices are in the set

$$\mathcal{I}_k = \{i : \mathbf{d}_k^T \mathbf{v}_i < 0\}$$

will affect the largest value of α that satisfies Eq. (16.113), and that value of α can be computed as

$$\bar{\alpha}_1 = \min_{i \in \mathcal{I}_k} \left(\frac{\mathbf{c}_k^T \mathbf{v}_i}{-\mathbf{d}_k^T \mathbf{v}_i} \right)$$

In order to satisfy the constraint in Eq. (16.110c), we solve the equation

$$\|\mathbf{c}_k + \alpha \mathbf{d}_k\|_2^2 = 1$$

to obtain the solution

$$\alpha = \bar{\alpha}_2 = \frac{[(\mathbf{c}_k^T \mathbf{d}_k)_2^2 - \|\mathbf{d}_k\|_2^2(\|\mathbf{c}_k\|_2^2 - 1)]^{1/2} - \mathbf{c}_k^T \mathbf{d}_k}{\|\mathbf{d}_k\|_2^2}$$

The value of $\bar{\alpha}$ can then be taken as $\min(\bar{\alpha}_1, \bar{\alpha}_2)$. In practice, we must keep the next iterate strictly inside the feasible region to ensure that the barrier function in Eq. (16.111a) is well defined. To this end we can use

$$\bar{\alpha} = 0.99 \min(\bar{\alpha}_1, \bar{\alpha}_2)$$

The above iterative optimization procedure is continued until the difference between two successive solutions is less than a prescribed tolerance. For a strictly feasible initial point, the Newton-barrier method described above always converges to the global minimizer for an arbitrary positive μ. However, the value of μ does affect the behavior of the algorithm. A small μ may lead to an ill-conditioned Hessian while a large μ may lead to slow convergence. A μ in the interval $[0.001, \ 0.1]$ would guarantee a well-conditioned Hessian and allow a fast convergence.

The BER performance of the CMBER detector is compared with that of the ML detector in Fig. 16.18 for a system with 10 equal-power users. As can be seen, the performance of the CMBER is practically the same as that of the ML detector.

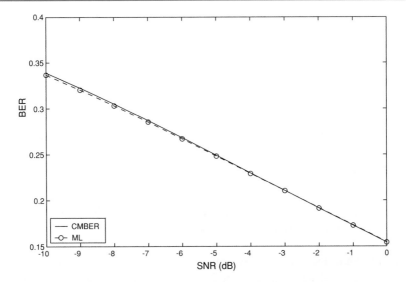

Fig. 16.18 Performance comparison of CMBER and ML detectors for a system with 10 equal-power users

Problems

16.1 Write a MATLAB program to implement the constrained optimization algorithm described in Sect. 16.2, and use it to obtain the design in Example 16.1.

16.2 Derive the expression in Eq. (16.12).

16.3 Show that the inequality in Eq. (16.13) holds if and only if matrix $D(\omega)$ in Eq. (16.14) is positive semidefinite.

16.4 *(a)* Show that the zeros of polynomial $z^K B(z)$ in Eq. (16.17c) are the eigenvalues of matrix $\mathbf{D}$ in Eq. (16.18).
   (b) Show that if the matrix inequality in Eq. (16.19) holds for some matrix $\mathbf{P} \succ \mathbf{0}$, then the largest modulus of the eigenvalues of matrix $\mathbf{D}$ in Eq. (16.18) is strictly less than one.

16.5 Show that the constraint in Eq. (16.23b) is equivalent to the matrix inequality in Eq. (16.24).

16.6 *(a)* Apply the SOCP formulation in Eqs. (16.29a)–(16.29b) to solve the problem in Example 16.2.
   (b) Apply the SOCP formulation sketched in Sect. 16.2.4 to solve the problem in Example 16.3.

16.7 Show that the matrix inequality in Eq. (16.42) is equivalent to the matrix inequalities in Eq. (16.43) or those in Eq. (16.44).

16.8 Using the results of Problem 16.7, show that matrix $\mathbf{W}$ in Eq. (16.41) is negative semidefinite if and only if the matrix inequality in Eq. (16.45) holds.

16.9 Assuming that matrix $\mathbf{S} \in R^{n \times n}$ is positive definite, $\mathbf{Y} \in R^{p \times n}$, and $\mathcal{E}$ is defined by Eq. (16.50), show that the formula in Eq. (16.53) is valid.

16.10 Using the result of Problem 16.7, show that the matrix inequality in Eq. (16.54) implies the inequality in Eq. (16.55).

16.11 Show that if there exists a symmetric matrix $\mathbf{X} \in R^{p \times p}$ such that the conditions in Eqs. (16.60a)–(16.60b) are satisfied, then $(\mathbf{YS}^{-1}\mathbf{Y}^T)_{j,j} \leq u_{j,\max}^2$.

16.12 *(a)* Show that Eq. (16.62) is equivalent to Eq. (16.72).
  (b) Show that the constraint in Eq. (16.64) assures the positive semidefiniteness of matrix $\mathbf{P}_i$ in Eq. (16.74).
  (c) Show that the constraints in Eqs. (16.65a)–(16.65b) assure the positive semidefiniteness of matrix $\mathbf{P}_i$ in Eq. (16.75).

16.13 *(a)* It has been shown that the smallest eigenvalue $\lambda_{\min}$ of matrix $\mathbf{P(c)}$ in Eq. (16.76c) can be viewed as a measure of the strictest friction and by how much the contact forces are away from slippage [18]. Modify the constraint in Eq. (16.76c) such that $\lambda_{\min}$ of $\mathbf{P(c)}$ is no less than a given threshold, say, ε, and the modified problem in Eqs. (16.76a)–(16.76c) remains an SDP problem.
  (b) Solve the optimal force distribution problem in Example 16.7 with the additional requirement that $\lambda_{\min}$ of $\mathbf{P(c)}$ be no less than $\varepsilon = 0.05$.

16.14 *(a)* Show that the objective function in Eq. (16.84a) can be expressed as

$$\mathrm{trace}(\mathbf{C}\hat{\mathbf{X}})$$

where

$$\hat{\mathbf{X}} = \begin{bmatrix} \mathbf{x}\mathbf{x}^T & \mathbf{x} \\ \mathbf{x}^T & 1 \end{bmatrix} \quad \text{and} \quad \mathbf{C} = \begin{bmatrix} \mathbf{H} & \mathbf{p}/2 \\ \mathbf{p}^T/2 & 1 \end{bmatrix}$$

  (b) Using the results obtained in part *(a)*, show that the optimization problem in Eqs. (16.84a)–(16.84b) can be reformulated as a problem that minimizes $\mathrm{trace}(\mathbf{C}\hat{\mathbf{X}})$ subject to the constraints in Eqs. (16.90a)–(16.90b).

16.15 *(a)* Show that any local minimizer of the BER cost function in Eq. (16.104) subject to the constraints in Eq. (16.106) is a global minimizer.

(a) Show that with an additional constraint $\|\mathbf{c}\|_2 = 1$, the global minimizer for the problem in part (a) is unique.

16.16 Show that if the signature vectors $\{\mathbf{s}_k : 1 \leq i \leq K\}$ in Eq. (16.103) are linearly independent, then set I defined by Eq. (16.107) contains an infinite number of elements.

16.17 Show that the constrained problem in Eqs. (16.110a)–(16.110c) is a CP problem.

References

1. A. Antoniou, *Digital Filters: Analysis, Design, and Signal Processing Applications*. New York: McGraw-Hill, 2018.
2. W.-S. Lu and A. Antoniou, *Two-Dimensional Digital Filters*. New York: Marcel Dekker, 1992.
3. J. W. Adams, "FIR digital filters with least-squares stopbands subject to peak-gain constraints," *IEEE Trans. Circuits Syst.*, vol. 38, pp. 376–388, Apr. 1991.
4. W.-S. Lu, "Design of nonlinear-phase FIR digital filters: A semidefinite programming approach," in *Proc. IEEE Int. Symp. Circuits and Systems*, vol. III, Orlando, FL, May 1999, pp. 263–266.
5. A. G. Deczky, "Synthesis of recursive digital filters using the minimum p-error criterion," *IEEE Trans. Audio Electroacoust.*, vol. 20, pp. 257–263, 1972.
6. A. T. Chottra and G. A. Jullien, "A linear programming approach to recursive digital filter design with linear phase," *IEEE Trans. Circuits Syst.*, vol. 29, pp. 139–149, Mar. 1982.
7. W.-S. Lu, S.-C. Pei, and C.-C. Tseng, "A weighted least-squares method for the design of stable 1-D and 2-D IIR filters," *IEEE Trans. Signal Processing*, vol. 46, pp. 1–10, Jan. 1998.
8. M. Lang, "Weighted least squares IIR filter design with arbitrary magnitude and phase responses and specified stability margin," in *IEEE Symp. on Advances in Digital Filtering and Signal Processing*, Victoria, BC, June 1998, pp. 82–86.
9. T. Kailath, *Linear Systems*. Englewood Cliffs, NJ: Prentice-Hall, 1980.
10. R. C. Nongpiur, D. J. Shpak, and A. Antoniou, "Improved design method for nearly linear-phase IIR filters using constrained optimization," *IEEE Trans. Signal Processing*, vol. 61, no. 4, pp. 895–906, Feb. 2013.
11. C. E. Garcia, D. M. Prett, and M. Morari, "Model predictive control: Theory and practice — a survey," *Automatica*, vol. 25, pp. 335–348, 1989.
12. M. V. Kothare, V. Balakrishnan, and M. Morari, "Robust constrained model predictive control using linear matrix inequalities," *Automatica*, vol. 32, no. 6, pp. 1361–1379, 1996.
13. H. Kwakernaak and R. Sivan, *Linear Optimal Control Systems*. New York: Wiley, 1972.
14. J. Kerr and B. Roth, "Analysis of multifingered hands," *Int. J. Robotics Research*, vol. 4, no. 4, pp. 3–17, 1986.
15. D. E. Orin and F.-T. Cheng, "General dynamic formulation of the force distribution equations," in *Proc. 4th Int. Conf. on Advanced Robotics*, Columbus, Ohio, June 13-15, 1989, pp. 525–546.
16. F.-T. Cheng and D. E. Orin, "Efficient algorithm for optimal force distribution — The compact-dual LP method," *IEEE Trans. Robotics and Automation*, vol. 6, pp. 178–187, 1990.
17. E. S. Venkaraman and T. Iberall, *Dextrous Robot Hands*. New York: Springer-Verlag, 1990.
18. M. Buss, H. Hashimoto, and J. B. Moore, "Dextrous hand grasping force optimization," *IEEE Trans. Robotics and Automation*, vol. 12, pp. 406–418, June 1996.

19. K. Shimoga, "Robot grasp synthesis algorithms: A survey," *Int. J. Robotics Research*, vol. 15, pp. 230–266, June 1996.

20. J. G. Proakis, *Digital Communications*, 3rd ed. New York: McGraw-Hill, 1995.

21. S. Verdú, *Multiuser Detection*. New York: Cambridge University Press, 1998.

22. S. Verdú, "Minimum probability of error for asynchronous Gaussian multiple-access channels," *IEEE Trans. Inform. Theory*, vol. 32, pp. 85–96, Jan. 1986.

23. X. M. Wang, W.-S. Lu, and A. Antoniou, "A near-optimal multiuser detector for CDMA channels using semidefinite programming relaxation," in *Proc. IEEE Int. Symp. Circuits and Systems*, Sydney, Australia, June 2001, pp. 525–546.

24. X. F. Wang, W.-S. Lu, and A. Antoniou, "Constrained minimum-BER multiuser detection," *IEEE Trans. Signal Processing*, vol. 48, pp. 2903–2909, Oct. 2000.

25. M. X. Geomans and D. P. Williamson, "Improved approximation algorithms for maximum cut and satisfiability problem using semidefinite programming," *J. ACM*, vol. 42, pp. 1115–1145, 1985.

26. M. X. Geomans and D. P. Williamson, ".878-approximation algorithm for MAX-CUT and MAX-2SAT," in *Proc. 26th ACM Symp. Theory of Computing*, 1994, pp. 422–431.

27. L. Vandenberghe and S. Boyd, "Semidefinite programming," *SIAM Review*, vol. 38, pp. 49–95, Mar. 1996.

28. H. Wolkowicz, R. Saigal, and L. Vandenberghe, *Handbook on Semidefinite Programming*. MA: Kluwer Academic Publishers, 2000.

29. G. W. Stewart, *Introduction to Matrix Computations*. New York: Academic Press, 1973.

30. K. C. Toh, R. H. Tütüncü, and M. J. Todd, "On the implementation of SDPT3 version 3.1 — a MATLAB software package for semidefinite-quadratic-linear programming," in *Proc. IEEE Conf. on Computer-Aided Control System Design*, Sept. 2004.

31. A. Nemirovski and P. Gahinet, "The projective method for solving linear matrix inequalities," *Math. Programming, Series B*, vol. 77, pp. 163–190, 1997.

Basics of Linear Algebra

A

A.1 Introduction

In this appendix we summarize some basic principles of linear algebra [1–5] that are needed to understand the derivation and analysis of the optimization algorithms and techniques presented in the book. We state these principles without derivations. However, a reader with an undergraduate-level linear-algebra background should be in a position to deduce most of them without much difficulty. Indeed, we encourage the reader to do so as the exercise will contribute to the understanding of the optimization methods described in this book.

In what follows, R^n denotes a vector space that consists of all column vectors with n real-valued components, and C^n denotes a vector space that consists of all column vectors with n complex-valued components. Likewise, $R^{m \times n}$ and $C^{m \times n}$ denote spaces consisting of all $m \times n$ matrices with real-valued and complex-valued components, respectively. Evidently, $R^{m \times 1} \equiv R^m$ and $C^{m \times 1} \equiv C^m$. Boldfaced uppercase letters, e.g., $\mathbf{A}, \mathbf{M}$, represent matrices, and boldfaced lowercase letters, e.g., $\mathbf{a}, \mathbf{x}$, represent column vectors. $\mathbf{A}^T$ and $\mathbf{A}^H = (\mathbf{A}^*)^T$ denote the transpose and complex-conjugate transpose of matrix $\mathbf{A}$, respectively. $\mathbf{A}^{-1}$ (if it exists) and $\det(\mathbf{A})$ denote the inverse and determinant of square matrix $\mathbf{A}$, respectively. The identity matrix of dimension n is denoted as $\mathbf{I}_n$. Column vectors will be referred to simply as vectors henceforth for the sake of brevity.

A.2 Linear Independence and Basis of a Span

A number of vectors $\mathbf{v}_1, \mathbf{v}_2, \ldots, \mathbf{v}_k$ in R^n are said to be *linearly independent* if

$$\sum_{i=1}^{k} \alpha_i \mathbf{v}_i = \mathbf{0} \tag{A.1}$$

© Springer Science+Business Media, LLC, part of Springer Nature 2021
A. Antoniou and W.-S. Lu, *Practical Optimization*, Texts in Computer Science,
https://doi.org/10.1007/978-1-0716-0843-2_A

only if $\alpha_i = 0$ for $i = 1, 2, \ldots, k$. Vectors $\mathbf{v}_1, \mathbf{v}_2, \ldots, \mathbf{v}_k$ are said to be *linearly dependent* if there exist real scalars α_i for $i = 1, 2, \ldots, k$, with at least one nonzero α_i, such that Eq. (A.1) holds.

A subspace S is a subset of R^n such that $\mathbf{x} \in S$ and $\mathbf{y} \in S$ imply that $\alpha\mathbf{x} + \beta\mathbf{y} \in S$ for any real scalars α and β. The set of all linear combinations of vectors $\mathbf{v}_1, \mathbf{v}_2, \ldots, \mathbf{v}_k$ is a subspace called the *span* of $\{\mathbf{v}_1, \mathbf{v}_2, \ldots, \mathbf{v}_k\}$ and is denoted as span$\{\mathbf{v}_1, \mathbf{v}_2, \ldots, \mathbf{v}_k\}$.

Given a set of vectors $\{\mathbf{v}_1, \mathbf{v}_2, \ldots, \mathbf{v}_k\}$, a subset of r vectors $\{\mathbf{v}_{i_1}, \mathbf{v}_{i_2}, \ldots, \mathbf{v}_{i_r}\}$ is said to be a maximal linearly independent subset if (*a*) vectors $\mathbf{v}_{i_1}, \mathbf{v}_{i_2}, \ldots, \mathbf{v}_{i_r}$ are linearly independent, and (*b*) any vector in $\{\mathbf{v}_1, \mathbf{v}_2, \ldots, \mathbf{v}_k\}$ can be expressed as a linear combination of $\mathbf{v}_{i_1}, \mathbf{v}_{i_2}, \ldots, \mathbf{v}_{i_r}$. In such a case, the vector set $\{\mathbf{v}_{i_1}, \mathbf{v}_{i_2}, \ldots, \mathbf{v}_{i_r}\}$ is called a *basis* for span$\{\mathbf{v}_1, \mathbf{v}_2, \ldots, \mathbf{v}_k\}$ and integer r is called the *dimension* of the subspace. The dimension of a subspace S is denoted as dim(S).

Example A.1 Examine the linear dependence of vectors

$$\mathbf{v}_1 = \begin{bmatrix} 1 \\ -1 \\ 3 \\ 0 \end{bmatrix}, \quad \mathbf{v}_2 = \begin{bmatrix} 0 \\ 2 \\ 1 \\ -1 \end{bmatrix}, \quad \mathbf{v}_3 = \begin{bmatrix} 3 \\ -7 \\ 7 \\ 2 \end{bmatrix}, \quad \text{and} \quad \mathbf{v}_4 = \begin{bmatrix} -1 \\ 5 \\ -1 \\ -2 \end{bmatrix}$$

and obtain a basis for span$\{\mathbf{v}_1, \mathbf{v}_2, \mathbf{v}_3, \mathbf{v}_4\}$.

Solution We note that

$$3\mathbf{v}_1 + 2\mathbf{v}_2 - 2\mathbf{v}_3 - 3\mathbf{v}_4 = \mathbf{0} \tag{A.2}$$

Hence vectors $\mathbf{v}_1, \mathbf{v}_2, \mathbf{v}_3$, and $\mathbf{v}_4$ are linearly dependent. If

$$\alpha_1\mathbf{v}_1 + \alpha_2\mathbf{v}_2 = \mathbf{0}$$

then

$$\begin{bmatrix} \alpha_1 \\ -\alpha_1 + 2\alpha_2 \\ 3\alpha_1 \\ -\alpha_2 \end{bmatrix} = \mathbf{0}$$

which implies that $\alpha_1 = 0$ and $\alpha_2 = 0$. Hence $\mathbf{v}_1$ and $\mathbf{v}_2$ are linearly independent. We note that

$$\mathbf{v}_3 = 3\mathbf{v}_1 - 2\mathbf{v}_2 \tag{A.3}$$

and by substituting Eq. (A.3) into Eq. (A.2), we obtain

$$-3\mathbf{v}_1 + 6\mathbf{v}_2 - 3\mathbf{v}_4 = \mathbf{0}$$

i.e.,

$$\mathbf{v}_4 = -\mathbf{v}_1 + 2\mathbf{v}_2 \tag{A.4}$$

Thus vectors $\mathbf{v}_3$ and $\mathbf{v}_4$ can be expressed as linear combinations of $\mathbf{v}_1$ and $\mathbf{v}_2$. Therefore, $\{\mathbf{v}_1, \mathbf{v}_2\}$ is a basis of span$\{\mathbf{v}_1, \mathbf{v}_2, \mathbf{v}_3, \mathbf{v}_4\}$. ∎

A.3 Range, Null Space, and Rank

Consider a system of linear equations

$$\mathbf{A}\mathbf{x} = \mathbf{b} \tag{A.5}$$

where $\mathbf{A} \in R^{m \times n}$ and $\mathbf{b} \in R^{m \times 1}$. If we denote the ith column of matrix $\mathbf{A}$ as $\mathbf{a}_i \in R^{m \times 1}$, i.e.,

$$\mathbf{A} = [\mathbf{a}_1 \ \mathbf{a}_2 \ \cdots \ \mathbf{a}_n]$$

and let

$$\mathbf{x} = [x_1 \ x_2 \ \ldots \ x_n]^T$$

then Eq. (A.5) can be written as

$$\sum_{i=1}^{n} x_i \mathbf{a}_i = \mathbf{b}$$

It follows from the above expression that Eq. (A.5) is solvable if and only if

$$\mathbf{b} \in \text{span}\{\mathbf{a}_1, \mathbf{a}_2, \ldots, \mathbf{a}_n\}$$

The subspace span$\{\mathbf{a}_1, \mathbf{a}_2, \ldots, \mathbf{a}_n\}$ is called the *range* of $\mathbf{A}$ and is denoted as $\mathcal{R}(\mathbf{A})$. Thus, Eq. (A.5) has a solution if and only if vector $\mathbf{b}$ is in the range of $\mathbf{A}$.

The dimension of $\mathcal{R}(\mathbf{A})$ is called the *rank* of $\mathbf{A}$, i.e., $r = \text{rank}(\mathbf{A}) = \dim[\mathcal{R}(\mathbf{A})]$. Since $\mathbf{b} \in \text{span}\{\mathbf{a}_1, \mathbf{a}_2, \ldots, \mathbf{a}_n\}$ is equivalent to

$$\text{span}\{\mathbf{b}, \mathbf{a}_1, \ldots, \mathbf{a}_n\} = \text{span}\{\mathbf{a}_1, \mathbf{a}_2, \ldots, \mathbf{a}_n\}$$

we conclude that Eq. (A.5) is solvable if and only if

$$\text{rank}(\mathbf{A}) = \text{rank}([\mathbf{A} \ \mathbf{b}]) \tag{A.6}$$

It can be shown that $\text{rank}(\mathbf{A}) = \text{rank}(\mathbf{A}^T)$. In other words, *the rank of a matrix is equal to the maximum number of linearly independent columns or rows.*

Another important concept associated with a matrix $\mathbf{A} \in R^{m \times n}$ is the *null space* of $\mathbf{A}$, which is defined as

$$\mathcal{N}(\mathbf{A}) = \{\mathbf{x} : \mathbf{x} \in R^n, \ \mathbf{A}\mathbf{x} = \mathbf{0}\}$$

It can be readily verified that $N(\mathbf{A})$ is a subspace of R^n. If $\mathbf{x}$ is a solution of Eq. (A.5) then $\mathbf{x} + \mathbf{z}$ with $\mathbf{z} \in N(\mathbf{A})$ also satisfies Eq. (A.5). Hence Eq. (A.5) has a unique solution only if $N(\mathbf{A})$ contains just one component, namely, the zero vector in R^n. Furthermore, it can be shown that for $\mathbf{A} \in R^{m \times n}$

$$\text{rank}(\mathbf{A}) + \dim[N(\mathbf{A})] = n \qquad (A.7)$$

(see [1, 3]). For the important special case where matrix $\mathbf{A}$ is square, i.e., $n = m$, the following statements are equivalent: (a) there exists a unique solution for Eq. (A.5); (b) $N(\mathbf{A}) = \{\mathbf{0}\}$; ($c$) $\text{rank}(\mathbf{A}) = n$.

A matrix $\mathbf{A} \in R^{m \times n}$ is said to have full column rank if $\text{rank}(\mathbf{A}) = n$, i.e., the n column vectors of $\mathbf{A}$ are linearly independent, and $\mathbf{A}$ is said to have full row rank if $\text{rank}(\mathbf{A}) = m$, i.e., the m row vectors of $\mathbf{A}$ are linearly independent.

Example A.2 Find the rank and null space of matrix

$$\mathbf{V} = \begin{bmatrix} 1 & 0 & 3 & -1 \\ -1 & 2 & -7 & 5 \\ 3 & 1 & 7 & -1 \\ 0 & -1 & 2 & -2 \end{bmatrix}$$

Solution Note that the columns of $\mathbf{V}$ are the vectors $\mathbf{v}_i$ for $i = 1, 2, \ldots, 4$ in Example A.1. Since the maximum number of linearly independent columns is 2, we have $\text{rank}(\mathbf{V}) = 2$. To find $N(\mathbf{V})$, we write $\mathbf{V} = [\mathbf{v}_1\ \mathbf{v}_2\ \mathbf{v}_3\ \mathbf{v}_4]$; hence the equation $\mathbf{V}\mathbf{x} = \mathbf{0}$ becomes

$$x_1\mathbf{v}_1 + x_2\mathbf{v}_2 + x_3\mathbf{v}_3 + x_4\mathbf{v}_4 = \mathbf{0} \qquad (A.8)$$

Using Eqs. (A.3) and (A.4), Eq. (A.8) can be expressed as

$$(x_1 + 3x_3 - x_4)\mathbf{v}_1 + (x_2 - 2x_3 + 2x_4)\mathbf{v}_2 = \mathbf{0}$$

which implies that

$$x_1 + 3x_3 - x_4 = 0$$
$$x_2 - 2x_3 + 2x_4 = 0$$

i.e.,

$$x_1 = -3x_3 + x_4$$
$$x_2 = 2x_3 - 2x_4$$

Hence any vector $\mathbf{x}$ that can be expressed as

$$\mathbf{x} = \begin{bmatrix} x_1 \\ x_2 \\ x_3 \\ x_4 \end{bmatrix} = \begin{bmatrix} -3x_3 + x_4 \\ 2x_3 - 2x_4 \\ x_3 \\ x_4 \end{bmatrix} = \begin{bmatrix} -3 \\ 2 \\ 1 \\ 0 \end{bmatrix} x_3 + \begin{bmatrix} 1 \\ -2 \\ 0 \\ 1 \end{bmatrix} x_4$$

with arbitrary x_3 and x_4 satisfies $\mathbf{Ax} = \mathbf{0}$. Since the two vectors in the above expression, namely,

$$\mathbf{n}_1 = \begin{bmatrix} -3 \\ 2 \\ 1 \\ 0 \end{bmatrix} \quad \text{and} \quad \mathbf{n}_2 = \begin{bmatrix} 1 \\ -2 \\ 0 \\ 1 \end{bmatrix}$$

are linearly independent, we have $\mathcal{N}(\mathbf{V}) = \text{span}\{\mathbf{n}_1, \mathbf{n}_2\}$. ∎

A.4 Sherman-Morrison Formula

The Sherman-Morrison formula [5] states that given matrices $\mathbf{A} \in C^{n \times n}$, $\mathbf{U} \in C^{n \times p}$, $\mathbf{W} \in C^{p \times p}$, and $\mathbf{V} \in C^{n \times p}$, such that $\mathbf{A}^{-1}$, $\mathbf{W}^{-1}$, and $(\mathbf{W}^{-1} + \mathbf{V}^H \mathbf{A}^{-1} \mathbf{U})^{-1}$ exist, then the inverse of $\mathbf{A} + \mathbf{UWV}^H$ exists and is given by

$$(\mathbf{A} + \mathbf{UWV}^H)^{-1} = \mathbf{A}^{-1} - \mathbf{A}^{-1} \mathbf{U} \mathbf{Y}^{-1} \mathbf{V}^H \mathbf{A}^{-1} \tag{A.9}$$

where

$$\mathbf{Y} = \mathbf{W}^{-1} + \mathbf{V}^H \mathbf{A}^{-1} \mathbf{U} \tag{A.10}$$

In particular, if $p = 1$ and $\mathbf{W} = 1$, then Eq. (A.9) assumes the form

$$(\mathbf{A} + \mathbf{uv}^H)^{-1} = \mathbf{A}^{-1} - \frac{\mathbf{A}^{-1} \mathbf{uv}^H \mathbf{A}^{-1}}{1 + \mathbf{v}^H \mathbf{A}^{-1} \mathbf{u}} \tag{A.11}$$

where $\mathbf{u}$ and $\mathbf{v}$ are vectors in $C^{n \times 1}$. Eq. (A.11) is useful for computing the inverse of a rank-one modification of $\mathbf{A}$, namely, $\mathbf{A} + \mathbf{uv}^H$, if $\mathbf{A}^{-1}$ is available.

Example A.3 Find $\mathbf{A}^{-1}$ for

$$\mathbf{A} = \begin{bmatrix} 1.04 & 0.04 & \cdots & 0.04 \\ 0.04 & 1.04 & \cdots & 0.04 \\ \vdots & \vdots & & \vdots \\ 0.04 & 0.04 & \cdots & 1.04 \end{bmatrix} \in \mathcal{R}^{10 \times 10}$$

Solution Matrix $\mathbf{A}$ can be treated as a rank-one perturbation of the identity matrix:

$$\mathbf{A} = \mathbf{I} + \mathbf{pp}^T$$

where $\mathbf{I}$ is the identity matrix and $\mathbf{p} = [0.2 \; 0.2 \; \cdots \; 0.2]^T$. Using Eq. (A.11), we can compute

$$
\mathbf{A}^{-1} = (\mathbf{I} + \mathbf{p}\mathbf{p}^T)^{-1} = \mathbf{I} - \frac{\mathbf{p}\mathbf{p}^T}{1 + \mathbf{p}^T\mathbf{p}} = \mathbf{I} - \frac{1}{1.4}\mathbf{p}\mathbf{p}^T
$$

$$
= \begin{bmatrix} 0.9714 & -0.0286 & \cdots & -0.0286 \\ -0.0286 & 0.9714 & \cdots & -0.0286 \\ \vdots & \vdots & & \vdots \\ -0.0286 & -0.0286 & \ldots & 0.9714 \end{bmatrix}
$$

■

A.5 Eigenvalues and Eigenvectors

The *eigenvalues* of a matrix $\mathbf{A} \in C^{n \times n}$ are defined as the n roots of its so-called *characteristic equation*

$$
\det(\lambda\mathbf{I} - \mathbf{A}) = 0 \tag{A.12}
$$

If we denote the set of n eigenvalues $\{\lambda_1, \; \lambda_2, \; \ldots, \; \lambda_n\}$ by $\lambda(\mathbf{A})$, then for a $\lambda_i \in \lambda(\mathbf{A})$, there exists a nonzero vector $\mathbf{x}_i \in C^{n \times 1}$ such that

$$
\mathbf{A}\mathbf{x}_i = \lambda_i\mathbf{x}_i \tag{A.13}
$$

Such a vector is called an *eigenvector* of $\mathbf{A}$ associated with eigenvalue λ_i.

Eigenvectors are not unique. For example, if $\mathbf{x}_i$ is an eigenvector of matrix $\mathbf{A}$ associated with eigenvalue λ_i and c is an arbitrary nonzero constant, then $c\mathbf{x}_i$ is also an eigenvector of $\mathbf{A}$ associated with eigenvalue λ_i.

If $\mathbf{A}$ has n distinct eigenvalues $\lambda_1, \; \lambda_2, \; \ldots, \; \lambda_n$ with associated eigenvectors $\mathbf{x}_1, \; \mathbf{x}_2, \; \ldots, \; \mathbf{x}_n$, then these eigenvectors are linearly independent; hence we can write

$$
\mathbf{A}[\mathbf{x}_1 \; \mathbf{x}_2 \; \cdots \; \mathbf{x}_n] = [\mathbf{A}\mathbf{x}_1 \; \mathbf{A}\mathbf{x}_2 \; \cdots \; \mathbf{A}\mathbf{x}_n] = [\lambda_1\mathbf{x}_1 \; \lambda_2\mathbf{x}_2 \; \cdots \; \lambda_n\mathbf{x}_n]
$$

$$
= [\mathbf{x}_1 \; \mathbf{x}_2 \; \cdots \; \mathbf{x}_n] \begin{bmatrix} \lambda_1 & & \mathbf{0} \\ & \ddots & \\ \mathbf{0} & & \lambda_n \end{bmatrix}
$$

In effect,

$$
\mathbf{A}\mathbf{X} = \mathbf{X}\boldsymbol{\Lambda}
$$

or

$$
\mathbf{A} = \mathbf{X}\boldsymbol{\Lambda}\mathbf{X}^{-1} \tag{A.14}
$$

with

$$\mathbf{X} = [\mathbf{x}_1 \ \mathbf{x}_2 \ \cdots \ \mathbf{x}_n] \quad \text{and} \quad \Lambda = \text{diag}\{\lambda_1, \ \lambda_1, \ \ldots, \ \lambda_n\}$$

where $\text{diag}\{\lambda_1, \ \lambda_2, \ \ldots, \ \lambda_n\}$ represents the diagonal matrix with components λ_1, $\lambda_2, \ \ldots, \ \lambda_n$ along its diagonal. The relation in (A.14) is often referred to as an *eigendecomposition* of $\mathbf{A}$.

A concept that is closely related to the eigendecomposition in Eq. (A.14) is that of similarity transformation. Two square matrices $\mathbf{A}$ and $\mathbf{B}$ are said to be *similar* if there exists a nonsingular $\mathbf{X}$, called a *similarity transformation*, such that

$$\mathbf{A} = \mathbf{X}\mathbf{B}\mathbf{X}^{-1} \tag{A.15}$$

From Eq. (A.14), it follows that if the eigenvalues of $\mathbf{A}$ are distinct, then $\mathbf{A}$ is similar to $\Lambda = \text{diag}\{\lambda_1, \ \lambda_2, \ \ldots, \ \lambda_n\}$ and the similarity transformation involved, $\mathbf{X}$, is composed of the n eigenvectors of $\mathbf{A}$. For arbitrary matrices with repeated eigenvalues, the eigendecomposition becomes more complicated. The reader is referred to [1–4] for the theory and solution of the eigenvalue problem for the general case.

Example A.4 Find the diagonal matrix Λ, if it exists, that is similar to matrix

$$\mathbf{A} = \begin{bmatrix} 4 & -3 & 1 & 1 \\ 2 & -1 & 1 & 1 \\ 0 & 0 & 1 & 2 \\ 0 & 0 & 2 & 1 \end{bmatrix}$$

Solution From Eq. (A.12), we have

$$\begin{aligned}
\det(\lambda\mathbf{I} - \mathbf{A}) &= \det\begin{bmatrix} \lambda - 4 & 3 \\ -2 & \lambda + 1 \end{bmatrix} \cdot \det\begin{bmatrix} \lambda - 1 & -2 \\ -2 & \lambda - 1 \end{bmatrix} \\
&= (\lambda^2 - 3\lambda + 2)(\lambda^2 - 2\lambda - 3) \\
&= (\lambda - 1)(\lambda - 2)(\lambda + 1)(\lambda - 3)
\end{aligned}$$

Hence the eigenvalues of $\mathbf{A}$ are $\lambda_1 = 1$, $\lambda_2 = 2$, $\lambda_3 = -1$, and $\lambda_4 = 3$. An eigenvector $\mathbf{x}_i$ associated with eigenvalue λ_i satisfies the relation

$$(\lambda_i\mathbf{I} - \mathbf{A})\mathbf{x}_i = \mathbf{0}$$

For $\lambda_1 = 1$, we have

$$\lambda_1\mathbf{I} - \mathbf{A} = \begin{bmatrix} -3 & 3 & -1 & -1 \\ -2 & 2 & -1 & -1 \\ 0 & 0 & 0 & -2 \\ 0 & 0 & -2 & 0 \end{bmatrix}$$

It is easy to verify that $\mathbf{x}_1 = [1 \ 1 \ 0 \ 0]^T$ satisfies the relation

$$(\lambda_1\mathbf{I} - \mathbf{A})\mathbf{x}_1 = \mathbf{0}$$

Similarly, $\mathbf{x}_2 = [3\ 2\ 0\ 0]^T$, $\mathbf{x}_3 = [0\ 0\ 1\ -1]^T$, and $\mathbf{x}_4 = [1\ 1\ 1\ 1]^T$ satisfy the relation

$$(\lambda_i \mathbf{I} - \mathbf{A})\mathbf{x}_i = \mathbf{0} \quad \text{for } i = 2,\ 3,\ 4$$

If we let

$$\mathbf{X} = [\mathbf{x}_1\ \mathbf{x}_2\ \mathbf{x}_3\ \mathbf{x}_4] = \begin{bmatrix} 1 & 3 & 0 & 1 \\ 1 & 2 & 0 & 1 \\ 0 & 0 & 1 & 1 \\ 0 & 0 & -1 & 1 \end{bmatrix}$$

then we have

$$\mathbf{AX} = \mathbf{\Lambda X}$$

where

$$\mathbf{\Lambda} = \text{diag}\{1,\ 2,\ -1,\ 3\}$$

∎

A.6   Symmetric Matrices

The matrices encountered most frequently in numerical optimization are symmetric. For these matrices, an elegant eigendecomposition theory and corresponding computation methods are available. If $\mathbf{A} = \{a_{ij}\} \in R^{n \times n}$ is a symmetric matrix, i.e., $a_{ij} = a_{ji}$, then there exists an orthogonal matrix $\mathbf{X} \in R^{n \times n}$, i.e., $\mathbf{XX}^T = \mathbf{X}^T\mathbf{X} = \mathbf{I}_n$, such that

$$\mathbf{A} = \mathbf{X\Lambda X}^T \tag{A.16}$$

where $\mathbf{\Lambda} = \text{diag}\{\lambda_1,\ \lambda_2,\ \ldots,\ \lambda_n\}$. If $\mathbf{A} \in C^{n \times n}$ is such that $\mathbf{A} = \mathbf{A}^H$, then $\mathbf{A}$ is referred to as a *Hermitian matrix*. In such a case, there exists a so-called *unitary matrix* $\mathbf{U} \in C^{n \times n}$ for which $\mathbf{UU}^H = \mathbf{U}^H\mathbf{U} = \mathbf{I}_n$ such that

$$\mathbf{A} = \mathbf{U\Lambda U}^H \tag{A.17}$$

In Eqs. (A.16) and (A.17), the diagonal components of $\mathbf{\Lambda}$ are eigenvalues of $\mathbf{A}$, and the columns of $\mathbf{X}$ and $\mathbf{U}$ are corresponding eigenvectors of $\mathbf{A}$.

The following properties can be readily verified:

(a) A square matrix is nonsingular if and only if all its eigenvalues are nonzero.
(b) The magnitudes of the eigenvalues of an orthogonal or unitary matrix are always equal to unity.

(c) The eigenvalues of a symmetric or Hermitian matrix are always real.

(d) The determinant of a square matrix is equal to the product of its eigenvalues.

A symmetric matrix $\mathbf{A} \in R^{n \times n}$ is said to be *positive definite, positive semidefinite, negative semidefinite, negative definite* if $\mathbf{x}^T \mathbf{A} \mathbf{x} > 0$, $\mathbf{x}^T \mathbf{A} \mathbf{x} \geq 0$, $\mathbf{x}^T \mathbf{A} \mathbf{x} \leq 0$, $\mathbf{x}^T \mathbf{A} \mathbf{x} < 0$, respectively, for all nonzero $\mathbf{x} \in R^{n \times 1}$.

Using the decomposition in Eq. (A.16), it can be shown that matrix $\mathbf{A}$ is positive definite, positive semidefinite, negative semidefinite, negative definite, if and only if its eigenvalues are positive, nonnegative, nonpositive, negative, respectively. Otherwise, $\mathbf{A}$ is said to be *indefinite*. We use the shorthand notation $\mathbf{A} \succ, \succeq, \preceq, \prec \mathbf{0}$ to indicate that $\mathbf{A}$ is positive definite, positive semidefinite, negative semidefinite, negative definite throughout the book.

Another approach to the characterization of a square matrix $\mathbf{A}$ is based on the evaluation of the *leading principal minor determinants*. A *minor determinant*, which is usually referred to as a *minor*, is the determinant of a submatrix obtained by deleting a number of rows and an equal number of columns from the matrix. Specifically, a minor of order r of an $n \times n$ matrix $\mathbf{A}$ is obtained by deleting $n - r$ rows and $n - r$ columns. For example, if

$$
\mathbf{A} = \begin{bmatrix} a_{11} & a_{12} & a_{13} & a_{14} \\ a_{21} & a_{22} & a_{23} & a_{24} \\ a_{31} & a_{32} & a_{33} & a_{34} \\ a_{41} & a_{42} & a_{43} & a_{44} \end{bmatrix}
$$

then

$$
\Delta_3^{(123,123)} = \begin{vmatrix} a_{11} & a_{12} & a_{13} \\ a_{21} & a_{22} & a_{23} \\ a_{31} & a_{32} & a_{33} \end{vmatrix}, \quad \Delta_3^{(134,124)} = \begin{vmatrix} a_{11} & a_{12} & a_{14} \\ a_{31} & a_{32} & a_{34} \\ a_{41} & a_{42} & a_{44} \end{vmatrix}
$$

and

$$
\Delta_2^{(12,12)} = \begin{vmatrix} a_{11} & a_{12} \\ a_{21} & a_{22} \end{vmatrix}, \quad \Delta_2^{(13,14)} = \begin{vmatrix} a_{11} & a_{14} \\ a_{31} & a_{34} \end{vmatrix}
$$

$$
\Delta_2^{(24,13)} = \begin{vmatrix} a_{21} & a_{23} \\ a_{41} & a_{43} \end{vmatrix}, \quad \Delta_2^{(34,34)} = \begin{vmatrix} a_{33} & a_{34} \\ a_{43} & a_{44} \end{vmatrix}
$$

are third-order and second-order minors, respectively. An nth-order minor is the determinant of the matrix itself and a first-order minor, i.e., if $n - 1$ rows and $n - 1$ columns are deleted, is simply the value of a single matrix component.[1]

If the indices of the deleted rows are the same as those of the deleted columns, then the minor is said to be a *principal minor*, e.g., $\Delta_3^{(123,123)}$, $\Delta_2^{(12,12)}$, and $\Delta_2^{(34,34)}$ in the above examples.

---

[1] The zeroth-order minor is often defined to be unity.

Principal minors $\Delta_3^{(123,123)}$ and $\Delta_2^{(12,12)}$ in the above examples can be represented by

$$\Delta_3^{(1,2,3)} = \det \mathbf{H}_3^{(1,2,3)}$$

and

$$\Delta_2^{(1,2)} = \det \mathbf{H}_2^{(1,2)}$$

respectively. An arbitrary principal minor of order i can be represented by

$$\Delta_i^{(l)} = \det \mathbf{H}_i^{(l)}$$

where

$$\mathbf{H}_i^{(l)} = \begin{bmatrix} a_{l_1 l_1} & a_{l_1 l_2} & \cdots & a_{l_1 l_i} \\ a_{l_2 l_1} & a_{l_2 l_2} & \cdots & a_{l_2 l_i} \\ \vdots & \vdots & & \vdots \\ a_{l_i l_1} & a_{l_i l_2} & \cdots & a_{l_i l_i} \end{bmatrix}$$

and $l \in \{l_1, l_2, \ldots, l_i\}$ with $1 \leq l_1 < l_2 < \cdots < l_i \leq n$ is the set of rows (and columns) retained in submatrix $\mathbf{H}_i^{(l)}$.

The specific principal minors

$$\Delta_r = \begin{vmatrix} a_{11} & a_{12} & \cdots & a_{1r} \\ a_{21} & a_{22} & \cdots & a_{2r} \\ \vdots & \vdots & & \vdots \\ a_{r1} & a_{r2} & \cdots & a_{rr} \end{vmatrix} = \det \mathbf{H}_r$$

for $1 \leq r \leq n$ are said to be the *leading principal minors* of an $n \times n$ matrix. For a 4×4 matrix, the complete set of leading principal minors is as follows:

$$\Delta_1 = a_{11}, \quad \Delta_2 = \begin{vmatrix} a_{11} & a_{12} \\ a_{21} & a_{22} \end{vmatrix}$$

$$\Delta_3 = \begin{vmatrix} a_{11} & a_{12} & a_{13} \\ a_{21} & a_{22} & a_{23} \\ a_{31} & a_{32} & a_{33} \end{vmatrix}, \quad \Delta_4 = \begin{vmatrix} a_{11} & a_{12} & a_{13} & a_{14} \\ a_{21} & a_{22} & a_{23} & a_{24} \\ a_{31} & a_{32} & a_{33} & a_{34} \\ a_{41} & a_{42} & a_{43} & a_{44} \end{vmatrix}$$

The leading principal minors of a matrix $\mathbf{A}$ or its negative $-\mathbf{A}$ can be used to establish whether the matrix is positive or negative definite whereas the principal minors of $\mathbf{A}$ or $-\mathbf{A}$ can be used to establish whether the matrix is positive or negative semidefinite. These principles are stated in terms of Theorem 2.9 in Chap. 2 and are often used to establish the nature of the Hessian matrix in optimization algorithms.

The fact that a nonnegative real number has positive and negative square roots can be extended to the class of positive semidefinite matrices. Assuming that matrix $\mathbf{A} \in R^{n \times n}$ is positive semidefinite, we can write its eigendecomposition in Eq. (A.16) as

$$\mathbf{A} = \mathbf{X}\boldsymbol{\Lambda}\mathbf{X}^T = \mathbf{X}\boldsymbol{\Lambda}^{1/2}\mathbf{W}\mathbf{W}^T\boldsymbol{\Lambda}^{1/2}\mathbf{X}^T$$

where $\boldsymbol{\Lambda}^{1/2} = \text{diag}\{\lambda_1^{1/2}, \lambda_2^{1/2}, \ldots, \lambda_n^{1/2}\}$ and $\mathbf{W}$ is an arbitrary orthogonal matrix, which leads to

$$\mathbf{A} = \mathbf{A}^{1/2}(\mathbf{A}^{1/2})^T \tag{A.18}$$

where $\mathbf{A}^{1/2} = \mathbf{X}\boldsymbol{\Lambda}^{1/2}\mathbf{W}$ and is called an *asymmetric square root* of $\mathbf{A}$. Since matrix $\mathbf{W}$ can be an arbitrary orthogonal matrix, an infinite number of asymmetric square roots of $\mathbf{A}$ exist. Alternatively, since $\mathbf{X}$ is an orthogonal matrix, we can write

$$\mathbf{A} = (\alpha\mathbf{X}\boldsymbol{\Lambda}^{1/2}\mathbf{X}^T)(\alpha\mathbf{X}\boldsymbol{\Lambda}^{1/2}\mathbf{X}^T)$$

where α is either 1 or -1, which gives

$$\mathbf{A} = \mathbf{A}^{1/2}\mathbf{A}^{1/2} \tag{A.19}$$

where $\mathbf{A}^{1/2} = \alpha\mathbf{X}\boldsymbol{\Lambda}^{1/2}\mathbf{X}^T$ and is called a *symmetric square root* of $\mathbf{A}$. Again, because α can be either 1 or -1, more than one symmetric square root exists. Obviously, the symmetric square roots $\mathbf{X}\boldsymbol{\Lambda}^{1/2}\mathbf{X}^T$ and $-\mathbf{X}\boldsymbol{\Lambda}^{1/2}\mathbf{X}^T$ are positive semidefinite and negative semidefinite, respectively.

If $\mathbf{A}$ is a complex-valued positive semidefinite matrix, then *non-Hermitian* and *Hermitian square roots of* $\mathbf{A}$ can be obtained using the eigendecomposition in Eq. (A.17). For example, we can write

$$\mathbf{A} = \mathbf{A}^{1/2}(\mathbf{A}^{1/2})^H$$

where $\mathbf{A}^{1/2} = \mathbf{U}\boldsymbol{\Lambda}^{1/2}\mathbf{W}$ is a non-Hermitian square root of $\mathbf{A}$ if $\mathbf{W}$ is unitary. On the other hand,

$$\mathbf{A} = \mathbf{A}^{1/2}\mathbf{A}^{1/2}$$

where $\mathbf{A}^{1/2} = \alpha\mathbf{U}\boldsymbol{\Lambda}^{1/2}\mathbf{U}^H$ is a Hermitian square root if $\alpha = 1$ or $\alpha = -1$.

Example A.5 Verify that

$$\mathbf{A} = \begin{bmatrix} 2.5 & 0 & 1.5 \\ 0 & \sqrt{2} & 0 \\ 1.5 & 0 & 2.5 \end{bmatrix}$$

is positive definite and compute a symmetric square root of $\mathbf{A}$.

Solution An eigendecomposition of matrix $\mathbf{A}$ is

$$\mathbf{A} = \mathbf{X}\boldsymbol{\Lambda}\mathbf{X}^T$$

with

$$\boldsymbol{\Lambda} = \begin{bmatrix} 4 & 0 & 0 \\ 0 & 2 & 0 \\ 0 & 0 & 1 \end{bmatrix} \quad \text{and} \quad \mathbf{X} = \begin{bmatrix} \sqrt{2}/2 & 0 & -\sqrt{2}/2 \\ 0 & -1 & 0 \\ \sqrt{2}/2 & 0 & \sqrt{2}/2 \end{bmatrix}$$

Since the eigenvalues of $\mathbf{A}$ are all positive, $\mathbf{A}$ is positive definite. A symmetric square root of $\mathbf{A}$ is given by

$$\mathbf{A}^{1/2} = \mathbf{X}\boldsymbol{\Lambda}^{1/2}\mathbf{X}^T = \begin{bmatrix} 1.5 & 0 & 0.5 \\ 0 & \sqrt{2} & 0 \\ 0.5 & 0 & 1.5 \end{bmatrix}$$

∎

A.7 Trace

The trace of an $n \times n$ square matrix, $\mathbf{A} = \{a_{ij}\}$, is *the sum of its diagonal components*, i.e.,

$$\text{trace}(\mathbf{A}) = \sum_{i=1}^{n} a_{ii}$$

It can be verified that the trace of a square matrix $\mathbf{A}$ with eigenvalues λ_1, λ_2, ..., λ_n is equal to the sum of its eigenvalues, i.e.,

$$\text{trace}(\mathbf{A}) = \sum_{i=1}^{n} \lambda_i$$

A useful property pertaining to the product of two matrices is that the trace of a square matrix $\mathbf{AB}$ is equal to the trace of matrix $\mathbf{BA}$, i.e.,

$$\text{trace}(\mathbf{AB}) = \text{trace}(\mathbf{BA}) \tag{A.20}$$

By applying Eq. (A.20) to the quadratic form $\mathbf{x}^T\mathbf{Hx}$, we obtain

$$\mathbf{x}^T\mathbf{Hx} = \text{trace}(\mathbf{x}^T\mathbf{Hx}) = \text{trace}(\mathbf{Hxx}^T) = \text{trace}(\mathbf{HX})$$

where $\mathbf{X} = \mathbf{xx}^T$. Moreover, we can write a general quadratic function as

$$\mathbf{x}^T\mathbf{Hx} + 2\mathbf{p}^T\mathbf{x} + \kappa = \text{trace}(\hat{\mathbf{H}}\hat{\mathbf{X}}) \tag{A.21}$$

where

$$\hat{\mathbf{H}} = \begin{bmatrix} \mathbf{H} & \mathbf{p} \\ \mathbf{p}^T & \kappa \end{bmatrix} \quad \text{and} \quad \hat{\mathbf{X}} = \begin{bmatrix} \mathbf{xx}^T & \mathbf{x} \\ \mathbf{x}^T & 1 \end{bmatrix}$$

A.8   Vector Norms and Matrix Norms

A.8.1   Vector Norms

The L_p norm of a vector $\mathbf{x} \in C^n$ for $p \geq 1$ is given by

$$\|\mathbf{x}\|_p = \left(\sum_{i=1}^{n} |x_i|^p \right)^{1/p} \tag{A.22}$$

where p is a positive integer and x_i is the ith component of $\mathbf{x}$. The most popular L_p norms are $\| \cdot \|_1$, $\| \cdot \|_2$, and $\| \cdot \|_\infty$, where the infinity norm $\| \cdot \|_\infty$ can easily be shown to satisfy the relation

$$\|\mathbf{x}\|_\infty = \lim_{p \to \infty} \left(\sum_{i=1}^{n} |x_i|^p \right)^{1/p} = \max_{i} |x_i| \tag{A.23}$$

For example, if $\mathbf{x} = [1 \ 2 \ \cdots \ 100]^T$, then $\|\mathbf{x}\|_2 = 581.68$, $\|\mathbf{x}\|_{10} = 125.38$, $\|\mathbf{x}\|_{50} = 101.85$, $\|\mathbf{x}\|_{100} = 100.45$, $\|\mathbf{x}\|_{200} = 100.07$, and, of course, $\|\mathbf{x}\|_\infty = 100$.

The important point to note here is that for an even p, the L_p norm of a vector is a *differentiable* function of its components but the L_∞ norm is *not*. So when the L_∞ norm is used in a design problem, we can replace it by an L_p norm (with p even) so that powerful calculus-based tools can be used to solve the problem. Obviously, the results obtained can only be *approximate* with respect to the original design problem. However, since

$$\lim_{p \to \infty} \|F(\omega)\|_p = \|F(\omega)\|_\infty = \max_{a \leq \omega \leq b} |F(\omega)|$$

where

$$\|F(\omega)\|_p = \left(\int_a^b |F(\omega)|^p \, d\omega \right)^{1/p}$$

with $p \geq 1$ (see Sect. B.9.3), the difference between the approximate and exact solutions becomes insignificant if p is made sufficiently large.

The inner product of two vectors $\mathbf{x}, \ \mathbf{y} \in C^n$ is a scalar given by

$$\mathbf{x}^H \mathbf{y} = \sum_{i=1}^{n} x_i^* y_i$$

where x_i^* denotes the complex-conjugate of x_i. Frequently, we need to estimate the absolute value of $\mathbf{x}^H \mathbf{y}$. There are two well-known inequalities that provide tight upper bounds for $|\mathbf{x}^H \mathbf{y}|$, namely, the *Hölder inequality*

$$|\mathbf{x}^H \mathbf{y}| \leq \|\mathbf{x}\|_p \|\mathbf{y}\|_q \tag{A.24}$$

which holds for any $p \geq 1$ and $q \geq 1$ satisfying the equality

$$\frac{1}{p} + \frac{1}{q} = 1$$

and the *Cauchy-Schwartz inequality*, which is the special case of the Hölder inequality with $p = q = 2$, i.e.,

$$|\mathbf{x}^H \mathbf{y}| \leq \|\mathbf{x}\|_2 \|\mathbf{y}\|_2 \tag{A.25}$$

If vectors $\mathbf{x}$ and $\mathbf{y}$ have unity length, i.e., $\|\mathbf{x}\|_2 = \|\mathbf{y}\|_2 = 1$, then Eq. (A.25) becomes

$$|\mathbf{x}^H \mathbf{y}| \leq 1 \tag{A.26}$$

A geometric interpretation of Eq. (A.26) is that for unit vectors $\mathbf{x}$ and $\mathbf{y}$, the inner product $\mathbf{x}^H \mathbf{y}$ is equal to $\cos \theta$, where θ denotes the angle between the two vectors, whose absolute value is always less than or equal to one.

Another property of the L_2 norm is its invariance under orthogonal or unitary transformation. That is, if $\mathbf{A}$ is an orthogonal or unitary matrix, then

$$\|\mathbf{A}\mathbf{x}\|_2 = \|\mathbf{x}\|_2 \tag{A.27}$$

The L_p norm of a vector $\mathbf{x}$, $\|\mathbf{x}\|_p$, is monotonically decreasing with respect to p for $p \geq 1$. For example, we can relate $\|\mathbf{x}\|_1$ and $\|\mathbf{x}\|_2$ as

$$\begin{aligned}
\|\mathbf{x}\|_1^2 &= \left(\sum_{i=1}^{n} |x_i| \right)^2 \\
&= |x_1|^2 + |x_2|^2 + \cdots + |x_n|^2 + 2|x_1 x_2| + \cdots + 2|x_{n-1} x_n| \\
&\geq |x_1|^2 + |x_2|^2 + \cdots + |x_n|^2 = \|\mathbf{x}\|_2^2
\end{aligned}$$

which implies that

$$\|\mathbf{x}\|_1 \geq \|\mathbf{x}\|_2$$

Furthermore, if $\|\mathbf{x}\|_\infty$ is numerically equal to $|x_k|$ for some index k, i.e.,

$$\|\mathbf{x}\|_\infty = \max_i |x_i| = |x_k|$$

then we can write

$$\|\mathbf{x}\|_2 = (|x_1|^2 + \cdots + |x_n|^2)^{1/2} \geq (|x_k|^2)^{1/2} = |x_k| = \|\mathbf{x}\|_\infty$$

i.e.,

$$\|\mathbf{x}\|_2 \geq \|\mathbf{x}\|_\infty$$

Therefore, we have

$$\|\mathbf{x}\|_1 \geq \|\mathbf{x}\|_2 \geq \|\mathbf{x}\|_\infty$$

In general, it can be shown that

$$\|\mathbf{x}\|_1 \geq \|\mathbf{x}\|_2 \geq \|\mathbf{x}\|_3 \geq \cdots \geq \|\mathbf{x}\|_\infty$$

A.8.2 Matrix Norms

The L_p norm of matrix $\mathbf{A} = \{a_{ij}\} \in C^{m \times n}$ is defined as

$$\|\mathbf{A}\|_p = \max_{\mathbf{x} \neq 0} \frac{\|\mathbf{A}\mathbf{x}\|_p}{\|\mathbf{x}\|_p} \quad \text{for } p \geq 1 \tag{A.28}$$

The most useful matrix L_p norm is the L_2 norm

$$\|\mathbf{A}\|_2 = \max_{\mathbf{x} \neq 0} \frac{\|\mathbf{A}\mathbf{x}\|_2}{\|\mathbf{x}\|_2} = \left[\max_i \left| \lambda_i (\mathbf{A}^H \mathbf{A}) \right| \right]^{1/2} = \left[\max_i \left| \lambda_i (\mathbf{A}\mathbf{A}^H) \right| \right]^{1/2} \tag{A.29}$$

which can be easily computed as the square root of the largest eigenvalue magnitude of $\mathbf{A}^H \mathbf{A}$ or $\mathbf{A}\mathbf{A}^H$. Some other frequently used matrix L_p norms are

$$\|\mathbf{A}\|_1 = \max_{\mathbf{x} \neq 0} \frac{\|\mathbf{A}\mathbf{x}\|_1}{\|\mathbf{x}\|_1} = \max_{1 \leq j \leq n} \sum_{i=1}^{m} |a_{ij}|$$

and

$$\|\mathbf{A}\|_\infty = \max_{\mathbf{x} \neq 0} \frac{\|\mathbf{A}\mathbf{x}\|_\infty}{\|\mathbf{x}\|_\infty} = \max_{1 \leq i \leq m} \sum_{j=1}^{n} |a_{ij}|$$

Another popular matrix norm is the Frobenius norm, which is defined as

$$\|\mathbf{A}\|_F = \left(\sum_{i=1}^{m} \sum_{j=1}^{n} |a_{ij}|^2 \right)^{1/2} \tag{A.30}$$

which can also be calculated as

$$\|\mathbf{A}\|_F = [\text{trace}(\mathbf{A}^H \mathbf{A})]^{1/2} = [\text{trace}(\mathbf{A}\mathbf{A}^H)]^{1/2} \tag{A.31}$$

Note that the matrix L_2 norm and the Frobenius norm are *invariant* under orthogonal or unitary transformation, i.e., if $\mathbf{U} \in C^{n \times n}$ and $\mathbf{V} \in C^{m \times m}$ are unitary or orthogonal matrices, then

$$\|\mathbf{U}\mathbf{A}\mathbf{V}\|_2 = \|\mathbf{A}\|_2 \tag{A.32}$$

and

$$\|\mathbf{U}\mathbf{A}\mathbf{V}\|_F = \|\mathbf{A}\|_F \tag{A.33}$$

Example A.6 Evaluate matrix norms $||\mathbf{A}||_1$, $||\mathbf{A}||_2$, $||\mathbf{A}||_\infty$, and $||\mathbf{A}||_F$ for

$$\mathbf{A} = \begin{bmatrix} 1 & 5 & 6 & 3 \\ 0 & 4 & -7 & 0 \\ 3 & 1 & 4 & 1 \\ -1 & 1 & 0 & 1 \end{bmatrix}$$

Solution

$$\|\mathbf{A}\|_1 = \max_{1 \le j \le 4} \left(\sum_{i=1}^{4} |a_{ij}| \right) = \max\{5, \ 11, \ 17, \ 5\} = 17$$

$$\|\mathbf{A}\|_\infty = \max_{1 \le i \le 4} \left(\sum_{j=1}^{4} |a_{ij}| \right) = \max\{15, \ 11, \ 9, \ 3\} = 15$$

$$\|\mathbf{A}\|_F = \left(\sum_{i=1}^{4} \sum_{j=1}^{4} |a_{ij}|^2 \right)^{1/2} = \sqrt{166} = 12.8841$$

To obtain $\|\mathbf{A}\|_2$, we compute the eigenvalues of $\mathbf{A}^T \mathbf{A}$ as

$$\lambda(\mathbf{A}^T \mathbf{A}) = \{0.2099, \ 6.9877, \ 47.4010, \ 111.4014\}$$

Hence

$$\|\mathbf{A}\|_2 = [\max_i |\lambda_i(\mathbf{A}^T \mathbf{A})|]^{1/2} = \sqrt{111.4014} = 10.5547$$

∎

A.9   Singular-Value Decomposition

Given a matrix $\mathbf{A} \in C^{m \times n}$ of rank r, there exist unitary matrices $\mathbf{U} \in C^{m \times m}$ and $\mathbf{V} \in C^{n \times n}$ such that

$$\mathbf{A} = \mathbf{U} \boldsymbol{\Sigma} \mathbf{V}^H \tag{A.34a}$$

where

$$\boldsymbol{\Sigma} = \begin{bmatrix} \mathbf{S} & \mathbf{0} \\ \mathbf{0} & \mathbf{0} \end{bmatrix}_{m \times n} \tag{A.34b}$$

and

$$\mathbf{S} = \text{diag}\{\sigma_1, \ \sigma_2, \ \ldots, \ \sigma_r\} \tag{A.34c}$$

with $\sigma_1 \ge \sigma_2 \ge \cdots \ge \sigma_r > 0$.

The matrix decomposition in Eq. (A.34a) is known as the *singular-value* decomposition (SVD) of $\mathbf{A}$. It has many applications in optimization and elsewhere. If $\mathbf{A}$ is a real-valued matrix, then $\mathbf{U}$ and $\mathbf{V}$ in Eq. (A.34a) become orthogonal matrices and $\mathbf{V}^H$ becomes $\mathbf{V}^T$. The positive scalars σ_i for $i = 1, \ 2, \ \ldots, \ r$ in Eq. (A.34c) are called the *singular values* of $\mathbf{A}$. If $\mathbf{U} = [\mathbf{u}_1 \ \mathbf{u}_2 \ \cdots \ \mathbf{u}_m]$ and $\mathbf{V} = [\mathbf{v}_1 \ \mathbf{v}_2 \ \cdots \ \mathbf{v}_n]$, vectors $\mathbf{u}_i$ and $\mathbf{v}_i$ are called the *left* and *right singular vectors* of $\mathbf{A}$, respectively. From Eqs. (A.34a)–(A.34c), it follows that

$$\mathbf{A}\mathbf{A}^H = \mathbf{U} \begin{bmatrix} \mathbf{S}^2 & \mathbf{0} \\ \mathbf{0} & \mathbf{0} \end{bmatrix}_{m \times m} \mathbf{U}^H \tag{A.35a}$$

and

$$\mathbf{A}^H\mathbf{A} = \mathbf{V} \begin{bmatrix} \mathbf{S}^2 & \mathbf{0} \\ \mathbf{0} & \mathbf{0} \end{bmatrix}_{n \times n} \mathbf{V}^H \tag{A.35b}$$

Therefore, the singular values of $\mathbf{A}$ are the positive square roots of the nonzero eigenvalues of $\mathbf{A}\mathbf{A}^H$ (or $\mathbf{A}^H\mathbf{A}$), the ith left singular vector $\mathbf{u}_i$ is the ith eigenvector of $\mathbf{A}\mathbf{A}^H$, and the ith right singular vector $\mathbf{v}_i$ is the ith eigenvector of $\mathbf{A}^H\mathbf{A}$.

Several important applications of the SVD are as follows:

(a) The L_2 norm and Frobenius norm of a matrix $\mathbf{A} \in C^{m \times n}$ of rank r are given, respectively, by

$$\|\mathbf{A}\|_2 = \sigma_1 \tag{A.36}$$

and

$$\|\mathbf{A}\|_F = \left(\sum_{i=1}^{r} \sigma_i^2 \right)^{1/2} \tag{A.37}$$

(b) The *condition number* of a nonsingular matrix $\mathbf{A} \in C^{n \times n}$ is defined as

$$\mathrm{cond}(\mathbf{A}) = \|\mathbf{A}\|_2 \|\mathbf{A}^{-1}\|_2 = \frac{\sigma_1}{\sigma_n} \tag{A.38}$$

(c) The range and null space of a matrix $\mathbf{A} \in C^{m \times n}$ of rank r assume the forms

$$\mathcal{R}(\mathbf{A}) = \mathrm{span}\{\mathbf{u}_1, \ \mathbf{u}_2, \ \ldots, \ \mathbf{u}_r\} \tag{A.39}$$

$$\mathcal{N}(\mathbf{A}) = \mathrm{span}\{\mathbf{v}_{r+1}, \ \mathbf{v}_{r+2}, \ \ldots, \ \mathbf{v}_n\} \tag{A.40}$$

(d) Properties and computation of Moore-Penrose pseudo-inverse:
The Moore-Penrose pseudo-inverse of a matrix $\mathbf{A} \in C^{m \times n}$ is defined as the matrix $\mathbf{A}^+ \in C^{n \times m}$ that satisfies the following four conditions:

(i) $\mathbf{A}\mathbf{A}^+\mathbf{A} = \mathbf{A}$
(ii) $\mathbf{A}^+\mathbf{A}\mathbf{A}^+ = \mathbf{A}^+$
(iii) $(\mathbf{A}\mathbf{A}^+)^H = \mathbf{A}\mathbf{A}^+$
(iv) $(\mathbf{A}^+\mathbf{A})^H = \mathbf{A}^+\mathbf{A}$

Using the SVD of $\mathbf{A}$ in Eqs. (A.34a)–(A.34c), the Moore-Penrose pseudo-inverse of $\mathbf{A}$ can be obtained as

$$\mathbf{A}^+ = \mathbf{V}\boldsymbol{\Sigma}^+\mathbf{U}^H \tag{A.41a}$$

where

$$\Sigma^+ = \begin{bmatrix} \mathbf{S}^{-1} & \mathbf{0} \\ \mathbf{0} & \mathbf{0} \end{bmatrix}_{n \times m} \tag{A.41b}$$

and

$$\mathbf{S}^{-1} = \text{diag}\{\sigma_1^{-1}, \; \sigma_2^{-1}, \; \ldots, \; \sigma_r^{-1}\} \tag{A.41c}$$

Consequently, we have

$$\mathbf{A}^+ = \sum_{i=1}^{r} \frac{\mathbf{v}_i \mathbf{u}_i^H}{\sigma_i} \tag{A.42}$$

(e) For an underdetermined system of linear equations

$$\mathbf{A}\mathbf{x} = \mathbf{b} \tag{A.43}$$

where $\mathbf{A} \in C^{m \times n}$, $\mathbf{b} \in C^{m \times 1}$ with $m < n$, and $\mathbf{b} \in \mathcal{R}(\mathbf{A})$, all the solutions of Eq. (A.43) are characterized by

$$\mathbf{x} = \mathbf{A}^+ \mathbf{b} + \mathbf{V}_r \boldsymbol{\phi} \tag{A.44a}$$

where $\mathbf{A}^+$ is the Moore-Penrose pseudo-inverse of $\mathbf{A}$,

$$\mathbf{V}_r = [\mathbf{v}_{r+1} \; \mathbf{v}_{r+2} \; \cdots \; \mathbf{v}_n] \tag{A.44b}$$

is a matrix of dimension $n \times (n-r)$ composed of the last $n-r$ columns of matrix $\mathbf{V}$ which is obtained by constructing the SVD of $\mathbf{A}$ in Eqs. (A.34a)–(A.34c), and $\boldsymbol{\phi} \in C^{(n-r) \times 1}$ is an *arbitrary* $(n-r)$-dimensional vector. Note that the first term in Eq. (A.44a), i.e., $\mathbf{A}^+ \mathbf{b}$, is a solution of Eq. (A.43) while the second term, $\mathbf{V}_r \boldsymbol{\phi}$, belongs to the null space of $\mathbf{A}$ (see Eq. (A.40)) . Through vector $\boldsymbol{\phi}$, the expression in Eqs. (A.44a)–(A.44b) parameterizes all the solutions of an underdetermined system of linear equations.

Example A.7 Perform the SVD of matrix

$$\mathbf{A} = \begin{bmatrix} 2.8284 & -1 & 1 \\ 2.8284 & 1 & -1 \end{bmatrix}$$

and compute $||\mathbf{A}||_2$, $||\mathbf{A}||_F$, and $\mathbf{A}^+$.

Solution To compute matrix $\mathbf{V}$ in Eq. (A.34a) , from Eq. (A.35b) we obtain

$$\mathbf{A}^T \mathbf{A} = \begin{bmatrix} 16 & 0 & 0 \\ 0 & 2 & -2 \\ 0 & -2 & 2 \end{bmatrix} = \mathbf{V} \begin{bmatrix} 16 & 0 & 0 \\ 0 & 4 & 0 \\ 0 & 0 & 0 \end{bmatrix} \mathbf{V}^T$$

where

$$V = \begin{bmatrix} 1 & 0 & 0 \\ 0 & 0.7071 & -0.7071 \\ 0 & -0.7071 & -0.7071 \end{bmatrix} = [\mathbf{v}_1 \ \mathbf{v}_2 \ \mathbf{v}_3]$$

Hence the nonzero singular values of $\mathbf{A}$ are $\sigma_1 = \sqrt{16} = 4$ and $\sigma_2 = \sqrt{4} = 2$. Now we can write Eq. (A.34a) as $\mathbf{U}\boldsymbol{\Sigma} = \mathbf{A}\mathbf{V}$, where

$$\mathbf{U}\boldsymbol{\Sigma} = [\sigma_1\mathbf{u}_1 \ \sigma_2\mathbf{u}_2 \ \mathbf{0}] = [4\mathbf{u}_1 \ 2\mathbf{u}_2 \ \mathbf{0}]$$

and

$$\mathbf{A}\mathbf{V} = \begin{bmatrix} 2.8284 & -1.4142 & 0 \\ 2.8284 & 1.4142 & 0 \end{bmatrix}$$

Hence

$$\mathbf{u}_1 = \frac{1}{4}\begin{bmatrix} 2.8284 \\ 2.8284 \end{bmatrix} = \begin{bmatrix} 0.7071 \\ 0.7071 \end{bmatrix}, \quad \mathbf{u}_2 = \frac{1}{2}\begin{bmatrix} -1.4142 \\ 1.4142 \end{bmatrix} = \begin{bmatrix} -0.7071 \\ 0.7071 \end{bmatrix}$$

and

$$\mathbf{U} = [\mathbf{u}_1 \ \mathbf{u}_2] = \begin{bmatrix} 0.7071 & -0.7071 \\ 0.7071 & 0.7071 \end{bmatrix}$$

On using Eqs. (A.36) and (A.37), we have

$$\|\mathbf{A}\|_2 = \sigma_1 = 4 \quad \text{and} \quad \|\mathbf{A}\|_F = (\sigma_1^2 + \sigma_2^2)^{1/2} = \sqrt{20} = 4.4721$$

Now from Eq. (A.42), we obtain

$$\mathbf{A}^+ = \frac{\mathbf{v}_1\mathbf{u}_1^T}{\sigma_1} + \frac{\mathbf{v}_2\mathbf{u}_2^T}{\sigma_2} = \begin{bmatrix} 0.1768 & 0.1768 \\ -0.2500 & 0.2500 \\ 0.2500 & -0.2500 \end{bmatrix}$$

∎

A.10 Orthogonal Projections

Let S be a subspace in C^n. Matrix $\mathbf{P} \in C^{n\times n}$ is said to be an orthogonal projection matrix onto S if $\mathcal{R}(\mathbf{P}) = S$, $\mathbf{P}^2 = \mathbf{P}$, and $\mathbf{P}^H = \mathbf{P}$, where $\mathcal{R}(\mathbf{P})$ denotes the range of transformation $\mathbf{P}$ (see Sect. A.3), i.e., $\mathcal{R}(\mathbf{P}) = \{\mathbf{y} : \mathbf{y} = \mathbf{P}\mathbf{x}, \ \mathbf{x} \in C^n\}$. The term 'orthogonal projection' originates from the fact that if $\mathbf{x} \in C^n$ is a vector outside S, then $\mathbf{P}\mathbf{x}$ is a vector in S such that $\mathbf{x} - \mathbf{P}\mathbf{x}$ is orthogonal to every vector in S and $\|\mathbf{x} - \mathbf{P}\mathbf{x}\|_2$ is the minimum distance between $\mathbf{x}$ and $\mathbf{s}$, i.e., $\min \|\mathbf{x} - \mathbf{s}\|_2$, for $\mathbf{s} \in S$, as illustrated in Fig. A.1.

Let $\{\mathbf{s}_1, \ \mathbf{s}_2, \ \ldots, \ \mathbf{s}_k\}$ be a basis of a subspace S of dimension k (see Sect. A.2) such that $\|\mathbf{s}_i\|_2 = 1$ and $\mathbf{s}_i^T\mathbf{s}_j = 0$ for $i, \ j = 1, \ 2, \ \ldots, k$ and $i \neq j$. Such a basis

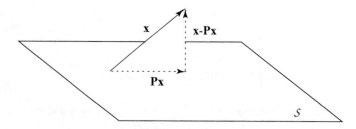

Fig. A.1 Orthogonal projection of $\mathbf{x}$ onto subspace S

is called *orthonormal*. It can be readily verified that an orthogonal projection matrix onto S can be explicitly constructed in terms of an orthonormal basis as

$$\mathbf{P} = \mathbf{SS}^H \tag{A.45a}$$

where

$$\mathbf{S} = [\mathbf{s}_1 \; \mathbf{s}_2 \; \cdots \; \mathbf{s}_k] \tag{A.45b}$$

It follows from Eqs. (A.39), (A.40), (A.45a), and (A.45b) that $[\mathbf{u}_1 \; \mathbf{u}_2 \; \cdots \; \mathbf{u}_r] \cdot [\mathbf{u}_1 \; \mathbf{u}_2 \; \cdots \; \mathbf{u}_r]^H$ is the orthogonal projection onto $\mathcal{R}(\mathbf{A})$ and $[\mathbf{v}_{r+1} \; \mathbf{v}_{r+2} \; \cdots \; \mathbf{v}_n] \cdot [\mathbf{v}_{r+1} \; \mathbf{v}_{r+2} \; \cdots \; \mathbf{v}_n]^H$ is the orthogonal projection onto $\mathcal{N}(\mathbf{A})$.

Example A.8 Let $S = \text{span}\{\mathbf{v}_1, \; \mathbf{v}_2\}$ where

$$\mathbf{v}_1 = \begin{bmatrix} 1 \\ 1 \\ 1 \end{bmatrix} \quad \text{and} \quad \mathbf{v}_2 = \begin{bmatrix} -1 \\ 1 \\ 1 \end{bmatrix}$$

Find the orthogonal projection onto S.

Solution First, we need to find an orthonormal basis $\{\mathbf{s}_1, \; \mathbf{s}_2\}$ of subspace S. To this end, we take

$$\mathbf{s}_1 = \frac{\mathbf{v}_1}{\|\mathbf{v}_1\|_2} = \begin{bmatrix} 1/\sqrt{3} \\ 1/\sqrt{3} \\ 1/\sqrt{3} \end{bmatrix}$$

Then we try to find vector $\hat{\mathbf{s}}_2$ such that $\hat{\mathbf{s}}_2 \in S$ and $\hat{\mathbf{s}}_2$ is orthogonal to $\mathbf{s}_1$. Such an $\hat{\mathbf{s}}_2$ must satisfy the relation

$$\hat{\mathbf{s}}_2 = \alpha_1 \mathbf{v}_1 + \alpha_2 \mathbf{v}_2$$

for some $\alpha_1, \; \alpha_2$ and

$$\hat{\mathbf{s}}_2^T \mathbf{s}_1 = 0$$

Hence we have

$$(\alpha_1 \mathbf{v}_1^T + \alpha_2 \mathbf{v}_2^T)\mathbf{s}_1 = \alpha_1 \mathbf{v}_1^T \mathbf{s}_1 + \alpha_2 \mathbf{v}_2^T \mathbf{s}_1 = \sqrt{3}\alpha_1 + \frac{1}{\sqrt{3}}\alpha_2 = 0$$

i.e., $\alpha_2 = -3\alpha_1$. Thus

$$\hat{\mathbf{s}}_2 = \alpha_1 \mathbf{v}_1 - 3\alpha_1 \mathbf{v}_2 = \alpha_1 \begin{bmatrix} 4 \\ -2 \\ -2 \end{bmatrix}$$

where α_1 is a parameter that can assume an arbitrary nonzero value.

By normalizing vector $\hat{\mathbf{s}}_2$, we obtain

$$\mathbf{s}_2 = \frac{\hat{\mathbf{s}}_2}{\|\hat{\mathbf{s}}_2\|_2} = \frac{1}{\sqrt{4^2 + (-2)^2 + (-2)^2}} \begin{bmatrix} 4 \\ -2 \\ -2 \end{bmatrix} = \frac{1}{\sqrt{6}} \begin{bmatrix} 2 \\ -1 \\ -1 \end{bmatrix}$$

It now follows from Eqs. (A.45a) and (A.45b) that the orthogonal projection onto S can be characterized by

$$\mathbf{P} = [\mathbf{s}_1 \ \mathbf{s}_2][\mathbf{s}_1 \ \mathbf{s}_2]^T = \begin{bmatrix} 1 & 0 & 0 \\ 0 & 0.5 & 0.5 \\ 0 & 0.5 & 0.5 \end{bmatrix}$$

∎

A.11 Householder Transformations and Givens Rotations

A.11.1 Householder Transformations

The *Householder transformation* associated with a nonzero vector $\mathbf{u} \in R^{n \times 1}$ is characterized by the symmetric orthogonal matrix

$$\mathbf{H} = \mathbf{I} - 2\frac{\mathbf{u}\mathbf{u}^T}{\|\mathbf{u}\|_2^2} \tag{A.46}$$

If

$$\mathbf{u} = \mathbf{x} - \|\mathbf{x}\|_2 \mathbf{e}_1 \tag{A.47}$$

where $\mathbf{e}_1 = [1 \ 0 \ \cdots \ 0]^T$, then the Householder transformation will convert vector $\mathbf{x}$ to coordinate vector $\mathbf{e}_1$ to within a scale factor $\|\mathbf{x}\|_2$, i.e.,

$$\mathbf{H}\mathbf{x} = \|\mathbf{x}\|_2 \begin{bmatrix} 1 \\ 0 \\ \vdots \\ 0 \end{bmatrix} \tag{A.48}$$

Alternatively, if vector $\mathbf{u}$ in Eq. (A.46) is chosen as

$$\mathbf{u} = \mathbf{x} + \|\mathbf{x}\|_2 \mathbf{e}_1 \tag{A.49}$$

then

$$\mathbf{H}\mathbf{x} = -\|\mathbf{x}\|_2 \begin{bmatrix} 1 \\ 0 \\ \vdots \\ 0 \end{bmatrix} \tag{A.50}$$

From Eqs. (A.47) and (A.49), we see that the transformed vector $\mathbf{H}\mathbf{x}$ contains $n-1$ zeros. Furthermore, since $\mathbf{H}$ is an orthogonal matrix, we have

$$\|\mathbf{H}\mathbf{x}\|_2^2 = (\mathbf{H}\mathbf{x})^T \mathbf{H}\mathbf{x} = \mathbf{x}^T \mathbf{H}^T \mathbf{H}\mathbf{x} = \mathbf{x}^T \mathbf{x} = \|\mathbf{x}\|_2^2$$

Therefore, $\mathbf{H}\mathbf{x}$ preserves the length of $\mathbf{x}$. For the sake of numerical robustness, a good choice of vector $\mathbf{u}$ between Eqs. (A.47) and (A.49) is

$$\mathbf{u} = \mathbf{x} + \text{sign}(x_1)\|\mathbf{x}\|_2 \mathbf{e}_1 \tag{A.51}$$

because the alternative choice, $\mathbf{u} = \mathbf{x} - \text{sign}(x_1)\|\mathbf{x}\|_2 \mathbf{e}_1$, may yield a vector $\mathbf{u}$ whose magnitude becomes too small when $\mathbf{x}$ is close to a multiple of $\mathbf{e}_1$.

Given a matrix $\mathbf{A} \in R^{n \times n}$, the matrix product $\mathbf{H}\mathbf{A}$ is called a *Householder update* of $\mathbf{A}$ and it can be evaluated as

$$\mathbf{H}\mathbf{A} = \left(\mathbf{I} - \frac{2\mathbf{u}\mathbf{u}^T}{\|\mathbf{u}\|_2^2}\right)\mathbf{A} = \mathbf{A} - \mathbf{u}\mathbf{v}^T \tag{A.52a}$$

where

$$\mathbf{v} = \alpha \mathbf{A}^T \mathbf{u}, \quad \alpha = -\frac{2}{\|\mathbf{u}\|_2^2} \tag{A.52b}$$

We see that a Householder update of $\mathbf{A}$ is actually a rank-one correction of $\mathbf{A}$, which can be obtained by using a matrix-vector multiplication and then an outer product update. In this way, a Householder update can be carried out efficiently without requiring matrix multiplication explicitly.

By successively applying the Householder update with appropriate values of $\mathbf{u}$, a given matrix $\mathbf{A}$ can be transformed into an upper triangular matrix. To see this, consider a matrix $\mathbf{A} \in R^{n \times n}$ and let $\mathbf{H}_i$ be the ith Householder update such that after $k-1$ successive applications of $\mathbf{H}_i$ for $i = 1, 2, \ldots, k-1$ the transformed matrix becomes

$$\mathbf{A}^{(k-1)} = \mathbf{H}_{k-1} \cdots \mathbf{H}_1 \mathbf{A} = \begin{bmatrix} * & * & \cdots & * & & & \\ 0 & * & & * & & * & \\ & & \ddots & & & & \\ 0 & 0 & & * & & & \\ \overline{0} & \overline{0} & \cdots & \overline{0} & & & \\ \vdots & \vdots & \cdots & & \mathbf{a}_k^{(k-1)} & \cdots & \mathbf{a}_n^{(k-1)} \\ 0 & 0 & \cdots & 0 & & & \end{bmatrix} \tag{A.53}$$

The next Householder update is characterized by

$$\mathbf{H}_k = \mathbf{I} - 2\frac{\mathbf{u}_k \mathbf{u}_k^T}{\|\mathbf{u}_k\|_2^2} = \begin{bmatrix} \mathbf{I}_{k-1} & \mathbf{0} \\ \mathbf{0} & \tilde{\mathbf{H}}_k \end{bmatrix} \qquad (A.54)$$

where

$$\mathbf{u}_k = \begin{bmatrix} 0 \\ \vdots \\ 0 \\ \mathbf{u}_k^{(k-1)} \end{bmatrix} \begin{matrix} \\ \}k-1 \\ \\ \end{matrix} , \quad \mathbf{u}_k^{(k-1)} = \mathbf{a}_k^{(k-1)} + \mathrm{sign}[\mathbf{a}_k^{(k-1)}(1)]\|\mathbf{a}_k^{(k-1)}\|_2 \mathbf{e}_1$$

$$\tilde{\mathbf{H}}_k = \mathbf{I}_{n-k+1} - 2\frac{\mathbf{u}_k^{(k-1)}(\mathbf{u}_k^{(k-1)})^T}{\|\mathbf{u}_k^{(k-1)}\|_2^2}$$

and $\mathbf{a}_k^{(k-1)}(1)$ represents the first component of vector $\mathbf{a}_k^{(k-1)}$.

Evidently, premultiplying $\mathbf{A}^{(k-1)}$ by $\mathbf{H}_k$ alters only the lower right block of $\mathbf{A}^{(k-1)}$ in Eq. (A.53) thereby converting its first column $\mathbf{a}_k^{(k-1)}$ to $[* \ 0 \ \cdots \ 0]^T$. Proceeding in this way, all the entries in the lower triangle will become zero.

Example A.9 Applying a series of Householder transformations, reduce matrix

$$\mathbf{A} = \begin{bmatrix} 1 & 0 & 3 & -1 \\ -1 & 2 & -7 & 5 \\ 3 & 1 & 7 & -1 \\ 0 & -1 & 2 & -2 \end{bmatrix}$$

to an upper triangular matrix.

Solution Using Eq. (A.51), we compute vector $\mathbf{u}_1$ as

$$\mathbf{u}_1 = \begin{bmatrix} 1 \\ -1 \\ 3 \\ 0 \end{bmatrix} + \sqrt{11} \begin{bmatrix} 1 \\ 0 \\ 0 \\ 0 \end{bmatrix} = \begin{bmatrix} 1+\sqrt{11} \\ -1 \\ 3 \\ 0 \end{bmatrix}$$

The associated Householder transformation is given by

$$\mathbf{H} = \mathbf{I} - \frac{2\mathbf{u}_1 \mathbf{u}_1^T}{\|\mathbf{u}_1\|_2^2} = \begin{bmatrix} -0.3015 & 0.3015 & -0.9045 & 0 \\ 0.3015 & 0.9302 & 0.2095 & 0 \\ -0.9045 & 0.2095 & 0.3714 & 0 \\ 0 & 0 & 0 & 1 \end{bmatrix}$$

The first Householder update is found to be

$$\mathbf{A}^{(1)} = \mathbf{H}_1 \mathbf{A} = \begin{bmatrix} -3.3166 & -0.3015 & -9.3469 & 2.7136 \\ 0 & 2.0698 & -4.1397 & 4.1397 \\ 0 & 0.7905 & -1.5809 & 1.5809 \\ 0 & -1.000 & 2.0000 & -2.0000 \end{bmatrix}$$

From Eq. (A.53), we obtain

$$\mathbf{a}_2^{(1)} = \begin{bmatrix} 2.0698 \\ 0.7905 \\ -1.0000 \end{bmatrix}$$

Using Eq. (A.54), we can compute

$$\mathbf{u}_2^{(1)} = \begin{bmatrix} 4.5007 \\ 0.7905 \\ -1.0000 \end{bmatrix}$$

$$\tilde{\mathbf{H}}_2 = \begin{bmatrix} -0.8515 & -0.3252 & 0.4114 \\ -0.3252 & 0.9429 & 0.0722 \\ 0.4114 & 0.0722 & 0.9086 \end{bmatrix}$$

and

$$\mathbf{H}_2 = \begin{bmatrix} 1 & \mathbf{0} \\ \mathbf{0} & \tilde{\mathbf{H}}_2 \end{bmatrix}$$

By premultiplying matrix $\mathbf{H}_1\mathbf{A}$ by $\mathbf{H}_2$, we obtain the required upper triangular matrix in terms of the second Householder update as

$$\mathbf{H}_2\mathbf{H}_1\mathbf{A} = \begin{bmatrix} -3.3166 & -0.3015 & -9.3469 & 2.7136 \\ 0 & -2.4309 & 4.8617 & -4.8617 \\ 0 & 0 & 0 & 0 \\ 0 & 0 & 0 & 0 \end{bmatrix}$$

■

A.11.2 Givens Rotations

Givens rotations are rank-two corrections of the identity matrix and are characterized by

$$\mathbf{G}_{ik}(\theta) = \begin{bmatrix} 1 & \cdots & 0 & \cdots & 0 & \cdots & 0 \\ \vdots & \ddots & \vdots & & \vdots & & \vdots \\ 0 & \cdots & c & \cdots & s & \cdots & 0 \\ \vdots & & \vdots & \ddots & \vdots & & \vdots \\ 0 & \cdots & -s & \cdots & c & \cdots & 0 \\ \vdots & & \vdots & & \vdots & \ddots & \vdots \\ 0 & \cdots & 0 & \cdots & 0 & \cdots & 1 \end{bmatrix} \begin{matrix} \\ \\ i \\ \\ k \\ \\ \\ \end{matrix}$$
$$\qquad\qquad i \qquad k$$

for $1 \leq i,\ k \leq n$, where $c = \cos\theta$ and $s = \sin\theta$ for some θ. It can be verified that $\mathbf{G}_{ik}(\theta)$ is an orthogonal matrix and $\mathbf{G}_{ik}^T(\theta)\mathbf{x}$ only affects the ith and kth components of vector $\mathbf{x}$, i.e.,

$$\mathbf{y} = \mathbf{G}_{ik}^T(\theta)\mathbf{x} \quad \text{with} \quad y_l = \begin{cases} cx_i - sx_k & \text{for } l = i \\ sx_i + cx_k & \text{for } l = k \\ x_l & \text{otherwise} \end{cases}$$

By choosing an appropriate θ such that

$$sx_i + cx_k = 0 \qquad\qquad (A.55)$$

the kth component of vector $\mathbf{y}$ is forced to zero. A numerically stable method for determining suitable values for s and c in Eq. (A.55) is described below, where we denote x_i and x_k as a and b, respectively.

    (a) If $b = 0$, set $c = 1$, $s = 0$.

    (b) If $b \neq 0$, then

        (i) if $|b| > |a|$, set

$$\tau = -\frac{a}{b}, \quad s = \frac{1}{\sqrt{1 + \tau^2}}, \quad c = \tau s$$

        (ii) otherwise, if $|b| \leq |a|$, set

$$\tau = -\frac{b}{a}, \quad c = \frac{1}{\sqrt{1 + \tau^2}}, \quad s = \tau c$$

Note that when premultiplying matrix $\mathbf{A}$ by $\mathbf{G}_{ik}^T(\theta)$, matrix $\mathbf{G}_{ik}^T(\theta)\mathbf{A}$ alters only the ith and kth rows of $\mathbf{A}$. The application of Givens rotations is illustrated by the following example.

Example A.10 Convert matrix $\mathbf{A}$ given by

$$\mathbf{A} = \begin{bmatrix} 3 & -1 \\ -3 & 5 \\ 2 & 1 \end{bmatrix}$$

into an upper triangular matrix by premultiplying it by an orthogonal transformation matrix that can be obtained using Givens rotations.

Solution To handle the first column, we first use $\mathbf{G}_{2,3}^T(\theta)$ to force its last component to zero. In this case, $a = -3$ and $b = 2$, hence

$$\tau = \frac{2}{3}, \quad c = \frac{1}{\sqrt{1 + \tau^2}} = 0.8321, \quad \text{and} \quad s = \tau c = 0.5547$$

Therefore, matrix $\mathbf{G}_{2,3}(\theta_1)$ is given by

$$\mathbf{G}_{2,3}(\theta_1) = \begin{bmatrix} 1 & 0 & 0 \\ 0 & 0.8321 & 0.5547 \\ 0 & -0.5547 & 0.8321 \end{bmatrix}$$

which leads to

$$\mathbf{G}_{2,3}^T(\theta_1)\mathbf{A} = \begin{bmatrix} 3.0000 & -1.0000 \\ -3.6056 & 3.6056 \\ 0 & 3.6056 \end{bmatrix}$$

In order to apply $\mathbf{G}_{1,2}^T(\theta_2)$ to the resulting matrix to force the second component of its first column to zero, we note that $a = 3$ and $b = -3.6056$; hence

$$\tau = \frac{3}{3.6056}, \quad s = \frac{1}{\sqrt{1 + \tau^2}} = 0.7687, \quad \text{and} \quad c = \tau s = 0.6396$$

Therefore, matrix $\mathbf{G}_{1,2}(\theta_2)$ is given by

$$\mathbf{G}_{1,2}(\theta_2) = \begin{bmatrix} 0.6396 & 0.7687 & 0 \\ -0.7687 & 0.6396 & 0 \\ 0 & 0 & 1 \end{bmatrix}$$

and

$$\mathbf{G}_{1,2}^T(\theta_2)\mathbf{G}_{2,3}^T(\theta_1)\mathbf{A} = \begin{bmatrix} 4.6904 & -3.4112 \\ 0 & 1.5374 \\ 0 & 3.6056 \end{bmatrix}$$

Now we can force the last component of the second column of the resulting matrix to zero by applying $\mathbf{G}_{2,3}^T(\theta_3)$. With $a = 1.5374$ and $b = 3.6056$, we compute

$$\tau = \frac{1.5374}{3.6056}, \quad s = \frac{1}{\sqrt{1 + \tau^2}} = 0.9199, \quad \text{and} \quad c = \tau s = 0.3922$$

Therefore, matrix $\mathbf{G}_{2,3}(\theta_3)$ is given by

$$\mathbf{G}_{2,3}(\theta_3) = \begin{bmatrix} 1 & 0 & 0 \\ 0 & -0.3922 & 0.9199 \\ 0 & -0.9199 & -0.3922 \end{bmatrix}$$

which yields

$$\mathbf{G}_{2,3}^T(\theta_3)\mathbf{G}_{1,2}^T(\theta_2)\mathbf{G}_{2,3}^T(\theta_1)\mathbf{A} = \begin{bmatrix} 4.6904 & -3.4112 \\ 0 & -3.9196 \\ 0 & 0 \end{bmatrix}$$

∎

A.12   QR Decomposition

A.12.1   Full-Rank Case

A QR decomposition of a matrix $\mathbf{A} \in R^{m \times n}$ is given by

$$\mathbf{A} = \mathbf{QR} \tag{A.56}$$

where $\mathbf{Q} \in R^{m \times m}$ is an orthogonal matrix and $\mathbf{R} \in R^{m \times n}$ is an upper triangular matrix.

In general, more than one QR decomposition exists. For example, if $\mathbf{A} = \mathbf{QR}$ is a QR decomposition of $\mathbf{A}$, then $\mathbf{A} = \tilde{\mathbf{Q}}\tilde{\mathbf{R}}$ is also a QR decomposition of $\mathbf{A}$ if $\tilde{\mathbf{Q}} = \mathbf{Q}\tilde{\mathbf{I}}$ and $\tilde{\mathbf{R}} = \tilde{\mathbf{I}}\mathbf{R}$ and $\tilde{\mathbf{I}}$ is a diagonal matrix whose diagonal comprises a mixture of 1's and -1's. Obviously, $\tilde{\mathbf{Q}}$ remains orthogonal and $\tilde{\mathbf{R}}$ is a triangular matrix but the signs of the rows in $\tilde{\mathbf{R}}$ corresponding to the -1's in $\tilde{\mathbf{I}}$ are changed compared with those in $\mathbf{R}$.

For the sake of convenience, we assume in the rest of this section that $m \geq n$. This assumption implies that $\mathbf{R}$ has the form

$$\mathbf{R} = \begin{bmatrix} \hat{\mathbf{R}} \\ \mathbf{0} \end{bmatrix} \begin{matrix} \}n \text{ rows} \\ \}m - n \text{ rows} \end{matrix}$$
$$\underbrace{}_{n \text{ columns}}$$

where $\hat{\mathbf{R}}$ is an upper triangular square matrix of dimension n, and that Eq. (A.56) can be expressed as

$$\mathbf{A} = \hat{\mathbf{Q}}\hat{\mathbf{R}} \tag{A.57}$$

where $\hat{\mathbf{Q}}$ is the matrix formed by the first n columns of $\mathbf{Q}$. Now if we let $\hat{\mathbf{Q}} = [\mathbf{q}_1 \ \mathbf{q}_2 \ \cdots \ \mathbf{q}_n]$, Eq. (A.57) yields

$$\mathbf{A}\mathbf{x} = \hat{\mathbf{Q}}\hat{\mathbf{R}}\mathbf{x} = \hat{\mathbf{Q}}\hat{\mathbf{x}} = \sum_{i=1}^{n} \hat{x}_i \mathbf{q}_i$$

In other words, if $\mathbf{A}$ has full column rank n, then the first n columns in $\mathbf{Q}$ form an orthogonal basis for the range of $\mathbf{A}$, i.e., $\mathcal{R}(\mathbf{A})$.

As discussed in Sect. A.11.1, a total of n successive applications of the Householder transformation can convert matrix $\mathbf{A}$ into an upper triangular matrix, $\mathbf{R}$, i.e.,

$$\mathbf{H}_n \ \cdots \ \mathbf{H}_2 \mathbf{H}_1 \mathbf{A} = \mathbf{R} \tag{A.58}$$

Since each $\mathbf{H}_i$ in Eq. (A.58) is orthogonal, we obtain

$$\mathbf{A} = (\mathbf{H}_n \ \cdots \ \mathbf{H}_2 \mathbf{H}_1)^T \mathbf{R} = \mathbf{Q}\mathbf{R} \tag{A.59}$$

where $\mathbf{Q} = (\mathbf{H}_n \ \cdots \ \mathbf{H}_2 \mathbf{H}_1)^T$ is an orthogonal matrix and, therefore, Eqs. (A.58) and (A.59) yield a QR decomposition of $\mathbf{A}$. This method requires $n^2(m - n/3)$ multiplications [4].

An alternative approach for obtaining a QR decomposition is to apply Givens rotations as illustrated in Sect. A.11.2. For a general matrix $\mathbf{A} \in R^{m \times n}$ with $m \geq n$, a total of $mn - n(n + 1)/2$ Givens rotations are required to convert $\mathbf{A}$ into an upper triangular matrix and this Givens-rotation-based algorithm requires $1.5n^2(m - n/3)$ multiplications [4].

A.12.2 QR Decomposition for Rank-Deficient Matrices

If the rank of a matrix $\mathbf{A} \in R^{m \times n}$ where $m \geq n$ is less than n, then there is at least one zero component in the diagonal of $\mathbf{R}$ in Eq. (A.56). In such a case, the conventional QR decomposition discussed in Sect. A.12.1 does not always produce an orthogonal basis for $\mathcal{R}(\mathbf{A})$. For such rank-deficient matrices, however, the

Householder-transformation-based QR decomposition described in Sect. A.12.1 can be modified as

$$AP = QR \qquad \text{(A.60a)}$$

where $\text{rank}(A) = r < n$, $Q \in R^{m \times m}$ is an orthogonal matrix,

$$R = \begin{bmatrix} R_{11} & R_{12} \\ 0 & 0 \end{bmatrix} \qquad \text{(A.60b)}$$

where $R_{11} \in R^{r \times r}$ is a triangular and nonsingular matrix, and $P \in R^{n \times n}$ assumes the form

$$P = [e_{s_1} \ e_{s_2} \ \cdots \ e_{s_n}]$$

where e_{s_i} denotes the s_ith column of the $n \times n$ identity matrix and index set $\{s_1, s_2, \ldots, s_n\}$ is a permutation of $\{1, 2, \ldots, n\}$. Such a matrix is said to be *a permutation matrix* [2].

To illustrate how Eqs. (A.60a) and (A.60b) are obtained, assume that $k - 1$ (with $k - 1 < r$) Householder transformations and permutations have been applied to A to obtain

$$R^{(k-1)} = (H_{k-1} \ \cdots \ H_2 \ H_1)A(P_1 \ P_2 \ \cdots \ P_{k-1})$$

$$= \begin{bmatrix} \mathbf{R}_{11}^{(k-1)} & \mathbf{R}_{12}^{(k-1)} \\ & \\ \mathbf{0} & \mathbf{R}_{22}^{(k-1)} \end{bmatrix} \begin{matrix} \}k-1 \\ \\ \}m-k+1 \end{matrix}$$

$$\underbrace{}_{k-1} \underbrace{}_{n-k+1} \tag{A.61}$$

where $\mathbf{R}_{11}^{(k-1)} \in R^{(k-1)\times(k-1)}$ is upper triangular and $\text{rank}(\mathbf{R}_{11}^{(k-1)}) = k - 1$. Since $\text{rank}(\mathbf{A}) = r$, block $\mathbf{R}_{22}^{(k-1)}$ is nonzero. Now we postmultiply Eq. (A.61) by a permutation matrix $\mathbf{P}_k$ which rearranges the last $n - k + 1$ columns of $\mathbf{R}^{(k-1)}$ such that the column in $\mathbf{R}_{22}^{(k-1)}$ with the largest L_2 norm becomes its first column. A Householder matrix $\mathbf{H}_k$ is then applied to obtain

$$\mathbf{H}_k \mathbf{R}^{(k-1)} \mathbf{P}_k = \begin{bmatrix} \mathbf{R}_{11}^{(k)} & \mathbf{R}_{12}^{(k)} \\ & \\ \mathbf{0} & \mathbf{R}_{22}^{(k)} \end{bmatrix} \begin{matrix} \}k \\ \\ \}m-k \end{matrix}$$

$$\underbrace{}_{k} \underbrace{}_{n-k}$$

where $\mathbf{R}_{11}^{(k)} \in R^{k\times k}$ is an upper triangular nonsingular matrix. If $r = k$, then $\mathbf{R}_{22}^{(k)}$ must be a zero matrix since $\text{rank}(\mathbf{A}) = r$; otherwise, $\mathbf{R}_{22}^{(k)}$ is a nonzero block, and we proceed with postmultiplying $\mathbf{R}^{(k)}$ by a new permutation matrix $\mathbf{P}_{k+1}$ and then premultiplying by a Householder matrix $\mathbf{H}_{k+1}$. This procedure is continued until the modified QR decomposition in Eqs. (A.60a) and (A.60b) is obtained where

$$\mathbf{Q} = (\mathbf{H}_r \mathbf{H}_{r-1} \cdots \mathbf{H}_1)^T \quad \text{and} \quad \mathbf{P} = \mathbf{P}_1 \mathbf{P}_2 \cdots \mathbf{P}_r$$

The decomposition in Eqs. (A.60a) and (A.60b) is called the *QR decomposition of matrix* $\mathbf{A}$ with *column pivoting*. It follows from Eqs. (A.60a) and (A.60b) that the first r columns of matrix $\mathbf{Q}$ form an orthogonal basis for the range of $\mathbf{A}$.

Example A.11 Find a QR decomposition of the matrix

$$\mathbf{A} = \begin{bmatrix} 1 & 0 & 3 & -1 \\ -1 & 2 & -7 & 5 \\ 3 & 1 & 7 & -1 \\ 0 & -1 & 2 & 2 \end{bmatrix}$$

Solution In Example A.9, two Householder transformation matrices

$$\mathbf{H}_1 = \begin{bmatrix} -0.3015 & 0.3015 & -0.9045 & 0 \\ 0.3015 & 0.9302 & 0.2095 & 0 \\ -0.9045 & 0.2095 & 0.3714 & 0 \\ 0 & 0 & 0 & 1 \end{bmatrix}$$

$$\mathbf{H}_2 = \begin{bmatrix} 1 & 0 & 0 & 0 \\ 0 & -0.8515 & -0.3252 & 0.4114 \\ 0 & -0.3252 & 0.9429 & 0.0722 \\ 0 & 0.4114 & 0.0722 & 0.9086 \end{bmatrix}$$

were obtained that reduce matrix $\mathbf{A}$ to the upper triangular matrix

$$\mathbf{R} = \mathbf{H}_2 \mathbf{H}_1 \mathbf{A} = \begin{bmatrix} -3.3166 & -0.3015 & -9.3469 & 2.7136 \\ 0 & -2.4309 & 4.8617 & -4.8617 \\ 0 & 0 & 0 & 0 \\ 0 & 0 & 0 & 0 \end{bmatrix}$$

Therefore, a QR decomposition of $\mathbf{A}$ can be obtained as $\mathbf{A} = \mathbf{QR}$ where $\mathbf{R}$ is the above upper triangular matrix.

$$\mathbf{Q} = (\mathbf{H}_2 \mathbf{H}_1)^{-1} = \mathbf{H}_1^T \mathbf{H}_2^T = \begin{bmatrix} -0.3015 & 0.0374 & -0.9509 & 0.0587 \\ 0.3015 & -0.8602 & -0.1049 & 0.3978 \\ -0.9045 & -0.2992 & 0.2820 & 0.1130 \\ 0 & 0.4114 & 0.0722 & 0.9086 \end{bmatrix}$$

∎

A.13 Cholesky Decomposition

For a symmetric positive-definite matrix $\mathbf{A} \in \mathcal{R}^{n \times n}$, there exists a unique lower triangular matrix $\mathbf{G} \in R^{n \times n}$ with positive diagonal components such that

$$\mathbf{A} = \mathbf{GG}^T \tag{A.62}$$

The decomposition in Eq. (A.62) is known as the *Cholesky decomposition* and matrix $\mathbf{G}$ as the *Cholesky triangle*.

One of the methods that can be used to obtain the Cholesky decomposition of a given positive-definite matrix is based on the use of the outer-product updates [2] as illustrated below.

A positive-definite matrix $\mathbf{A} \in R^{n \times n}$ can be expressed as

$$\mathbf{A} = \begin{bmatrix} a_{11} & \mathbf{u}^T \\ \mathbf{u} & \mathbf{B} \end{bmatrix} \tag{A.63}$$

where a_{11} is a positive number. It can be readily verified that with

$$\mathbf{T} = \begin{bmatrix} \frac{1}{\sqrt{a_{11}}} & \mathbf{0} \\ -\mathbf{u}/a_{11} & \mathbf{I}_{n-1} \end{bmatrix} \tag{A.64}$$

we have

$$\mathbf{TAT}^T = \begin{bmatrix} 1 & \mathbf{0} \\ \mathbf{0} & \mathbf{B} - \mathbf{uu}^T/a_{11} \end{bmatrix} \equiv \begin{bmatrix} 1 & \mathbf{0} \\ \mathbf{0} & \mathbf{A}_1 \end{bmatrix} \tag{A.65}$$

which implies that

$$\mathbf{A} = \begin{bmatrix} \sqrt{a_{11}} & \mathbf{0} \\ \mathbf{u}/\sqrt{a_{11}} & \mathbf{I}_{n-1} \end{bmatrix} \begin{bmatrix} 1 & \mathbf{0} \\ \mathbf{0} & \mathbf{B} - \mathbf{uu}^T/a_{11} \end{bmatrix} \begin{bmatrix} \sqrt{a_{11}} & \mathbf{u}/\sqrt{a_{11}} \\ \mathbf{0} & \mathbf{I}_{n-1} \end{bmatrix}$$

$$\equiv \mathbf{G}_1 \begin{bmatrix} 1 & \mathbf{0} \\ \mathbf{0} & \mathbf{A}_1 \end{bmatrix} \mathbf{G}_1^T \tag{A.66}$$

where G_1 is a lower triangular matrix and $A_1 = B - uu^T/a_{11}$ is an $(n-1) \times (n-1)$ symmetric matrix. Since A is positive definite and T is nonsingular, it follows from Eq. (A.65) that matrix A_1 is positive definite; hence the above procedure can be applied to matrix A_1. In other words, we can find an $(n-1) \times (n-1)$ lower triangular matrix G_2 such that

$$A_1 = G_2 \begin{bmatrix} 1 & 0 \\ 0 & A_2 \end{bmatrix} G_2^T \tag{A.67}$$

where A_2 is an $(n-2) \times (n-2)$ positive-definite matrix. By combining Eqs. (A.66) and (A.67), we obtain

$$A = \begin{bmatrix} \sqrt{a_{11}} & 0 \\ u/\sqrt{a_{11}} & G_2 \end{bmatrix} \begin{bmatrix} I_2 & 0 \\ 0 & A_2 \end{bmatrix} \begin{bmatrix} \sqrt{a_{11}} & u^T/\sqrt{a_{11}} \\ 0 & G_2^T \end{bmatrix}$$

$$\equiv G_{12} \begin{bmatrix} I_2 & 0 \\ 0 & A_2 \end{bmatrix} G_{12}^T \tag{A.68}$$

where I_2 is the 2×2 identity matrix and G_{12} is lower triangular. The above procedure is repeated until the second matrix at the right-hand side of Eq. (A.68) is reduced to the identity matrix I_n. The Cholesky decomposition of A is then obtained.

Example A.12 Compute the Cholesky triangle of the positive-definite matrix

$$A = \begin{bmatrix} 4 & -2 & 1 \\ -2 & 7 & -1 \\ 1 & -1 & 1 \end{bmatrix}$$

Solution From Eq. (A.66), we obtain

$$G_1 = \begin{bmatrix} 2 & 0 & 0 \\ -1 & 1 & 0 \\ 0.5 & 0 & 1 \end{bmatrix}$$

and

$$A_1 = \begin{bmatrix} 7 & -1 \\ -1 & 1 \end{bmatrix} - \frac{1}{4} \begin{bmatrix} -2 \\ 1 \end{bmatrix} [2 \ 1] = \begin{bmatrix} 6 & -0.50 \\ -0.50 & 0.75 \end{bmatrix}$$

Now working on matrix A_1, we get

$$G_2 = \begin{bmatrix} \sqrt{6} & 0 \\ -0.5/\sqrt{6} & 1 \end{bmatrix}$$

and

$$A_2 = 0.75 - (-0.5)^2/6 = 0.7083$$

In this case, Eq. (A.66) becomes

$$A = \begin{bmatrix} 2 & 0 & 0 \\ -1 & \sqrt{6} & 0 \\ 0.5 & -0.5/\sqrt{6} & 1 \end{bmatrix} \begin{bmatrix} 1 & 0 & 0 \\ 0 & 1 & 0 \\ 0 & 0 & 0.7083 \end{bmatrix} \begin{bmatrix} 2 & 0 & 0 \\ -1 & \sqrt{6} & 0 \\ 0.5 & -0.5/\sqrt{6} & 1 \end{bmatrix}^T$$

Finally, we use

$$\mathbf{G}_3 = \sqrt{0.7083} \approx 0.8416$$

to reduce $\mathbf{A}_2$ to $\mathbf{A}_3 = 1$, which leads to the Cholesky triangle

$$\mathbf{G} = \begin{bmatrix} 2 & 0 & 0 \\ -1 & \sqrt{6} & 0 \\ 0.5 & -0.5/\sqrt{6} & \sqrt{0.7083} \end{bmatrix} \approx \begin{bmatrix} 2 & 0 & 0 \\ -1 & 2.4495 & 0 \\ 0.5 & -0.2041 & 0.8416 \end{bmatrix}$$

∎

A.14 Kronecker Product

Let $\mathbf{A} \in R^{p \times m}$ and $\mathbf{B} \in R^{q \times n}$. The Kronecker product of $\mathbf{A}$ and $\mathbf{B}$, denoted as $\mathbf{A} \otimes \mathbf{B}$, is a $pq \times mn$ matrix defined by

$$\mathbf{A} \otimes \mathbf{B} = \begin{bmatrix} a_{11}\mathbf{B} & \cdots & a_{1m}\mathbf{B} \\ \vdots & & \vdots \\ a_{p1}\mathbf{B} & \cdots & a_{pm}\mathbf{B} \end{bmatrix} \tag{A.69}$$

where a_{ij} denotes the $(i,\ j)$th component of $\mathbf{A}$ [6]. It can be verified that

(i) $(\mathbf{A} \otimes \mathbf{B})^T = \mathbf{A}^T \otimes \mathbf{B}^T$
(ii) $(\mathbf{A} \otimes \mathbf{B}) \cdot (\mathbf{C} \otimes \mathbf{D}) = \mathbf{AC} \otimes \mathbf{BD}$ where $\mathbf{C} \in R^{m \times r}$ and $\mathbf{D} \in R^{n \times s}$
(iii) If $p = m, q = n$, and $\mathbf{A},\ \mathbf{B}$ are nonsingular, then

$$(\mathbf{A} \otimes \mathbf{B})^{-1} = \mathbf{A}^{-1} \otimes \mathbf{B}^{-1}$$

(iv) If $\mathbf{A} \in R^{m \times m}$ and $\mathbf{B} \in R^{n \times n}$, then the eigenvalues of $\mathbf{A} \otimes \mathbf{B}$ and $\mathbf{A} \otimes \mathbf{I}_n + \mathbf{I}_m \otimes \mathbf{B}$ are $\lambda_i \mu_j$ and $\lambda_i + \mu_j$, respectively, for $i = 1,\ 2,\ \ldots,\ m$ and $j = 1,\ 2,\ \ldots,\ n$, where λ_i and μ_j are the ith and jth eigenvalues of $\mathbf{A}$ and $\mathbf{B}$, respectively.

The Kronecker product is useful when we are dealing with matrix variables. If we use nvec($\mathbf{X}$) to denote the column vector obtained by stacking the column vectors of matrix $\mathbf{X}$, then it is easy to verify that for $\mathbf{M} \in R^{p \times m}$, $\mathbf{N} \in R^{q \times n}$ and $\mathbf{X} \in R^{n \times m}$, we have

$$\text{nvec}(\mathbf{NXM}^T) = (\mathbf{M} \otimes \mathbf{N})\text{nvec}(\mathbf{X}) \tag{A.70}$$

In particular, if $p = m = q = n$, $\mathbf{N} = \mathbf{A}^T$, and $\mathbf{M} = \mathbf{I}_n$, then Eq. (A.70) becomes

$$\text{nvec}(\mathbf{A}^T\mathbf{X}) = (\mathbf{I}_n \otimes \mathbf{A}^T)\text{nvec}(\mathbf{X}) \tag{A.71a}$$

Similarly, we have

$$\text{nvec}(\mathbf{XA}) = (\mathbf{A}^T \otimes \mathbf{I}_n)\text{nvec}(\mathbf{X}) \tag{A.71b}$$

For example, we can apply Eqs. (A.71a) and (A.71b) to the Lyapunov equation [6]

$$\mathbf{A}^T\mathbf{P} + \mathbf{PA} = -\mathbf{Q} \tag{A.72}$$

where matrices $\mathbf{A}$ and $\mathbf{Q}$ are given and $\mathbf{Q}$ is positive definite. First, we write Eq. (A.72) in vector form as

$$\text{nvec}(\mathbf{A}^T\mathbf{P}) + \text{nvec}(\mathbf{PA}) = -\text{nvec}(\mathbf{Q}) \tag{A.73}$$

Using Eqs. (A.71a) and (A.71b), Eq. (A.73)becomes

$$(\mathbf{I}_n \otimes \mathbf{A}^T)\text{nvec}(\mathbf{P}) + (\mathbf{A}^T \otimes \mathbf{I}_n)\text{nvec}(\mathbf{P}) = -\text{nvec}(\mathbf{Q})$$

which can be solved to obtain nvec($\mathbf{P}$) as

$$\text{nvec}(\mathbf{P}) = -(\mathbf{I}_n \otimes \mathbf{A}^T + \mathbf{A}^T \otimes \mathbf{I}_n)^{-1}\text{nvec}(\mathbf{Q}) \tag{A.74}$$

Example A.13 Solve the Lyapunov equation

$$\mathbf{A}^T\mathbf{P} + \mathbf{PA} = -\mathbf{Q}$$

for matrix $\mathbf{P}$ where

$$\mathbf{A} = \begin{bmatrix} -2 & -2 \\ 1 & 0 \end{bmatrix} \quad \text{and} \quad \mathbf{Q} = \begin{bmatrix} 1 & -1 \\ -1 & 2 \end{bmatrix}$$

Solution From Eq. (A.69), we compute

$$\mathbf{I}_2 \otimes \mathbf{A}^T + \mathbf{A}^T \otimes \mathbf{I}_2 = \begin{bmatrix} -4 & 1 & 1 & 0 \\ -2 & -2 & 0 & 1 \\ -2 & 0 & -2 & 1 \\ 0 & -2 & -2 & 0 \end{bmatrix}$$

Since

$$\text{nvec}(\mathbf{Q}) = \begin{bmatrix} 1 \\ -1 \\ -1 \\ 2 \end{bmatrix}$$

Equation A.74 gives

$$\text{nvec}(\mathbf{P}) = -(\mathbf{I}_2 \otimes \mathbf{A}^T + \mathbf{A}^T \otimes \mathbf{I}_2)^{-1}\text{nvec}(\mathbf{Q})$$

$$= - \begin{bmatrix} -4 & 1 & 1 & 0 \\ -2 & -2 & 0 & 1 \\ -2 & 0 & -2 & 1 \\ 0 & -2 & -2 & 0 \end{bmatrix}^{-1} \begin{bmatrix} 1 \\ -1 \\ -1 \\ 2 \end{bmatrix} = \begin{bmatrix} 0.5 \\ 0.5 \\ 0.5 \\ 3 \end{bmatrix}$$

from which we obtain

$$\mathbf{P} = \begin{bmatrix} 0.5 & 0.5 \\ 0.5 & 3 \end{bmatrix}$$

∎

A.15 Vector Spaces of Symmetric Matrices

Let S^n be the vector space of real symmetric $n \times n$ matrices. As in the n-dimensional Euclidean space where the inner product is defined for two vectors, the *inner product* for matrices $\mathbf{A}$ and $\mathbf{B}$ in S^n is defined as

$$\mathbf{A} \cdot \mathbf{B} = \text{trace}(\mathbf{AB})$$

If $\mathbf{A} = (a_{ij})$ and $\mathbf{B} = (b_{ij})$, then we have

$$\mathbf{A} \cdot \mathbf{B} = \text{trace}(\mathbf{AB}) = \sum_{i=1}^{n} \sum_{j=1}^{n} a_{ij} b_{ij} \tag{A.75}$$

The norm $\|\mathbf{A}\|_{S^n}$ associated with this inner product is

$$\|\mathbf{A}\|_{S^n} = \sqrt{\mathbf{A} \cdot \mathbf{A}} = \left[\sum_{i=1}^{n} \sum_{j=1}^{n} a_{ij}^2 \right]^{1/2} = \|\mathbf{A}\|_F \tag{A.76}$$

where $\|\mathbf{A}\|_F$ denotes the Frobenius norm of $\mathbf{A}$ (see Sect. A.8.2).

An important set in space S^n is the set of all positive-semidefinite matrices given by

$$\mathcal{P} = \{\mathbf{X} : \mathbf{X} \in S^n \text{ and } \mathbf{X} \succeq \mathbf{0}\} \tag{A.77}$$

A set $\mathcal{K}$ in a vector space is said to be a *convex cone* if $\mathcal{K}$ is a convex set such that $\mathbf{v} \in \mathcal{K}$ implies $\alpha \mathbf{v} \in \mathcal{K}$ for any nonnegative scalar α. It is easy to verify that set $\mathcal{P}$ forms a convex cone in space S^n.

Let matrices $\mathbf{X}$ and $\mathbf{S}$ be two components of $\mathcal{P}$, i.e., $\mathbf{X} \succeq \mathbf{0}$ and $\mathbf{S} \succeq \mathbf{0}$. The eigendecomposition of $\mathbf{X}$ gives

$$\mathbf{X} = \mathbf{U} \boldsymbol{\Lambda} \mathbf{U}^T \tag{A.78}$$

where $\mathbf{U} \in R^{n \times n}$ is orthogonal and $\boldsymbol{\Lambda} = \text{diag}\{\lambda_1, \lambda_2, \ldots, \lambda_n\}$. The decomposition in Eq. (A.78) can be expressed as

$$\mathbf{X} = \sum_{i=1}^{n} \lambda_i \mathbf{u}_i \mathbf{u}_i^T$$

where $\mathbf{u}_i$ denotes the ith column of $\mathbf{U}$. By using the property that

$$\text{trace}(\mathbf{AB}) = \text{trace}(\mathbf{BA})$$

(see Eq. (A.20)), we can compute the inner product $\mathbf{X} \cdot \mathbf{S}$ as

$$\mathbf{X} \cdot \mathbf{S} = \text{trace}(\mathbf{XS}) = \text{trace}\left(\sum_{i=1}^{n} \lambda_i \mathbf{u}_i \mathbf{u}_i^T \mathbf{S}\right) = \sum_{i=1}^{n} \lambda_i \, \text{trace}(\mathbf{u}_i \mathbf{u}_i^T \mathbf{S})$$

$$= \sum_{i=1}^{n} \lambda_i \, \text{trace}(\mathbf{u}_i^T \mathbf{S} \mathbf{u}_i) = \sum_{i=1}^{n} \lambda_i \mu_i \tag{A.79}$$

where $\mu_i = \mathbf{u}_i^T \mathbf{S} \mathbf{u}_i$. Since both $\mathbf{X}$ and $\mathbf{S}$ are positive semidefinite, we have $\lambda_i \geq 0$ and $\mu_i \geq 0$ for $i = 1, 2, \ldots, n$. Therefore, Eq. (A.79) implies that

$$\mathbf{X} \cdot \mathbf{S} \geq 0 \tag{A.80}$$

In other words, the inner product of two positive-semidefinite matrices is always nonnegative.

A further property of the inner product on set $\mathcal{P}$ is that if $\mathbf{X}$ and $\mathbf{S}$ are positive semidefinite and $\mathbf{X} \cdot \mathbf{S} = 0$, then the product matrix $\mathbf{XS}$ must be the zero matrix, i.e.,

$$\mathbf{XS} = \mathbf{0} \tag{A.81}$$

To show this, we can write

$$\mathbf{u}_i^T \mathbf{X} \mathbf{S} \mathbf{u}_j = \mathbf{u}_i^T \left(\sum_{k=1}^{n} \lambda_k \mathbf{u}_k \mathbf{u}_k^T\right) \mathbf{S} \mathbf{u}_j = \lambda_i \mathbf{u}_i^T \mathbf{S} \mathbf{u}_j \tag{A.82}$$

Using the Cauchy-Schwartz inequality (see Eq. (A.25)), we have

$$|\mathbf{u}_i^T \mathbf{S} \mathbf{u}_j|^2 = |(\mathbf{S}^{1/2} \mathbf{u}_i)^T (\mathbf{S}^{1/2} \mathbf{u}_j)|^2 \leq \|\mathbf{S}^{1/2} \mathbf{u}_i\|_2^2 \|\mathbf{S}^{1/2} \mathbf{u}_j\|_2^2 = \mu_i \mu_j \tag{A.83}$$

Now if $\mathbf{X} \cdot \mathbf{S} = 0$, then Eq. (A.79) implies that

$$\sum_{i=0}^{n} \lambda_i \mu_i = 0 \tag{A.84}$$

Since λ_i and μ_i are all nonnegative, Eq. (A.84) implies that $\lambda_i \mu_i = 0$ for $i = 1, 2, \ldots, n$; hence for each index i, either $\lambda_i = 0$ or $\mu_i = 0$. If $\lambda_i = 0$, Eq. (A.82) gives

$$\mathbf{u}_i^T \mathbf{X} \mathbf{S} \mathbf{u}_j = 0 \tag{A.85}$$

If $\lambda_i \neq 0$, then μ_i must be zero and Eq. (A.83) implies that $\mathbf{u}_i^T \mathbf{S} \mathbf{u}_j = 0$ which in conjunction with Eq. (A.82) also leads to Eq. (A.85). Since Eq. (A.85) holds for any i and j, we conclude that

$$\mathbf{U}^T \mathbf{X} \mathbf{S} \mathbf{U} = \mathbf{0} \tag{A.86}$$

Since $\mathbf{U}$ is nonsingular, Eq. (A.86) implies Eq. (A.81).

Given $p+1$ symmetric matrices $\mathbf{F}_0, \mathbf{F}_1, \ldots, \mathbf{F}_p$ in space $\mathcal{S}^n$, and a p-dimensional vector $\mathbf{x} = [x_1 \ x_2 \ \cdots \ x_p]^T$, we can generate a symmetric matrix

$$\mathbf{F}(\mathbf{x}) = \mathbf{F}_0 + x_1 \mathbf{F}_1 + \cdots + x_p \mathbf{F}_p = \mathbf{F}_0 + \sum_{i=1}^{p} x_i \mathbf{F}_i \tag{A.87}$$

which is said to be *affine* with respect to $\mathbf{x}$. Note that if the constant term $\mathbf{F_0}$ were a zero matrix, then $\mathbf{F(x)}$ would be a *linear* function of vector $\mathbf{x}$, i.e., $\mathbf{F(x)}$ would satisfy the condition $\mathbf{F}(\alpha\mathbf{x} + \beta\mathbf{y}) = \alpha\mathbf{F(x)} + \beta\mathbf{F(y)}$ for any vectors $\mathbf{x}$, $\mathbf{y} \in R^p$ and any scalars α and β. However, because of the presence of $\mathbf{F_0}$, $\mathbf{F(x)}$ in Eq. (A.87) is *not* linear with respect to $\mathbf{x}$ in a strict sense and the term 'affine' is often used in the literature to describe such a class of matrices. In effect, the affine property is a somewhat relaxed version of the linearity property.

In the context of linear programming, the concept of an *affine manifold* is sometimes encountered. A manifold is a subset of the Euclidean space that satisfies a certain structural property of interest, for example, a set of vectors satisfying the relation $\mathbf{x}^T\mathbf{c} = \beta$. Such a set of vectors may possess the affine property, as illustrated in the following example.

Example A.14 Describe the set of n-dimensional vectors $\{\mathbf{x} : \mathbf{x}^T\mathbf{c} = \beta\}$ for a given vector $\mathbf{c} \in R^{n \times 1}$ and a scalar β as an affine manifold in the n-dimensional Euclidean space E^n.

Solution Obviously, the set of vectors $\{\mathbf{x} : \mathbf{x}^T\mathbf{c} = \beta\}$ is a subset of E^n. If we denote $\mathbf{x} = [x_1 \ x_2 \ \cdots \ x_n]^T$ and $\mathbf{c} = [c_1 \ c_2 \ \cdots \ c_n]^T$, then equation $\mathbf{x}^T\mathbf{c} = \beta$ can be expressed as $F(\mathbf{x}) = 0$ where

$$F(\mathbf{x}) = -\beta + x_1c_1 + x_2c_2 + \cdots + x_nc_n \tag{A.88}$$

By viewing $-\beta$, c_1, c_2, $\ldots$, c_n as one-dimensional symmetric matrices, $F(\mathbf{x})$ in Eq. (A.88) assumes the form in Eq. (A.87), which is affine with respect to $\mathbf{x}$. Therefore the set $\{\mathbf{x} : \mathbf{x}^T\mathbf{c} = \beta\}$ is an affine manifold in E^n.

∎

Example A.15 Convert the following constraints

$$\mathbf{X} = (x_{ij}) \succeq \mathbf{0} \quad \text{for } i, j = 1, 2, 3 \tag{A.89a}$$

and

$$x_{ii} = 1 \quad \text{for } i = 1, 2, 3 \tag{A.89b}$$

into a constraint of the type

$$\mathbf{F(x)} \succeq \mathbf{0} \tag{A.90}$$

for some vector variable $\mathbf{x}$ where $\mathbf{F(x)}$ assumes the form in Eq. (A.87).

Solution The constraints in Eqs. (A.89a) and (A.89b) can be combined into

$$\mathbf{X} = \begin{bmatrix} 1 & x_{12} & x_{13} \\ x_{12} & 1 & x_{23} \\ x_{13} & x_{23} & 1 \end{bmatrix} \succeq \mathbf{0} \tag{A.91}$$

Next we write matrix $\mathbf{X}$ in Eq. (A.91) as

$$\mathbf{X} = \mathbf{F_0} + x_{12}\mathbf{F_1} + x_{13}\mathbf{F_2} + x_{23}\mathbf{F_3}$$

where $\mathbf{F}_0 = \mathbf{I}_3$ and

$$\mathbf{F}_1 = \begin{bmatrix} 0 & 1 & 0 \\ 1 & 0 & 0 \\ 0 & 0 & 0 \end{bmatrix}, \quad \mathbf{F}_2 = \begin{bmatrix} 0 & 0 & 1 \\ 0 & 0 & 0 \\ 1 & 0 & 0 \end{bmatrix}, \quad \mathbf{F}_3 = \begin{bmatrix} 0 & 0 & 0 \\ 0 & 0 & 1 \\ 0 & 1 & 0 \end{bmatrix}$$

Hence the constraint in Eqs. (A.89a) and (A.89b) can be expressed in terms of Eq. (A.90) with $\mathbf{F}_i$ given by the above equations and $\mathbf{x} = [x_{12} \; x_{13} \; x_{23}]^T$.

■

A.16 Polygon, Polyhedron, Polytope, and Convex Hull

A *polygon* is a closed plane figure with an arbitrary number of sides. A polygon is said to be convex if the region inside the polygon is a convex set (see Def. 2.7). A convex polygon with m sides can be described in terms of m linear inequalities which can be expressed in matrix form as

$$\mathcal{P}_y = \{\mathbf{x} : \mathbf{A}\mathbf{x} \geq \mathbf{b}\} \tag{A.92}$$

where $\mathbf{A} \in R^{m \times 2}$, $\mathbf{x} \in R^{2 \times 1}$, and $\mathbf{b} \in R^{m \times 1}$.

A *convex polyhedron* is an n-dimensional extension of a convex polygon. A convex polyhedron can be described by the equation

$$\mathcal{P}_h = \{\mathbf{x} : \mathbf{A}\mathbf{x} \geq \mathbf{b}\} \tag{A.93}$$

where $\mathbf{A} \in R^{m \times n}$, $\mathbf{x} \in R^{n \times 1}$, and $\mathbf{b} \in R^{m \times 1}$. For example, a 3-dimensional convex polyhedron is a 3-dimensional solid which consists of several polygons, usually joined at their edges such as that shown in Fig. 11.4.

A polyhedron may or may not be bounded depending on the numerical values of $\mathbf{A}$ and $\mathbf{b}$ in Eq. (A.93). A bounded polyhedron is called a *polytope*.

Given a set of points $\mathcal{S} = \{p_1, \; p_2, \; \ldots, \; p_L\}$ in an n-dimensional space, the *convex hull* spanned by $\mathcal{S}$ is defined as the smallest convex set that contains $\mathcal{S}$. It can be verified that the convex hull is characterized by

$$\text{Co}\{p_1, \; p_2, \; \ldots, \; p_L\} = \{p : \; p = \sum_{i=1}^{L} \lambda_i p_i, \; \lambda_i \geq 0, \; \sum_{i=1}^{L} \lambda_i = 1\}$$

In the above definition, each point p_i represents an abstract n-dimensional point. For example, if point p_i is represented by an n-dimensional vector, say, $\mathbf{v}_i$, then the convex hull spanned by the L vectors $\{\mathbf{v}_1, \mathbf{v}_2, \ldots, \mathbf{v}_L\}$ is given by

$$\text{Co}\{\mathbf{v}_1, \; \mathbf{v}_2, \; \ldots, \; \mathbf{v}_L\} = \{\mathbf{v} : \; \mathbf{v} = \sum_{i=1}^{L} \lambda_i \mathbf{v}_i, \; \lambda_i \geq 0, \; \sum_{i=1}^{L} \lambda_i = 1\}$$

Alternatively, if point p_i is represented by a pair of matrices $[\mathbf{A}_i \ \mathbf{B}_i]$ with $\mathbf{A}_i \in R^{n \times n}$ and $\mathbf{B}_i \in R^{n \times m}$, then the convex hull spanned by $\{[\mathbf{A}_i \ \mathbf{B}_i] \text{ for } i = 1, 2, \ldots, L\}$ is given by

$$\text{Co}\{[\mathbf{A}_1 \ \mathbf{B}_1], \ [\mathbf{A}_2 \ \mathbf{B}_2], \ldots, \ [\mathbf{A}_L \ \mathbf{B}_L]\} =$$

$$\{[\mathbf{A} \ \mathbf{B}] : \ [\mathbf{A} \ \mathbf{B}] = \sum_{i=1}^{L} \lambda_i [\mathbf{A}_i \ \mathbf{B}_i], \ \lambda_i \geq 0, \ \sum_{i=1}^{L} \lambda_i = 1\}$$

References

1. G. Strang, *Introduction to Linear Algebra*, 5th ed. Wellesley, MA: Wellesley Cambridge Press, 2016.
2. G. H. Golub and C. F. Van Loan, *Matrix Computations*, 4th ed. Baltimore, MD: Johns Hopkins University Press, 2013.
3. R. A. Horn and C. R. Johnson, *Matrix Analysis*, 2nd ed. New York: Cambridge University Press, 2013.
4. G. W. Stewart, *Introduction to Matrix Computations*. New York: Academic Press, 1973.
5. P. E. Gill, W. Murray, and M. H. Wright, *Numerical Linear Algebra and Optimization, Vol. I*. Reading: Addison-Wesley, 1991.
6. S. Barnett, *Polynomials and Linear Control Systems*. New York: Marcel Dekker, 1983.

Basics of Digital Filters

B

B.1 Introduction

Several of the unconstrained and constrained optimization algorithms described in this book have been illustrated in terms of examples taken from the authors' research on the application of optimization algorithms for the design of digital filters. To enhance the understanding of the application of the algorithms presented to the design of digital filters, we provide in this appendix a concise introduction to the basic concepts and principles of digital filters as well as typical design problems associated with these systems. A detailed treatment of the subject can be found in [1].

B.2 Characterization

Digital filters are digital systems that can be used to process discrete-time signals. A single-input single-output digital filter can be represented by a block diagram as shown in Fig. B.1a where $x(nT)$ and $y(nT)$ are the excitation (input) and response (output), respectively. The excitation and response are sequences of numbers such as those illustrated in Fig. B.1b and c. In the most general case, the response of a digital filter at instant nT is a function of a number of values of the excitation $x[(n+K)T]$, $x[(n+K-1)T], \ldots, x(nT), x[(n-1)T], \ldots, x[(n-M)T]$ and a number of values of the response $y[(n-1)T], y[(n-2)T], \ldots, y[(n-N)T]$ where M, K, and N are positive integers, i.e.,

$$y(nT) = \mathcal{R}x(nT) = f\{x[(n+K)T], \ldots, x[(n-M)T],$$
$$y[(n-1)T], \ldots, y[(n-N)T]\} \tag{B.1}$$

where $\mathcal{R}$ is an operator that can be interpreted as "*is the response produced by the excitation*".

© Springer Science+Business Media, LLC, part of Springer Nature 2021
A. Antoniou and W.-S. Lu, *Practical Optimization*, Texts in Computer Science,
https://doi.org/10.1007/978-1-0716-0843-2_B

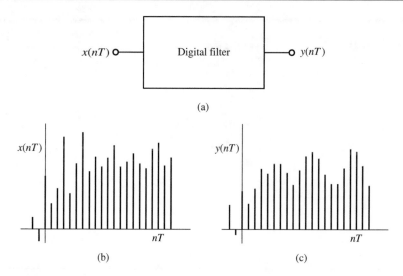

Fig. B.1 **a** Block diagram representation of a digital filter, **b** excitation, **c** response

Digital filters can be linear or nonlinear, time invariant or time dependent, causal or noncausal. In a linear digital filter, the response of the filter to a linear combination of two signals $x_1(nT)$ and $x_2(nT)$ satisfies the relation

$$\mathcal{R}[\alpha x_1(nT) + \beta x_2(nT)] = \alpha \mathcal{R} x_1(nT) + \beta \mathcal{R} x_2(nT)$$

In an initially relaxed time-invariant digital filter, we have

$$\mathcal{R} x(nT - kT) = y(nT - kT)$$

This relation states, in effect, that a delayed excitation will produce a delayed but otherwise unchanged response. In this context, an *initially relaxed* digital filter is one whose response $y(nT)$ is zero for $nT < 0$ if $x(nT) = 0$ for all $nT < 0$. A causal digital filter, on the other hand, is a filter whose response at instant nT depends only on values of the input at instants $nT, (n-1)T, \ldots, (n-M)T$, i.e., it is independent of future values of the excitation.

In a general linear, time-invariant, causal digital filter, Eq. (B.1) assumes the form of a linear recursive difference equation, i.e.,

$$y(nT) = \sum_{i=0}^{N} a_i x(nT - iT) - \sum_{i=1}^{N} b_i y(nT - iT) \qquad \text{(B.2)}$$

where a_i for $0 \leq i \leq N$ and b_i for $1 \leq i \leq N$ are constants. Some of these constants can be zero. A digital filter characterized by Eq. (B.2) is said to be *recursive*, since the response depends on a number of values of the excitation as well as a number of values of the response. Integer N, namely, the order of the difference equation, is said to be the *order* of the digital filter.

If the response of a digital filter at instant nT is independent of all the previous values of the response, i.e., $y(nT - T)$, $y(nT - 2T)$, ..., then Eq. (B.2) reduces to the nonrecursive equation

$$y(nT) = \sum_{i=0}^{N} a_i x(nT - iT) \qquad (B.3)$$

which characterizes an Nth-order *nonrecursive filter* (see Sects. 2.2 and 2.3 of [1]). The number of coefficients in the difference equation, namely $N + 1$, is said to be the *length* of the nonrecursive filter.

B.3 Time-Domain Response

The time-domain response of a digital filter to some excitation is often required and to facilitate the evaluation of time-domain responses, a number of standard signals are frequently used. Typical signals of this type are the unit impulse, unit step, and unit sinusoid, which are defined in Table B.1 (see Sect. 2.5 of [1]).

From Eq. (B.3), the impulse response of an arbitrary Nth-order nonrecursive filter, denoted as $h(nT)$, is given by

$$h(nT) \equiv y(nT) = \mathcal{R}\delta(nT)$$

$$= \sum_{i=0}^{N} a_i \delta(nT - iT).$$

Now from the definition of the unit impulse in Table B.1, we can readily show that ..., $h(-2T) = 0$, $h(-T) = 0$, $h(0) = a_0$, $h(T) = a_1$, ..., $h(NT) = a_N$, $h[(N + 1)T] = 0$, $h[(N + 2)T] = 0$, ..., i.e.,

$$h(nT) = \begin{cases} a_i & \text{for } n = i \\ 0 & \text{otherwise} \end{cases} \qquad (B.4)$$

In effect, the impulse response in nonrecursive digital filters is of finite duration and for this reason these filters are also known as *finite-duration impulse response*

Table B.1 Discrete-time standard signals

Function	Definition
Unit impulse	$\delta(nT) = \begin{cases} 1 & \text{for } n = 0 \\ 0 & \text{for } n \neq 0 \end{cases}$
Unit step	$u(nT) = \begin{cases} 1 & \text{for } n \geq 0 \\ 0 & \text{for } n \leq 0 \end{cases}$
Unit sinusoid	$u(nT) \sin \omega nT$

(FIR) filters. On the other hand, the use of Eq. (B.2) gives the impulse response of a recursive filter as

$$y(nT) = h(nT) = \sum_{i=0}^{N} a_i \delta(nT - iT) - \sum_{i=1}^{N} b_i y(nT - iT)$$

and if the filter is initially relaxed, we obtain

$$y(0) = h(0) = a_0\delta(0) + a_1\delta(-T) + a_2\delta(-2T) + \cdots$$
$$- b_1 y(-T) - b_2 y(-2T) - \cdots = a_0$$
$$y(T) = h(T) = a_0\delta(T) + a_1\delta(0) + a_2\delta(-T) + \cdots$$
$$- b_1 y(0) - b_2 y(-T) - \cdots = a_1 - b_1 a_0$$
$$y(2T) = h(2T) = a_0\delta(2T) + a_1\delta(T) + a_2\delta(0) + \cdots$$
$$- b_1 y(T) - b_2 y(0) - \cdots$$
$$= a_2 - b_1(a_1 - b_1 a_0) - b_2 a_0$$
$$\vdots$$

Evidently, in this case the impulse response is of infinite duration since the response at instant nT depends on previous values of the response, which are always finite. Hence recursive filters are also referred to as *infinite-duration impulse response (IIR) filters.*[2]

Other types of time-domain response are the unit-step and the sinusoidal responses. The latter is of particular importance because it leads to a frequency-domain characterization for digital filters.

Time-domain responses of digital filters of considerable complexity can be deduced by using the z transform. The z transform of a signal $x(nT)$ is defined as

$$\mathcal{Z}x(nT) = X(z) = \sum_{n=-\infty}^{\infty} x(nT)z^{-n} \tag{B.5}$$

where z is a complex variable. The conditions for the convergence of $X(z)$ can be found in Sect. 4.3 of [1].

---

[2]Note that FIR and IIR filters are often said to be finite impulse response and infinite impulse response filters, respectively, but these designations are misnomers as both kinds of digital filters actually have finite impulse responses. Digital filters with infinite impulse responses are unstable by definition and are, therefore, of no practical use as filters. It is the duration of the impulse response that is finite or infinite.

B.4   Stability Property

A digital filter is said to be *stable* if any bounded excitation will produce a bounded response. In terms of mathematics, a digital filter is stable if and only if any input $x(nT)$ such that

$$|x(nT)| \leq P < \infty \qquad \text{for all } n$$

will produce an output $y(nT)$ that satisfies the condition

$$|y(nT)| \leq Q < \infty \qquad \text{for all } n$$

where P and Q are positive constants.

A necessary and sufficient condition for the stability of a causal digital filter is that its impulse response be absolutely summable over the range $0 \leq nT \leq \infty$, i.e.,

$$\sum_{n=0}^{\infty} |h(nT)| \leq R < \infty \tag{B.6}$$

(see Sect. 2.7 of [1] for proof).

Since the impulse response of FIR filters is always of finite duration, as can be seen in Eq. (B.4), it follows that it is absolutely summable and, therefore, these filters are always stable.

B.5   Transfer Function

The analysis and design of digital filters is greatly simplified by representing the filter in terms of a transfer function. This can be derived from the difference equation or the impulse response and it can be used to find the time-domain response of a filter to an arbitrary excitation or its frequency-domain response to an arbitrary linear combination of sinusoidal signals.

B.5.1   Definition

The transfer function of a digital filter can be defined as the ratio of the z transform of the response to the z transform of the excitation, i.e.,

$$H(z) = \frac{Y(z)}{X(z)} \tag{B.7}$$

From the definition of the z transform in Eq. (B.5), it can be readily shown that

$$\mathcal{Z}x(nT - kT) = z^{-k}X(z)$$

and

$$\mathcal{Z}[\alpha x_1(nT) + \beta x_2(nT)] = \alpha X_1(z) + \beta X_2(z)$$

Hence if we apply the z transform to both sides of Eq. (B.2), we obtain

$$Y(z) = \mathcal{Z}y(nT) = \mathcal{Z}\sum_{i=0}^{N} a_i x(nT - iT) - \mathcal{Z}\sum_{i=1}^{N} b_i y(nT - iT)$$

$$= \sum_{i=0}^{N} a_i z^{-i} \mathcal{Z}x(nT) - \sum_{i=1}^{N} b_i z^{-i} \mathcal{Z}y(nT)$$

$$= \sum_{i=0}^{N} a_i z^{-i} X(z) - \sum_{i=1}^{N} b_i z^{-i} Y(z) \tag{B.8}$$

Therefore, from Eqs. (B.7) and (B.8), we have

$$H(z) = \frac{Y(z)}{X(z)} = \frac{\sum_{i=0}^{N} a_i z^{-i}}{1 + \sum_{i=1}^{N} b_i z^{-i}} = \frac{\sum_{i=0}^{N} a_i z^{N-i}}{z^N + \sum_{i=1}^{N} b_i z^{N-i}} \tag{B.9}$$

The transfer function happens to be the z transform of the impulse response, i.e., $H(z) = \mathcal{Z}h(nT)$ (see Sect. 5.2 of [1]).

In FIR filters, $b_i = 0$ for $1 \leq i \leq N$ and hence the transfer function in Eq. (B.9) assumes the form

$$H(z) = \sum_{i=0}^{N} a_i z^{-i}$$

Since coefficients a_i are numerically equal to the impulse response values $h(iT)$ for $0 \leq i \leq N$, as can be seen in Eq. (B.4), the transfer function for FIR filters is often expressed as

$$H(z) = \sum_{i=0}^{N} h(iT)z^{-i} \quad \text{or} \quad \sum_{n=0}^{N} h_n z^{-n} \tag{B.10}$$

where h_n is a simplified representation of $h(nT)$.

B.5.2 Zero-Pole Form

By factorizing the numerator and denominator polynomials in Eq. (B.9), the transfer function can be expressed as

$$H(z) = \frac{A(z)}{B(z)} = \frac{H_0 \prod_{i=1}^{Z}(z - z_i)^{m_i}}{\prod_{i=1}^{P}(z - p_i)^{n_i}} \tag{B.11}$$

where $z_1, z_2, \ldots, z_Z$ and $p_1, p_2, \ldots, p_P$ are the zeros and poles of $H(z)$, m_i and n_i are the orders of zero z_i and pole p_i, respectively, $\sum_i^Z m_i = \sum_i^P n_i = N$, and H_0 is a multiplier constant. Evidently, the zeros, poles, and multiplier constant describe the transfer function and, in turn, the digital filter, completely. Typically, the zeros and poles of digital filters are simple, i.e., $m_i = n_i = 1$ for $1 \leq i \leq N$, and in such a case $Z = P = N$.

From Eq. (B.10), the transfer function of an FIR filter can also be expressed as

$$H(z) = \frac{1}{z^N} \sum_{n=0}^{N} h_n z^{N-n}$$

and, in effect, all the poles in an FIR filter are located at the origin of the z plane.

B.6 Time-Domain Response Using the Z Transform

The time-domain response of a digital filter to an arbitrary excitation $x(nT)$ can be readily obtained from Eq. (B.7) as

$$y(nT) = \mathcal{Z}^{-1}[H(z)X(z)] \tag{B.12}$$

i.e., we simply obtain the inverse-z transform of $H(z)X(z)$ (see Sect. 5.4 of [1]).

B.7 Z-Domain Condition for Stability

The stability condition in Eq. (B.6), namely, the requirement that the impulse response be absolutely summable over the range $0 \le nT \le \infty$ is difficult to apply in practice because it requires complete knowledge of the impulse response over the specified range. Fortunately, this condition can be converted into a corresponding z-domain condition that is much easier to apply as follows: *A digital filter is stable if and only if all the poles of the transfer function are located strictly inside the unit circle of the z plane.* In mathematical terms, a digital filter with poles

$$p_i = r_i e^{j\psi_i}$$

where $r_i = |p_i|$ and $\psi_i = \arg p_i$ for $1 \le i \le N$ is stable if and only if

$$r_i < 1$$

The z-plane areas of stability and instability are illustrated in Fig. B.2.

From the above discussion we note that for a stable digital filter, the denominator of the transfer function, $B(z)$, must *not* have zeros on or outside the unit circle $|z| = 1$ and, therefore, an alternative way of stating the z-domain stability condition is

$$B(z) \neq 0 \quad \text{for } |z| \ge 1 \tag{B.13}$$

The poles of the transfer function are the zeros of polynomial $B(z)$ in Eq. (B.11) and it can be easily shown that these are numerically equal to the eigenvalues of matrix

$$\mathbf{D} = \begin{bmatrix} -b_1 & -b_2 & \cdots & -b_{N-1} & -b_N \\ 1 & 0 & \cdots & 0 & 0 \\ 0 & 1 & \cdots & 0 & 0 \\ \vdots & \vdots & \ddots & \vdots & \vdots \\ 0 & 0 & \cdots & 1 & 0 \end{bmatrix}$$

Fig. B.2 Z-plane areas of stability and instability in IIR digital filters

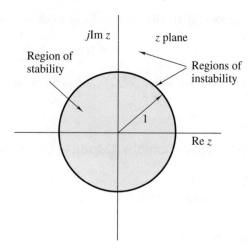

(see Problem 16.4(a)). Consequently, *an IIR filter is stable if and only if the moduli of the eigenvalues of matrix* **D** *are all strictly less than one.*

B.8   Frequency, Amplitude, and Phase Responses

A most important type of time-domain response for a digital filter is its steady-state sinusoidal response, which leads to a frequency-domain characterization for the filter.

As shown in Sect. 5.5.1 of [1], the steady-state sinusoidal response of a stable digital filter can be expressed as

$$\tilde{y}(nT) = \lim_{n \to \infty} \mathcal{R}[u(nT) \sin \omega nT]$$
$$= M(\omega) \sin[\omega nT + \theta(\omega)]$$

where

$$M(\omega) = |H(e^{j\omega T})| \quad \text{and} \quad \theta(\omega) = \arg H(e^{j\omega T}) \tag{B.14}$$

Thus, the steady-state effect of a digital filter on a sinusoidal excitation is to introduce a *gain* $M(\omega)$ and a phase shift $\theta(\omega)$, which can be obtained by evaluating the transfer function $H(z)$ on the unit circle $z = e^{j\omega T}$ of the z plane. $H(e^{j\omega T})$, $M(\omega)$, and $\theta(\omega)$ in Eq. (B.14) as functions of ω are known as the *frequency response, amplitude response,* and *phase response,* respectively.

The frequency response of a digital filter characterized by a transfer function of the form given in Eq. (B.11) can be obtained as

$$H(z)\big|_{z \to e^{j\omega T}} = H(e^{j\omega T}) = M(\omega)e^{j\theta(\omega)} \tag{B.15}$$

$$= \frac{H_0 \prod_{i=1}^{Z}(e^{j\omega T} - z_i)^{m_i}}{\prod_{i=1}^{P}(e^{j\omega T} - p_i)^{n_i}} \tag{B.16}$$

and by letting

$$e^{j\omega T} - z_i = M_{z_i} e^{j\psi_{z_i}} \tag{B.17}$$

$$e^{j\omega T} - p_i = M_{p_i} e^{j\psi_{p_i}} \tag{B.18}$$

Equations B.14–B.18 give

$$M(\omega) = \frac{|H_0| \prod_{i=1}^{Z} M_{z_i}^{m_i}}{\prod_{i=1}^{P} M_{p_i}^{n_i}} \tag{B.19}$$

$$\theta(\omega) = \arg H_0 + \sum_{i=1}^{Z} m_i \psi_{z_i} - \sum_{i=1}^{P} n_i \psi_{p_i} \tag{B.20}$$

where $\arg H_0 = \pi$ if H_0 is negative. Thus the gain and phase shift of a digital filter at some frequency ω can be obtained by calculating the magnitudes and angles of the complex numbers in Eqs. (B.17) and (B.18) and then substituting these values in Eqs. (B.19) and (B.20). These calculations are illustrated in Fig. B.3 for the case of a transfer function with two zeros and two poles. The vectors from the zeros and poles to point B represent the complex numbers in Eqs. (B.17) and (B.18), respectively.

The amplitude and phase responses of a digital filter can be plotted by evaluating the gain and phase shift for a series of frequencies $\omega_1, \omega_2, \ldots, \omega_K$ over the frequency range of interest.

Point A in Fig. B.3 corresponds to $\omega = 0$, i.e., zero frequency, and one complete revolution of vector $e^{j\omega T}$ about the origin corresponds to an increase in frequency of $\omega_s = 2\pi/T$ rad/s; this is known as the *sampling frequency*. Point C, on the other hand, corresponds to an increase in frequency of π/T, i.e., half the sampling frequency, which is often referred to as the *Nyquist frequency*. In the design of digital filters, a normalized sampling frequency of 2π rad/s is usually used, for the sake of simplicity, which corresponds to a Nyquist frequency of π and a normalized sampling period, $T = 2\pi/\omega_s$, of 1 s.

If vector $e^{j\omega T}$ in Fig. B.3 is rotated k complete revolutions starting at some arbitrary point, say, point B, it will return to its original position and the values of $M(\omega)$ and $\theta(\omega)$ will be the same as before according to Eqs. (B.19) and (B.20). Therefore,

$$H(e^{j(\omega+k\omega_s)T}) = H(e^{j\omega T})$$

In effect, the *frequency response* of a digital filter is a *periodic function* of frequency with a period ω_s. It, therefore, follows that knowledge of the frequency response of a digital filter over the base period $-\omega_s/2 \le \omega \le \omega_s/2$ provides a complete frequency-domain characterization for the filter. This frequency range is often referred to as the *baseband*.

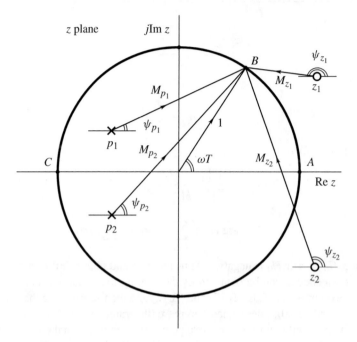

Fig. B.3 Calculation of gain and phase shift of a digital filter

Assuming real transfer-function coefficients, the amplitude response of a digital filter can be easily shown to be an even function and the phase response an odd function of frequency, i.e.,

$$M(-\omega) = M(\omega) \quad \text{and} \quad \theta(-\omega) = -\theta(\omega)$$

Consequently, a frequency-domain description of a digital filter over the positive half of the baseband, i.e., $0 \le \omega \le \omega_s/2$, constitutes a complete frequency-domain description of the filter.

Digital filters can be used in a variety of applications, for example, to pass low and reject high frequencies (lowpass filters), to pass high and reject low frequencies (highpass filters), or to pass or reject a range of frequencies (bandpass or bandstop filters). In this context, low and high frequencies are specified in relation to the positive half of the baseband, e.g., frequencies in the upper part of the baseband are deemed to be high frequencies. A frequency range over which the digital filter is required to pass or reject frequency components is said to be a *passband* or *stopband* as appropriate.

In general, the amplitude response is required to be close to unity in passbands and approach zero in stopbands. A constant passband gain close to unity is required to ensure that the different sinusoidal components of the signal are subjected to the same gain. Otherwise, so-called *amplitude distortion* will occur. The gain is required to be as small as possible in stopbands to ensure that undesirable signals are as far as possible rejected.

Note that the gain of a filter can vary over several orders of magnitude and for this reason it is often represented in terms of decibels as

$$M(\omega)_{dB} = 20 \log_{10} M(\omega)$$

In passbands, $M(\omega) \approx 1$ and hence we have $M(\omega)_{dB} \approx 0$. On the other hand, in stopbands the gain is a small fraction and hence $M(\omega)_{dB}$ is a negative quantity. To avoid this problem, stopbands are often specified in terms of *attenuation*, which is defined as the reciprocal of the gain in decibels, i.e.,

$$A(\omega) = 20 \log_{10} \frac{1}{M(\omega)} \tag{B.21}$$

Phase shift in a signal is associated with a delay and the delay introduced by a digital filter is usually measured in terms of the *group delay*, which is defined as

$$\tau(\omega) = -\frac{d\theta(\omega)}{d\omega} \tag{B.22}$$

If different sinusoidal components of the signal with different frequencies are delayed by different amounts, a certain type of distortion known as *phase distortion* (or *delay distortion*) is introduced, which is sometimes objectionable. This type of distortion can be minimized by ensuring that the group delay is as far as possible constant in passbands, and a constant group delay corresponds to a linear phase response as can be readily verified by using Eq. (B.22) (see Sect. 5.7 of [1]).

B.9   Design

The design of digital filters involves four basic steps as follows:

- Approximation
- Realization
- Implementation
- Study of the effects of roundoff errors.

The *approximation* step is the process of deducing the transfer function coefficients such that some desired amplitude or phase response is achieved. *Realization* is the process of obtaining a digital network that has the specified transfer function. *Implementation* is the process of constructing a system in hardware or software form based on the transfer function or difference equation characterizing the digital filter. Digital systems constructed either in terms of special- or general-purpose hardware are implemented using finite arithmetic and there is, therefore, a need to proceed to the fourth step of the design process, namely, the study of the effects of roundoff errors on the performance of the digital filter (see Sect. 9.1 of [1]). In the context of optimization, the design of a digital filter is usually deemed to be the solution of just the approximation problem.

B.9.1 Specified Amplitude Response

The design of an FIR or IIR filter that would satisfy certain amplitude-response requirements starts with an idealized amplitude response such as those depicted in Figs. B.4a, b and B.5a, b for lowpass, highpass, bandpass, and bandstop filters, respectively. In lowpass and highpass filters, parameters ω_p and ω_a are the *passband* and *stopband edges*. In bandpass and bandstop filters, on the other hand, ω_{p1} and ω_{p2} are the lower and upper passband edges, and ω_{a1} and ω_{a2} are the lower and upper stopband edges. The frequency bands between passbands and stopbands, e.g., the range $\omega_p < \omega < \omega_a$ in a lowpass filter, are called *transition bands*, for obvious reasons. The idealized passband gain is usually assumed to be unity but some other value could be used if necessary.

The objective of design in digital filters is to find a set of transfer function coefficients that would yield an amplitude response which would fall within the passband

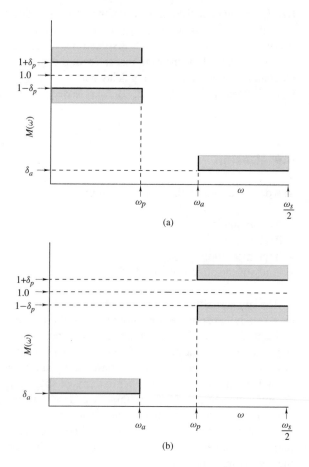

Fig. B.4 Idealized amplitude responses for lowpass and highpass filters

Fig. B.5 Idealized amplitude responses for bandpass and bandstop filters

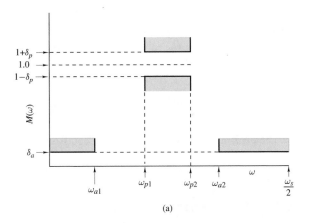

(a)

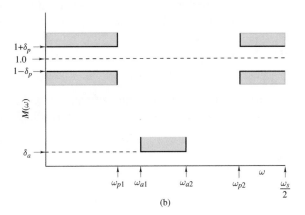

(b)

and stopband bounds shown in Figs. B.4a, b and B.5a, b. For example, in the case of a lowpass filter, we require that

$$1 - \delta_p \leq M(\omega) \leq 1 + \delta_p$$

with respect to the passband and

$$M(\omega) \leq \delta_a$$

with respect to the stopband.

The peak-to-peak passband error in Figs. B.4a, b and B.5a, b is often expressed in decibels as

$$A_p = 20 \log_{10}(1 + \delta_p) - 20 \log_{10}(1 - \delta_p) = 20 \log_{10} \frac{1 + \delta_p}{1 - \delta_p} \qquad \text{(B.23)}$$

which is often referred to as the *passband ripple*. The peak error δ_p can be deduced from Eq. (B.23) as

$$\delta_p = \frac{10^{0.05A_p} - 1}{10^{0.05A_p} + 1}$$

Since the maximum stopband gain δ_a is a small fraction, which corresponds to a large negative quantity when expressed in decibels, the stopband specification is often expressed in terms of the minimum stopband attenuation, A_a, which can be obtained from Eq. (B.21) as

$$A_a = 20 \log_{10} \frac{1}{\delta_a}$$

The peak stopband error can be deduced from A_a as

$$\delta_a = 10^{-0.05 A_a}$$

B.9.2   Linear Phase Response

A linear phase response is most easily obtained by designing the filter as an FIR filter. It turns out that a linear phase response can be achieved by simply requiring the impulse response of the filter to be symmetrical or antisymmetrical with respect to its midpoint (see Sect. 10.2 of [1]), i.e.,

$$h_n = h_{N-n} \qquad \text{for } n = 0, \ 1, \ \dots, \ N \tag{B.24}$$

Assuming a normalized sampling frequency of 2π, which corresponds to a normalized sampling period of 1 s, the use of Eqs. (B.10) and (B.24) will show that

$$H(e^{j\omega}) = e^{-j\omega N/2} A(\omega)$$

where

$$A(\omega) = \sum_{n=0}^{N/2} a_n \cos n\omega$$

$$a_n = \begin{cases} h_{N}/2 & \text{for } n = 0 \\ 2h_{N/2-n} & \text{for } n \neq 0 \end{cases}$$

for even N and

$$A(\omega) = \sum_{n=0}^{(N-1)/2} a_n \cos[(n + 1/2)\omega]$$

$$a_n = 2h_{(N-1)/2-n}$$

for odd N. The quantity $A(\omega)$ is called the *gain function* and, in fact, its magnitude is the gain of the filter, i.e.,

$$M(\omega) = |A(\omega)| \tag{B.25}$$

B.9.3   Formulation of Objective Function

Let us assume that we need to design an Nth-order IIR filter with a transfer function such as that in Eq. (B.9) whose amplitude response is required to approach one of the idealized amplitude responses shown in Figs. B.4a, b and B.5a, b. Assuming a normalized sampling frequency of 2π rad/s, the amplitude response of such a filter can be expressed as

$$|H(e^{j\omega})| = M(\mathbf{x},\ \omega) = \left| \frac{\sum_{i=0}^{N} a_i e^{-j\omega i}}{1 + \sum_{i=1}^{N} b_i e^{-j\omega i}} \right|$$

where

$$\mathbf{x} = [a_0\ a_1\ \cdots\ a_N\ b_1\ b_2\ \cdots b_N]^T$$

is the parameter vector. An error function can be constructed as

$$e(\mathbf{x},\ \omega) = M(\mathbf{x},\ \omega) - M_0(\omega) \qquad (B.26)$$

where $M_0(\omega)$ is the required idealized amplitude response, for example,

$$M_0(\omega) = \begin{cases} 1 & \text{for } 0 \le \omega \le \omega_p \\ 0 & \text{otherwise} \end{cases}$$

in the case of a lowpass filter. An objective function can now be constructed in terms of one of the standard norms of the error function, e.g., the L_2 norm

$$F = \int_{\omega \in \Omega} |e(\mathbf{x},\ \omega)|^2 d\omega \qquad (B.27)$$

where Ω denotes the positive half of the normalized baseband $[0,\ \pi]$. Minimizing Eq. (B.27) would yield a *least-squares* solution. Alternatively, we can define the objective function as

$$F = \max_{\omega \in \Omega} |e(\mathbf{x},\ \omega)| = \lim_{p \to \infty} \int_{\omega \in \Omega} |e(\mathbf{x},\ \omega)|^p d\omega \qquad (B.28)$$

where p is a positive integer, which would yield a minimax solution.

A more general design can be accomplished by forcing the frequency response of the filter, $H(e^{j\omega})$, to approach some desired idealized frequency response $H_d(\omega)$ by minimizing the least-pth objective function

$$F = \max_{\omega \in \Omega} |e(\mathbf{x},\ \omega)| = \lim_{p \to \infty} \int_{\omega \in \Omega} |H(e^{j\omega T}) - H_d(\omega)|^p d\omega \qquad (B.29)$$

As before, we can assign $p = 2$ to obtain a least-squares solution or let $p \to \infty$ to obtain a minimax solution.

Discretized versions of the objective functions in Eqs. (B.27)–(B.29) can be deduced by sampling the error in Eq. (B.26), $e(\mathbf{x},\ \omega)$, at frequencies $\omega_1,\ \omega_2, \ldots,$ ω_K, and thus the vector

$$\mathbf{E}(\mathbf{x}) = [e_1(\mathbf{x})\ e_2(\mathbf{x})\ \ldots\ e_K(\mathbf{x})]^T$$

can be formed where

$$e_i(\mathbf{x}) = e(\mathbf{x}, \omega_i)$$

for $i = 1, 2, \ldots, K$. At this point, an objective function can be constructed in terms of the L_p norm of $\mathbf{E}(\mathbf{x})$ as

$$F = ||\mathbf{E}(\mathbf{x})||_p = \left[\sum_{i=1}^{K} |e_i(\mathbf{x})|^p \right]^{1/p}$$

where we can assign $p = 2$ for a least-squares solution or $p \to \infty$ for a minimax solution.

The above objective functions can be readily applied for the design of FIR filters by setting the denominator coefficients of the transfer function, b_i for $1 \leq i \leq N$, to zero. If a linear phase response is also required, it can be readily achieved by simply forcing the coefficients $a_i \equiv h_i$ for $1 \leq i \leq N$ to satisfy the symmetry property in Eq. (B.24) and this can be accomplished by using the amplitude response given by Eq. (B.25).

Reference

1. A. Antoniou, *Digital Filters: Analysis, Design, and Signal Processing Applications*. New York: McGraw-Hill, 2018.

Index

A

Active
  constraint matrix, 349
  constraints, 18
    inequality, 290
Active-set methods for strictly convex
    quadratic-programming problems,
    429
  advantages of, 434
  dual active-set method, 434–435
  dual problem, 435
  primal active-set method, 430–434
    active-set, 430
    algorithm, 432
    example, 432
Additive white Gaussian noise in wireless
    communication channels, 615
ADMM, *see* Alternating direction method of
    multipliers
Affine
  function in sequential convex programming,
    540
  manifold, 670
    example, 670
  property of a matrix, 670
Algorithms
  accelerated alternating direction method of
    multipliers, 524
  accelerated alternating minimization
    algorithm, 530
  alternating convex optimization algorithm,
    547
  alternating minimization algorithm, 530
  backtracking line search, 505

barrier algorithm for general convex
    programming problems, 510
basic alternating direction method of
    multipliers, 522
basic quasi-Newton algorithm, 187
basic SQP algorithm, 554
block diagram, 64
CCP algorithm, 559
Charalambous minimax algorithm, 215
  modified version, 222
closed, 66
computational efficiency, 73
conjugate-gradient algorithm, 155
  advantages and disadvantages, 155
  solution of systems of linear equations,
    174
continuous, 64
convergence, 8
coordinate-descent algorithm, 146
cubic interpolation search, 96
Davies, Swann, and Campey algorithm,
    97–101
descent, 64
fast iterative shrinkage-thresholding
    algorithm, 515
  for the minimization of composite convex
    functions, 515
Fibonacci search, 86
Fletcher-Reeves algorithm, 159
Fletcher's inexact line search algorithm, 107
Gauss–Newton algorithm, 139
general structure, 8
global convergence, 69
golden-section search, 89

© Springer Science+Business Media, LLC, part of Springer Nature 2021
A. Antoniou and W.-S. Lu, *Practical Optimization*, Texts in Computer Science,
https://doi.org/10.1007/978-1-0716-0843-2

initialization, 8
interior-point
    primal-dual path-following algorithm
        for convex quadratic-programming
        problems, 439
iterative, 64
L_1 algorithms, 23
L_2 algorithms, 23
least-pth minimax algorithm, 214
    modified version, 221
L_∞ algorithms, 23
Matthews and Davies algorithm for the
    modification of the Hessian to achieve
    positive definiteness, 131
Mehrotra's predictor-corrector algorithm,
    418
minimax algorithms, 23
modified primal-dual path-following
    algorithm, 410
near-optimal multiuser detector using
    semidefinite-programming relaxation,
    623
Nesterov's accelerated gradient descent
    algorithm, 250
Newton algorithm, **128**, 259
    alternative, 182
Newton algorithm for convex problems with
    equality constraints, 504
    nonfeasible initial point, 505
Newton algorithm for unconstrained
    minimization of convex functions, 502
nonfeasible-initialization interior-point
    primal-dual path-following algorithm
        for convex quadratic-programming
        problems, 440
        for linear complementarity problems, 445
        for linear-programming problems, 413
over-relaxed alternating direction method of
    multipliers, 523
PCCP-based algorithm, 560
point-to-point mapping, 64
point-to-set mapping, 65
Powell's algorithm, 164
practical quasi-Newton algorithm, 202–205
practical SQP algorithm, 556
predictor-corrector algorithm, 418
    for semidefinite problems, 461
primal
    active-set algorithm for quadratic-
        programming problems with inequality
        constraints, 432

affine scaling linear programming
    algorithm, 401
    Newton barrier algorithm, 406
primal-dual
    interior-point algorithm for second-order
        cone programming problems, 474
    path-following algorithm for linear-
        programming problems, 409
    path-following algorithm for semidefinite
        programming problems, 457
proximal-point algorithm, 513, 514
quadratic interpolation search, 92
robustness, 9
rule of correspondence, 64
scaled ADMM algorithm, 564
scaled alternating direction method of
    multipliers, 523
sequential convex programming algorithm,
    543
    based on an exact penalty function, 545
simplex algorithm for alternative-form
    linear-programming problem
    degenerate vertices, 372
    nondegenerate vertices, 365
simplex algorithm for standard-form
    linear-programming problem, 380
steepest descent, 119
    without line search, 121
technique for updating the size of the trust
    region, 545
Zangwill's algorithm, 169
Alternating
    accelerated minimization algorithm, 530
    convex optimization, 546–550
        algorithm, 547
    minimization algorithm, 529–530
    updates, 521
Alternating direction
    method of multipliers, 519–525, 539
        accelerated algorithm, 524
        application to general constrained convex
            problem, 526–529
        basic algorithm, 522
        heuristic technique for nonconvex
            problems, 562–565
        over-relaxed algorithm, 523
        scaled algorithm, 523
    methods, 519–530
Alternative-form linear-programming problem,
    340
AMA, *see* Alternating minimization algorithm

Amplitude
   distortion in digital filters, 682
   response error in IIR filters, 589
   response in digital filters, 212, **680**
Analytical optimization methods, 2
Analytic center
   example, 506
   in linear programming, 421
Application of alternating direction method of
      multipliers to general constrained
      convex problem, 526–529
Applications
   constrained optimization
      introduction to, 571
Approximation
   error, 5
   methods (one-dimensional), 78
      cubic interpolation, 94–97
      Davies, Swann, and Campey algorithm,
      97
      quadratic interpolation, 91–94
   step in digital filters, 683
Asymmetric square root, 645
Attenuation in digital filters, 683
Augmented
   convex problem, 510
   dual, 520
   dual problem, 521
   Lagrangian, 520, 521, 523
      merit function, 553
   objective function, 307
Average
   convergence ratio, 75
   linear convergence, 75
   order of convergence, 74
      example, 74
AWGN, *see* Additive white Gaussian noise in
      wireless communication channels

B
Backtracking line search, **135**, 501, 505
   relation with Fletcher's inexact line search,
      136
   use in Gauss–Newton algorithm, 140
Ball
   in linear algebra, 292
   sequential convex programming, 540
Bandpass digital filters, 682
Bandstop digital filters, 682
Barrier

algorithm for general convex programming
   problems, 510
function in linear programming, 402
function methods, 296
parameter in
   linear programming, 402
Barzilai–Borwein
   formulas, 123
   Step-size estimation formulas, 122–124
   two-point formulas
      efficiency of, 123
Baseband in digital filters, 681
Basic
   conjugate-directions method, 148–151
      convergence, 149
      orthogonality of gradient to a set of
      conjugate directions, 150
   convex-concave procedure, 558–559
   optimization problem, 4
   quasi-Newton
      algorithm, 187
      method, 180
   SQP algorithm, 552–554
   variables in linear programming, 374
Basic SQP algorithm, 554
Basis for a span, 636
   example, 636
BER, *see* Bit error rate
BFGS, *see* Broyden–Fletcher–Goldfarb–
      Shanno, *see* Broyden-Goldfarb-
      Fletcher-Shanno method
   modified formula, 554
Binary classification
   use of logistic regression for, 243–245
Bit-error rate in multiuser detection, 615
Block diagram for an algorithm, 64
Boundary point, 18
Bounded sets, 70
Bracket, 77
Branches of mathematical programming:
   dynamic programming, 23
   integer programming, 22
   linear programming, 21
   nonlinear programming, 23
   quadratic programming, 22
Broyden–Fletcher–Goldfarb–Shanno method,
      197–199
   comparison with Davidon–Fletcher–Powell
      method, 205
   example, 198
   updating formula, 197

Broyden method, 200
   equivalence with Fletcher–Reeves method,
     200
   updating formula, 200

C

Cauchy-Schwartz inequality, 648
Causality property in digital filters, 674
CCP, *see* Convex-concave procedure
   algorithm, 559
CDMA, *see* Code-division multiple access in
     multiuser detection
Center for Biological and Computational
     Learning at MIT, 548
Centering
   condition in semidefinite programming, 452
   direction in linear programming, 423
   parameter in linear programming, 415
Central path in
   linear programming, 396–398
     example, 397
   quadratic programming, 436
   semidefinite programming, 451
Channel model in wireless communications,
     615–617
Characteristic
   equation, 640
   polynomial
     roots of, 129
Characterization of
   nonrecursive (FIR) digital filters in terms of
     a difference equation, 675
   recursive (IIR) digital filters in terms of a
     difference equation, 674
   symmetric matrices, 42–50
     examples, 44–50
Charalambous
   method
     design of lowpass IIR filters, 230
   minimax algorithm, 215
     modified version, 222
Cholesky
   matrix decomposition, 664
     example, 665
   triangle, 664
Class Γ_0
   definition, 495
Classes of nonlinear optimization problems, 77
   one-dimensional, 77
Classification of
   constrained optimization problems, 292–296

     example, 295
   handwritten digits, 240–253
     using softmax regression, 249–253
     using the MNIST data set, 250–253
   stationary points, 40–42
Closed
   algorithms, 66
     examples, 67–68
   kinematic chains in robotic systems, 602
   sets, 70
CMBER, *see* Constrained minimum BER
Code-division multiple access in multiuser
     detection, 614
Code sequences in multiuser detection, 614
Column
   pivoting in QR decomposition, 663
   rank, 638
   vector, 4
Compact sets, 70
Complementarity in Karush-Kuhn-Tucker
     conditions, 316
Complex L_1-norm approximation problem
     formulated as a second-order cone
     programming problem, 469
Computational
   complexity
     in minimax methods, 219
     of practical quasi-Newton algorithm, 203
     in predictor-corrector method, 418
     in simplex method, 385–387
   efficiency, 73
   effort
     dichotomous search, 79
     Fibonacci search, 83
Concave functions, 50–57
   definition, 51
Concepts and properties of convex functions,
     483–500
Conditional probability
   maximization of, 247
   use in logistic regression for binary
     classification, 245
Condition number
   of a Hilbert matrix, 174
   of a matrix, 174, 651
Conjugate
   definition of a proper function, 496
   functions, 494–498
Conjugate directions
   in Davidon–Fletcher–Powell method, 193
   definition, 146

generation of conjugate directions in
Powell's method, 161–162
methods
convergence, 149
introduction to, 145
orthogonality of gradient to a set of, **150**, 152
Conjugate-gradient
algorithm, 155
advantages and disadvantages, 155
comparison with steepest-descent
algorithm, 157
example, 156
solution of systems of linear equations,
174
trajectory of solution, 156
method, 152–156
convergence, 152
solution of systems of linear of equations,
173
Constrained optimization
application of singular-value decomposition,
288
applications
introduction to, 571
augmented objective function, 307
barrier function methods, 296
basic assumptions, 286
classification of constrained problems,
292–296
example, 295
constraint qualification, 335
convexity, 326
convex programming, 295
duality, 328–331
gap, 330
dual problem, 328, **330**
equality constraints, 286
example, 289–290, 307
example, 318, 543, 545, 556, 561, 564
Farkas lemma, 336
feasible region, 286
example, 287
general constrained problem, 295
geometrical interpretation of gradient,
310–311
example, 311
global constrained minimizer, 292
inequality constraints, 290
active, 290
inactive, 290
interior-point methods, 292

introduction to, 393
introduction to, 285
Lagrange multipliers
equality and inequality constraints, 315
equality constraints, 308
example, 306
Lagrangian, 307
linear programming, 293
local constrained minimizer, 292
nonnegativity bounds, 300
normal plane, 310
notation, 286
primal problem, 328, **330**
problems, 9
quadratic programming, 294
regular point, 287
relation between linearity of inequality
constraints and convexity of the feasible
region, 291
sequential quadratic-programming method,
296
simplex method, 292
slack variables, 291
strong local constrained minimizer, 293
sufficiency of Karush-Kuhn-Tucker
conditions in convex problems, 327
tangent plane, 309
transformation methods, 296
unconstrained minimizer, 307
variable elimination method, 296–303
example, 298
linear equality constraints, 296
nonlinear equality constraints, 299
variable transformations, 300–303
example, 301
interval-type constraints, 301
Constraints
active, 18
blocking, 372
constraint matrix, 340
equality, 9, **286**
example, 289–290
inequality, 9, **290**
active, 290
inactive, 290
interval-type, 301
qualification, 335
relation between linearity of inequality
constraints and convexity of the feasible
region, 291
working set of active, 371

Contact forces in dextrous hands, 602
Continuous algorithms, 64
Contour, 2
Convergence, 8
  average
    convergence ratio, 75
    linear convergence, 75
    order of, 74
  conjugate-directions method, 149
  conjugate-gradient method, 152
  global, 69
  of inexact line search, 104
  linear, 74
    in steepest-descent method, 125
  order of, 73
    in Newton method, 128
  quadratic rate of convergence in conjugate-
    direction methods, 157
  quadratic termination, 157
  rates, 73
  ratio, 73
    example, 73
    in steepest-descent method, 125
  of a sequence, 64
  steepest-descent method, 124–126
  superlinear, 74
Convex
  cone, 446, **668**
  functions, 50–57
    convexity of linear combination of convex
     functions, 52
    definition, 51
    existence of a global minimizer, 58
    optimization, 57–59
    properties of strictly convex functions, 56
    property relating to gradient, 54
    property relating to Hessian, 55
    relation between local and global
     minimizers, 57
    with Lipschitz-continuous gradients,
     488–491
  hull, 671
    in model predictive control of dynamic
     systems, 592
  optimization problems with linear matrix
    inequality constraints, 447
  polygon, 671
  polyhedron, 291, 671
  programming, 295
  quadratic problems with equality constraints,
    426–429
    example, 427
  sets
    definition, 50
    theorem, 53
Convex-concave procedure, 542, 557–562
Convexity
  in constrained optimization, 326
    theorem, 326
  relation between linearity of inequality
    constraints and convexity of the feasible
    region, 291
Coordinate-descent algorithm, 146
Coordinate directions, 146
Correction matrix, 183
Corrector direction, 461
Cost function, 4
CP, *see* Convex programming
CPU time
  average, 124
  evaluation of, 124
  comparison of basic quasi-Newton and
    Fletcher–Reeves algorithms, 190
  comparison of conjugate-gradient and
    steepest-descent algorithms, 157
  comparison of different versions of
    steepest-decent algorithm, 124
  comparison of Fletcher–Reeves and
    steepest-descent algorithms, 161
  normalized, 124
  pseudocode for the evaluation of, 124
Cubic
  approximation of Taylor series, 30
  interpolation search, 94–97
    algorithm, 96
Cutting plane, 486
Cycling in simplex method, 371
  least-index rule, 371

D

Davidon–Fletcher–Powell method, 190–196
  advantage relative to rank-one method, 190
  alternative updating formula, 196
  comparison with Broyden–Fletcher–
    Goldfarb–Shanno method, 205
  example, 195, 196
  positive definiteness of S matrix, 191
  updating formula, 191
Davies, Swann, and Campey algorithm, 97–101
Deconvolution
  of distorted and noise-contaminated signals,
    example, 516–519

problem, 516
Decrement
    of Newton, 135
Definitions
    class Γ_0, 495
    indicator function, 495
    closed algorithm, 66
    concave function, 51
    condition number of a matrix, 174
    conjugate of a proper function, 496
    conjugate vectors, 146
    convex
        cone, 446
        function, 51
        set, 50
    degenerate vertex, 347
    descent function, 69
    epigraph, 494
    feasible direction, 33
    global
        constrained minimizer, 292
        minimizer, 31
    Lagrange
        dual function, 328
        dual problem, 329
    local constrained minimizer, 292
    lower-semicontinuous function, 494
    nonconvex set, 51
    nondegenerate vertex, 347
    point-to-set mapping, 65
    proper function, 494
    regular point, 287
    residue of a system of linear equations, 173
    saddle point, 40
    second-order cone, 465
    strictly
        concave function, 51
        convex function, 51
    strong
        local constrained minimizer, 293
        minimizer, 31
    strong and weak duality, 330
    strongly convex functions, 491
    subgradient, 485
    weak
        global minimizer, 31
        local minimizer, 31
Degenerate linear-programming problems, 403
Delay distortion in digital filters, 683
Denavit-Hartenberg parameters, 253
Descent

algorithms, 64
    functions, 69
        example, 69
Design of
    digital filters
        discretized objective function, 687
        formulation of objective function,
            687–688
        introduction to, 261–262
        least-squares objective function, 687
        minimax objective function, 687
    FIR filters with linear phase response, 686
    FIR or IIR filters with a specified amplitude
        response, 684–686
    linear-phase FIR filters using quadratic
        programming, 572–574
        constraint on passband error, 572
        constraint on stopband gain, 573
        discretization, 573
        example, 574
        optimization problem, 574
    linear-phase FIR filters using unconstrained
        optimization, 264–267
        example, 265–267
    minimax FIR filters using semidefinite
        programming, 574–578
        discretization, 577
        example, 577
        frequency response, 575
        minimax problem, 576
        Remez exchange algorithm, 574
        squared weighted error, 576
        weighted-Chebyshev method, 574
    minimax FIR filters using unconstrained
        optimization, 267–275
        algorithm, 269
        direct and sequential optimization, 270
        example, 270, 273
        gradient and Hessian, 269
        low-delay filters, 268
        objective function, 268
        Toeplitz matrix, 270
    minimax IIR filters using semidefinite
        programming, 578
        example, 584
        formulation of design problem, 581
        frequency response, 581
        introduction to, 578
        iterative SDP algorithm, 584
        linear matrix inequality constraint, 580
        Lyapunov equation, 583

optimization problem, 582
  stability constraint, 583
  weighting function, 582
weighted least-squares FIR filters using
  unconstrained optimization, 262–267
  linear phase response, 264–267
  specified frequency response, 263–264
  weighting, 263
Determinant of a square matrix in terms of its
  eigenvalues, 643
Dextrous hands in robotic systems, 602
DFP, *see* Davidon–Fletcher–Powell190
DH, *see* Denavit-Hartenberg parameters253
Diagonalization of matrices, 44
Dichotomous search, 78–79
  computational effort, 79
Digital filters
  2-band, 588
  3-band, 590
  amplitude
    distortion, 682
    response, 589, 680
  approximation step, 683
  attenuation, 683
  bandpass, 682
  bandstop, 682
  baseband, 681
  causality property, 674
  characterization of
    nonrecursive (FIR) filters in terms of a
      difference equation, 675
    recursive (IIR) filters in terms of a
      difference equation, 674
  definition, 673
  delay distortion, 683
  design of FIR filters with linear phase
    response, 686
  design of FIR or IIR filters with a specified
    amplitude response, 684–686
  design of IIR filters satisfying prescribed
    specifications, 587–591
  design of linear-phase FIR filters using
    quadratic programming, 572–574
    constraint on passband error, 572
    example, 574
  design of lowpass filters using
    modified Charalambous method, 230
    modified least-pth method, 224
  design of lowpass IIR filters using
    Charalambous method, 230
    least-pth method, 224

design of minimax FIR filters using
  semidefinite programming
  Remez exchange algorithm, 574
  weighted-Chebyshev method, 574
desired amplitude response, 212
discretized objective function, 687
excitation (input), 673
finite-duration impulse response (FIR) filters,
  676
formulation of objective function, 687–688
frequency response, 680
gain, 680
  in decibels, 683
group delay, 683
highpass, 682
impulse response, 675
infinite-duration impulse response (IIR)
  digital filters, 676
initially relaxed, 674
least-squares objective function, 687
length of a nonrecursive (FIR) digital filter,
  675
linearity property, 674
lower
  passband edge, 684
  stopband edge, 684
lowpass, 682
L_p norm, 589
maximum stopband gain, 686
minimax design of FIR filters using
  semidefinite programming, 574–578
  discretization, 577
  example, 577
  frequency response, 575
  minimax problem, 576
  squared weighted error, 576
minimax design of FIR filters using SOCP,
  586
minimax design of IIR filters satisfying
  prescribed specifications, 587
minimax design of IIR filters using
  semidefinite programming
  example, 584
  formulation of design problem, 581
  frequency response, 581
  introduction to, 578
  iterative SDP algorithm, 584
  linear matrix inequality constraint, 580
  Lyapunov equation, 583
  optimization problem, 582
  stability constraint, 583

weighting function, 582
minimax design of IIR filters using SOCP, 587
minimax objective function, 687
minimum stopband attenuation, 686
normalized
    sampling frequency, 681
    sampling period, 681
Nyquist frequency, 681
order, 674
passband, 682
    edge, 684
    ripple, 685
passband error function in IIR filters, 589
peak stopband error, 686
peak-to-peak passband error, 685
periodicity of frequency response, 681
phase
    distortion, 683
    response, 680
    shift, 680
$\mathcal{R}$ operator, 673
sampling frequency, 681
Schur polynomials, 302
stability
    condition imposed on eigenvalues, 680
    condition imposed on impulse response, 677
    condition imposed on poles, 679
    property, 677
stability margin, 583, 587
stabilization technique, 226
standard signals
    unit impulse, 675
    unit sinusoid, 675
    unit step, 675
steady-state sinusoidal response, 680
stopband, 682
    edge, 684
system properties, 674
time-domain response, 675–676
    using the z transform, 679
time-invariance property, 674
transfer function
    definition, 677
    derivation from difference equation, 678
    multiplier constant, 678
    poles, 678
    in zero-pole form, 678
    zeros, 678
transition band, 684

upper
    passband edge, 684
    stopband edge, 684
z transform, 676
Dimension of a subspace, 636
Direction
    centering direction in linear programming, 423
    coordinate, 146
    corrector, **417**, 461
    matrix **S**, 179
        generation of, 181
        positive definiteness of, 191
        relation with the Hessian, 181
    Newton direction, 128
    orthogonal direction, 119
    primal affine scaling, 400
    projected steepest-descent, 398
    quasi-Newton direction, 183
    steepest
        ascent direction, 116
        descent direction, 116
Direct-sequence CDMA channels, 615
Discretization, 11
Disjoint feasible region, 20
DNB, *see* Dual Newton barrier421
Dot product, 116
Double inverted pendulum, 10
DS-CDMA, *see* Direct-sequence CDMA channels615
Dual
    active-set method for convex quadratic-programming problems, 434–435
    ascent iterations, 520
    Newton barrier, 421
    problem in
        constrained optimization, 328
        linear programming, 394
        quadratic programming, 435
        second-order cone programming, 466
    problem in constrained optimization, 330
    residual, 504, 522
    semidefinite problem, 446
Duality
    in constrained optimization, 328–331
    of DFP and BFGS updating formulas, 197
    gap in
        constrained optimization, 330
        linear programming, 396
        primal-dual method for second-order cone programming problems, 472

quadratic programming, 436
semidefinite-programming problems, 447,
451
of Hoshino formula, 199
strong and weak, 330
Dynamic programming, 23

E

Edge of a convex polyhedron, 346
Efficient multiuser detector based on duality,
621
Eigendecomposition, 641
example, 641
Eigenvalues, 45, **640**
ratio of smallest to the largest eigenvalue in
steepest-descent method, 125
Eigenvectors, 147, **640**
Elemental error, 212
Elementary
matrices, 44
transformations, 44
Elimination of
spikes in the error function, 219
End-effector, 253
Epigraph
definition, 494
of a convex function, 494
Equality constraints, 9, **286**
example, 289–290
Equation of forward kinematics, 255
Error
approximation, 5
elemental, 212
residual, 7
Euclidean
norm, 7
space, 4
Exact penalty formulation, 544–546
Examples
Broyden–Fletcher–Goldfarb–Shanno
method, 198
active constraints, 376
affine manifold, 670
alternative-form linear-programming
problem, 345
bounded solution, 366
degenerate vertex, 373
unbounded solution, 369
analytic center problem, 506
average order of convergence, 74
basic quasi-Newton algorithm, 188, 189

basis for a span, 636
central path in linear programming, 397
characterization of symmetric matrices,
44–50
Cholesky decomposition, 665
classification of constrained optimization
problems, 295
closed algorithms, 67–68
conjugate-gradient algorithm, 156
constrained optimization, 318
equality constraints, 307
problem, 543, 545, 556, 561, 564
convergence ratio, 73
convex quadratic-programming problems
with equality constraints, 427
Davidon–Fletcher–Powell method, 195, 196
deconvolution of distorted and noise-
contaminated signals, 516–519
descent functions, 69
design of
bandpass FIR digital filter using
unconstrained optimization, 273
linear-phase FIR filter using quadratic
programming, 574
linear-phase FIR filter using unconstrained
optimization, 265–267
lowpass digital filters using least-pth
method, 224
lowpass digital filters using modified
Charalambous method, 230
lowpass digital filters using modified
least-pth method, 224
lowpass FIR digital filter using
unconstrained optimization, 270
lowpass IIR filters using Charalambous
method, 230
minimax FIR filters using semidefinite
programming, 577
minimax IIR filters using semidefinite
programming, 584
double inverted pendulum, 10
eigendecomposition, 641
equality constraints, 289–290
feasible region, 287
first-order necessary conditions for a
minimum in constrained optimization,
313
Fletcher–Reeves algorithm, 159
geometrical interpretation of gradient, 311
Givens rotations, 659
global convergence, 71–73

graphical optimization method, 18–20
Householder transformation, 657
investment portfolio, 15
joint angles in three-link manipulator, 258
$L_1 - L_2$ problem, 524–525
Lagrange
    dual problem, **330**, 331
    multipliers, 306
linear dependence of vectors, 636
Lipschitz continuous gradient, **489**, 490
LP problem, 510
Lyapunov equation, 667
matrix norms, 649
modified primal-dual path-following method, 411
multilayer thin-film system, 13
near-optimal multiuser detector using semidefinite-programming relaxation, 624
nonfeasible-initialization interior-point primal-dual path-following method for
    convex quadratic-programming problems, 441
    linear-programming problems, 414
nonlinear equality constraints in constrained optimization, 300
nonnegative matrices, 548
null space, 638
optimal force distribution in multifinger dextrous hands
    using linear programming, 609
    using semidefinite programming, 612
orthogonal projections, 654
position and orientation of the robot tip in a manipulator, 255
predictor-corrector method, 418
primal
    active-set method for quadratic-programming problems with inequality constraints, 432
    affine scaling linear-programming method, 401
    Newton barrier method, 406
primal-dual
    interior-point method for second-order cone programming problems, 474
    path-following method for semidefinite programming problems, 459
QR decomposition, 663
rank of a matrix, 638
robust

constrained model predictive control using semidefinite programming, 601
unconstrained model predictive control of dynamic systems using semidefinite programming, 596
second-order
    necessary conditions for a minimum, 321
    Schur polynomial, 302
    sufficient conditions for a minimum, 323
    sufficient conditions for a minimum in a general constrained problem, 326
Sherman-Morrison formula, 639
singular-value decomposition, 652
solution of overdetermined system of linear equations, 216
sparse solution of an undetermined system of linear equations, 527–529
standard-form linear-programming problem, 342, 380
step response of a control system, 5
subdifferential, 486
symmetric square root, 645
system of linear equations, 174
system of nonlinear equations, 139
transportation problem, 14
variable elimination method in constrained optimization, 298
variable transformations in constrained optimization, 301
vertex of a convex polyhedron, 352, 353
Excitation (input) in digital filters, 673
Existence of
    primal-dual solution in linear programming, 395
    a vertex minimizer in
        alternative-form linear-programming problem, 360
        standard-form linear-programming problem, 361
Experimental methods of optimization, 2
Extension of Newton method
    barrier method for general convex programming problems, 507–511
    convex constrained and unconstrained problems, 500–511
    minimization of smooth convex functions subject to equality constraints, 502–504
        with a nonfeasible initial point, 504–506
    minimization of smooth convex functions without constraints, 500–502
Exterior point, 18

Extrapolation formula for inexact line search,
        103
Extrema, 31

F
Face of a convex polyhedron, 346
Facet of a convex polyhedron, 346
Facial image representation, 548
Farkas lemma, 336
Fast
    algorithm for the minimization of composite
        convex functions, 514–519
    iterative shrinkage-thresholding algorithm,
        515
        for the minimization of composite convex
            functions, 515
FDMA, *see* Frequency-division multiple access
Feasible
    descent directions in linear programming,
        348
    direction, 33
    domain, 17
    linear-programming problem, 394
    point, 17
        in linear programming, 356
    region, **17**, 286
        disjoint, 20
        example, 287
        relation between linearity of inequality
            constraints and convexity of the feasible
            region, 291
        simply connected, 20
Fenchel inequality, 496
Fibonacci
    search, 79–86
        algorithm, 86
        comparison with golden-section search,
            89
        computational effort, 83
        sequence, 82
Finding
    a feasible point, 356
    a linearly independent normal vector, 357
    a vertex minimizer, 360–362
FIR digital filters
    amplitude response, 267
    design of linear-phase FIR filters using
        quadratic programming, 572–574
    design of linear-phase FIR filters using
        unconstrained optimization, 264–267
        example, 265–267

design of minimax FIR filters using
    unconstrained optimization, 267–275
    low-delay filters, 268
    objective function, 268
design of weighted least-squares FIR filters
    using unconstrained optimization
    linear phase response, 264–267
finite-duration impulse response (FIR)
    digital filters, 676
high order, 264
introduction to, 261–262
linear phase, 262
minimax design of FIR filters using
    unconstrained optimization
    algorithm, 269
    bandpass filter, 273
    direct and sequential optimization, 270
    example, 270, 273
    gradient and Hessian, 269
    lowpass filter, 271
    Toeplitz matrix, 270
objective function, 263
using unconstrained optimization, 239, 261
weighted least-squares design of FIR
    filters using unconstrained optimization,
    262–275
    specified frequency response, 263–264
weighting in the design of least-squares FIR
    filters, 263
First-order conditions for a minimum
    necessary conditions in constrained
        optimization, 312–319
        example, 313
    necessary conditions for equality constraints,
        312
        theorem, 312
    necessary conditions for inequality
        constraints, 314
    necessary conditions in unconstrained
        optimization, 34–35
        theorem, 35
First-order gradient methods, 115
First-order optimality condition for nondif-
            ferentiable unconstrained convex
            problems
    theorem, 487
FISTA, *see* Fast iterative shrinkage-
            thresholding algorithm
Fixed-point of proximal-point operator, 513
Fletcher
    inexact line search algorithm, 107

switch method, 200
Fletcher–Reeves method, 158–159
  advantages, 158
  algorithm, 159
    comparison with steepest-descent
      algorithm, 161
    example, 159
Formulas for step-size estimation due to
      Barzilai and Borwein, 122–124
FR, *see* Fletcher-Reeves
Frame in robotics, 253
Frequency-division multiple access in
      communications, 614
Frequency response in digital filters, 680
Friction
  cone in dextrous hands, 606
  force in dextrous hands, 606
  limits in dextrous hands, 607
Frobenius norm, 649
  use in softmax regression, 249

G
Gain
  in decibels, 683
  in digital filters, 680
Gaussian elimination in Matthews and Davies
      method for the modification of the
      Hessian, 129
Gauss–Newton method, 137–140
  algorithm, 139
    use of backtracking line search, 140
    solution of a system of nonlinear equations,
      139
GCO, *see* General constrained optimization
General
  constrained optimization problem, **9**, 286
  structure of optimization algorithms, 8
Generation of conjugate directions in Powell's
      method, 161–162
Geometrical interpretation of gradient in a
      constrained optimization problem,
      310–311
  example, 311
Geometry of a linear-programming problem,
      346–360
  degenerate vertex, 347
  edge, 346
  face, 346
  facet, 346
  method for finding a vertex, 350–356
  nondegenerate vertex, 347

vertex, 346
  example, 352, 353
Givens rotations, 658
  example, 659
Global
  constrained minimizer, 292
  convergence, 69
    examples, 71–73
    theorem, 69
  minimizer, 31
GN, *see* Gauss-Newton
Golden ratio, 89
Golden-section
  golden ratio, 89
  search, 88–89
    algorithm, 89
    comparison with Fibonacci search, 89
  sequence, 82
Goldstein
  conditions, 103
  tests, 102
Gradient
  geometrical interpretation
    constrained optimization problem,
      310–311
    example, 311
  information, 27
  in least-pth method, 214
  vector, 27
Gradient methods, 28
  conjugate directions
    introduction to, 145
  first-order, 115
  Gauss–Newton method, 137–140
    algorithm, 139
  introduction to multidimensional, 115
  Newton, 126–134
    direction, 128
    modification of the Hessian to achieve
      positive definiteness, 128–134
    order of convergence, 128
    relation with steepest-descent method,
      129
  second-order, 115
  steepest descent, 116–126
    algorithm, 119, 121
    convergence, 124–126
    estimation of step size, 120–122
    relation with Newton method, 129
    scaling, 126
Graphical optimization method, 2

examples, 18–20
Group delay in digital filters, 683

H

Handwritten-digit recognition
  applications, 240
  gradient, 242
  image blocks, 241
  magnitude and angle of gradient, 242
  partial derivatives, 242
  problem, 240
  training data, 240
  use of two-dimensional discrete convolution,
    242
Heavisine test signal, 517
Hermitian
  matrix, 642
  square root, 645
Hessian matrix, 28
  computation, 135
  modification to achieve positive definiteness,
    128–134
    examples, 132–134
  positive definite approximation, 554–555
Highpass digital filters, 682
Hilbert matrix, 174
  condition number of, 174
Histogram of oriented gradients, 240–243
Hölder inequality, 647
Homogeneous transformation, 255
Hoshino
  method, 199
  updating formula, 199
Householder
  transformation, 655
    example, 657
  update, 656
Huang family
  of quasi-Newton methods, 201
  updating formula, 201
Hull, 671
HWDR, see Handwritten-digit recognition
Hypersurface, 286

I

I.i.d., see Independent and identically
    distributed
Identity matrix, 46
IIR digital filters:
  design of lowpass IIR filters using
    least-pth method, 224

infinite-duration impulse response filters,
    676
L_p norm, 589
minimax design of IIR filters using
    semidefinite programming, 578–584
  example, 584
  formulation of design problem, 581
  frequency response, 581
  introduction to, 578
  iterative SDP algorithm, 584
  linear matrix inequality constraint, 580
  Lyapunov equation, 583
  optimization problem, 582
  stability constraint, 583
  weighting function, 582
passband error function in IIR filters, 589
Impulse response in digital filters, 675
Inactive inequality constraints, 290
Independent and identically distributed (i.i.d.)
    random variables
  use in logistic regression for binary
    classification, 245
Indicator function
  definition, 495
Inequality constraints, 9, **290**
  active, 290
  inactive, 290
  relation between linearity of inequality
    constraints and convexity of the feasible
    region, 291
Inequality of Fenchel, 496
Inexact line searches, 101–108
  choice of parameters, 108
  convergence theorem, 104
  Fletcher's algorithm, 107
    modified version, 202
  Goldstein
    conditions, 103
    tests, 102
  interpolation formula, 103
Infimum, 328
Infinite-duration impulse response filters, see
    IIR digital filters
Initialization of an algorithm, 8
Inner product
  for matrices, 668
  for vectors, 647
Integer programming, 22
Interior point, 18
Interior-point methods, 292
  introduction to, 393

Interior-point methods for convex quadratic-
        programming problems, 435
    central path, 436
    dual problem, 435
    duality gap, 436
    linear complementarity problems, 442
        algorithm, 445
    mixed linear complementarity problems, 442
    monotone linear complementarity problems,
        442
    nonfeasible-initialization interior-point
        primal-dual path-following method,
        439–442
        algorithm, 440
        example, 441
    primal-dual path-following method, 437–439
        algorithm, 439
        example, 441
        iteration complexity, 438
        potential function, 438
Interpolation formula
    for inexact line search, 103
    polynomial, 221
    simplified three-point quadratic formula, 93
    three-point quadratic formula, 92
    two-point quadratic formula, 94
Interval-type constraints in optimization, 301
Inverse Hilbert matrix, 175
Inverse kinematics for robotic manipulators,
        253–261
    Denavit-Hartenberg parameters, 253
    end-effector, 253
    equation of forward kinematics, 255
    frame in robotics, 253
    homogeneous transformation, 255
    joint angles in three-link manipulator
        example, 258
    optimization problem, 256
    position and orientation of a manipulator,
        253–256
        example, 255
    robotic manipulator, 253
    solution of, 257–261
Investment portfolio, 15
IP, *see* Integer programming
Iteration, 8
    complexity in primal-dual path-following
        method, 438
Iterative algorithms, 64

J
Jacobian, 137
Joint angles in three-link manipulator
    example, 258

K
Karush-Kuhn-Tucker conditions, 304
    complementarity, 316
    for nondifferentiable constrained convex
        problems,
        theorem, 487
    for second-order cone programming
        problems, 472
    for semidefinite-programming problems, 450
    for standard-form linear-programming
        problems, 341
    sufficiency conditions in convex problems,
        327
    theorem, 316
KKT, *see* Karush-Kuhn-Tucker
Kronecker product, 666

L
L_1
    norm
        of a vector, 647
    algorithms, 23
    norm, 7
        of a matrix, 649
L_1 merit function, 553
$L_1 - L_2$ minimization problem, 549
    example, 524–525
L_2
    algorithms, 23
    norm, 7
        design of FIR filters using unconstrained
            optimization, 262–267
        invariance under orthogonal or unitary
            transformation, 648
        of a matrix, 649
        of a vector, 647
Lagrange
    dual, 519
    dual function, 328
    dual problem, **329**
        example, 330, 331
    multipliers, 307, 328, 502
        equality and inequality constraints, 315
        equality constraints, 305–308
        example, 306
        introduction to, 303

Lagrangian, **307**, 322, 328, 329, 509, 519, 552
  augmented, 520, 521, 523
Large-scale convex problems, 519
Leading principal minor, 644
Least-pth minimax
  algorithm, 214
    modified version, 221
  method, 213–214
    choice of parameter μ, 214
    design of lowpass IIR filters, 224
    gradient, 214
    numerical ill-conditioning, 214
Least-squares
  minimization problem with quadratic
    inequality constraints, 480
  problem, 7
Left singular vector, 651
Length of a nonrecursive (FIR) digital filter,
    675
L_∞
  algorithms, 23
  norm, 7
    design of FIR filters using unconstrained
      optimization, 267–275
    of a function of a continuous variable,
      262, 647
    of a matrix, 649
    of a vector, 647
Limit of a sequence, 64
Line searches
  backtracking, 135
Linear
  complementarity problems in convex
    quadratic programming, 442
    algorithm, 445
  convergence, 74
  dependence in Powell's method, 165
  dependence of vectors
    example, 636
  equations, *see* Systems of linear equations
  fractional problem formulated as a second-
    order cone programming problem,
    470
  independence of columns in constraint
    matrix, theorem, 374
  matrix inequality, 447
Linear algebra,
  affine
    manifold, 670
    property of a matrix, 670
  asymmetric square root, 645

ball, 292
Cauchy-Schwartz inequality, 648
characteristic equation, 640
Cholesky
  decomposition, 664
  triangle, 664
column rank, 638
condition number of a matrix, 651
convex
  cone, 668
  hull, 671
  polygon, 671
  polyhedron, 291, **671**
determinant of a square matrix in terms of its
  eigenvalues, 643
dimension of a subspace, 636
eigendecomposition, 641
  example, 641
eigenvalues, 640
eigenvectors, 640
Frobenius norm, 649
Givens rotations, 658
Hermitian
  matrix, 642
  square root of, 645
Hölder inequality, 647
Householder
  transformation, 655
  update, 656
hypersurface, 286
inner product, 647
  for matrices, 668
introduction to, 635
Kronecker product, 666
L_1 norm
  of a matrix, 649
  of a vector, 647
L_2 norm
  invariance under orthogonal or unitary
    transformation, 648
  of a matrix, 649
  of a vector, 647
left singular vector, 651
L_∞ norm
  of a matrix, 649
  of a vector, 647
L_p norm
  of a matrix, 649
  of a vector, 647
manifold, 670
maximal linearly independent subset, 636

Moore-Penrose pseudo-inverse, 651
  evaluation of, 651
non-Hermitian square root of, 645
nonsingular matrices, 642
normal
  plane, 310
  vector, 340
null space, 637
orthogonal
  matrices, 642
  projection matrix, 653
orthonormal basis of a subspace, 654
polygon, 671
polytope, 671
QR decomposition
  with column pivoting, 663
  for full-rank case, 660
  mathematical complexity, 661
  for rank-deficient case, 662
range of a matrix, 637
rank of a matrix, 637
right singular vector, 651
row rank, 638
Schur polynomials, 302
Sherman-Morrison formula, 639
similarity transformation, 641
singular
  value decomposition, 651
  values, 651
span, 636
  basis for, 636
subspace, 636
symmetric
  matrices, 642
  square root, 645
tangent plane, 309
trace of a matrix, 646
vector spaces of symmetric matrices,
  668–671
Linear approximation, 30
Linear programming, 15, **21**, 293, 425
  active
    constrained matrix, 349
    constraints, example, 376
  alternative-form LP problem, 340
    example, 345
    necessary and sufficient conditions for a
      minimum, 344, 350
  analytic center, 421
  centering direction, 423
  central path, 396–398

constraint matrix, 340
degenerate LP problems, 403
duality gap, 396
dual problem, 394
existence of a vertex minimizer in
  alternative-form LP problem,
  theorem, 360
existence of a vertex minimizer in
  standard-form LP problem,
  theorem, 361
existence of primal-dual solution, 395
feasible
  descent directions, 348
  LP problem, 394
finding a
  feasible point, 356
  linearly independent normal vector, 357
  vertex minimizer, 360–362
geometry of an LP problem, 346–360
  degenerate vertex, 347
  edge, 346
  face, 346
  facet, 346
  nondegenerate vertex, 347
  vertex, 346
interior-point methods
  introduction to, 393
introduction to, 339
Karush-Kuhn-Tucker conditions
  theorem, 341
linear independence of columns in constraint
  matrix,
  theorem, 374
Mehrotra's predictor-corrector algorithm,
  418
modified primal-dual path-following method,
  410
nondegenerate LP problems, 403
nonfeasible-start primal-dual path-following
  algorithms, 412
normal vector, 340
optimality conditions, 341–346
primal
  LP problem, 394
primal-dual
  solutions, 394–396
projected steepest-descent direction, 398
relation between
  alternative-form linear programming and
    semidefinite programming problems, 448

standard-form linear-programming and
semidefinite-programming problems, 445
scaling, 398
affine scaling transformation, 399
primal affine scaling direction, 400
simplex method
for alternative-form LP problem, 363–374
basic and nonbasic variables in
standard-form LP problem, 374
blocking constraints, 372
computational complexity, 385–387
cycling, 371
least-index rule, 371
pivot in tabular form, 383
for standard-form LP problem, 374–382
tabular form, 383–384
standard-form LP problem, 339
example, 342
necessary and sufficient conditions for a
minimum, 350
strict feasibility of primal-dual solutions, 395
strictly feasible LP problem, 394
uniqueness of minimizer of alternative-form
LP problem
theorem, 361
uniqueness of minimizer of standard-form
LP problem
theorem, 362
Linearity property in digital filters, 674
Linearly independent vectors, 636
Lipschitz
convex functions with Lipschitz-continuous
gradients, 488–491
constant, 488, 512
continuous gradient, 488, 512, 514
example, 489
continuous gradient example, 490
$L_1 - L_2$ minimization problem, 549
LMI, *see* Linear matrix inequality
Local constrained minimizer, 292
Location of maximum of a convex function, 58
Logarithmic barrier function, 402, **508**
Logistic regression for binary classification,
243–245
independent and identically distributed
(i.i.d.) random variables in, 245
use of conditional probability in, 245
use of logistic sigmoid function in, 244
Logistic sigmoid function
use in logistic regression for binary
classification, 244

Lorentz
cone in second-order cone programming,
465
Low-delay FIR filters using unconstrained
optimization, 268
Lower
passband edge in digital filters, 684
stopband edge in digital filters, 684
triangular matrix, 129
Lower-semicontinuous function
definition, 494
Lower-upper factorization, 385
Lowpass digital filters, 682
LP, *see* Linear programming
problem example, 510
L_p norm, 6
amplitude response in IIR digital filers, 589
of a function of a continuous variable, 262,
647
of a matrix, 649
of a vector, 647
passband error function in IIR filers, 589
LSQI, *see* Least-squares minimization problem
with quadratic inequality
LU, *see* Lower-upper
Lyapunov equation, 583, 667
example, 667

M
Machine-learning techniques, 239, **240**
MAI, *see* Multiple access
Manifold, 670
Mathematical
complexity
QR decomposition, 661
programming, 3
introduction to nonlinear, 27
Matrices
active constraint matrix, 349
affine property, 670
asymmetric square root, 645
characteristic equation, 640
characterization of symmetric matrices,
42–50
via diagonalization, 43
Cholesky
decomposition, 664
triangle, 664
column rank, 638
computation of the Hessian matrix, 135
condition number, 174, **651**

constraint matrix, 340
correction matrix, 183
determinant of a square matrix in terms of its
    eigenvalues, 643
diagonalization, 44
direction matrix **S**, 179
    generation of, 181
eigendecomposition, 641
    of symmetric matrices, 45
eigenvalues, 45, **640**
eigenvectors, 147, **640**
elementary, 44
Frobenius norm, 649
Gaussian elimination, 129
Givens rotations, 658
Hermitian, 642
    square root of, 645
Hilbert matrix, 174
identity, 46
inner product for matrices, 668
inverse Hilbert matrix, 175
Jacobian, 137
Kronecker product, 666
L_1 norm, 649
L_2 norm, 649
leading principal minor, 644
L_∞ norm, 649
L_p norm, 649
minor determinant (or minor), 643
modification of the Hessian to achieve
    positive definiteness, 128–134
    examples, 132–134
Moore-Penrose pseudo-inverse, 651
    evaluation of, 651
negative part of a symmetric matrix, 541
non-Hermitian square root of, 645
nonnegative, 548
nonsingular, 642
norms
    example, 649
notation, 635
orthogonal, 45, **642**
    projection, 653
positive definite, positive semidefinite,
    negative definite, negative semidefinite,
    43, 643
    notation, 643
positive definiteness of **S** matrix, 191
principal minor, 643
properties, 46
QR decomposition

with column pivoting, 663
    example, 663
    for full-rank case, 660
    mathematical complexity, 661
    for rank-deficient case, 662
range, 637
rank, 637
relation between direction matrix **S** and the
    Hessian, 181
roots of the characteristic polynomial, 129
row rank, 638
Schur complement matrix, 456
Sherman-Morrison formula, 639
similarity transformation, 641
singular
    value decomposition, 651
    values, 651
sparse, 386
strictly feasible, 450
symmetric, 642
    square root, 645
Toeplitz matrix, **270**, 516
trace, 646
unitary matrix, 45, **642**
unit lower triangular matrix, 129
upper triangular matrix, 131
vector spaces, 668–671
working set, 371
Matthews and Davies
    algorithm for the modification of the Hessian
        to achieve positive definiteness, 131
    method, 129–132
Maximal linearly independent subset, 636
Maximization of conditional probability, 247
Maximizer, 31
Maximum, 1
    stopband gain in digital filters, 686
Maximum-likelihood
    estimation
        gradient, 245
        Hessian, 245
        unconstrained problem, 245
        use in logistic regression for binary
            classification, 245
    multiuser detector, 615, 617
McCormick updating formula, 202
Mean-value theorem for differentiation, 30
Mehrotra's predictor-corrector linear-
        programming algorithm, 418
Memoryless BFGS updating formula, 208
Merit function, 553

augmented Lagrangian merit function, 553
L_1 merit function, 553
Minimax
  algorithms, 23, 213
    Charalambous, 215
    improved, 219–224
    least-pth, 214
    modified Charalambous, 222
    modified least-pth, 221
  design
    stability of IIR filters, 587
  design of
    FIR and IIR filters using SOCP, 586–587
    IIR filters satisfying prescribed
      specifications, 587–591
    IIR filters using SOCP, 587
  methods
    computational complexity in, 219
    elimination of spikes, 219
    introduction to, 211
    nonuniform variable sampling technique,
      219
    objective function, 211
    use of interpolation in, 221
  multipliers, 215
  problem, 7
Minimization of
  composite convex functions, 512–519
    fast algorithm, 514–519
    proximal-point algorithm, 512
  nonquadratic functions, 127
    using conjugate-directions methods,
      157–158
  a sum of L_2 norms formulated as a second-
    order cone programming problem,
    468
Minimizer, **8**, 31
  global, 31
  strong, 31
  uniqueness of minimizer of alternative-form
    linear-programming problem,
    theorem, 361
  uniqueness of minimizer of standard-form
    linear-programming problem,
    theorem, 362
  weak
    global, 31
    local, 31
Minimum, 1
  point, 8
  stopband attenuation in digital filters, 686

value, 8
Minor determinant (or minor), 643
Mixed linear complementarity problems, 442
ML, *see* Maximum-likelihood
MLE, *see* Maximum-likelihood estimation
MNIST, *see* Modified National Institute for
    Standards and Tehnology
Model predictive control of dynamic systems,
    591–614
  convex hull, 592
  introduction to, 591
  introduction to robust MPC, 592, 592–594
  minimax optimization problem, 592
  polytopic model for uncertain dynamic
    systems, 591–592
  robust constrained MPC using semidefinite
    programming, 597–614
    componentwise input constraints, 600
    Euclidean norm constraint, 599
    example, 601
    invariant ellipsoid, 599
    L_2-norm input constraint, 597
    modified SDP problem, 601
    SDP problem, 600
  robust unconstrained MPC using semidefinite
    programming, 594–597
    example, 596
    optimization problem, 595
Modification of the Hessian to achieve positive
    definiteness, 128–134
  examples, 132–134
  Matthews and Davies algorithm, 131
Modified BFGS formula, 554
Modified National Institute for Standards and
    Tehnology
  database, 249
    testing data set, 249
    training data set, 249
  data set
    use for classification of handwritten digits,
      250–253
Monotone linear complementarity problems,
    442
Moore-Penrose pseudo-inverse, 510, 651
  evaluation of, 651
MPC, *see* Model predictive control
Multiclass classification
  using softmax regression, 243–249
Multidimensional
  optimization
    introduction to, 115

unconstrained problems, 115
Multilayer thin-film system, 13
Multilegged vehicles as robotic systems, 602
Multimodal problems, 275
Multiple
    access interference, 617
    manipulators as robotic systems, 602
Multiplier constant in digital filters, 678
Multiuser access interference in wireless
        communications, 617
Multiuser detection in wireless communication
        channels, 614–630
    additive white Gaussian noise, 615
    bit-error rate in multiuser detection, 615
    channel model, 615–617
    code-division multiple access, 614
    code sequences, 614
    constrained minimum-BER multiuser
        detector, 625–630
        formulation as a convex-programming
            problem, 628
        problem formulation, 625–628
        solution based on a Newton-barrier
            method, 629–630
    frequency-division multiple access, 614
    introduction to, 614
    maximum-likelihood multiuser detector, 617
    multiuser access interference, 617
    multiuser detector, 617
    near-optimal multiuser detector using
        semidefinite-programming relaxation,
        617–625
        algorithm, 623
        binary solution, 620
        efficient detector based on duality, 621
        example, 624
        optimization problem, 620
        relaxation of MAX-CUT problem,
            618–619
        SDP-relaxation-based multiuser detector,
            619
        solution suboptimality, 621
    spreading
        sequence, 615
        gain, 615
    time-division multiple access, 614
    transmission delay, 615
Multiuser detector, 617

N
NAG, *see* Nesterov's accelerated gradient
        descent algorithm
Necessary and sufficient conditions
    for a minimum in alternative-form
        linear-programming problem
        theorem, 344, 350
    for a minimum in standard-form linear-
        programming problem
        theorem, 350
    for local minima and maxima, 33–40
Negative definite
    matrix, **43**, 643
    notation, 643
    quadratic form, 36
Negative part of a symmetric matrix, 541
Negative semidefinite
    matrix, **43**, 643
    notation, 643
    quadratic form, 36
Nesterov's accelerated gradient descent
        algorithm, 250
Newton
    algorithm, **128**, 259
        alternative, 182
        backtracking line search, 135
        comparison with Fletcher–Reeves
            algorithm, 190
        error after convergence, 135
        example, 188, 189
    algorithm for convex problems with equality
        constraints, 504
        nonfeasible initial point, 505
    algorithm for unconstrained minimization of
        convex functions, 502
    decrement, 135, 501
    direction, 128, 501
        independence with respect to linear
            changes in variables, 136
    method, 126–134
        barrier method for general convex
            programming problems, 507–511
        extension to convex constrained and
            unconstrained problems, 500–511
        minimization of smooth convex functions
            subject to equality constraints, 502–504
        minimization of smooth convex functions
            subject to equality constraints with a
            nonfeasible initial point, 504
        minimization of smooth convex functions
            without constraints, 500–502

modification of the Hessian to achieve
positive definiteness, 128–134
order of convergence, 128
relation with steepest-descent method,
129
NNF, *see* Nonnegative matrix factorization
Nonbasic variables in linear programming, 374
Nonconvex sets
definition, 51
Nondegeneracy assumption in simplex method,
363
Nondegenerate linear programming problems,
403
Nonfeasible-initialization interior-point
primal-dual path-following method
for convex quadratic-programming
problems, 439–442
algorithm, 440
example, 441
for linear-programming problems, 412–414
algorithm, 413
example, 414
Nonfeasible point, 17
Nonfeasible-start interior-point primal-dual
path-following algorithms, 412
Non-Hermitian square root, 645
Nonlinear
equality constraints in constrained
optimization, 299
example, 300
programming, 23
introduction to, 27
Nonnegative
matrices, 548
example, 548
matrix factorization, 548
Nonnegativity bounds in constrained
optimization, 300
Nonquadratic functions
minimization, 127
using conjugate-directions methods,
157–158
Nonsingular matrix, 642
Nonuniform variable sampling technique, 219
segmentation of frequency axis, 220
virtual sample points, 219
Normal
plane, 310
vector, 340
Normalized
CPU time, 124

sampling
frequency in digital filters, 681
period in digital filters, 681
Norms:
Euclidean, 7
Frobenius, 649
L_1 norm, 7
of a matrix, 649
of a vector, 647
L_2 norm, 7
of a matrix, 649
of a vector, 647
L_∞ norm, 7
of a function of a continuous variable,
262, 647
of a matrix, 649
of a vector, 647
L_p norm, 6
of a function of a continuous variable,
262, 647
of a matrix, 649
of a vector, 647
Null space, 637
example, 638
Numerical
ill-conditioning in least-pth minimax
method, 214
methods of optimization, 3
Nyquist frequency in digital filters, 681

O

Objective
function, 4
augmented, 307
in a minimax problem, 211
One-dimensional optimization
approximation methods, 78
cubic interpolation, 94–97
quadratic interpolation, 91–94
Davies, Swann, and Campey algorithm,
97–101
inexact line searches, 101–108
problems, 77
range of uncertainty, 78
search methods, 77
dichotomous, 78–79
Fibonacci, 79–86
golden-section, 88–89
Optimal force distribution for robotic systems,
602–614
closed kinematic chains, 602

dextrous hands, 602
force distribution problem in multifinger
    dextrous hands, 602–614
    compact linear programming method, 610
    contact forces, 602
    example, 609, 612
    friction cone, 606
    friction force, 606
    friction limits, 607
    point-contact model, 606
    soft-finger contact model, 606
    torsional moment, 606
    using linear programming, 608–610
    using semidefinite programming, 611–614
    introduction to, 602
    multilegged vehicles, 602
    multiple manipulators, 602
Optimality conditions for linear programming,
    341–346
Optimization,
    by analytical methods, 2
    basic problem, 4
    boundary point, 18
    classes of nonlinear optimization problems,
        77
    of constrained problems, 9
    of convex functions, 57–59
    cost function, 4
    by experimental methods, 2
    exterior point, 18
    feasible
        domain, 17
        point, 17
        region, 17
    of a function of a continuous independent
        variable, 5
    gradient vector, 27
    by graphical methods, 2
    Hessian matrix, 28
    interior point, 18
    introduction to, 1
        multidimensional, 115
    nonfeasible point, 17
    by numerical methods, 3
    objective function, 4
    optimum, 1
    practice, 1
    as a process, 1
    saddle point, 40
    of several functions, 4
    stationary point, 40

theory, 1
    tolerance, 9
    types of optimization problems, 1
    unconstrained, 10
Optimum, 1
    investment portfolio, 15
    point, 8
    value, 8
Order of
    a digital filter, 674
    convergence, 73
Orthogonal
    direction, 119
    matrices, 45, **642**
    projections
        example, 654
        matrix, 653
    vectors, 147
Orthogonality
    condition, 147
    of gradient to a set of conjugate directions,
        150, 152
Orthonormal basis of a subspace, 654
Overdetermined system of linear equations,
        216

P
Partan
    algorithm
        solution trajectory, 171
    method, 169–172
Particle method, 541–542
PAS, *see* Primal afifine scaling
Passband
    in digital filters, 682
    edge in digital filters, 684
    ripple in digital filters, 685
Passband error function in IIR filters, 589
PCCP, *see* Penalty convex-concave procedure
PCCP-based algorithm, 560
PCM, *see* Predictor-corrector method
Peak stopband error in digital filters, 686
Peak-to-peak passband error in digital filters,
        685
Penalty convex-concave procedure, 559–562
Periodicity of frequency response in digital
        filters, 681
Phase
    distortion in digital filters, 683
    response in digital filters, 680
    shift in digital filters, 680

Phase-one optimization problem, 510
PNB, *see* Primal Newton barrier
Point-contact model in multifinger dextrous
        hands, 606
Point-to-point mapping, 64
Point-to-set mapping, 65
Poles in digital filters, 678
Polygon, 671
Polyhedron, 671
Polynomial interpolation, 221
Polytope, 671
Polytopic model for uncertain dynamic
        systems, 591–592
Position and orientation of a manipulator,
        253–256
Positive definite
    matrix, **43**, 643
        notation, 643
    quadratic form, 36
Positive definite approximation of Hessian,
        554–555
Positive semidefinite
    matrix, **43**, 643
        notation, 643
    quadratic form, 36
Powell's
    method, 161–169
        advantages and disadvantages, 165
        algorithm, 164
        alternative approach, 166
        generation of conjugate directions,
            161–162
        linear dependence in, 165
        solution trajectory, 165
        Zangwill's technique, 166–169
Practical quasi-Newton algorithm, 202–205
    choice of line-search parameters, 205
    computational complexity, 203
    modified inexact line search, 202
    positive definiteness condition in, 205
    termination criteria in, 205
Practical SQP algorithm, 556–557
Predictor-corrector method, 415–419
    algorithm, 418
    centering parameter, 415
    computational complexity, 418
    corrector direction, 417
    example, 418
    for semidefinite-programming problems,
        460–464
        algorithm, 461

    corrector direction, 461
    predictor direction, 460
    Mehrotra's algorithm, 418
    predictor direction, 415
Prescribed specifications
    in minimax design of IIR filters, 587–591
Primal
    active-set method for convex quadratic-
        programming problems, 430–434
    affine scaling method for linear-programming
        problems, 398–402
        affine scaling transformation, 399
        algorithm, 401
        example, 401
        primal affine scaling direction, 400
        projected steepest-descent direction, 398
        scaling, 398
    linear-programming problem, 394
    Newton barrier method for linear-
        programming problems, 402–407
        algorithm, 406
        barrier function, 402
        barrier parameter, 402
        example, 406
        logarithmic barrier function, 402
    problem in
        constrained optimization, 328, 330
        second-order cone programming, 465
        semidefinite programming, 445
    residual, 504, 522
Primal-dual
    interior-point methods, 292
    path-following method, 407–412
        algorithm, 409
        example, 411
        modified algorithm, 410
        nonfeasible initialization, 412–414
        nonfeasible-initialization algorithm, 413
        nonfeasible-initialization example, 414
        short-step algorithms, 410
    path-following method for convex
        quadratic-programming problems,
        437–439
        algorithm, 439
        example, 441
        potential function, 438
    path-following method for semidefinite
        programming, 453–460
        algorithm, 457
        example, 459
        reformulation of centering condition, 453

Schur complement matrix, 456
symmetric Kronecker product, 454
solutions in
    linear programming, 394–396
    semidefinite programming, 451
Principal minor, 643
Projected steepest-descent direction in linear
    programming, 398
Projection, 563, 565
Proper function definition, 494
Properties of
    Broyden method, 200
    conjugate functions, 496–498
    convex functions, 483–500
    proximal operators, 498–500
    semidefinite programming, 450–453
Proximal operators, 498–500
    fixed point, 513
    properties, 498–500
Proximal-point algorithm, 512–514

Q
QP, *see* Quadratic programming
QR decomposition
    with column pivoting, 663
    example, 663
    for full-rank case, 660
    for rank-deficient case, 662
Quadratic
    approximation of Taylor series, 30
    cone in second-order cone programming,
        465
    form, 36
        positive definite, positive semidefinite,
            negative definite, negative semidefinite,
            36
    interpolation search
        algorithm, 92
        simplified three-point formula, 93
        three-point formula, 91–94
        two-point formula, 94
    programming, 22, 294
    rate of convergence in conjugate-direction
        methods, 157
    termination, 157
Quadratic programming
    central path, 436
    convex quadratic-programming problems
        with equality constraints, 426–429
        example, 427
    dual problem, 435

duality gap, 436
introduction to, 425
mixed linear complementarity problems, 442
monotone linear complementarity problems,
    442
primal-dual potential function, 438
problem formulated as a second-order cone
    programming problem, 468
relation between convex quadratic-
    programming problems and semidefinite-
    programming problems, 448
relation between convex quadratic-
    programming problems with
    quadratic constraints and semidefinite-
    programming problems, 449
Quasi-Newton
direction, 183
algorithm
    basic, 187
    practical, 202–205
methods
    advantage of DFP method relative to
        rank-one method, 190
    basic, 180
    Broyden method, 200
    Broyden–Fletcher–Goldfarb–Shanno
        method, 197–199
    choice of line-search parameters, 205
    comparison of DFP and BFGS methods,
        205
    correction matrix, 183
    Davidon–Fletcher–Powell method,
        190–196
    disadvantages of rank-one method, 187
    duality of DFP and BFGS formulas, 197
    duality of Hoshino formula, 197
    equivalence of Broyden method with
        Fletcher–Reeves method, 200
    Fletcher switch method, 200
    generation of inverse Hessian, 186
    Hoshino method, 199
    introduction to, 179
    McCormick updating formula, 202
    memoryless BFGS updating formula, 208
    positive definiteness condition, 205
    rank-one method, 185
    relation between direction matrix **S** and
        the Hessian, 181
    termination criteria, 205
    updating formula for BFGS method, 197

updating formula for Broyden method, 200

updating formula for DFP method, 191

updating formula for Hoshino method, 199

updating formula for rank-one method, 185

R

Range of
a matrix, 637
uncertainty, 78
Rank of a matrix, 637
example, 638
Rank-one method, 185
disadvantages, 187
updating formula, 185
Rates of convergence, 73
Regular
index assignment, 547
point, 287
Regularization parameters, 514, 548
Relation between second-order cone
programming problems and
linear-programming, quadratic-
programming, and semidefinite-
programming problems, 466
Relation between semidefinite-programming
problems
and alternative-form linear-programming
problems, 448
and convex quadratic-programming
problems, 448
and convex quadratic-programming
problems with quadratic constraints, 449
and standard-form linear-programming
problems, 445
Relaxation of MAX-CUT problem, 618–619
Remainder of Taylor series, 29
Residual
dual, **504**, 522
error, 7
primal, **504**, 522
Response (output) in digital filters, 673
Right singular vector, 651
Robotic manipulator, 253
Robustness and solvability of QP subproblem, 555–556
Robustness in algorithms, 9
Roots of characteristic polynomial, 129
$\mathcal{R}$ operator in digital filters, 673

Rosenbrock function, **176**, 189, 196, 198
Row rank, 638
Rule of correspondence, 64

S

Saddle point, 40
in steepest-descent method, 125
Sampling frequency in digital filters, 681
Scalar product, 116
Scaled ADMM algorithm, 564
Scaling in
linear programming, 398
affine scaling transformation, 399
steepest-descent method, 126
Schur polynomials, 302
example, 302
SCP, *see* Sequential convex programming
SD, *see* Steepest-descent
SDP, *see* Semidefinite programming
SDP relaxation dual, 624
SDP relaxation primal, 624
SDPR-D, *see* SDP relaxation dual
SDPR-P, *see* SDP relaxation primal
Search
methods (multidimensional)
introduction to, 115
methods (one-dimensional), 77
dichotomous, 78–79
Fibonacci, 79–86
golden-section, 88–89
Second-order
cone, 465
gradient methods, 115
Second-order conditions for a maximum
necessary conditions in unconstrained
optimization, 39
sufficient conditions in unconstrained
optimization, 40
sufficient conditions in unconstrained
optimization, theorem, 40
Second-order conditions for a minimum
in constrained optimization, 319–323
necessary conditions
constrained optimization, 320
equality constraints, 321
example, 321
general constrained problem, 322
necessary conditions in unconstrained
optimization, 36
sufficient conditions
equality constraints, 323

example, 323, 326
general constrained problem, 324
sufficient conditions in unconstrained
optimization, theorem, 39
Second-order cone programming, 465–475
application for minimax design of FIR and
IIR filters, 586–587
application for minimax design of IIR filters,
587
complex L_1-norm approximation problem
formulated as an SOCP problem, 469
definitions, 465
dual problem, 466
introduction to, 425
linear fractional problem formulated as an
SOCP problem, 470
Lorentz cone, 465
minimization of a sum of L_2 norms
formulated as an SOCP problem, 468
notation, 465
primal problem, 465
primal-dual method, 472–475
assumptions, 472
duality gap in, 472
example, 474
interior-point algorithm, 474
Karush-Kuhn-Tucker conditions in, 472
quadratic cone, 465
quadratic-programming problem with
quadratic constraints formulated as an
SOCP problem, 468
relations between second-order cone
programming problems and linear-
programming, quadratic-programming,
and semidefinite-programming problems,
466
second-order cone, 465
stability of IIR filters, 587
Semidefinite programming
assumptions, 450
centering condition, 452
central path in, 451
convex
cone, 446
optimization problems with linear matrix
inequality constraints, 447
definitions, 445
dual problem, 446
duality gap in, **447**, 451
introduction to, 425
Karush-Kuhn-Tucker conditions in, 450

notation, 445
primal problem, 445
primal-dual
solutions, 451
properties, 450–453
primal-dual path-following method,
453–460
relation between semidefinite programming
and alternative-form linear programming
problems, 448
relation between semidefinite-programming
and convex quadratic-programming
problems, 448
and convex quadratic-programming
problems with quadratic constraints, 449
and standard-form linear-programming
problems, 445
Semidefinite-programming relaxation-based
multiuser detector, 619
Sequential convex programming, 540–550
affine
approximation, 541–543
functions, 540
algorithm, 543
based on an exact penalty function, 545
ball, 540
convex approximations, 541–543
principle, 540
trust region, 540
Sequential quadratic programming, 550–557
introduction to, 296
Sets,
bounded, 70
closed, 70
compact, 70
Sherman-Morrison formula, 639
example, 639
Signal-to-noise ratio, 518, 519, 525, 624
Similarity transformation, 641
Simplex method, 292
computational complexity, 385–387
for alternative-form linear-programming
problem, 363–374
algorithm, degenerate vertices, 372
algorithm, nondegenerate vertices, 365
blocking constraints, 372
cycling, 371
degenerate case, 371
example, bounded solution, 366
example, degenerate vertex, 373
example, unbounded solution, 369

least-index rule, 371
nondegeneracy assumption, 363
nondegenerate case, 362, 363–370
working index set, 371
working set of active constraints, 371
working-set matrix, 371
for standard-form linear-programming
problem, 374–382
algorithm, 380
basic and nonbasic variables, 374
example, 380
tabular form, 383–384
pivot, 383
Simply connected feasible region, 20
Singular
value decomposition, 651
application to constrained optimization,
288
example, 652
values, 651
Slack variables in constrained optimization,
291
Snell's law, 12
SNR, *see* Signal-to-noise ratio
SOCP, *see* Second-order cone programming
Soft-finger model in multifinger dextrous
hands, 606
Softmax regression
for use in multiclass classification, 243–249
use for classification of handwritten digits,
249–253
use of Frobenius norm, 249
weight decay, 249
Soft shrinkage, 514
Solution
point, 8
Span, 636
basis for, 636
example, 636
Sparse
matrices, 386
solution of an undetermined system of linear
equations, 527–529
Spreading
gain in multiuser detection, 615
sequence in multiuser detection, 615
SQP, *see* Sequential quadratic programming
Stability
condition imposed on
eigenvalues, 680
impulse response, 677

poles, 679
margin in digital filters, 583
property in digital filters, 677
Stability margin in digital filters, 587
Stabilization technique for digital filters, 226
Standard-form linear-programming problem,
339
Standard signals in digital filters
unit impulse, 675
unit sinusoid, 675
unit step, 675
Stationary points, 40
classification, 40–42
Steady-state sinusoidal response in digital
filters, 680
Steepest-ascent direction, 116
Steepest-descent
algorithm, 119
without line search, 121
solution trajectory, 170
direction, 116
formula with a constant line search step, 490
method, 116–126
convergence, 124–126
relation with Newton method, 129
saddle point, 125
scaling, 126
solution trajectory, 118
step-size estimation, 120–122
Step response of a control system, 5
Step-size estimation
for steepest-descent method, 120–122
using the Barzilai–Borwein formulas,
122–124
Stopband
in digital filters, 682
edge in digital filters, 684
Strict feasibility
of primal-dual solutions in linear
programming, 395
Strictly
concave functions,
definition, 51
convex functions,
definition, 51
theorem, 56
feasible
linear-programming problem, 394
matrices, 450
Strong
local constrained minimizer, 293

minimizer, 31
Strongly convex functions, 491–493
  definition, 491
Subdifferential, 485
  example, 486
  notation, 485
Subgradient, 484–488
  definition, 485
  uniqueness, 485
Subspace, 636
  dimension of, 636
Superlinear convergence, 74
Suppression of spikes in the error function, 219
SVD, *see* Singular value decomposition
Symmetric
  matrices, 642
  square root, 645
    example, 645
Systems of linear equations
  example, 174
  residue of, 173
  solution of, 173–175
  use of the conjugate-gradient method for the solution of, 173

T

Tabular form of simplex method, 383–384
  pivot, 383
Tangent plane, 309
Taylor series, 29
  cubic approximation of, 30
  higher-order exact closed-form expressions for, 30
  linear approximation of, 30
  quadratic approximation of, 30
  remainder of, 29
TDMA, *see* Time-division multiple access
Techniques:
  updating the size of the trust region, 545
    algorithm, 545
Theorems:
  characterization of
    matrices, 43
    symmetric matrices via diagonalization, 43
  conjugate directions in Davidon–Fletcher–Powell method, 193
  convergence of
    conjugate-directions method, 149
    conjugate-gradient method, 152
    inexact line search, 104

convex sets, 53
convexity of linear combination of convex functions, 52
eigendecomposition of symmetric matrices, 45
equivalence of Broyden method with Fletcher–Reeves method, 200
existence of
  a global minimizer in convex functions, 58
  a vertex minimizer in alternative-form linear-programming problem, 360
  a vertex minimizer in standard-form linear-programming problem, 361
  primal-dual solution in linear programming, 395
first-order necessary conditions for a minimum
  equality constraints, 312
  unconstrained optimization, 35
first-order optimality condition for nondifferentiable unconstrained convex problems, 487
fixed-point of proximal-point operator, 513
generation of
  conjugate directions in Powell's method, 161
  inverse Hessian, 186
global convergence, 69
globalness and convexity of minimizers in convex problems, 326
Karush-Kuhn-Tucker conditions, 316
  for standard-form linear-programming problem, 341
Karush-Kuhn-Tucker conditions for nondifferentiable constrained convex problems, 487
linear independence of
  columns in constraint matrix, 374
  conjugate vectors, 147
location of maximum of a convex function, 58
mean-value theorem for differentiation, 30
necessary and sufficient conditions for a minimum in
  alternative-form linear-programming problem, 344, 350
  standard-form linear-programming problem, 350
optimization of convex functions, 57–59

orthogonality of gradient to a set of conjugate
directions, **150**, 152
positive definiteness of **S** matrix, 191
properties of
Broyden method, 200
matrices, 46
strictly convex functions, 56
property of convex functions relating to
gradient, 54
Hessian, 55
relation between local and global minimizers
in convex functions, 57
second-order necessary conditions for a
maximum
unconstrained optimization, 39
second-order necessary conditions for a
minimum
equality constraints, 321
general constrained problem, 322
unconstrained optimization, 36
second-order sufficient conditions for a
maximum
unconstrained optimization, 40
second-order sufficient conditions for a
minimum
equality constraints, 323
general constrained problem, 324
unconstrained optimization, 39
strict feasibility of primal-dual solutions, 395
strictly convex functions, 56
sufficiency of Karush-Kuhn-Tucker
conditions in convex problems, 327
uniqueness of minimizer of
alternative-form linear-programming
problem, 361
standard-form linear-programming
problem, 362
upper bound of a convex function with
Lipschitz continuous gradients, 489
Weierstrass theorem, 70
Time-division multiple access in communications, 614
Time-domain response
in digital filters, 675–676
using the z transform, 679
Time invariance property in digital filters, 674
Toeplitz matrix, **270**, 516
Torsional moment in multifinger dextrous
hands, 606
Trace of a matrix, 646
Training data

in handwritten-digit recognition, 240
Trajectory of solution in
conjugate-gradient algorithm, 156
partan algorithm, 171
Powell's algorithm, 165
steepest-descent algorithm, 118, 170
Transfer function of a digital filter
definition, 677
derivation from difference equation, 678
in zero-pole form, 678
Transformations,
affine scaling, 399
in constrained optimization, 296
elementary, 44
homogeneous, 255
Transition band in digital filters, 684
Transmission delay in communication
channels, 615
Transportation problem, 14
Trust region
sequential convex programming, 540
Two-point Barzilai–Borwein formulas, *see*
Barzilai–Borwein formulas

U

Unconstrained minimizer, 307
Unconstrained optimization
applications
introduction to, 239
multidimensional problems, 115
problems, 10
Undetermined system of linear equations,
example, 527–529
Unimodal, 77
Uniqueness of minimizer of
alternative-form linear-programming
problem, 361
standard-form linear-programming problem,
362
Unitary matrix, 45, 642
Unit lower triangular matrix, 129
Updating formulas:
alternative formula for Davidon–Fletcher–
Powell method, 196
Broyden–Fletcher–Goldfarb–Shanno
formula, 197
Broyden formula, 200
DFP formula, 191
duality of DFP and BFGS formulas, 197
Hoshino formula, 199
Huang formula, 201

McCormick formula, 202
memoryless BFGS updating formula, 208
rank-one formula, 185
Upper
  passband edge in digital filters, 684
  stopband edge in digital filters, 684
  triangular matrix, 131
Upper bound of a convex function with
        Lipschitz continuous gradients
  theorem, 489

V

Variable elimination methods in constrained
        optimization, 296
  example, 298
  linear equality constraints, 296
  nonlinear equality constraints, 299
    example, 300
Variable elimination methods in constrained
        optimization, 296
Variable transformations in constrained
        optimization, 300, 303
  example, 301
  interval-type constraints, 301
  nonnegativity bounds, 300
Vector spaces of symmetric matrices, 668–671
Vectors:
  conjugate, 146
  eigenvectors, 147
  inner product, 647
  L_1 norm, 647
  L_2 norm, 647
  left singular, 651
  L_∞ norm, 647
  linear independence, 147, 636
  L_p norm, 647
  notation, 635

orthogonal, 147
right singular, 651
Vertex minimizer
  existence of vertex minimizer in alternative-
        form linear programming problem,
        theorem, 360
  existence of vertex minimizer in standard-
        form linear programming problem,
        theorem, 361
  finding a, 360–362
Vertex of a convex polyhedron, 346
  degenerate, 347
  example, 352, 353
  method for finding a vertex, 350–356
  nondegenerate, 347
Virtual sample points, 219

W

Weak
  global minimizer, 31
  local minimizer, 31
Weierstrass theorem, 70
Weight decay
  in softmax regression, 249
Weighting in the design of least-squares FIR
        filters, 263
Weights, 7
Working
  index set, 371
  set of active constraints, 371

Z

Zangwill's
  algorithm, 169
  technique, 166–169
Zeros in digital filters, 678